# 23rd International Conference on High Pressure Science and Technology 2011



# (AIRAPT-23)

Journal of Physics: Conference Series Volume 377

## Mumbai, India
## 25-30 September 2011

ISBN: 978-1-62276-811-0
ISSN: 1742-6588

**Printed from e-media with permission by:**

Curran Associates, Inc.
57 Morehouse Lane
Red Hook, NY 12571

**Some format issues inherent in the e-media version may also appear in this print version.**



Printed by Curran Associates, Inc. (2013)

For permission requests, please contact the Institute of Physics at the address below.

Institute of Physics
Dirac House, Temple Back
Bristol BS1 6BE UK

Phone: 44 1 17 929 7481
Fax: 44 1 17 920 0979

techtracking@iop.org

**Additional copies of this publication are available from:**

Curran Associates, Inc.
57 Morehouse Lane
Red Hook, NY 12571 USA
Phone: 845-758-0400
Fax: 845-758-2634
Email: curran@proceedings.com
Web: www.proceedings.com

## High Pressure Instrumentation



## Experimental Static High Pressure Phase Transitions, Superconductivity Melting, EOS and Strength Properties

## Experimental Shock Compression of Materials



## High Pressure Biosceinces

**High Pressure Geophysics**



**High Pressure Theory, Modelling and Simulations**

# 23rd International Conference on High Pressure Science and Technology (AIRAPT-23)

Satish C Gupta

## Abstract Metrics

### PREFACE

The 23rd AIRAPT International Conference on High Pressure Science and Technology was held at Bhabha Atomic Research Centre, Mumbai, from 25–30 September 2011. This conference is part of the series of AIRAPT International Conferences which are held biennially. AIRAPT is an acronym for the French title which translates as 'International Association for the Advancement of High Pressure Science and Technology'. This was the second time the AIRAPT Conference was organized in India. The first was held 20 years ago at the National Aeronautical Laboratory, Bangalore in 1991.

The 23rd Conference covered many important topics in the area of both static and dynamic high pressures including theoretical and experimental investigations on the response of materials under high pressures, new developments using neutron and synchrotron sources, investigations on superconductivity under high pressure, studies of geophysical and planetary sciences, biosciences, and the synthesis of new materials.

The conference program included Bridgman award lecture, Jemieson award lecture, seven plenary talks, 85 invited talks, 83 oral presentations and about 195 posters. In all there were 372 presentations. 285 scientists from 19 countries participated in the conference. The countries represented included Austria, Canada, China, Estonia, France, Germany, India, Israel, Italy, Japan, Nepal, New Zealand, Poland, Russia, South Korea, Spain, Sweden, Switzerland, Turkey, UK, Ukraine and USA.

Many new developments were presented, for example, measurement techniques using the new generation synchrotron sources, more powerful neutron sources and much brighter laser sources; integration of gas-gun with synchrotron source; the achievement of multi-megabar pressures in shock-less dynamic compressions; and capabilities to synthesize centimeter size diamonds with better quality. All these developments have opened up new opportunities for understanding the physics of materials under high pressures.

I would like to thank all those who have made valuable contributions to the success of the conference, which include the members of the AIRAPT executive committee, the International Advisory Committee and National Advisory Committee, the plenary speakers, invited speakers, the chairmen of various sessions, all the participants, and the authors of the papers in this volume.

All the papers accepted for the proceedings have been reviewed by two independent referees. I am extremely thankful to all the anonymous referees, who have spent their valuable time to ensure the quality of the papers of this volume.

I wish to express my gratitude to the members of the Local Organizing Committee for their help and hard work for the success of the conference. Finally, I convey my special thanks to Dr T C Kaushik and Dr K D Joshi, who worked tirelessly and enthusiastically towards making this conference a success.

I am confident that this volume of the Conference proceedings will provide an excellent source of information on the current trends in the field of High Pressure Science and Technology.

**Satish C Gupta**
*Conference Chairman*
25–30 September 2011

# Peer review statement

**Abstract** Metrics

All papers published in this volume of *Journal of Physics: Conference Series* have been peer reviewed through processes administered by the proceedings Editors. Reviews were conducted by expert referees to the professional and scientific standards expected of a proceedings journal published by IOP Publishing.

Journal of Physics: Conference Series **377** (2012) 012001 doi:10.1088/1742-6596/377/1/012001

# Development of the Kawai-type Multi-anvil Apparatus (KMA) and Its Application to High Pressure Earth Science

**E Ito**

Institute for Study of the Earth's Interior Okayama University Misasa Tottori-ken 682-0193 Japan

E-mail: eiito645@gmail. com

**Abstract**. Since Birch's prediction on the structure of the Earth's interior high pressure Earth science has rapidly been grown by the Kawai-type multi-anvil apparatus (KMA) and the diamond anvil cell (DAC). An important advantage of the KMA is its large specimen volume which makes it possible to conduct experiments under precisely controlled P-T conditions. A typical application of the KMA to the Earth science might be determination of the post-spinel phase equilibria in the system $Mg_2SiO_4$-$Fe_2SiO_4$. By adopting sintered diamond (SD) as the anvil material accessible pressure of the KMA has substantially increased. Melting experiments of mantle materials up to 35 GPa opened a new paradigm on the mantle fractionation in early Earth. Combining the KMA equipped with SD anvils with the synchrotron radiation phase equilibria of Fe GaN and $Fe_2O_3$ were determined by means of in situ X-ray diffraction. Special attention was paid to define the phase boundaries between perovskite and post-perovskite in $MgGeO_3$ and $MnGeO_3$ as analogues of $MgSiO_3$. We also observed the spin transition of $Fe^{2+}$ in $(Mg_{0.87}Fe_{0.17})O$ at 300 and 700 K. Recently the maximum attainable pressure is reaching 100 GPa and high P-T experiments up to 90 GPa are our ordinary jobs. In order to produce still higher pressure however innovation of SD such as NPD is indispensable.

## 1. Beginning of the high pressure Earth science

In the early 20th century the Earth's interior became an object of scientific research due to the information provided by seismology [1] which made it possible to determine the density and thereby pressure as functions of depth from the surface to the center [2]. Following these pioneering works Birch [3] analyzed homogeneity of the subdivided layers of the Earth's interior [2] using the equation of state based on finite strain theory and Bridgman's compression data and concluded that minerals constituting the uppermost mantle such as olivine pyroxenes and garnet successively transform into closed-packed oxides similar to corundum rutile spinel or perovskite in structure. As his analyses were so concrete and convincing that the prediction had been the guiding principle for exploration of the Earth's interior for long time and strongly stimulated the high pressure experiments.

Figure 1 shows a standard Earth's model PREM [4]. The mantle is characterized by two sharp increases in seismic velocities at depths of 410 and 660 km which divides the mantle into three parts the upper mantle (B) the transition zone (C) and the lower mantle (D). Following Birch [3] high pressure phase equilibria of the system $Mg_2SiO_4$-$Fe_2SiO_4$ were extensively studied by many workers represented by Ringwood and Major [5] and Akimoto and Fujisawa [6] because olivine $(Mg_{0.9}Fe_{0.1})_2SiO_4$ is the most dominant (> 60%) constituent of the upper mantle. And the 410 km discontinuity was reasonably assigned to the transformation of the olivine ($\alpha$) into the modified spinel (β) structure up to the early 1970s. As capability of the apparatuses employed in these studies were limited to ca. 15 GPa neither synthesis of the end member $Mg_2SiO_4$ spinel (γ) nor acquisition of insight into the post-spinel transformation were successful at that time.

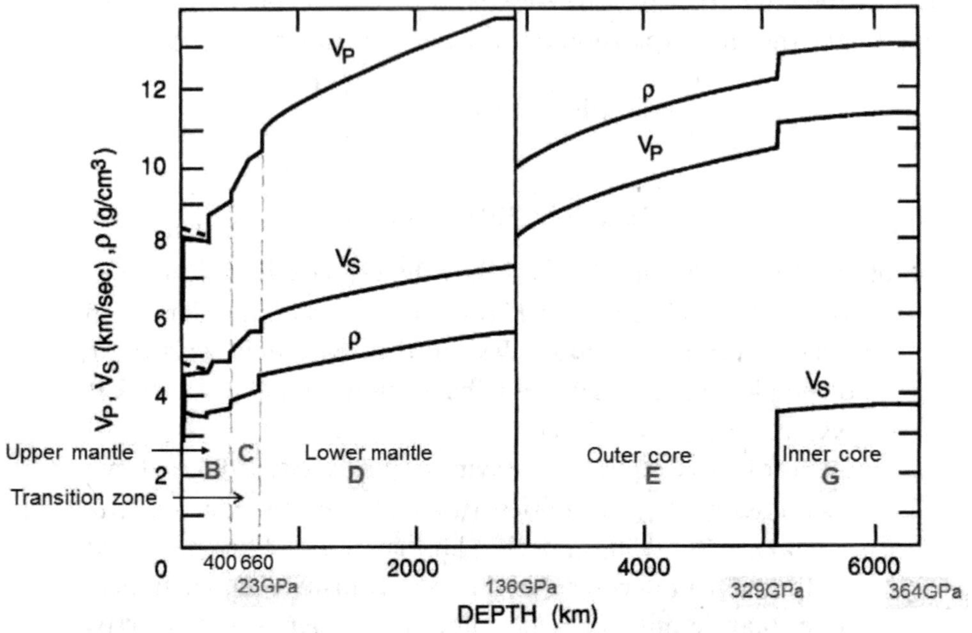


**Figure 1.** A standard Earth's model (PREM) by Dziewonski and Anderson (1981). Layers A-G are correspond to division by Bullen [2].

## 2. Devices employed in the high pressure Earth sciences Beginning of the high pressure Earth science

Soon after however γ–Mg2SiO4 was synthesized by Suito [7] in the Kawai-type multi-anvil apparatus (KMA) and the decomposition of γ-Fe2SiO4 into an assemblage of FeO (wüstite) and $SiO_2$ (stishovite) was found by Mao and Bell [8] by adopting the diamond anvil cell (DAC). Hereafter the high pressure Earth science has been extensively developed by adopting both the devices. In the DAC it is possible to compress a small amount of sample to multi Mbar heat it to thousands of Kelvin and observe the state of the sample under these extreme conditions by means of X-ray optical and other measurements. A marked advantage of the KMA over the DAC on the other hand is the much larger sample volume which makes it possible to conduct experiments under precisely controlled P-T conditions. Therefore the KMA has been adopted in various researches determination of phase equilibria synthesis of high pressure phases and measurements of physical properties. Development of the KMA was carried out mainly during 1965-1973 at Osaka University under the direction of the late N. Kawai [9]. Photos of the KMA are shown in Figure 2. The cubic assemblage of eight cubes of tungsten carbide (WC) an octahedral pressure medium and gaskets so-called the Kawai-cell (left) is squeezed along the [111] direction in the split sphere guide blocks with the aid of a uniaxial press (right).

**Figure 2.**The Kawai-type multi anvil apparatus (KMA). The Kawai-cell composed of cubic WC anvils MgO octahedral pressure medium pyrophyllite pre-gaskets etc. (upper photo) is squeezed in the split sphere guide block with the aid of uniaxial press (lower photo).

The easiest way to convince the performance of the KMA may be to show "huge (over 1 mm)" as grown single crystals of $MgSiO_3$ perovskite [10] (see Figure 3) which is stable only at pressures higher than 23 GPa and could be the most abundant material in our planet. The single crystals of the deep mantle material thus synthesized have been used to measure various kinds of physical properties.

**Figure 3.** "Huge" (over 1mm) single crystals of MgSiO3 perovskite which is stable only at pressures higher than 23 GPa. After [10].

## 3. The post-spinel transformation in the system $Mg_2SiO_4$-$Fe_2SiO_4$: An application of the KMA to phase equilibrium study

Synthesis of $MgSiO_3$ perovskite (Pv) and confirmation of the dissociation of $\gamma$–$Mg_2SiO_4$ into an assemblage of $MgSiO_3$Pv and MgOpericlase were first carried out by Liu [11] using DAC coupled with the laser heating system. Later the post-spinel transformation was examined in detail all over the system $Mg_2SiO_4$-$Fe_2SiO_4$ by Ito and Takahashi [12] using the KMA. Pseudobinary diagrams at 1100 and 1600 ℃ determined by quenching method are reproduced in Figure 4. The diagrams indicate that the mantle spinel with composition $(Mg_{0.9}Fe_{0.1})_2SiO_4$ dissociates into an assemblage of $(Mg_{0.97}Fe_{0.03})SiO_3$ perovskite (Pv) and $(Mg_{0.83}Fe_{0.17})O$ magnesiowüstite (Mw) (or ferropericlase Fp) at pressures of 23.1-24.5 GPa and at 1100-1600 ℃. As the dissociation brings about 10 % increases in density and seismic velocities [13] the dissociation must be responsible for the 660 km discontinuity. It is realized that the silicate Pv's and Fp are by far dominant constituents of the lower mantle taking the high pressure transformations in pyroxenes and garnet into consideration as well.

The post-spinel dissociation possesses two important characteristics. First it completes within a quite narrow pressure interval less than 0.15 GPa (or 4 km in depth) for the mantle spinel. Therefore by making the dissociation boundary correspond to the depth of the 660 km discontinuity we have a temperature fixed point of 1600 ± 100 ℃ at the depth and can construct the temperature profile through the transition zone [14].

Secondly the phase boundary has a negative slope dP/dT~ -3 MPa/deg. Seismic tomography has revealed that some of descending slabs stagnate around the 660 km discontinuity before collapsing down into the lower mantle [15]. On the other hand  it is known that a phase boundary with a negative slope in the mantle suppresses material transport through the boundary [16]. Therefore  it is highly likely that the spinel dissociation substantially contributes to the stagnation of slabs.

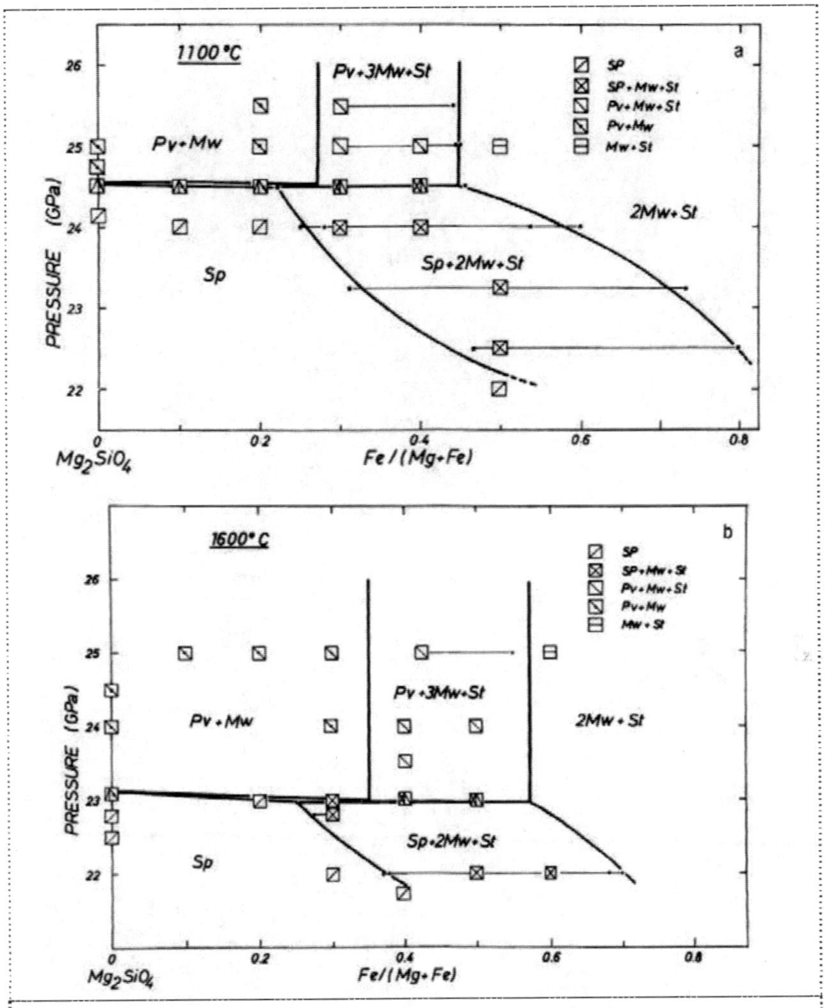


**Figure 4.** Pseudobinary diagrams of the post-spinel transformation in the system $Mg_2SiO_4$-$Fe_2SiO_4$ at 1200 and 1600°C. After [12].

## 4. Adoption of sintered diamond (SD) for anvils of the KMA

Decisive disadvantage of the KMA in comparison with the DAC was that the maximum attainable pressure was limited to 28 GPa so far as WC was used as the anvil material. Fortunately sintered diamond (SD) which was twice harder than WC [17] became available for anvils of the KMA in the late 1980s. Figure 5 shows SD cubes of Co-binder. We have employed cubes of an edge length 14 mm with truncated corners of 2 to 0.75 mm.

We first adopted the split-sphere system [9] to squeeze the Kawai-cell of SD cubes and compared the performances with those of WC cubes. Generated pressure was calibrated using several fixed points and plotted versus oil pressure for both the SD and WC cubes as Figure 6. Superiority of SD to WC is overwhelming  suggesting high potential of SD in pressure generation. Following these calibration runs we carried out high P-T quenching experiments. The followings are summaries of studies which had never been achieved by using WC anvils.

**Figure 5.**Sintered diamond cubes with Co binder.

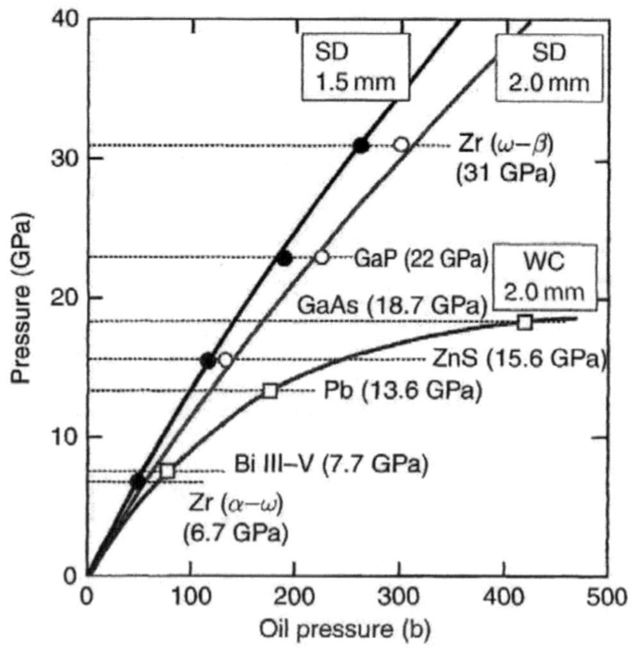


**Figure 6.**Comparison of pressure generation using SD with WC anvils. Superiority of SD to WC anvils is overwhelming.

*4.1. Synthesis of perovskite with pyrope ($Mg_3Al_2Si_3O_{12}$)*

As the system $MgSiO_3$-$Al_2O_3$ represents the compositions of pyroxene and garnet in the mantle high pressure phase equilibria of the system are also indispensable to understand mineralogy of the lower mantle. Systematic study by Irifune et al. [18] had shown that pyrope ($Mg_3Al_2Si_3O_{12}$) broke-down into solid solutions of the $MgSiO_3$-rich perovskite and $Al_2O_3$-rich corundum structures at ca. 27 GPa and that the binary loop between the two phases would persist to higher pressures.

We examined the stability of pyrope composition using SD anvils and confirmed formation of the orthorhombic perovskite phase at 37 GPa and 1600 ℃ [19]. The result indicates closing of the binary loop and the silicate perovskite can accommodate certain amount of $Al_2O_3$. Ubiquitous coexistence of tiny amount of SiO2 stishovite suggested slight nonstoichiometry of $Al_2O_3$-bearing perovskite.

*4.2Melting experiments of the primordial mantle materials*

In order to simulate possible material fractionation in a deep magma ocean formed on the early Earth melting phase relations of the Earth material and element portioning between liquidus phases and melt are indispensable. Therefore melting experiments of fertile peridotite [20] and CI chondritic mantle material (CMM) [21] had been carried out up to 25 GPa. Nevertheless a clear model for the mantle fractionation had been postponed.

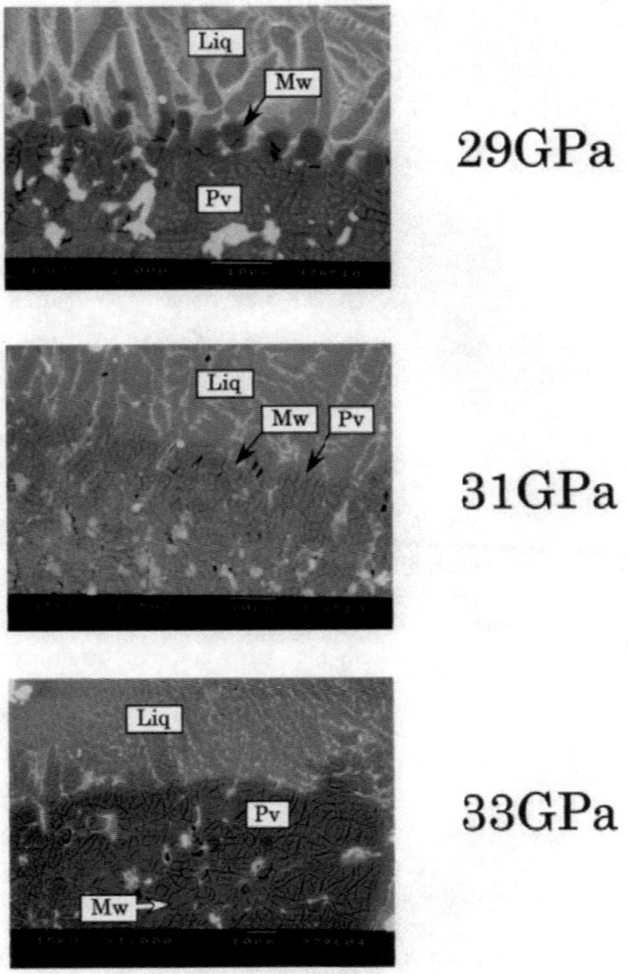


**Figure 7.** Back scattered electron images of quenched charges of peridotite at 29 31 and 33 GPa.

We extended melting experiment on both the materials to 35 GPa [22]. In periditite the first liquidus phase changed from ferropericlase (Fp) to Mg-perovskite (Mg-Pv) at 31 GPa and at 33 GPa liquidus Mg-Pv was successively followed down temperature by Fp and Ca-perivskite (Ca-Pv) within a small temperature range. The change of liquidus phase with pressure is clearly demonstrated in a series of pictures shown in Figure 7. In the CMM Mg-Pv was the liquidus phase which was followed down temperature by Ca-Pv at pressures higher than 28 GPa and Fp was absent in super solidus conditions. Differentiation in a deep magma ocean was examined by crystal fractionation of Mg-Pv Fp and Ca-Pv for CMM and peridotitic bulk silicate Earth models. Mass balance calculation indicated that subtraction of about 40% Mg-Pv and 2% Ca-Pv from CMM yielded a residual melt with characteristics of fertile upper mantle composition. The fractionated Mg-Pv and Ca-Pv would pile up to a depth ca. 1400 km from bottom of the mantle and might be characterized as an enriched and possibly heat producing reservoir by the higher capability of Ca-Pv to accommodate large cations such as rare earth elements and alkaline elements than melt. The observed effect of pressure on element partitioning suggested that better mass balance solutions might be obtained for higher pressure liquidus composition.

## 5. In situ X-ray observation using the KMA

In situ X-ray observation using synchrotron radiation has rapidly pushed the high pressure Earth science towards the exact sciences because the pressure value is determined continuously and uniquely from volume of a pressure marker via its equation of state. The pressure marker is usually mixed with the sample or put next to the sample. By being interfaced with synchrotron radiation versatility of the KMA has been expanded in various research fields such as phase equilibria reaction kinetics equation of state radiographic measurements etc. At the synchrotron facility SPring-8 two DIA-type presses [23] were installed on beam line BL04B1 to squeeze the Kawai-cell. One of them SPEED mkII [24] has exclusively been used to squeeze the Kawai-cell of SD cubes. Experimental methodology of in situ X-ray observation using the Kawai-cell were described to some extent elsewhere [25].

### *5.1. In situ X-ray diffraction studies on Fe GaN and $Fe_2O_3$*

We have first studied high pressure polymorphism of iron by means of in situ X-ray diffraction. Our aim of the study was to clarify the stability of β-phase the 5th polymorph of iron which was claimed to be stable under the conditions higher than 35 GPa/1500 K by several groups from the experiments using the DAC [26]. Our experiments up to 44 GPa and 2100 K confirmed progression of the $\varepsilon \rightarrow \gamma$and the reverse transitions on increasing and decreasing temperature respectively and the presence of β-phase was thus excluded [27].

We successively examined phase relations in GaN [28] and $Fe_2O_3$ [29]. In GaN onset of the wurtzite→rocksalt transformation was observed at conditions of 54 GPa/300 K and 51.5 GPa/ 750 K.suggesting a negative dP/dT slope. However it was observed that the rocksalt phase persisted for 90 min at 48.9 GPa/850 K. Therefore the reverse transformation is rather sluggish and more examination is needed to decide the equilibrium phase boundary. Electrical resistance of GaN was tens of mega ohms up to ca. 62 GPa at 300 K and no remarkable change in resistance was observed on the wurtzite – rocksalt transformation.

Phase equilibria of $Fe_2O_3$ were examined up to 58 GPa and 1400 K. Hematite (phase I) transformed successively into the $Rh_2O_3$ II type structure (II) and an orthorhombic phase (III) with increasing pressure. The phase boundaries for the I - II and II - III were defined to be $P(GPa) = -0.015\, T(K) + 44.2$ and $P(GPa) = -0.005\, T(K) + 48.7$ respectively by observing both the forward and backward reactions in situ. It should be noted that the transformations were observed only at temperatures higher than 500 K and hematite persisted metastably at least up to 58 GPa at 300 K. The electrical resistanceof hematite on the other hand decreased monotonically from tens of mega ohms at 10 GPa to a few ohms with increasing pressure and then plateaus at ca. 60 GPa as shown in Figure 8. The isostructural electrical resistance reduction strongly suggests the occurrence of a Mott transition. The plateau point of electric resistance 60 GPa is usable as a pressure fixed point at room temperature.

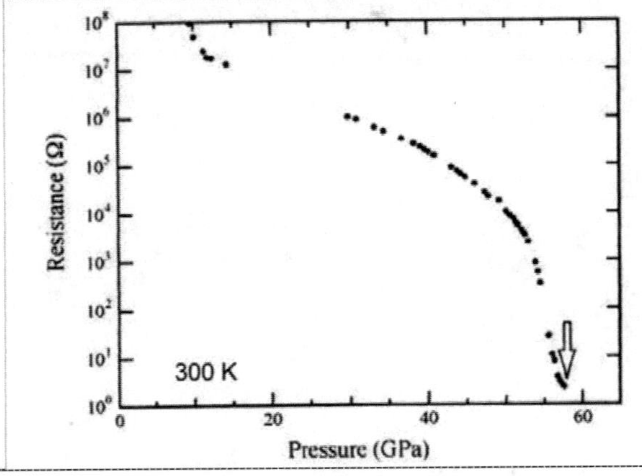


**Figure 8.** Change in electrical resistance of $Fe_2O_3$ (hematite) with pressure at 300 K.

*5.2. Post-perovskite transformation in meta-germanates*

The lowermost 200-300 km region of the mantle called the D" layer is characterized by various aseismic anomalies such as velocity jumps [30] anisotropy [31] anti-correlation between shear velocity VS and bulk sound velocity VC [32] and presence of ultra-low velocity zone at the base [33]. These enigmatic features would be refection of active exchange of energy and material between the mantle and the core. Therefore understanding of the D" layer should be of essential importance to clarify dynamics and evolution of the Earth.

In 2004 Murakami et al. [34] and Oganov and Ono [35] discovered the transformation of $MgSiO_3$ perovskite (Pv) into the $CaIrO_3$ structure (the post-perovskite (PPv) transformation) at ca. 120 GPa and 2200 K using the laser-heated DAC. The post-perovskite transformation should play an important role in formation and structure of the D" layer. Therefore detailed knowledge on the transformation is of urgent need. Especially the slope of the phase boundary dP/dT would give some insight to the thermal structure through the D" layer and to estimate of thermal flux from the core [36].

As the P-T conditions for the PPv transformation in $MgSiO_3$ are out of the ability of the KMA at present we examined the PPv transformation in $MgGeO_3$ [37] and MnGeO3 [38] which are regarded as the best analogs of $MgSiO_3$.

Cross section of a sample assembly adopted in $MgGeO_3$ experiments is shown in Figure 9. Fine mixture of $MgGeO_3$ilmenite + 0.1Au (in weight) was put in a cylindrical TiB2 heater which is set at the center of octahedron of MgO + 5% $Cr_2O_3$. Temperature was measured by a thin W3%Re/W25%Re thermocouple whose junction was at the center of the sample. By adopting the $TiB_2$ heater we could see through the sample and the thermocouple by the CCD camera. Pressure value was determined from the volume of gold based on the Anderson et al.'s [39] Au scale.

The transformation from Pv to PPv was first confirmed at 63.2 GPa and 1325 K as shown by the diffraction profiles in Figure 10(A). Then we tried to observe the reverse transformation by decreasing pressure by reducing the applied load and keeping temperature constant at 1323 K. Growth of Pv was not recognized until pressure went down to 60.5 GPa as clarified by the successive profiles in Figure 10 (B). We carried out more few runs to recognize growth either PPv or Pv from the counter phases over the conditions up to 74 GPa and 2200 K. Figure 11 shows P-T plots of the X-ray observations relevant to define the phase boundary between Pv and PPv. The phase boundary between Pv and PPv was defined by passing through the Pv-growth PPv-growth and the coexistence points and expressed by T(K) = 177P(GPa) – 9677 with dP/dT = 5.6 MPa/K.

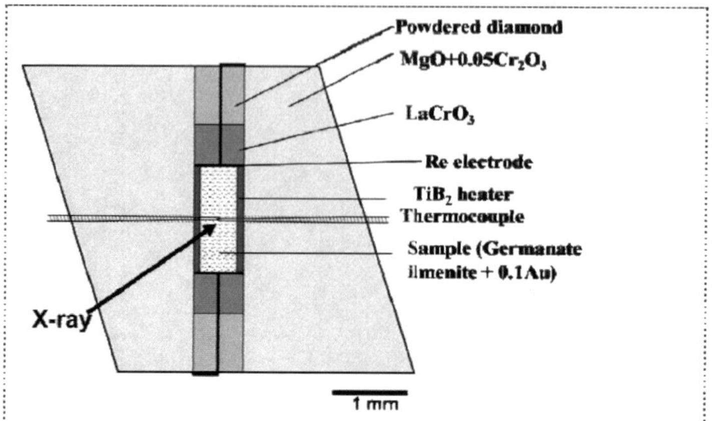


**Figure 9.** A schematic drawing of the sample assembly for experiments for the post-perovskite transformation in $MgGeO_3$.

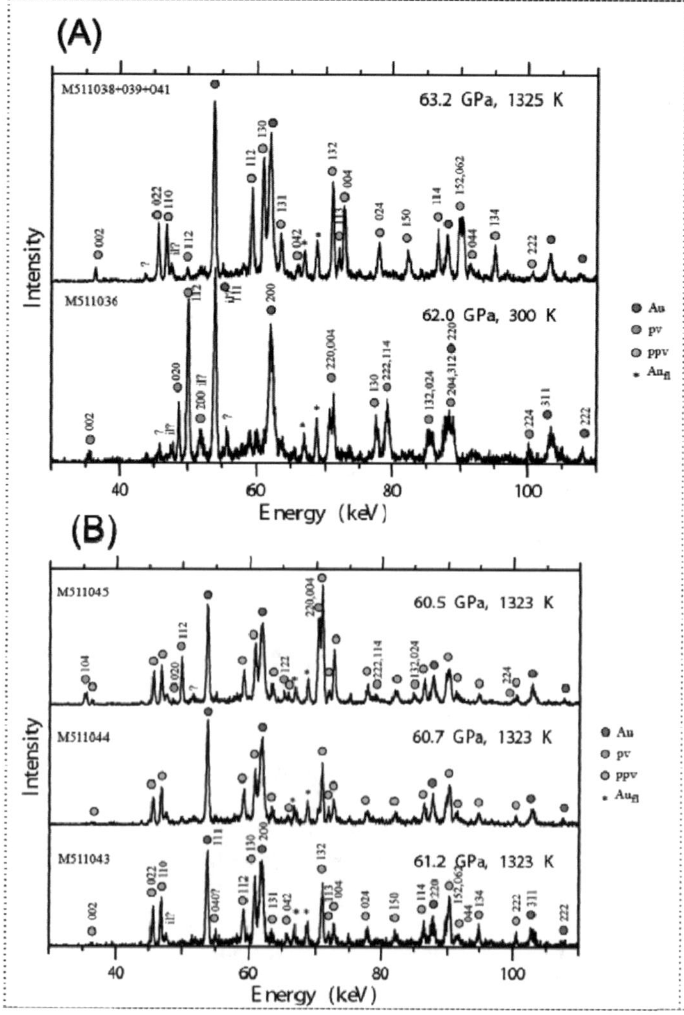


**Figure 10.** Diffraction profiles of run M511. Profiles in (A) demonstrate the growth of post-perovskite phase from the perovskite at 63.2 GPa and 1325 K and a series of profiles from the bottom to top in (B) do the reverse transformation at 60.5 GPa and 1323 K.

The dP/dT value is slightly smaller than those estimated for $MgSiO_3$ [36]. The smaller value sets the D" discontinuity deeper and so called PPv lens formed by the double-crossing of the phase boundary and the geotherm [40] would be present only in relatively cold region such as beneath mantle downwellings. The slope of 5.6 MPa/K suggests a large heat flux of 16 TW across the D" layer assuming thermal conductivity of 10 Wm-1K-1 [41]. Nevertheless if we adopt other Au pressure scale such as Tsuchiya [42] instead of Anderson et al.'s [38] our phase boundary shifts towards higher pressure by 3.5 GPa and the dP/dT increases to 8.7 MPa/K. Therefore establishment of reliable pressure scale is fatally important.

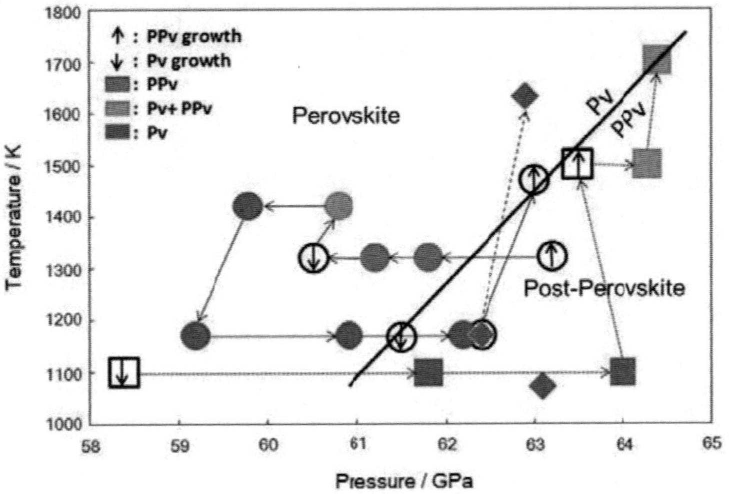


**Figure 11.** P-T plots of the X-ray observations relevant to define the phase boundary between Pv and PPv (see text). Circles squares and diamonds denote the different runs. Growing phases and phases present are indicated by the arrows and color markers. After [37].

The Pv-PPv phase equilibria in $MnGeO_3$ [38] were also examined by in situ X-ray diffraction method as in $MgGeO_3$. The phase boundary was determined to be P(GPa) = 39.2 + 0.013T(K) based on Tsuchiya's Au scale [42] which is located around ca. 10 GPa lower pressures than that of $MgGeO_3$ with a lager dP/dT value [38].

*5.3.High spin-low spin transition in ferropericlase ($Mg_{0.83}Fe_{0.17}$)O*

High spin-low spin transition of $Fe^{2+}$ in $(Mg_{1-x}Fe_x)$OFp occurring under lower mantle conditions has been attracted special attention because the transition is considered to substantially affect geophysical and geochemical processes [43]. The transition can be detected by observing an anomaly in the compression curve due to a drastic reduction of effective ionic radius of Fe2+ accompanied with the transition [44]. We selected Fp with a composition $(Mg_{0.83}Fe_{0.17})$O because the composition is a counter part of magnesian perovskite in the assemblage produced by dissociation of $(Mg_{0.9}Fe_{0.1})_2SiO_4$ spinel [12].

We acquired pressure (P)-volume (V) data at 300 and 700 K and up to 90 GPa employing the sample assembly shown Figure 9. Pressure was determined from volume of Au mixed with the $(Mg_{0.83}Fe_{0.17})$OFp powder based on Anderson et al.'s scale [39]. The acquired P-V data are shown in

Figure 12. Errors in pressure determination caused those of Au volume and temperature fluctuation were within ± 0.4 GPa and those in volume determination were typically within ± 0.05 %.

We tried to fit the 3rd order Birch-Murnaghan equation of state (B-M EoS) with $K_0$' = 4 to the data up to 40 GPa which were successfully performed as shown by the solid curves. And the resultant zero pressure values of volume V0 and bulk modulus $K_0$ were in good agreement with the literature values [45] which are noted as HS values in the insertions of the figures. However volume data higher than 50 GPa at 300 K and those higher than 55 GPa at 700 K show gradual downward

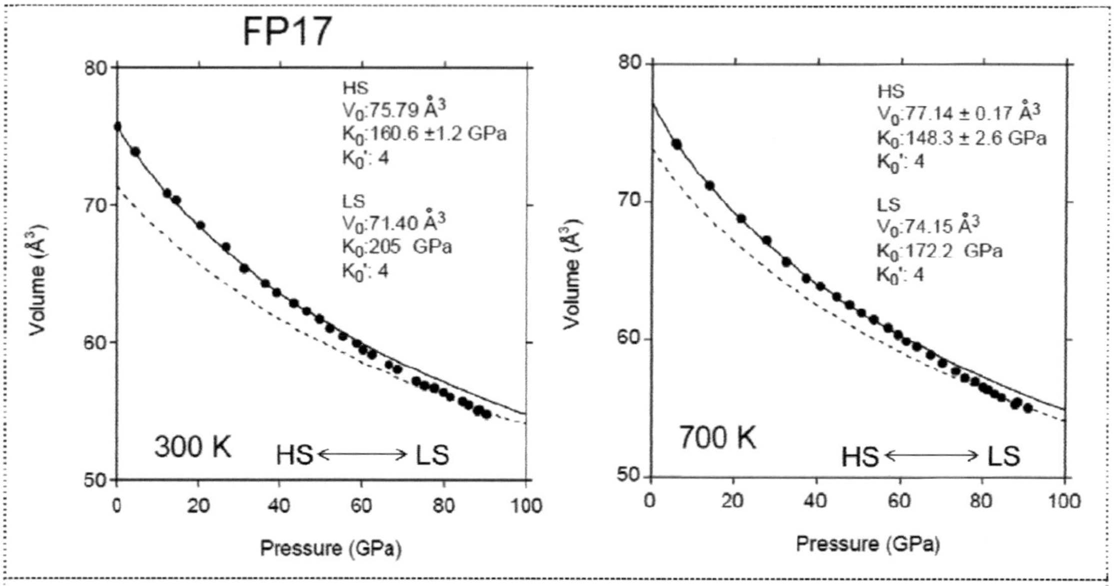


**Figure 12.**Compression data of $(Mg_{0.83}Fe_{0.17})O$Fp and fitting the 3rd order Birch-MunaghanEoS's. Fitting up to 40 GPa assuming K0' = 4.0 (solid curves) yields reasonable $V_0$ and $K_0$ values for both 300 and 700 K data. The data for higher pressures clearly deviate toward lower direction indicating occurrence of the spin transition. For low spin regime of 70-90 GPa at 300 K and 80-90 GPa at 700 K we fit EoS's with parameters noted as LS in the insertions which are shown by the broken lines.

deviation from the solid curves with increasing pressure indicating onset of the spin transition. Assuming that the transition completes at 70 GPa at 300 K and 80 GPa at 700 K we tried to fit the B-M EoS with $K_0$' = 4 for the data higher than these pressures. The resultant $V_0$'s and $K_0$'s are noted in the insertions as LS values in the figures and the corresponding EoS's are shown by broken curves. The $V_0$ and $K_0$ at 300 K for the low spin regime are larger and smaller than those of Lin et al. [46] respectively in the same composition. Our data suggest that the mixed spin region shifts toward higher pressure and expands with increasing temperature in accord with theoretical prediction [47]. Nevertheless it is evident that acquisition of LS data up to higher pressure is required to specify the LS state in more detail.

## 6. Recent pressure generation and future perspectives

By adopting SD anvils the attainable pressure of the KMA has increased year by year since 1998 [19]. Figure 13 shows our recent performance of pressure generation carried out by means of in situ X-ray observation at SPring-8 using anvils with 1.0 mm truncation. The maximum attainable pressure is reaching 100 GPa and high pressure and temperature experiments up to 90 GPa have become our ordinary jobs.

One of our urgent objects is to clarify the PPv transformation in detail just like done in the post-spinel transformation. In order to accomplish the object we have to extend the accessible pressure at least by 20 GPa. The systematics between the hardness of anvil materials and the maximum attainable pressures for WC SD and single crystal diamond and the trends shown in Figure 13 suggest possibility to produce pressures high than 100 GPa. However innovation of SD is of essential importance to produce still higher pressures. In this context binder-free nano-polycrystalline diamond (NPD) recently developed by Irifune et al. [48] expands our dream because itsKnoop hardness reaches 140 GPa

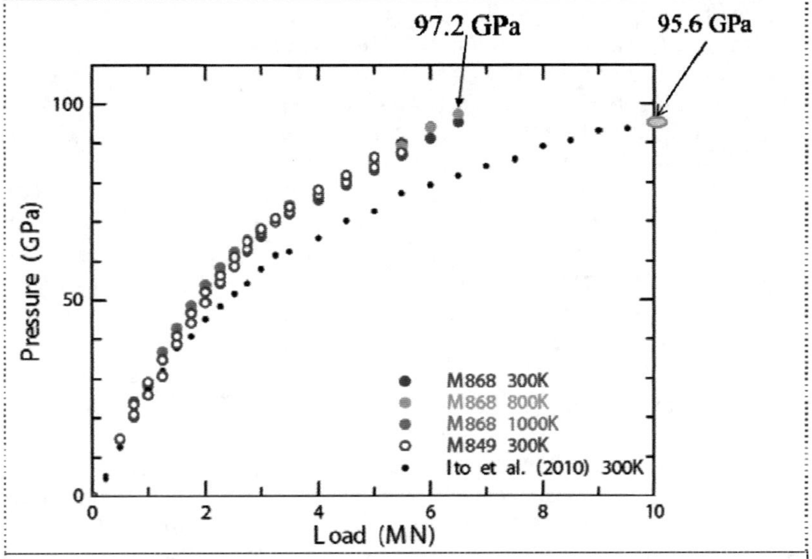


**Figure 13.** Plots of generated pressure vs. applied load. Truncation was 1.0 mm and pressure values are based on Tsuchiya's Au scale [42].

**Acknowledgements**

I am very grateful to all those who have been in contact with me on the high-pressure research for their guidance help and collaboration. I would like to dedicate this article to the late professors N. Kawai S. Akimoto and T. Matsumoto and also to the late super technician K. Tanaka .

**References**

[1] Jeffreys H 1939 *Mon. Not. R. Astr. Soc. Geophys. Sup.* **4** 498.
[2] Bullen E 1940 *Bull. Seis. Soc. Amer.* **30** 235.
[3] Birch F P 195) *J. Geophys. Res.* **57** 227.
[4] Dziewonski A E and Anderson D L 1981 *Phys. Earth Planet. Inter.* **25** 297.
[5] Ringwood A E and Major A 1970 *Phys. Earth Planet Inter.* **3** 89.
[6] Akimoto S and Fujisawa H 1968 *J. Geophys Res.* **73** 1467.
[7] Suito K 1972 *J. Phys. Earth* **20** 225.
[8] Mao H-K and Bell P M 1971 *Carnegie Institute Year Book* **70** 176.
[9] Kawai N 1966 *Proc. Jpn. Acad.* **42** 385; Kawai N and Endo S 1970 *Rev. Sci. Instr.* **41** 425; Kawai N Togaya M and Onodera A 1973 *Proc. Jpn. Acad.* **49** 623.
[10] Shatskiy A Fukui H Matsuzaki T et al. 2007 *Am. Mineral.* **92** 1744.
[11] Liu L 176 Nature 262 770-772.
[12] Ito E and E. Takahashi E 1989 *J. Geophys. Res.* **94** 10637
[13] Sawamoto H Weidner D J Sasaki S and Kumazawa M 1984 *Science* **224** 749; Yeganeh-Haeri A Weidner D J and Ito E 1989 *Science* **243** 787.

[14] Ito E and Katsura T 1989 *Geophys. Res. Lett.* **16** 425.
[15] Fukao Y Widiyantoro S and Obayashi M 2001 *Rev. Geophys.* **39** 291.
[16] Verhoogen J 1965 *Phil. Trans. R. Soc. London* **A258** 276.
[17] Sung C.-M. and Sung M 1996 *Mat. Chem. Phys.* **43** 1.
[18] Irifune T Koizumi T and Ando J 1996 *Phys. Earth Planet. Inter.* **96** 147.
[19] Ito E Kubo A Katsura T Akaogi M and Fujita T 1998 *Geophys. Res. Lett.* **25** 821.
[20] Trønes R G and Frost D J 2002 *Earth Planet Sci. Lett.* **197** 117.
[21] Ohtani E and Sawamoto H 1987 *Gephys. Res. Lett.* **14** 733.
[22] E. Ito E Kubo A Katsura T and Walter M J 2004 *Phys. Earth Planet. Inter.* **143-144** 397.
[23] Osugi J Shimizu K Inoue K et al. 1964 *Rev. Phys. Chem. Jpn.* **34** 1.
[24] Katsura T Funakoshi K Kubo A et al. 2004 *Phys. Earth Planet. Inter.* **143-144** 497.
[25] Ito E 2007 Theory and Practice-Multianvil Cells and High-Pressure Experimental Methods In: Treatises on Geophysics 2 edited by G. D. Price and G. Schubert Elsevier B. V. 197.
[26] Boehler R 1993 *Nature* **363** 534; Andrault D Fiquet G Charpin T and le Bihan T 2000 *Am. Mineral.* **85** 364; Saxena S K and Dubrovinsky L S 2000 *Am. Mineral.* **85** 372.
[27] Kubo A Ito E Katsura T et al. 2003 *Geophys. Res. Lett.* **30** 1126 doi: 1029/200GL06394.
[28] Ito E Katsura T Aizawa Y et al. 2005 In: Advanced in High-Pressure Technology for Geophysical Applications edited by J. Chen Y. Wang T. Duffy and L. F. Dobrhinetskaya Elsevier B.V. 451.
[29] Ito E Fukui H Katsura T et al. 2009 *Am. Mineral.* **94** 205.
[30] Wysession M E Lay T Revenaugh J et al. 1998 In: The D" discontinuity and its implications edtied. M.E. Wysession E. Knittle and B. A. Buffet 28 273 AGU Washington D.C.
[31] Pinning M and Romanowicz B 2004 *Science* **303** 351.
[32] Masters G Laske G Bolton H Dziewonski A 2000 In: Erath's deep interior Mineral physics and tomography From the Atomic to the Global Scale edited by S. Karato A. M. Forte R. C. Liebermann G. Masters and L. Stixrude 117 Geophys. Mono.AGU Washington DC.63.
[33] Lay T Williams Q and Garnero J 1998 *Nature* **392** 641.
[34] Murakami M Hirose K Kawamura K et al. 2004 *Science* **304** 855.
[35] Oganov A R and Ono S 2004 *Nature* **430** 445.
[36] van der Hilst R. D. de Hoop M V Wang P et al. 2007 *Science* **315** 1813.
[37] Ito E Yamazaki D Yoshino T et al. 2010 *Earth Planet. Sci. Lett.* **293** 84.
[38] Yamazaki D Ito E Katsura T et al. 2011 *Am. Mineral.* **96** 89.
[39] Anderson O Isaak D G and Yamamoto S 1989 *J. Appl. Phys.* **65** 1534.
[40]Herunlund J W Thomas C and Tackley P J 2005 *Nature* **434** 882.
[41] Stacey F 1992 Physics of the Earth 3rd ed. Brookfield Brisbane Australia p. 513.
[42] Tsuchiya T 2003 *J. Geophys. Res.* **108** dio:10.1029/3002JB002446.
[43] Lin J-F and Tsuchiya T 2008 *Earth Planet. Sci. Lett.* **170** 248.
[44] Burns R G Mineralogical Applications of Crystal Field Theory Cambridge Univ. Press p. 224.
[45] Anderson O L and Isaak D G 1995 Elasticity of Minerals Glasses and Melts In: Mineral Physics & Crystallography A Handbook of Physical Constants edited by T. J. Ahrens AGU Reference Shelf 2 64.
[46] Lin J-F Struzhkin V V Jacobsen S D et al. 2005 *Nature* **436** doi:10.1038/nature03825.
[47] Tsuchiya T Wentzcovitch R M da Silva C R S et al. 2006 *Phys. Rev. Lett.* **96** 198501-198504.
[48] Irifune T Kurio S Sakamoto S et al. 2003 *Nature* **421** 599; Irifune T 2011 ABSTRACTS AIRAPT-23 52.

Journal of Physics: Conference Series **377** (2012) 012002 doi:10.1088/1742-6596/377/1/012002

# Transparent Large Anvil Press for In-situ Raman and Laser Heating*

**Amartya Sengupta[1,†], Young-Jay Ryu[2] and Choong-Shik Yoo[1,2,‡]**

[1] Institute for Shock Physics, Washington State University Pullman, WA 99164, USA
[2] Department of Chemistry, Washington State University, Pullman, WA 99164, USA

**Abstract:** It has been shown that high-pressure polymorphs of simple materials often display novel thermal/mechanical/optical properties and can be recovered at ambient conditions with significant stability (or metastability). Therefore, high-pressure synthesis of such materials offers an avenue toward novel materials applications. However, the present high-pressure method of diamond-anvil cell (DAC) or large volume press (LVP) has a fairly limited use in this regard because of a small amount of sample or the lack of in-situ optical diagnostic, respectively. Hybridizing the merits of DAC and LVP, we have recently developed a simple yet unique device, called a Transparent Large Anvil Press (TLAP), which consists of an opposed diamond anvil cell and a Paris-Edinburgh press. Coupling with a wide range of optical diagnostics such as Raman spectroscopy and laser heating, the TLAP is capable of *in-situ* investigation and synthesis of a milligram quantity of cryogenic samples at high pressures and temperatures.

## 1. Introduction

Application of high pressure on materials has resulted in considerable change in its physical and chemical properties. Recent experiments have also demonstrated that high pressure and temperature is a very powerful tool to synthesize new stable and metastable phases of low Z elements [1-3]. For many materials, such high-pressure modifications can be conserved in metastable form at ambient conditions for a significant amount of time [4]. Discovery of novel properties like optical nonlinearity [5], high energy density [6] and super-hardness [7] of the high pressure polymers of several low Z compounds, specifically, $CO_2$, $N_2$, CO to name a few, have opened up new prospects of their applications in science and technology [8,9].

Almost all of the high pressure-high temperature synthesis experiments reported in the literature has been performed by using microscopic quantities of samples in diamond anvil cells (DACs) at formidable high pressure-high temperature conditions [10,11]. This restricts the characterization of the novel properties of the materials by using conventional techniques like differential scanning calorimetry, solid state NMR or other mechanical testing tools, which typically require large milligram quantities of samples.

* The work has been done at Washington State University in support of DARPA (W911NF-10-1-0081) and DTRA (HDTRA1-09-1-0041).
† Current Address: Department of Geosciences, Princeton University, Princeton, NJ 08544, USA, E-mail: amartyas@princeton.edu
‡ Author to whom all correspondence should be addressed: csyoo@wsu.edu

Large-volume presses (LVPs) such as a multi-anvil press or the more compact Paris-Edinburgh cell (PEC) even though more suitable for compressing milligram to gram quantities of sample to few tens of GPa, are mainly used for X-Ray and neutron diffraction measurements [12-14]. The main reason being they typically use opaque WC or sintered diamond anvil assemblies which severely restricts in-situ optical or spectroscopic diagnostics. Additionally, the extraordinarily large mass of the press components prevent efficient cryogenic loading of many gaseous samples at ambient conditions. This has also limited the types of samples that can be studied using LVPs to those which are either solids or liquids at ambient conditions.

There have been recent efforts which have tried to address the above issues in LVPs systematically; however in most of the cases, optical information from the sample chamber were limited to ruby fluorescence measurements. The techniques used either a sequence of NaCl window plugs [15] or optical fibers fixed into a ceramic tube with molten glass [16]. There have been other approaches too; using multi-anvil apparatus [17] or large sapphire ball cells using anvils ranging from 0.5 to 1 inch in diameter [18] and even large synthetic sapphire anvils [19]. However, these devices cannot be considered to be LVPs in the strict sense of the term, they were more of the pressure cell type.

On the other hand, there have been very few efforts to address the issue of cryogenic loading of samples inside LVPs; the most notable ones being that of Klotz et al. [20] where he successfully loaded deuterated $CH_4$ and $NH_3$ in a PEC at liquid nitrogen temperatures even though the technique had its safety concerns due to the amount of energy stored in the hydraulic fluid system. The other method was pioneered by Lipp et al. [21] where they thermally isolated the sample/gasket/anvil area and incorporated a leak-tight sample containment chamber which they used to load CO. However, this whole procedure required a long time because of the multiple steps involved and also used several resistive heaters to warm the LVP up to ambient temperature before the actual experiments could be started. Another point to note in both the above works is none of the LVP assemblies allowed optical access to the sample chamber.

In this work, we present a relatively simple, portable and unique approach to using the a modified version of the PEC using a separate cell assembly with either WC or sintered diamond (SD) seats and opposing diamond anvils in the Bridgman configuration. A key feature of this type of design is it allows efficient and prompt loading of samples under cryogenic conditions as well as provides optical access to the sample chamber for in-situ laser heating, Raman and other spectroscopy measurements. We demonstrate the application of this modified PEC, or Transparent Large Anvil Press (TLAP) by loading liquid $N_2$ and liquid $CO_2$ samples and by performing Raman spectroscopy measurements on them. Additionally, results of in-situ laser heating on $CO_2$ samples at pressures greater than 15 GPa using this assembly and efforts to recover novel phases of tetracyanoethylene (TCNE) are also presented. Both the experimental arrangement for optical measurements and the mechanical design of the complete PEC assembly are very general and can be adapted to a wide variety of different LVP configurations currently in use among numerous high pressure and geo-physical research communities.

## 2. Design of the TLAP & optical arrangement

In our mechanical design of the cell assembly, we use the systematics of the PEC, which operates by injecting hydraulic fluid into the reservoir of a piston/cylinder cavity which, in turn, applies the force on the sample that is studied. The applied load is determined by the hydraulic fluid pressure and the cross-sectional area of the piston surface of the reservoir.

Following is the modification that we incorporated in a 200 ton VX2 type PE press for our experimental measurements. We introduce a separate cell assembly between the steel spacer of both the breech and the piston sides of the PEC (see Figs. 1, 2a) that allows the sample chamber to be portable and eases the procedure of cryogenic loading. This separate assembly which is made of Vascomax maraging steel and subsequently heat-treated after final testing is carried out, houses the

WC (supported by Vascomax maraging steel) or GE Compax SD (supported by WC) seats with opposing diamond anvils in the Bridgman configuration (see Figs. 2 b,c).

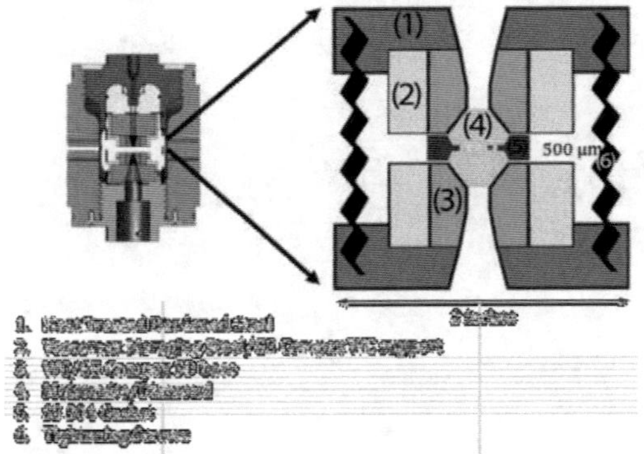


**Figure 1.** Internal schematics of the TLAP. The portable arrangement on the right has been designed in-house at WSU for portability and optical access inside the TLAP. The center portion of the PEC (white area between the WC supports [13] in the left figure) is where the portable arrangement on the right is placed after loading the sample. After this, the hydraulic arrangement of the PEC is used to increase the pressure. For further details, please see text.

The outer dimensions of this portable cell assembly is custom made to fit inside our cryogenic gas loading device, that has been demonstrated to be very effective in loading gaseous samples in DACs by condensing them to a liquid under appropriate P-T conditions [22]. The WC (SD) seats have funnel shaped access holes drilled in the center which merge into a 1 mm straight bore before opening up into an inverted cone that had been precisely cut to accept the diamond (moissanite) anvils (see Fig. 2d). The diamond (moissanite) anvils having a height of 2 mm are 45° crown angled and are obtained from Almax Industries (Charles and Colvard).

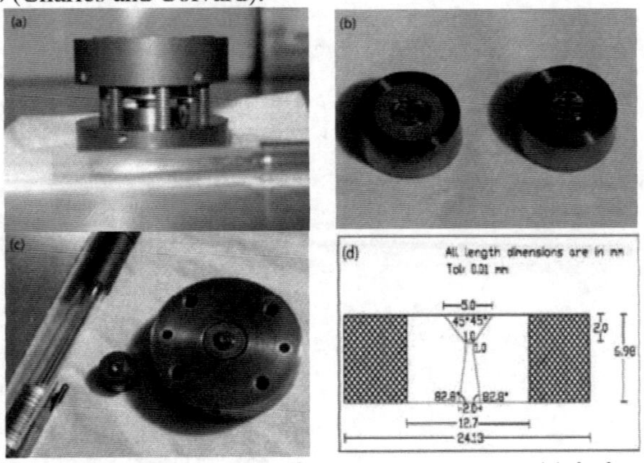


**Figure 2.** Images and drawings of the cell and seat arrangements. (a) in-house designed portable, small cell assembly; (b) one of the typical seats used in the measurements, WC supported by maraging steel (pressure-fitted and heat-treated in-house) with the mounted diamonds; (c) commercial SD seats supported by WC used for $CO_2$ measurements; (d) the diagram of the cone arrangement that was drilled in WC or SD for housing the diamond anvils. For further details, see text.

Finally, after the alignment of the diamonds is complete so that they are centered with the entire seat arrangement, STYCAST 2850T is used as an epoxy to hold them in place even under cryogenic conditions. Initially to close the cell after the sample has been loaded, the screws are used to apply the pressure. To avoid any ambiguities, there is a notch that is machined in into the top and bottom parts of this portable cell assembly so that it can be perfectly aligned with the backing plates both on the piston and the breech side of the PEC that would allow us to evenly distribute the hydraulic load across the sample area during actual measurements.

We employ a standard confocal micro-Raman system in back-scattering geometry to collect the Raman signal with some differences as mentioned below. For ease of incorporation of the TLAP system in our current optical table, we use 200 μm core multi-mode optical fibers to collect and deliver the Raman signal to the Princeton Instruments Acton 2500 spectrometer (see Fig. 3).

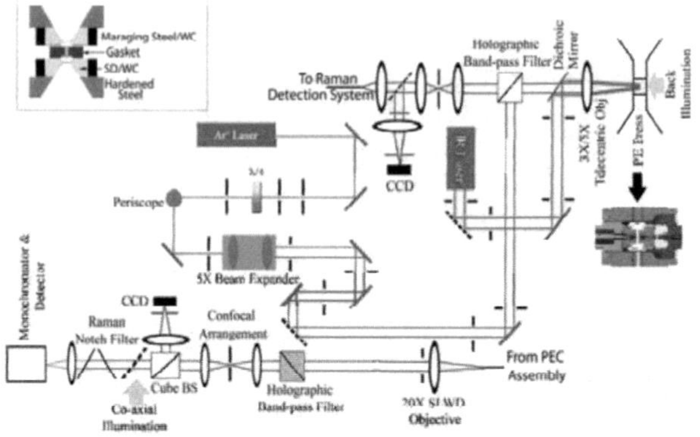

**Figure 3.** Experimental arrangement used for in-situ laser heating and Raman spectroscopy inside the TLAP. The inset shows the internal schematics of the small cell that is used in the measurements.

Stainless steel gaskets having initial thicknesses ~ 300 μm were used to contain the samples ($N_2$ or $CO_2$ with multiple ruby chips) in a hole of diameter 2.5 mm. Long WD Mitutoyo objectives were used to deliver (collect) the optical signal to (from) the sample chamber which was combined with a lens system to image the sample chamber onto a camera (magnification: 3×, WD = 75 mm).

### 3. Experimental Results

Fig. 4 shows a typical plot of correlation of the applied load with the pressure inside the sample chamber (for TCNE samples) as obtained from the ruby fluorescence technique. It is seen that we are operating the PEC in the linear regime and so the stresses are distributed evenly.

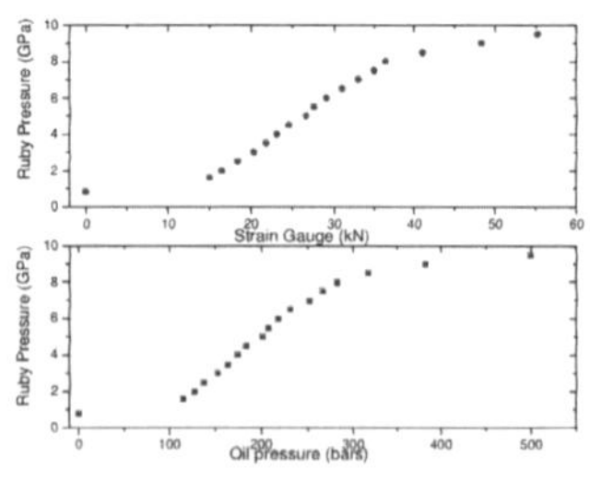

**Figure 4.** Plot of load and oil pressure vs Ruby fluorescence measurements. The calculated ratio F/A (assuming the area of compression to be that of the diamond/moissanite table) shows the pressure multiplication factor, $m$ ~ 2.5; which is in stark contrast to the assumption of a parabolic pressure profile.

*3.1 Nitrogen ($N_2$ measurements)*

A 300 μm thick spring steel gasket was pre-indented using the PEC to about 280 μm and a hole of diameter 1.5 mm was drilled using a standard drilling machine. A few micrometer sized ruby chips were scattered inside the cell for in-situ pressure measurements. Liquid $N_2$ was then cryogenically loaded onto the maraging steel-WC-diamond anvil cell assembly by immersing the cell into a bath of liquid $N_2$. All the four screws were used to close the cell inside the liquid and the loading pressure after the cell had warmed up to room temperature was ~ 3kbars. The cell assembly was then inserted into the PEC and the pressure was increased using the hydraulic mechanism. The oil pressure and

Journal of Physics: Conference Series **377** (2012) 012002 doi:10.1088/1742-6596/377/1/012002

subsequently the load on the sample chamber could be controlled with a high degree of accuracy, which allowed in-situ ruby fluorescence measurements to estimate the pressure inside the sample chamber.

Fig. 5 shows a succession of the in-situ Raman spectra of Nitrogen starting from 3 GPa to about 14 GPa. It shows the fact that nitrogen freezes around 5 GPa in the δ phase which shows the two stretch modes ($\nu_1$, $\nu_2$) unique to this phase [23]. Their frequency increases steadily as the pressure is increased as does the relative spacing between the two Raman peaks. The sample size as is shown in the sample microphotograph (see Fig. 4) is about 1.5 mm in diameter and we estimate that the thickness of the sample at the highest pressures to be of the order of 100 μm. Because of the large area of the sample chamber, we are also able to make highly accurate spatially resolved ruby fluorescence data to know the actual pressure inside the sample chamber.

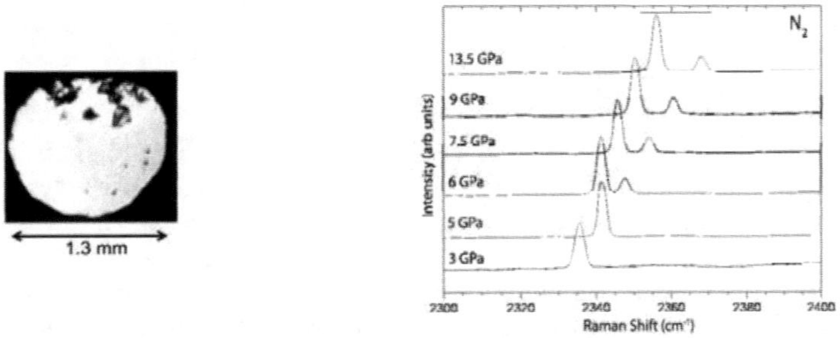


**Figure 5.** Sample microphotographs for $N_2$ inside the modified PEC assembly at a pressure of around 6 GPa. The scattered particles are ruby crystals. The right panel shows the successive Raman spectra of $N_2$ starring from a liquid at the loading conditions to about 14 GPa.

Further increase of pressure beyond 14 GPa resulted in fracturing of the anvils accompanied by a significant loss of pressure. The post-experimental observation showed that the WC seats have fractured radially which is not unexpected since WC anvils are known to fail around 10 GPa.

*3.2 Carbon Dioxide ($CO_2$) measurements*

To overcome the limitations of using WC seats, the measurements on $CO_2$ were taken using the WC-SD-diamond anvil cell assembly. After preparing a gasket on the same lines as above, $CO_2$ samples were loaded in the small cell assembly from a liquid by condensing $CO_2$ gas to -35 °C and 15 atmospheres. Besides ruby chips, a thin Pt foil (10 μm) was placed in the sample chamber to heat the $CO_2$ sample using laser powers around 10-30 W. The time taken to complete the entire process from starting the condensation of the gas to a liquid followed by subsequent loading of the liquid sample inside the cell and then warming up the cell to room temperature takes about 30-35 minutes.

The cell assembly was then inserted into the PEC and the pressure was increased using the hydraulic mechanism as in the earlier measurements. Figs. 6b and 6a show the microphotograph of the sample chamber and results of Raman measurements on this sample starting from as low as 3 kbars which show the onset of transformation of $CO_2$ from a liquid to the 'dry ice' form ($CO_2$-I) starting around 13 kbars and finally transforming to a mix of $CO_2$-I and III above 130 kbars [24, 25]. The pressure was measured using the ruby fluorescence technique and it was found that there is no appreciable pressure gradient across the sample at these pressures.

We also did in-situ laser heating of $CO_2$ inside the modified PEC at pressures around 16 GPa when the mix of $CO_2$-I and III transforms to $CO_2$-IV and II [26] along the direction of decreasing temperature gradient. Fig 6c shows the Raman spectrum of this transformation. Due to the large sample chamber in the TLAP, measuring the Raman spectrum along the temperature gradient was trivial as compared with our earlier laser heating measurements inside DACs [22].

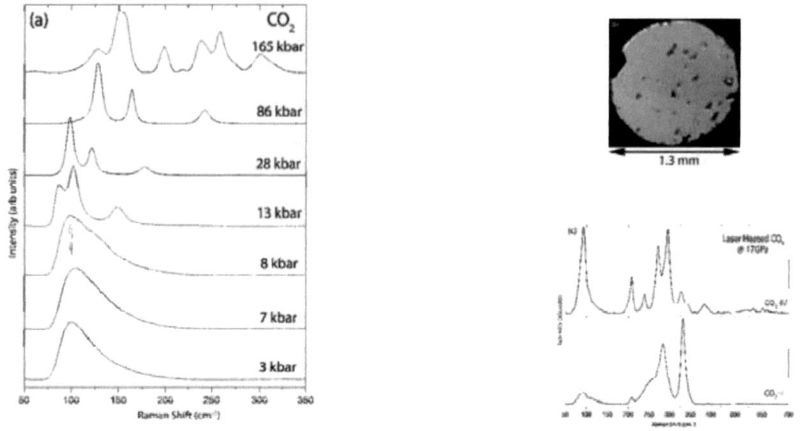


**Figure 6.** Experimental results using $CO_2$ as the sample in the modified PEC assembly. (a) Successive Raman spectra of $CO_2$ using our modified PEC assembly. (b) $CO_2$ sample microphotograph at 3 kbars, (c) Raman spectra of laser heated $CO_2$ around 17 GPa inside the TLAP.

### *3.3 Tetracyanoethylene (TCNE) measurements and sample recovery*

Owing to its high electron affinity, TCNE molecule forms many charge transfer complexes with a variety of electron donors. These charge transfer complexes exotic electrical, magnetic and other chemical properties [27, 28]. Interesting TCNE-based compounds have also been synthesized [29] wherein the magnetization states can be switched using visible light of different wavelengths. TCNE is also known to polymerize under different physical conditions, which can find applications in several electronic devices [30]. Figure 7 summarizes our results on Raman measurements using WC anvil assembly and our efforts on recovering the novel polymeric phase. The measurements show that TCNE undergoes amorphous transformations on pressure release to give very broad graphitic-like D-G band Raman spectra.

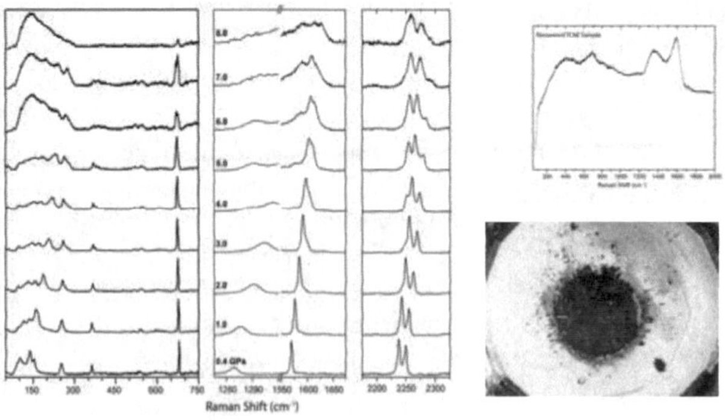


**Figure 7.** TCNE sample recovery. (Left) Raman spectra of TCNE from ambient conditions to about 8 GPa, when it starts polymerizing. The ruby was at the edge of the sample and so the sample pressures would be at least 15-20% higher than labeled. (Right Top) Raman spectra of recovered TCNE sample after pressure cycling, indicating a possible amorphization. (Right Bottom) Sample microphotograph of the recovered TCNE sample outside the TLAP. The sample is transparent to begin with, however, it becomes black on polymerization and remains black on subsequent sample recovery.

Journal of Physics: Conference Series **377** (2012) 012002 doi:10.1088/1742-6596/377/1/012002

## 4. Conclusions

We have successfully demonstrated an optically transparent anvil design assembly that could be used for in-situ laser heating, optical spectroscopy measurements and also efficiently load condensed gas samples inside the sample chamber for use inside LVPs like the PEC. Our TLAP design permits high quality imaging of the samples both at ambient conditions and under pressure inside the modified PEC, which is necessary while doing laser heating or photo-catalysis measurements. Pressures in excess of 15 GPa have been achieved using our current design of WC-SD-diamond anvil assembly with macroscopic quantities of samples several mm in diameter, which will allow the convenient use of routine characterization tools like DSC and NMR spectroscopy on the recovered samples.

## 5. References


1. Yoo C S and Nicol M J 1986 *J. Phys. Chem.* **90** 6726
2. Sihachakr D and Loubeyre P in *Abstracts of Joint 19th AIRAPT – 41st EHPRG Int.Conf. on High Pressure Science and Technology*, S8P144 (Bordeaux Univ.,Bordeaux, 2003).
3. Eremets M I *et al. Abstracts of Joint 19th AIRAPT – 41st EHPRG Int. Conf. on High Pressure Science and Technology*, S10-11 (Bordeaux Univ.,Bordeaux, 2003).
4. McMillan P F 2002 *Nature Materials* **1** 19
5. Iota V, Yoo C S and Cynn H 1999 *Science* **283** 1510
6. Barbee T W III *et al.* 1997 *Phys. Rev. B* **56** 5148
7. Yoo C S *et al.* 1999 *Phys. Rev. Lett.* **83** 5527
8. Lipp M J *et al.* 2005 *Nature Materials* **4** 211
9. Mailhiot C, Yang L H and McMahan A 1992 *Phys. Rev. B* **46** 14419
10. Jayaraman A 1986 *Rev. Sci. Instrum.* **57** 1013
11. Eremets M I *et al.* 2001 *Nature* **411** 170
12. Besson J M *et al.* 1992 *Phys. Rev. B* **45** 2613
13. Klotz S, Hamel G and Frelat J 2004 *High Press. Res.* **24** 219
14. Klotz S *et al.* 1995 *Appl. Phys. Lett.* **66** 1735
15. Drickamer H G and Balchan A S in *Modern Very High Pressure Techniques*, edited by R.H. Wentorf (Butterworths,Washington, 1962).
16. Eremets M in *High Pressure Experimental Methods* (Oxford University Press, Oxford, 1996).
17. Secco R A and Balog P S 1998 *Rev. Sci. Instrum.* **69** 2802
18. Daniels, W B *et al.*, in *Recent Trends in High Pressure Research* (Proceedings of the XIII AIRAPT International Conference on High Pressure Science and Technology), edited by A.K. Singh (Oxford & IBH Publishing Co., New Delhi, 1992)
19. Goncharenko I N 2004 *High Press. Res.* **24** 193
20. Klotz S *et al.* 1995 *Appl. Phys. Lett.* **67** 1188
21. Lipp, M J, Evans W J and Yoo C S 2005 *Rev. Sci. Instrum.* **76** 053903
22. Sengupta A and Yoo C S 2009 *Phys. Rev. B* **80** 014118
23. Bini R *et al.* 2000 *J. Chem. Phys.* **112** 8522
24. Schmidt J Wand Daniels W B 1980 *J. Chem. Phys.* **73** 4848
25. Hanson R C 1985 *J. Phys. Chem.* **89** 4499
26. Iota V and Yoo C S 2001 *Phys. Rev. Lett.* **86** 5922
27. Chetkina L A and Bel'skii V K 2002 *Crystallogr. Rep.* **47** 581
28. Dixnon D A and Miller J S 1987 *J. Am. Chem. Soc.* **109** 3656
29. Pejakovic D A, Kitamora C, Miller J S and Epstein A J 2002 *J. Appl. Phys.* **91** 7176
30. Yanagi H, Itoh H, Ueda Y and Ashida M 1994 *Appl. Phys. Lett.* **65** 1909

Journal of Physics: Conference Series **377** (2012) 012003 doi:10.1088/1742-6596/377/1/012003

# Explosive Laboratory Devices for Dynamic Shock Loading of Materials in Mbar Pressure Region

**Manjit Singh**
Terminal Ballistics Research Laboratory, DRDO, Ministry of Defense,
Sector 30, Chandigarh -160 030, INDIA

E-mail: manjitsingh78@yahoo.com

**Abstract.** Dynamic methods utilizing strong shock waves are used for investigating the properties of condensed matter at very high pressures and temperatures. Explosive driven plate impact tests have been conducted to find out the shock Hugoniot of materials up to pressure of 2 Mbar. Explosive cascading utilizing overdriven detonation waves in high explosives produced by flyer impact from the first stage to accelerate comparatively thin flyer plates to very high velocities have been demonstrated. Numerical simulations using Autodyn 2D/3D have been performed to optimize the various parameters in two stage explosive assemblies to accelerate flyer to velocities exceeding 10 km/s. Shock pressure up to 20 Mbar has been successfully measured using explosive assemblies in a convergent flow. Expertise and infrastructure available in TBRL for launching metal flyers to high velocity and monitoring its in-flight velocity and profile have been discussed in this paper.

## 1. Introduction

The study of material behaviour under extreme conditions of pressure and temperature is a fascinating subject for scientific community all around the globe for years and this quest is still not over. This helps to reveal the truth of natural occurrences from core of earth to the Sun and other planetary bodies. Although static methods using Diamond Anvil Cells and other such devices are employed to generate the high pressure in the materials, but the range of achievable conditions by these method is limited. In order to generate high temperature in DAC external heating devices such as lasers or conventional heaters are required, however, in shock compression the generation of temperature is part of the process itself. For experiments at pressures higher than those obtainable by the static methods, resort has to be taken to the inertial forces set up in the material itself by the acceleration accompanying strong shock waves.

The dynamic methods utilizing strong shock waves are used for investigating the properties of condensed matter at very high pressures and temperatures and are of particular interest to scientists and technologists working in the various disciplines of science and engineering. The shock waves which propagate in the materials at supersonic velocities, typically at several km/s, subject the material to high pressure from few hundred kbar to Mbar pressure region. The contact explosive detonation [1] produces triangular pressure pulse in target materials in the range of 200-500 Kbar, which is limited by the availability of high power explosives and shock impedance of materials. Shock waves can be generated in number of ways and the most common is the impact onto a target specimen by the projectiles accelerated by a variety of systems such as gas guns, electromagnetic rail guns, plasma guns, laser and explosives etc. The impact of explosively accelerated flyer on target produce uniaxial strain states and is widely used for the determination of shock hugoniot of the materials and equation of state [2, 3] of different materials at high pressure. The terminal velocity of flyer is a function of explosive to metal mass ratio and the Gurney energy of the explosive and a flyer can be launched to a maximum velocity of approx 5 km/s. An impact of a steel plate moving at 5km/s with the stationary steel plate generates a shock pressure of the order of 1.5 Mbar.

To generate still high pressures in materials, we have used explosive cascading technique [5], which allows to utilize overdriven detonation waves in high explosives produced by impact from the first stage flyer (driver plate)to accelerate comparatively thin flyer plates (driven plate) to very high velocities. The thicknesses of driver and driven plates, flight distances, thickness of the second stage explosive and the separation of the two plates from the explosives to maintain planarity are the few issues which need to be addressed for designing a cascaded plate impact experiment. Numerical simulation is the best tool to optimize the different dimensions of a cascaded explosive system. We have demonstrated two stage explosive system to launch the metal flyer to 7 -8 km/s in planer configuration and up to 15 km/s in a spherically convergent system [6]. Shock pressure up to 20 Mbar has successfully been measured experimentally across a converging shock wave.

Dynamic methods of investigating the properties of condensed matter at high pressure and temperature are based on the kinematics parameters of the shock waves. For experimental determination of different shock parameters the experimental techniques such as electrical shock arrival pins, fibre optical pins and streak photography have been used in the present studies.

## 2. Conventional Plate Impact Experiments

In this type of experimental setup a flat metal plate is launched to different velocities using the explosives, gas gun etc. On impact at target, this generates pressures much higher than those achievable by contact explosive detonation. The control over the dimensions of the flyer and careful design of target restricts the strain state in target to 1-D, which simplifies the use of theoretical models and mathematical calculations for determining the shock hugoniot of the materials. Even the pressure pulse in the target is approximately rectangular in shape, thus allowing broader area for deployment of sensors to monitor the shock propagation. We have established explosive driven plate impact technique to launch the 3-5 mm thick metal flyers of diameter 120 mm to velocities from 1-5 km/s. The explosive-metal assemblies were finalized after inputs from theoretical models, which predict the terminal flyer velocity to a reasonable degree of accuracy. These assemblies were used to in the laboratory to determine the shock hugoniot of the metals, ceramics, polymers and heavy metal alloys.

### 2.1 Theoretical Models for Terminal Flyer Velocity

The use of an explosive charge detonation to drive the metal fragments was first studied by RW Gurney [7, 8] in early 1940's. The theoretical model relied on the partition of an explosive energy between the KE of metal and the explosive gases driving it. This model yielded a simple relationship between the final metal velocity, the explosive energy, and the ratio of the mass of driven metal to that of loaded explosive. For a metal launched by unconfined explosive charge, the final flyer plate velocity is given as

$$V_p = \sqrt{2E}\left[\frac{1+\left(2\frac{M}{C}\right)^3}{6\left(1+\frac{M}{C}\right)}+\frac{M}{C}\right]^{\frac{-1}{2}}$$

where $\sqrt{2E}$ is Gurney velocity coefficient, which is specific to a particular explosive and is determined experimentally for each explosive.

Aziz et al [9] considered the metal flyer as a rigid piston driven by expanding detonation gases and predicted the final plate velocity as

$$V_p = \frac{D(Z-1)}{(Z+1)} \qquad \text{where} \qquad Z = \sqrt{1 + \frac{32}{27}\frac{C}{M}}$$

These two models ignored the effect of the side rarefaction losses and shock characteristics of the materials, which affect the energy coupling due to decrease in impulse and different shock impedances of materials and hence the plate velocities. These models predict the same terminal flyer velocity for different explosive-metal combinations with same charge to metal mass (c/m) ratio.

Backofen and Weickert [10] incorporated the forward momentum of detonation products and modified the Gurney relation as

$$V_p = \sqrt{(V_P)^2 + (u_P)^2}$$

where Vp is the plate velocity given by Gurney relation and up the initial particle velocity imparted to metal surface. The particle velocity has been found to be dependent on nature and thickness of explosive metal combination through a relation

$$u_p = \frac{\rho_e D}{1 + \frac{M}{C}} * \frac{1}{3 \cdot \left\{1 + \frac{9}{16}\left[\frac{(\rho_p C_o)}{(\rho_e D)}\right]\right\}}$$

where $\rho_p$, $\rho_e$ are densities of metal plate and explosive respectively, D, the detonation velocity and c0 is the sound velocity in the plate material.

These models do not account for the deformity of the flyer plate during motion, caused due to the side rarefactions so the experimentally measured velocities are different from those predicted by Gurney, Aziz or Backofen. Further flyer plate assembly design with introduction of air gap between the flyer and explosive were analysed and introduced by HS Yadav [11], which are presently used in our studies.

## 2.2 Design of Explosive-Metal Assemblies and Determination of Shock Hugoniot

The above discussed models were used to determine the dimensions of explosive assembly to achieve a particular flyer velocity. Plane wave initiated octol explosive was used to accelerate flyer plates of diameter 120 mm and thickness 3-5 mm to velocities from 1-5 km/s. Further refinement in the preliminary explosive assembly design was done through modelling and simulation using AUTODYN. Experiments were conducted to validate the results of simulation and to determine the velocity and planarity of the flyer prior to impact. Experimental techniques deploying optical, electrical and photographic methods have been established to record the flyer plate velocity and profile and post impact shock velocity in the target material.

The figure 1(a) below illustrates the use of optical pins to record the flyer plate velocity and profile. The explosive was initiated through a plane wave lens consisting of fast and slow octol and baratol explosives respectively, which accelerates the flyer to its terminal velocity. This flyer impacts the optical pins, producing a light flash, which has been recorded by the streak camera in fig 1(b). The pins at different height w.r.t target record velocity, whereas pins at the same height are used to calculate the impact symmetry. The flyer velocity and profile prior to impact was also recorded using electrical shock arrival pins and is shown in figure 1(c).

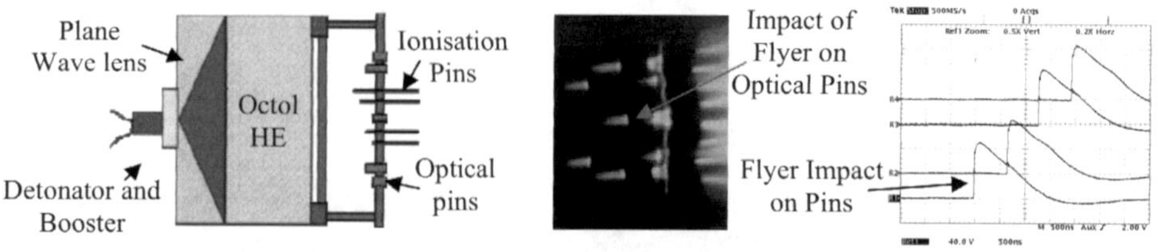


(a) (b) (c)

**Fig.1** .(a)Explosive assembly with optical and electrical pins to record velocity and planarity of flyer (b) Streak camera record and (c) oscilloscope record showing arrival of flyer prior to impact

These assemblies were used to determine the shock hugoniot of metallic, Ceramics, Rock Samples etc. The figure 2(a) shows the experimental assembly using the symmetric impact technique used to determine the shock hugoniot of materials wherein electrical and optical shock arrival pins were used to record the shock velocity in the target materials The thickness of the explosive was changed along with the air spacing in between explosive and flyer to achieve flyer velocities from 1.2 to around 5 km/s. Shock Hugoniot between pressure and particle velocity found for the metal sample has been shown in fig 2(b).

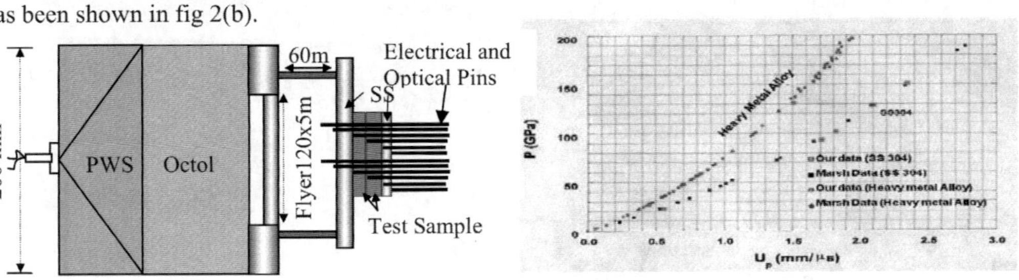


(a) (b)

**Fig.2** .(a) Experimental assembly to determine shock hugoniot and (b) Comparison of our experimental data with literature data for SS 304 and a heavy metal alloy

## 3. Cascaded Plate Impact Experiment

There is an upper limit to the velocity of flyer, which can be achieved by single stage explosive device as the velocity depends principally on the Gurney Energy of explosive (a constant) and explosive to metal mass ratio. The velocity increases with charge to metal mass ratio and reaches an asymptotic value. So there is an upper limit on velocity and hence the pressure which can be generated in the target. But quest to realize the higher pressure states goes on. The technique of using explosive and impedance cascading or staging is popular among the shock physics community. We have established the concept of cascaded explosive system to launch metal flyers to velocities of the order of 7-8 km/s and even more.

In staged flyer launch systems, impedance cascading and explosive cascading techniques can be used o accelerate the flyer to very high velocity. In the impedance cascading, buffer plates of different materials and varying thicknesses are deployed, in order of decreasing shock impedances as target to first stage. A striker plate is explosively accelerated to terminal velocity, which then hits a set of buffer plates. In this manner the driven plate can be launched to a velocity equal to the free surface velocity of the last buffer plate. The concept of achieving high flyer velocity by this method was given by Balchan and Cowan [12] and was demonstrated by the authors in their previously published studies [13].

### 3.1 Explosive Cascading

In the explosive cascading arrangement, the first stage explosive launches a metal flyer to its terminal velocity. At the terminal end it impacts another explosive and produces overdriven detonation in it. The overdriven detonation in the second charge decays rapidly, so it is essential to use optimum thickness of it to drive the second stage flyer. This flyer is relatively very thin, but gets accelerated to very high velocities.

In an two stage explosive system, if the velocity $u_0$ of first flyer is more than the CJ particle velocity of explosive, overdriven detonation travels in the second stage explosive and the strong detonation speed D can be deduced as

$$D = \frac{D_J^2 + (\gamma+1)^2 {u_0}^2}{2(\gamma+1)u_0}$$

where D is overdriven detonation velocity, Dj is normal detonation velocity and $\gamma$ is the isentropic expansion co-efficient of detonation products. The final velocity of the second stage flyer can be estimated by relations given by Wen Shanggang et al [4]. In the present studies explosive cascading technique has been utilized to launch the metal flyer to extremely high flyer velocities.

### 3.2 Simulation and Experimental Validation

In a cascaded system the velocity and thickness of the driver and driven plate, second stage explosive thickness and flight distances need to be optimized to get maximum efficiency. The numerical simulation using AUTODYN 2D was performed to finalize the various parameters. A two stage explosive assembly with Octol explosive is shown in the figure 3 (a). The experimental assembly used to propel a metal flyer, launched a nearly flat flyer which made an impact on the second stage explosive, in turn accelerating the second flyer to high velocity. The results predicted through the AUTODYN simulation are shown in the figure 3(b).

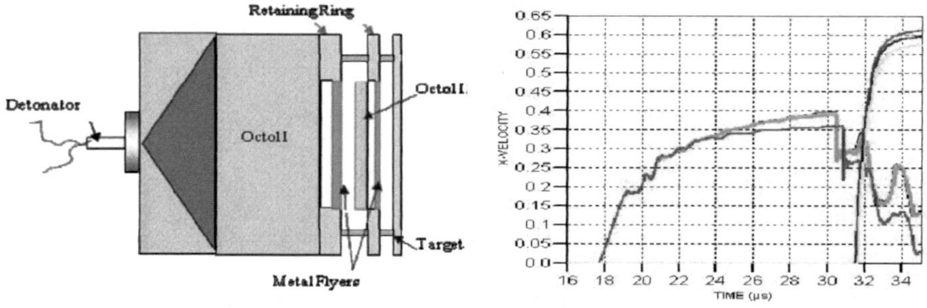


(a) (b)

Figure 3(a). Two Stage Experimental Setup 3(b) Simulation Result of

The experimental assemblies were evaluated to monitor the second stage flyer velocity and planarity. The ionisation and optical pins were deployed along with the high streak photography to monitor the motion of second stage flyer.

This method has successfully been demonstrated by us to generate pressures or the order of 3Mbar in the different target materials. Beyond this pressure we can use the three stage arrangement, but there are practical limitations which introduce mis-alignment problems, limiting the use of this method. Even the thickness of explosive and flyers decreases in each stage, which are very difficult to fabricate with adequate strength, flatness and within acceptable tolerances. Still higher pressures have been achieved by translating the planer arrangement to spherical one, thus introducing the concept of convergence to realize pressures in excess of 10 Mbar.

## 4 Converging Detonation and Shock Waves for Mega bar Pressure Generation

In order to realize the conditions of pressure and temperature inside the cores of planetary bodies and to simulate the conditions of impact of terrestrial bodies, converging detonation systems along with the two stage explosive cascading systems can be used. Either a cylindrical system converging on axis or a spherical system with peak pressures at centre is generally used.

The phenomenon of convergence of a spherical detonation wave at the centre and the cylindrical wave on the axis was studied by LD Landau and KP Stanyukovich in 1944[14]. They found out that the pressure rise in convergence as $r^{-1.13}$ in spherical cases and as $r^{-0.47}$ in the cylindrical case. If the effect of converging detonation waves is augmented by a converging shock wave, tremendous rise in the pressures occurs. Practical use of the convergence of detonation waves and shock waves has been demonstrated by Al'tshuller et al [4] to achieve very high pressures in the target materials.

In this technique the explosives used are in hollow cylinders or spheres. The outer surface is initiated simultaneously to generate converging detonation waves in the explosives propagating towards the cylinder axis or sphere centre. The converging detonation is further used to accelerate the metal liners of matching shape. The theoretical relations to find the collapsing liner velocity have been given by R. Cheret [15] and Hirsh [16]. The liners can be accelerated to very high velocities generating pressures upto 5 Mbar in single stage arrangements. If the concept of explosive cascading discussed earlier is extended to the spherical geometry, pressures of the order of 10-20 Mbar or even more can be achieved.

In one such experimental assembly we used Octol explosive to accelerate the first stage 4 mm thick liner. This liner impacted the second stage 10 mm thick explosive, which was used to accelerate the second liner. The velocity of the second liner was estimated using simulation nearing 14 km/s at a radial location where target was placed. The results of AUTODYN simulation are presented in figure 4(a) showing the predicted value of the collapsing liner in the range 13.6-14 km/s at different locations. This liner impacted the hollow metal target in the form of concentric metal shells. Different types of optical, electrical and piezoelectric pins were deployed at the interfaces between target shells at possible azimuth locations to record the converging shock wave. Shock velocity of the order of 20-22 km/s was recorded with calculated pressures reaching nearly 20 Mbar. In the two experiments, the average pressures were measured at $r/r_0$ of 0.024 and 0.029 from centre of convergence and are compared with the simulation results in figure 4(b).

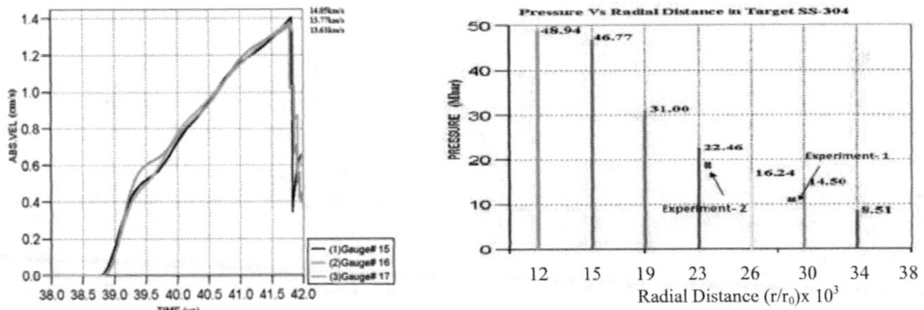


(a) (b)

**Fig.4** (a) Flyer velocity-time history in two stage explosive cascaded system (b) Comparison of experimental data with simulation results

## 5 Comparison of Two Stage with Single Stage Explosive Cascaded Systems

This flyer velocity can be increased considerably using two stage systems wherein higher pressure in excess of normal C-J detonation pressure is developed in the inner pad due to an over-driven detonation, and hence a higher velocity for the inner flyer plate. The numerical simulation shows the comparison of a single stage and two stage systems within same dimensions. Cascaded system allows to achieve detonation pressure of the order of 1 Mbar in second stage explosive whereas in the single stage system the convergent flow allows detonation pressure of 500 kbar or more with large duration. The predictions of the explosive impulse in the two systems are shown in figure 5(a).As the pressure pulse is very sharp in the two stage systems; the inner flyer plate has to be very thin. Thickness of this flyer must be much smaller than width of the pressure pulse. The velocity time history of the flyer in a single stage vs two stage explosive system is shown in fig 5(b).

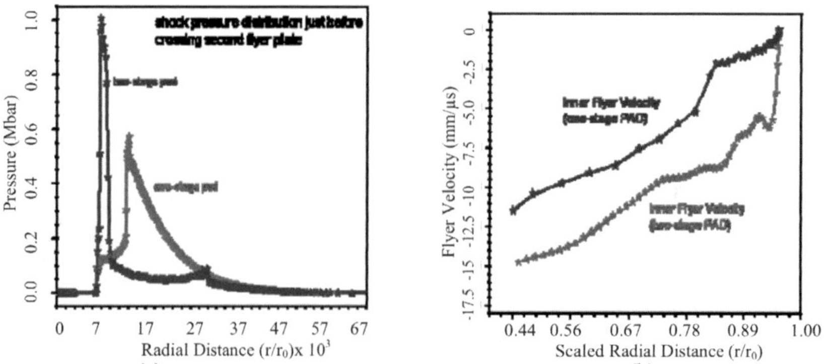


(a) (b)

Fig 5(a) Flyer velocity history in single and two stage systems (b) Impulse delivered to the liner in staged system

## 6. Conclusions

A two stage explosive system both in the planer and converging geometry has been designed and successfully tested. Materials have been shock loaded to pressure upto 2.5 Mbar in planer assembly and upto 20 Mbar in converging system.

## 7. Acknowledgments

The author wish to acknowledge the help rendered by A C Sharma and Bhupinder Sewak in carrying out the numerical simulations using AUTODYN. This work have been made possible by the sincere efforts of Rajwant Rai, PC Gautam, Ritu Khurana, Ruchi Joshi, and Anil Kumar in designing and carrying out the experiments.

## 8. References

[1] C.H. Johanson and P.A. Pearson, 1970 Detonics of High Explosives, Academic Press, London
[2] J.M. Walsh, M.H. Rice, R.G. McQueen and F.L. Yarjer 157 Phy. Rev 87 108
[3] R.G. McQueen and S.P. Marsh 1960 J. Appl. Phy.31 1253
[4] Wen Shanggang Snezhinsk, 2003 VII-th Zababakhin Scientific Talks, A study of hypervelocity flyer driven by strong detonation products
[5] Zhi-yue Liu 2001 International Journal of Impact Engineering, Numerical study on hypervelocity acceleration of flyer plate by overdriven detonation of high explosive 26 443
[6] L.V. Al'tshuller, R.F. Trunin, K.K. Kvapnikov & N.V. Panov, 1996 Physics – Uspekhi 39 539
[7] P.W. Cooper, 1996 Fundamentals of Explosive Engineering. VCH, New York
[8] R.W. Gurney 1943 The initial velocities of fragments from bombs, shells and grenades'. Army Ballistic Research Laboratory
[9] A.R, Aziz, H. Hurwitz, H.M. Sten Berg 1961 Phys Fluids Energy transferred to a rigid piston under detonation loading 4(3) 380-384
[10] J.F. Backofen, C.A. Weickert 1993 Proc of 14th International symposium on ballistics A Gurney formula for forward projection from the end of explosive charge vol, page no
[11] H.S. Yadav, N.K. Gupta 1988 International Journal of Impact Engineering, Flyer plate motion and deformation during the flight Vol 1, 71-83
[12] A.S. Balchan & G.R Cowan 1964 Rev. Sci. Instruments 35 937
[13] Manjit Singh, H.R. Suneja, A.K. Madan, M.S. Bola 1997 Advances in high pressure research in condensed matter Explosive accelerators for hypervelocity impact studies. 466
[14] K.P. Stanyukovich, 1955 Gostekhizdat, Unsteady motion of a continuous medium,
[15] R. Cheret, F. Chaisse, J. Zoe, Some Results on the converging spherical detonation in a solid explosive, Centre d'Etudes de Vaujours Sevran-FRANCE
[16] E. Hirsch,1986 Propellants, Explosives, Pyrotechnics, Simplified and Extended Gurney Formulas for Imploding Cylinders and Spheres, Vol 11, Issue 1 6

# Raman spectroscopic study of pressure induced amorphization in Cavansite

**T. R. Ravindran[A*], A.K. Arora[A] and G. Parthasarathy[B]**
[A]Condensed Matter Physics Division, Indira Gandhi Centre for Atomic Research, Kalpakkam 603102, India
[B]National Geophysical Research Institute, Hyderabad 500007, India

E-mail: trr@igcar.gov.in

**Abstract.** High pressure phase transformations in calcium aluminum silicate (cavansite), a rare mineral, are reported up to a pressure of 15 GPa using Raman spectroscopy. Structural transformations are indicated at 1, 4, 6 and 8 GPa followed by amorphization above 15 GPa. The amorphous phase is quenchable to ambient pressure.

## 1. Introduction

Hydrated calcium vanadium silicate ($Ca(VO)Si_4O_{10} \cdot 4H_2O$) - mnemonically known as cavansite - is a rare mineral that crystallizes in an orthorhombic structure with four formula units per unit cell. This brilliant blue (or greenish blue) coloured mineral was first discovered in the year 1960 in Oregon, USA. Excellent crystals of cavansite have been obtained from the zeolite mines at Wagholi, near Pune, India. Its structural, mineralogical and thermal properties have been reported [1,2]. The structure of cavansite consists of silicate tetrahedra linked by vanadyl groups resulting in an infinite framework of vanadium silicates. This open structure is analogous to zeolites and other materials with negative thermal expansion (NTE) that also exhibit pressure induced amorphization [3]. Many complex silicate minerals have been found to turn amorphous at high pressure [4]. Here we report the first high pressure Raman spectroscopic study of cavansite.

## 2. Experimental

Crystalline samples were obtained from afore mentioned mines at Wagholi. X-ray diffraction pattern recorded at ambient conditions confirmed that the crystals are of good quality. Crystalline fragments of dimensions about 50-100 microns were loaded into a compact, symmetric diamond anvil cell with ruby pressure markers of a few microns size; 4:1 mixture of methanol and ethanol was used as pressure transmitting medium. Pressure measurement and sample spectra were recorded from the same spot without moving the DAC to avoid error in pressure measurement. This was possible because of the small size of the ruby chips and the high intensity of its luminescence signal. Raman spectra were recorded using a Renishaw Invia micro-Raman system with 514 nm laser excitation.

Journal of Physics: Conference Series **377** (2012) 012004 doi:10.1088/1742-6596/377/1/012004

### 3. Results and discussion:

Cavansite crystallizes in a centro-symmetric orthorhombic unit cell (space group $D_{2h}^{16}$) with four formula units per primitive cell. Factor group analysis (Bhagavantam and Venkatrayudu method) results in the following irreducible representation

$$\Gamma_{Total} = 46A_g + 41B_{1g} + 46B_{2g} + 41B_{3g} + 41A_u + 46B_{1u} + 41B_{2u} + 46B_{3u}$$

The first four modes are Raman active, Au are optically inactive and the others are IR active, so there are 174 Raman active and 133 IR active modes expected from the crystal. However, at ambient conditions cavansite exhibits 36 Raman bands, some of which could be overtones and combinations. Less number of modes than predicted by group theory is not uncommon and could be due to accidental degeneracy or low Raman scattering cross section. The modes below 300 $cm^{-1}$ are lattice modes; those in the region of 500-800 $cm^{-1}$ are due to Si-O bonds and 900-1000 $cm^{-1}$ due to V-O bonds. A band about 3546 $cm^{-1}$ with three shoulders arises from the four molecules of water of crystallization.

As pressure is increased to 1.1 GPa (Figs. 1 - 3) there are qualitative changes in the spectrum: the bands at 90 and 935 $cm^{-1}$ disappear and the most prominent internal mode at 981 cm-1 undergoes significant softening. Even more dramatic is the appearance of new Raman bands in the region 2800-3400 $cm^{-1}$, above the second order features of the diamond anvils (not shown in the figure). These new bands - that could be due to hydrogen atoms of water making weak bonds with other atoms such as Si or V - appear at 2856 and 2960 $cm^{-1}$; the water bands get closer to each other and become less distinguishable from one another. As the pressure is increased to 1.6 GPa the spectral features change again: new modes appears at 337 and 957 $cm^{-1}$, the water bands around 3544 $cm^{-1}$ intensify before splitting into three distinct modes at higher pressures. Most of the Raman modes show the usual behavior of hardening as pressure is increased; a few modes such as 236 and 958 $cm^{-1}$ show softening. Above 14 GPa the lattice modes disappear; internal modes broaden and merge into four broad modes, at 332, 640, 717 and 973 $cm^{-1}$, indicating a possible transformation into an amorphous state. The transition into the amorphous state is irreversible.

Examination of the ω vs. P plot (fig. 4) shows also the following: There are large changes at 1.1 GPa: there is a down-shift of frequency from 90 to 76 $cm^{-1}$, 112 to 106 $cm^{-1}$, and an up shift from 133 to 141 $cm^{-1}$, and disappearance of the mode at 164 $cm^{-1}$. Additionally, new modes appear at 2856, 2960, 3236 and 3424 $cm^{-1}$. At 3 GPa the mode at 600 $cm^{-1}$ splits into two and a new mode at 964 $cm^{-1}$ appears. In the pressure range 4-5 GPa, a new mode appears at 453 $cm^{-1}$ and the modes at 254 and 270 $cm^{-1}$ merge into a single one at 259 $cm^{-1}$; as the pressure is increased further, the mode at 215 $cm^{-1}$ splits into 3-4 modes at 6 GPa; similar splitting of modes occurs at 8 GPa for the 221 $cm^{-1}$ mode; the mode at 1050 $cm^{-1}$ disappears between 6 and 7 GPa; a new mode at 1116 $cm^{-1}$ appears and the slope of the 989 $cm^{-1}$ mode changes above 8 GPa; all the modes disappear and merge into broad, weak modes about 250, 340, 630, 700 and 925 $cm^{-1}$ at 15 GPa. These results indicate possible phase transformations around 1, 4, 6 and 8 GPa followed by amorphization above 15 GPa.

Studies of phase transformations of calcium silicates at high pressure are scarce. There are only a few Ca silicates whose high pressure behaviour is well documented. Anorthite ($CaAl_2Si_2O_8$) disproportionates into a mixture of grossularite ($Ca_3Al_2$-$Si_3O_{12}$) + kyanite ($Al_2SiO_5$) + quartz ($SiO_2$) at pressures above 3 GPa [5]. At 15 GPa the quartz component transforms into stishovite ($SiO_2$) [6]. A nonquenchable phase was observed at pressures greater than 22 GPa [4]. Both $CaSiO_3$ • 10% $Al_2O_3$ and $CaMgSi_2O_6$ • 10% $Al_2O_3$ crystallize in a garnet-type structure in the pressure range 100-150 kbar [7]. $CaSiO_3$ transforms to a cubic perovskite-type structure at about 15 GPa as detected by in situ X-ray diffraction.

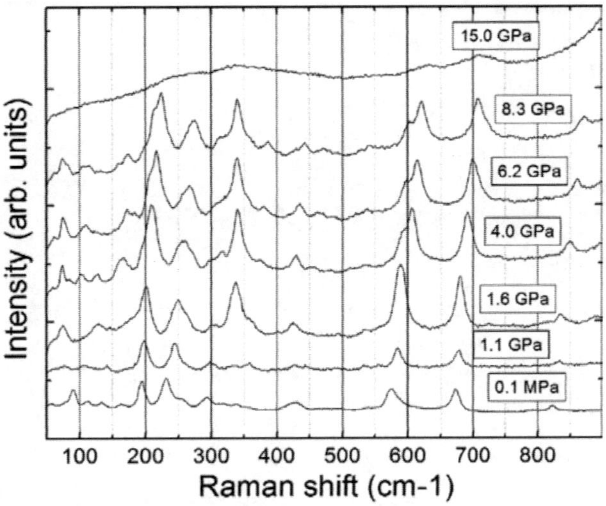


**Fig. 1.** Raman spectra of cavansite at different pressures. Changes are clearly seen around 100, 340, 600 $cm^{-1}$

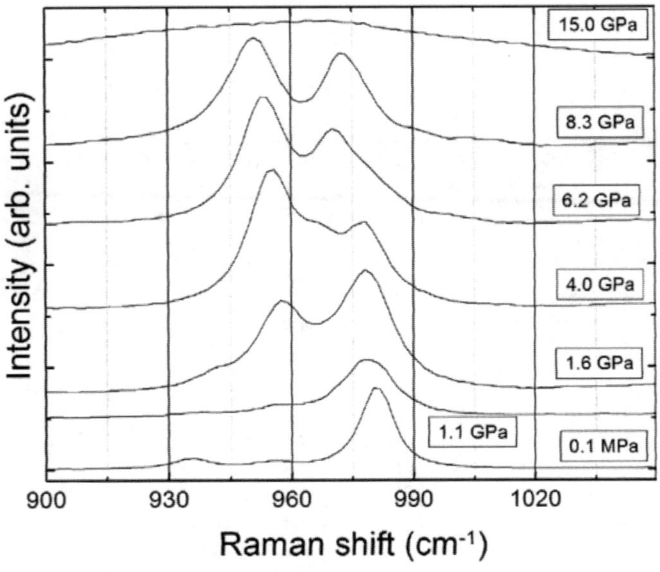


**Fig. 2.** Raman spectra of cavansite at different pressures. Changes can be seen about 935, 950 $cm^{-1}$

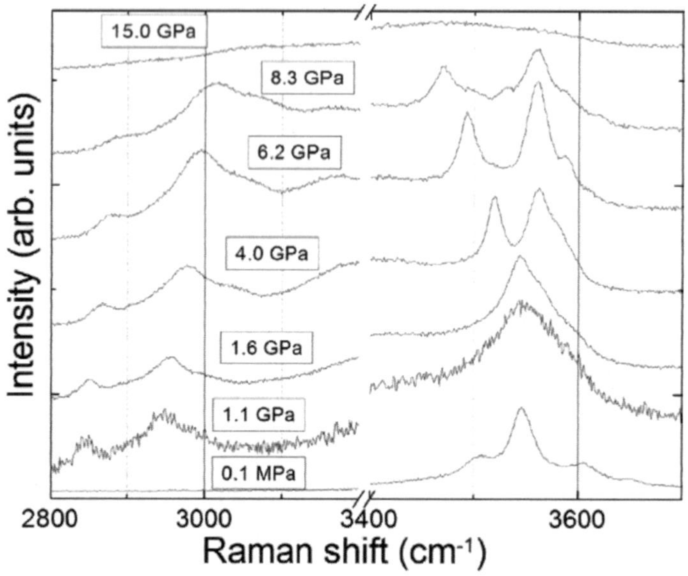


**Fig. 3.** Raman spectra of cavansite at different pressures. New bands appear about 2856 and 2960 cm$^{-1}$ at 1.1 GPa; water bands around 3546 cm$^{-1}$ undergo drastic changes

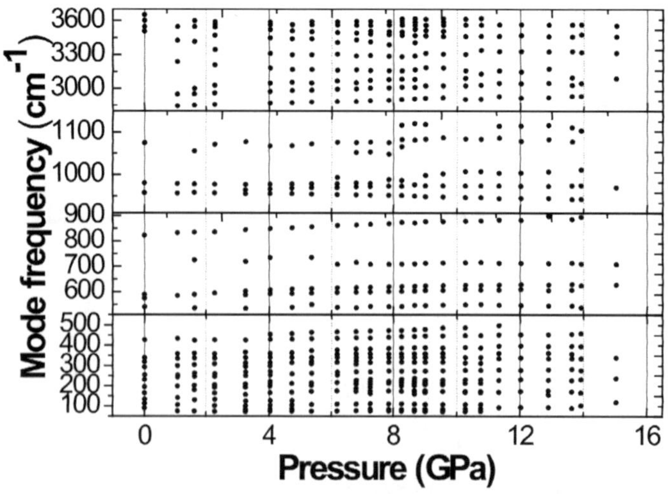


**Fig. 4.** ω vs. P for cavansite. Changes are noticeable at 1, 4, 6, 8 and 15 GPa

A possible decomposition of cavansite into its isochemical mixed oxides is as follows: $Ca(VO)Si_4O_{10} \Rightarrow CaO + VO_2 + 4SiO_2$. It can be expected that at high temperature and pressure, cavansite would decompose into these oxides if the combined molar volumes of the daughter phases is lower than the original cavansite. It is possible that amorphization of cavansite takes place at 15 GPa due to insufficient kinetics for decomposition: at a sufficiently high temperature, cavansite may decompose into its daughter phases above (CaO, $VO_2$ and $SiO_2$) rather than amorphizing. Since at ambient temperature the mobility of the constituent atoms is not enough for the daughter phases to nucleate and grow, the structure settles into an amorphous state that is of the lowest energy. It is noteworthy that recent, extensive investigations on the structure and chemical composition of cavansite by Ishida et al [8] have revealed that it contains $H_3O^+$ and $OH^-$ species in addition to $H_2O$, and the correct structural formula is $Ca(VO)(Si_4O_{10})\cdot(H_2O)_{4-2x}(H_3O)_x(OH)_x$ rather than $Ca(VO)Si_4O_{10}\cdot 4H_2O$. The latter applies only to pentagonite, a mineral that coexists with cavansite but crystallizes with a different space group motif (*Ccm2₁*) in the orthorhombic structure [2]. The difference between the structures of the two minerals is essentially in the manner in which some of the $H_2O$ units are split as $H_3O^+$ and $OH^-$ in the lattice. This does not, however, affect the present analysis, since water molecules are not considered in the above decomposition; molecules of water of crystallization do seem to play a role in the phase transformations as seen from the Raman spectra, and this aspect needs further investigation.

## References


[1] Staples L W, Evans H T, Jr. and Lindsay J R 1973 *Amer. Mineral.* **58** 405

[2] Evans H T 1973 *Amer. Mineral.* **58** 412

[3] Havenga E A, Huang Y and Secco R A 2003 *Mater. Res. Bull.* **38** 381

[4] Arora A K 2000 *Solid State Commun.* **115** 665

[5] Boyd F R and England J L 1961 *Melting of silicates at high pressures*, Carnegie Inst. Washington Year book **60** 113

[6] Ringwood A E 1970 *Phys. Earth Planet. Inter.* **3** 109

[7] Ringwood A E and Major A 1966 *Earth Planet. Sci. Lett.* **1** 351

[8] N. Ishida, M. Kimata, N. Nishida, T. Hatta, M. Shimizu and T. Akasara 2009 Journal of Mineralogical and Petralogical Sciences **104** 241

Journal of Physics: Conference Series **377** (2012) 012005 doi:10.1088/1742-6596/377/1/012005

# Polyamorphism in network structures

**A K Arora**
Condensed Matter Physics Division, Indira Gandhi Centre for Atomic Research, Kalpakkam, 603102, India

E-mail: aka@igcar.gov.in

**Abstract.** Reports of high-density amorphous forms of ice, quartz and Si have intensified the search for amorphous polymorphs in other systems. The network structures that exhibit pressure-induced amorphization are interesting from the point of view of polyamorphism. Recent *in-situ* high pressure Raman scattering and synchrotron x-ray diffraction studies on negative thermal expansion material zirconium tungstate have revealed an amorphous to amorphous transition at 19 GPa. The transition was identified for the discontinuities in the structural parameters such as position and height of the first peak of the structure factor and that in the tungstate internal vibrational mode frequency. Another compound, vanadium pentaoxide that forms a glass during melt-quenching, is also found to turn amorphous at 40 GPa, pointing to a possibility of polyamorphism.

## 1. Introduction

Instead of exhibiting a conventional crystal-crystal transition, a few compounds turn amorphous at high pressure [1]. This is believed to arise due to kinetic hindrance of transformation to final equilibrium state [2] that could be either a new high-density polymorph or a mixture of dense-packed daughter compounds [3, 4]. In analogy with the crystalline solids, it likely that an amorphous solid (or a glass) could also undergo a structural transition from one non-crystalline state to another, i.e. may exhibit distinct amorphous phases under suitable pressure and temperature conditions. The amorphous phases could be different in terms of density, entropy, local coordination, intermediate range order, and also composition [5, 6]. Consequently the electronic structure and the vibrational properties are also expected to be different. It is much more challenging to distinguish between different amorphous polymorphs than in crystalline polymorphs. The early evidences of distinct amorphous states were found in ice [7] and quartz [8] and later in silicon [9]. A ~10% change in density was found between low-density amorphous (LDA) and high-density amorphous (HDA) ice. The coordination number is reported to be higher in the HDA phase of silica. However, their structure and other properties have not been investigated in detail. It has also been argued that one could, in principle, expect amorphous phases corresponding to each of the crystalline polymorphs; however, few studies have been done to support this. In this context the network structures that undergo pressure-induced amorphization [10] are of interest from the point of view of exploring the possibility of polyamorphism. Recent *in-situ* high pressure Raman scattering and synchrotron X-ray diffraction studies on negative thermal expansion material zirconium tungstate have revealed an amorphous to amorphous transition at 19 GPa [11]. Here the transition was identified for the discontinuities in the structural parameters such as position and height of the first peak of the structure factor and that in the tungstate internal vibrational mode frequency. Another compound, vanadium pentaoxide that forms a glass during melt-quenching

[12], is also found to turn amorphous at 40 GPa, pointing to a possibility of polyamorphism. This article reviews recent results on the exploration of polyamorphism in network structures.

## 2. Experimental details

Powdered samples were loaded into a 160 μm hole of a stainless-steel gasket (preindented to 70 μm thickness) of a Mao-Bell-type high-pressure diamond anvil cell (DAC). *In-situ* x-ray diffraction measurements were made using synchrotron radiation ($\lambda = 0.4131(2)$ Å) from NE1 beam-line at AR-Photon Factory, KEK. A 30-μm collimator was used and a y-z translation stage was used for aligning the sample in the hole of the gasket with the collimator. A thin gold foil of 30 μm size was also placed within the gasket hole and the pressure was obtained from equation of state of gold [13]. The DAC was translated to bring either the sample area free from gold foil or that containing it in front of the collimator. An image plate was used as the detector. 514.5 nm line of an argon ion laser was used for exciting the sample in the DAC and the spectra were recorded using a micro-Raman set up. For Raman experiments pressure was measured using ruby fluorescence method.

## 3. Results and discussion

The cubic phase of $Zr(WO_4)_2$ is metastable and transforms to an orthorhombic structure at 0.3 GPa before turning amorphous irreversibly at 2.1 GPa [14]. The amorphous phase is about ~26% denser than the cubic phase. *a*-$Zr(WO_4)_2$ is reported to decompose into mixture of simple oxides ($ZrO_2$ + $2WO_3$), when its temperature is raised *in-situ* at high pressure [15]. On the other hand, it crystallizes back to the cubic phase if it is heated at ambient pressure [16]. Based on the closeness of the amorphous density with that of a mixture of simple oxides, the amorphous structure has been argued to be a precursor to decomposition [17]. Studies of recovered *a*-$Zr(WO_4)_2$ have suggested W-coordination to be five [18, 19].

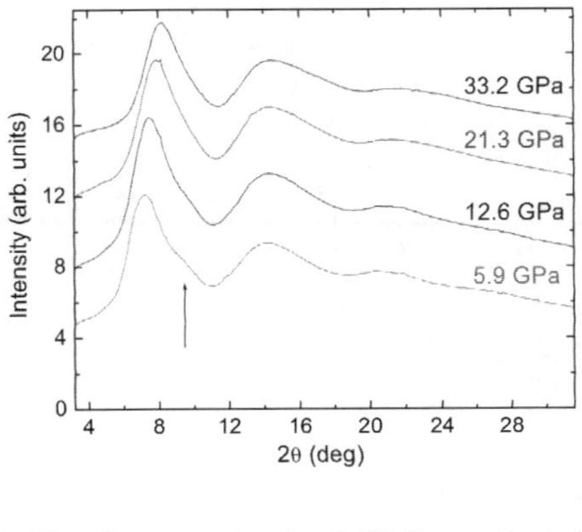


**Figure 1.** Diffraction patterns of *a*-$ZrW_2O_8$ at selected pressures.

Figure 1 shows the diffraction patterns at selected pressures. Four broad amorphous diffraction peaks were found up to a scattering angle of 32°. One can see that at high pressure the shoulder of the first amorphous diffraction peak (ADP) (between 9 and 11°) is absent. All the patterns below 17 GPa exhibited this shoulder while those above 21 GPa did not. The diffraction patterns were analyzed quantitatively to obtain the structure factor $S(Q)$ using procedure described earlier [19]. One can also see a systematic shift of the first ADP to higher 2θ and a consequent shift of the centre $Q_1$ of the first peak in $S(Q)$. This is shown in Fig. 2 as a function of pressure. In analogy with the shift of diffraction peaks in crystalline solids at high pressure due to compression, within a given amorphous phase the position of first ADP is expected to shift monotonically to higher 2θ (larger $Q_1$). On the other hand, the data shows a change of slope around 19 GPa.

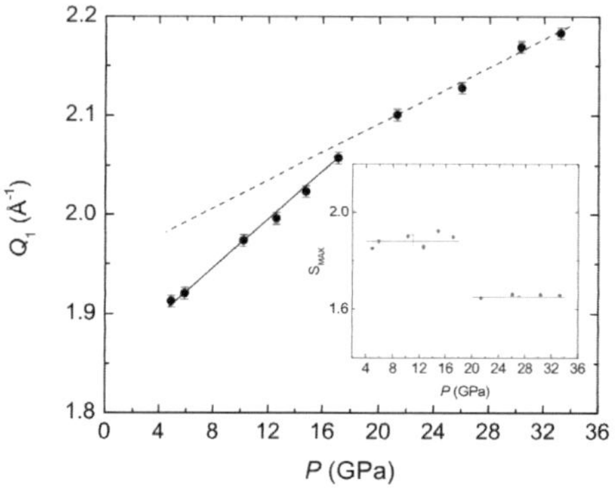


**Figure 2.** Scattering vector $Q_1$ corresponding to first ADP as a function of pressure. The inset shows the pressure dependence of $S_{MAX}$.

The height $S_{MAX}$ of the first ADP of the $S(Q)$ as a function of pressure is shown as an the inset of Fig. 2. $S_{MAX}$ is a measure of the magnitude of the structural correlations in an amorphous phase [20, 21]. One can see a large discontinuous change above 19 GPa. In absence of an amorphous-amorphous (structural) transformation, the $S_{MAX}$ is expected to show only a constant or monotonic behaviour. Thus this discontinuity also supports occurrence of a transition. It is well known that the densities of amorphous solids and liquids scale like $(Q_1)^3$ in analogy with the diffraction in crystalline solids [22, 23]. The $Q_1$ of the samples recovered from 17 GPa is found to be smaller than that for samples recovered from 33 GPa, suggesting that the two recovered amorphous phases have different densities. The high-density amorphous (HDA) phase appears to be about 12±1 % more dense than the low-density amorphous (LDA) phase.

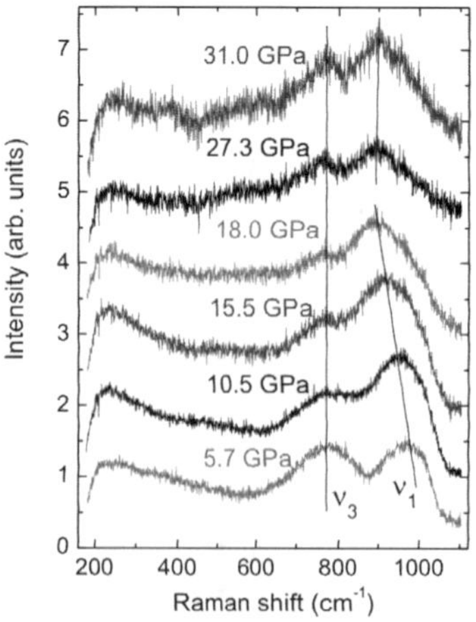


**Figure 3.** Raman spectra of *a*-$ZrW_2O_8$ at different pressures.

One can also find corresponding changes in the phonon spectra across amorphous-amorphous transition. Figure 3 shows the Raman spectra at different pressures. Because of the disorder (lack of periodicity) all the lattice and external modes disappear and only the two broad peaks corresponding to the stretching and bending vibrations of $WO_4$ polyhedra are found. The spectra were fitted to a sum of two Voigt functions along with a suitable background. The symmetric stretching mode ($\nu_1$) shows a softening up to 18 GPa, above which it exhibits hardening. On the other hand, the asymmetric stretching mode ($\nu_3$) shows only a weak dependence on pressure. It may be pointed out that the symmetric stretching vibration of tetrahedral ions such as $SO_4$ and $WO_4$ depends strongly on the bond-length and the mode frequency is known to increase at high pressure due to compression of the bond [24, 25]. A decrease in $\nu_1$ is anomalous [3] and would correspond to increase in the mean bond-distance. The observed softening of the $\nu_1$ mode in *a*-$Zr(WO_4)_2$ at high pressure suggests an increase of the mean bond length upon compression. This can occur only when oxygen atoms are displaced to facilitate a higher average oxygen coordination of W. Thus the present Raman results suggest a gradual increase of coordination number within the LDA phase. The discontinuous change in $\nu_1$ mode frequency above 19 GPa supports the occurrence of an amorphous-amorphous transformation. The normal hardening behaviour of the stretching mode at still higher pressures implies that the oxygen coordination remains stable in the HDA phase.

We now discuss the possibility of polyamorphism in another compound ($V_2O_5$) with a network structure. Several high-pressure and high-temperature studies have been carried out on $V_2O_5$ and two structural phase transitions to high pressure polymorphs have been reported [26, 27]. The ambient pressure α-phase has layered orthorhombic structure with corner and edge sharing $VO_5$ square pyramids. Above 7 GPa the high-pressure monoclinic (β) phase consists of chains of quadruple units of edge-sharing $VO_6$ octahedra along *b*-axis [26]. A transition to another monoclinic phase (δ) has also been identified above 15 GPa [27]. Diamond-anvil cell based Raman and X-ray diffraction measurements have revealed only the α-β transition occurring around 6-7 GPa [28, 29]. It may also be pointed out that in $V_2O_5$ a glassy/ amorphous state, formed by splat-cooling [30] (bulk) and physical vapor deposition [31] (thin-film) is well established. However, glassy $V_2O_5$ is known to recrystallize at relatively low temperature of 185 °C [12].

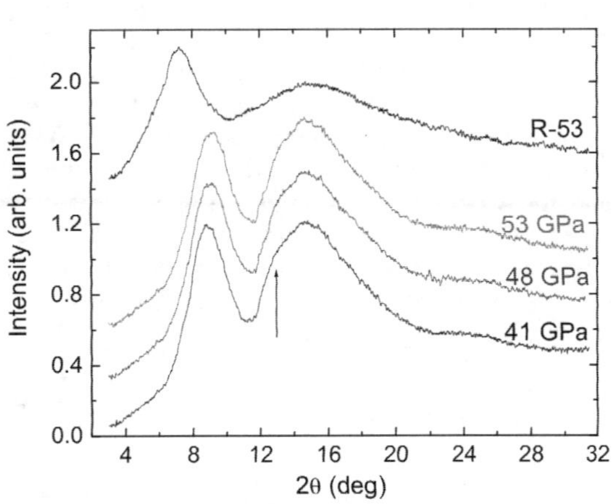


**Figure 4.** The diffraction patterns of pressure-amorphized $V_2O_5$. The pattern labelled R-53 is that of the recovered sample at ambient pressure after subjecting it to 53 GPa.

The diffraction patterns of $V_2O_5$ at high pressure are shown in Fig. 4. These unambiguously correspond to amorphous patterns with broad peaks centered at 8.8 and 14.6° 2θ. The first peak shifts to higher 2θ as pressure is increased further due to the compression of the amorphous phase. Figure 4 also includes the diffraction pattern measured after releasing the pressure from 53 GPa. The persistence of the broad peaks in the recovered pattern suggests irreversible pressure-induced amorphization of $V_2O_5$ above 40 GPa, similar to other network structures. Thus it is of interest to examine the high-pressure amorphous phase in more detail and compare with those obtained using other methods. Unfortunately high quality diffraction patterns of glassy and vapor-deposited amorphous films with good signal to noise ratio have not been reported.

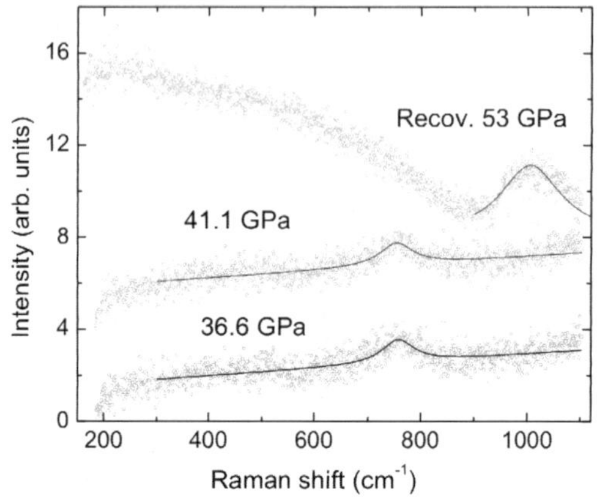


**Figure 5.** Raman spectra of $V_2O_5$ at high pressure and after recovering from 53 GPa.

Figure 5 shows the Raman spectra in the high-pressure amorphous phase. Here only one broad peak is observed at 755 cm$^{-1}$. This represents average V-O stretching vibration frequency corresponding to average V-O bond distance in the amorphous state. The mode frequency, when extrapolated to the ambient, is around 757 cm$^{-1}$, which is close to the doubly-coordinated (V-O-V) vibrations of the oxygen. This suggests that the high-pressure $a$-$V_2O_5$ phase has predominantly $V_2$-O-type of bonds. Upon releasing the pressure the recovered $a$-$V_2O_5$ exhibits a broad Raman peak at a wave number close to that for V=O stretching vibration at ambient. This implies restoration or reformation of V=O-like bonds in the recovered amorphous state. This is understandable as a relaxation of the amorphous state upon release of pressure. A comparison of the vibrational spectra of the samples recovered from high pressure with that reported [31] for vapor-deposited $a$-$V_2O_5$ is in order. The vapor-deposited $a$-$V_2O_5$ exhibited prominent peaks at 520, 650, 932 and 1027 cm$^{-1}$. The peaks at 520, 650 and 1027 cm$^{-1}$ have been assigned to vibrations of triply-, doubly- and singly-coordinated oxygen atoms respectively, in analogy with those of the α-phase. It may be pointed out that the V=O mode frequency in the recovered $a$-$V_2O_5$ (1006 cm$^{-1}$) and in vapor-deposited $a$-$V_2O_5$ film (1027 cm$^{-1}$) are close to the that of V=O bond in the α-phase (995 cm$^{-1}$). Thus the recovered $a$-$V_2O_5$ appears to have random arrangement of $VO_5$-like structural units similar to those of α-$V_2O_5$. The difference in the vibrational frequencies in the two amorphous phases could be due to the differences in the intermediate-range order as the vibrational frequency of any given bond is strongly influenced by its immediate neighborhood.

## 4. Conclusions

The network structures provide unique possibilities for exploration of polyamorphism. Discontinuities in the pressure dependence of structural parameters such as position and height of the first peak of the structure factor and that in the internal vibrational mode frequencies of polyhedral units can be used for the identification of amorphous-amorphous transitions. Occurrence of pressure-induced amorphization in $V_2O_5$ at 40 GPa and formation of glassy state during melt-quenching provide compelling reasons for further exploration of polyamorphism in this system.

**Acknowledgments**
It is a pleasure to acknowledge a very fruitful collaboration with Prof. Takehiko Yagi, Dr. Tomoko Sato and Dr. Taku Okada.

**References**

[1] Ravindran T R, Arora A K and Gopalakrishnan R 2002 *J. Phys.: Condens. Matter* **14** 6579
[2] Arora AK 2000 *Solid State Commun.* **115** 665
[3] Arora A K, Nithya R, Yagi T, Miyajima N, and Mary T A 2004 *Solid State Commun.* **129** 9
[4] Arora AK, Yagi T, Miyajima N and Mary T A 2005 *J. Appl. Phys.* **97** 013508
[5] Aasland S and McMillan P F 1994 *Nature (London)* **369** 633
[6] Poole P H, Grande T, Angell C A, and McMillan P F 1997 *Science* **275** 322
[7] Mishima O, Calvert L D and Whalley E 1985 *Nature (London)* **314** 76
[8] Mukherjee G D, Vaidya S N and V. Sugandhi V 2001 *Phys. Rev. Lett.* **87** 195501
[9] Deb S K, Wilding M, Somayazulu M and McMillan P F 2001 *Nature (London)* **414** 528
[10] Sakuntala T, Arora A K, Sivasubramanian V, Rao R, Kalavathy S, and Deb S K 2007 *Phys. Rev. B* **75**, 174119
[11] Arora AK, Sato T and Yagi T 2011 *J. Phys.: Condens. Matter* **23** 112207
[12] Suzuki T, Saito S and Arakawa W 1977 *J. Non-Cryst. Solids* **24** 355
[13] Anderson O L, Isaak D G and Yamamoto S 1989 *J. Appl. Phys.* **65** 1534
[14] Ravindran T R, Arora A K and Mary T A 2000 *Phys. Rev. Lett.* **84** 3879
[15] Gallardo-Amores J M, Amador U, Moran E, and Franco M A A 2000 *Int. J. Inorg. Mater.* **2** 123
[16] Ravindran T R, Arora A K, Sastry V S, and Sahu P C 2009 *J. Non Crystall. Solids* **355** 2289
[17] Arora A K, Sastry V S, Sahu P C, and Mary T A 2004 *J. Phys.: Condens. Matter* **16** 1025
[18] Keen D A, Goodwin A W, Tucker M G, Dove M T, Evans J S O, Crichton W A, and Brunelli M 2007 *Phys. Rev. Lett.* **98** 225501
[19] Arora A K, Okada T and Yagi T 2010 *Phys. Rev. B* **81** 134103
[20] Kesavamoorthy R, Tata B V R, Arora A K, and Sood A K 1989 *Phys. Lett. A* **138** 208
[21] Hansen J P and Verlet L 1969 *Phys. Rev.* **184** 151
[22] Daisenberger D, Wilson M, McMillan P F, Cabrera R Q, Wilding M C, and Machon D 2007 *Phys. Rev. B* **75** 224118
[23] Kesavamoorthy R and Arora A K 1985 *J. Phys. A: Math. Gen.* **18** 3389
[24] Ravindran T R and Arora A K 1999 *High Pressure Research* **16** 233
[25] Lacomba-Perales R, Errandonea D, Garcia D M, Hernandez P R, Radescu S, Mujica A, Munoz A, Chervin J C and Polian A 2009 *Phys. Rev. B* **79** 094105
[26] V.P. Filonenko, Sundberg M, Werner P E, Zibrov I P 2004 *Acta Crystallogr. B* **60** 375
[27] Balog P, Orosel D, Cancarevik Z, Schon C and Jansen M 2007 *J. Alloy Compounds* **429** 87
[28] Loa I, Grzechnik A, Schwarz U, Syassen K, Hanfland M and Kremer R K 2001 *J. Alloy Compounds* **317-318** 103
[29] Grzechnik A 1998 *Chem. Mater.* **10** 2505
[30] Nabavi M, Sanchez C and Livage J 1991 *Philos. Mag. B* **63** 941
[31] Lee S H, Cheong H M, Seong M J, Liu P, Tracy E, Mascarenhas A, Pitts J R and Deb S K 2003 *Solid State Ionics* **165** 111

# High pressure phase transition in $Nd_2O_3$

**K. K. Pandey[1], Nandini Garg, A. K. Mishra, Surinder M. Sharma**
Bhabha Atomic Research Centre, Mumbai, India

[1]E-mail: kkpandey@barc.gov.in

**Abstract.** High pressure structural and spectroscopic investigations have been carried out on rare-earth sesquioxide, $Nd_2O_3$ using energy dispersive x-ray diffraction and Raman scattering techniques. Our measurements show that ambient hexagonal structure of $Nd_2O_3$ is stable up to 27 GPa. Beyond this pressure, it undergoes a structural phase transition to a lower symmetry phase. On release of pressure, it reversibly transforms back to the hexagonal phase suggesting the nature of phase transition to be of displacive kind.

## 1. Introduction

Rare-Earth (RE) sesquioxides ($R_2O_3$-R-rare earth element) are important materials, both scientifically as well as technologically due to their existence in various polymorphic forms and wide variety of applications such as laser rods, phosphors, refractory and abrasive materials etc.[1] Like the RE metals, the physical properties of the RE sesquioxides are also affected by the partial filling of their inner 4f-electron shells leading to the well known effect, lanthanide contraction.[2] However 4f electrons do not participate in bonding and are core like in nature. The s and d valence electrons thus dominate in bonding and leave their respective atomic cores' in a trivalent state except for Eu, Yb, Ce, and Pr where divalent and tetravalent states are also observed. The screening of the 4f electrons by the Xe core serves to leave the bulk properties of the lanthanides relatively unchanged as electrons are added to the 4f shell. It is because of this reason that the properties of the trivalent lanthanides are expected to vary in a gradual manner as one traverse the series. As a consequence, depending on $RE^{+3}$ cation radii, there exist three polymorphic modifications of RE sesquioxide under ambient conditions *viz.* hexagonal (A-type), monoclinic (B-type) and cubic structures (C-type).[3] The rare earth sesquioxides were first investigated systematically by Goldschmidt and his co-workers in 1925.[4] His original designations for the phases (A,B and C) are still in use today. Larger $RE^{+3}$ cation radii sesquioxides are found to be stabilized in hexagonal structure whereas for smaller ones, cubic structure is more stable. In-between sesquioxides (*viz.* Sm, Eu, Gd) are found to have monoclinic structure at elevated temperatures.[5] The hexagonal phase (A-type) has been found to be denser than the monoclinic as well as cubic phase. It has been observed that pressure also affects the stability of various polymorphs in RE sesquioxides. This is also corroborated by the fact that molar volumes of hexagonal and monoclinic phases are relatively lower than that of cubic phase. As a consequence, under high pressures B-type structure is favored over the C-type and the hexagonal (A) type over both of the other structure types. Pressure is also known to enhance the stability regime of monoclinic phase as a function of lanthanide atomic number at elevated temperatures. Under ambient temperature, cubic and monoclinic structured sesquioxides are known to undergo polymorphous transition to hexagonal at under high pressures. However, the role of high pressure on hexagonal structured sesquioxide is relatively less studied. Since high pressure should lead 4f electrons to become more delocalized or hybridized, it would be interesting to investigate high pressure behavior of some partially 4f filled rare-earth sexquioxide. With this motivation we have investigated high pressure behavior of $Nd_2O_3$ using energy dispersive x-ray diffraction (EDXRD) and Raman scattering techniques.

## 2. Experimental details

High pressure EDXRD measurements were carried out on $Nd_2O_3$ up to 40 GPa at BL-11 (EDXRD beam line) of Indian synchrotron source, INDUS-II (RRCAT, Indore). A- type $Nd_2O_3$ was obtained commercially from Sigma-Aldrich and had a stated purity of 99.999%. Since $Nd_2O_3$ is highly hygroscopic, several milligrams of sample were annealed in inert environment at ~700°C for about 24 hours prior to measurements in order to get rid of carbonate and hydroxide impurities. Mao-bell kind of diamond anvil cell (DAC) was used for high pressure measurements. Sample was loaded in ~100 μm hole in pre-indented tungsten gasket. For hydrostatic compression, 1:1 pentane iso-pentane mixture was used as pressure transmitting medium. It also helped in getting rid of any type of contamination of sample during experiments. Since diffraction peaks of standard elements, generally used as pressure calibrants e.g. gold, copper etc. are nuisance in diffraction measurements, ruby chip (~20 μm) was used as pressure calibrant in diamond anvil cell (DAC). Pressure calibration at different pressures was carried out using offline ruby pressure measurement setup configured around Ocean optics Maya2000 spectrometer. Further, to avoid diffraction peaks of gasket material in EDXRD measurements, we used collimated white beam of synchrotron radiation cut down to a size of ~50 μm. Raman spectroscopic measurements were carried out in a similar manner using Mao-bell kind of DAC on our laboratory based micro-focus confocal Raman setup up to 47 GPa. 532 nm frequency doubled Nd:YAG diode pumped solid state laser was used as an excitation source. Same setup was used for pressure calibration using ruby chip which was also loaded along with sample and 1:1 pentane iso-pentane mixture.

## 3. Results and discussion

EDXRD measurements (Fig. 1a) show that the ambient phase of $Nd_2O_3$ has the symmetry of A –type hexagonal phase. Most of the prominent peaks viz. 100,011,110 112 etc. of hexagonal phase are visible even at the highest measured pressure (40.1 GPa). Assuming hexagonal phase to be stable up to the highest measured pressure we have determined *a* and *c* axis at each measured pressure.

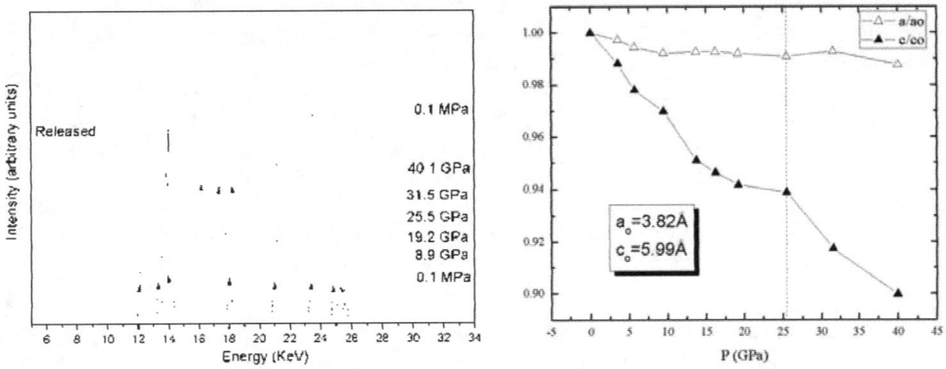


(a) (b)

**Figure 1.** (a) EDXRD data at representative pressures, (b) lattice parameter as function of pressure

As shown in Fig. 1b, *c* axis has been found to be more compressible than the *a* axis and *c/a* ratio changes from 1.57 to 1.48 at the highest pressure of our measurement. However we observed enhanced compressibility of *c* axis beyond 25.5 GPa. This is also concomitant with the emergence of a few additional diffraction peaks between 011 and 012 peaks (Fig. 1a). This is suggestive of a structural phase transition to a lower symmetry phase. Other diffraction peaks present earlier have been found to broaden beyond 25.5 GPa. Since the diffraction pattern has not changed much except a few additional peaks, the transformed phase might be structurally very close to the initial phase and the diffraction peaks of new phase are appearing at similar positions to that of initial phase.

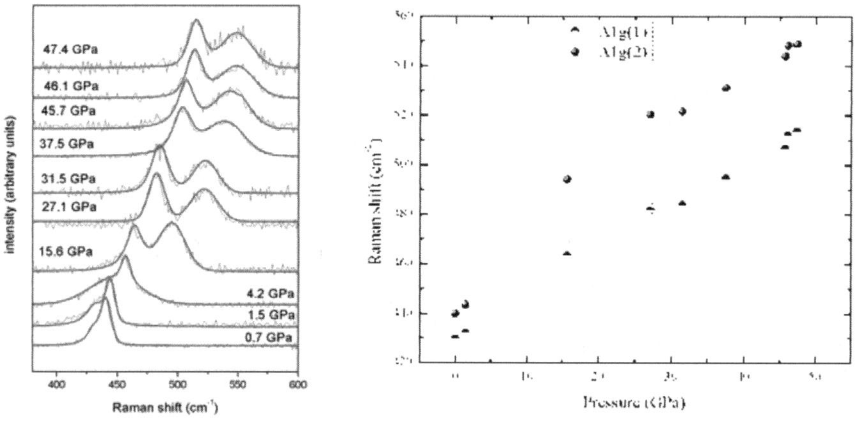


(a) (b)

**Figure 2.** $A_{1g}$ Raman mode as a function of pressure

Our Raman scattering measurements show that the most intense $A_{1g}$ mode at 440 $cm^{-1}$ splits with pressure (Fig. 2 a). J. P. McClure has explained this splitting in his thesis work[6] in terms of asymmetric change in the bond lengths of Ln cation (in present study Ln is Nd) and O atoms along the *c* axis which introduces addition Raman mode. The gap widens as a function of pressure. However, there is $d\omega/dp$ slop change above 27 GPa (Fig. 2b). This reduced stiffening of Raman modes suggests structural relaxation above this pressure as this change is concomitant with the emergence of new diffraction peaks in EDXRD measurements. On release of pressure, the hexagonal phase (A-type) reappears suggesting the phase transition to be of displacive kind. This new phase could have been similar to the novel hexagonal type super lattice structure observed in $La_2O_3$ obtained by cell doubling along *c* axis.[6] However; we did not observe any super-lattice peak in our diffraction measurements.

This new phase also seems to be different from monoclinic (B-type) structure of $Nd_2O_3$. Even though the known monoclinic phase has diffraction peaks between the (002) and (012) peaks of hexagonal phase (A type), there are other prominent diffraction peaks of monoclinic phase which are absent in our measurements. In order to get the symmetry of new phase we considered all the diffraction peaks observed at 31.5 GPa and analysed using CRISFIRE and Chekcell software. Our analysis shows the best solution to be a monoclinic phase with P21/m (space group: 11)

**Table 1.** Lattice parameter of the monoclinic phase using CRISFIRE and Chekcell software

| Lattice parameter | Value |
|---|---|
| $a$(Å) | 6.922 |
| $b$(Å) | 6.286 |
| $c$(Å) | 5.003 |
| $\beta$(Degree) | 113.91° |

It is interesting to note that the high pressure phase transition in A-type $La_2O_3$ was reported to be at ~18 GPa whereas our investigation shows phase transition in case of $Nd_2O_3$ takes place at ~27 GPa. The bulk modulus of the hexagonal phase of $Nd_2O_3$ has been found to be 135.6 GPa using Birch-Murnaghan equation of state ( Fig. 3) which is also higher than that of $La_2O_3$ (113 GPa). This suggests that the stability region of hexagonal phase increases with the increasing Ln atomic number or in other words with decreasing cation radii. This observation is similar to the behaviour of cubic to hexagonal phase transition observed in other rare-earth sesquioxides with further smaller cation radii.

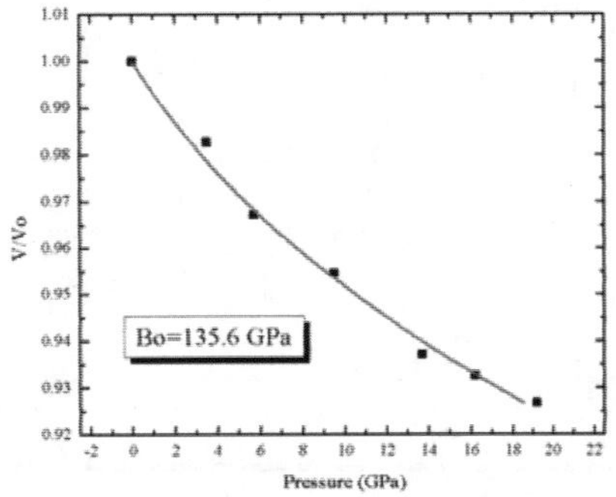


Figure 3. EOS of hexagonal (Type-A) phase of Nd2O3

## 4. Conclusion

Our EDXRD and Raman measurements show that $Nd_2O_3$ transforms to a lower symmetry phase at ~27 GPa. As observed in $La_2O_3$, this could be due to the relatively different bond length compressibility of the two RE-O bonds in the hexagonal structure. However the high pressure structure may not be similar to the superlattice hexagonal structure conjectured in case of $La_2O_3$. Our structural analysis suggests the high pressure phase of $Nd_2O_3$ to be of monoclinic symmetry. Further high pressure experiments based on high resolution angle dispersive x-ray diffraction measurements might give further insight into the new phase of $Nd_2O_3$.

## References

[1] Dayana Lonappan et. al, 2010 *Mat. Chem. Phys.* **120**, 65
[2] S. Cotton 2006 *Lanthanide and Actinide Chemistry*
[3] M. Zinkevich 2007 *Prog. In Mat. Sci.* **52,** 597
[4] V. M. Goldschmidt, F. Ulrich and T. Barth 1925, *Skrifter Norske Videnskaps-Akad, Oslo, I : Mat. Naturev, Kl.* **5**
[5] Henry R. Hoekstra 1965, *Inorganic Chemistry* **5**, 754
[6] J. P. McClure 2009 Thesis, Univ. of Nevada Las Vegas

Journal of Physics: Conference Series **377** (2012) 012007 doi:10.1088/1742-6596/377/1/012007

# Strength of solid helium under high pressure

**A K Singh** [1]
[1]Materials Science Division, National Aerospace Laboratories, Council of Scientific and Industrial Research (CSIR), Bangalore 560017, India

E-mail: aksingh@nal.res.in

**Abstract.** Helium is commonly used as a pressure transmitting medium to render the sample pressure hydrostatic in high pressure experiments with diamond anvil cells. On solidification under pressure helium begins to develop strength that is characteristic of any solid. The estimation of the compressive strength of helium as a function of pressure is important as it helps in estimating the nonhydrostatic stresses that can develop in the sample even when it remains immersed in helium pressure transmitting medium. The x-ray diffraction data on polycrystalline samples obtained with the radial diffraction geometry are commonly used to determine the compressive strength of solids. This method fails in the case of helium because of its low scattering power. The nonhydrostatic stresses that develop in crystalline solids immersed in helium pressure transmitting medium were used to estimate the strength of helium. The diffraction data available in the literature were selected for this study. It was important for this study to consider the data from the experiments that were conducted ensuring that the samples did not come in contact with the anvils. The analysis of these data suggests that the strength of helium remains low (< 0.1 GPa) at pressures below 20 GPa and increases to ~2 GPa at 100 GPa.

## 1. Introduction

X-ray diffraction experiments using diamond anvil cells are routinely conducted to study a variety of pressure-induced effects in solids. For a meaningful interpretation of the high-pressure data and its comparison with the theoretical predictions, it is important to carry out the measurements under hydrostatic pressure. The pressure on the sample is rendered hydrostatic by containing the sample and a fluid pressure-transmitting medium in a metal gasket. The performance of a fluid as a pressure medium depends on its inability to support shear stresses. Ideally, the pressure medium should remain in fluid state and fully surround the sample. With increasing pressure, the fluids exhibit increase in viscosity and ultimately solidify at high enough pressures [1]. This leads to the appearance of nonhydrostatic stresses that increase with increasing pressure. Invariably, a pressure medium is much more compressible than the solid sample. This can result in the sample coming in contact with the anvils at pressures much below the pressure range where appreciable viscosity increase or solidification of the pressure medium occurs. The presence of the pressure medium in such cases becomes redundant. This problem has to be dealt with through careful assembly of the high-pressure cell. Helium, the best-known pressure medium, crystallizes under pressure [2, 3] but retains fluid-like flow property to much higher pressures than any other fluid. Eventually, it is expected to develop high mechanical stiffness to be able to support detectable nonhydrostatic stresses. These stresses are transmitted to the sample and the pressure on the sample begins to deviate from hydrostatic. It is therefore important to estimate the compressive strength of helium as a function of pressure. This will define the lower limit of nonhydrostaticity of the sample pressure that will be present even when helium pressure medium is used.

Several investigators have used helium pressure-medium and followed as function of pressure the parameter like diffraction line-width or the quality of Brillouin scattering signals [4–8] to detect the onset of nonhydrostaticity. These studies suggest a hydrostatic-pressure limit that varies between 35 and 55 GPa. Such a wide range in the estimates, obviously, is due to the finer details of the loading of the sample and pressure medium in the metal gasket and the sensitivity to nonhydrostatic stresses of the parameter used for detection. In any case, these estimates are qualitative. The measurement of stresses has to be done through the measurement of strains these stresses produce. These strains (lattice strains) are derived from the shift of the diffraction-peak positions. Of course, appropriate elastic moduli are required to convert the measured strains to the stresses that produce them. Since the samples are in the form of polycrystalline aggregates, estimation of the appropriate elastic moduli is not straightforward. This necessitated the modelling of stresses in the sample compressed between the anvils [9, 10] and study of the effect of such stresses on the strains measured by x-ray diffraction [11–17]. The equations connecting the x-ray measured strains and stresses are commonly known as the lattice strain equations. These equations have been extensively used in deriving from the diffraction data the strength and elasticity of crystalline solids as a function of pressure. We present a brief account of this theory before discussing the strength of solid helium.

## 2. Lattice strain equations

The stress state of a solid sample under nonhydrostatic compression is axially symmetric about the load axis of the diamond anvil cell and is described by two equal stress components $\sigma_{11}$ along the two orthogonal set of axes parallel to the anvil face and a component $\sigma_{33}$ along the bad axis [9, 10]. The off-diagonal terms in the stress tensor are assumed to be zero. The off-diagonal terms are often present in actual experiments. The effect of these terms on the lattice strain equations has been discussed in earlier publications [18, 19]. The difference between axial and radial stress components is termed uniaxial stress component or differential stress denoted by $t$. The maximum value of $t$ is a measure of the compressive strength of the sample material at a pressure that equals the equivalent hydrostatic pressure (mean normal stress). The following relations can be written down

$$(\sigma_{33} - \sigma_{11}) = t \le \sigma_Y \quad (2.1)$$

$$\sigma_P = (2\sigma_{11} + \sigma_{33})/3 \quad (2.2)$$

The term $\sigma_Y$ denotes the yield strength of the solid sample at a confining pressure (equivalent hydrostatic pressure) $\sigma_P$. The $d$-spacing measured under such a stress state is given by [13–15]

$$d_m(hkl) = d_P(hkl)[1 + (1 - 3\cos^2\psi)Q(hkl)] \quad (2.3)$$

$$Q(hkl) = (t/6)\{\alpha[G_R^X(hkl)]^{-1} + (1-\alpha)(G_V)^{-1}\} \quad (2.4)$$

The subscripts m and P denote the quantities under nonhydrostatic (measured) and hydrostatic compression, respectively, and $\psi$ is the angle between the reflecting-plane normal and load axis of the diamond anvil cell. The term $G_R^X(hkl)$ is the diffraction shear modulus computed from the single-crystal elastic moduli under assumption of stress continuity across the boundaries separating the crystallites in the aggregate, $G_V$ is the aggregate shear modulus computed under the assumption of strain continuity, and $\alpha$ is a weight factor [13–15, 18]. Equations (2.1–2.4) are general and valid for all crystal systems. It is seen from equation (2.3) that, in presence of nonhydrostatic stresses, $d_m(hkl) > d_P(hkl)$ even though both are at same pressure $\sigma_P$. The differential stress is given by the following relation [15]

$$t = 6G\langle Q(hkl)\rangle \quad (2.5)$$

The term <*Q*(*hkl*)> represents the average of *Q*(*hkl*) over all observed reflections. The parameter $t$ is compressive and is taken to be positive by convention. The symbol $G$ denotes the pressure dependent shear modulus. The measurement of $t$ offers a unique method of determining the compressive strength of crystalline solids as a function of pressure. The use of equation (2.5) to determine $t$ requires $d_m(hkl)$ versus $\psi$ data. Such data can be obtained with the radial diffraction geometry [17]. The strength of solid argon under pressure has been measured using radial diffraction technique, which gives the strength without any ambiguity. The strength of argon is small at low pressures but increases rapidly with pressure [20]. Helium too is expected to show similar behaviour. However, because of low scattering power for x-rays, recording of radial diffraction patterns of polycrystalline helium is challenging. Because of this difficulty we use the high-pressure x-ray diffraction data from polycrystalline solids recorded with helium pressure medium. The strength of helium is derived from the nonhydrostatic stresses it transmits to the sample. Such data have been recorded using the conventional diffraction geometry wherein the incident x-ray beam passes parallel to the load axis of the diamond anvil cell. In case of the cubic system, $a_m(hkl)$ versus $3(1\text{-}3\sin^2\theta)\Gamma(hkl)$ plot (gamma plot) is a straight line given by [17],

$$a_m(hkl) \cong M_0 + M_1[3(1-3\sin^2\theta)G(hkl)] \quad (2.6)$$

$$M_0 = a_P\{1+(\alpha t/3)(1-3\sin^2\theta)[(S_{11}-S_{12})-(1-\alpha^{-1})(2G_V)^{-1}]\} \quad (2.7)$$

$$M_1 = -a_P \alpha t S/3 \quad (2.8)$$

$$G(hkl) = (h^2k^2+k^2l^2+l^2h^2)/(h^2+k^2+l^2)^2 \quad (2.9)$$

$$S = (S_{11} - S_{12} - S_{44}/2) \quad (2.10)$$

$$t \cong -3M_1/[\alpha M_0 S] \quad (2.11)$$

Equation (2.11) suggests that the estimation of $t$ from the diffraction data recorded using the conventional geometry depends on $\alpha$. The value of $\alpha$ is not known *a priori*. In the early work in this field, $\alpha$ was assumed to lie between 0 and 1. However, it has been shown recently [18, 19] that this limit is valid only for solids with positive $S$, and $\alpha = 0.5$ is a good approximation in this case. For solids with negative $S$, $\alpha$ is greater than 1.

### 3. Strength of solid helium

For the present discussion we use the published x-ray diffraction data on niobium [21, 22], gold [8, 23, 24] and silver [25]. The data from these studies are chosen as the investigators ensured that the sample did not come in contact with anvils. Both $\sigma_P$ and $t$ in such cases are transmitted to the sample by the helium pressure medium. These studies [21–25] were conducted to measure the equation of state under hydrostatic pressure. The data were analyzed in these studies to assess the hydrostaticity of pressure by estimating $t$ in the polycrystalline samples using equation (2.11) and assuming $\alpha = 1$. In view of the recent studies [18, 19] it turns out that $t$ derived with $\alpha = 1$ for gold was underestimated while for niobium overestimated [21–25]. We recalculated $t$ for gold with $\alpha = 0.5$. The $t$-values for niobium published earlier [22] were recalculated using the $\alpha$-values as a function of pressure obtained in a recent work [19]. The data on niobium in helium pressure medium from another study [21] show abnormally large values and are not considered here. The recalculated $t$-values for gold and niobium are shown in figure 1. Also shown in figure 1 are the $t$-values of silver derived from the diffraction line-width analysis [25].

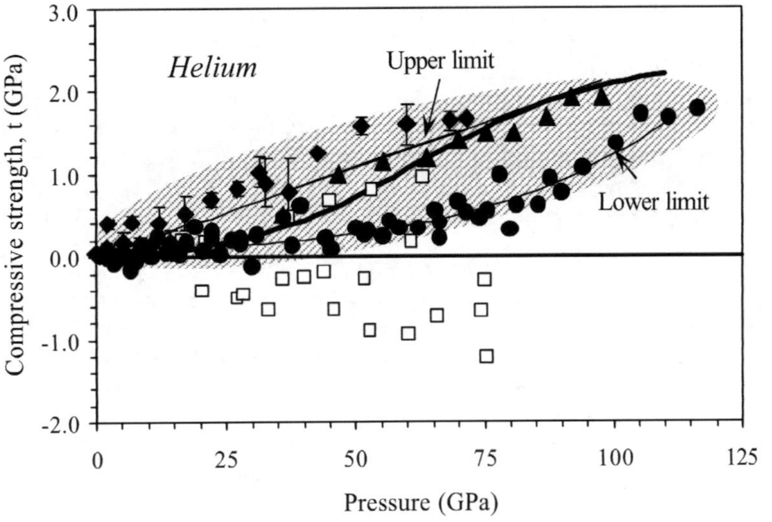


**Figure 1.** Recalculated values of $t$ in niobium [22] – filled diamonds, gold [23, 24] – dots (ESRF data), and unfilled squares (Photon factory data). For details see [24]. Line-width analysis of silver data [25] – filled triangles. Suggested pressure dependence of strength of helium – thick line.

In case the samples are fully immersed in the helium pressure medium, the measured $t$ should be independent of the sample material as it is $t$ of the solid helium that is expected to be transmitted to the crystalline sample. Of course, implicit in this argument is the assumption that niobium, gold and silver have higher compressive strength than helium. Contrary to this expectation, the $t$–$P$ data show considerable spread. Large spread is observed even among the data on the same material. For example, the $t$-values of gold exhibit a spread over negative to positive values. What is surprising is the occurrence of negative $t$. One set of data on niobium [22] also gave large negative $t$-values (not shown in figure 1). This situation does not arise when a solid sample is directly compressed between the anvils but is often observed in the experiments with helium pressure medium. Since $t = (\sigma_{33} - \sigma_{11})$, negative $t$ would imply $\sigma_{11} > \sigma_{33}$. It is likely that highly compressible helium present in the cell as the pressure medium allows the metal gasket to collapse inward and leads to a situation where $\sigma_{11} > \sigma_{33}$. We term this as 'abnormal compression' of the gasket. The magnitude of the measured $t$ in such cases depends on the deformation characteristic of the helium filled gasket. This possibly can explain rather large scatter observed in the $t$–$P$ data shown in figure 1. If it were possible to fine-tune the deformation of the gasket, one could possibly achieve a condition with $\sigma_{33} = \sigma_{11}$, which would imply $t = 0$. From practical point of view this situation is of great interest as the unit cell parameters measured under this condition would be free from the systematic effect of nonhydrostatic stresses defined by equation (2.3). This, however, does not imply that pressure is truly hydrostatic. The shear stresses in the form of micro-stresses can still be present and the signatures of the resulting micro-strains could be seen in the diffraction line-broadening. The strength of the sample in such cases could still be determined from the line width analysis [26]. It may be pointed out that equation (2.11) could not be used to measure $t$ in aluminium because it has small $S$ but the line-width analysis could be used successfully [27]. It is important to point out that $M_1 = 0$ in the gamma plot occurs when either $S$ or $t$

equals zero. As it is seen from equation (2.6), the systematic effect of nonhydrostatic stresses on the unit cell parameter still persists in case $S = 0$ but $t \neq 0$.

We exclude from the present discussion the negative $t$-values and consider only the data bound by the shaded region. The positive $t$-values show a lower limit predominantly determined by the $t$-$P$ data (dots) of gold. The $t$-$P$ data of silver and niobium mark the upper limit. Noting that $t$-values of gold also could be influenced by the abnormal compression of the gasket, it is tempting to conclude that the upper limit represents the strength of solid helium. However, we give credence to the data close to the lower limit and obtain a weighted average of the two limits. The $t$-values for niobium are influenced by the larger uncertainty in $\alpha$ for niobium in the low pressure range [19]. We assign arbitrarily higher weight (2:1) to the data lying close to the lower limit for pressures below 30 GPa. At higher pressure the weights are reversed. The averaged data are smoothened by fitting a function containing a three-parameter sigmoidal function and a linear term. The resulting curve is shown by the thick line in figure 1. To express the strength data as a fraction of the pressure-dependent shear modulus, we used the experimental data between 13 and 32 GPa [28]. We fitted a two-parameter logarithmic equation to the measured data and used this equation to compute shear moduli at required pressures. The commonly used finite strain equation [29] could not be used because only six data points were available for this fit. The strength of solid helium remains small ($< 0.005G$) at pressures below 20 GPa and increases to $0.03G$ at 100 GPa. These data should be considered only rough estimates because of the uncertainties in the extrapolation of the shear modulus data over a large pressure range.

## 4. Discussion

The analysis of diffraction data recorded under nonhydrostatic compression gives the uniaxial stress component $t$. For $t$ to be equal to the yield strength it is important that the maximum value of $t$, which the sample material can support, should develop in the experiments. In most cases, this condition is achieved by compressing the solid sample directly between the anvils without any pressure medium. In such cases, the gasket begins to deform upon compression by outward flow. Such deformation is essential for the development of maximum $t$ in the sample. The situation is different when the gasket contains highly compressible material like helium. The deformation of the gasket may begin with inward flow of the gasket. This gives rise to the $t$-values that are lower than the maximum and, in the extreme case, can result in negative $t$. There is no way of ensuring the development of maximum $t$ in these experiments. For this reason, the strength of helium obtained here may have been underestimated. The $t$-values derived from the diffraction data recorded with the conventional geometry is influenced by the choice of $\alpha$. Uncertainty in $t$ arising from the uncertainty in $\alpha$ could be as high as ±20%. The uncertainty in $t$ from this source could be eliminated if the radial diffraction data become available but the effect of abnormal compression of the gasket will continue to persist.

Tempere and Silvera [30] recently analyzed the pressure distribution inside the cell containing solid sample (pressure marker) and helium pressure medium using finite element analysis. These investigators show that the pressures of the sample and pressure medium differ. This difference is a function of the elastic properties of the sample and pressure medium. Further, pressure gradients exist in the pressure medium in the vicinity of the sample. In the present study, the pressure of the medium is inferred from the pressure of solid sample. The difference between the two pressures is expected to be only a few gigapascal at the highest pressure. The second factor does not affect the diffraction line-shift and, therefore, will not affect $t$ determined from the gamma plots. However, the strength of silver inferred from the line-width analysis may be in error. In view of the overall uncertainty in estimating the strength of solid helium, the influence of both the factors is not significant.

The strength of argon, another inert gas, was determined as a function of pressure from the radial diffraction data in an earlier study [20]. The strengths of argon and helium are compared in figure 2. The strength of argon is nearly zero at low pressure and increases steeply as the pressure is increased. The strength, however, is unlikely to continue increasing at this rate and is bound to exhibit a gradually decreasing slope such that the strength asymptotically approaches infinity at infinite pressure. In comparison, helium also has near-zero strength at low pressures and increases with increasing pressure but not as steeply as for argon. At 50 GPa, the strength of argon is nearly seven times the strength of helium.

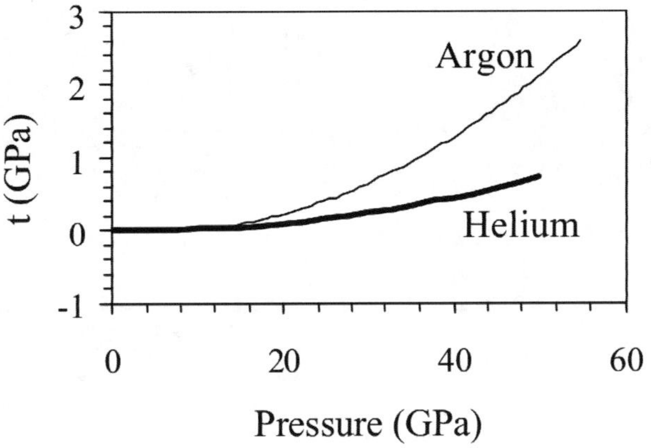


**Figure 2.** A comparison of the compressive strengths of helium and argon. Thick line – helium (this study). Thin line – argon. The strength of argon was determined from the radial x-ray diffraction data [20].

## 5. References


[1] Angel R J, Bujak M, Zhao J, Diego Gatta G D and Jacobsen S D 2007 *J. Appl. Cryst.* **40** 26–32
[2] Loubeyre P, LeToullec R, Pinceaux J P, Mao H K, Hu J and Hemley R J 1993 *Phys. Rev. Lett.* **71** 2272–5
[3] Mao H K, Hemley R J, Jephcoat A P, Finger L W, Zha C S and Bassett W A 1988 *Phys. Rev. Lett.* **60** 2649–52
[4] Bell P M and Mao H K 1981 *Carnegie Inst. Washington Publ.* **80** 404
[5] Alexandrov L V, Gonchrov A F, Zosman A N, and Stishov S M 1987 *Sov. Phys. JET* **66** 384
[6] Zhang L, Ahsbahs H, and Kuteglu A 1998 *Phys. Chem. Miner.* **25** 301
[7] Zha C S, Mao H K, and Hemley R J 2000 *Proc. Natl. Acad. Sci. U.S.A.* **97** 13494–9
[8] Takemura K 2001 *J. Appl. Phys.* **89** 662–8
[9] Singh A K and Kennedy G C 1974 *J. Appl. Phys.* **45** 4686–91
[10] Ruoff A L 1975 *J. Appl. Phys.* **46** 1389–92
[11] Singh A K 1993 *J. Appl. Phys.* **73** 4278–86: Erratum 1993 *J. Appl. Phys.* **74** 5920
[12] Uchida T, Funamori N and Yagi T 1996 *J. Appl. Phys.* **80** 739–6
[13] Singh A K, Balasingh C, Mao H-k, Shu J and Hemley R J 1998 *Rev. High Pressure Sci. Technol.* **7** 205–10
[14] Singh A K, Mao H-k, Shu J and Hemley R J 1998 *Phys. Rev. Lett.* **80** 2157–60
[15] Singh A K, Balasingh C, Mao H-k, Hemley R J and Shu J 1998 *J. Appl. Phys.* **83** 7567–75
[16] Matthies S, Priesmeyer G and Daymond M R 2001 *J. Appl. Cryst.* **34** 585–601
[17] Singh A K 2004 *J. Phys. Chem. Solids* **65** 1589–96
[18] Singh A K 2009 *J. Appl. Phys.* **106** 043514
[19] Singh A K and Liermann H-P 2011 *J. Appl. Phys.* **109** 113539
[20] Mao H-k, Badro J, Shu J, Hemley R J and Singh A K 2006 *J. Phys.: Condens. Matter* **18** S963–8
[21] Singh A K and Takemura K 2001 *J. Appl. Phys.* **90** 3269–75: Erratum Singh A K and Takemura K 2002 *J. Appl. Phys.* **91** 5507
[22} Takemura K and Singh A K 2006 Phys. Rev. B **73** 224119
[23] Dewaele A, Loubeyre P and Mezouar M 2004 *Phys. Rev.* B **70** 094112
[24] Takemura K and Dewaele A 2008 *Phys. Rev*. B **78** 104119

[25] Dewaele A and Loubeyre P 2007 *High Press. Res.* **27** 419–29
[26] Singh A K, Liermann H-P and Saxena S K 2004 *Solid State Commun.* **132** 795–8
[27] Singh A K, Liermann H-P, Akahama Y and Kawamura H 2007 *J. Appl. Phys.* **101** 123526
[28] Zha C-S, Mao H-k and Hemley R J 2004 *Phys. Rev.* B **70** 174107
[29] Birch, F 1978 *J. Geophys. Res.* **83**, No. B3 1257–68
[30] Tempere J and Silvera I F 2011 *J. Appl. Phys.* (under publ.)

# Strength of rhenium from x-ray diffraction experiments under nonhydrostatic compression to 250 GPa

**A K Singh [1*], J Hu [2], J Shu [2], H-k Mao [2] and R J Hemley [2]**

[1]Materials Science Division, National Aerospace Laboratories, Council of Industrial and Scientific Research (CSIR), Bangalore 560017, India
[2]Geophysical Laboratory, Carnegie Institution of Washington, 5251 Broad Branch Road, NW, Washington, DC 20015, USA

E-mail: aksingh@nal.res.in

**Abstract.** Polycrystalline rhenium in the form of a powder contained in a two-stage gasket was compressed in a diamond anvil cell. No pressure medium was used to ensure that maximum nonhydrostatic stresses that rhenium can support were produced. The pressure was increased in steps to a maximum of 250 GPa and the diffraction pattern recorded at each pressure using energy dispersive technique. The differential stress $t$, a measure of the compressive strength, was determined at any given pressure from the offset between the measured unit cell volume and volume computed from the pressure-volume relation under hydrostatic pressure. The data suggest that $t$ is 2.5 GPa at a pressure of 5 GPa and increases linearly to 26 GPa at 250 GPa. The present data agree well with those obtained from the radial diffraction data to 37 GPa in an earlier study but differ significantly from the results of another study. Though comparable with the theoretical strength of ideal solids, the observed strengths are extremely large for a polycrystalline aggregate of a solid. The pure pressure effect on the strength described by the shear modulus scaling is inadequate to account for such a large increase of strength with pressure. It is suggested that the major contribution to strength comes from the strain hardening of rhenium that arises due to plastic deformation of the sample during nonhydrostatic compression in a diamond anvil cell.

## 1. Introduction

Rhenium, a group-VII transition metal, crystallizes in the hexagonal-close-packed (hcp) structure. Measurements of the volume change produced by static [1] and dynamic [2] pressures indicate that Re is highly incompressible. High-pressure x-ray diffraction studies [3] do not indicate any pressure-induced phase transition up to 216 GPa and suggest a nearly pressure independent $c/a$-ratio. Ultrasonic velocity measurements give high values of the single-crystal elastic moduli confirming the low compressibility of Re [4]. First-principles computations of the elastic moduli [5, 6] and their pressure derivatives are found to be in reasonable agreement with the ultrasonic data. From the consideration of the Gibbs free energies of the hcp and body-centered-cubic (bcc) structures, Kaufman [7] suggested

[*] To whom any correspondence should be addressed.

that Re might transform to the bcc phase at high pressure. However, first-principles calculations predict that hcp Re is stable to pressures in the range of a terapascal [8].

The increase in the yield (flow) stress of Re with increasing plastic deformation, termed strain or work hardening, is found to be the largest of all metallic materials [9, 10]. The successful use of Re as a gasket material in high-pressure studies with a diamond anvil cell is due to the fact that it develops, on compression, high enough compressive strength to retain appreciable thickness even at several hundred gigapascal without any "punch-through" at the edges of the diamond anvils. Two studies [11, 12] carried out in the past reported the strength of Re as a function of high pressure. The first study [11] combined the pressure-volume data measured under nonhydrostatic compression and that under hydrostatic pressure to estimate the strength under pressure up to 120 GPa The second study [12] used radial diffraction data to determine the strength to 37 GPa. In this study, we derive the compressive strength of Re as a function of pressure up to 250 GPa using an approach that is essentially same as the one used in the earlier study [11]. However, the equations we use to analyze the data are based on the lattice strain theory [13–19] and differ from that used in the earlier study [11].

## 2. Experimental details

X-ray diffraction experiments were carried out at the wiggler beam line X17C of the National Synchrotron Light Source of Brookhaven National Laboratory, using the energy dispersive x-ray diffraction method. A diamond anvil cell with a pair of bevel diamonds was used. The culet size of one of the diamonds was 0.17 mm flat face and 0.48 mm at the beveled edge and that of the second diamond differed slightly (0.19 mm and 0.49 mm). A two-stage gasket was used to contain the Re sample. The outer gasket consisted of a pre-indented 0.025mm thick stainless steel disk with a 0.090 mm central hole. A 0.025 mm-thick Re disk with a 0.040 mm sample hole was inserted in the hole of the stainless steel gasket. Re powder (99.99%) mixed with Pt-powder was loaded in the 0.040 mm hole. Pt was used as a pressure standard. The primary incident beam slit was $0.1\times0.1\text{mm}^2$ and the collimator was $0.010\times0.010\text{ mm}^2$. The first receiving slit was $0.15\times0.35\text{ mm}^2$ and the second receiving slit was $0.1\times5$ mm. An intrinsic solid state Ge-detector was used to record the diffraction patterns. A fixed $2\theta_0 = 20°$ was used in the experiments.

## 3. Lattice strain equations

The stress state of a solid sample under nonhydrostatic compression is axially symmetric about the load axis of the diamond anvil cell and is described by two equal stress components $\sigma_{11}$ along the two orthogonal set of axes parallel to the anvil face and a component $\sigma_{33}$ along an axis parallel to the load axis. The off-diagonal terms in the stress tensor are assumed to be zero. This assumption is not always valid in practice. The effect of the breakdown of this assumption on the lattice strain theory has been discussed in detail elsewhere [20, 21]. The difference $(\sigma_{33} - \sigma_{11})$ is denoted by $t$, and its maximum value is limited by the compressive yield strength at the confining pressure that equals the mean normal stress $(2\sigma_{11} + \sigma_{33})/3$. At a given pressure, the unit cell volume $V_m$ measured under nonhydrostatic compression is larger than $V_P$, the unit cell volume under hydrostatic compression. As shown earlier [14], $t$ is given by the relation

$$t = \frac{V_m - V_P}{V_P} \frac{3K}{\langle(1-3\sin^2\theta_0)\rangle} \left[\frac{1-2v}{1+v}\right] \tag{3.1}$$

In terms of the shear modulus, equation (3.1) can be expressed as

$$t = \frac{V_m - V_P}{V_P} \frac{2\mu}{\langle(1-3\sin^2\theta_0)\rangle} \tag{3.2}$$

$K, \mu$ and $v$ denote the aggregate bulk modulus, shear modulus and Poisson's ratio, respectively. The negative sign in Eq. (33) of the earlier paper [14] appears because the convention of taking compressive stress as negative was used. Noting that the pressure offset is given by $\Delta P = K\,[(V_m - V_P)/V_P]$, equation (3.1) can be rewritten as

$$t = \frac{3\Delta P}{\langle(1-3\sin^2\theta_0)\rangle}\left[\frac{1-2\nu}{1+\nu}\right] \qquad (3.3)$$

Here, $\Delta P = (P_m - P_h)$. The terms $P_m$ and $P_h$ are pressures computed from the $P$-$V$ relation measured under nonhydrostatic and hydrostatic compressions, respectively, at a fixed $V/V_0$. Equations (3.1), (3.2) and (3.3) are equivalent if $P$ at which $V_m$ and $V_P$ in (3.1) are computed is chosen at $(P_m + P_h)/2$. It may be noted that the elastic moduli in these equations are to be taken at the relevant pressure.

**4. Results**

The diffraction pattern recorded at 251 GPa is shown in figure 1. The pressure in each run was determined by using the measured volume compression of Pt in the equation of state proposed by Holmes et al [22]. The cell volumes $V_m$ were determined from the measured $d$-spacings. These values together with the data obtained by Jeanloz et al [11] are shown figure 2. To determine $P$-$V_P$ data under hydrostatic pressure the fourth order Birch equation given below was used

$$F = \frac{P}{3f(1+2f)^{5/2}} = K_0[1 + a_1 f + a_2 f^2] \qquad (4.1)$$

$$f = [(V_0/V)^{2/3} - 1]/2 \qquad (4.2)$$

Here, $F$, and $f$, are the normalized pressure and Eulerian strain, respectively. $K_0$ is the zero-strain bulk modulus, and $a_1$, and $a_2$ are functions of $K_0$, $K_0'$, and $K_0''$ [23]. The single and double primes indicate the first and second pressure derivatives of the bulk modulus at zero strain. Equation (4.1) was fitted to the pressure-volume data obtained from shock wave measurements [2]. This gave $K_0$=364.97(±0.64), $a_1$=0.48(±0.05) and $a_2$=−2.03(±0.33). The $P$-$V_P$ data computed from equation (4.1) with these constants are shown in figure 2 by continuous lines. It may be noted that the constants in equations (4.1) used by Jeanloz et al [11] are significantly different from those used in this study. However, the $P$-$V_P$ data computed from the constants used by Jeanloz et al [11] do not show any significant difference with the present data up to 150 GPa. The strength $t$ was determined from equation (3.2). The pressure dependent $\mu$ required for these calculations was determined using $\mu_0$= 179 GPa and $\mu_0'$ = 1.8 in the Birch equation [24] viz.:

$$\mu = \mu_0 x^{5/3}[1 + \tfrac{1}{2}(3K_0\mu_0'/\mu_0 - 5)(x^{2/3} - 1)] \qquad (4.3)$$

The term $(1 - 3\sin^2\theta_0)$ appearing in equation (3.2) equals 0.91 here and is constant for all reflections. The compressive strength $t$ determined in this study as a function of pressure is shown in figure 3. The error bars indicate standard errors arising only from the uncertainty in $V_m$. Uncertainties in $\mu$ and $V_P$ are hard to estimate. The strength data from the earlier studies [11, 12] are also shown in figure 3.

**5. Discussion**

The pressure-volume data from this study and an earlier measurements to 119 GPa [11] are shown in figure 2. To compare the data from two studies we fitted the third-order Birch-Murnaghan equation to each set of data. The computed $V_0/V_m$ from this study is larger than that from [11] by 0.3% at 20 GPa and increases to 0.6% at 120 GPa. Though not relevant to this comparison, the difference begins to decrease above ~150 GPa and vanishes at ~225 GPa. The trend is reversed above 225 GPa. These differences are small in comparison with the combined errors of measurements in the two studies. It is seen that at any given pressure, $V_m > V_P$ due to the effect of nonhydrostatic compression.

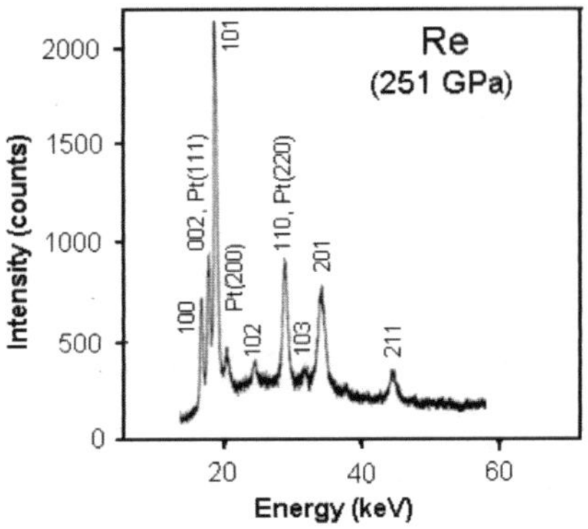


**Figure 1.** Diffraction pattern of Re.

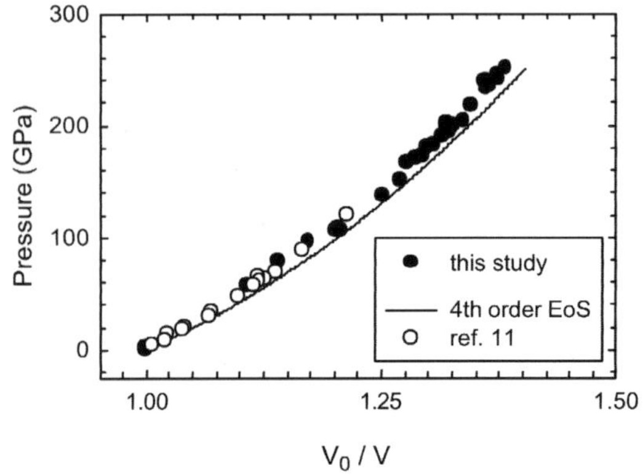


**Figure 2.** Volume compressions data under nonhydrostatic (symbols) and hydrostatic pressure.

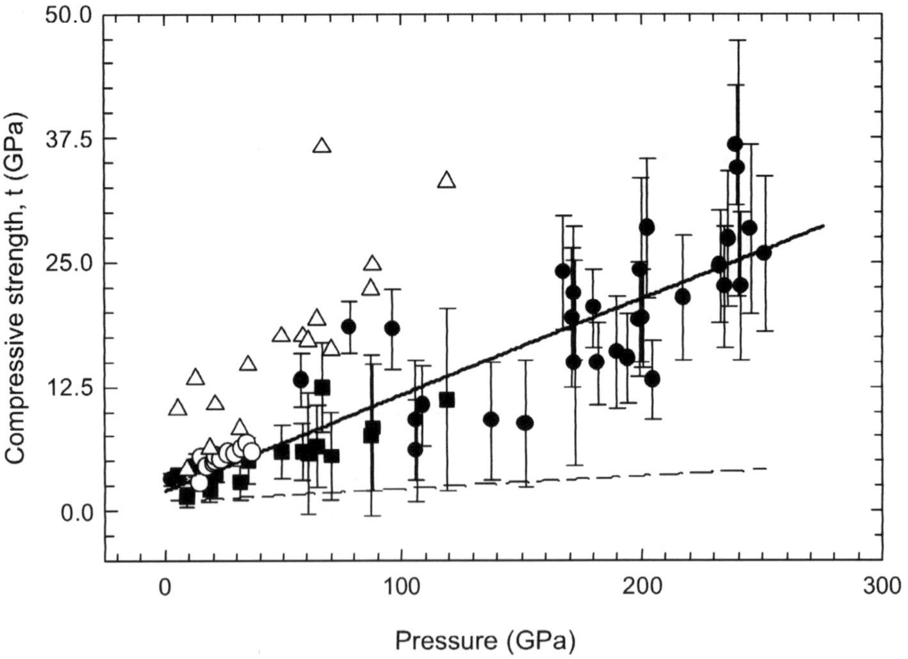


**Figure 3.** Strength of Re as function of pressure. Dots – present data, circles – Duffy et al. [12], triangles – Jeanloz et al. [11], and squares – recomputed from equation (3.2) using measured $V_m$ from [11]. Solid line – straight line fit through dots, circles, and squares gives $t = 2 + 0.096P$. Dashed line – shear modulus scaling taking ambient pressure strength as 1 GPa.

Duffy et al [12] used radial diffraction data to determine the strength. It is seen from figure 3 that the strength data of Duffy et al [12] are in good agreement with the present data. This agreement is not surprising as both the present set of formulae and those used by Duffy et al are based on the same lattice strain theory. The results of Jeanloz et al. [11], however, show large differences. Since the measured unit cell volumes under nonhydrostatic compression and the computed $V_P$ in [11] are in excellent agreement with those in this study, the differences in estimated strength solely arise due to the fact that the formula used to derive the strength differed in the two studies. It can be easily verified that equation (4) of [11] gives $t$ that is ~3-times larger than $t$ obtained from equation (3.3) for any given set of experimental data.

The pure pressure effect on yield strength is given by shear modulus scaling [25]. The ambient pressure yield strength of Re reported in the literature varies in the range 0.7-1.3 GPa [9]. Taking the ambient-pressure shear modulus as 179 GPa and yield strength as 1 GPa, the shear-modulus scaling suggests that the strength of Re under pressure is given by $0.006\mu$. Considering the pressure dependence of $\mu$, this effect predicts strength of ~5 GPa at 250 GPa. The strength measured in this study varies from $0.01\mu$ at 2.5 GPa to $0.04\mu$ at 250 GPa. These magnitudes, though comparable with the strength of ideal solids [26–28] are extremely large for the polycrystalline aggregate of a real solid. The plastic deformation during nonhydrostatic compression of the sample in a diamond anvil cell is an important factor in the interpretation of strength derived from the diffraction data. It changes the state of the sample in two ways: The crystallite size [29–31] and the sample thickness decrease with increasing compression. This decrease in grain size results in an increase in strength following the Hall-Petch relation [32, 33]

$$\sigma_Y = \sigma_0 + K_{HP} / D^{1/2} \quad (5.1)$$

Here, $\sigma_0$ is the yield strength at one atmosphere and $D$ is the grain size, which is taken as the crystallite size determined from diffraction line-width analysis [29–31]. The term $K_{HP}$ is a material-dependent constant called strengthening coefficient or the Hall-Petch parameter. In absence of the information on the grain-size reduction under nonhydrostatic compression and $K_{HP}$ of Re, the quantitative estimate of the grain-size contribution to strength is not possible. However, the magnitude of strengthening from this source can be estimated by assuming the value of $K_{HP}$ for pure Re as 12 GPa/√nm, a value close to those for hexagonal metals. With this value of $K_{HP}$, a grain size in the range of 10-20 nm at 250 GPa would account for a strength increase in the range of 3–4 GPa. Thus, the contribution to strength from both the shear modulus scaling and grain-size factor together is ~10 GPa at 250 GPa. The major contribution to strength of Re comes from the effect of plastic deformation on the flow (yield) stress known as strain or work hardening. The strain hardening effect is described by the relation [34],

$$\sigma_{fY} = k\varepsilon^n \quad (5.2)$$

Here, $\sigma_{fY}$ is the flow or yield stress, $k$ and $n$ are material constants, and $\varepsilon$ is the plastic strain. With the experimenta values [10], $k$ = 28.7 GPa and $n$ = 0.78 [10], Re shows the highest strain hardening effect of all metals. Equation (5.2) suggests that plastic strain of ~50% at 250 GPa can result in a strengthening of ~16 GPa. Such large plastic strains are not inconceivable at high nonhydrostatic compression. The constants $k$ and $n$ are expected to be pressure dependent [35].

In conclusion, the strength of Re at high pressure determined by x-ray diffraction is found to be extremely large. The increased strength under pressure cannot be explained by shear-modulus scaling. It is pointed out that the nonhydrostatic compression, which is essential for the strength determination by the diffraction method, causes severe plastic deformation of the sample. This results in the reduction of the crystallite size and sample thickness leading to strengthening due to grain size and work hardening. In the case of Re, work hardening effect appears to be the dominant factor leading to strengthening under pressure.

## 6. References

[1] Bridgman P W 1955 *Proc. Am. Acad. Arts Sci.* **84** 112–29
[2] McQueen R G, Marsh S P, Tayler J W, Fritz J N and Carter W J 1970 *High Velocity Impact Phenomenon*, (New York: Academic), p. 293
[3] Vohra Y K, Duclos S J and Ruoff A L 1987 *Phys. Rev.* B **36** 9790–2
[4] Manghnani M H, Katahara K and Fisher E. S 1974 *Phys. Rev.* B **9** 1421–31
[5] Fast L, Willis J M, Johansson B and Eriksson O 1995 *Phys. Rev.* B **51** 17431–8
[6] Steinle-Neumann G, Stixrude L and Cohen R E 1999 *Phys. Rev.* B **60** 791–9. Erratum: 2004 *Phys. Rev.* B **69**, 219903(E)
[7] Kaufman L 1967 *Phase Stability in Metals and Alloys*, edited (New York: McGraw -Hill) p.125
[8] Verma A K, Ravindran P, Rao S R, Godwal B K and Jeanloz R 2003 *Bull. Mater. Sci.* **26** 183–7
[9] Schwartz M M 2002 *Encyclopedia of Materials, Parts, and Finishes*, Second Edition, (Boca Raton: CRC Press) p 651
[10] Mannheim R L and Garin J L 2000 *Z. Metallkd.* **91** 848–53
[11] Jeanloz R, Godwal B K and Meade C 1991 *Nature* **349** 687–9
[12] Duffy T S, Shen G, Heinz D L, Shu J, Ma Y, Mao H-k, Hemley R J and Singh A K 1999 *Phys. Rev.* B **60** 156063–73
[13] Singh A K 1993 *J. Appl. Phys.* **73** 4278–86: Erratum 1993 *J. Appl. Phys*. **74** 5920
[14] Singh, A K and Balasingh C 1994 *J. Appl. Phys.* **75** 4956–62
[15] Uchida T, Funamori N and Yagi T 1996 *J. Appl. Phys.* **80** 739–46
[16] Matthies S, Priesmeyer G and Daymond M R 2001 *J. Appl. Cryst.* **34** 585–601
[17] Singh A K, Mao H-k, Shu J and Hemley R J 1998 *Phys. Rev. Lett*. **80** 2157–60
[18] Singh A K, Balasingh C, Mao H-k, Hemley R J and Shu J 1998 *J. Appl. Phys*. **83** 7567–75
[19] Singh A K 2004 *J. Phys. Chem. Solids* **65** 1589–96
[20] Singh A K 2009 *J. Appl. Phys.* **106** 043514
[21] Singh A K and Liermann H-P 2011 *J. Appl. Phys.* **109** 113539
[22] Holmes, N C Moriarty, J A Gathers, G R and Nellis, W J 1989 *J.Appl. Phys.* **66** 2962–7
[23] Jeanloz R 1989 *J. Geophys. Res.* **94** 5873–86
[24] Birch, F 1978 *J. Geophys. Res*. **83**, No. B3 1257–68
[25] Chua J O and Ruoff A L 1975 *J. Appl. Phys.* **49** 4659–63
[26] Kelly A and Macmillan N H 1986 *Strong Solids*, 3rd ed. (Oxford: Clarendon)
[27] Krenn C R, Roundy D, Cohen M L, Chrzan D C and Morris Jr. J W 2002 *Phys. Rev.* B **65** 134111
[28] Ogata, S Li, J Shibutani, Y and Yip, S 2004 *IUTAM Symposium on Mesoscopic Dynamics of Fracture Process and Material Strength* (Dordrecht: Kluwer) pp. 1-13
[29] Singh A K, Liermann H P and Saxena S K 2004 *Solid State Commun*. **132** 795–9
[30] Singh A K, Liermann H P and Saxena S K, Mao H K and Usha Devi S 2006 *J. Phys.: Condens. Matter* **18** S969–78
[31] Liermann H P, Jain A, Singh A K and Saxena S K 2010 *J. Phys. Chem. Solids* **71** 1088–93
[32] Hall E O 1951 *Proc. Phys. Soc. (London) B* **64** 747–53
[33] Petch N J 1953 *J. Iron Steel Inst.* **174** 25–8
[34] Dieter G E 1988 *Mechanical Metallurgy – SI Metric Edition* (Singapore: McGraw-Hill) p. 520
[35] Abey A E 1970 *J. Appl. Phys.* **41** 2537–41

Journal of Physics: Conference Series **377** (2012) 012009 doi:10.1088/1742-6596/377/1/012009

# Behaviour of NTE Material $Ag_3[Co(CN)_6]$ under Pressure

**Pallavi Ghalsasi[1,3*], H.K. Poswal[2], Ravi Tripathi[1], Nandini Garg[2], M.N. Deo[2] and Surinder M. Sharma[2]**

[1]Department of Physics, Indian Institute of Technology Gandhinagar, Ahmedabad-382424, Gujarat, India
[2]High Pressure and Synchrotron Radiation Physics Division, Bhabha Atomic Research Center, Mumbai-400085, India
[3](At Present), Department of Physics, The M.S. University of Baroda, Vadodara- 390 002, Gujarat, India

E-mail: pallavi.teredesai@gmail.com

**Abstract.** Recent discovery of colossal negative thermal expansion (NTE) behaviour in Silver Hexacyanocobaltate $Ag_3[Co(CN)_6]$ has triggered interest among researchers to understand the basic mechanism causing such an unusual behaviour. This report presents our results on the behaviour of $Ag_3[Co(CN)_6]$ using *in-situ* high pressure Raman and FTIR spectroscopy. The reported trigonal-monoclinic phase transition occurring in this compound is observed at 0.2 GPa. Upon increasing the pressure further, $Ag_3[Co(CN)_6]$ becomes irreversibly amorphous above 13 GPa.

## 1. Introduction

Negative thermal expansion materials have got practical applications in making composites where zero or very low coefficient of thermal expansion (CTE) is required. To make advancement in this applicability, however it is very important to understand the behaviour of these materials with respect to changes from ambient conditions. Research has progressed in this area after the recent discoveries of isotropic NTE property in framework materials like $ZrW_2O_8$ [1], Zn $(CN)_2$ [2] along with many dicyanometallates [3, 4] and many metal-organic frameworks (MOFs). These structures have many low energy rigid unit modes (RUMs) which can be observed spectroscopically as soft modes with negative Grüneissen parameters [2,5]. It is found that in these structures, bridging arrangement of atoms offers greater flexibility as the constraint is only on the metal atom distances. Many of these materials exhibit pressure-temperature dependent phase transitions and also pressure induced amorphization (PIA) [6, 7].

Recently silver hexacyanocobaltate $Ag_3[Co(CN)_6]$ was shown to have colossal anisotropic negative thermal expansion with a CTE having magnitude much larger than that of many NTE materials including $ZrW_2O_8$ [8,9]. $Ag_3[Co(CN)_6]$ is a framework material assembled from highly under constrained Co-CN-Ag-NC-Co linkages. Its crystal structure consists of alternating layers of $Ag^+$ and $[Co(CN)_6]^{3-}$ ions, stacked parallel to the unique axis of its trigonal $P\bar{3}1m$ unit cell. Within any given silver containing layer, the Ag atoms are arranged at the vertices of a Kagome lattice with the octahedral $[Co(CN)_6]^{3-}$ ions positioned above and below the hexagonal Kagome holes. This compound behaves like a garden fencing where a contraction along c-direction is coupled to an expansion along a-direction. The compound also exhibits a negative linear compression (NLC) [9]. It is understood that the flexible framework structure along with the Ag...Ag argentophillic interactions give rise to such an unusual NTE effect [10, 11]. Present work is aimed for getting better insight for the high pressure behaviour of the compound pertaining to its NLC property and also to look for possible pressure induced amorphization (PIA). Hence *in-situ* high pressure Raman, FT infrared spectroscopic investigations have been carried out and the data is presented in accordance to our current understanding.

## 2. Experimental Details

White polycrystalline $Ag_3[Co(CN)_6]$ was synthesized by mixing aqueous solutions of $K_3[Co(CN)_6]$ and $AgNO_3$ following precipitation as described in literature [8]. The sample was characterized by infrared spectroscopy as well as X-ray diffraction.

*In-situ* high pressure micro-Raman scattering measurements were done using a Jobin Yvon HR460 single stage monochromator and diode pumped frequency doubled solid state laser (532 nm) was used for excitation. Non-hydrostatic pressure Raman measurements were carried out without any pressure transmitting medium. High pressure was generated using a Mao Bell type diamond anvil cell (Diacell Lever DAC-Mega). *In-situ* quasi-hydrostatic pressure transmission spectra were recorded using Bruker's VERTEX 80V Fourier transform infrared spectrometer attached with a HYPERION 2000 infrared microscope and a mid-infrared globar source. High pressure was generated using a symmetric diamond anvil cell equipped with CVD grown synthetic diamonds. The sample was prepared by making a homogeneous mixture of 2% $Ag_3[Co(CN)_6]$ in CsI. The spectra were recorded using liquid nitrogen cooled HgCdTe (MCT) detector. The pressure calibration in both Raman as well as IR experiments was done by ruby fluorescence technique [12].

## 3. Results and Discussion

At ambient conditions $Ag_3[Co(CN)_6]$ exists in trigonal phase $Ag_3[Co(CN)_6]$-I in space group $P\bar{3}1m$. This phase has centre of symmetry and consists of three interpenetrating distorted cubic networks having $CoC_6$ octahedra at the cube vertices and edges formed by approximately linear –CN–Ag–NC– linkages. At 0.2 GPa, the compound is shown to undergo a structural phase transition from trigonal to monoclinic phase (space group *C2/m*), $Ag_3[Co(CN)_6]$-II accompanied by a very large volume change $\Delta V/V$=16.25% [9].

Owing to its centrosymetric structure, Raman and infrared active modes are mutually exclusive. To our best knowledge, there are no reports on group theoretical calculations for $Ag_3[Co(CN)_6]$ to explain the normal modes of vibration. Since $H_3[Co(CN)_6]$ is isostructural [13] to the compound of present study, direct correlation has been drawn from the vibrational modes of $H_3[Co(CN)_6]$ [14] and with comparison to that of $K_3[Co(CN)_6]$ [15]. Figure 1(a) shows infrared spectrum and 1(b) shows Raman spectrum of $Ag_3[Co(CN)_6]$ recorded at ambient conditions. The infrared spectrum shows a strong mode at 2184 $cm^{-1}$ corresponding to C≡N stretching vibrations. Also, there are modes observed at 581 $cm^{-1}$ corresponding to Co-C≡N bending and two sharp modes in 4000 $cm^{-1}$ region which are mainly combination of C≡N stretching Raman and IR active modes. Apart from these modes, there are several modes observed between 760-950 and 2300-2950 $cm^{-1}$, some of which may be combination modes. Ambient Raman spectrum of $Ag_3[Co(CN)_6]$ shows strong modes at 139, 165, 475 and 522 $cm^{-1}$ corresponding to C-Co-C deformation, N-Ag-N bending, Ag-N stretching and Co-C stretching respectively. Two very strong modes appear at around 2183 $cm^{-1}$ and 2201 $cm^{-1}$ corresponding to C≡N stretching. All these observed mode frequencies match well with the expected mode frequencies and with those reported in recent study [16], (Table 1). The evolution of Raman modes with increasing non-hydrostatic pressure up to 17 GPa is shown in Figure 2. The compound becomes amorphous around 13 GPa and the process of amorphization is found to be irreversible. There are significant changes in the spectra around 0.2 GPa at which $Ag_3[Co(CN)_6]$-II is formed.

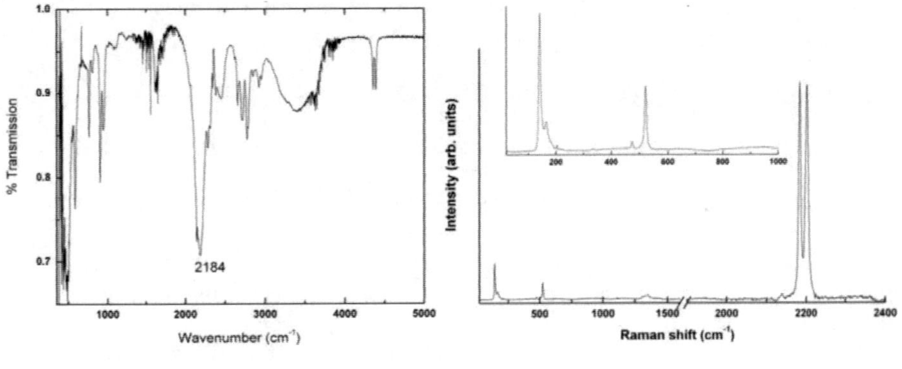


**(a)** **(b)**

**Figure 1 :** Vibrational Spectra of $Ag_3[Co(CN)_6]$ at ambient conditions (a) Infrared spectrum (b) Raman spectrum

The mode at 139 $cm^{-1}$ could not be traced beyond 0.2 GPa whereas the one at 165 $cm^{-1}$ increases in frequency with pressure. The Co-C stretching 522 $cm^{-1}$ mode disappears at 0.2 GPa and a new mode at 481 $cm^{-1}$ appears. The two C≡N stretching modes at 2183 $cm^{-1}$ disappear at 0.2 GPa and two new modes appear at lower frequencies at around 2140 $cm^{-1}$ and 2158 $cm^{-1}$ which harden with pressure. One more peak appears at 2162 $cm^{-1}$ around 7.5 GPa. This may be associated with puckering of silver atoms in the 2-D lattice leading to a triangular network [9]. To our observation, none of these modes show negative pressure dependence. It was difficult to trace a peak at 335 $cm^{-1}$ due to Co-C≡N bending in our data due to poor signal to noise ratio. Interestingly, this peak has been observed to show negative pressure dependence by a recent study [16]. The variation of Raman mode frequencies with pressure are shown in Figure 2 and the values of the slopes $d\omega/dP$ for each of the modes are mentioned alongside each mode frequency variation shown with linear fits.

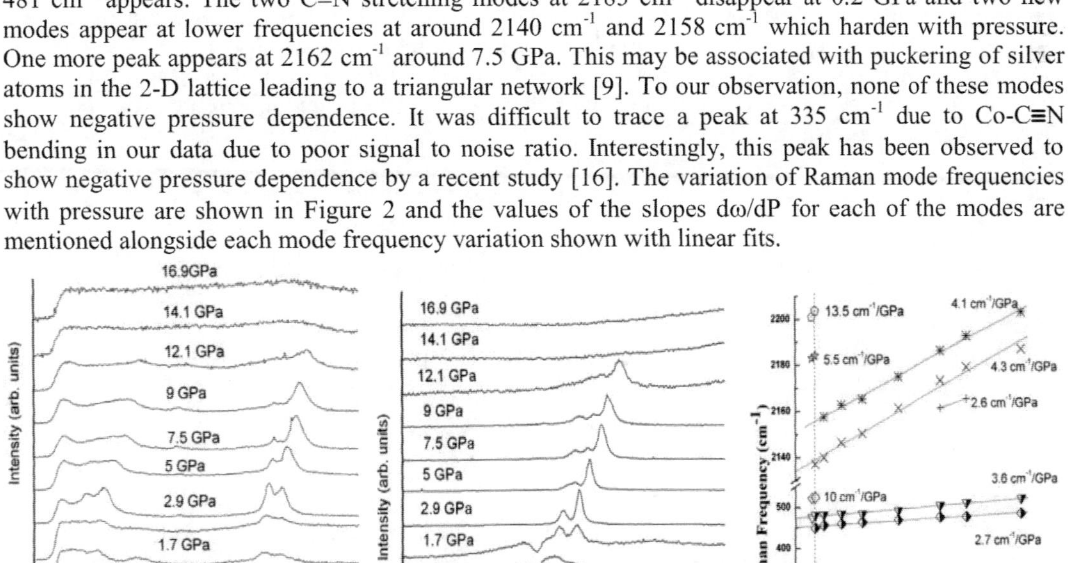


**Figure 2:** Raman spectra of $Ag_3[Co(CN)_6]$ recorded under non-hydrostatic pressure environment and variation of Raman mode frequency with pressure

| ν (cm$^{-1}$) | Vibration | ν (cm$^{-1}$) | Vibration |
|---|---|---|---|
| (R)139 | C-Co-C deformation | (IR) 760-950 | Combination modes |
| (R) 165 | N-Ag-N bending | (IR) 2184 | C≡N stretching |
| (R) 475 | Ag-N stretching | (R) 2183, 2201 | C≡N stretching |
| (R) 522 | Co-C stretching | (IR)2300-2950 | Combination modes |
| (IR) 581 | Co-C≡N bending | (IR) 4000 | Combination of C≡N stretching R & IR modes |

**Table 1:** Observed vibrational frequencies in $Ag_3[Co(CN)_6]$ at ambient conditions

Before we discuss the results of in-situ high pressure infrared spectroscopy, we would like to point out that the C≡N stretching vibrational mode around 2185 cm$^{-1}$ does get shifted to around 2128 cm$^{-1}$ along with red shift of all the other IR modes at ambient conditions if the sample is prepared by mixing of $Ag_3[Co(CN)_6]$ with other transparent matrix like KBr [17, 18]. There are two possibilities to understand this: (1) the pressure of 0.2 GPa required to transform $Ag_3[Co(CN)_6]$ –I to $Ag_3[Co(CN)_6]$ –II can easily be achieved while preparing the sample for IR experiments. If this phase transition is irreversible then it is hard to study this compound using such matrix materials for IR measurements. (2) In case of KBr matrix, small traces of water in the title compound can facilitate exchange of Ag and K producing AgBr and $K_3[Co(CN)_6]$ for which the C≡N stretching mode appears at 2128 cm$^{-1}$. Note that our high pressure IR measurements refer to evolution of the spectrum with pressure for which ambient IR spectrum does show the C≡N peak shift to 2128 cm$^{-1}$. CsI, being less reactive than KBr, the second of the above possibility seems too rare. So presently, we are inclined to conclude that the effect is related to the formation of $Ag_3[Co(CN)_6]$ –II due to hand driven pressure in mortar and pestle. In order to completely rule out the second possibility, we would like to carry out our experiments with $CaF_2$ as a matrix which is negligibly reactive as compared to KBr and CsI. The ambient IR spectrum of $Ag_3[Co(CN)_6]$ does not show the peak shift to 2128 cm$^{-1}$ as it is measured on nascent $Ag_3[Co(CN)_6]$ compound in CsI (without any grinding) by DRIFT method. Figure 3 shows the behaviour of vibrational modes with pressure as studied by micro FTIR experiments. Modes only above 2700 cm$^{-1}$ could be followed due to limitation of MCT detector and strong absorption in 2000 cm$^{-1}$ region by the diamonds.

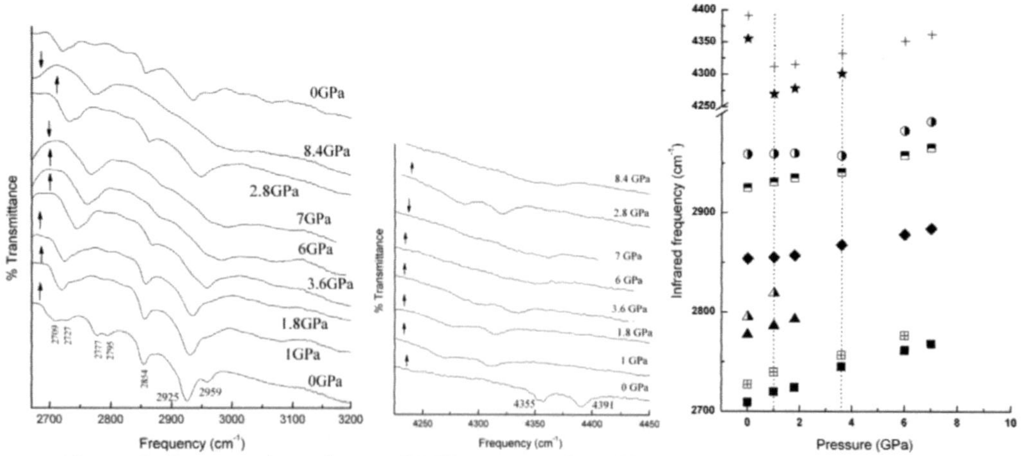


**Figure 3:** Pressure dependence of FTIR spectra of $Ag_3[Co(CN)_6]$ and IR mode frequencies

The pressure in IR run was increased to 7 GPa. The compound was then decompressed to 2.8 GPa and again pressurized to 8.4 GPa. Finally the pressure was completely released. To our surprise, we found that the structure of $Ag_3[Co(CN)_6]$ –II is quite flexible and shows almost complete reversibility except some hysteresis in the mode frequencies. With pressure, all the modes in 2700-3200 cm$^{-1}$ region show positive pressure dependence. The original modes in 4000 $cm^{-1}$ region which are combinations of IR and Raman active C≡N stretching vibrations still exist in the sample filled inside DAC at lowest pressure and two new modes appear on lower frequency side which are very weak may be due to diamond absorption. With increase in pressure, the original modes lose their intensity and the new modes exhibit positive pressure dependence. Figure 3 shows pressure variation of IR mode frequencies. The rest of the modes show a marked change in pressure dependence dω/dP, at around 3.6 GPa. The reason for this slope change needs to be understood.

A detailed study of our *in-situ* high pressure infrared as well as X-ray diffraction results is in progress.

## 4. Conclusion

We have studied behaviour of polycrystalline $Ag_3[Co(CN)_6]$ under non-hydrostatic pressure environment using Raman and under quasi-hydrostatic pressure environment using FTIR spectroscopy spanning the pressure range across both the phases of $Ag_3[Co(CN)_6]$. This compound shows irreversible amorphization above 13 GPa. We didn't notice negative pressure dependence for any of the observed vibrational modes and more detailed study needs to be done in this regard. We also observed flexibility in the structure of $Ag_3[Co(CN)_6]$ –II from our high pressure FTIR data. It would be of importance to know if any of the observed modes can be related to Ag...Ag argentophillic interactions.

## 5. Acknowledgement

PG would like to thank IIT Gandhinagar for financial support.

## 6. References

[1] Mary T A, Evans J S O, Vogt T and Sleight A W 1996 *Science* **272** 90
[2] Goodwin A L and Kepert C J 2005 *Phys. Rev.* B **71** 140301(R)
[3] Goodwin A L, Kennedy B J and Kepert C J 2009 *J. Am. Chem. Soc.* **131** 6334
[4] Korcok J L, Katz M J and Leznoff D B 2009 *J. Am. Chem. Soc.* **131** 4866
[5] Zwanziger J W 2007 *Phys. Rev.* B **76** 052102
[6] Perottoni C A and da Jornada J A H 1998 *Science* **280** 886
[7] Catafesta J, Haines J, Zorzi J E, Pereira A S and Perottoni C A 2008 *Phys. Rev.* B **77** 064104
[8] Goodwin A L, Calleja M, Conterio M J, Dove M T, Evans J S O, Keen D A, Peters L and Tucker M G 2008 *Science* **319** 794
[9] Goodwin A L, Keen D A and Tucker M G 2008 *Proc. Natl. Acad. Sci.* USA **48(105)** 18708
[10] Goodwin A L ,Keen D A, Tucker M G, Dove M T, Peters L and Evans J S O 2008 *J. Am. Chem. Soc.* **130** 9660
[11] Conterio M J, Goodwin A L, Tucker M G, Keen D A, Dove M T, Peters L and Evans J S O 2008 J. Phys.: Conden. Matter **20** 255225
[12] Forman R A, Piermarini G J, Barnet J D and Block S 1972 *Science* **176**, 284
[13] Pauling L and Pauling P 1968 *Proc. Natl. Acad. Sci.* USA **60** 363
[14] Haser R, Bonnet Band Rozere J 1977 J. Mol. Struct. **40** 177
[15] Jones L H 1962 *J. Chem. Phys.* **36** 1209
[16] Rao R, Achary S N, Tyagi A K and Sakuntala T 2011 *Phys. Rev.* B **84** 054107
[17] Kob N E and House Jr J E 1994 *Trans. Met. Chem.* **19** 31
[18] Reguera E, Quintana G and Fernandez-Bertran J 1997 *Trans. Met. Chem.* **22** 527

Journal of Physics: Conference Series **377** (2012) 012010 doi:10.1088/1742-6596/377/1/012010

# Raman spectroscopic studies on $CeVO_4$ at high pressures

**Rekha Rao[1], Alka B.Garg[2] and B.N.Wani[3]**
[1] Solid State Physics Division, [2] High Pressure & Synchrotron Radiation Physics Division, [3]Chemistry Division,
Bhabha Atomic Research Centre, Mumbai - 400085, India

E-mail: rekhar@barc.gov.in

**Abstract.** Raman spectroscopic studies of zircon $CeVO_4$ is carried out at high pressures upto 13 GPa. Further pressurization leads to vanishing of Raman intensity. Changes in the Raman spectra across the zircon to monoclinic transition are reported. In addition, evolution of Raman spectra of the recovered monazite phase is reported.

## 1. Introduction

Rare- earth orthovanadates (*Re*$VO_4$) crystallize in zircon structure at ambient conditions [1] and follow a sequence of phase transitions from zircon→scheelite→fergusonite as a function of pressure in general [2,3]. Unlike the other members of this series, zircon $CeVO_4$ (which has a larger cationic radius) follows a different sequence of transitions from zircon → monazite→scheelite under pressure [4]. Apart from basic research interest in systematics of phase transitions, orthovanadates are technologically important as thermo-phosphors, scintillators, in photocatalysis and in lithium batteries. Though there are many structural and spectroscopic studies [2,3] on general high pressure behavior of orthovanadates, the trend in $CeVO_4$ is unusual. Recently XRD and Raman spectroscopic investigations have been reported in $CeVO_4$ at high pressure across the zircon to monazite transition [5,6]. Here we report Raman scattering investigations of zircon as well as recovered monazite phases $CeVO_4$ at high pressures. In addition, results of non-hydrostatic experiments which produced scheelite phase are also reported. Raman spectra of various phases of $CeVO_4$ are compared with that of other *Re*$VO_4$.

## 2. Experimental Details

Polycrystalline $CeVO_4$ was prepared by solid state reaction of $CeO_2$ and$V_2O_5$ and characterized by powder XRD [1]. High pressure Raman spectroscopic measurements were carried out as per experimental details given in reference [2].

## 3. Results and Discussion

### *3.1 High pressure behaviour of the zircon phase*

In the ambient zircon phase, (space group $I4_1$/amd, Z = 4), of the 33 expected modes, group theory predicts 12 distinct Raman active modes $2A_{1g}+4B_{1g}+B_{2g}+5E_g$ . Raman spectra of $CeVO_4$ agrees with that reported [7]. Figure 1 shows Raman spectra of $CeVO_4$ compared with that of typical *Re*$VO_4$ like $LuVO_4$ in the zircon phase. In the spectra of $CeVO_4$, the strongest mode at 860 $cm^{-1}$ is the symmetric stretching mode ($\nu_1$) of $A_{1g}$ symmetry which is much lower than the symmetric stretching frequency(899 $cm^{-1}$) in $LuVO_4$ because of the larger ionic radius (114.3 pm) of $Ce^{3+}$ as compared to that of $Lu^{3+}$ (97.7pm) resulting in reduced cation potential in the case of $CeVO_4$. The other two modes in $CeVO_4$ at 797 and 785 $cm^{-1}$ are the asymmetric stretching modes ($\nu_3$) of symmetry $E_g$ and $B_{1g}$ respectively. The V-O bending modes appear in the range 250-500 $cm^{-1}$.The mode at 376 $cm^{-1}$ is the symmetric bending ($\nu_2$) mode of $A_{1g}$ symmetry and that at 466 $cm^{-1}$ is the asymmetric bending mode

($\nu_4$) of $B_{1g}$ symmetry; the other asymmetric bending mode of $E_g$ symmetry could not be detected. The mode at 263 $cm^{-1}$ is the $B_{2g}$ mode due to symmetric bending of the $VO_4$ tetrahedra. The external modes due to Ce-$VO_4$ vibrations appear at frequencies below 250 $cm^{-1}$.

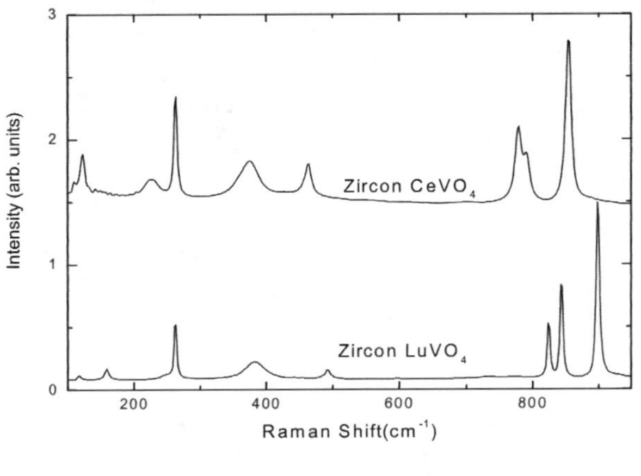


**Figure1.**Raman spectra of $CeVO_4$ in comparison with that of $LuVO_4$

With increase in pressure, all the modes except the $B_{2g}$ mode at 263 $cm^{-1}$ increase in frequency. At around 4 GPa, abrupt changes in the Raman spectra are observed; around the same pressure as the observed zircon to monazite transition by XRD measurements [4]. Discontinuities in mode frequencies and appearance of new modes indicate first order nature of the transition. Softening of the $B_{2g}$ mode in $CeVO_4$ is similar to that of $B_{1g}$ mode seen in other vanadates also across the zircon to scheelite transition [2,3]. Co-existence of the two phases was found in a small pressure range 3.9 to 4.9 GPa. While the bending mode region shows a different set of modes, there is a small discontinuity in the stretching mode frequency. Overall there is a marked increase in the number of modes consistent with the transition to a low symmetry phase. Figure 2 shows the evolution of Raman spectra of $CeVO_4$ with pressure and the pressure dependence of mode frequencies. The Raman spectra were recorded upto 13 GPa beyond which all the peaks were very weak and only the symmetric stretching mode could be followed. The sample became dark and the signal became poor beyond 13 GPa probably due to transition involving reduction in bandgap. It may be noted that $NdVO_4$ which also follows the trend of $CeVO_4$; undergoes zircon to monazite transition [8] followed by transition to scheelite phase involving discontinuous decrease in bandgap across the transitions. A similar effect like band gap reduction is expected in $CeVO_4$ also which could be the reason for loss of intensity of the Raman modes.

When the pressure is released, $CeVO_4$ remains in the high pressure monazite phase. In vanadates like $TbVO_4$, $DyVO_4$ Elliott et al [9] have identified the softening of $B_{2g}$ and $B_{1g}$ modes to a distortion of the primitive cell along [110] or [100] directions, respectively. The softening of the $B_{2g}$ mode could be indicative of similar structural distortion in the system leading to monazite phase.

*3.2 High pressure behaviour of the monazite phase*

Figure 3 shows the evolution of Raman spectra of the pressure quenched $CeVO_4$ which is in monazite phase at ambient conditions. It may be noted that one of the modes exhibits further softening with pressure. As in the previous case, complete loss of Raman intensity was observed above 13 GPa.

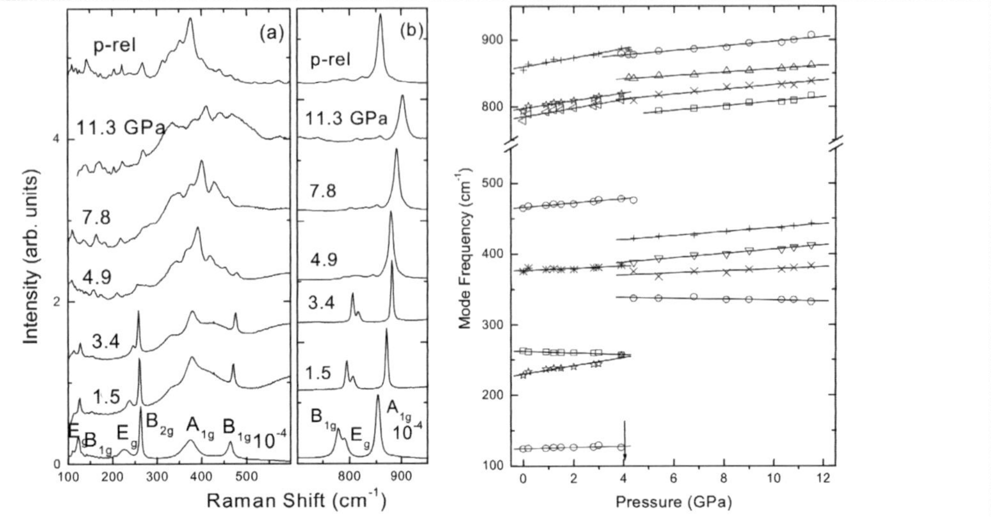


**Figure 2(a)and (b)** Raman spectra of $CeVO_4$ at various pressures showing the changes across the zircon to monazite transition. **(c)** Pressure dependence of mode frequencies. Arrow indicates the zircon to monazite transition pressure. Solid lines denote the linear fit of the data within a phase. Only the prominent mode frequencies in the high pressure phase are plotted.

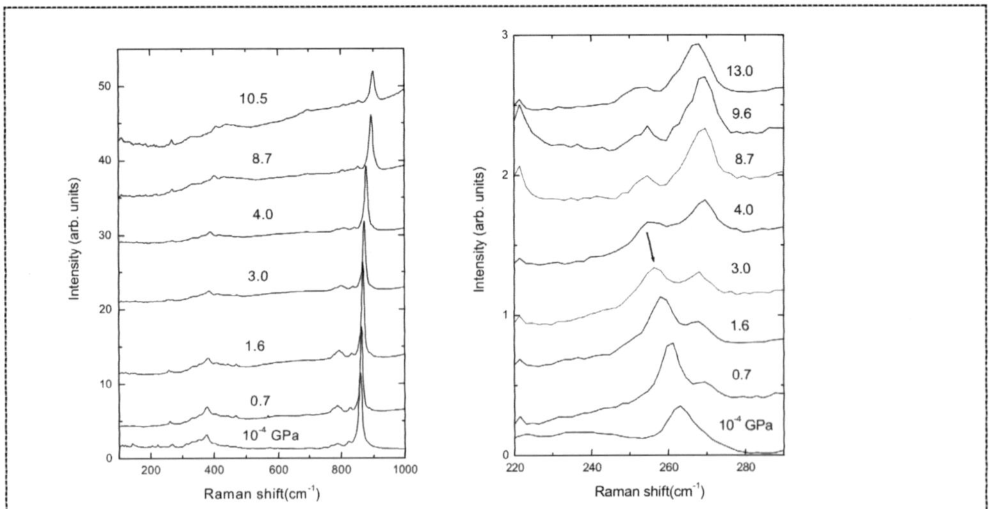


**Figure 3.**Pressure evolution of monazite phase of $CeVO_4$. Arrow mark indicates continuous softening of a mode in the monazite phase

*3.3 Non-hydrostatic experiments*

In order to recover other phases, non-hydrostatic pressure was generated in a diamond anvil cell without any pressure transmitting medium. Pressurizing to 8 GPa without any pressure transmitting medium resulted in a mixture of zircon and scheelte phase as discussed in reference [5]. Figure 4 shows the Raman spectra of $CeVO_4$ in the three phases along with that of scheelite $LuVO_4$ for comparison. Though the zircon $CeVO_4$ stretching frequencies are very much different from that of $LuVO_4$, in the scheelite interstingly, the V-O stretching frequencies are nearly same. This indicates the interatomic bond strength is similar in spite of larger ionic radius of $Ce^{3+}$ which implies that the scheelite phase of $CeVO_4$ has shorter V-O bonds compared to that of $LuVO_4$. While the V-O interaction strength in zircon phase in $CeVO_4$ is less than that of $LuVO_4$, it is same in the scheelite phase in $CeVO_4$ and $LuVO_4$.Panchal *et al* in their calculations of band structure have shown that bandgap depends on the V-O distances in this class of orthovanadates [8]. It would be possible to estimate the band gap using V-O stretching frequency. It appears that $CeVO_4$ which has the lowest bandgap in the zircon phase among the orthovanadates, undergoes transitions involving discontinuous decrease in bandgap similar to $NdVO_4$, resulting in loss of Raman intensity above 13 GPa. In analogy with $NdVO_4$, the loss intensity above 13 GPa could be due to a phase transition.

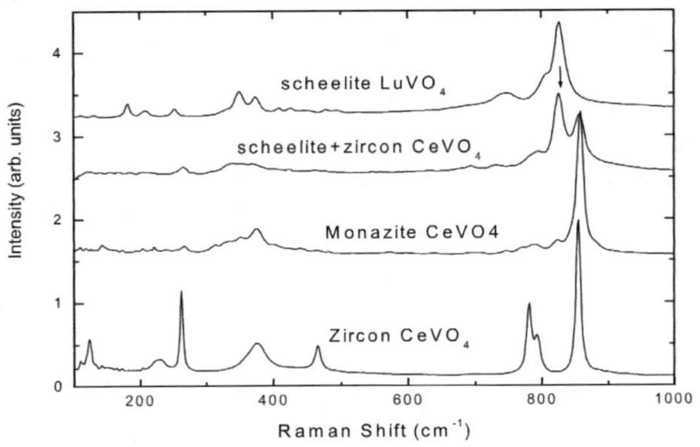


**Figure 4.** Raman spectra of $CeVO_4$ in zircon, monazite and scheelite phases. Arrow mark indicates the symmetric stretching mode in scheelite $CeVO_4$

## 4. Conclusion

Raman spectroscopic studies have been carried out on zircon phase and monazite phases of $CeVO_4$ at high pressures upto 13 GPa beyond which there is a loss of Raman intensity. Raman spectra of $CeVO_4$ in various phases is compared with other *Re*$VO_4$.

## 5. References

[1] Varma S, Wani B N and Gupta N M 2002 *Mater. Res. Bull.* **37** 2117
[2] Garg A B, Rao R, Sakuntala T, Wani B N and Vijayakumar V 2009 *J Appl Phys* **106** 063513
[3] Rao R, Garg A B, Sakuntala T, Achary S N and Tyagi A K 2009 *J Solid State Chem* **182**1879
[4] Range K.-J, Meister H and Klement, U 1990 *Z. Naturforsch.* **B45** 598
[5] Panchal V, Lopez-Moreno S, Santamarıa-Perez D, Errandonea D, Manjon F J, Rodrıguez-Hernandez P, Munoz A, Achary S N and Tyagi A K 2011 *Phys. Rev B* **84** 024111
[6] Errandonea D, Kumar R S, Achary S N and Tyagi A K 2011 *Phys. Rev B* **84** 224121
[7] Hirata T and Watanabe A 2001 *J Solid State Chem.* **158** 264
[8] Panchal V, Errandonea D, Segura A, Rodrıguez-Hernandez P, Munoz A, Lopez-Moreno S, and Bettinelli M 2011 *J. App. Phys.* **110** 043723
[9] Elliott R J, Harley R T, Hayes W and Smith S P R 1972 *Proc. R. Soc.* **A 328** 217

Journal of Physics: Conference Series **377** (2012) 012011 doi:10.1088/1742-6596/377/1/012011

# High pressure experiments at the XAFS Beamline, INDUS-2

**Nitya Ramanan[1], Debdutta Lahiri[1,*], Nandini Garg[1], D Bhattacharyya[2], S N Jha[2], N K Sahoo[2] and Surinder M Sharma[1]**

[1]High Pressure and Synchrotron Radiation Physics Division, [2]Applied Spectroscopy Division, Bhabha Atomic Research Centre, Mumbai-400085, India.

Email: debduttalahiri@yahoo.com, dlahiri@barc.gov.in

**Abstract.** The dispersive XAFS beamline BL-08 at the INDUS-2 synchrotron radiation source, RRCAT, Indore uses a bent Si (111) crystal as dispersive-cum-focusing element and a position sensitive CCD detector to enable instantaneous measurement of the whole XAFS spectrum around the absorption edge of interest. This beamline is ideal for characterisation of materials under high pressure using Diamond Anvil Cell with ~50μm spot size. For this setup, the theoretically determined spot size (Horizontal × Vertical) varies between 17 × 137 μm and 37 × 142 μm for the x-ray energy range 5 keV-20 keV. To reduce the vertical spot size to <50 μm, we have designed an additional focusing mirror between the polychromator and sample position. The mirror, procured from SESO (France), will be installed shortly. Meanwhile, we have developed a dummy mirror bender setup at CDM (BARC) and have carried out feasibility tests to confirm reduction in spot size using the same. We have also conducted preliminary XAFS experiments (at BL-08) on $SrRuO_3$ at ~16 keV, under ambient conditions and inside diamond anvil cell, in order to assess the signal intensity and quality. We have obtained reasonably good signal.

## 1. Introduction:

X-ray Absorption Fine Structure (XAFS) beamline BL-08 (5-20 keV) has been recently commissioned at INDUS-2 Synchrotron Radiation Source, RRCAT (India)[1]. The main optical component of the beamline is a bent crystal polychromator, which serves as both energy dispersive and focusing optical element. White synchrotron radiation originating from the bending magnet undergoes Bragg reflection from the crystal at preset angle, with a certain bandwidth (determined by the crystal curvature) around the desired photon energy. This pseudo-monochromatic (pink) incident beam is focused at the sample position and the beam transmitted by the sample is detected by a position-sensitive CCD detector (Andron DH 432 FO). One of the proposed activities with the beamline is the XAFS characterization of materials under high pressure. In situ studies of materials under high pressure and temperature are one of the major applications of energy dispersive XAFS studies. XAFS with synchrotron radiation is used to correlate changes in the local and electronic structures [2].

The main requirements for high pressure EXAFS studies (involving Diamond Anvil Cell (DAC)) are (i) small focal spot (~20-30μm), (ii) position stability of focal spot in the whole energy range, (iii) high flux (to override the absorption by the diamonds in transmission configuration) and

* To whom any correspondence should be addressed

(iv) easy removal of diamond diffraction peaks in the energy range of interest (Bragg diffraction from the diamonds is inevitable, and leads to intense peaks superposed on the absorption spectrum, at energy values that match the Bragg condition on the intercepted crystal planes of the two diamond anvils)[3]. The energy-dispersive setup broadly caters to these requirements due to the sharp focusing by the elliptically bent crystal and absence of any mechanical movement of the crystal (during XAFS recording), leading to required focal spot stability.

For the present setup, the focal spot size at the sample position was calculated to be between [17 μm(H)×137 μm(V)] and [37 μm(H)×142 μm(V)] for the photon energy range 5-20 keV. The vertical spot size needs to be reduced to conform to DAC sample size. SHADOW calculations [4] show that inserting an additional focusing mirror prior to the sample position (Figure 1) reduces the spot size to [16 μm(H)×20 μm(V)] and [37 μm(H)×27 μm(V)] for the photon energy range 5-20 keV [5]. The radius of curvature of the mirror should vary between 277m and 401m correspondingly. For the above purpose, an elliptically bendable mirror has been procured from SESO, France and due to be installed in March 2012. The mirror is 150mm long, and has useful width and thickness 15mm and 15mm respectively. It has a U-shape and is Rhodium coated (coating thickness=500Å). Roughness of the mirror is less than 3 Å.rms. It has a meridional radius of 0.44197-0.401071 km and a bending resolution of 0.05%. An actuator is used to bend the mirror and is driven by a stepper motor. The actuator applies two equal forces on the legs of the U-mirror allowing it to apply the same moment on each leg of the mirror, causing it to bend.

2.1. *Feasibility test to check the reduction in spot size:*

To test the feasibility of the reduction in spot size, a dummy mirror bender was developed at the Centre for Design and Manufacture (CDM), Bhabha Atomic Research Centre (BARC) [6]. The focusing properties of the bendable mirror were simulated using divergent laser light. The dummy bender setup consists of a copper mirror plate supported on four support pins located at the two ends of the system. (Fig.3a) Two bender plates are supported on two pins and are rotated by the translating movements of two shafts. When the bender plate rotates, the copper plate is subjected to mechanical torques at its two ends, which results in the bending of the mirror plate. The translating movement to the shafts is realized by a special nut and worm wheel assembly fitted in the system. The worm wheel is rotated by a PC controlled Performax 2EX-SA two axis controller [7], the PC interface for which was developed by Integrated Electronics.

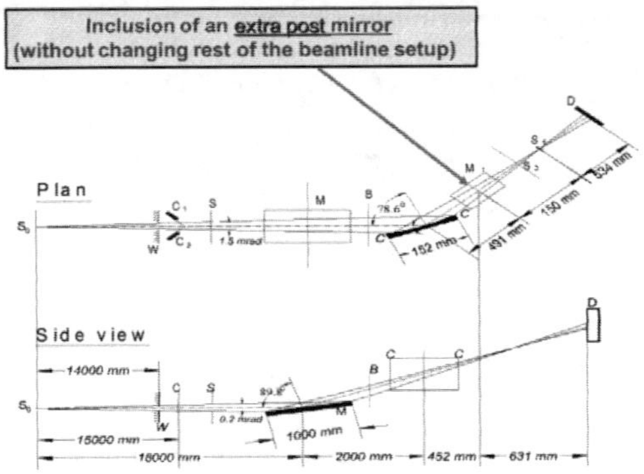


**Figure 1.** Modified optical layout of the beamline.

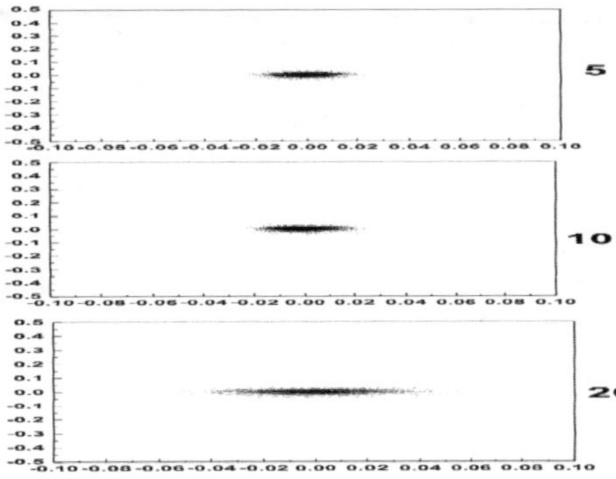


**Figure 2.** Focus spot at the image position for photon energies 5 keV, 10 keV and 20 keV

Figure 3b shows the schematic experimental arrangement for testing the focusing properties of the bendable mirror using laser light. He-Ne laser beam is focused at a pinhole by the microscope objective (MO). Light diverging from the pinhole is initially focused at the point $S_1$ by the focusing lens to form the primary image of the pinhole. Subsequently, the bent mirror focuses the laser light at the final image point $S_2$. There was a significant improvement of the focusing in the vertical direction due to reflection of laser light by the bendable mirror used in the grazing incidence configuration. The size of the focused laser spot also varied with the radius of curvature of the mirror substrate (varied by PC controlled stepper motor)

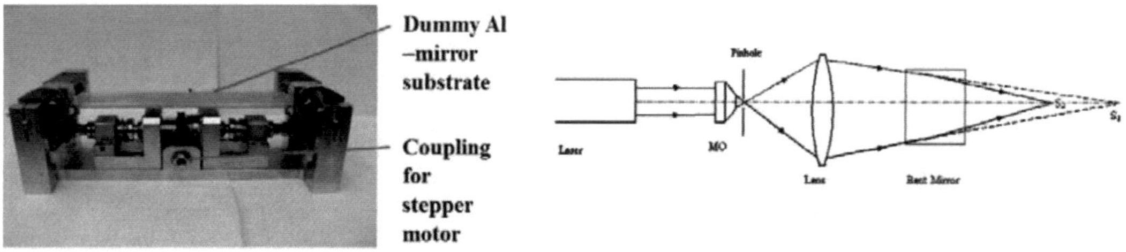


a b

Figure 3: a. Photograph of the mirror bender developed at CDM, BARC, b. Schematic Experimental arrangement for testing focusing properties of the bent mirror.

*2.2. Feasibility test to check the transmission of synchrotron radiation through diamond:*

For XAFS experiments at BL-08( INDUS-2), *$SrRuO_3$* sample was ground to powder of average particle size ~ 5 μm and pasted uniformly on tape [8]. XAFS at *Sr* K edge (~16 keV) was measured both within and without the diamond anvil cell (DAC) (2 diamonds, each having thickness = 2mm) at 60mA beam current and 2GeV ring energy. *Sr* K edge was selected since at this energy (~16 keV), diamond absorption is less [9]. A Mao Bell type DAC, with anvil culet of diameter 400μm, was used for this experiment. The optimized spot size in this experiment was 200μm × 500μm (H×V). Note that the spot size is bigger than the calculated value (see Introduction) due to crystal imperfection and bigger source size from the design parameters. We have obtained a reasonable intensity signal for the DAC setup. The decrease in transmitted signal intensity is by ~40% when DAC is used (Figure 5). EXAFS spectra extraction and conversion to energy for this *raw* data is currently in progress. It may be noted that this feasibility test was conducted *without* the extra elliptical mirror. We expect the intensity to improve after the mirror is installed, due to reduced spot size (<50 μm V). The theoretically calculated energy resolution, photon flux and spectral intensity at the final image point upon inclusion of the extra elliptical mirror are summarized in Table 1. Further, for high pressure studies at lower edges, perforated diamonds from Almax industries are in consideration.

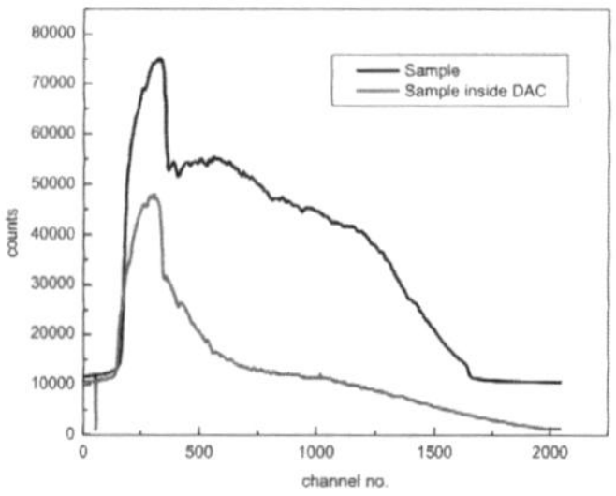

Figure 5: Transmitted intensity through the sample without DAC (black curve) and with DAC (red curve)

Table 1. Energy Resolution and Spectral Intensity

| E (keV) | $\phi_o(\lambda)$ Photons/ sec/mrad/0.1% $\Delta\lambda$ | $\Delta E/E$ | $\Delta E$ (eV) | $\phi(\lambda)$ Photons / sec | I ($\lambda$) Photons / sec/ mm$^2$ |
|---|---|---|---|---|---|
| 5 | $1.21 \times 10^{13}$ | $1.44\times 10^{-4}$ | 0.72 | $5.37\times 10^{11}$ | $2.68\times 10^{13}$ |
| 10 | $5.62 \times 10^{12}$ | $1.39\times 10^{-4}$ | 1.39 | $3.42\times 10^{11}$ | $1.71\times 10^{13}$ |
| 20 | $9.98 \times 10^{11}$ | $1.31\times 10^{-4}$ | 2.62 | $0.87\times 10^{11}$ | $0.22\times 10^{12}$ |

### 3. Conclusion:

In summary, we have evaluated the possibility of high pressure experiments at the XAFS Beamline, RRCAT, Indore. Preliminary tests conducted at CDM (BARC), using an indigenously developed bender setup, confirm reduction in focal spot size on addition of an elliptically bent mirror prior to the sample position.

We have recorded XAFS of $SrRuO_3$ (powder tape sample) under ambient conditions, both within and without the DAC, in order to assess signal intensity and quality. We have obtained reasonable intensity signal. Following the installation of an elliptically bent mirror prior to the sample position, we expect the intensity to improve due to reduced spot size.

[1] Bhattacharyya D, Poswal A K, Jha S N, Sangeeta and Sabharwal S C 2009 *Nuc. Insts. Meth. in Phys. Res. A* **609** 286.
[2] Zeng Q, Ding Y, Mao W L, Yang W, Sinogeikin S V, Shu J, Mao H K and Jiang J Z 2010, *Phys Rev Lett*, **104** 105702; Giefers H, Porsch F and Wortmann G 2005 *Physica Scripta* **T115** 538; Miyauchi K, Qiu J, Shojiya M, Kawamoto Y, Kitamura N, Fukumi K, Katayama Y, Nishihata Y 2002 *Sol. State Comm.* **124** 189; Jackson W E, Leon J M, Brown G E.Jr, Waychunas G A, Conradson S D, Combes J M 1993, *Science* **262** 29; Sadoc A, Itie J P, Polian A, Lefebvre S 1997 *J Phys IV France* **7** C2-991; Vaccari M, Aquilanti G, Pascarelli S, Mathon O 2009 *J Phy: Cond Matt.* **21** 145403
[3] Pascarelli S, Mathon O, Aquilanti G 2004 *J. Alloys and Compds* **362** 33
[4]http://www.esrf.eu/computing/scientific/raytracing/PDF/shadow_basics.pdf, http://www.esrf.eu/computing/scientific/raytracing/PDF/primer.pdf
[5] Das N C, Bhattacharyya D, Lahiri D, Sharma S M, "Studies on the possibilities of high pressure X-ray absorption experiments using XAFS beamline at INDUS-2 SRS"2007 *BARC report* E/001
[6] Ramanan N, Lahiri D, Sharma S M, Das N C, Udupa D V, Patil A D, "Development of a bendable mirror for studying materials under high pressure using EXAFS beamline at INDUS-2 SRS"2010 *BARC report* E/012
[7] http://www.arcus-technology.com/products/24.html
[8] Lahiri D 2006 *Physica C* **436** 32
[9] Dadashev A, Pasternak M P, Rozenberg G K, Taylor R D 2001 *Rev. Sci. Inst.* **72** 2633

Journal of Physics: Conference Series **377** (2012) 012012 doi:10.1088/1742-6596/377/1/012012

# Effect of the surfactant CTAB on the high pressure behavior of CdS nano particles

**A. K. Mishra**[1], **Nandini Garg, K. K. Pandey and Vineet Singh**[+]
High Pressure & Synchrotron Radiation Physics Division,
Bhabha Atomic Research Centre, Trombay, Mumbai-40085, India
[+] Dept. of Physics, University of Allahabad - 211002, India

E-mail: akmishra@barc.gov.in

**Abstract.** Angle dispersive x-ray diffraction studies on CTAB capped cadmium sulphide (CdS) nano particles of sizes 70 nm, 44 nm and 10 nm have been carried out up to ~39, ~30 and ~28 GPa respectively. All the three types of nano particles transform to the rocksalt phase at pressures higher than that of bulk CdS. Our analysis of the x-ray diffraction data indicates that the phase transition pressure as well as the bulk modulus of the rocksalt phase of CdS exhibits a non monotonous behaviour as a function of the size of the nano particles. Our high pressure Raman scattering studies on the 10 nm CdS nano particles are in agreement with the x-ray diffraction studies. The presence of the Raman mode in the high pressure rocksalt phase may be attributed to a disorder activated scattering in this phase or to the formation of a core shell structure beyond the transformation pressure.

## 1. Introduction

Studies of II-VI or III-VI binary semiconductors like ZnS, ZnO CdS GaN, InP etc are of fundamental interest because of their technological importance and also from a basic physics point of view. It has been observed that reducing particle size to the nano meters either modifies their physical properties or introduces some novel properties making them much sought after materials for applications like disease detection, quantum computing etc [1]. Reduction of size results in an increase in the surface to volume ratio and therefore the surface energy plays an important role in deciding the stability, structure and properties of the nano materials. It is often found that depending on the size of the nano particle a particular compound can exhibit a myriad of structures even at ambient temperature and pressure [2]. For eg. CdS nano particles $< 2$ nm are amorphous whereas 2 nm and $\geq 4$ nm particles of CdS crystallize in the rocksalt and zinc blende/wurtzite structures respectively.

Knowledge of the phase stability of these nano semiconductors is important and has been a central issue in nano science and technology. In nano particles the phase stability, not only depends on stimuli of pressure, temperature and composition of the specimen but also on the size of these particles [3]. There are several studies where it has been demonstrated that solid-solid phase transition pressures or the high pressure behavior itself is affected by the size of the nano particles. It has been shown that the phase stability of nano particles also depends upon the capping surfactant as the surface energy of these particles is altered by different surfactants.

Cadmium sulphide (CdS), a direct band gap semiconductor, is extensively used in the electronic industry especially in the nano crystalline form [4]. As in the nano form it is commercially used as a phosphor, hetrojunction laser, field effect transistor, photovoltaic cell, fabrication of optoelectronic devices, it is important to understand its structural behaviour under different thermodynamic conditions. There are several studies on CdS which have shown that the wurtzite or zinc blende form

[1] To whom any correspondence should be addressed.

of bulk CdS transforms to the rocksalt form at ~ 3.1 GPa. It has been shown by Alivisatos et al. that size reduction enhances the stability range of the initial structure, thus delaying the transformation to the rocksalt phase [5].

These authors have also shown that surfactants play a vital role in the high pressure structural behavior of CdS. They observed that CdS nano particles (4.5 nm) stabilized by sodium polyphosphate transformed to rocksalt structure at 8.0 GPa, where as tetrabutylammonium-EDTA (4 nm) stabilized particles transformed at 6.5 GPa . However, since the effect of CTAB (cetyl trimethyl ammonium bromide) surfactant on the high pressure behavior of CdS nano particles has not been studied so far , we have carried out angle dispersive x-ray diffraction and Raman scattering studies on these nano particles.

## 2. Method

Nano-sized particles of CdS (in CTAB matrix) were prepared using wet chemical routes [6]. The samples yielded different particle sizes when annealed from different temperatures. We have studied three samples annealed at 500°C (A), 300° C (B) and 20°C (C). The x-ray diffraction pattern of 'A' type and 'B' type sample could be indexed to pure wurtzite phase while that of 'C' to a mixture of wurtzite and zinc blend phase. The lattice parameters of the different phases were determined by carrying out a Rietveld refinement using GSAS [7] and have been tabulated in table 1.

Table 1.

| Type of CdS nano particles | Phase (at ambient) | Space group | Lattice parameter |
|---|---|---|---|
| 'A' type | Wurtzite | $P6_3mc$ | a = 4.147 (2) Å, c = 6.733 (3) Å |
| 'B' type | Wurtzite | $P6_3mc$ | a = 4.141 (4) Å, c = 6.713 (3) Å |
| 'C' type | Wurtzite + Zinc blende | $P6_3mc$ + $F\bar{4}3m$ | a = 4.181(3) Å, c = 6.843 (2) Å<br>a = 5.838 (2) Å |

The particle size of these three types samples was determined by using the Deby-Scherrer formula given below

Particle size (t) = $K \times \lambda / (W \times Cos\ (\theta_B))$

where K is a dimensionless constant with typical value of 0.94, W is FWHM of the x-ray diffraction peak used to determine the particle size and $\theta_B$ is the corresponding Bragg angle. The Scherrer equation relying on FWHM of diffraction peaks is generally used as a quick and reliable method for the particle size determination in case of bulk as well as nano crystalline samples using x-ray diffraction pattern. Since other factors such as instrumental broadening, strain broadening etc. also contribute to the peak width, the FWHM needs to be deconvoluted in order to get reliable particle size. In this study we have obtained particle size by factoring out instrumental broadening only as other contributions are negligible under ambient conditions.

However the contribution to FWHM of x-ray diffraction peak due to instrument broadening was factored out during determination of the particle size. The particle size of the A, B and C type samples were determined to be 70nm, 44 nm and 10nm respectively.

For the high pressure experiments, nano crystalline samples of CTAB capped CdS along with a few specs of gold were loaded in a hole of ~100 μm diameter drilled in a pre-indented (~ 80 micron thick) tungsten gasket of a diamond-anvil cell (DAC). Methanol: ethanol (4:1) mixture was used as the pressure transmitting medium. The pressure was determined from the known equation of state of gold [8]. High-pressure angle dispersive x-ray-diffraction experiments, were carried out at 5.2 R (XRD1) beamline of Elettra Synchrotron source using monochromatized x-rays of λ = 0.6888 Å. The

Journal of Physics: Conference Series **377** (2012) 012012 doi:10.1088/1742-6596/377/1/012012

diffraction patterns were recorded using the MAR345 imaging plate detector kept at a distance of ~ 20 cm from the sample. Two-dimensional x-ray diffraction patterns were transformed to one-dimensional diffraction profiles by the radial integration of diffraction rings using the FIT2D software [9].

High pressure Raman scattering studies were carried out using our indigenously developed micro Raman set up with confocal optics. The Raman scattered light, excited by 532nm DPSS laser, is collected using CCD based single stage spectrograph and a supernotch filter. 4:1 methanol:ethanol mixture was used as pressure transmitting medium and ruby fluorescence lines were used for monitoring the pressure [10].

## 3. Results and Discussion

Figure 1(a, b &c) show the x-ray diffraction patterns of 70 nm (A), 44nm (B) and 10 nm (C ) CdS nano particle respectively stacked at a few representative pressures. The x-ray diffraction (xrd) peaks marked with G and Au represents the xrd peaks from gasket and gold pressure marker respectively.

In case of 70 nm CdS particles the transformation to the rocksalt (face centered cubic) phase takes place at 6.7 GPa which is higher than that of bulk CdS. The rocksalt phase was found to be stable up to the highest pressure of measurement (39 GPa) of this study. On release of pressure, the diffraction peaks of the rocksalt phase became very broad and were hardly discernible below 15 GPa. This could be an indicative of amorphization of the 70 nm CTAB capped CdS nano particles on release of pressure.

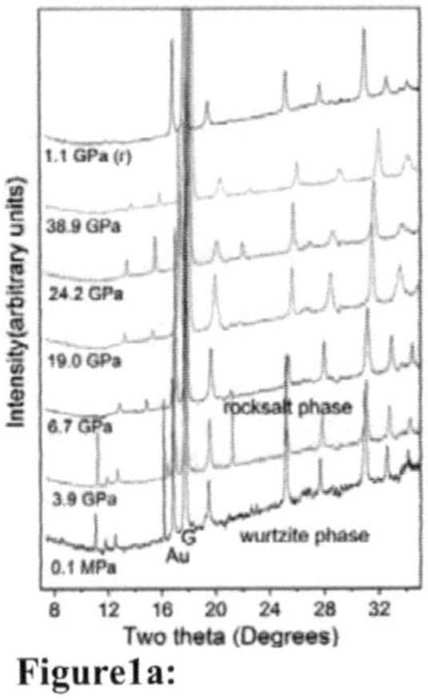


**Figure1a:**

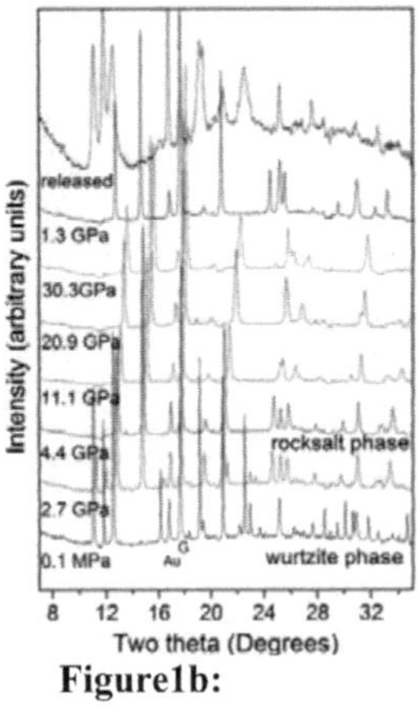


**Figure1b:**

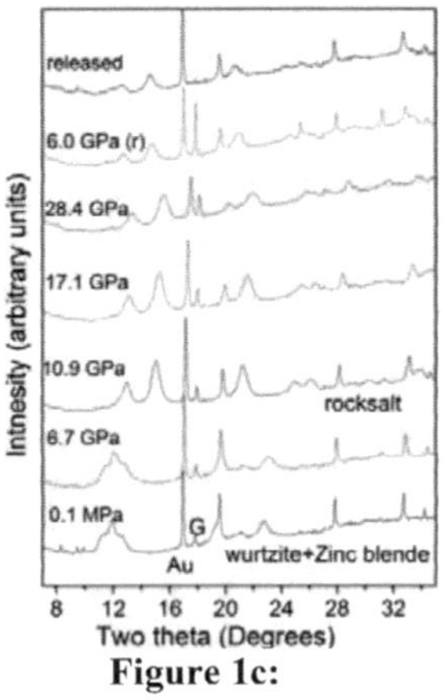


**Figure 1c:**

**Figure1:**X-ray diffraction patterns of 70 nm, 44 nm and 10 nm CdS nano particles at few representative pressures as shown in Figure1a, 1b and 1c respectively.

However, in the case of the 44 nm sized particles the transformation to the rocksalt phase takes place at ~ 4.4 GPa whereas for the 10 nm particles the transformation takes place at pressures beyond~ 6.7 GPa as shown in Fig. 1b &1c. As in the case of the 44 nm particle, the rocksalt phase was found to be stable upto ~ 30 GPa and on release it reverts back to wurtzite phase but in contrast to the 70 nm particles the high pressure phase was retained on release of pressure for 10 nm particles. This shows that the high pressure behavior of the 70 nm particles of CdS is distinct from that of the 44 and 10 nm particlesThis implies that the CTAB surfactant tends to increase the stability range of the CdS nano particles when compared to that of tetrabutylammonium-EDTA surfactant but decreases with respect to the sodium polyphosphate surfactant [5].

Journal of Physics: Conference Series **377** (2012) 012012 doi:10.1088/1742-6596/377/1/012012

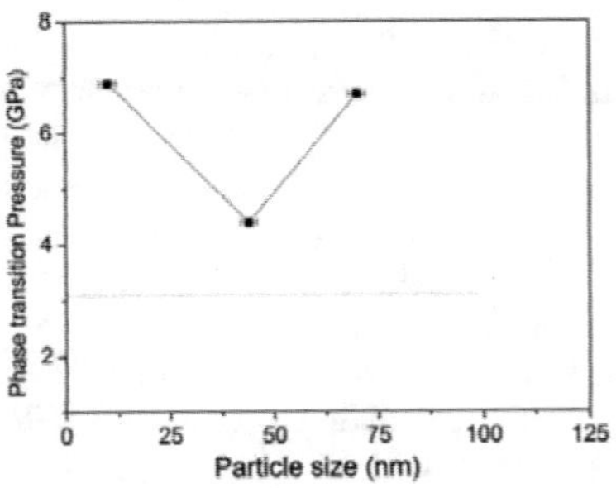


**Figure2:** Variation of phase transition pressure with particle size.

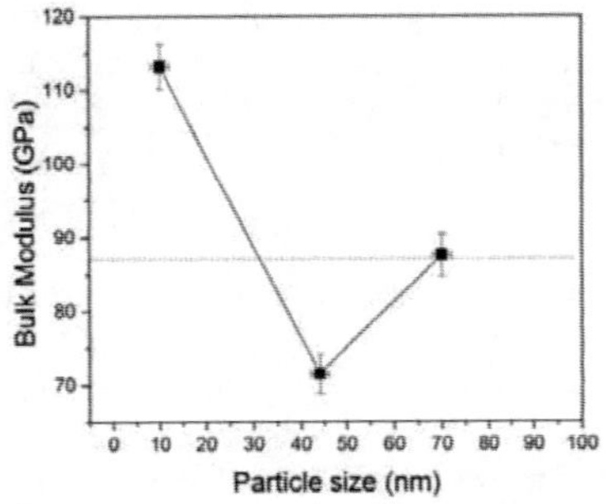


**Figure 3**: Variation of bulk modulus of rocksalt phase with particle size.

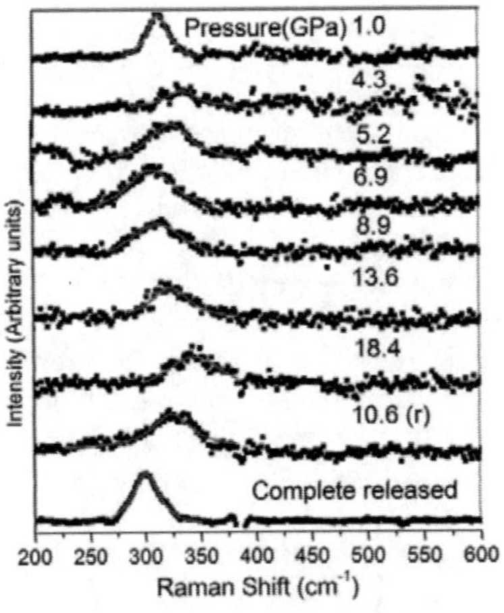


**Figure4:** Raman spectrum of 10 nm CdS nano particles at a few representative pressures.

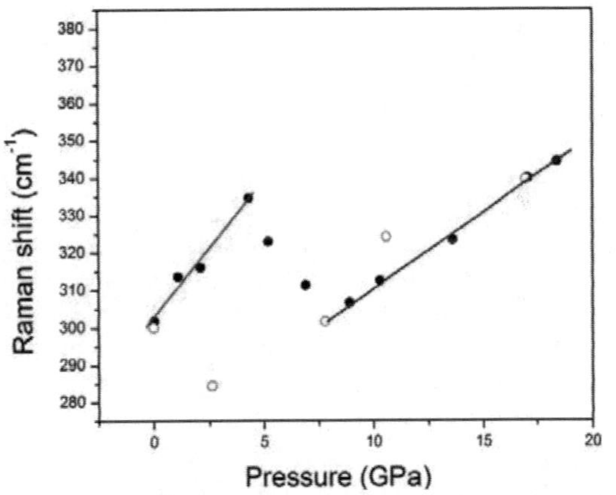


**Figure5:** Variation of Raman shift for A1 (LO) mode with pressure.

Figure 2 shows the variation of transition pressure with particle size. Here dotted line corresponds to the phase transition pressure of bulk CdS sample. It is interesting to note that transition pressure decreases as the particle size increases from 10 to 44 nm. However it increases again for 70 nm crystals, for which it becomes comparable to that of 10nm. It is noteworthy that, annealing of the samples at different temperatures does affect the particle size. However, effect of the annealing temperature on the state of the surfactant cannot be ruled out. Therefore the variation in transition pressures of the nano crystalline CdS particles may be attributed to both particle size and state of the surfactant.

The observed pressure induced variations of volume has been fitted with third order Birch Murnaghan equation of state to determine the bulk modulus of rocksalt phase for all the three types of CdS nano particles. Figure 3 shows the plot for the bulk modulus of rock salt phase as a function of particle size. The dotted line corresponds to the bulk modulus of rock salt phase of bulk CdS sample.

The bulk modulus for rock salt phase is lowest for the 44 nm particles size thus the variation of bulk modulus with particle size is also non monotonous.

Raman scattering studies were carried out on the 10 nm particles upto 18.4 GPa. Representative Raman spectra at a few pressures are shown in Fig. 4. The black square symbols represent the observed data while the red line corresponds to the Lorentzian fit for the observed Raman mode. The letter (r) adjacent to some pressure values represents the release run. As expected, we have observed a Raman mode at ~302 $cm^{-1}$ assigned as $A_1$ (LO) corresponding to the cation-anion (Cd-S) longitudinal vibration. On application of pressure the observed Raman mode at 302 $cm^{-1}$ stiffens and broadens and at 6.9 GPa it appears to soften by ~23 $cm^{-1}$which is indicative of a phase transition. As our x-ray diffraction measurements show that the high pressure phase has rocksalt structure and it does not have any Raman active mode, this softened Raman mode could be due to the coexisting surface wurtzite motifs and the rocksalt phase as core beyond this pressure. Softening of the A1(LO) Raman mode of wurtzite at the phase transformation could be due to surface relaxation caused by the large volume collapse (~18%) of the core . Though our x-ray diffraction studies do not show coexistence of the wurtzite and rocksalt phases, this possibility cannot be ruled out as Raman scattering is a local probe and more sensitive to local coordination whereas x-ray diffraction probes the long range ordering. However, to ascertain this further surface sensitive studies on the released samples should be carried out.

Other possibility is that this Raman mode beyond transformation pressure could be a disorder activated, symmetry forbidden Raman mode of rock salt phase as was observed in case of GaN [11]. For Bulk samples wurtzite to rocksalt phase transition has been proposed to proceed via an orthorhombic cell. One of the notable aspects of this phase transition is the large reduction in volume that occurs. During this transition the (001) plane of rocksalt structure could have a high degree of disorder as there is an arbitrariness in the relationship between the initial crystallographic a/b plane of wurtzite and (001) plane of rocksalt. This poor long range order in the lattice would allow relaxation of k-selection rules in the Raman scattering; resulting in disorder activated Raman scattering.

This mode was found to stiffen on further application of pressure as shown in figure 5. Here the closed symbol corresponds to the compression while the open ones are from the released run. Slope of this Raman mode is 7.4±0.8 $cm^{-1}$/GPa for the ambient phase while for high pressure phase it is 3.9±0.2 $cm^{-1}$/GPa. It is interesting to note that the $d\omega/dP$ for bulk CdS is 5.0 ± 0.2 $cm^{-1}$/GPa.

## 4. Conclusions

Our high pressure studies on CTAB capped CdS nano particles of different sizes show that the range of stability of the ambient phase can be controlled by using different surfactants like CTAB, EDTA [5] etc. The detailed x-ray diffraction analysis suggests that the size variation of CdS nano-particles has a non monotonous effect on the pressure of transformation to the rocksalt phase. However, further work needs to be carried out with several particle sizes to ascertain this non monotonous behavior. Our Raman results are in agreement with the x-ray diffraction results. The presence of the Raman mode in the high pressure phase may be attributed to the coexisting wurtzite shell and rocksalt core as interpreted in the earlier reports or it could be due to the disorder activated scattering from the rocksalt phase itself. Our measurements also suggest that the 70 nm particles of CTAB capped CdS amorphize on release of pressure.

## 5. References

[1] Buffat P. H., Borel J. P. 1976 *Phys. Rev. A* **13** 2287.
[2] Owen N. B. et al. 1963 *J. Phys. Chem. Solids* **24** 1519.
[3] Tolbert S. H. and Alivisatos A. P. 1995 Annu. Rev. Phys. Chem. **46** 595.
[4] Sowa H. 2005 *Solid State Sciences* **7**(2005) 73.
[5] Hasse M. and Alivisatos A.P. 1992 *J. Phys. Chem.* **96** 6756.
[6] Singh V., Sharma P. K., Chauhan P. 2010 *Materials Chemistry and Physics* **121** 202.

[7] Larson A. C. and Von Dreele R. B., GSAS: General Structure Analysis System (Los Alamos National Laboratory USA 2000.
[8] Dewaele A., Loubeyre P. and Mezouar M. 2004 *Phys. Rev. B* **70** 094112.
[9] Hammersely A. P., Svensson S. O., Hanfland M., Fitch A. N. and Hausermann D. 1996 *High Press. Res.* **14** 235.
[10] Piermarini G.J., et al. 1973 *J. Appl. Phys.* **44** 5377.
[11] Besson J. M., Itie J. P., Polian A., Weill G., Mansot J. L. and Gonzalez J. 1991 *Phys. Rev. B* **44** 4214.

# Neutron powder diffraction of small-volume samples at high pressure using compact opposed-anvil cells and focused beam

**T. Okuchi[1], S. Sasaki[2], Y. Ohno[2], J. Abe[3], H. Arima[4], T. Osakabe[3], T. Hattori[3], A. Sano-Furukawa[3], K. Komatsu[5], H. Kagi[5], W. Utsumi[3], S. Harjo[4], T. Ito[4] and K. Aizawa[4]**

[1]Institute for Study of the Earth's Interior, Okayama University, 827 Yamada, Misasa, Tottori 682-0193 Japan
[2]Department of Materials Science and Technology, Gifu University, Gifu 501-1193, Japan
[3]Quantum Beam Science Directorate, Japan Atomic Energy Agency, Tokai, Ibaraki 319-1195, Japan
[4]J-PARC Center, Japan Atomic Energy Agency, Tokai, Ibaraki 319-1195, Japan
[5]Geochemical Research Center, Graduate School of Science, University of Tokyo, Tokyo 113-0033, Japan

E-mail: okuchi@misasa.okayama-u.ac.jp

**Abstract**. Neutron powder diffraction techniques of small-volume samples at high pressure using compact opposed-anvil cells were developed at J-PARC pulsed neutron source. For this purpose we apply a few types of super-hard materials as opposed anvils with culet diameters between 3 to 5 mm. Generated pressures with these anvils were up to 9 GPa for 2 to 4 $mm^3$ and up to 14 GPa for 0.7 $mm^3$ sample volumes, which not only depends on the anvil geometry and material but even more depends on the metallic gasket geometry and material. A representative anvil geometry with 4 mm in culet diameter, along with TiZr "null alloy" metallic gasket containing varying sample volumes, were then applied to time-of-flight neutron powder diffraction experiments, where methane hydrate of 4 $mm^3$ volume and lead of 0.7 $mm^3$ volume were separately measured and their signal-to-background ratios were evaluated. A neutron-focusing optics was used to concentrate the neutron beam into these small-volume samples to increase the intensity of diffraction. Although spurious diffraction peaks from the anvils were prominent, more than seven diffraction peaks are clearly observed from both of the samples. In spite of the smaller sample capacity than previous standard high-pressure apparatus for neutron, it is concluded that the opposed-anvil cells will become alternative apparatuses for neutron scattering at strong pulsed neutron sources where sufficient neutron intensity was granted.

## 1. Introduction

Neutron plays the complementary role to x-ray in solid state physics and chemistry, geophysics, and material science at high pressures, because it is the unique probe for structural analysis of hydrogen-bearing materials which works effectively even at high pressures. Japan Proton Accelerator Research Complex (J-PARC) at Tokai, Japan now has become one of the strongest pulsed neutron sources which enable us to utilize this unique potential of the neutron beam. To make the best use of the strongest pulsed neutron beam, we are developing the experimental techniques of neutron powder diffraction of small-volume samples of the order of a few $mm^3$ at high pressures, which is compressed by compact opposed-anvil cells. We are working at a state-of-the-art engineering difftractometer TAKUMI (BL19) at J-PARC Materials and Life Science Experimental Facility (MLF), where time-of-flight (TOF) neutron diffraction experiments at ambient condition and high pressure are conducted with L1=40m, L2=2m and 90-degree scattering geometry [1,2]. The compact anvil cells are designed to work together with a neutron focusing optics previously designed and installed at TAKUMI [3]. Here we present our preliminary results using these apparatuses and facility.

## 2. Experimental

We expect that by using compact anvil cells with their best-optimized manner, detectable sample volume for TOF neutron diffraction at high pressures will be significantly reduced from that of the current standard Paris-Edinburgh (P-E) apparatus [4]. This is because the compact cells have not only much wider opening angle from sample to see neutron detectors, but also equipped with much more transparent anvils for the neutron beam than the P-E press. Therefore we expect a considerable increase of signal-to-background ratio per unit sample volume which will compensate the intrinsically-limited sample capacity of the compact cells. To evaluate this expectation, we have conducted neutron powder diffraction experiments for sample volumes of one to two orders of magnitude smaller than that of the P-E press.

Before touching the neutron, we applied several types of super-hard anvil materials as opposed anvils with culet diameter of 3 to 5 mm, which is considerably larger than those used for previous compact anvil cell experiments [5,6]. We thus evaluated pressure generation efficiency of the-largest-order volume samples for the compact cells. First of all, culet sizes of single crystal diamond anvils at ≤1 mm are too much small, so that their application was not considered. Instead, we applied single crystal moissanite (SiC) [5], sintered polycrystalline diamond (SD) [7], and nano-polycrystalline diamond (NPD) [8]. The SiC anvil was made of synthetic gem-quality silicon carbide single crystal commercially supplied from Charles & Colvard. Its size was 10 mm in outside diameter, 11 mm in height, and 5 mm in culet diameter. The SD anvils were cut from a die commercially supplied from Sumitomo Electric Industries. Its geometry was 13 mm in outside diameter, 12 mm in length, and 4 to 5 mm in culet diameter. The NPD anvils were cut from test pieces supplied from Sumitomo Electric Industries, LTD (Itami, Osaka, Japan). Their sizes were 5 to 6 mm in outside diameter, 4 to 5 mm in length, and 3 to 4 mm in culet diameter, which were precisely cut and shaped into the double-conical geometry by laser-machining scheme [9]. These anvils and their surrounding supports were designed to accept full lateral support while compressive force of the order of 100 kilonewtons (~10 tons) were applied to compress the increased sample volume.

As the anvil culet size increased, our earliest designs of compact anvil cells for neutrons [6,10] are required to be improved to generate even more forces, so that the cell designs were revised [11,12]. Using these cells and anvils, we conducted a series of pressure generation experiments, and found that the generated maximum pressure strongly depends on the sample volume. We generated 9 GPa pressure with moissanite or NPD on one side and SD on the other side of the sample, which was 2 to 4 $mm^3$ in volume between the 4 to 5 mm anvil culet diameters. We also generated 5 to 14 GPa pressures using a pair of NPD with 4 mm in culet diameter; 5 GPa was for 3 $mm^3$ sample volume enclosed in a $Ti_{52.5}Zr_{47.5}$ "TiZr null-alloy" gasket, while 14 GPa was for 0.7 $mm^3$ sample volume in a fully-hardened Be-Cu 25 alloy gasket. Therefore, it was experimentally demonstrated that such a large difference in generated pressure using the same anvil material and geometry came exclusively from the difference in gasket geometry (especially in the sample volume) [13] and selection of its material. We note that for measuring the sample pressures, microscope-based ruby fluorescence systems worked effectively, which collected good fluorescence from the pressure sensors in sample chambers through transparent moissanite and NPD anvils [6].

Considering the results of these pressure generation experiments, we selected a pair of the 4 mm culet NPD anvils and the TiZr gasket material as the working standard. Two representative samples with different volumes, 4 $mm^3$ of gas hydrate and 0.7 $mm^3$ of lead, were prepared and gently compressed with these anvils and gasket (Fig. 1).

At TAKUMI beamline, the incident neutron beam is shielded and collimated by the combination of boron rubber plate and hBN (boron nitride) rod collimators which stop the direct neutron beam, and also by cadmium foil surrounding the cell which stops scattered neutrons from the cell body (Fig. 1). A number of hBN collimators with different inner diameters were prepared to best-fit into the size of the sample. Fine tuning of the neutron focusing device works not only to increase the beam intensity at the sample position but also to reduce the background neutron intensity scattered from other materials than the sample. This device produces twice higher intensity of neutron beam at the sample position while keeping the resolution Δd/d at around 0.6% which is only 0.1% larger than the original performance of TAKUMI [3].

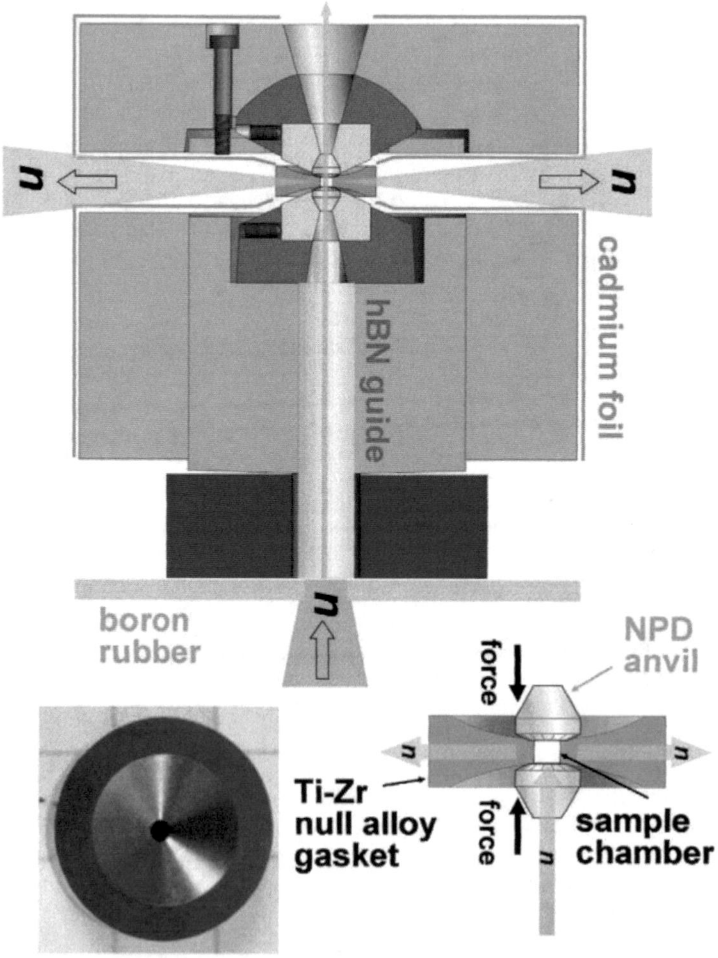


**Figure 1.** An example of compact anvil cell and neutron beam path geometry. Boron rubber plate, hBN guide, and cadmium foil simultaneously shield and collimate the incident neutron beam from the bottom of the figure and reduce the background scattered from the cell body.

## 3. Results and Summary

Figs. 2 and 3 show the anvil cell setup and observed diffraction patterns for the two representative samples. Some of the strongest diffraction peaks shown with asterisks came from the two anvils directly contacting with the samples. These anvils were polycrystalline which intrinsically produce strong diffraction. Other indexed peaks were all coming from the samples.

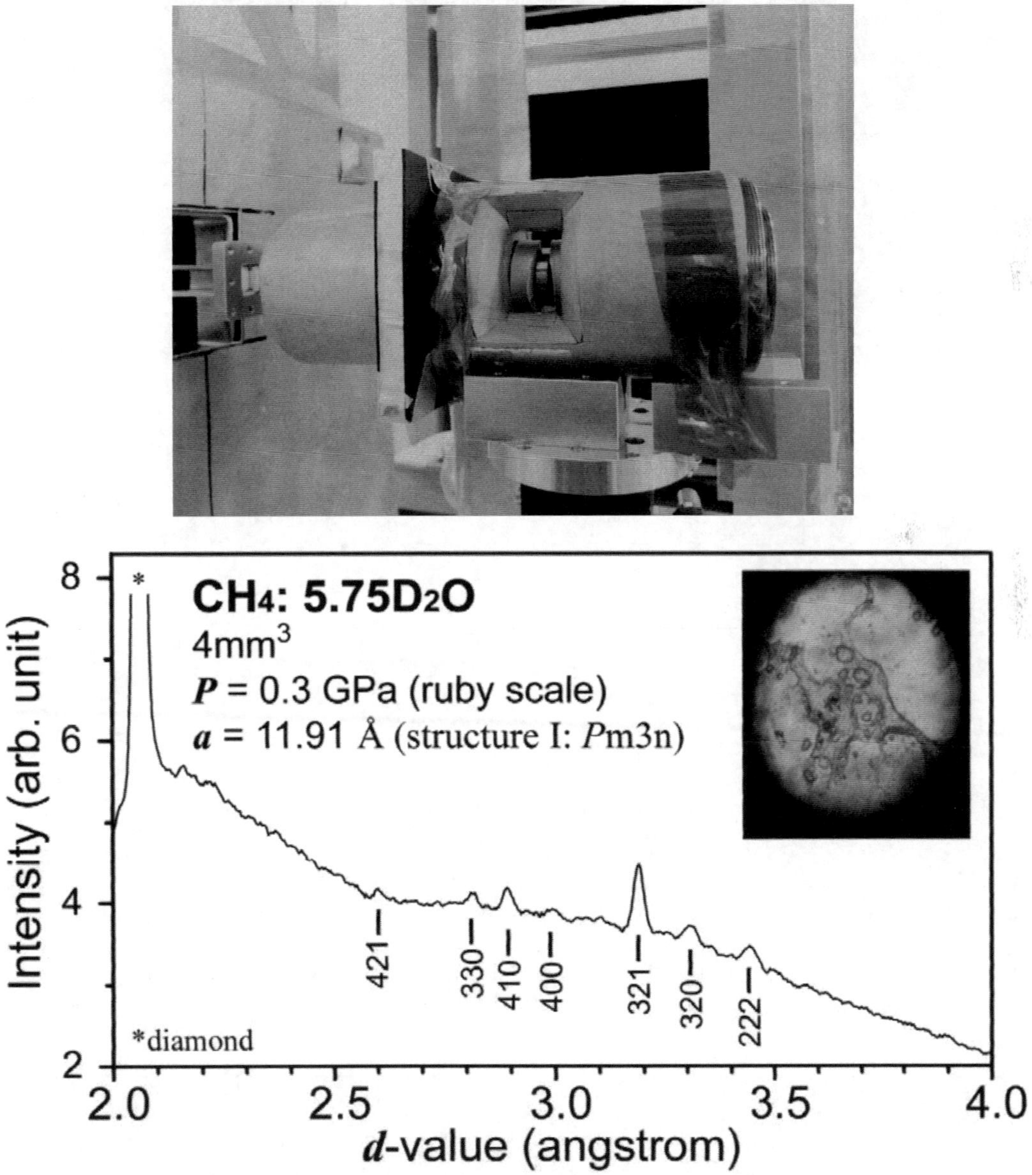


**Figure 2.** Setup and observed diffraction pattern of partially-deuterated methane hydrate with 4 $mm^3$ in volume before compression. The neutron beam came from the left of the cell. The small photo is the sample of ~2mm in diameter observed through the anvil by microscope. Experimental conditions were 120 kW proton power, 6.5 hours of accumulation, and radial collimator of 2 mm gauge volume installed only at the single side (north) detector bank. This pattern was constructed from that bank only.

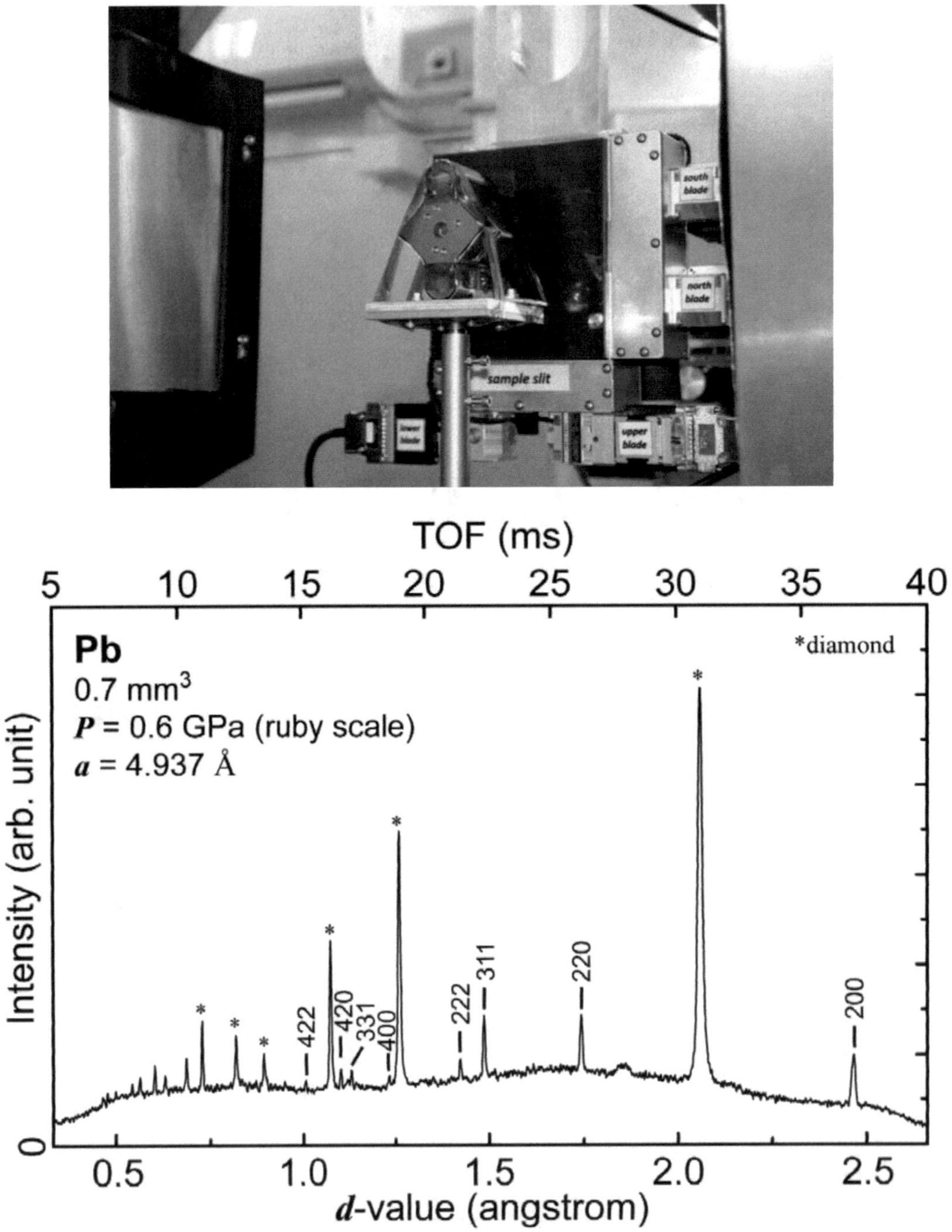


**Figure 3.** Setup and observed diffraction pattern of lead with 0.7 $mm^3$ in volume before compression. The neutron beam came from the back of the cell at the centre of the photo. Experimental conditions were 100 kW proton power, 7 hours of accumulation, and radial collimator of 2 mm gauge volume installed only at the single side (north) detector bank. This pattern was constructed from that bank only.

It is noteworthy that more than seven diffraction peaks from both samples are clearly observed. Especially in Fig. 3, only 0.7 $mm^3$ sample volume gave enough diffraction intensity to accurately determine the lattice constant at high pressure by Le Bail fitting [14]. Therefore, in spite of the small sample capacity, the opposed-anvil cells gave very high signal-to-noise ratio per unit volume, and will become alternative apparatuses for powder neutron diffraction at high pressures at strong pulsed neutron sources where enough neutron intensity was granted.

Acknowledgments
This work is supported by JSPS Grant-in-Aid for Creative Scientific Research (19GS0205) and Grant-in-Aid for Scientific Research on Innovative Areas (20103001).

**References**

[1] Harjo S, Moriai A, Torii S, Suzuki H, Suzuya K, Morii Y, Arai M, Tomota Y, Akita K, Akiniwa Y 2006 *Mater. Sci. Forum* **524–525** 199
[2] Hattori T, Fukazawa H, Okuchi T, Komatsu K, Arima H *MLF Annual Report*, in press
[3] Arima H, Komatsu K, Ikeda K, Hirota K, Kagi H 2009 *Nucl. Instr. and Meth. A* **600** 71
[4] Besson J M, Nelmes R J, Hamel G, Loveday J S, Weill G, Hull S 1992 *Physica B* **180&181**, 907
[5] Xu J, Mao H K, Hemley R J, Hines E 2004 *Rev. Sci. Instrum.* **75** 1034
[6] Okuchi T, Sasaki S, Osakabe T, Ohno Y, Odake S, Kagi H 2010 *J. Phys. Conf. Ser.* **215** 012188
[7] Klotz S, Besson J M, Hamel G, Nelmes R J, Loveday J S, Marshall W G, Wilson R M 1995 *Appl. Phys. Lett.* **66** 1735
[8] Irifune T, Kurio A, Sakamoto S, Inoue T and Sumiya H 2003 *Nature* **421**, 599
[9] Okuchi T, Ohfuji H, Odake S, Kagi H, Nagatomo S, Sugata M and Sumiya H 2009 *Appl. Phys. A: Mater. Sci. Process.* **96** 833
[10] Okuchi T, Sasaki S, Ohno Y, Komatsu K, Kagi H, Arima H, Abe J, Hattori, T, Sano A, Osakabe T, Utsumi W, Harjo S, Ito T, Aizawa K, Irifune T 2010 *Rev. High Press. Sci. Tech.* **20** 175 (in Japanese with english abstract)
[11] Okuchi T, Sasaki S, Ohno Y, Komatsu K, Kagi H, Arima H, Abe J, Hattori, T, Sano A, Osakabe T, Utsumi W, Irifune T 2010 S*pecial Issue of Rev. High Press. Sci. Tech.* **51** 2B01 (in Japanese)
[12] Okuchi T, *in preparation*
[13] Ohno Y, Fujii T, Sasaki S, Osakabe T, Okuchi T, Kagi H 2010 S*pecial Issue of Rev. High Press. Sci. Tech.* **51** 2P09 (in Japanese)
[14] Le Bail A, Duroy H, Fourquet J L 1988 *Mat. Res. Bull.* **23** 447

Journal of Physics: Conference Series **377** (2012) 012014 doi:10.1088/1742-6596/377/1/012014

# Twin chamber sample assembly in DAC and HPHT studies on GaN nano-particles

**Balmukund Shukla*, N V Chandra Shekar, N R Sanjay Kumar, T R Ravindran, P Sahoo[@], S Dhara[@], and P Ch Sahu**
Condensed Matter Physics Division, [@]Surface and Nanoscience Division, Materials Science Group, Indira Gandhi Centre for Atomic Research, Kalpakkam, Tamil Nadu, India

*E-mail: bshukla@igcar.gov.in

**Abstract**: In this paper, we have suggested a novel idea of twin chamber sample assembly for separating ruby from the sample to overcome certain problems during high pressure–high temperature experiments using diamond anvil cell. Two holes of diameter 70μm were drilled symmetrically about the centre of the preindented area (500 μm diameter) in a stainless steel gasket using EDM. Using ruby pressure calibration, good pressure correlation between these two holes was established up to about 15GPa. Also, high pressure - high temperature (HPHT) experiments on the III-V compound semiconductor GaN were performed up to ~ 3GPa and 2000K using the two chamber sample assembly and a laser heated diamond anvil cell facility. NaCl, both hydrous and anhydrous, were used as the pressure transmitting media in two separate experiments. Micro-Raman spectroscopy was used to characterise the HPHT treated samples. While GaN remained stable in its wurtzite phase when heated in anhydrous NaCl medium, it transformed to $Ga_2O_3$ with rocksalt structure in the hydrous NaCl medium.

## 1. Introduction

The diamond anvil cells (DACs) are well suited for *in-situ* high pressure study of various forms of samples ranging from solids and liquids to gases. Merely keeping the sample in between the diamonds in a DAC does not serve the purpose of achieving high pressure, because of various problems like the sample squeezing out at high pressures and anisotropic transmission of pressure, *etc.* This has been elegantly solved by using hard gaskets for containing the sample. Stainless steel or rhenium gasket is used as a container for the sample. The maximum static pressure attainable nowadays using the diamond anvil cell (DAC) technique has reached several mega bars (100 GPa=1 mbar) [1,2]. Simultaneously, high temperatures ~ 5000°C can be generated by using the laser heated DAC (LHDAC) facility [3]. During laser heating experiments, the pressure calibrant is placed along with the sample. In certain experiments, this configuration is not desirable. For example, while working with porous silicon under high pressure, we can not measure the pressure using ruby in the same chamber, as their photoluminescence spectra overlap [4]. Also during synthesis of metal nitrides under HPHT condition, there should not be any oxygen source in the sample chamber. We know that ruby ($Al_2O_3:Cr^{3+}$) which decomposes above ~2000K becomes a source of oxygen and hence can not be used in the same chamber in the above experiment [5]. To overcome this situation, a two-hole sample assembly is desired, one for mounting the sample and the other for mounting Ruby pressure calibrant.

In DAC experiments, a gasket for holding the sample is prepared by pre-indenting it and drilling a central hole of desired size. For example, in a stainless steel gasket pre-indented to thickness of ~ 50 μm, 200-250μm diameter hole is drilled for carrying out experiments up to ~50-60GPa using 500μm culet diamond anvils. To attain higher pressure (~100GPa), smaller culet sizes (100-300μm) are used. The common method for drilling a hole larger than 100μm is by mechanical drilling using tungsten carbide drill bits. This process is very tedious and time consuming. Also, this process involves breaking of expensive drill bits and formation of burr while drilling. Drilling two holes in a single gasket by this method is difficult. To achieve this we have deployed a technique by modifying a low cost electric discharge machine. A 50μm tungsten wire was used as one of the electrodes in the discharge machine and the gasket as the other electrode. A two hole sample assembly was made for studying the HPHT behaviour of GaN nanoparticles by loading GaN in one hole and the Ruby in the other. This twin chamber technique also can be used to study two samples, a reference sample and the actual sample under HPHT conditions.

## 2. Experiment

***Drilling Holes by EDM:*** For drilling holes in SS gaskets, we have used a modified electric discharge machine (JOEMARS, Korea model no. TR 100) which supports voltages ranging from 100 to 240 V with max power consumption of 450 VA. The power was supplied through a variac transformer. The gasket was mounted on an x-y-z stage (Fig. 1) [6]

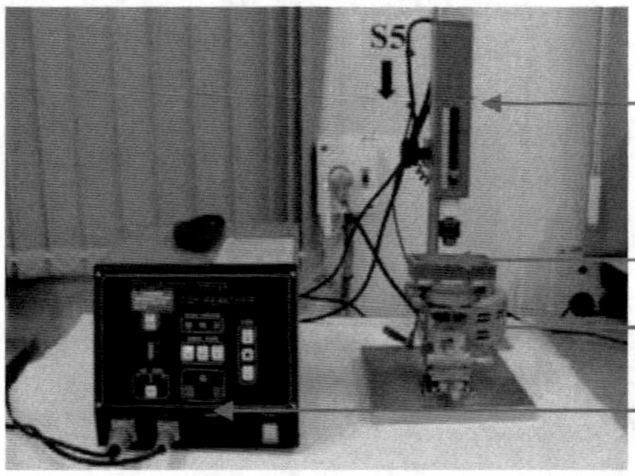


**Figure 1.** Electric discharge machine (EDM) for drilling DAC gasket holes

Micro EDM involves a thermal process in which a spark is utilised to erode a conducting material [7,8]. As there is no contact between the tip of the tungsten wire and SS gasket, there is no force acting between them. An x-y-z stage is used for fixing the SS gasket and positioning of the tungsten wire at the desired point is done using a stereo zoom microscope. Both the wire and the SS gasket are submerged in dielectric fluid. De-ionized water is used as a dielectric. It acts as insulation in the gap, flushes out eroded particles, and also cools the electrodes. Depending upon the gap between the wire and gasket, a suitable voltage is applied to establish an electric field. The voltage applied must be enough to create an electric field higher than the dielectric rigidity of the fluid used in the process. The SS gasket stage is then slowly moved upwards toward the tungsten wire until the spark gap (the nearest distance between both electrodes) is small enough to ionize the dielectric. As the electric field is established between the wire and the gasket, the plasma channel is formed between them. The electrical resistance of such plasma channel would be very less. Thus all of a sudden, a large number of electrons flow from the wire to the gasket and ions from the gasket to the wire. This is called avalanche motion of electrons. Such movement of electrons and ions can be visually seen as a spark. Thus the electrical energy is dissipated as the thermal energy of the spark.

The high speed electrons then impinge on the gasket and ions on the wire, and create a localized heat flux. Such intense localized heat flux leads to instantaneous rise in temperature. Such localized extreme rise in temperature leads to material removal. Material removal occurs due to instant vaporization of the material as well as due to melting. If the potential difference is lowered below the threshold value, the plasma channel is no longer sustained. As the plasma channel collapses, it generates pressure or shock waves, which evacuates the molten material forming a crater of removed material around the site of the spark. Due to the sudden decrease of internal pressure of the gas ball, the dielectric fluid breaks it making the ball to implode. As a consequence of this implosion, the molten metal gets ejected out. Since tungsten is a hard material, there is very little erosion of the tungsten wire. For high precision applications, the energy must be reduced and it will take longer time to make a hole. In our experiment, we used a 50μm diameter tungsten wire as electrodes for drilling two equidistant holes of diameter 70μm on a preindented SS gasket. We used a Mao-Bell DAC with 500 μm diamond culets.

***Synthesis of GaN Nanoparticles***

GaN nanoparticles were synthesized in the atmospheric pressure chemical vapour deposition (APCVD) technique using Ga (99.999%, Aldrich) as precursor and $NH_3$ as reacting gas for about 1.5 hrs. The growth temperature was kept at 900 °C. As a final product, powder like GaN matrix with light yellowish colour was formed around the source material within the alumina. Field emission scanning electron microscopy (FESEM, Zeiss Supra 40) image of GaN nanoparticles (Fig. 2), reveals an average particle size of ~ 100 nm with faceted morphology.

**Figure 2.** SEM image of GaN nano-particles

***Laser heating system:*** HPHT studies of GaN nanoparticles by the two hole method was carried out using the LHDAC facility. A $CO_2$ laser was used for laser heating the sample. The diameter of laser spot on sample was ~40 μm.

***Micro-Raman Set up:*** A high throughput Renishaw micro-Raman spectrometer (model Invia) was employed to record the spectra using the 514 nm laser excitations. The laser was focused to a spot size of about 2 μm with a 20× objective and the signal was detected by a CCD detector. Collection time for each spectrum was typically 100 seconds.

**3. Results and Discussion:**

***Pressure correlation in two chambers:***

Using the modified EDM, two holes of diameter ~ 70 μm equispaced from the centre of the culet impression of the diamond were made to separate sample and pressure calibrant. The schematic of the gasket is shown in Fig. 3a and the photo of the actual gasket containing two holes is shown in Fig 3b.

Journal of Physics: Conference Series **377** (2012) 012014 doi:10.1088/1742-6596/377/1/012014

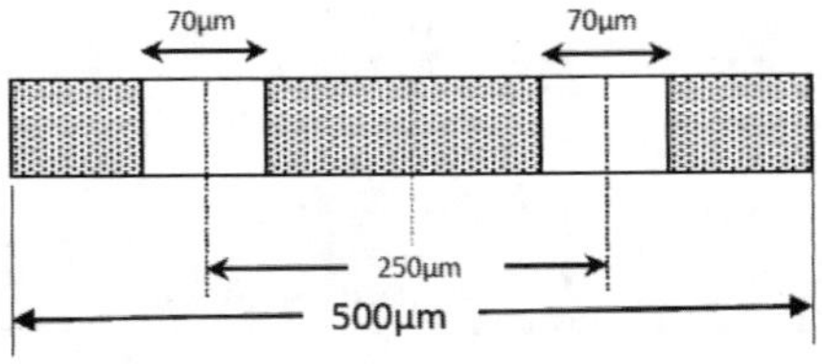


**Figure 3a.** Hole positions in the culet impression on SS gasket

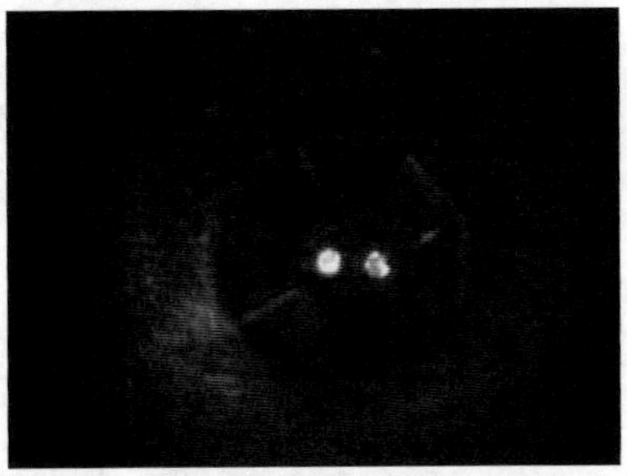

**Figure 3b.** Two hole gasket mounted on piston cylinder of the DAC with the both chambers filled with ruby

To the pressure correlation between the two holes, we put ruby particles in both the holes along with the sodium chloride as the pressure transmitting medium. Pressure was estimated using ruby fluorescence technique. It was seen that the pressure values in the two holes were almost same up to 15 GPa (Fig. 4), demonstrating that the two holes were equidistant from the centre of the preindented area. At higher pressures, there are chances for the sample chambers to be deformed disturbing the above correlation. Therefore, this process is well suited for the low pressure experiments.

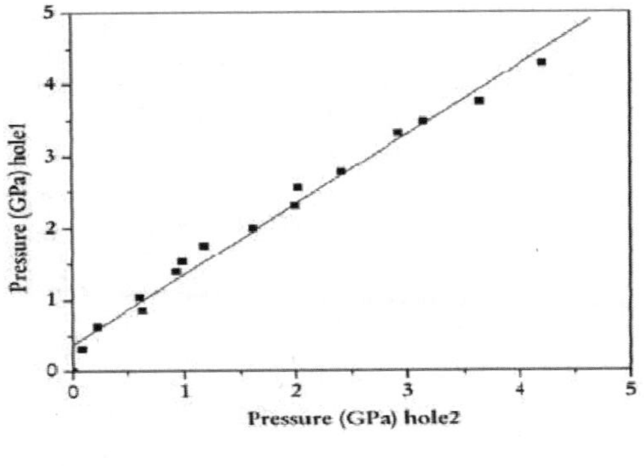


**Figure 4.** Pressure correlation in two holes of the gasket

***HPHT study of GaN nanoparticles using two chamber sample assembly:***

GaN is an III-V semiconductor with a band gap of 3.4eV and exists in wurtzite structure. To our knowledge, there are no reports on HPHT behaviour of GaN in the literature. Micro-Raman spectrum of GaN before heating certifies the existence of wurtzite structure at ambient pressure and temperature. The cell was pressurised to ~3GPa and laser heating was carried out for ~10 min and temperature was estimated to be ~2000K. Figure 5 shows the $CO_2$ laser heating of GaN inside DAC. Raman spectra of the laser heated sample do not show any change with respect to the spectrum of GaN obtained before heating. It is reported in literature that GaN converts to gallium oxide at ambient pressure at a temperature of ~900°C. The probable reason for not observing oxide formation may be due to the inert atmosphere present in the sample chamber. This is so because in the above sample assembly for laser heating, the NaCl was pre-heated before loading in DAC for up to 3 hours to remove the moisture.

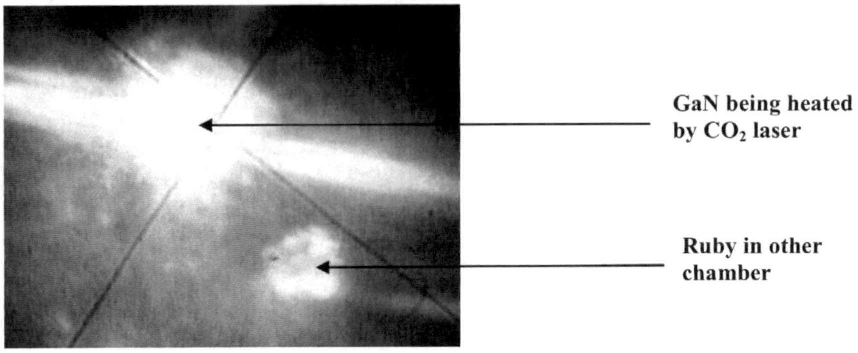


**Figure 5**. Laser heating of GaN under high pressure using $CO_2$ laser

For verifying the above conclusion, another experiment was carried out, wherein NaCl was used without pre-heating treatment for removal of moisture. Raman spectra before laser heating contains modes at 273 $cm^{-1}$, 430 $cm^{-1}$, 574 $cm^{-1}$ and 737 $cm^{-1}$ after laser heating it is observed that all the modes of GaN have disappeared along with appearance of new modes at 201 $cm^{-1}$,358 $cm^{-1}$, 449 $cm^{-1}$, 483 $cm^{-1}$ and 780 $cm^{-1}$. The Raman spectra before and after laser heating are shown in Fig 6. All the Raman modes observed after laser heating can be ascribed to $Ga_2O_3$ confirming the conversion of GaN into $Ga_2O_3$ under HPHT conditions in presence of moisture.

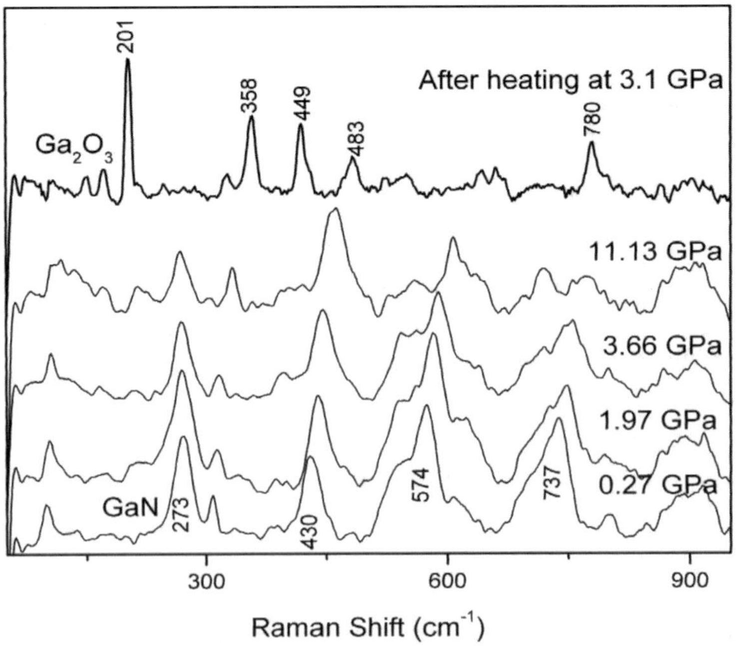


**Figure 6.** Raman modes of GaN before and after heating at various pressures. Complete transformation to $Ga_2O_3$ is seen in the top most pattern. Bottom four graphs are before heating in anhydrous NaCl and top graph is after laser heating in hydrous NaCl.

**Acknowledgment:**
The authors thank Mrs. L Dayana, Mr. L M Sundaram, Dr. N Subramanian, Mr. M Sekar for the help during various stages of this work; Dr. A K Arora and Dr. C S Sundar for their encouragement and support.

**References:**

1. Hemley R J and Ashcroft W 1998 *Phys. Today* **51** 26
2. Sahu P Ch, Chandra Shekar N V 2007 *Resonance* **12** 49; ibid *2007 Resonance* **12** 10
3. Subramanian N, Chandra Shekar N V, Sanjay Kumar N R and Sahu P Ch 2006 *Curr. Sci.* **91** 176
4. Huang W Q, Zhang R T, Wang H X, Jin F, Xu Li, Qin S J, Wu K Y, Liu S R, Qin C J 2008 *Optical Communications* **281** 5229–5233
5. Horvath-Bordon E, Riedel R, Zerr A, McMillan P F, Auffermann G, Prots Y, Bronger W, Kniep R and Kroll P 2006 *Chem. Soc. Rev.* 35 987-1014.
6. Dayana Lonappan, Chandra Shekar N V, Sundaram L. M, Edwin T and Sahu P Ch, Solid State Physics: Proceedings of the55th DAE Solid State Physics Symposium 2010, Part A, Part B, edited by Garg A, Mittal R and Mukhopadhyay R, AIP Conf. Proc. No. 1349 (AIP, New York, 2011) p.461.
7. Driver C, Atkinson J, Helml H J, Li L 2004 *J. Mat. Proc. Tech.* **149** 296-303
8. Yap Ber-Chin and Fromm L 2003 *Electrical Overstress/Electrostatic Discharge Symposium EOS/ESD '03* pp.1-9.

# High Pressure Structural Studies on Rare-Earth Sesquioxides

**P. Ch. Sahu*, Dayana Lonappan[@] and N.V. Chandra Shekar**
Condensed Matter Physics Division, Materials Science Group
Indira Gandhi Centre for Atomic Research, Kalpakkam 603102, Tamil Nadu, India
[@]Crystal Growth Centre, Anna University, Chennai 600025, Tamil Nadu, India

*E-mail: pcsahu@igcar.gov.in

**Abstract.** The Rare-Earth sesquioxides ($RE_2O_3$) exhibit interesting physical and chemical properties. Some of these oxides have tremendous technological applications. At ambient conditions, the $RE_2O_3$ systems exist in three polymorphic forms, namely: hexagonal A- type, monoclinic B- type, and cubic C- type. The structural stability of three $RE_2O_3$ systems: RE = Gd, Ho, Tm and the $Tb_4O_7$ system have been studied under high pressure. All the four systems exhibited interesting pressure induced phase transitions. $Gd_2O_3$ exhibited C-A transition at ~ 12 GPa, whereas $Ho_2O_3$ and $Tm_2O_3$ exhibited C-B transition at pressures of 9.5 and 7 GPa respectively. $Tb_4O_7$ transformed from its cubic fluorite structure to probably the cotunnite phase at a relatively higher pressure of 27 GPa. The unusual large pressure stability of $Tb_4O_7$ was attributed to the presence of $Tb^{4+}$ ions. The bulk moduli show systematic increase with higher cations, probably due to the enhanced influence of the 4*f* electron states and increased nature of covalency. However, the mechanisms of the structural transitions C-B-A with increased pressure and increased cation radii are yet to be understood.

## 1. Introduction

The Rare-Earth sesquioxides ($RE_2O_3$) exhibit interesting structural, physical and chemical properties and also have important technological applications [1]. Some of these oxides have potential application as nuclear waste host materials[2]. Therefore, it is crucial to understand their structural stability at extreme conditions of pressure and temperature. At ambient conditions, the $RE_2O_3$ systems exist in three polymorphic forms, namely: hexagonal A- type with space group SG: P-3m1 and Z=1, monoclinic B- type with SG: C2/m and Z=6, and cubic C- type with SG: Ia-3 and Z=16 [3]. The lighter $RE_2O_3$ systems with larger cation sizes (La-Pm) exist in A-type structure. The cubic structure is characteristic of heavier $RE_2O_3$ with smaller cation sizes (Tb-Lu). The $RE_2O_3$ systems with medium cation radii (Sm-Gd) may exist in both B and C type structures. Thus the $RE_2O_3$ systems adopt the structural sequence A-B-C with the increasing atomic number or decreasing ionic radii. Two more high temperature polymorphs, the cubic X (SG: Im-3 m ) and hexagonal H (SG: $P6_3$/ mmc) phases also have been reported above 2000°C [4]. The molar volumes of these compounds decrease with increasing cation coordination in the sequence C-B-A [4,5]. Hence, the phase transition sequence of C-B-A is usually expected for most of the $RE_2O_3$ systems under pressure [4-6].

In this investigation, the structural stability of three $RE_2O_3$ systems: RE=Gd, Ho, Tm and the $Tb_4O_7$ system have been studied under high pressure using X-ray diffraction (XRD) and laser Raman. The systematic structural transition data have been compared and contrasted with the results available for other $RE_2O_3$ systems and plausible reasons have been suggested for the observed trends.

## 2. Experimental Details

All the oxide samples of purity 99.99% were characterised by XRD using a high precision Guinier diffractometer[7] and were found to be in single phase. High pressure X-ray diffraction (HPXRD) experiments were carried out in the angle dispersive mode by using a Mao-Bell type Diamond Anvil Cell (DAC). Either the PSD based Guinier diffractometer [7] or image plate based mar345dtb diffractometer was used for carrying out the HPXRD experiments. A mixture of methanol, ethanol and water (MEW) in the volume ratio 16:3:1 was used as the pressure transmitting medium. Either the equation of state (EOS) of Ag [8] or the ruby fluorescence method [9] was employed for determining the sample pressure. The incident Mo X-ray was obtained from Rigaku 18 kW rotating anode X-ray generator. The incident X-ray was monochromatised by a curved crystal monochromator in the case of Guinier diffractometer and a flat graphite monochromator in the case of mar345dtb diffractometer. Some samples in their high pressure metastable phase were laser heated using the LHDAC facility developed in our laboratory [10] to check the reversibility of the high pressure phase.

## 3. Results and Discussion

HPXRD experiments on the three $RE_2O_3$ systems (RE=Gd, Ho and Tm) and the $Tb_4O_7$ were carried out up to the maximum pressures of 25, 17, 30 and 35 GPa respectively [11-14]. All the four systems exhibited interesting pressure induced structural transitions. $Gd_2O_3$ transformed from its C-type cubic structure to the A-type hexagonal structure at ~ 12 GPa with a 13% volume collapse (Figure 1)[11]. On releasing the pressure from 25 GPa, the daughter phase was retained at ambient. The pressure quenched sample at NTP was laser heated to about 500°C using the LHDAC facility[10] to check the reversibility of the above transition. Although a small fraction of the cubic phase reappeared, the structure remained mostly hexagonal indicating the irreversible nature of the transition. A recent study on Er-doped $Gd_2O_3$ up to ~40 GPa indicated the C→A transition at 8.57-12.5 GPa [15]. However, it

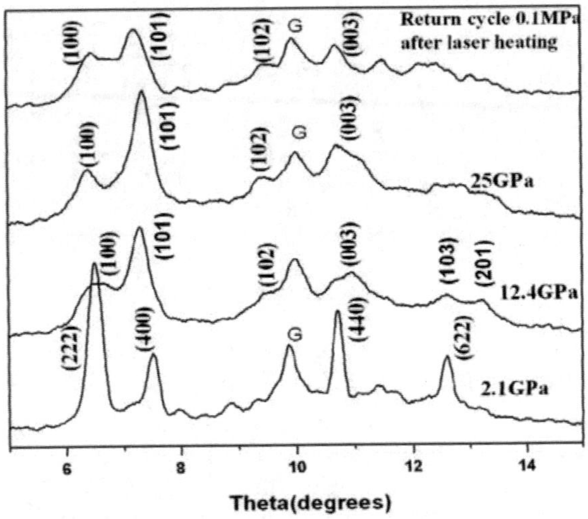


**Figure 1.** High pressure XRD patterns of $Gd_2O_3$ showing the C-A transition above 12 GPa. The top pattern is at 0.1MPa after return cycle[11].

transformed to the B phase during the return cycle. Similar results have been observed by other investigators in nano-crystalline Eu-doped $Gd_2O_3$ [16] and pure $Gd_2O_3$ [17,18]. The differences in the phase transition behaviour during the return cycle may be due to the difference in the sample quality and the thermal history. The P–V data up to 12 GPa were fitted to the Birch–Murnaghan equation of state [19] and the bulk modulus and its pressure derivative were found to be: $B_0 = 118 \pm 21$ GPa and $B_0' = 13.9 \pm 7.7$ respectively [11]. This value of bulk modulus is smaller compared to the reported values of 164, 134 and 188 GPa [15,17,18] with $B_0'$ fixed at 4. These values of the bulk moduli are comparable with those reported for other $RE_2O_3$ systems [4, 20-22].

$Ho_2O_3$ on the other hand transformed from the C-type cubic phase to B-type monoclinic phase at 9.5 GPa[12]. The high pressure B-phase remained stable up to 17 GPa. High pressure Raman studies on $Ho_2O_3$ indicated a Ho–O stretching vibration peak at wave number 372 $cm^{-1}$. This peak slowly broadened and moved towards higher wave number on compression, and finally disappeared above 7 GPa, signifying the above phase transition. A recent study on $Ho_2O_3$ up to 63 GPa indicated C→B transition at 8.9 GPa and B→A at 14.8 GPa [23]. During the return cycle, it transformed to the B phase. In our experiment, since the B-phase was stable up to 17 GPa, the same B-phase was obtained on releasing the pressure. In case of $Tm_2O_3$, the irreversible C-B transition was observed at 7 GPa [13]. The P–V data for $Ho_2O_3$ up to 9.5 GPa and that of $Tm_2O_3$ up to 7 GPa were fitted to the Birch–Murnaghan equation of state[19] and the values of the bulk modulus and its pressure derivative obtained from the above fit were: $B_0 = 178 \pm 7$ GPa , $B_0' = 4$ and $B_0 = 154.5 \pm 7.5$ GPa , $B_0' = 4$ respectively. The $B_0$ and $B_0'$ values for $Ho_2O_3$ reported by Jiang *et al* [23] are 206 GPa and 4.8 respectively.

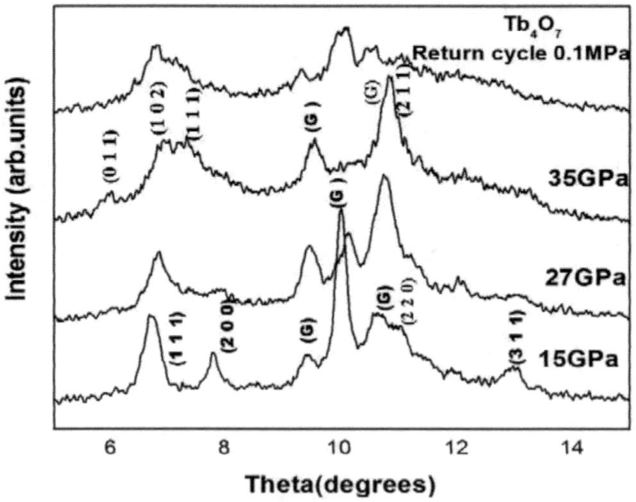


**Figure 2.** High pressure XRD patterns of $Tb_4O_7$ showing the structural transition above 27 GPa [14].

Terbium oxide $Tb_4O_7$ transformed from its fluorite cubic structure to probably cotunnite type (orthorhombic) structure at ~27 GPa (Figure 2) [14]. The large structural stability of the compound under pressure is thus unusual, when compared with other $RE_2O_3$ systems. The P–V data was fitted to the Birch–Murnaghan equation of state [19]. The bulk modulus and its pressure derivative of the cubic phase were found to be $B_0 = 195.7 \pm 3.8$ GPa and $B_0' = 4$ respectively. This bulk modulus is comparable to that observed in other rare earth oxides [18, 20, 21, 23]. However, the transition pressure of 27 GPa for $Tb_4O_7$ is much higher than that generally found in other rare-Earth oxide systems (<15 GPa) [18, 20-24]. This has been attributed to the presence of $Tb^{4+}$ ions in $Tb_4O_7$. The

XPS spectra on the sample confirm the presence of $Tb^{4+}$ ions [14]. In $Tb_4O_7$, the different cation sites in the fluorite lattice are occupied by trivalent and tetravalent Tb ions, which are coordinated by 6 and 8 oxygen anions respectively. X-ray absorption spectroscopy studies reveal that the *f8* ground state configuration of Tb is well localized in the trivalent oxide, while in the case of tetravalent oxide, the last *4f* electron delocalizes in the valence band. The remaining seven- *4f* electrons of the cation remain localized. This indicates that the last *f* electron participates in a partial type covalent bonding [25]. In other sesquioxides, the bonding is mostly ionic in nature with the cations in the 3+ state. Since our sample mostly consisted of $Tb^{4+}$ ions due to the presence of $Tb_4O_7$, this may probably be one of the reasons for the large structural stability seen here.

The summary of our phase transition data along with that reported in literature are shown in figure 3. It can be seen that the C-A transition pressure increases with increasing Z of the cation [20,24]. The bulk moduli of these systems along with that of the other C-type $RE_2O_3$ systems reported in the literature [18, 20-24] are shown in figure 4. The increase in $B_0$ with increasing atomic number Z of the cation can be clearly seen. These anomalous phase transition and compressibility behaviour of the $RE_2O_3$ systems have to be understood in terms of their electronic structure.

In the C-type structure, the cations have 6 fold coordination, that is each cation is surrounded by 6 anions. There are two types of cation sites- 3/4th in 24*d* and 1/4th 8*a* positions. In the B-type structure, 2/3rd cations are surrounded by 7 anions and 1/3rd by 6 anions. In A-type structure, each cation is surrounded by 7 anions. Although the cation radii increase in the sequence C-B-A, the molar volumes decrease ( density increase) and the cation coordination numbers increase in the same sequence[4, 5, 20, 26]. This contrasting behaviour can be understood from the following arguments. The $RE_2O_3$ systems mostly exhibit ionic bonding with the cations mostly in the 3+ state and the *4f* electrons localized [4,20, 21, 26]. Cations with 4+ state, in which one *4f*–electron participates in bonding, have tendency to form higher oxides like $REO_2$ [20]. Electronic structure calculations have clearly indicated that the nature of covalency between the O-p and RE-*pd* as well as the influence of *4f* states increase in the sequence A-B-C [4,20,26]. The increased nature of directional covalent bonding keep the ions apart leading to their higher molar volume and lower cation coordination, and hence the increased molar volume in the sequence A-B-C. Considering the molar volume difference, the structural sequence C-B-A is expected under high pressure.

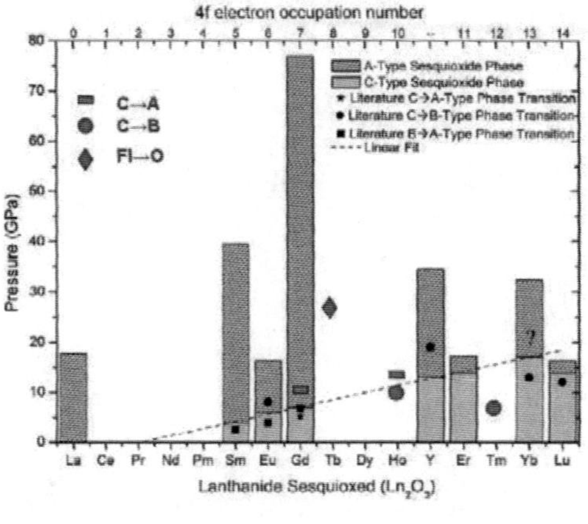


**Figure 3.** Summary of structural transitions in $Re_2O_3$ systems [24] appended with our data.

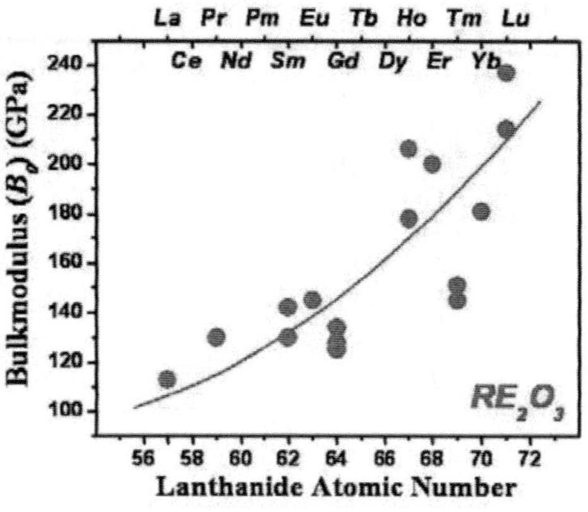


**Figure 4.** Summary of bulk modulii data on C-type $Re_2O_3$ systems.

Further, the free energy difference between C and B type phases are very large compared to that of the A and B phases [18, 20, 21,26]. The C→B transitions are accompanied with large scale rearrangement of atoms, breaking of bonds and large volume collapse (`~10%) [4,5,18, 27]. That is why the C→B transitions are reconstructive type and irreversible. The B→A transitions involve minor displacement of atoms, small volume collapse (~ 2-5%), displacive in nature and are reversible [4, 6, 18, 21, 28]. Depending on the sample history and the experimental conditions ( such as hydrostaticity, rate of pressurization, temperature, *etc.*), either the sequence C → B → A ($Sm_2O_3$, $Yb_2O_3$, $Ho_2O_3$, $Lu_2O_3$) or direct C → A ($Gd_2O_3$, $Dy_2O_3$, $Er_2O_3$) transitions are observed [4-6, 18, 21, 23, 28, 29]. Considering the very small free energy difference between the B and A phases[21], missing of the intermediate B phase during the C→A transition is not surprising. However, during the return cycle, the systems mostly land up in the B phase ($Dy_2O_3$, $Ho_2O_3$, $Er_2O_3$,$Yb_2O_3$). This is due to the fact that there is a large kinetic barrier in the reconstructive type transition from B→C, whereas the A→B displacive transition proceeds without any hindrance.

Now returning to the compressibility behaviour, the bulk moduli of these systems along with that of the other C-type $RE_2O_3$ systems reported in the literature [18,20-24] are shown in figure 4. The increase in $B_0$ with increasing atomic number Z of the cation can be clearly seen. The theoretical estimated values of bulk moduli show very little difference between the A and C type $RE_2O_3$ systems [20]. Thus it can be safely stated that the lighter $RE_2O_3$ systems (with lower molar volume and higher density) have lower bulk moduli compared to that of the heavier $RE_2O_3$ systems (with higher molar volume and lower density). Considering the higher density (lower molar volume) of the lighter $RE_2O_3$ systems, the bulk moduli of these are expected to be higher than that of the heavier $RE_2O_3$ systems. However, as can be seen from figure 4, the observation is just the opposite, that is, the heavier sesquioxides show higher bulk moduli compared to that of the lighter sesquioxides. This behaviour again can be understood from the increased covalent bonding in the heavier sesquioxides compared to the mostly ionic bonding in the lighter sesquioxides [20,26].

In summary, the $RE_2O_3$ systems exhibit the same structure sequence C-B-A on increasing cation radii as well as increasing pressure. The origin of this strange structural behavior may be due to the steric effects. A systematic structural study has to be carried out to understand this behaviour. For this purpose, systematic high resolution synchrotron based X-ray diffraction experiments and Rietveld analysis along with micro-Raman studies have to be carried out. The compressibility behaviour may be attributed to the increased covalent bonding in the heavier sesquioxides (less compressible) compared to the mostly ionic bonding in the lighter sesquioxides (more compressible).

## Acknowledgements

The authors thank Dr. N Subramanian, Dr. Anthony Arulraj, Dr. T.R. Ravindran, Shri M Sekar, Shri N R Sanjay Kumar and Shri B. Shukla for useful discussions. They thank Shri L M Sundaram for his help in sample preparation and characterization, Shri V Kathirvel and Shri Y A Sorb for their help in carrying out experiments, Dr. A.K. Bandyopadhyay, Dr. B.V. Kumarasamy and Prof. M. Rajagopalan for fruitful collaborations. They thank Dr. A K Arora and Dr. C S Sundar for their encouragement and support. DL expresses her gratitude to the IGCAR authorities for granting permission to carry out her Ph.D. work at the Condensed Matter Physics Division of IGCAR.

## References

[1] Hong M, Kwo J, Kortan A R , Mannaerts J P, and Sergent A M 1999 *Science* **283** 1897.

[2] Ewing R C, and Weber W J 1995 *J. Mater. Res.* **10** 243; Ewing R C, Weber W J and Clinard F W Jr 1995 *Prog. Nucl. Energy* **29** 63.

[3] Eyring L 1979 *Handbook on the Physics and Chemistry of Rare Earths* Vol. 3, ed K A Gschneidner Jr and L Eyring (Amsterdam : North-Holland) p. 337.

[4] Zinkevich M 2007 *Prog. Mater. Sci.* **52** 597.
[5] Hoekstra H, and Gingerich K 1964 *Science* **146** 1163.
[6] Atou T, Kusaba K, Tsuchida Y, Utsumi W, Yagi T, and Syono Y 1989 *Mater. Res. Bull.* **24** 1171.
[7] Sahu P Ch, Chandra Shekar N V , Subramanian N, Yousuf M, and Govinda Rajan K 1995 *Rev. Sci. Instrum.* **66** 2599.
[8] Liu L and Bassett W A 1973 *J. Appl. Phys.* **44** 1475; Mao H K, Bell P M, Shaner J W, and Steinberg D J 1978 *J. Appl. Phys.* **49** 3276
[9] Piermarini G J, Block S, Barnett J D, and Forman R A 1975 *J. Appl. Phys.* **46** 2774; Chijioke A D, Nellis W J, Soldatov A, and Silvera I F 2005 *J. Appl. Phys.* **98** 114905.
[10] Subramanian N, Chandra Shekar N V, Sanjay Kumar N R, and Sahu P Ch 2006 *Current Science* **91** 175.
[11] Dayana Lonappan, Chandra Shekar N V, Sahu P Ch, Kumarasamy B V, Bandyopadhyay A K, and Rajagopalan M 2008 *Phil. Mag. Lett.* **88** 473.
[12] Dayana Lonappan, Chandra Shekar N V, Ravindran T R and Sahu P Ch 2010 *Mat. Chem. Phys.* **120** 65.
[13] Chandra Shekar N V, Sanjay Kumar N R, and Sahu P Ch 2011 *AIRAPT-23 –Abstracts* (Mumbai: BARC) p.189.
[14] Dayana Lonappan, Chandra Shekar N V, Sahu P Ch, Kumar J, Ranita Paul and Pinaki Paul 2010 *J. Alloys Compds.* **490** 47.
[15] Zou Xu *et al.* 2011 *Phys. Stat. Solidi B* **248** 1123.
[16] Chen Hai-Yong *et al.* 2007 *Chin. Phys. Lett.* **24** 158.
[17] Bai L *et al.* 2009 *J. Appl. Phys.* **106** 073507.
[18] Zhang F X, Lang M, Wang J W, Becker U, and Ewing R C 2008 *Phys. Rev. B* **78** 064114.
[19] Birch F 1978 *J. Geophys. Res.* **83** 1257.
[20] Rahm M and Skorodumova N V 2009 *Phys. Rev. B* **80** 104105.
[21] Wu Bo, Zinkevich Matvei, Aldinger Matvei, Wen Dingzhong, and Chen Lu 2007 *J. Solid State Chem.* **180** 3280.
[22] Mikami M and Nakamura S 2006 *J. Alloys and Comp.* **408** 687.
[23] Jiang Sheng *et al.* 2011 *J. Appl. Phys.* **110** 013526.
[24] McClure J P 2009 *High Pressure Phase Transitions in the Lanthanide Sesquioxides, Ph.D. Thesis* (Las Vegas: University of Nevada).
[25] Esteva J M, Karnatak R C, Dexpert H, Gasgnier M, Caro P E, and Albert L 1986 *J. De Phys.* **12** C8–955.
[26] Petit L, Svane A, Szotek Z, and Temmerman W M 2005 *Phys. Rev. B 72* 205118.
[27] Schweda E 1992 *Key Eng. Mater.* **68** 187.
[28] Atou T, Kusaba K, Syono Y, Kikegawa T, and Iwasaki H 1992 *High - Pressure Research: Application to Earth and Planetary Sciences, Geophysical Monograph Series* vol.67, ed Y Syono and M H Manghnani (Washington: AGU) p. 469.
[29] Sharma N D, Singh J, Dogra S, Varandani D, Poswal H K, Sharma S M, and Bandyopadhyay A K 2011 *J. Raman Spectroscopy* **42** 438.

Journal of Physics: Conference Series **377** (2012) 012016 doi:10.1088/1742-6596/377/1/012016

# High-pressure structural phase transitions in chromium-doped $BaFe_2As_2$

**W O Uhoya[1], J M Montgomery[1], G K Samudrala[1], G M Tsoi[1], Y K Vohra[1], S T Weir[2], A S Sefat[3]**

[1]Department of Physics, University of Alabama at Birmingham (UAB), Birmingham, AL 35294, USA
[2]Mail Stop L-041, Lawrence Livermore National Laboratory (LLNL) Livermore, CA 94550, USA
[3]Oak Ridge National Laboratory (ORNL), Oak Ridge, TN 37831, USA

**Abstract.** We report on the results from high pressure x-ray powder diffraction and electrical resistance measurements for hole doped $BaFe_{2-x}Cr_xAs_2$ (x=0, 0.05, 0.15, 0.4, 0.61) up to 81 GPa and down to 10 K using a synchrotron source and diamond anvil cell (DAC). At ambient temperature, an isostructural phase transition from a tetragonal (T) phase (*I4/mmm*) to a collapsed tetragonal (CT) phase is observed at 17 GPa. This transition is found to be dependent on ambient pressure unit cell volume and is slightly shifted to higher pressure upon increase in the Cr-doping. Unlike $BaFe_2As_2$ which superconduct under high pressure, we have not detected any evidence of pressure induced superconductivity in chromium doped samples in the pressure and temperature range of this study. The measured equation of state parameters are presented for both the tetragonal and collapsed tetragonal phases for x= 0.05, 0.15, 0.40 and 0.61.

## 1. Introduction

The discovery of pressure enhanced superconductivity in chemically doped iron arsenide compounds in 2008 [1] has attracted extensive research activities on these materials in the field of high pressure physics [2]. One of the most intensively investigated iron-based compound is $BaFe_2As_2$ with the tetragonal $ThCr_2Si_2$ (122) type crystal structure [3, 4]. At ambient pressure, pure $BaFe_2As_2$ is known to undergo a crystallographic phase transition from a tetragonal phase (*I4/mmm*) to an orthorhombic phase (*Fmmm*) phase and a concurrent antiferromagnetic ordering (spin-density wave, SDW) at ~140 K [3]. The transition is suppressed by either chemical doping or application of high pressure resulting in the appearance of superconductivity [3, 4]. Ambient pressure studies show that hole-doping using chromium leads to suppression of the magnetic and structural transitions, but does not lead to superconductivity [5]. Studies under pressure have shown that $BaFe_2As_2$ interestingly undergoes an isostructural phase transition from tetragonal (T) phase to an isostructural collapsed tetragonal (CT) phase which is accompanied by anomalous compressibility effect and a sharp drop in axial ratio *c/a* [6-8]. In the present work, we present high pressure electrical resistivity and x-ray diffraction measurements on hole doped $BaFe_{2-x}Cr_xAs_2$ up to 81 GPa and down to 10 K using a synchrotron source and diamond anvil cell. We investigate the influence of chemical doping and high pressure on the tetragonal crystal structure and the possibility of the pressure induced superconductivity.

## 2. Experimental Methods

Single crystals of $BaFe_{2-x}Cr_xAs_2$ were grown out of a mixture of FeAs and CrAs flux and characterized by x-ray diffraction and electron probe micro-analysis as described in the original publication [5]. The crystals were ground and loaded into an 80-micron hole of a spring-steel gasket that was first pre-indented to a ~ 50-micron thickness and mounted between a matched pair of beveled diamond anvil cell ready for high-pressure experiments.

The synchrotron x-ray powder diffraction experiments were performed at the high pressure beam-line 16-ID-B of the Advanced Photon Source (APS) in Argonne National Laboratory. An angle dispersive technique with a MAR345 image-plate area detector was employed using a focused monochromatic beam with x-ray wavelength, $\lambda = 0.4246$ Å. The image plate x-ray diffraction patterns were recorded with a focused x-ray beam of 5μm by 5μm on an 80 μm diameter sample mixed with powdered copper pressure marker. Experimental geometric constraints and the sample to image plate detector distance were calibrated using $CeO_2$ diffraction pattern. The lattice parameters of the copper pressure marker obtained from the Rietveld refinement of the x-ray diffraction patterns of copper-powder mixed with the sample in the DAC were employed for the calibration of pressure [9]. At ambient temperature, the Birch-Murnaghan equation [10] as shown by equation (1) was fitted to the available equation of state data on copper pressure standard [9] to determine sample pressure.

$$P=\tfrac{3}{2}B_0[(\tfrac{V_0}{V})^{7/3}-(\tfrac{V_0}{V})^{5/3}]\{1+\tfrac{3}{4}(B_0'-4)[(\tfrac{V_0}{V})^{2/3}-1]\} \quad (1)$$

Where $B_0$ is the bulk modulus, $B_0'$ is the first derivative of bulk modulus at ambient pressure, and $V_0$ is the ambient pressure volume. The fitted values for the copper pressure standard are $B_0 = 121.6$ GPa, $B' = 5.583$, and $V_0 = 11.802$ Å$^3$/atom [9]. We employed the software package FIT2D [11] to integrate the collected MAR3450 image plate diffraction patterns. The integrated patterns were analyzed by EXPGUI/GSAS software [12] using full-pattern Rietveld refinements and LeBail fit techniques to extract structural parameters.

## 3. Results and discussion

At room temperature and nearly ambient pressure (P~ 0.1±0.2 GPa), $BaFe_{2-x}Cr_xAs_2$ samples are all identified as $ThCr_2Si_2$ (*I4/mmm*) with the following measured lattice parameters: $a$ =3 .962Å(18), $c$ = 13.0760(20)Å and $V_0$= 205.3 Å$^3$ for $BaFe_{1.95}Cr_{0.05}As_2$; $a$ = 3.9615(21) Å, c = 13.1731(13)Å and $V_0$= 206.7Å$^3$ for $BaFe_{1.85}Cr_{0.15}As_2$; $a$ = 3.9682(15) Å, $c$ = 13.1671(8)Å and $V_0$ = 207.3Å$^3$ for $BaFe_{1.6}Cr_{0.4}As_2$; $a$ = 3.9705(21) Å, $c$ = 13.2397(22)Å and $V_0$ = 208.7Å$^3$ for $BaFe_{1.39}Cr_{0.61}As_2$. The incorporation of Cr in to Fe site is observed to increase the unit cell volume without any evidence of a structural phase transition and this is in good agreement with previous ambient pressure studies [5].

Figure 1(a) shows the measured axial ratios (*c/a*) as a function of pressure at ambient temperature for hole doped $BaFe_{2-x}Cr_xAs_2$. For x = 0.05, *c/a* ratio show a rapid decrease (~21% decrease) with increasing pressure up to ~ 16.9 GPa and a gradual decrease above this pressure. The *c/a* ratio variation as a function of pressure can be divided into two linear regions and the fits for the two linear regions are shown as solid lines in the figure. The intersection of the two linear regions defines the phase transition pressure from the ambient pressure T-phase to the CT-phase. The T-CT phase transition is defined by the present experiments to occur at ~17 GPa. We made similar observations at ambient temperature for the case of x = 0.15, x = 0.4 and x = 0.61 with the T-CT phase transitions shifting slightly to ~19.8 GPa, ~14.8 GPa and ~20.3 GPa respectively.

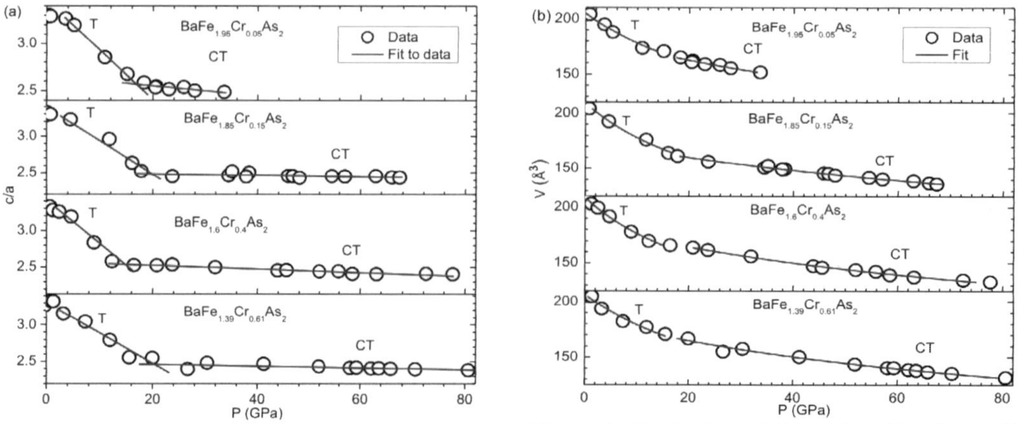


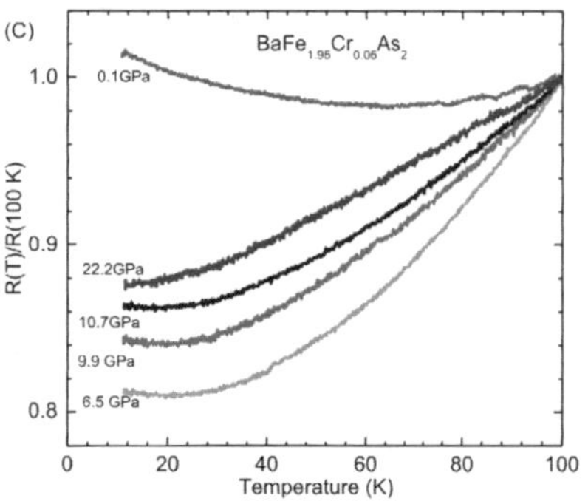


Figure 1. Evolution of the unit cell volume (V) and axial ratio (*c/a*) for the tetragonal (T) and collapsed tetragonal (CT) phases of chromium doped $BaFe_{2-x}Cr_xAs_2$ (x=0.05, 0.15, 0.4 and 0.61) as a function of pressure (P) up to 81 GPa. Intersections of the linear fit to *c/a* data shown in figure 1 (a) indicate the T-CT phase transition pressure. The solid lines in (b) indicate Birch Murnaghan equation of state fit to data in T and CT phases. Figure 1 (c) shows the normalized temperature dependence of electrical resistance for pressures up to 22.2 GPa and temperatures below 100 K.

Collected in figure 1 (b) are the measured pressure-volume (P-V) curves for $BaFe_{2-x}Cr_xAs_2$ up to 81 GPa and at ambient temperature. Also shown are the Birch Murnaghan equation of state fits as described by equation (1). The fit parameters for the tetragonal and collapsed tetragonal phases are summarized in table 1. The fitted ambient pressure volume ($V_0$) for the tetragonal phases are consistent with the measured values of $V$ = 205.3 Å$^3$ for $BaFe_{1.95}Cr_{0.05}As_2$, $V$ = 206.7Å$^3$ for $BaFe_{1.85}Cr_{0.15}As_2$, $V$ = 207.3Å$^3$ for $BaFe_{1.6}Cr_{0.4}As_2$ and $V$ = 208.7Å$^3$ for $BaFe_{1.39}Cr_{0.61}As_2$. Bishop et al [13] have recently shown that the T-CT transition pressure for $BaFe_2As_2$ compounds can be predicted by equation 2.

$$P_{CT} = 0.388V_0(\pm 0.034)GPa\text{Å}^{-3} - 63.259(\pm 6.2)GPa \quad (2)$$

Using equation (2) and the measured volume $V_0$ given above, we predicted the T-CT transitions for the series of $BaFe_{2-x}Cr_xAs_2$ to occur at $P_{CT}$=16.4 GPa, 17.0 GPa, 17.2 GPa and 17.7 GPa for x=0.05, 0.15, 0.4 and 0.61 respectively and these are in consistent with the measured values shown in table 1. The measured equations of state show considerable stiffening at the tetragonal to collapsed tetragonal phase transition and the T-CT transition shift slightly to higher pressure value as evidenced by the change in slope of the volume-pressure curves as depicted in figure 1(b). The fitted value of bulk modulus ($B_0$), pressure derivative of bulk modulus ($B_0'$) and unit cell volume ($V_0$) at ambient conditions for both T and CT phases are summarized in table 1.

The high-pressure temperature dependence electrical resistance measurements were performed down to 10 K using four-probe method in an eight-probe designer diamond anvil cell as described in our earlier publication [14]. Our resistance data do not show any evidence of superconductivity up to a pressure of 22.2 GPa and down to 10 K as shown in figure 1(c) for $BaFe_{1.95}Cr_{0.05}As_2$. Instead, we observe an upturn in resistance below 80 K and at 0.1 GPa. The resistance decreases with increase in pressure up to 6 GPa. Further increase in pressure above 6 GPa leads to increase in the magnitude resistivity up to 22 GPa. This is in contrast to the results on pure $BaFe_2As_2$ in which superconducting transition temperature $T_C$ with a peak of 34 K at 0.1 GPa was shown to be induced by pressure [6].

Table 1. Equation of state parameters for the Cr-doped $BaFe_2As_2$ in the tetragonal (T) and collapsed tetragonal (CT) phase at ambient temperature.

| Phase and hole-dope (x) value of $BaFe_{2-x}Cr_xAs_2$ | Pressure range used in fit | Unit cell volume ($V_0$) at ambient conditions | bulk modulus ($B_0$) | ($B_0'$) Pressure derivative of bulk modulus |
|---|---|---|---|---|
| T-phase, x=0.05 | 0 < P < 17 GPa | 204.9± 0.9 Å$^3$ | 57.4 ± 2.9 GPa | 4.0 |
| CT-phase, x=0.05 | 17 < P < 33.8 GPa | 186.8 ± 3.5 Å$^3$ | 117.3 ± 16.7 GPa | 3.1 |
| T-phase, x=0.15 | 0 < P < 19.8 GPa | 209.1 ± 3.3 Å$^3$ | 48.5 ± 2.8 GPa | 4.0 |
| CT-phase, x=0.15 | 19.8 < P < 67.4 GPa | 176.4 ± 1.2 Å$^3$ | 178.6 ± 7.9 GPa | 3.1 |
| T-phase, x=0.4 | 0 < P < 16.5 GPa | 208.7± 1.7 Å$^3$ | 46.5 ± 2.8 GPa | 4.0 |
| CT-phase, x=0.4 | 16.5 < P < 77.6 GPa | 186.2± 2.0 Å$^3$ | 134.5 ± 7.4 GPa | 3.1 |
| T-phase, x=0.61 | 0 < P < 20.3 GPa | 206.7 ± 3.6 Å$^3$ | 57.2 ± 8.0 GPa | 4.0 |
| CT-phase, x=0.61 | 20.3 < P < 80.7 GPa | 186.5 ± 2.4 Å$^3$ | 134.9 ± 8.8 GPa | 3.1 |

In conclusion, our high pressure studies have shown absence of superconductivity and presence of a collapsed tetragonal phase that is correlated to ambient pressure volume of $BaFe_{2-x}Cr_xAs_2$. The T-CT phase transitions reported here are similar to those observed previously in undoped parents (e.g. $BaFe_2As_2$) of Fe-based superconductors. The T-CT transition is dependent on the ambient pressure unit cell volume which increases with concentration of Cr incorporated in Fe site and can be predicted using the equation (2) developed for the parent compounds. The equation of state for T and CT phases show distinct bulk moduli and ambient unit cell volume ($V_0$) for all the studied $BaFe_{2-x}Cr_xAs_2$ samples.

**Acknowledgments**

W O Uhoya acknowledges support from the Carnegie/Department of Energy (DOE) Alliance Center (CDAC) under Grant No. DE-FC52- 08NA28554. J M Montgomery acknowledges support from the Department of Education Grant No. P200A090143. Research at ORNL is sponsored by the Division of Materials Sciences and Engineering, Office of Basic Energy Sciences, U.S. Department of Energy. Portions of this work were performed at HPCAT (Sector 16), APS, Argonne National Laboratory.

**References**

[1] Takahashi H, Igawa K, Arii K, Kamihara Y, Hirano H and Hosono H 2008 *Nature*.**453**, 376.
[2] Sefat A S, 2011 *Rep. Prog. Phys.* **74** 124502.
[3] Rotter M, Tegel M and Johrendt D 2008 *Phys. Rev.lett.***101**, 107006.
[4] Alireza P L, Ko Y T C, Gillett J, Petrone C M, Cole J M, Lonzarich G G and Sebastian S E 2008 *J. Phys.: Condens. Matter* **21** 012208.
[5] Sefat A S, Singh D J, VanBebber L H, Mozharivskyj Y, McGuire M A, Jin R Y, Sales B C, Keppens V and Mandrus D 2009 *Phys. Rev. B* **79** 224524.
[6] Uhoya W, Stemshorn A, Tsoi G, Vohra Y K, Sefat A S, Sales B C, Hope K M and Weir S T 2010 *Phys. Rev.* B **82** 144118.
[7] Mittal R, Mishra S K, Chaplot S L, Ovsyannikov S V, Greenberg E, Trots D M, Dubronvinsky L, Su Y, Brueckel T, Matsuishi S, Hosono H and Garbarino G 2011 *Phys. Rev.* B **83** 054503.
[8] Colonna N, Profeta G, Continenza A, and Massidda S 2011 *Phys. Rev.* B **83**, 094529.
[9] Velisavljevic N and Vohra Y K 2004 *High Press. Res.* **24** 295.
[10] Birch F 1947 *Phys. Rev.* **71** 809.
[11] Hammersley A P 1995 *Report No. EXP/AH/*95-01.
[12] Toby B H 2001 *J. Appl. Crystallogr.* **34** 210–3.
[13] Bishop M, Uhoya W, Tsoi G, Vohra Y K, Sefat A S and Sales B 2010 *J. Phys.: Condens. Matter* **22** 425701.
[14] Tsoi G, Stemshorn A, Vohra Y K, Wu P M, Hsu F C, Huang Y L, Wu M K, Yeh K W and Weir S T 2009 *J. Phys.: Condens. Matter* 21 232201.

# High Pressure Studies of as Grown $WX_{2-x}$ Single Crystals

**G K Solanki[1], Y A Patel[2] and M K Agarwal[1]**

[1] Department of Physics, Sardar Patel University, Vallabh Vidyanagr- 388 120, Gujarat, India.

[2] Department of Physics, N.V.Patel College of Pure and Applied Sciences, Sardar Patel University, Vallabh Vidyanagr- 388 120, Gujarat, India.

E-mail: solankigunvant@yahoo.co.in

**Abstract.** Single crystals of $WX_{2-x}$ (X=S, Se and x=0, 0.1) have been grown by PVT technique. The grown crystals have been characterized by XRD. The results of simultaneous thermoelectric power and resistivity measurements upto 8GPa on these crystals have been reported. The decrease in resistivity and thermoelectric power S beyond 2 GPa suggests a possibility of metallization in these crystals. A possible explanation for this decrease has been provided. The variation of thermoelectric power factor with pressure has been thoroughly studied and discussed. The implications of the results have been pointed out.

## 1.Introduction

Littleton et al. [1] have studied the variation of power factor with pressure in pentatelluride materials and have observed an enhancement in its value by an order of magnitude with increase in pressure. To author's knowledge, there are very few reports on the high pressure studies of tungsten dichalcogenides $WX_2$ single crystals. Webb et al. [2] studied the effect of pressure ($\leq 0.65$ GPa) on resistivity and structure of $WS_2$. Mahalawy and Evans [3] carried out experiments on the dependence of electrical conductivity in p-type 2H $WSe_2$ at pressures upto 0.30 GPa. The derived values for pressure dependent activation energy were interpreted in terms of the existing band models. Vaidya et al. [4] have studied the effect of pressure on electrical resistance of $WSe_2$ upto pressure of 27GPa. There was no clear indication of any phase transition till the highest pressure reached in their studies. Agarwal et al. [5] made thermoelectric and electrical resistance measurements for p and n-type $WS_2$ crystals grown by CVT as a function of pressure upto 0.80 GPa. All these measurements indicated the presence of a phase transition at 0.43 and 0.41 GPa for p and n-type $WS_2$ single crystals respectively.

## 2.Experimental

Single crystals of $WX_{2-x}$ (X=S, Se and x=0, 0.1) were grown by a PVT (Physical Vapour Transport) technique [6]. Stoichiometric amounts of constituent elements Tungsten [99.99% purity], Selenium [99.999% purity] and Sulphur [99.99 % purity] were sealed in an evacuated quartz ampoule. Compound synthesis was carried out by slowly heating the ampoule upto 1023K in case of $WSe_{2-x}$ and upto 1123K in case of $WS_{2-x}$ and maintaining it at these temperatures for 72 hours. Complete details of the 240 hours growth procedure for both the compounds $WS_{2-x}$ and $WSe_{2-x}$ are summarized in Table 1. The lattice parameters obtained by X-ray diffraction are listed in Table 2.

**Table 1** Growth conditions for $WX_{2-x}$ (X=S, Se and x=0, 0.1) grown using Physical Vapour Transport (PVT) technique.

| Crystal | Temperature distribution | | Physical characteristics of crystals | | | |
|---|---|---|---|---|---|---|
| | Hot Zone(K) | Cold Zone(K) | Growth Time(h) | Plate area (mm×mm) | Thickness (μm) | Colour |
| $WSe_2$ | 1223 | 1213 | 240 | 16×16 | 25 | Black |
| $WSe_{1.9}$ | 1223 | 1213 | 240 | 20×10 | 20 | Black |
| $WS_2$ | 1353 | 1323 | 240 | 15×10 | 20 | Gray |
| $WS_{1.9}$ | 1353 | 1323 | 240 | 18×19 | 50 | Gray |

**Table 2** Crystallographic data of $WX_{2-x}$ (X=S, Se and x=0, 0.1) single crystals grown using PVT technique.

| Parameter | $WSe_2$ | $WSe_{1.9}$ | $WS_2$ | $WS_{1.9}$ |
|---|---|---|---|---|
| a (Å) | 3.240 ± 0.025 | 3.226±0.006 | 3.158±0.004 | 3.16±0.005 |
| c (Å) | 12.427±0.005 | 12.41±0.005 | 12.34±0.001 | 12.3±0.008 |
| c/a | 1.918 | 1.925 | 1.954 | 1.951 |
| Unit Cell Volume $(Å)^3$ | 112.973 | 111.927 | 106.59 | 106.91 |
| X-ray density $(g/cm^3)$ | 10.046 | 9.906 | 7.725 | 7.695 |

For the high pressure studies the Bridgemen opposed anvil cell was used. The cell assembly consisted of two pyrophyllite gaskets, 12.7 mm outer diameter, 3 mm inside diameter and 0.25mm thick, and the specimen was located between two talc disks, 3 mm in diameter and 0.2 mm thick, at the centre of the gasket ring. A small temperature gradient was produced across the specimen by means of a nicrome heater at one end, and two pairs of chromel/alumel wires were used as probes. The potential drop across the wires, $E_C$ and $E_A$ (where C refers to chromel and A to alumel), were measured with a Keithley 150 Null Detector and the thermoelectric power was calculated from the equation [1]:

$$S = S_A \frac{(1 + S_C E_A / S_A E_C)}{1 - E_A / E_C} \tag{1}$$

where $S_C$ and $S_A$ are the Seebeck coefficients of the chromel and alumel respectively. The four probe assembly was used for the determination of the conductivity of the specimen.

*2.1 Variation of resistivity with pressure*

The results on the pressure dependence of resistivity for $WX_{2-x}$ (X=S, Se and x=0, 0.1) crystals are shown in Figure 1 and 2. The crystals of $WS_{2-x}$ are more resistive than $WSe_{2-x}$. Also in both the cases the off-stoichiometry leads to a reduction in the overall resistivity of the samples. In all samples of $WX_{2-x}$ resistivity is independent of frequency and all the materials show an exponential fall of resistivity (ρ) on increase in pressure (P). At pressures upto 2 GPa, the resistivity decreased at rates described by different hydrostatic piezoelectric coefficients $\frac{\partial \rho}{\partial P}$ ,19.8, 15.75, 100, 101.5 and 4.4, 3.75, 7.2 and 52.79 for $WSe_2$, $WSe_{1.9}$, $WS_2$ and $WS_{1.9}$ respectively. At pressures beyond 2.2 GPa in $WSe_{2-x}$ and 2.55 GPa in $WS_{2-x}$ , the slopes of the curves ρ (P) decrease substantially and they almost coincide with pressure axis with $\frac{\partial \rho}{\partial P}$ , equal to zero.

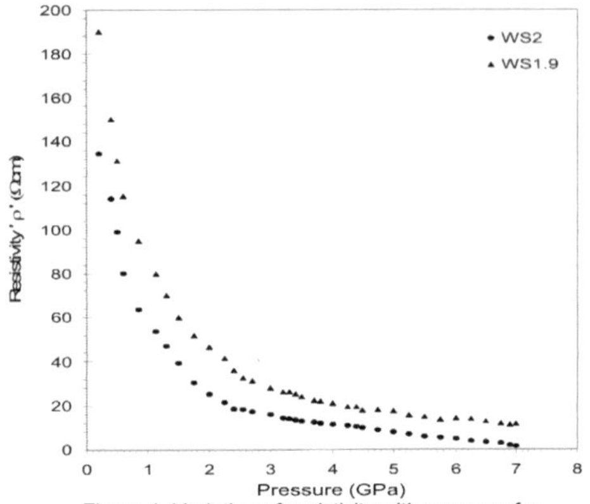

Figure 1 Variation of resistivity with pressure for $WS_2$ and $WS_{1.9}$ single crystal

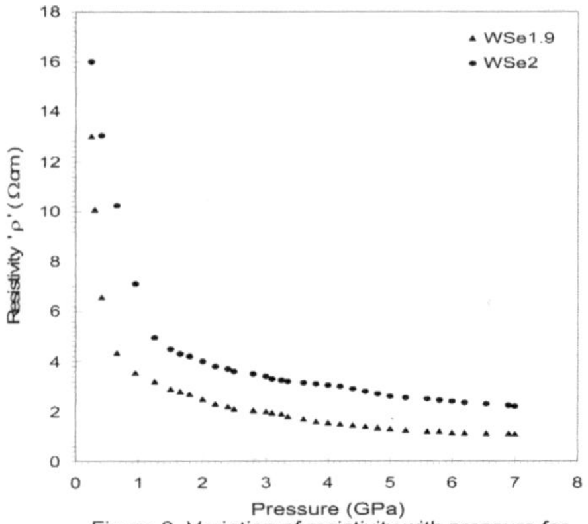

Figure 2 Variation of resistivity with pressure for $WSe_2$ and $WSe_{1.9}$ single crystal

*2.2Variation of Thermoelectric power factor with pressure*

The thermoelectric power factor variation with pressure for specimens of $WSe_{2-x}$ (x=0 and 0.1) and $WS_{2-x}$ (x=0 and 0.1) which are representative of all samples is shown in Figures 3 and 4 respectively.

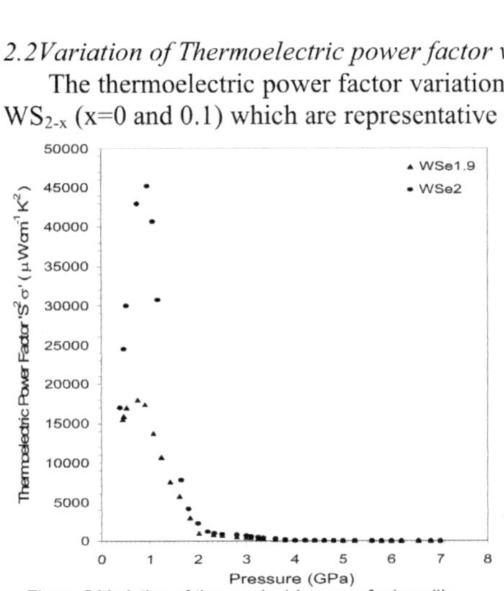

Figure 3 Variation of thermoelectric power factor with pressure for $WSe_2$ and $WSe_{1.9}$ single crystal

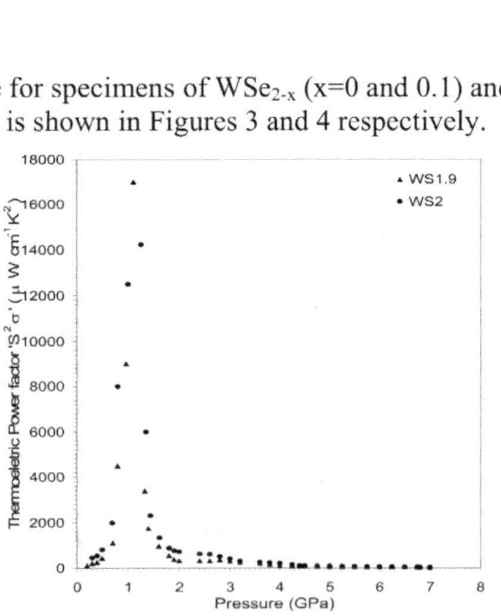

Figure 4 Variation of thermoelectric power factor with pressure for $WS_2$ and $WS_{1.9}$ single crystal

In both the cases the power factor increases steadily with increase in pressure and attains a maximum and thereafter decreases monotonically with further rise in pressure. The values of pressure at which power factor attains a maximum value are 0.95GPa, 0.75 GPa, 1.25 GPa and 1.1 GPa for $WSe_2$, $WSe_{1.9}$, $WS_2$ and $WS_{1.9}$ respectively. The value of thermoelectric power factor is more for stoichiometrically perfect crystals of $WSe_2$ and $WS_2$ as compared to off-stoichiometric crystals of $WSe_{1.9}$ and $WS_{1.9}$. This trend is maintained at all pressures upto ~2.2 GPa. In comparison to $WS_{2-x}$ crystals, the value of thermoelectric power factor at each pressure is more for $WSe_{2-x}$ crystals.

## 3.Discussion

When the layered material is subjected to pressure the interlayer spacing will change more rapidly than the intralayer spacing since the interlayer bonds are much weaker than the intralayer bonds. For such materials compression therefore forces the layers closer together without affecting the nearest neighbour distances. Lattice vibration pressure Raman studies of layer crystals reveals that change in $r_0$ (the covalent bond length) i.e. $\Delta$ $(r_0)$ is one to two orders of magnitude smaller than the change in $r_1$ (the intralayer spacing) i.e. $\Delta$ $(r_1)$. Accordingly, pressure induced modifications in the electron energy levels of the layered semiconductors like $WX_2$ should show a decrease in the resistance with increasing pressure. The present data on $WX_{2-x}$ upto 7GPa are consistent with the above explanation. In addition to these effects, the structural defects, like stacking faults in layer crystals and defects introduced due to off-stoichiometry also influence the electron transport under pressure, which probably the reason for the higher values of conductivity in off –stoichimetric samples. In all samples of $WX_{2-x}$ the monotonous decrease in resistivity with pressure to very low values is similar to that observed by other research workers in layered compounds and clearly shows a tendency of the materials towards metallization at higher pressure. Animalu [8] has discussed the variation of TEP as a function of carrier concentration. He has clearly shown that TEP decreases with an increase in carrier concentration. The sharp drop in TEP in this sense suggests the setting of metallization in $WX_{2-x}$ with increasing pressure. This result supports the conclusions drawn from the variation of resistivity with pressure in all the samples of $WX_{2-x}$ above. The results on variation of thermoelectric power factor with pressure clearly lead to an enhancement in the values of $S^2\sigma$ with increase in pressure. This type of enhancement of power factor with pressure has recently been observed by Littleton et al. [1]. The thermoelectric power factor of a material is given by $S^2\sigma$ where S is the Seebeck coefficient and $\sigma$ is the conductivity of the material. In order to increase the power factor, the material should be such that its conductivity is high and at the same time the Seebeck coefficient should not be low. In case of $WX_{2-x}$, it is only $WSe_2$ single crystals which meet this requirement. Although off-stoichiometric crystals of $WSe_2$ (i.e. $WSe_{1.9}$) possess higher conductivity as compared to $WSe_2$ and their TEP values are comparatively lower, which leads to an all over reduction in $S^2\sigma$. This is probably due to the fact that the range of off-stoichiometry in the present work is maximum. Efforts should therefore be made to grow $WSe_{2-x}$ crystals with different values of x, so that the off-stoichiometry could be adjusted to provide materials with higher values of thermoelectric power factor.

## 4.Conclusion

The decrease in resistivity and thermoelectric power S in all samples of $WX_{2-x}$ beyond 2 GPa suggest the setting up of metallization in these crystals with increasing pressure. The variation of thermoelectric power factor with pressure has been thoroughly studied. In order to get tungsten diselenide with higher values of $S^2\sigma$ its off-stoichiometry should be adjusted to provide a still superior thermoelectric material.

**Acknowledgement**
Authors are grateful to Director, Bhabha Atomic Research Centre (B.A.R.C.), Mumbai for proving the facilities of high pressure work.

## References

[1] Littleton R T, Tritt T M, Feger C R, Kolis J, Wilson M L, Marone M, Payne J, Verebeli D and Levy F 1998 *Appl. Phys. Lett,* **72** (16) 2056-2058
[2] Webb A W, Towle L C, Skelton E F, Feldman J L, Liu C Y and Spain I L 1975 *Euro. Phys. Conf. Abstr.* **1(A)** 103
[3] El Mahalawy L H and Evans B L 1978 *Phys. Stat. Solidi* **86b** 151
[4] Vaidya R, Neha Bhatt, Patel S G, Jani A R, Garg A B, Vijaykumar V and Godwal B 2003 *Pramana* **61** (1) 183
[5] Agarwal A, Goyal S, Agarwal M K, Vaidya S N and Karunakaran C 1995/1996 *High Temp. High Press.* **27/28** 163-169
[6] Al-Hilli A A and Evans B L 1972 *J. Crystal Growth* **15** 93
[7] Patel Y A 2005 *Ph.D. Thesis, Sardar Patel University, Vallabh Vidyanagar, Gujarat, India*
[8] Animalu A O E 1978 *Intermediate Quantum Theory of Crystalline Solids*, (Prentice-Hall of India Pvt. Limited, New Delhi) 243

# Pressure dependent luminescence properties of $Sm^{3+}$ ions in fluorophosphate glass

**Ch Srinivasa Rao[1], V Venkatramu[2], C K Jayasankar[1*], Th Tröster[3], W Sievers[4] and G Wortmann[4]**

[1] Department of Physics, Sri Venkateswara University, Tirupati-517 502, India
[2] Department of Physics, Yogi Vemana University, Kadapa- 516 003, India
[3] Fakultät für Maschinenbau, Universität Paderborn, D-33098 Paderborn, Germany
[4] Department Physik, Universität Paderborn, D-33095 Paderborn, Germany

*Corresponding author email address: ckjaya@yahoo.com

**Abstract:** Pressure dependent luminescence spectra and decay curves for the $^4G_{5/2}$ level of $Sm^{3+}$ ions in fluorophosphate glass of composition 55.975 $P_2O_5$ + 14 $K_2O$ + 6 KF + 14.975 BaO + 9 $Al_2O_3$ + 0.05 $Sm_2O_3$ (referred as PKFBASm005) have been measured as a function of pressure up to 40.2 GPa. In this pressure range, the red-shift (during increase in pressure), the blue-shift (while releasing pressure) and enhanced crystal-field splitting with increase in pressure are observed for $^4G_{5/2} \rightarrow {}^6H_J$ (J = 9/2,7/2 and 5/2) multiplet transitions of PKFBASm005 glass. These pressure induced shifts as well as crystal-field splittings were explained by the interaction of $Sm^{3+}$ ions with its surrounding ligands under pressure. The decay curves for the $^4G_{5/2}$ level are found to be non-exponential in the entire pressure range. The lifetime for the $^4G_{5/2}$ level decreases from 3.23 to 1.62 ms with increase in pressure from ambient to 40.2 GPa. The generalized Yokota-Tanimoto (YT) model has been used to explain non-exponential behaviour of decay curves.

*Keywords*: $Sm^{3+}$ ions in glass; High pressure; Luminescence properties; Crystal-field.

## 1. Introduction

Spectroscopic investigations on rare-earth (RE) doped glasses have considerable interest to design and develop new optical devices [1] due to desirable energy level structure for optical pumping, sharp lines observed in the optical absorption and emission spectra, relatively longer lifetimes of excited energy levels, easier fabrication compared to crystals, etc. The interest in selection of phosphate glass as a host matrix for $RE^{3+}$ ions is due to its unique characteristics such as high gain density that is due to high solubility of $RE^{3+}$ ions, high transparency, low refractive index and dispersion, low melting point and good thermal stability. But, the complexity with the use of phosphate glasses as laser ion host is that they have larger thermal expansion coefficient and lower fracture toughness than silicates [2,3].

The importance of preparing fluorophosphate glasses by introducing suitable metal fluorides into phosphate glass composition is of their smaller fluorescence linewidth, low melting temperature, higher stimulated emission cross-section, longer lifetime and reduction in phonon energy. The presence of fluorides in glass medium strongly reduces OH absorption and shifts the IR cutoff edge towards longer wavelengths making them highly suitable for fiber amplifiers [4,5] in telecommunication systems. In phosphate glasses, $K_2O$ reduces the melting point and BaO favors the mechanical properties which are important for fiber production. On the other hand, $Al_2O_3$ alters the glass structure that can improve the physical properties and chemical stability of the glass. The trivalent samarium ($Sm^{3+}$) ion is one of the most interesting $RE^{3+}$ ions to study the luminescence properties because its excited $^4G_{5/2}$ level shows relatively high quantum efficiency and also shows different quenching emission channels. The present work reports luminescence spectral and decay curve properties of 0.05 mol % $Sm_2O_3$ –doped fluorophosphate glass as a function of pressure up to 40.2GPa.

## 2. Experimental

The composition (in mol %) of the glass studied in the present work 55.975 $P_2O_5$ + 14 $K_2O$ + 6 KF + 14.975 BaO + 9 $Al_2O_3$ + 0.05 $Sm_2O_3$ (referred as PKFBASm005) was prepared by conventional melt quenching technique [6]. High-pressure photoluminescence spectra were recorded with a double monochromator equipped with a photomultiplier tube. The 476.5 nm line of an Ar+ laser was used as an excitation source. The glass under investigation, together with a small piece of ruby, which acts as a pressure calibrant, was placed in a hole of 100 µm diameter of 200 µm thick stainless-steel gasket, which was located between the parallel faces of specially made miniature diamond anvil cells (DAC) [6-9]. A mixture of methanol-ethanol-water 16:3:1 was used as a hydrostatic pressure-transmitting fluid. Decay curves were obtained using a mechanical chopper with a multichannel scalar interfaced to a personal computer that recorded and averaged the signal. All measurements were carried out at room temperature.

## 3. Results and discussion

The luminescence spectra of the PKFBASm005 glass at ambient conditions for successive increasing and decreasing pressures are shown in figure 1, which were measured at room temperature (RT). As seen from figure 1, the peak positions corresponding to the $^4G_{5/2} \rightarrow {}^6H_J$ (J = 9/2, 7/2 and 5/2) transitions are shifted towards lower energy side (red-shift) with increase in pressure (figure 1(a)) whereas it has shifted towards higher energy side (blue-shift) while release of pressure (figure 1(b)). The energy shifts are found to be -5.19, -5.04 and -5.17 $cm^{-1}$/GPa for the $^4G_{5/2} \rightarrow {}^6H_J$ (J = 9/2,7/2 and 5/2) transitions, respectively, which are more or less similar to the corresponding shifts of -5.3, -4.7 and -5.2 $cm^{-1}$/GPa observed in PKFBASm01 glass [10]. The peak shift of these transitions towards lower energy side with increase in pressure is attributed to the changes in Coulomb and spin-orbit coupling interactions due to changes in the overlap of the ligand orbitals with the $4f$– wave functions of the $Sm^{3+}$ ions [11].

In figure 1, the $^4G_{5/2} \rightarrow {}^6H_J$ (J = 9/2,7/2 and 5/2) transitions are split into two components and the magnitude of the crystal-field (CF) splitting increases with increase in pressure as well as decreases while release of pressure (see Table 1) for all these transitions. For instance, the splitting of the $^4G_{5/2} \rightarrow {}^6H_{7/2}$ and $^4G_{5/2} \rightarrow {}^6H_{5/2}$ transitions increase with a slope of 5.04 and 5.17 $cm^{-1}$/GPa, respectively, for the present glass and also similar changes were observed in other $Sm^{3+}$:glasses [10,12]. These changes are mainly due to decrease in average inter-ionic distance between $Sm^{3+}$ ions and ligands with increase of pressure [13]. However, it is not possible to estimate CF parameters because of complex electronic structure of the $Sm^{3+}$ ion in $Sm^{3+}$-doped systems [14,15].

The luminescence decay curves for the $^4G_{5/2} \rightarrow {}^6H_{7/2}$ transition of $Sm^{3+}$ ions in PKFBASm005 glass have been measured at various pressures by excitation with 476.5 nm line of $Ar^+$ laser. In general, the decay curves exhibit single exponential behaviour at lower concentrations and turn into non-exponential for higher concentrations [16,17]. In the present work, the decay curves are found to be non-exponential in the entire pressure range. The effective lifetimes ($\tau_{eff}$) for non-exponential decay curves were evaluated using the following expression [18].

$$\tau_{eff} = \frac{\int t\, I(t)dt}{\int I(t)dt}$$

where $I(t)$ is the fluorescence intensity at time $t$ starting from excitation.

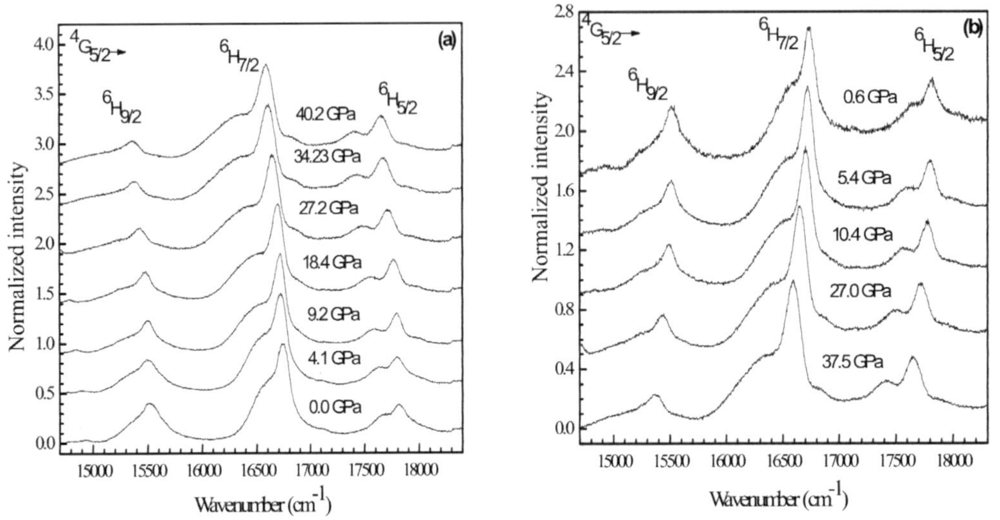


**Figure 1.** Luminescence spectra of PKFBASm005 glass with (a) increasing and (b) decreasing pressures.

The $\tau_{eff}$ for the $^4G_{5/2}$ level of $Sm^{3+}$ ion decreases from 3.23 to 1.62 ms with increase in pressure from ambient to 40.2 GPa and are collected in Table 1. These non-exponential decay curves observed beyond 4.1GPa are well-fitted to the generalized Yokota-Tanimoto (YT) model [19] for S = 6. This generalized YT model fit is used to explain non-exponential behaviour of decay curves. The analysis indicate that the dominant interaction mechanism involved in the energy transfer processes through cross-relaxation among $Sm^{3+}$ ions is of dipole-dipole interaction. The calculated energy transfer (Q), donor-acceptor interaction ($C_{DA}$) and diffusion (D) parameters were shown in table 1.

The decrease in lifetimes with increase in pressure, associated with increasing non-exponential behaviour of the decay curves, has been attributed to the gradual increase in CF strengths around the $Sm^{3+}$ ions as well as energy transfer processes through cross-relaxation channels ($^4G_{5/2}$, $^6H_{5/2}$) → ($^6F_{5/2}$, $^6F_{11/2}$) and ($^4G_{5/2}$, $^6H_{5/2}$) → ($^6F_{9/2}$, $^6F_{7/2}$) in the $Sm^{3+}$ ion [13,16]. This cross-relaxation is due to the energy transfer from the $Sm^{3+}$ ion in an excited $^4G_{5/2}$ state to a nearby $Sm^{3+}$ ion in the ground $^6H_{5/2}$ state. Moreover, while releasing pressure the peak positions corresponding to the $^4G_{5/2} \rightarrow {}^6H_J$ (J = 9/2,7/2 and 5/2) transitions are shifted towards higher energy side (blue-shift) with a small hysteresis but not any permanent deformation of structure around $Sm^{3+}$ ions. Similar results have also been noticed in our earlier studies [10,12].

**Table 1.** Effective lifetimes ($\tau_{eff}$, ms) of the $^4G_{5/2}$ level, energy transfer parameter (Q), peak positions (cm$^{-1}$) of the $^4G_{5/2} \rightarrow {}^6H_{9/2,7/2,5/2}$ emission bands, crystal-field splitting (CFS, cm$^{-1}$) for the $^4G_{5/2} \rightarrow {}^6H_{7/2,5/2}$ emission bands, donor-acceptor interaction ($C_{DA}$, $\times 10^{-41}$ cm$^6$ s$^{-1}$) and diffusion (D, $\times 10^{-14}$ cm$^2$ s$^{-1}$) parameters of $Sm^{3+}$ ions in PKFBASm005 glass at (a) increasing and (b) decreasing pressures.

| Pressure (GPa) | $\tau_{eff}$ | Q | Peak positions $^6H_{9/2}$ | Peak positions $^6H_{7/2}$ | Peak positions $^6H_{5/2}$ | CFS $^6H_{7/2}$ | CFS $^6H_{5/2}$ | $C_{DA}$ | D |
|---|---|---|---|---|---|---|---|---|---|
| (a) Increasing pressure | | | | | | | | | |
| 0 | 3.23 | 0 | 15511 | 16732 | 17809 | 180 | 145 | - | - |
| 4.1 | 2.89 | 0.011 | 15499 | 16718 | 17798 | 199 | 176 | 0.9 | 2.4 |
| 9.2 | 2.57 | 0.021 | 15488 | 16707 | 17787 | 221 | 211 | 3.2 | 4.8 |
| 18.4 | 2.26 | 0.032 | 15469 | 16685 | 17765 | 228 | 223 | 8.0 | 7.4 |
| 27.2 | 1.96 | 0.036 | 15422 | 16644 | 17710 | 236 | 245 | 9.7 | 11.8 |
| 34.2 | 1.76 | 0.041 | 15378 | 16606 | 17666 | 275 | 249 | 14.0 | 15.1 |
| 40.2 | 1.62 | 0.043 | 15361 | 16588 | 17654 | 279 | 255 | 13.2 | 17.8 |
| (b) Decreasing pressure | | | | | | | | | |
| 37.5 | 1.67 | 0.049 | 15373 | 16592 | 17650 | 273 | 254 | 18.7 | 15.0 |
| 27.0 | 1.99 | 0.027 | 15431 | 16646 | 17715 | 235 | 246 | 5.7 | 13.5 |
| 10.4 | 2.30 | 0.024 | 15491 | 16699 | 17780 | 223 | 221 | 4.7 | 4.9 |
| 5.4 | 2.58 | 0.022 | 15512 | 16718 | 17801 | 212 | 192 | 3.8 | 3.9 |
| 0.6 | 3.00 | 0.015 | 15521 | 16736 | 17812 | 188 | 168 | 1.7 | 0.8 |

## 4. Conclusions

The pressure dependent luminescence and decay properties of $Sm^{3+}$ ions in fluorophosphate glass have been studied. A red-shift as well as increase in crystal-field splitting for the $^4G_{5/2} \rightarrow {}^6H_J$ (J = 9/2,7/2 and 5/2) transitions have been observed with increase in pressure and are attributed to the changes in Coulomb and spin-orbit coupling interactions besides changes in the overlap of the ligand orbitals with the $4f$- wave functions of the $Sm^{3+}$ ions. The blue-shift has also been noticed for $^4G_{5/2} \rightarrow {}^6H_J$ transitions while releasing pressure, The non-exponential behaviour of the decay curves increased with increase in pressure and is well-fitted to the generalized Yokota-Tanimoto model indicating that the energy transfer among $Sm^{3+}$ ions is of dipole-dipole interaction. The decrease in lifetime of the $^4G_{5/2}$ level with increase in pressure is due to enhancement in energy transfer efficiency. It has also been observed that there is no permanent deformation of structure around $Sm^{3+}$ ions when pressure is released.

## Acknowledgements

This work has been supported through a Major Research Project funded by DAE-BRNS (No.2009/34/36/BRNS/3174, dt.12-02-2010), Government of India. One of the authors (CKJ) is grateful to Department Physik, Universität Paderborn, Germany for his stay as a Guest Scientist.

**References**

[1] 1993 *Rare Earth Doped Fibre Lasers and Amplifiers* ed M J Digonnet Dekker, New York
[2] Suratwala T I, Steele R A, Wike G D, Campbell J H and Takeuchi K 2000 *J. Non-Cryst. Solids* **263-264** 213
[3] Jahn W 1961 *Glasstechn. Bar* **34** 107
[4] Naftaly M and Jha A 2000 *J. Appl. Phys.* **87** 2098
[5] Marion J E and Weber M J 1991 *Eur. J. Solid. State. Inorg. Chem.* **28** 271
[6] Jayasankar C K, Venkatramu V, Babu P, Tröster Th, Sievers W, Wortmann G and Holzapfel W B 2005 *J. Appl. Phys.* **97** 093523-1
[7] Von der Ahe W, *Diploma thesis* 1989 Universitat Paderborn
[8] Piermarini G J, Block S, Barnett J D and Forman R A 1975 *J. Appl. Phys.* **46** 2774
[9] Holzapfel W B 2003 *J. Appl. Phys.* **93** 1813
[10] Jayasankar C K, Jyothi L, Basavapoornima Ch, Tröster Th, Sievers W and Wortmann G 2010 *High Press. Res.* **30** 424
[11] Chen G, Stump N A, Haire R G, Burns J B and Peterson J R 1994 *High Press. Res.* **12** 83
[12] Jayasankar C K, Upendra Kumar K, Venkatramu V, Babu P, Tröster Th, Sievers W and Wortmann G 2008 *J. Lumin.* **128** 718
[13] Qiuping W, Lijun L, DingZheng Z, Yuanbin C and Lizhong W 1992 *J. Phys. Condens. Matter* **4** 6491
[14] Renuka Devi A, Jayasankar C K and Reid M F 1994 *Phys. Rev. B* **49** 12551
[15] Garcia D and Faucher M 1998 *Handbook on the Physics and Chemistry of Rare Earths* vol 21, ed K A Gschneidner Jr. and L Eyring (Elsevier Science, New York) Chapter 144
[16] Venkatramu V, Babu P, Jayasankar C K, Tröster Th, Sievers W and Wortmann G 2007 *Opt. Mater.* **29** 1429
[17] Praveena R, Venkatramu V, Babu P and Jayasankar C K 2008 *Physica B* **403** 3527
[18] Nakazawa E 1999 *Phosphor Handbook* ed Shionoya S and Yen W M (Boca Raton, FL:CRC Press) p. 104.
[19] Yokota M and Tanimoto O 1967 *J. Phys. Soc. Jpn.* **22** 779.

Journal of Physics: Conference Series **377** (2012) 012019 doi:10.1088/1742-6596/377/1/012019

# Pressure Induced Metallization in $Zn_{1-x}Be_xSe$ Ternary Mixed Crystals

**Gopal K. Pradhan,[1] Chandrabhas Narayana,[1] S. K. Deb,[2] O. Pagès,[3] F. Firszt,[4] W. Paszkowicz[5]**

[1]Chemistry and Physics of Materials Unit, Jawaharlal Nehru Centre for Advanced Scientific Research (JNCASR), Jakkur P.O., Bangalore 560064, India

[2]Indus Synchrotron Utilization Division – RRCAT, Indore 452013, India

[3]Laboratoire de chimie et physique approche Multi-échelles Milieux Complexes – IJB, Université de Lorraine, Metz 57078, France

[4]Institute of Physics, N. Copernicus University, 87-100 Toruń, Poland

[5]Institute of Physics, Polish Academy of Sciences, 02-668 Warsaw, Poland

**Abstract.** We use high pressure Raman scattering to investigate the structural stability and the pressure effects on the semiconductor-metal phase transition of $Zn_{1-x}Be_xSe$ (x = 0.112, 0.16 and 0.24) mixed crystals up to about 25 GPa. We find that introduction of Be in Zn site results in an increase in the semiconductor-metal phase transition pressure as compared to pure ZnSe. We also determine the mode Gruneisen parameters ($\gamma$) and find a decrease in the ratio of the mode Gruneisen parameter ($\gamma_{TO}/\gamma_{LO}$) values with increasing Be content suggesting an increased covalency in the system with Be incorporation.

## 1. Introduction

Wide band gap II–VI compounds are promising for short-wavelength optoelectronic applications in laser diodes and light emitting diodes which display from green to ultraviolet spectral region. They find potential applications in optoelectronics as these lasers could revolutionize areas such as optical information storage and color displays. In this direction, the developments of ZnSe based semiconducting compounds and heterostructures are of particular interest [1, 2]. However, due to defect formation favored by the high ionicity and the smaller bond energies of conventional II-VI materials, devices based on these materials have shorter lifetime compared to III-V based devices. Degradation of devices based on these materials because of shorter lifetimes has been a major challenge. Use of Be in ZnSe based mixed crystals (i.e. $Zn_{1-x}Be_xSe$ ternary alloy) has been suggested [3,4] to improve the device lifetime because of dominant covalent bonding and high cohesive energy of BeSe. The high degree of covalent bonding of Be chalcogenides leads to an increase of their shear modulus [5]. Therefore, mixing of Be chalcogenides with ZnSe would impart rigidity to the lattice and increase the resistance of the structure to defect generation and propagation. The variation of the composition in $Zn_{1-x}Be_xSe$ crystals induces significant changes of the physical properties such as bond polarity, electronic band structure and lattice dynamics. The effect of pressure is of fundamental interest in altering the structural, electrical and optical properties of semiconductors. The pressure dependence of the vibrational modes provides additional insight into the nature of the lattice vibrations. Such information is important in characterizing strained-layer superlattices and heterostructures. High pressure Raman spectroscopy is a very powerful technique to study the lattice dynamics and phase transition of semiconductor materials related to their structural stability. However, the structural stability of $Zn_{1-x}Be_xSe$ with varying Be concentration and the pressure-induced semiconductor (zincblende)-metal(rocksalt) transition as observed in ZnSe [6-9] and related systems [9-11] remained unexplored until a recent x-ray diffraction study [12]. Under pressure, ZnSe transforms to the six-fold coordinated cubic NaCl rocksalt (RS) phase (metallic) whereas BeSe transforms to the six-fold coordinated hexagonal NiAs structure [13,14]. Moreover, the BeSe transition pressure (~56 GPa) is much higher than that of ZnSe (~13 GPa). In these circumstances, an interesting question to address is that what happens to the nature of the pressure induced semiconductor-metal transition in the ZnBeSe mixed crystals with increased Be content. In this paper, we use Raman scattering to investigate the pressure effects on the semiconductor-metal phase transition of $Zn_{1-x}Be_xSe$ (x= 0.112, 0.16 and 0.24) mixed crystals at high pressures up to about 25 GPa. The effect of Be on the bonding and the semiconductor-metal phase transition is discussed.

## 2. Experimental Details

$Zn_{1-x}Be_xSe$ ($x$ = 0.11, 0.16, 0.24) single crystals were grown by the high-pressure Bridgman method. Samples, mechanically polished to optical quality, were loaded with ruby chips in a Mao/Bell-type diamond anvil cell. Methanol-ethanol-water (16:3:1) was used as the pressure transmitting medium, the pressure being determined via the ruby fluorescence linear scale [15]. Unpolarized Raman spectra were recorded in backscattering geometry by a custom built in Raman spectrometer details of which is described elsewhere [16].

## 3. Results and Discussion

Unpolarized Raman spectra for $Zn_{1-x}Be_xSe$ mixed crystals at room temperature are shown in Figure 1. The dashed lines are Lorentzian fit to the experimental data. Pages *et al* [17-18] have done a very detailed and systematic room temperature Raman investigations on these systems. Hence the origin of the various Raman modes is briefly discussed here. According to Pages *et al*, in the ZnBeSe ternary alloy, there is formation of a percolating pseudo-continuous chain of Be–Se bonds throughout the alloy which consists of randomly formed hard-like (h) Be-rich bounded clusters embedded in a soft-like (s) Zn-rich matrix. Due to the different mechanical properties of the two host media, the Be–Se bonds from each of them vibrate at two separate frequencies giving rise to a so-called percolation doublet of the transverse optic (TO) modes (i.e. $TO^{Zn}_{Be-Se}$ and $TO^{Be}_{Be-Se}$). In labelling the modes, the subscript and superscript refer to the bond species and to the host region, respectively. A basic rule of the percolation model [18] is that *bonds are longer when they stay in the environment of the short species (thereby vibrating at a lower frequency), and vice versa.* This is because the Be–Se bonds in the h-region undergo a larger internal tensile strain to match the surrounding lattice parameter than those dispersed within the much softer ZnSe-like host matrix.With this, the TO modes from the BeSe-like region vibrate at a lower frequency than those from the ZnSe-like region, in both the Zn-Se and Be-Se spectral ranges and at any alloy composition. This explains the origin of the $TO^{Zn}_{Be-Se}$ - $TO^{Be}_{Be-Se}$ splitting (as shown in Figure 1). A similar splitting has also been evidenced for the Zn-Se bonds, but shows up clearly only at high pressure [19]. Furthermore, in these mixed crystals, the wave vector conservation selection rule is relaxed as a result of the violation of translational invariance, and some of the Raman features correspond to phonons from regions of the Brillouin zone (BZ) where the phonon density of state has a maximum, that is, at the edge of the BZ. Therefore, apart from the zone centre (Γ) modes, a disorder activated $LA_L$ mode assigned as longitudinal acoustic modes at L-point of the BZ is also seen. Combinational overtones ($2TA + TO_{Zn\text{-}Se}$) are also seen.

Figure 2 shows the pressure evolution of the Raman spectra at a few representative pressures in the increasing pressure run. It can be seen (marked by arrows in Figure 2) that new modes start appearing at ~ 8 and 9 GPa. Origin of these modes has been explained by a pressure-induced Fano-decoupling of the TO mode from the $2TA_K$ and $(TA+LA)_L$ two-phonon continua at the BZ edge [19]. When starting from the pressure-induced TO splitting pattern in pure ZnSe [20], these two new modes can be explained simply by postulating the existence of an extra TO ( $TO^{Be}_{Zn-Se}$ ), which substitutes for the main TO ( $TO^{Zn}_{Zn-Se}$ ) for the anharmonic decay into the 2TA continuum. With this, the 2TA-decoupling occurs at 8 GPa in ZnBeSe, in reference to the first new mode, while the second new mode at ~ 9 GPa is due to the transient appearance of the TA+LA continuum just after decoupling from the main TO and just before re-coupling with the additional minor TO ( $TO^{Be}_{Zn-Se}$ ) [19].

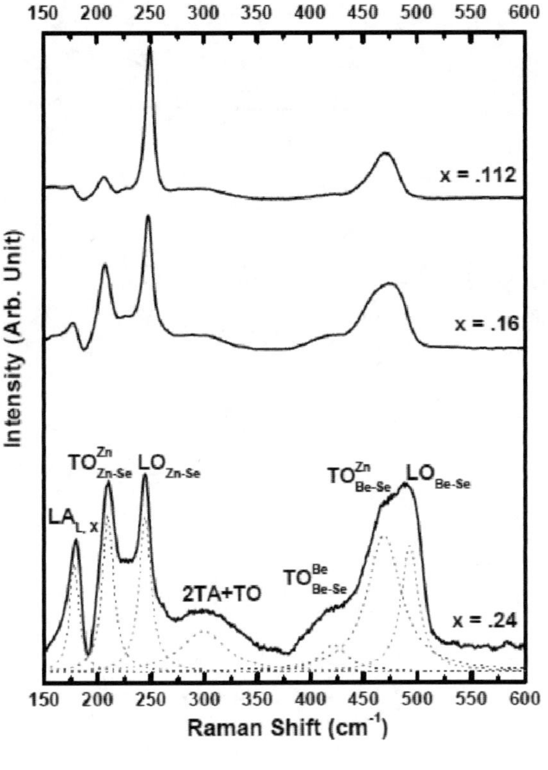


**Figure 1.** Raman spectra for $Zn_{1-x}Be_xSe$ mixed crystals at ambient conditions

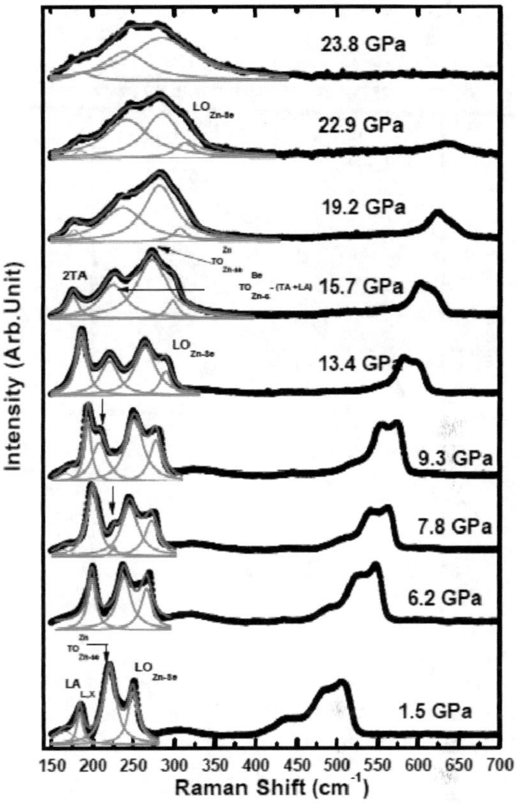


**Figure 2.** Pressure evolution of the Raman spectra of $Zn_{1-x}Be_xSe$ (x = 24%).

With increasing pressure, the intensity of the Raman modes decreases gradually. For the 24% Be sample, at ~23.8 GPa, the sample turned opaque (black) with a sudden decrease in Raman intensity. Therefore, at this pressure the spectra was recorded for 45 minutes and the disappearance of $LO_{ZnSe}$, $LO_{BeSe}$, and $TO_{BeSe}$ modes was observed. For both $Zn_{0.84}Be_{0.16}Se$ and $Zn_{0.888}Be_{0.112}Se$, similar behavior was observed but at a lower pressure. Both the sample became opaque along with the disappearance of the $LO_{ZnSe}$, $LO_{BeSe}$ and $TO_{BeSe}$ modes at 21.1 and 19.5 GPa, respectively for $Zn_{0.84}Be_{0.16}Se$ and $Zn_{0.888}Be_{0.112}Se$, respectively. We attribute these changes to the semiconductor (zinc blende) → metal (rock salt) phase transition as observed in ZnSe [6-9] and $Zn_{1-x}Fe_xSe$ [9-10] but at different pressure values. It can be recalled that ZnSe also exhibits similar optical behavior [6] (i.e. transparent to opaque transformation) at 13.5 GPa and turns metallic when the structure changes from zinc blende to rocksalt structure [7-9]. From the temperature dependence of resistivity, Galit Itkin *et al* [8] have suggested that in the pressure range of 13.5–17 GPa, the semiconducting and the metallic rock salt phase of ZnSe coexist. Beyond that, there is a complete metallization due to gap closure. In contrast, the semiconductor to metal transition pressure of $Zn_{1-x}Fe_xSe$ is lowered with increased Fe concentration [10]. However, the TO modes remain visible till the highest experimental pressure of 35 GPa is reached. Furthermore, it has been calculated [10] that the skin depth to be only about several tens of an Å. If it is to be a skin depth issue then the LO modes will be of extremely weak intensity. This is because of the transverse nature of the incident light which can couple rather strongly to the TO modes. In our experiments, the TO modes were found till the highest achieved pressure of about ~25 GPa and the $LO_{ZnSe}$, $TO_{BeSe}$, $LO_{BeSe}$ modes disappear at 23.8, 21.1 and 19.5 GPa for $Zn_{0.76}Be_{0.24}Se$,

$Zn_{0.84}Be_{0.16}Se$ and $Zn_{0.888}Be_{0.112}Se$, respectively, marking the onset of the semiconductor to metallic transition.

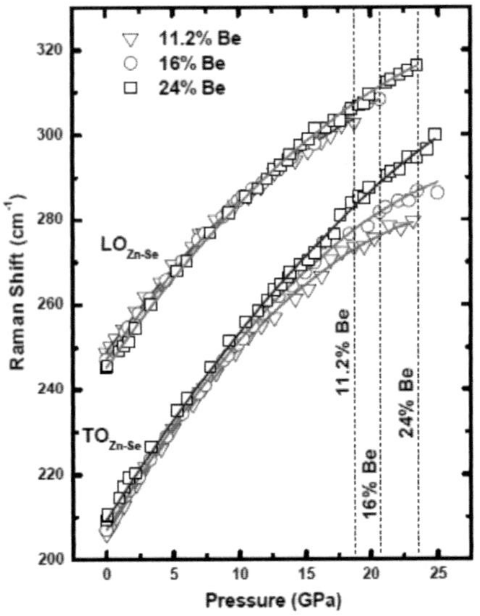


**Figure 3.** The frequency~pressure variations of the TO and LO modes in the Zn-Se region. The dashed vertical lines indicate transition pressure.

**Table 1.** Fitting parameter for various modes and calculated mode Gruneisen parameters for different Be concentration.

| Be content ($x$) | Phonon Modes | $d\omega/dP$ | $\gamma_i$ | Transition Pressure |
|---|---|---|---|---|
| .112 | $TO^{Zn}_{Zn-Se}$ | 5.3 | 1.6 | 19.5 |
| | $LO^{Zn}_{Zn-Se}$ | 4.3 | 1 | |
| .16 | $TO^{Zn}_{Zn-Se}$ | 5.2 | 1.5 | 21.1 |
| | $LO^{Zn}_{Zn-Se}$ | 4.5 | 1.1 | |
| .24 | $TO^{Zn}_{Zn-Se}$ | 4.9 | 1.4 | 23.8 |
| | $LO^{Zn}_{Zn-Se}$ | 4.5 | 1.1 | |

The pressure dependencies of all the mode frequencies can be obtained by a quadratic ($\omega_P = \omega_i + AP + BP^2$) fitting equation, where $\omega_P$ is the mode frequency at pressure $P$, $\omega_i$ is the mode frequency at ambient pressure, $A$ and $B$ are constants. The frequency versus pressure variations of the TO and LO modes in the Zn-Se region are shown in Figure 3. The dotted vertical lines in Figure 3 indicate the semiconductor-metal transition for each sample. Solid lines are quadratic fit to the experimental data (open symbols). Table 1 lists all the fitting parameters, mode Gruneisen parameter ($\gamma_i$) and the metallic transition pressure values for each Be concentration. $\gamma_i$ is calculated using the equation $\gamma_i = (B_0/\omega_i)(d\omega/dP)$, where $B_0$ is the bulk modulus of ZnSe, and was taken as 62.4 GPa [7]. We have used the same $B_0$ value for all values of $x$ in $Zn_{1-x}Be_xSe$ system. From table I, the $\gamma_{TO}/\gamma_{LO}$ values for the ZnSe –like modes are found to be 1.2, 1.3, and 1.6 for $Zn_{0.76}Be_{0.24}Se$, $Zn_{0.84}Be_{0.16}Se$ and $Zn_{0.888}Be_{0.112}Zn_{0.84}$ respectively whereas the same for ZnSe is 2.082 [9]. One observes a decrease of $\gamma_{TO}/\gamma_{LO}$ for the ZnSe–like modes with increase in Be concentration implying an increasing covalency [21]. This again shows that incorporation of Be indeed change the bonding properties by increasing the covalency (reducing the ionicity) thus imparting strength to the mixed crystal lattice. This increased covalency because of Be incorporation increases the transition pressure. It is worth mentioning that in case of $Zn_{1-x}Fe_xSe$ an opposite behavior is seen because of the increased ionicity due to Fe incorporation.

## 4. Conclusion

High-pressure Raman scattering measurements were performed on $Zn_{0.76}Be_{0.24}Se$, $Zn_{0.84}Be_{0.16}Se$ and $Zn_{0.888}Be_{0.112}$ mixed crystals up to about 25 GPa. The semiconductor-metal phase transition pressure was found to be 23.8, 21.1 and 19.5 GPa respectively. Our studies reveal that introduction of Be in Zn site results in an increase in the semiconductor-metal phase transition pressure as compared to pure ZnSe. The disappearance of most of the phonon modes along with the sample turning opaque marks the onset of this transition. Decrease in $\gamma_{TO}/\ \gamma_{LO}$ values with increasing Be content proves the increased covalency in the system with Be incorporation.

**Acknowledgements:** The authors would like to thank IFCPAR for the financial support under project No.3204-1 and Prof. K Rustagi for several discussions.

**References**

[1] F Vigue, E Tournie, and J P Faurie, 2001, *IEEE J. Quant. Electr.* **37**, 1146.
[2] F Fischer, 1997, *J. Cryst. Growth* **175-176**, 532.
[3] A Waag, F Fischer, H J Lugauer, Th Litz, J Laubender, U Lunz, U Zehnder, W Ossau, T Gerhardt, M Muller, and G Landwehr, 1996, *J. App. Phys.* **80**, 792.
[4] V Wagner, J J Liang, R Kruse, S Gundel, M Keim, A Waag, and J Geurts, 1999, *Phys. Stat. Sol.* B **215**, 87.
[5] C Vèrié, 1998, *J. Cryst. Growth* **184-185**, 1061.
[6] S Ves, K. Strossner, N. E. Christensen, C. K. Kim and M. Cardona, 1985, *Sol. St. Comm.* **56**, 479.
[7] R G Greene, H A Luo, and A L Ruoff 1995, *J. Phys. Chem. Sol.* **56**, 521.
[8] G Itkin, Gr Hearne, E Sterer, Mp Pasternak, and W Potzel 1995, *Phys. Rev.* B **51**, 3195.
[9] C-M Lin, D. S. Chuu, T. J. Yang, W. C. Chou, J. Xu and E. Huang 1997, *Phys. Rev.* B **55**, 13641.
[10] C-M Lin and D.S. Chuu,1999, *J. Raman Spectr.* **30**, 951.
[11] S. Qadri, E. Skelton, A. Webb, N. Moulton, J. Hu, and J. Furdyna 1992, *Phys. Rev.* B **45**, 5670.
[12] T. Ganguli, J. Mazher, A. Polian, S. K. Deb, F. Villain, O. Pages, W. Paszkowicz F. Firsz 2010, *J. App. Phys.* **108**, 083539.
[13] H. Luo, K. Ghandehari, R. Greene, A. Ruoff, S. Trail, and F. DiSalvo 1995, *Phys. Rev.* B **52**, 7058.
[14] A. Mujica, Angel Rubio, a. Muñoz, and R. Needs 2003, *Rev. Mod. Phys.* **75**, 863.
[15] H. K. Mao, J. Xu, and P. M. Bell 1986, *J. Geophys. Res.* **91**, 4673.
[16] G. K. Pradhan 2011, "*Inelastic Light Scattering Investigations on Novel Materials*". *Ph. D. Thesis*, Jawaharlal Nehru Centre for Advanced Scientific Research, India.
[17] O Pagès, M Ajjoun, and T Tite 2004, *Phys Rev* B **70**, 155319.
[18] O Pagès, A V Postnikov, M Kassem, A Chafi, A Nassour, and S Doyen 2008, *Phys. Rev.* B, **77**, 125208.
[19] G K Pradhan, C Narayana, O Pages, A Breidi, J Souhabi, A V Postnikov, S K Deb, F Firszt, W Paszkowicz, A Shukla, and F El Haj Hassan 2010, *Phys. Rev.* B **81**, 115207.
[20] W. C. Chou, C. M. Lin, R. S. Ro, C. S. Ho, D. Y. Hong, C. S. Yang, D. S. Chuu, T. J. Yang, J. Xu, and E. Huang 1997, *Chin. J. Phys. (Taipei)* **35**, 266.
[21] S. S. Mitra, O. Brafman, W. B. daniels and R. K. Crawford 1969, *Phys. Rev.* **186**, 942.

# Raman spectroscopic investigations on delafossite $CuLaO_2$ at high pressures

**Nilesh P. Salke**[1], **Rekha Rao**[1,*], **S. N. Achary**[2], **A. K. Tyagi**[2]

[1]Solid State Physics Division, Bhabha Atomic Research Center, Mumbai – 400085, India
[2]Chemistry Division, Bhabha Atomic Research Center, Mumbai – 400085, India

E-mail: rekhar@barc.gov.in

**Abstract.** Vibrational properties of delafossite $CuLaO_2$ are investigated using Raman spectroscopy at ambient conditions and at high pressures. Changes are observed in the Raman spectrum at around 1.8 GPa which could be related to a phase transition. The changes observed are reversible from 7 GPa.

## 1. Introduction

Transparent conducting oxides (TCO) of the delafossite family $AMO_2$ ( $A$ = Cu, Ag ; $M$ = Al, Ga, In) have many technological applications in solar cells, touch panels, as electrodes in flat panel displays, ultraviolet-emitting diodes [1]. While there are many $n$-type TCOs, $p$- type conductivity found in $CuMO_2$ is rare and useful in pn junctions [2]. Recently, Copper delafossites $CuMO_2$ have also been used as photo-catalysts for hydrogen production by water splitting [3].

Delafossite type crystals have layered structure with space group $R\bar{3}m$ with general formula $AMO_2$, in which monovalent cations $A$ ( $A$ = Cu, Ag) are linearly coordinated with two oxygen ions along the $c$-axis and the trivalent cations $M$ are octahedrally coordinated to oxygen atoms [4]. Earlier high pressure investigations of delafossite compounds have revealed interesting phase transitions in which Cu environment is affected whereas environment of $M$ is unchanged [5]. Both $CuAlO_2$ and $CuGaO_2$ have been reported to transform to unresolved structures above 34 and 26 GPa respectively [1,6]. $CuScO_2$ is reported to have structural transitions with pressure above 13 GPa [7]. Recent first-principles calculations determined the critical pressure of transition for $CuAlO_2$ to be 60 GPa [8].

Recently, Negative thermal expansion (NTE) has been reported in some members of this family like $CuScO_2$, and $CuLaO_2$ [4,9]. NTE is attributed to the anharmonicity of linear O-Cu-O bond along the $c$-axis, and this effect increases with increase in the size of the atom at $M$ site. Interesting fact is that $CuLaO_2$ exhibits NTE in both $a$ and $c$ edges at low temperature and it shows nearly isotropic thermal expansion though its structure is anisotropic. Neutron diffraction studies [4] have reported that the cell parameter $a$ shows positive expansion above 100 K, while $c$- parameter contracts in the temperature range 30-200 K and expands at higher temperatures upto 600 K.

High pressure Raman investigation of NTE materials often exhibits interesting phenomena like softening of phonons, phase transitions and also amorphization. Raman spectroscopic studies at high

* To whom any correspondence should be addressed.

pressures gives information about the mode Grüneissen parameter, which is useful for understanding the thermal expansion behavior. It has been observed that in most of the NTE materials, strong vibrations perpendicular to some interatomic bond plays an important role; such as O-Cu-O bond in $Cu_2O$. Though Raman spectroscopy gives information about only the zone-center phonons, it is quite useful in understanding the contribution of different vibrations to the thermal expansion. Here we report Raman spectroscopic studies of high pressure behavior of $CuLaO_2$.

## 2. Experimental details

The compound $CuLaO_2$ was synthesized by solid state reaction of $Cu_2O$ and $La_2O_3$. Raman spectroscopic measurements at high pressure were carried out from inside a diamond anvil cell Diacell B-05 with 4:1 methanol-ethanol mixture as pressure transmitting medium. Spectrum of polycrystalline sample was excited using 532 nm line of power ~15 mW. Pressure was measured using the ruby fluorescence technique. Scattered light was analyzed using a home built 0.9 m single monochromator, coupled with an edge filter and detected by a cooled CCD. Entrance slit was kept at 50 μm, which gives a spectral band pass of 3 $cm^{-1}$.

## 3. Results and Discussions

In the delafossite structure $R\bar{3}m$, primitive unit cell consists of 4 atoms resulting in 12 normal modes which transform as $\Gamma = A_{1g} + E_g + 3A_{2u} + 3E_u$. Movement in the direction of Cu-O bonds along the hexagonal $c$ axis is represented by $A$ modes, whereas $E$ modes correspond to vibrations in the perpendicular direction. $3A_{2u} + 3E_u$ are IR active modes which include acoustic modes ($A_{2u} + E_u$). $E_g$ and $A_{1g}$ are Raman active modes. In the absence of single crystals, Raman modes were identified using the Raman spectra of analogous compounds [1,6].

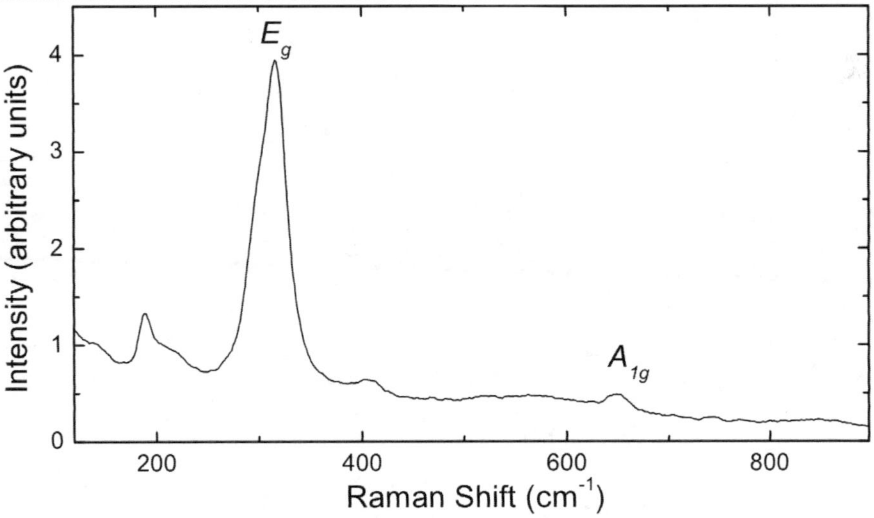


**Figure 1.** Raman spectra of $CuLaO_2$ at ambient conditions

Figure 1 shows the Raman spectra of $CuLaO_2$ at ambient conditions (Phase PI). It consists of two modes at 318 and 652 $cm^{-1}$ identified as $E_g$ and $A_{1g}$ respectively. Both the frequencies are lower than that of other delafossites due to higher La mass and weaker La-O bonds. $E_g$ mode is more intense than $A_{1g}$. In addition, several features of intensity comparable to that of $A_{1g}$ mode were also observed. Such features in other delafossites were attributed to defect induced modes (due to Cu vacancies or interstitial oxygen) which appear due to relaxation of selection rules and were identified by *abinitio* calculations[1,6].

Figure 2 shows the Raman spectra at various pressures. Both the Raman modes increase monotonically with pressure. Above 1.8 GPa, new modes appear in the low frequency region and the $E_g$ mode shows splitting. With further pressurization one of the split components of $E_g$ showed softening. Appearance of new modes and changes in the pressure dependencies of mode frequencies indicate a phase transition to a phase PII. Above 1.8 GPa, the $A_{1g}$ mode could not be followed. With further pressurization, Raman intensity drastically reduced above 7 GPa. Pressure coefficients of the Raman modes is 5.5 cm$^{-1}$/GPa for $E_g$ and 9.5 cm$^{-1}$/GPa for $A_{1g}$ mode in phase PI. Appearance of many new modes in PII indicates lower symmetry of the high pressure phase. Nature of changes observed at 1.8 GPa is similar to the transition observed in other delafossite members $CuAlO_2$ and $CuGaO_2$ which transform at 34 and 26 GPa respectively [1,6].

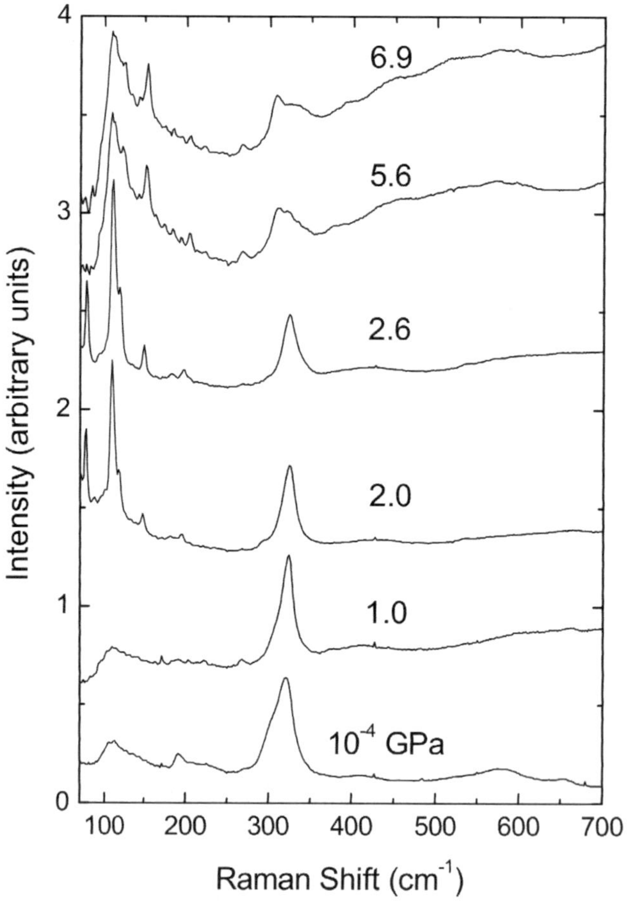


**Figure 2.** Raman spectra of $CuLaO_2$ at various pressures. Note the appearance of new modes above 1.8 GPa.

The Cu-O bond length decreases and unit cell edge $a$ increases with ionic radius of the $M^{3+}$ ion in this class of compounds. The variation in the transition pressure can be correlated to the ionic radii of trivalent cation which is coordinated octahedrally to oxygen atoms. We observe that the transition pressure $P_c$ is inversely proportional to the ionic radius of $M$ in this class of delafossites $CuMO_2$. Figure 3 gives the relation between transition pressure of the delafossites to the ionic radii of octahedrally coordinated trivalent cation.

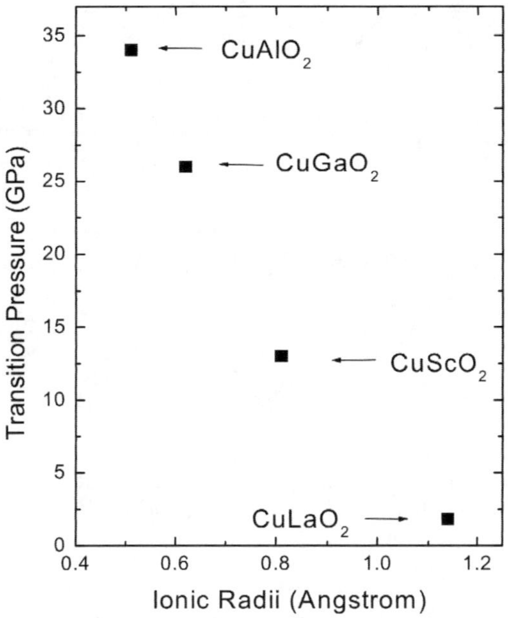


**Figure 3.** Onset pressure of the first phase transition plotted as a function of Ionic radii of trivalent cation $M^{3+}$ of delafossite Cu$M$O$_2$ [1,6,7]; $CuLaO_2$ is from the present study.

## 4. Conclusion

High pressure behaviour of $CuLaO_2$ is investigated upto 7 GPa, beyond which Raman modes could not be followed due to abrupt decrease in intensity. The compound exhibits a phase transition at 1.8 GPa similar to the phase transitions observed in $CuAlO_2$ and $CuGaO_2$ at 34 and 26 GPa respectively.

## 5. References

[1] Pellicer-Porres J, Martinez-Garcia D, Segura A, Rodriguez-Hernandez P, Munoz A, Chervin J C, Garro N and Kim D 2006 *Phys. Rev. B* **74** 184301
[2] Kawazoe H, Yasukawa M, Hyodo H, Kurita M, Yanagi H and Hosono H 1997 *Nature (London)* **389** 939
[3] Huda M N, Yan Y and Al-Jassim M M 2011 *J App. Phys* **109** 113710
[4] Li J, Sleight A W, Jones C Y and Toby B H 2005 *J. Solid State Chem.* **178** 285
[5] Pellicer-Porres J, Segura A, Ferrer-Roca Ch, Martinez-Garcia D, Sans J A, Martinez E, Itie J P, Polian A, Baudelet F, Munoz A, Rodriguez-Hernandez P and Munsch P 2004 *Phys Rev B* **69** 024109
[6] Pellicer-Porres J, Segura A, Martinez E, Saitta A M, Polian A, Chervin J C and Canny B 2005 *Phys. Rev. B* **72** 064301
[7] Gilliland S, Pellicer-Porres J, Segura A, Muñoz A, Rodríguez-Hernández P, Kim D, Lee M S and Kim T Y 2007 *phys. stat. sol. (b)* **244 No. 1** 309–314
[8] Nakanishi A and Katayama-Yoshida H 2011 *J Phys Soc Japan* **80** 024706
[9] Ahmed S I, Dalba G, Fornasini P, Vaccari M, Rocca F, Sanson A, Li J and Sleight A W 2009 *Phys. Rev. B* **79** 104302

**Acknowledgement**

Nilesh P. Salke acknowledges Department of Atomic Energy, India for financial support. The authors thank Dr. S. L. Chaplot for support and encouragement.

Journal of Physics: Conference Series **377** (2012) 012021 doi:10.1088/1742-6596/377/1/012021

# Rearrangement of electronic structure of $Pb_{1-x-y}Sn_xV_yTe$ under pressure

**E P Skipetrov[1], A N Golovanov[1], B B Kovalev[1], A M Mousalitin[2], E I Slyn'ko[3] and V E Slyn'ko[3]**
[1] Faculty of Physics, Moscow State University, 119991 Moscow, Russia
[2] Moscow Institute of Steel and Alloys, 119049 Moscow, Russia
[2] Institute of Materials Science Problems, 58001 Chernovtsy, Ukraine

E-mail: skip@mig.phys.msu.ru

**Abstract.** The galvanomagnetic properties in weak magnetic fields ($4.2 \leq T \leq 300$ K, $B \leq 0.07$ T) and the Shubnikov-de Haas effect (T=4.2 K, $B \leq 7$ T) in the single crystal $Pb_{1-x-y}Sn_xV_yTe$ (x=0.05-0.20, $y \leq 0.01$) alloys at atmospheric pressure and under hydrostatic compression up to 15 kbar have been investigated. We found that the increase of the vanadium impurity content leads to the p-n-conversion, transition to the insulating phase and to the pinning of Fermi level by the deep impurity level, lying under the bottom of the conduction band. Under pressure a decrease of the activation energy of the vanadium level, the n-p-inversion of the conductivity type at low temperatures and an insulator-metal transition occur. The pressure and the temperature coefficients of vanadium deep level energy are determined and the diagram of the electronic structure rearrangement for $Pb_{1-x-y}Sn_xV_yTe$ under pressure is proposed.

## 1. Introduction

Doping of lead telluride-based semiconductors with transition elements (Ti, V, Cr) creates donor-type deep impurity levels, stabilizing Fermi level in strongly doped samples near the bottom of conduction band [1]. In this case impurity level is partially filled with electrons and densities of occupied and empty states correspond to concentrations of impurity ions in 2+ and 3+ states, possessing different electrical and magnetic activities. Thus, these materials are diluted magnetic semiconductors whose electrical and magnetic properties should be dependent on their electronic structure (energy position of the impurity level relative to the band edges and occupancy of the level with electrons) [2,3].

It was known, that in PbTe vanadium induces appearance of a deep impurity level, situated under the bottom of conduction band ($E_V \approx E_c$-20 meV) [4,5]. In $Pb_{1-x}Sn_xTe$ alloys activation energy of the level rises with the increase of the tin content and it approaches the middle of the gap [6]. To investigate the rearrangement of the electronic structure of vanadium doped alloys under pressure in the present work the galvanomagnetic properties (T=4.2-300 K, $B \leq 7$ T) of $Pb_{1-x-y}Sn_xV_yTe$ (x=0.05-0.20, $y \leq 0.01$) at atmospheric pressure and under hydrostatic pressure $P \leq 15$ kbar have been studied.

## 2. Experimental details

Single crystal $Pb_{1-x-y}Sn_xV_yTe$ ingot with nominal concentrations x=0.08 and y=0.01 was grown by the vertical Bridgman technique and cut into slices $\approx 1.5$ mm thick perpendicular to the growth direction.

The crystal structure was controlled at room temperature by powder high-resolution X-ray diffraction using Rigaku X-ray diffractometer. It is shown that all peaks in the diffraction patterns correspond to those of the undoped PbTe, indicating that all samples are single-phase. The chemical composition was determined by the X-ray fluorescent microanalysis using scanning electron microscope LEO Supra 50 VP with Oxford Instruments INCA Energy+ system. It was found that, in accordance with the exponential distribution of substitution components, typical for $A^4B^6$ solid solutions [7], the tin content grows monotonously from the origin to the end of ingot (x=0.05–0.20), where as the vanadium impurity concentration rises up to $y \approx 0.010$ and then remains almost constant.

At the atmospheric pressure temperature dependencies of the resistivity ρ and the Hall coefficient $R_H$ (T=4.2-300 K, B≤0.07 T) were measured by the four-probe technique. Then the sample with maximum tin content (x=0.20, $y \approx 0.01$) was studied under pressure. For this sample dependencies ρ(T) and $R_H$(T), as well as Shubnikov-de Haas effect (T=4.2 K, B≤7 T) under pressure were measured. Pressures up to 15 kbar were obtained in the beryllium bronze chamber with kerosene-oil-pentane pressure transmitting media. The actual pressure in the chamber at helium temperatures was controlled by measuring the superconducting transition temperature in the pure tin as a function of pressure.

## 3. Galvanomagnetic properties under pressure

It is found that at atmospheric pressure lightly doped with vanadium samples from the origin of the ingot are p-type. Temperature dependencies of resistivity ρ and Hall coefficient $R_H$ exhibit metal-like behaviour due to the high density of free holes, typical for the undoped PbTe-based alloys. As the vanadium content in the samples grows, the hole concentration decreases, p-n-conversion and transition to the insulating phase at $x \approx 0.08$, $y \approx 0.008$ take place. In this phase dependencies ρ(1/T) and $R_H$(1/T) have activation character due to the pinning of the Fermi level by the deep vanadium level $E_V$, situated under the bottom of the conduction band. With the increase of the tin content, the slope of ρ(1/T) and $R_H$(1/T) dependencies grows gradually, indicating an increase of activation energy $\Delta E_V = E_c - E_V$ from 14 meV to 32 meV and slow shift of the level towards the middle of the gap [6].

Under pressure the resistivity ρ at helium temperature decreases monotonously by approximately seven orders of magnitude, while the slope of ρ(1/T), associated with the impurity ionization,

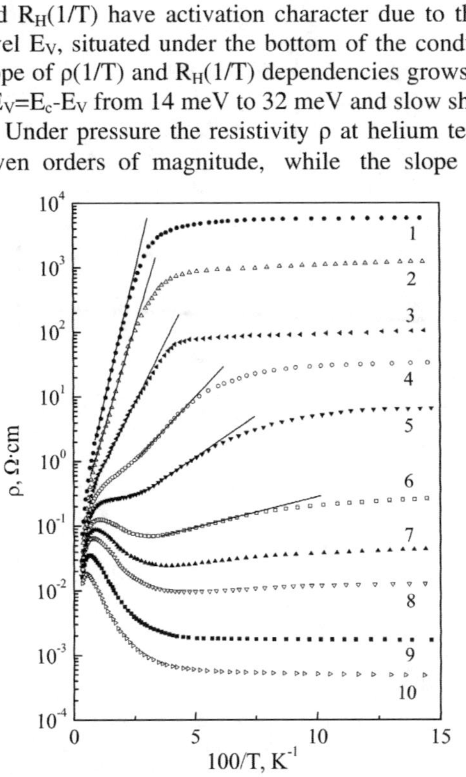


**Figure 1.** Temperature dependencies of the resistivity in $Pb_{1-x-y}Sn_xV_yTe$ (x=0.20, $y \approx 0.01$). P, kbar: 1-0, 2-1.8, 3-3.7, 4-5.3, 5-5.7, 6-6.6, 7-7.5, 8-8.7, 9-11.0, 10-14.2.

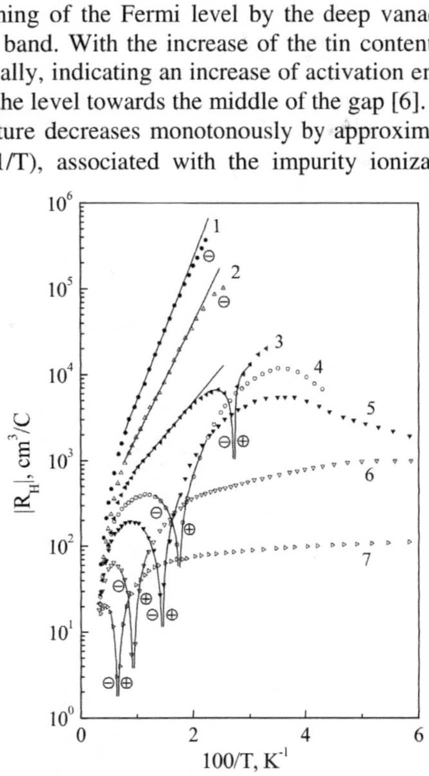


**Figure 2.** Temperature dependencies of the Hall coefficient in $Pb_{1-x-y}Sn_xV_yTe$ (x=0.20, $y \approx 0.01$). P, kbar: 1-0, 2-1.8, 3-3.7, 4-5.3, 5-5.7, 6-8.7, 7-14.2.

decreases up to the critical pressure P* ≈ 7 kbar (figure 1). Then the character of ρ(1/T) dependencies changes to the metal-type one and an insulator-metal transition takes place. At the same time, on the temperature dependencies of the Hall coefficient $R_H(1/T)$ the n-p-inversion of the $R_H$ sign at pressures P≥3.7 kbar is observed (figure 2). With the increase of pressure the sign-reversal temperature $T_{inv}$ grows at a rate $dT_{inv}/dP$=(7-16) K/kbar and the low-temperature Hall coefficient decreases gradually.

In the metallic phase (P>8.7 kbar), the distinct Shubnikov-de Haas (SdH) oscillations are observed. The period of SdH oscillations in inverse magnetic field Δ(1/B) decreases, indicating an increase of the free hole concentration under pressure. Experimental observation of quantum SdH oscillations under pressure may be considered as a direct confirmation of the high crystal perfection of the sample and of the uniformity of vanadium distribution in the investigated $Pb_{1-x-y}Sn_xV_yTe$ alloy.

## 4. Energy level diagram under pressure

Obtained experimental results allow us to construct the following model of the electronic structure rearrangement for $Pb_{1-x-y}Sn_xV_yTe$ (x=0.2) under pressure (figure 3). At atmospheric pressure, the deep vanadium level stabilises the Fermi level in the upper half of the gap, slightly above its centre. Under pressure, the gap decreases linearly at a rate $dE_g/dP \approx -7.4$ meV/kbar [8], whereas the vanadium level intersects the middle of the gap $E_i$, moving almost parallel to the bottom of the conduction band. So, at pressures P ≥ 3.7 kbar the p-n-inversion of the Hall coefficient on the $R_H(1/T)$ dependencies appears and the activation of electrons from the valence band to the level becomes the main thermal activation mechanism at low temperatures. Then the vanadium level approaches the top of the valence band and crosses it at the critical pressure P* ≈ 7 kbar, inducing an insulator-metal transition. In the metallic phase (P>P*), the vanadium level moves down into the valence band and the concentration of free holes increases due to the flow of electrons from the band to empty states of the level.

Experimental values of the vanadium level energy $E_V$ (points in figure 3) were obtained from the analysis of the ρ(1/T) dependencies and SdH data, supposing the pinning of the Fermi level by the impurity level in the whole pressure range. In the insulating phase $E_V$ values were estimated using the activation energy of the level $\Delta E_V$, while in the metallic phase – using the Fermi energy, calculated

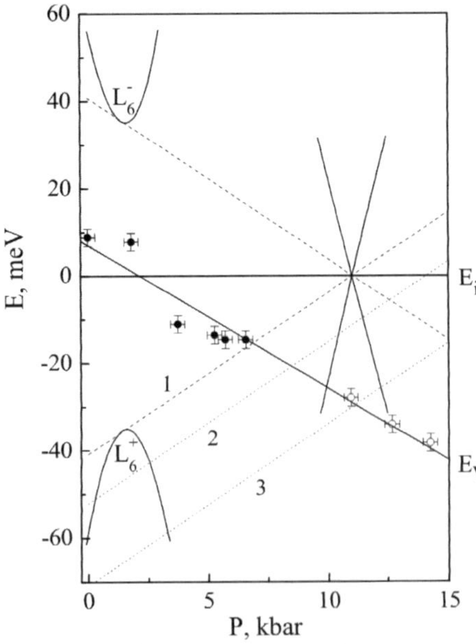


**Figure 3.** The model of electronic structure rearrangement for $Pb_{1-x-y}Sn_xV_yTe$ (x=0.20) under pressure. T, K: 1-4.2, 2-50, 3-100.

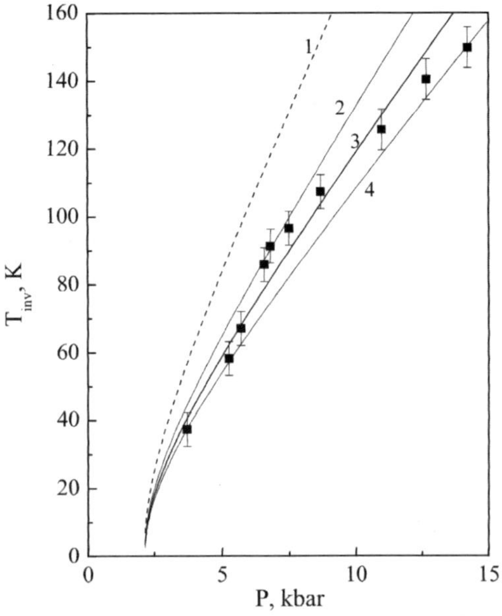


**Figure 4.** Pressure dependence of the $R_H$ sign-reversal temperature in $Pb_{1-x-y}Sn_xV_yTe$ (x=0.20). α, meV/K: 1-0, 2-0.08, 3-0.12, 4-0.16.

from the hole concentration in the framework of the two-band Kane dispersion law for $A^4B^6$ semiconductors [8] with parameters, determined for undoped $Pb_{1-x}Sn_xTe$ ($x \approx 0.2$) [9]. One can see that over the entire investigated pressure range, the vanadium level $E_V$ moves almost parallel to the term $L_6^-$, slowly approaching the term $L_6^-$ at a rate about 0.5 meV/kbar.

To estimate the temperature coefficient of the motion of the vanadium level with respect to band edges, we plot the experimental pressure dependence of the sign-reversal temperature $T_{inv}$ (figure 4). A monotonous shift of the p-n-inversion point to higher temperatures under pressure indicates that the vanadium level moves with respect to the middle of the gap when the temperature is increased. Anticipating that the energy of the level, measured from the bottom of the conduction band, is only slightly dependent on temperature and that the vanadium level crosses the middle of the gap at $T=T_{inv}$, we have calculated theoretical dependence $T_{inv}(P)$, using well-known temperature dependence of the gap for $Pb_{1-x}Sn_xTe$ alloys [8] (lines in figure 4). The best fit (full line in figure 4) is achieved assuming that as the temperature increases the vanadium impurity level moves with respect to the middle of the gap and slowly approaches the bottom of the conduction band at a rate $\alpha=d(E_V-E_c)/dT \approx 0.12$ meV/K.

## 5. Conclusion

Obtained experimental results allow us to conclude that doping of p-$Pb_{1-x}Sn_xTe$ alloys with vanadium leads to the p-n-conversion and introduces a deep impurity level under the bottom of the conduction band ($\Delta E_V=E_c-E_V=14$-$32$ meV for $x=0.09$-$0.20$). Under pressure, the deep vanadium level in $Pb_{1-x-y}Sn_xV_yTe$ ($x=0.20$) alloy moves almost parallel to the conduction band bottom and subsequently intersects the middle of the gap and the top of the valence band, inducing the n-p-conversion and the insulator-metal transition at critical pressure $P^* \approx 7$ kbar at T=4.2 K. In the metallic phase further increase of pressure leads to the increase of the hole concentration due to the redistribution of electrons between the valence band and the impurity level. With the increase of temperature at fixed pressure the opposite movement of the level, inducing the intersection of the level with the middle of the gap and the p-n-inversion of the conductivity type, takes place. In the framework of this model, the pressure and the temperature coefficients of the vanadium level energy, counted from the bottom of the conduction band, are estimated as $d(E_V-E_c)/dP \approx 0.5$ meV/kbar and $d(E_V-E_c)/dT \approx 0.12$ meV/K.

## Acknowledgments

The authors would like to thank Dr. S.A. Ibragimov and Dr. A.V. Knotko (Faculty of Chemistry, Moscow State University) for performing the X-ray diffraction study and the X-ray fluorescent microanalysis. This research was supported by the Russian Foundation for Basic Research (Grant N 11-02-01298) and by the Ministry of Education and Science of the Russian Federation (State contract No 14.740.11.0051).

## References

[1] Volkov B A, Ryabova L I and Khokhlov D R 2002 *Physics-Uspekhi* **45** 819
[2] Story T 1996 *Acta Phys. Pol.* A **92** 663
[3] Story T, Grodzicka E, Witkowska B, Gorecka J and Dobrowolski W 1992 *Acta Phys. Polon.* A **82** 879
[4] Vinokurov A A, Artamkin A I, Dorofeev S G, Kuznetsova T A and Zlomanov V P 2008 *Inorg. Mater.* **44** 576
[5] Artamkin A I, Dobrovolsky A A, Vinokurov A A, Zlomanov V P, Gavrilkin S Y, Ivanenko O M, Mitzen K V, Ryabova L I and Khokhlov D R 2010 *Semiconductors* **44** 1543
[6] Skipetrov E P, Golovanov A N, Zvereva E A, Slyn'ko E I and Slyn'ko V E 2009 *Physica* B **404** 5262
[7] Slyn'ko V E 2001 *Vestn. L'vovsk. Univ., Ser. Fiz.* **34** 291
[8] Dornhaus R, Nimtz G and Schlicht B 1983 *Narrow-Gap Semiconductors* (Berlin: Springer-Verlag) pp 35-49
[9] Akimov B A, Vadkhva R S and Chudinov S M 1978 *Sov. Phys. Semiconductors* **12** 1146

# Synchrotron based XRD study on nano crystalline $SnO_2$ under pressure

**Alka B. Garg[1*], P. Thangadurai[2], S. Meenakshi[1] and S. Ramasamy[2]**

[1]High Pressure and Synchrotron Radiation Physics Division,
Bhabha Atomic Research Centre, Mumbai 400085, India
[2]Department of Nuclear Physics, University of Madras,
Guindy Campus, Chennai 600025

[*]E-mail: alkagarg@barc.gov.in

**Abstract.** Results of in-situ high pressure x ray diffraction measurements on 5 nm $SnO_2$ nanoparticles are reported. The ambient rutile phase remains stable upto the pressure of 18.1 GPa. However, beyond 19 GPa, the diffraction pattern shows the new diffraction peaks indicating the presence of structural phase transition in the material. These new peaks could be assigned to cubic structure. The coexistence of both the phases could be seen over a large pressure region indicating the sluggishness of the transition.

## 1. Introduction

In the last few years the scientific community has cherished the technical potential of nano sized (1-100 nm) clusters or particles of various semiconductors, elemental solids and compounds. These entities often have different physical and chemical properties than their bulk counterpart as the number of atoms or molecules on their surface can be comparable to that inside the particles. For example, number of semiconductors prepared in nano size exhibit particle size dependence of optical and electronic properties. The studies on phase transitions in nanocrystalline materials under pressure can reveal new types of nucleation and growth dynamics and transition paths as compared to the bulk. The issue of their structural stability against pressure is of considerable interest from fundamental point of view also. There are number of research papers on the studies of phase stability of nano structured semiconductors such as CdSe [1], Si[2], CdS [3], ZnS [4], ZnO [5], PbS [6] etc. under pressure. Results of these measurements clearly show a trend that, with the decrease in the particle size the transition pressure increases in all these semiconductors, and the compressibility increases with decreasing particle size. However, there are few compounds *i.e.* $Fe_2O_3$ [7], $TiO_2$ [8], $CeO_2$ [9] where the phase transition pressure is lowered than their bulk counterpart. The reason for the decrease in the transition pressure was suggested as the large volume collapse at the transition. The thermodynamic considerations regarding the phase stability of these materials reveals that mainly three factors namely volume collapse, surface energy difference and the difference in the internal energy govern the transition pressure in nanocrystals as compared to their bulk counterpart. Tin dioxide $SnO_2$, yet another technologically important semiconductor crystallizes in the rutile-type structure. Rutile is tetragonal with space group $P4_2/mnm$ and Z=2. A basic building block of the structure is the $SnO_6$ octahedra; edge-linked chains of octahedral extend parallel to the *c* axis. Each chain is linked to four others by a corner sharing of anions. There are few reports on the high pressure investigations on nano $SnO_2$. The electrical resistivity and dielectric measurements under pressure shows the presence of phase transition below 3 GPa [10]. However the Raman scattering measurements did not reveal the pressure induced structural phase transition in this pressure region. The reports on the high pressure x ray diffraction on nano $SnO_2$ reveals the vast difference in the transition pressure [11, 12]. In this work we report the results of high quality synchrotron based angle dispersive x-ray diffraction measurements on 5nm $SnO_2$ to map out the transition pressure and other structural details.

## 2. Experimental

### i Sample Preparation

Nanocrystalline $SnO_2$ was prepared by chemical precipitation method with different hydrolysis time. The precursor material used in this work was a 0.1 N aqueous solution of tin chloride (SnCl2.H2O). The particle size of 5nm was obtained with hydrolysis time of 144 hours. The prepared sample was characterized by powder x-ray diffraction (XRD) using a rotating anode generator (Rigaku-Make) with the Mo K$\alpha$ ($\lambda$=0.7107 Å) radiation. The average particle size estimated using the Scherrer formula is ~ 5 nm

### ii ADXRD Measurements

Angle dispersive x-ray diffraction (ADXRD) measurements were carried out at the powder x-ray diffraction beam line of ELETTRA synchrotron source, Trieste, Italy. For high-pressure experiments fine powdered samples of $SnO_2$ were loaded in a Mao–Bell type diamond anvil cell (DAC). Hardened stainless steel gasket with a central hole of 100 μm diameter and thickness 50 μm contained the sample. The x-ray beam was collimated by 80 μm diameter pinhole. A mixture of methanol–ethanol (4:1) was used as pressure transmitting medium. The pressure was determined in-situ by using platinum powder mixed with the sample as pressure calibrant. The wavelength of the x-ray employed and the sample to image plate distance were calibrated employing $LaB_6$ diffraction pattern. X-ray powder patterns at various pressures were collected employing x-rays of wavelength 0.6704 Å. Images of the powder diffraction rings were collected on a Mar345 image plate detector and read with a resolution of 100x100 pixel size. Typical exposure times of 20 to 30 minutes were employed for data collection at each pressure point. The powder rings thus obtained were integrated using the versatile FIT2D [13] software and converted to the one dimensional 2θ *vs.* intensity plot.

## 3. Results and Discussion

Figure 1(a) shows the two dimensional diffraction image of as prepared nano $SnO_2$ and figure 1(b) depicts the Rietveld refined one dimensional intensity vs. 2θ plot. The x ray diffraction data shows the single phase formation of the compound with rutile structure. The ambient lattice parameter obtained for as prepared sample from Rietveld refined data is a= 4.7427(6) Å; c = 3.1851(6) Å with Sn position at (0, 0, 0) and refined oxygen position at (0.3056, 0.3056, 0.3056). The average grain size estimated using the Scherrer formula gives the particle size to be 5 nm.

Pressure evolution of few selected x- ray diffraction patterns is shown in figure 2. The ambient rutile phase remains stable against pressure upto 18.1 GPa. However the pattern collected at 19.8 GPa shows a weak peak developing between (110) and (101) lines of rutile phase indicating the onset of structural phase transition in this material. The new peak position is marked by an arrow in figure 2. The intensity of this new peaks increases with further increase in pressure while that of rutile phase goes down.

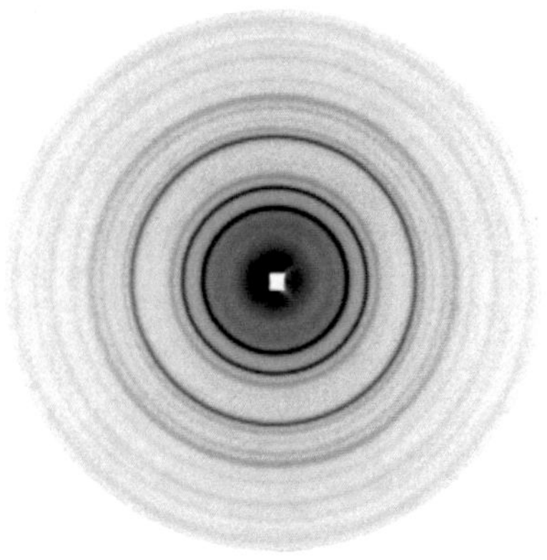

Figure 1(a) 2-D diffraction rings

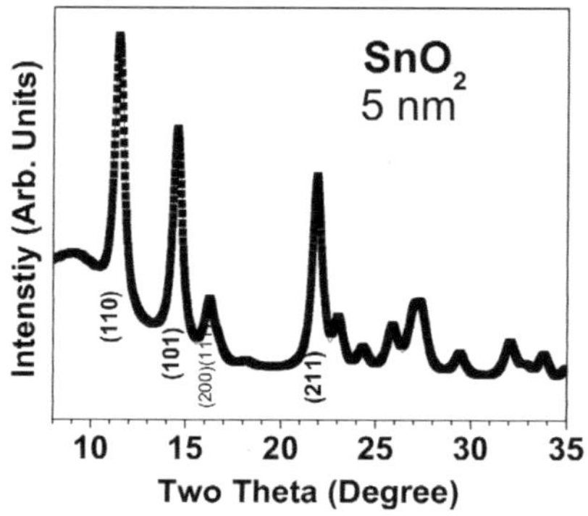


Figure 1(b) One dimensional Intensity vs. 2θ plot. Solid line is the Rietveld refined data whereas the solid squares are the experimental points.

The new peaks could be assigned to cubic structure as described in the literature [12]. The fitted lattice parameter for the new cubic phase is 4.9214 Å at 22 GPa. The coexistence of both phases was seen upto 30.8 GPa, the highest pressure of the present investigations indicating the sluggishness of this transition. The reported transition pressure for 3 nm particle size $SnO_2$ is 30 GPa whereas for 8 nm $SnO_2$ it is around 18 GPa. Defects and lattice strain may be responsible for difference in transition pressure. It is to be noted here that the bulk $SnO_2$ shows structural phase transition from rutile to $CaCl_2$ at 11.8 GPa and then to orthorhombic $PbO_2$ type phase. Beyond 21 GPa the structure change to cubic structure [14].

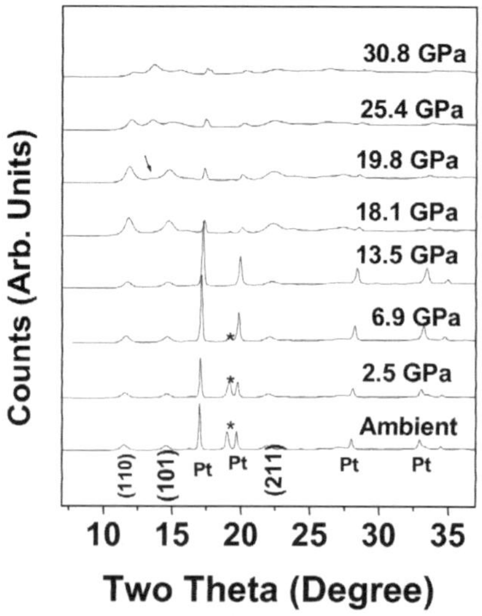


Figure 2 Pressure evolutions of x ray diffraction patterns. Appearance of new peak has been marked by an arrow. Peaks marked by an asterisk are from the gasket. Pt indicates the diffraction peaks from the pressure calibrant.

## Conclusion

The high pressure x ray diffraction measurements on 5 nm particle size of $SnO_2$ prepared by chemical precipitation method has been carried out. The data shows stability of ambient rutile phase against pressure upto 18 GPa. However beyond 19 GPa the rutile phase becomes unstable and material transforms to cubic structure.

## Acknowledgement

One of the authors (ABG) acknowledges the Indian DST for travel support and Italian DST for local hospitality. The help received from Dr. V. Vijaykumar during XRD data collection at Elettra is highly acknowledged.

## References

[1] Haase M, Alivisatos A P 1992 *Journal of Physical Chemistry* 96 p 6756.
[2] Tolbert S H, Alivisatos A P 1995 *Journal of Chemical Physics* 102 p 4642.
[3] Tolbert S H, Herhold A B, Brus L E, Alivisatos A P 1996 *Physical Review Letters* 76, p 4384.
[4] Jiang J Z, Gerward L, Frost D, Secco R, Peyronneau J, Olsen J S 1999 *J. Appl. Phys.* 86 p 6608.
[5] Jianga J Z, Olsen J S, Gerward L, Frost D, Rubie D, Peyronneau J 2000 *Europhys. Lett.* 50 p 48.
[6] Jianga J Z, Gerward L, Secco R, Frost D, Olsen J S, Truckenbrodt J 2000 *J. Appl. Phys.* 2658.
[7] Jianga J Z, Olsen J S, Gerward L, Morup S 1998 *Europhys. Lett* 44 p 620
[8] Olsen J S, Gerward L, Jianga J Z 1999 *J. Phys. Chem. Solids* 60 p 229
[9] Rekhi A, Saxena S K, Lazor P 2001 *J. Appl. Physics* 89 p 2968
[10] Thangadurai P, Bose A C, Ramasamy S, Kesavamoorthy R, Ravindran T R 2005 *J. Phys. Chem. Solids* 66 p 1621
[11] Jianga J Z, Gerward L, Olsen J S 2001 *Scripta Mater.* 44 p 1983
[12] He Y, Liu J F, Chen W, Wang y, Wang H, Zeng Y W, Zhang G Q, Wang L N, Liu J, Hu T D, Hahm H, Gleiter H and Jiang J Z 2005 *Phys. Rev. B* 72 p 212102
[13] Hammersley A P, Svensson S O, Hanfland M, Fitch A N and Hausermann D 2008 *High Pressure Res.* 14 p 235
[14] Sheih S R, Kubo A, Duffy T S, Prakapenka V B and Shen Guoyin 2006 *Phys. Rev. B* 73 p 014105

# High pressure investigation of ductile intermetallic HoCu

**S Meenakshi and Alka B Garg**
High Pressure and Synchrotron Radiation Physics Division,
Bhabha Atomic Research Centre, Mumbai-400085, INDIA

E-mail: msunder@barc.gov.in

**Abstract** The results of high-pressure angle dispersive X-ray diffraction measurements up to 27.8 GPa on the ductile intermetallic HoCu is presented. The ambient CsCl structure (*SG:Pm-3m*) is found to be stable upto the highest pressure of the present measurements. The pressure-volume data fitted to second order Birch-Murnaghan equation of state yields the bulk modulus and its pressure derivative as 76.8 GPa and 4 respectively.

## 1. Introduction

Intermetallic compounds have always attracted considerable attention as an engineering material due to their interesting combination of mechanical, electrical, magnetic, thermal and elastic properties that are often superior to ordinary alloys. Many of them possess desirable combinations of low density, good oxidation resistance and high strength at elevated temperature. However, their engineering application has been limited by the generally poor ductility and low fracture toughness and brittle behaviour at ambient temperature [1]. Recently however a large family of fully ordered stoichiometric binary rare earth intermetallic compounds with high ductility and high fracture toughness at room temperature have been reported [2]. Combining a rare earth element "R" (Y, Dy, Ho, Er, Ce, and Nd) with metals "M" like Cu and Ag forms these equiatomic binary compounds "RM" with CsCl type structure (*SG: Pm* $\bar{3}m$) as shown in Figure 1 wherein the corners of the unit cell cube are occupied by the "M" element and the atom in the cube centre is the "R" element. These RM compounds are "line-compounds" with the exact 1:1 stoichiometry. These ductile materials are promising materials for coatings that are highly resistant to corrosion or that maintain strength at high temperature. Many theoretical and experimental investigations have been carried out to reveal the mechanism of the high room temperature ductility in these RM intermetallics because it may provide insights into improving ductility in other intermetallics [3-6].

A number of features make the observation of ductility even more striking. Calculations by Morris *et al* [3] have shown that the RM ductile materials are significantly different than previously studied intermetallic B2 compounds. The elastic constants are significantly more isotropic, and the Poisson's ratio is smaller. The bonding in these materials is less covalent than, for example, NiAl, which shows significant ductility only in single crystals oriented for single slip modes. The fracture toughness value for the RM compounds is two to five times larger than that of NiAl (which also has the B2 structure) but about the same as that of aluminium alloys. Also, the tensile strength and elongation values for the RM phases and commercial grade aluminium alloys are comparable, but the elongations are five to ten times larger than that of the brittle NiAl intermetallic compound [4]. The high fracture toughness observed in these RM compounds has been attributed to their multiple slip systems, fine grain size and to the possible "R" element counteracting environmental embrittlement [5]. Gschneidner *et al* [1] have indicated that these materials possess much lower unstable stacking fault energies due to which it is easier for them to plastically deform instead of fracturing at the grain boundaries. Furthermore theoretical calculations have indicated the Fermi energy to be near a local minimum for some of the RM systems and the hybridization in these systems has been found to be stronger than that of NiAl alloy indicating a better ductility in these intermetallics than the NiAl alloy [6].

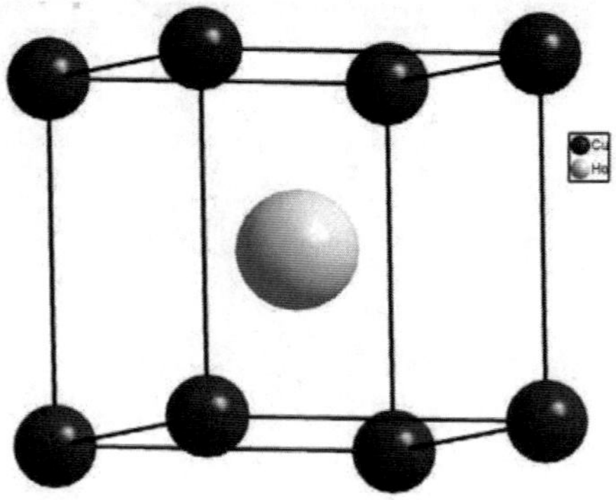

**Figure 1**. Crystal Structure of HoCu.

HoCu belongs to this class of ductile intermetallics. Neutron diffraction [7] studies on a related compound GdCu have indicated a diffusionless martensitic transformation from the ambient CsCl to orthorhombic FeB structure over a broad temperature range near the Neel temperature. This phase transition is characterized by pronounced anomalies in unit cell volume and electrical resistivity. Heat capacity [8] and thermal expansion [9] measurements on HoCu and DyCu have indicated anomaly near 26 K and 61K respectively. These low temperature anomalies are related to spin reorientations and antiferromagnetic transition [10]. The structural instability in this series of compounds have been correlated with the magnitude of the metallic radii of the rare earth ion, which produces small variation in the density of states near the Fermi level [9]. To get a better understanding of the anomalous ductility of RM intermetallics, fundamental investigations of their structural properties are desirable. Application of intensive variables *e.g.,* pressure, is known to influence the structure and properties of materials. The present synchrotron based high pressure angle dispersive X-ray diffraction (ADXRD) on HoCu up to 27.8 GPa is aimed at investigating its structural stability under compression.

## 2. Experimental Details

HoCu was prepared from pure metal constituents by argon arc-melting in a water-cooled copper hearth. The sample characterization using XRD indicated a cubic symmetry (*SG: $Pm\bar{3}m$*) with a lattice parameter of 3.440 Å. High pressure ADXRD measurements were carried out on powdered HoCu at the synchrotron radiation source at Elettra, Trieste, Italy. A Mao-Bell type diamond anvil cell (DAC) with diamonds of culet size 400 μm was employed. Fine powder of HoCu along with silver as the internal pressure calibrant was loaded into a hardened stainless steel gasket of hole 150 μm in diameter with 4:1 methanol:ethanol as the pressure-transmitting medium. The experimental station was based on an image plate area detector (Marresearch). The sample to image plate distance and wavelength were calibrated using $LaB_6$ as standard. The X-ray beam wavelength was calibrated to be 0.652 Å. The data was collected up to a pressure of 27.8 GPa with a typical exposure time of about 200 secs depending on the beam current. The scanned two-dimensional diffraction patterns were corrected for image plate tilt and converted to intensity *vs* 2θ through radial integration using FIT2D

Journal of Physics: Conference Series **377** (2012) 012023 doi:10.1088/1742-6596/377/1/012023

software [11]. The lattice parameters of the sample and silver were determined by carrying out full profile refinement using GSAS software [12]. The pressure was then determined from the equation of state of silver [13].

### 3. Results and Discussion

The evolution of the diffraction pattern at various pressures as seen in Figure 2 shows that the ambient cubic phase is retained up to the highest pressure of 27.8 GPa attained in the present measurements. There was no indication of any pressure-induced phase transition in the investigated pressure range.

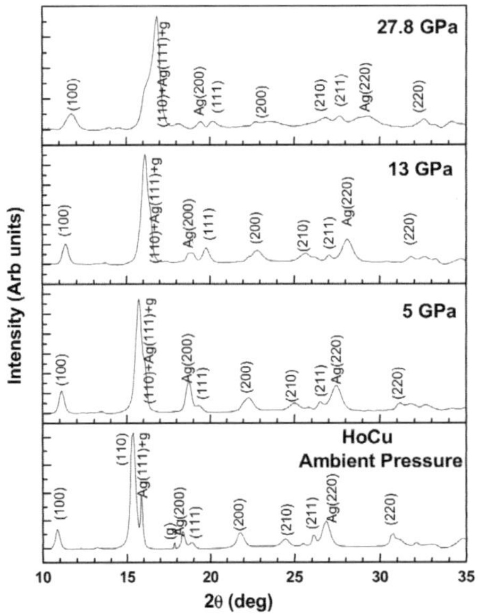


**Figure 2.** Evolution of ADXRD pattern with Pressure for HoCu

The pressure volume (P-V) data obtained at various pressures is shown in Figure 3. A second order Birch-Murnaghan equation of state fit to the pressure volume (P-V) data yielded a bulk modulus of 76.8 GPa with its pressure derivative fixed at 4.

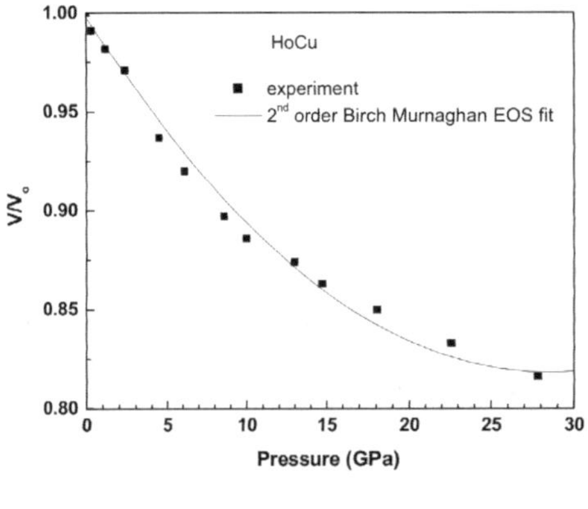


**Figure 3.** Equation of state for HoCu

The results of the present investigation are also consistent with the recent study on the deformation behavior of DyCu ductile intermetallic compound under compression which showed no indications that DyCu had undergone stress-induced phase transformation or twinning [14].

## 4. Conclusions

The ductile intermetallic HoCu has been investigated by high-pressure synchrotron powder diffraction for its structural stability up to 27.8 GPa. A second order Birch-Murnaghan equation of state fit to the pressure volume (P-V) data yielded a bulk modulus of 76.8 GPa with its pressure derivative fixed at 4.

## Acknowledgements

Measurements were carried out under the Indo-Italian Program of Cooperation in Science and Technology. We acknowledge the unfailing help received from Dr. P. Maurizio and Dr. V. Vijayakumar during the measurements at Elettra. The DST (India) and the Italian foreign ministry have provided the financial support for travel to Elettra and accommodation respectively.

## References

[1] Fleischer R L and Zabala R 1990 *Met. Mat. Trans* A **21** 2709
[2] Gschneidner Jr K A, Russell A, Pecharsky A, Morris J, Zhang Z, Lograsso T, Hsu D, Chester Lo C H, Ye Y,Slager A and Kesse D 2003 *Nature Materials* **2** 587
[3] Morris J R, Ye Y, Lee Y B, Harmon B N ,Gschneidner Jr. K A, Russell A M 2004 *Acta. Materialia.* **52** 4849
[4] Gschneidner Jr. K A 2009 *Acta Materialia* **57** 18
[5] Zhang Z, Russell A M, Biner S B, Gschneidner Jr. K A, Lo C C H 2005 *Intermetallics* **13** 559
[6] Wu Y, Hu W, Han S 2008 *Physica* B **403** 3792
[7] Blanco J A, Espeso J I, Soldevilla J G, Sal J C G, Ibarra M R, Marquina C, Fischer H E 1999 *Phys. Rev.* B **59** 512
[8] Ho J C , He S ,Wu C, Chien T S and Pavlovic A S 1993 *Chinese J. Phys.* **29** 147
[9] Ibarra R M, Chien T S and Pavlovic A S 1989 *J. Less Common Met.* **153** 233
[10] Amara M, Mourin P and Bourdarot F 1997 *J. Phys. Chem. Solids* 9, 7441
[11] Hammersley A P, Svensson S O, Hanfland M, Fitch A N and Hausermann D 1996 *High Press. Res.* **14** 235
[12] Larson A C, Von Dreele R B 1994 *General Structure Analysis System (GSAS), Los Alamos National Laboratory Report LUAR* **86** 748
[13] Dewaele A, Torrent M, Loubeyre P and Mezouar M 2008 *Phys. Rev. B* **78** 104102
[14] Cao G H, Yu Z, Russell A M 2011 *Mat. Sci. and Engg.* A **528** 7173

Journal of Physics: Conference Series **377** (2012) 012024 doi:10.1088/1742-6596/377/1/012024

# Pressure induced phase transition in defect chalcopyrite compounds

**S Meenakshi**
High Pressure and Synchrotron Radiation Physics Division,
Bhabha Atomic Research Centre, Mumbai-400085, INDIA

E-mail: msunder@barc.gov.in

**Abstract** The large band gap semiconducting compounds of the type $A^{II}B_2^{III}C_4^{VI}$ crystallizing in the defect chalcopyrite (DCP) tetragonal structure are of great technological interest and have potential applications in non linear optical and photovoltaic devices. These compounds contain a crystallographically ordered array of vacancies (stoichiometric voids) in the cation sublattice and exhibit several interesting physical and chemical properties. The presence of the vacancies facilitates the doping of these compounds by impurities and has stimulated investigations about occurrence of order–disorder effects in the cation sublattice. In this presentation our recent experimental results on the high pressure investigations on some of the defect chalcopyrite compounds would be discussed.

## 1. Introduction

The large band gap ternary chalcopyrite compounds have attracted considerable attention from fundamental and application point of view as a new generation of semiconducting materials. Their crystal structure is common to compounds of chemical formula $A^{II}B^{IV}C_2^{V}$ and $A^{I}B^{III}C_2^{VI}$. These can be thought of as being derived from the parent III-V and II-VI compounds respectively, by replacing the group-III element by alternating a group-II and group-IV element or the group II by a group-I and group-III element. The presence of nonequivalent tetrahedrally coordinated cations in these compounds because of the displacement of the anion from the sphalerite-like position results in a distortion of the crystal lattice from the cubic to tetragonal symmetry.

Another class of tetrahedrally bonded materials, known as the defect chalcopyrites can be derived by successive substitution from the sphalerite (zinc blende) structure. For instance, the formula unit of the $A^{I}B^{III}C_2^{VI}$ compound may be doubled and subsequently one of the group-I elements could be removed, thus creating a vacancy. In order to maintain the valency the other group I element could be replaced by a group-II element thus forming the $A^{II}B_2^{III}C_4^{VI}$ compound. Most of the $A^{II}B_2^{III}C_4^{VI}$ semiconductors crystallize in the thiogallate structure, also called defect chalcopyrite (DCP), {*SG:* $I\bar{4}$}; while some of them crystallize in the defect famatinite (DF), also known as defect stannite, with higher symmetry {*SG:* $I\bar{4}2m$}. The difference between the two structures according to the partial cation ordering model lies in a partial disordering between the A and B cations for the $I\bar{4}2m$ structure. Table 1 and Table 2 summarizes the structural parameters for a representative of both the DCP and DF structures respectively [1].

**Table 1.** Atomic positions and site occupancy for $CdGa_2S_4$ [1] in the defect-chalcopyrite structure, {*SG: I* $\bar{4}$, *Z*=2} at ambient pressure.

| *Atom* | *Site* | *x* | *y* | *z* | *site occupancy* |
|---|---|---|---|---|---|
| Cd | 2a | 0 | 0 | 0 | 1 |
| Ga(1) | 2b | 0 | 0 | 0.5 | 1 |
| Ga(2) | 2c | 0 | 0.5 | 0.25 | 1 |
| Vacancy | 2d | 0 | 0.5 | 0.75 | 0 |
| S | 8g | 0.271(8) | 0.261(8) | 0.140(4) | 1 |

**Table 2.** Atomic positions and site occupancy for $ZnGa_2Se_4$ [1] in the defect-stannite structure, {*SG: I* $\bar{4}$ *2m*, *Z*=2} at ambient pressure.

| *Atom* | *Site* | *x* | *y* | *z* | *site occupancy* |
|---|---|---|---|---|---|
| Zn | 4d | 0 | 0.5 | 0.25 | 0.5 |
| Ga(1) | 4d | 0 | 0.5 | 0.25 | 0.5 |
| Ga(2) | 2a | 0 | 0 | 0 | 1 |
| Vacancy | 2b | 0 | 0 | 0.5 | 0 |
| S | 8i | 0.264(8) | 0.264(8) | 0.117(4) | 1 |

It may be noted that these structures contain a crystallographically ordered array of vacancies (stoichiometric voids or vacancies) in the cation sublattice and hence they can be considered as ordered vacancy compounds [2]. The crystal structure of a typical DCP compound is shown in Figure 1. The vacancy in these compounds is an ordered defect and various attempts to completely fill up the vacancy are unsuccessful indicating that the defect is an inherent one. Hence this class of compounds form a natural bridge between the impurity physics and crystal physics and they can help to understand the role played by vacancies in the physical and chemical properties of solids. The tetrahedral coordination of atoms and the stoichiometric vacancies also result in a low packing efficiency in the crystal lattice. This facilitates considerably the doping of the DCP compound by impurities and also the formation of solid solution and the occurrence of order-disorder effects in the cation sublattice [3, 4]. Furthermore, this family of semiconductors are of great technological interest and are promising optoelectronic materials due to their high nonlinearity, intense luminescence and high photosensitivity [5]. In particular, tunable filters of $CdGa_2S_4$ and UV photodetectors of $CdAl_2S_4$ are already used as devices [5, 6].

High pressure investigation of the DCP compound could be of great interest for establishing the general systematics of phase transitions in these ternary compounds. Furthermore, the understanding of the role played by the vacancies in bonding and in pressure induced phase transition is of current theoretical and experimental interest. The knowledge of these data is also important from material design consideration.

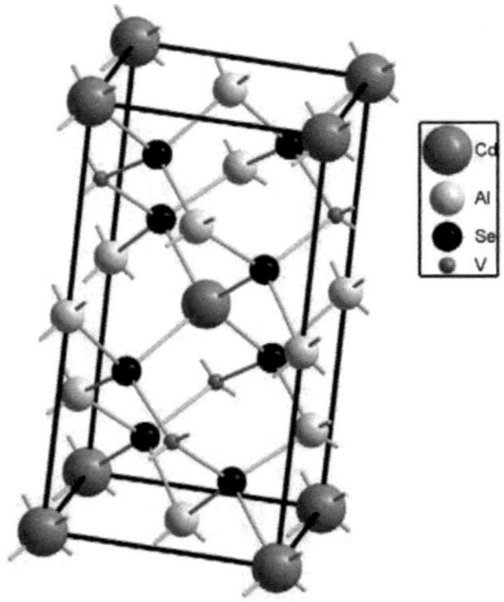


**Figure 1.** Crystal structure of the defect chalcopyrite compound $CdAl_2Se_4$. Atom denoted by V is actually the ordered vacancy position.

Several $A^{II}B_2^{III}C_4^{VI}$ compounds have been extensively investigated both theoretically and experimentally for their high pressure behaviour [1,7-14]. High pressure angle dispersive X-ray diffraction (ADXRD) on $CdGa_2Se_4$ by Gresechnik *et al* [9] has shown an order-disorder phase transition at 21 GPa to the rock-salt (SG: $Fm\bar{3}m$, $Z=4$) structure which upon decompression transforms into disordered zinc-blende (SG:$F\bar{4}3m$, $Z=4$). High pressure Raman studies on $CdGa_2Se_4$ [8, 9] has indicated an irreversible disappearance of Raman signal around 20.9 GPa due to transition to the Raman inactive rock salt phase. High pressure ADXRD measurements on $ZnGa_2Se_4$ and $CdGa_2S_4$ [1] have revealed a pressure induced phase transition from the defect tetragonal to a defect cubic NaCl type structure around 15.5 GPa and 17 GPa respectively. Marquina *et al* [7] have observed a tetragonal to cubic NaCl type structure at about 14 GPa with a two phase region (defect tetragonal + NaCl type) between 12 and 14 GPa in $MnGa_2Se_4$.

In view of these interesting results we have carried out high pressure angle dispersive X-ray diffraction and Raman investigation of the DCP compounds $AAl_2Se_4$ (A = Cd, Zn, Hg) and $CdAl_2S_4$ to investigate their structural evolution and vibrational properties under compression. In this paper the results of some of these would be reviewed.

## 2. Experimental Details

The DCP compounds for the study were grown using the chemical vapour transport (CVT) with iodine as transporting agent [15]. High pressure ADXRD measurements were carried out at the synchrotron radiation source at Elettra, Trieste, Italy using a Mao-Bell type of diamond anvil cell (DAC) with diamonds of culet size 400 μm. A 150 μm hole in a hardened stainless steel gasket pre-indented to a thickness of ~50 μm constituted the sample chamber. Silver or platinum was used as the internal pressure calibrant. The experimental station was based on an image plate area detector (Marresearch). The sample to image plate distance and wavelength were calibrated using $LaB_6$ as standard. The scanned two-dimensional diffraction patterns were corrected for image plate tilt and converted to intensity *vs* $2\theta$ through radial integration using FIT2D software [16]. The lattice parameters of the sample and pressure calibrant were determined by carrying out full profile refinement using GSAS software [17]. The pressure was then determined from the equation of state of the pressure calibrant.

The Raman studies at high pressure were carried out using a 0.25 m Czerny-Turner spectrometer (SpectraPro 275, Acton Res. Corp.) in backscattering geometry with pressure calibration using ruby fluorescence technique [15]. For all the experiments 4:1 methanol: ethanol served as pressure transmitter except in the case of chemically reactive $HgAl_2Se_4$, where paraffin oil was used as the pressure medium.

## 3. Results and Discussion

High pressure ADXRD measurements on $AAl_2Se_4$ (A = Cd, Zn, Hg) and $CdAl_2S_4$ indicated a structural phase transition from the ambient tetragonal (*SG: I* $\bar{4}$) structure [15, 18, 19]. Table 3 lists the phase transition pressures for all these compounds. The high pressure phase was characterized by a broadening of the diffraction peaks in addition to the appearance of new diffraction peaks and could be fitted to a disordered NaCl type (*SG: Fm* $\bar{3}m$) structure using full profile refinement. The typical $R_p$ and $wR_p$ factor of the fit was about 2.8% and 4.3 % respectively. The disordered NaCl phase was stable up to the highest pressure of the current measurements. Furthermore a mixed phase region where the sharp lines of the tetragonal phase coexist with the broad lines of the NaCl phase was observed in the case of $CdAl_2S_4$. The structural evolution for $CdAl_2Se_4$ is shown in Figure 2 as a typical case. Upon unloading, the diffraction pattern in all the four compounds was different from that of the initial phase as well as from the high-pressure phase implying that the structural phase transition is irreversible. This pattern with broad diffraction lines could be fitted to a disordered zinc blende (*SG: F* $\bar{4}3m$) structure. A second order Birch-Murnaghan equation of state [20] fit to the pressure volume (P-V) data yielded a bulk modulus and volume discontinuity as indicated in Table 3. The presence of a volume discontinuity reaffirms the first order nature of the structural transition.

High pressure Raman measurements for these compounds indicated a decreasing intensity of the spectra with pressure and an irreversible vanishing of the spectrum above a certain transition pressure [15]. This pressure correlates with the structural phase transition pressure to the Raman inactive rock salt structure observed in the X-ray measurements. Table 3 lists the transition pressure for all the four compounds investigated. In addition, the line widths of the Raman peaks was found to change at intermediate pressures between 4 and 6 GPa. This is attributable to the pressure induced two stage order–disorder transition proposed by Bernard *et al* [2]. According to this model in the first stage A and B cations substitute each other leading to a partially cation disordered phase. In the second stage, mutual disorder of both cations and vacancies is established, resulting in a disordered zinc blende phase. If pressure is increased further, this structure transforms to a high symmetry rock salt type structure.

**Table 3.** Transition pressures in X-ray and Raman experiments, bulk modulus and volume discontinuity for the investigated DCP compounds [15, 18, 19].

| Compound | X-ray transition pressure (GPa) | Raman transition pressure (GPa) | Bulk Modulus (GPa) | Volume discontinuity (%) |
|---|---|---|---|---|
| $CdAl_2Se_4$ | 9.1 | 9.6 | 52.1 | 10.0 |
| $ZnAl_2Se_4$ | 10.6 | 14.4 | 43.7 | 7.6 |
| $HgAl_2Se_4$ | 13.0 | 12.7 | 66.0 | 2.1 |
| $CdAl_2S_4$ | 14.4 | 13.5 | 44.6 | 7 |

The phase transition from the defect chalcopyrite to the disordered NaCl type structure implies an increase in the symmetry and increase in the coordination from tetrahedral to octahedral. Interestingly, a similar pressure induced phase transition to the cubic rock salt type structure has also been observed in the normal cubic zinc-blende and chalcopyrite ($A^{I}B^{III}C_2^{VI}$) compounds [21-23].The similitude between the high-pressure behaviour of defect chalcopyrite and pure chalcopyrite and II-VI semiconductors is not surprising since the DCP structure could be considered to be derived from the zinc-blende (ZnSe) structure with the unit cell nearly doubled along [001] axis. On comparing the bulk modulus of the DCP compounds with the pure chalcopyrite and the II-VI compounds it is noted that the DCP compounds are more compressible than the latter.

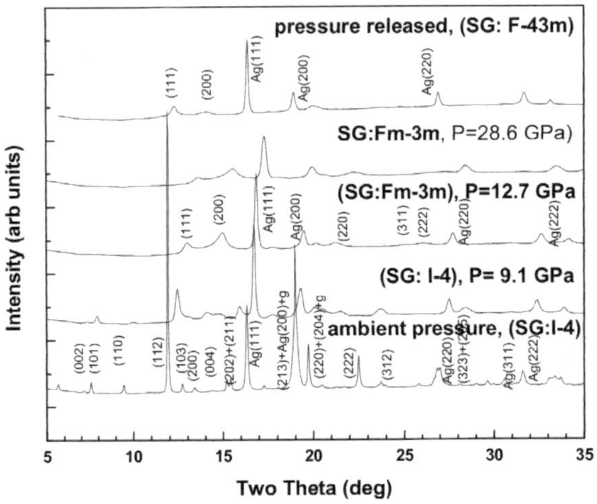


**Figure 2.** Structural evolution of $CdAl_2Se_4$ under compression [15].

## 4. Conclusions

High pressure ADXRD measurements on the defect chalcopyrites $AAl_2Se_4$ (A = Cd, Zn, Hg) and $CdAl_2S_4$ indicate an irreversible structural phase transition to a disordered rock salt phase which upon pressure release goes to a disordered zinc blende phase with broad X-ray diffraction lines.

## Acknowledgements

The high pressure X-ray Measurements were carried out under the Indo-Italian Program of Cooperation in Science and Technology. The DST (India) and the Italian foreign ministry have provided the financial support for travel to Elettra and accommodation respectively. The contributions of my collaborators Dr. V. Vijayakumar, Dr. B.K. Godwal, Prof. H.D. Hochheimer, Dr.A. Eifler, Dr. I. Orgzall and Dr. S. Tkachev in the work presented here are gratefully acknowledged.

## References

[1] Errandonea D, Kumar R S, Manjon F J, Ursaki V V and Tiginyanu I M 2008 *J. Appl. Phys.* **104** 063524
[2] Bernard J E and Zunger A 1988 *Phys. Rev. B* **37** 6835
[3] Kim W T, Cho G J and Kim C S *Phys. Rev. B* **43** 14265
[4] Park T Y, Kim C D, Yoon C S, Yang D I, Song H J, Kim W T 1998 *J. Phys. Chem. Solids* **59** 645
[5] Radautsan and I.M. Tiginyanu I M 1993 *Jpn. J. Appl. Phys. Suppl.* **32** 5
[6] Badikov V V, Matveev I N, Pshenichnikov S M, Rychik O V, Trotsenko N K, and Ustinov N D 1981 *Sov. J. Quantum Electron.* **11** 548

[7] Marquina J, Power C, Grima P, Morocoima M, Quintero M, Couzinet B, Chervin J C, Munsch P and Gonzalez J 2006 *J. Appl. Phys.* ***100*** 093513

[8] Ursaki V V, Burlakov I I, Tiginyanu I M, Raptis Y S, Anastassakis E and Anedda A 1999 *Phys. Rev. B* **59** 257

[9] Mitani T, Naitou T, Matsuishi K, Onari S, Allakhverdiev K, Gashimzade F and Kerimova T 2003 *Phys. Stat. Solidi. B,* **235** 321

[9[ Grzechnik A, Ursaki V V, Syassen K, Loa I, Tigiyanu I M and Hanfland M 2001 *J. Solid State Chem.* **160** 205

[11] Allakhverdiev K, Gashimzade F, Kerimova T, Mittani T, Naitou T, Matsuishi K and Onari S 2003 *J. Phys. Chem. Solids* **64** 1597

[12] Ursaki V V, Burlakov I I, Tiginyanu I M, Raptis Y S, Anastassakis E, Aksenov I and Sato K 1998 *Jpn. J. Appl. Phys.* **37** 135

[13] Cabrera M F and Sankey O F 2001 *J. Phys. Condens. Matter* **13** 1669

[14] Waskowska A, Gerward L, Olsen J S, Feliz M, Liusar R, Gracia L, Marques M and Recio J M 2004 *J. Phys. Condens. Matter* **16** 53

[15] Meenakshi S, Vijayakumar V, Godwal B K, Eifler A, Orgzall I, Tkachev S and Hocheimer H D 2006 *J. Phys. Chem. Solids* **67** 1660

[16] Hammersley A P, Svensson S O, Hanfland M, Fitch A N and Hausermann D 1996 *High Press. Res.* **14** 235

[17] Larson A C and Von Dreele R B 1994 *General Structure Analysis System (GSAS), Los Alamos National Laboratory Report LUAR* **86** 748

[18] Meenakshi S, Vijayakumar V, Eifler A and Hocheimer H D 2010 *J. Phys. Chem. Solids* **71** 832

[19] Meenakshi S, Vijayakumar V, Eifler A and Hocheimer H D 2010 *Azerbaijan J of Phys.* **16** 10

[20] Birch F 1978 *J. Geophys. Res.* **83** 1257

[21] McMahon M I and Nelmes R J 1995 *J. Phys. Chem. Solids* **56** 485

[22] Werner A, Hocheimer H D and Jayaraman A 1981 *Phys. Rev. B* **23** 3836

[23] Arora A K, Sakuntala T and Artus L 1993 *J. Phys. Chem. Solids* **54** 381

# High - pressure synchrotron x-ray diffraction study of $RMnO_3$ (R = Eu, Gd, Tb and Dy) upto 50 GPa

**D V S Muthu[1,*], A E Midgley[2], P R Scott[2], M B Kruger[2], J R Sahu[3], A K Sood[1] and C N R Rao[3]**

[1]Department of Physics, Indian Institute of Science, Bangalore – 560012. India
[2]Department of Physics, University of Missouri, Kansas City 64110 USA
[3]Jawaharlal Nehru Centre for Advanced Scientific Research, Jakkur, Bangalore – 560064. India

[*]E-mail: victor@physics.iisc.ernet.in

**Abstract**. We have carried out synchrotron based high-pressure x-ray diffraction study of orthorhombic $EuMnO_3$, $GdMnO_3$, $TbMnO_3$ and $DyMnO_3$ up to 54.4, 41.6, 47.0 and 50.2 GPa, respectively. The diffraction peaks of all the four manganites shift monotonically to higher diffraction angles and the crystals retain the orthorhombic structure till the highest pressure. We have fitted the observed volume versus pressure data with the Birch-Murnaghan equation of state and determined the bulk modulus to be 185 ± 6 GPa, 190 ± 16 GPa, 188 ± 9 GPa and 192 ± 8 GPa for $EuMnO_3$, $GdMnO_3$, $TbMnO_3$ and $DyMnO_3$, respectively. The bulk modulus of $EuMnO_3$ is comparable to other manganites, in contrast to theoretical predictions.

## 1. Introduction

Rare-earth manganites $RMnO_3$ (R = Eu, Gd, Tb, and Dy) have attracted a lot of interest due to their coupling between ferroelectric and anti-ferromagnetic properties [1]. $RMnO_3$ compounds which contain large R ions crystallize in the orthorhombic structure while those with smaller *R* ions crystallize in hexagonal structure. The boundary between these structures is present between Dy and Ho. At room temperature, R = Eu, Gd, Tb and Dy are paraelectric and paramagnetic with a distorted perovskite structure (due to tilting and deformation of $MnO_6$ octahedra) of orthorhombic symmetry with space group Pbnm [2 - 4]. In comparison to many studies under temperature and magnetic field, only a few experimental studies have been reported under pressure in perovskite $RMnO_3$, R = Tb and Dy [5, 6] and R = La, Pr and Sm [7-10]. Theoretical calculations [11] have shown that the bulk moduli in $RMnO_3$ (R = Dy-La) decreases with increasing molar volume with an anomalous dip for $EuMnO_3$. This motivated us to perform high pressure synchrotron based x-ray diffraction measurements on Eu, Gd, Tb and $DyMnO_3$ to determine the equation of state and look for possible phase transitions.

## 2. Experimental

High-pressure x-ray experiments were performed at room temperature at beamline X17C of the National Synchrotron Light Source at Brookhaven National Laboratory. Powder samples were loaded along with ruby chips, into a hole of ~100 μm diameter drilled in a pre-indented steel gasket. A Mao-Bell type diamond-anvil cell (DAC) was used with methanol:ethanol (4:1) mixture as pressure-transmitting medium. Pressure was monitored using the ruby fluorescence technique [12]. The experiments were carried out up to ~50 GPa using monochromatic x-rays ($\lambda$ = 0.4066 Å) and a two-dimensional (2D) imaging plate. The sample to imaging plate distances were 209.22 mm for $EuMnO_3$, 185.02 mm for Gd (Tb) $MnO_3$ and 185.09 mm for $DyMnO_3$. The 2D images were transformed into plots of intensity versus 2θ by radial integration using the FIT2D software package. Since the ambient structure is known, the lattice parameters and unit cell volumes were determined at each pressure by profile refinement of all four samples using Jana2000 package [13] without doing structural refinement. The equation of state was determined from the observed volume versus pressure using Igor Pro software.

## 3. Results and discussion

Eu, Gd, Tb and $DyMnO_3$ crystallize in the orthorhombic structure (space group Pbnm -no. 62). With increasing pressure the diffraction peaks shifts monotonically to higher diffraction angle value till the highest pressures without any phase transition. The lattice parameters and unit cell volumes at each pressure were determined. The zero pressure bulk modulus is found to show a small increase with decreasing ionic radius (molar volume also decreases).

### *3.1. $EuMnO_3$*

Figure 1a shows x-ray diffraction patterns, after background subtraction at selected pressures of $EuMnO_3$. With increasing pressure the diffraction peaks shift to higher angular values (decrease in d-spacing). The decompressed sample (from 54.4 GPa) with little broadening of peaks compared to the lowest starting pressure diffraction peaks. Profile fitting of the diffraction peak taken at 0.2 GPa yields the lattice parameters a = 5.3479 Å, b = 5.8459 Å, and c = 7.4578 Å and the unit cell volume V = 233.2 $Å^3$ which matches with an earlier report [2]. The cell parameters and the unit cell volume were subsequently determined for each pressure. The pressure versus unit cell volume is shown in figure 1b. The experimental data were fitted using the Birch-Murnaghan equation of state (EOS) ($P = 1.5K_0[(V/V_0)^{-7/3} - (V/V_0)^{-5/3}]\,[1 + 3/4\,(K_0'-4)\,[(V/V_0)^{-2/3}-1]]$, where $V_0$ is the unit cell volume, $K_0$ is the bulk modulus and $K_0'$ is the first pressure derivative of bulk modulus under ambient pressure) [14]: yields a bulk modulus of $K_o$ = 185 ± 6 GPa, and its first pressure derivative $K_0'$ = 3.7 ± 0.5.

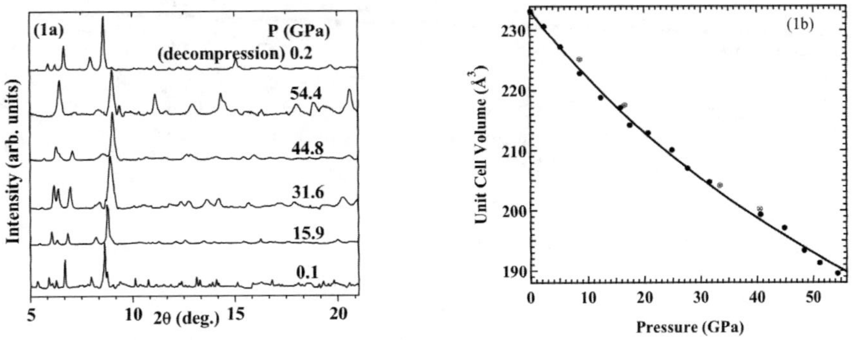


Figure 1(a) Diffraction patterns of $EuMnO_3$ at a few representative pressures, (b) Pressure versus unit cell volume. Solid (open) circles represent unit cell volumes for compression (decompression) pressure. The solid line represents the Birch-Murnaghan equation of state.

### *3.2 $GdMnO_3$*

The lattice parameters and the unit cell volume of $GdMnO_3$ at the lowest pressure were determined to be: a = 5.3202 Å, b = 5.8627 Å, c = 7.4291 Å and V = 231.7 $Å^3$ which are in good agreement with the reported values [2]. Figure 2a shows the variation of the unit cell volume with pressure. The solid line represents the Birch-Murnaghan EOS fit to the experimental data. This yields a bulk modulus of 190 ± 16 GPa and a first pressure derivative of 3.2 ± 1.6.

*3.3 $TbMnO_3$*

Profile refinement of the diffraction peaks collected at 0.2 GPa yields the lattice parameters: a = 5.2939 Å, b = 5.8407 Å, c = 7.3886 Å and V = 228.5 $Å^3$ which match with previously reported values [3]. Chen et al. have reported pressure-induced variations of the electronic state as well as structural distortion of Tb and $DyMnO_3$ but the equation state was not measured [5, 6]. The pressure dependence of the unit cell volume for $TbMnO_3$ is shown in Figure 2b. The bulk modulus and its pressure derivative were determined by fitting the Birch-Murnaghan EOS to the data. The results are $K_o$ = 188 ± 9 GPa and $K_0$' = 8.9 ± 1.2.

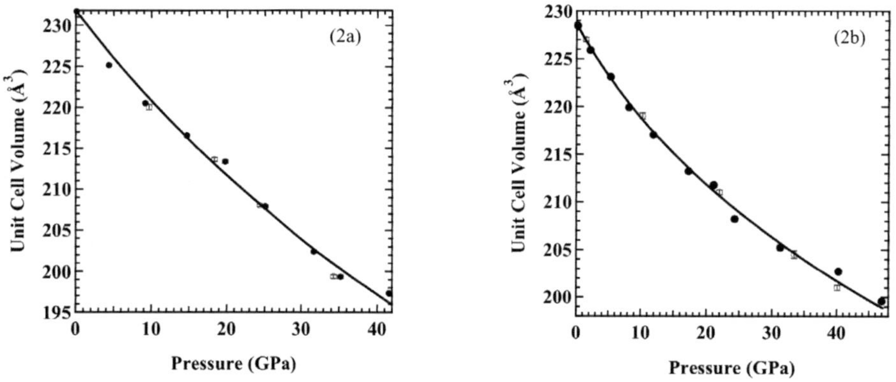


Figure 2 (a) Pressure dependence of unit cell volume for $GdMnO_3$. (b) Pressure dependence of unit cell volume for $TbMnO_3$. Solid (open) circles represent unit cell volumes for compression (decompression) pressure. The solid line represents the Birch-Murnaghan EOS.

*3.4 $DyMnO_3$*

The lattice parameters and unit cell volume (a = 5.2713 Å, b = 5.8335 Å, c = 7.3786 Å and V = 226.9 $Å^3$) were determined from the profile refinement of the diffraction pattern recorded at the lowest pressure (0.7 GPa). Figure 3 shows the pressure dependence of the unit cell volume. The Birch-Murnaghan EOS fit to the pressure versus volume data yields a bulk modulus of $K_o$ = 192 ± 8 GPa and its first pressure derivative $K_0'$ = 1.6 ± 0.6.

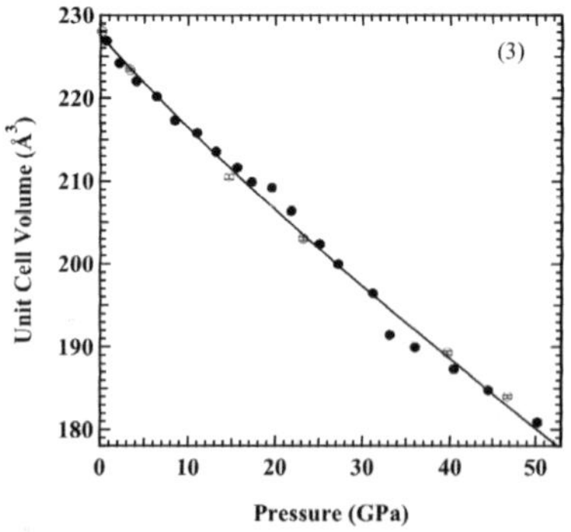


Figure 3. Pressure dependence of unit cell volume for $DyMnO_3$. Solid (open) circles represent unit cell volumes for compression (decompression) pressure. The solid line represents the Birch-Murnaghan EOS.

Figure 4 shows the molar volume dependence of bulk modulus of $RMnO_3$. As the ionic radius decreases from La-Dy more distortion is introduced because the buckling of Mn-O-Mn angle increases and the structure deviates from orthorhombic to distorted orthorhombic. The unit cell volume decreases as the distortion increases correspondingly the bulk modulus increases. A noteworthy observation from figure 4 is that the experimental value of the bulk modulus for $EuMnO_3$ does not show an anomalous decrease as compared to the theoretical predictions [11].

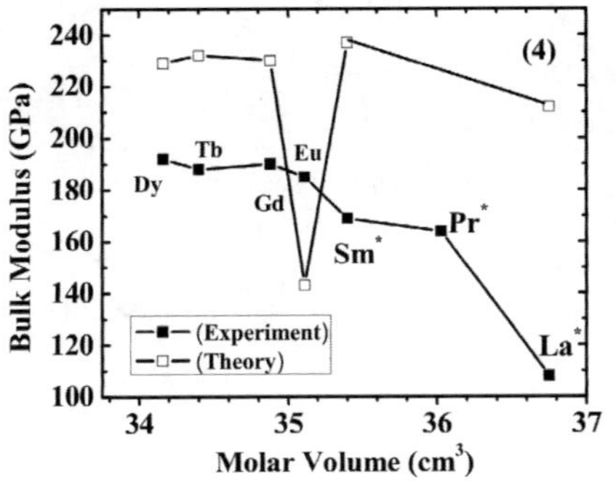


Figure 4. Molar volume dependence of bulk modulus for various $RMnO_3$. Filled squares represent experimental values. The values for La, Pr and $SmMnO_3$ are taken from Ref. 7 and 8. Open squares represent the theoretical values from Ref. 11.

## 4. Conclusion

In conclusion, we have performed a synchrotron based high-pressure study on rare-earth manganites, $RMnO_3$ (R = Eu, Gd, Tb and Dy) up to 54.4, 41.6, 47.0 and 50.2 GPa, respectively. Our studies do not reveal any phase transition, and structure remains orthorhombic till the highest pressures. The fit to the experimental data using the Birch-Murnaghan EOS yields the zero pressure bulk modulus to be 185, 190, 188 and 192 GPa in Eu, Gd, Tb and $DyMnO_3$, respectively. These values are compared with the theoretical predictions [11].

## 5. Acknowledgments

AKS thanks Department of Science and Technology, India for financial assistance. This work was partially supported by NSF Contract Number DMR-0605493, NSF COMPRES EAR 10-43050. Use of the National Synchrotron Light Source, Brookhaven National Laboratory, was supported by the U.S. Department of Energy, Office of Science, Office of Basic Energy Sciences, under Contract No. DE-AC02-98CH10886.

## References

[1] Goto T et al., 2004 *Phys. Rev. Lett.* **92** 257201
[2] McCarthy G J et al., 1973 *Mater. Res. Bull.,* **8** 1277
[3] Alonso J A et al., 2000 *Inorg. Chem,* **39** 917
[4] Kimura T et al., 2003 *Nature* **419** 818
[5] Chen J M et al., 2009 *Phys. Rev. B* **79** 165110
[6] Chen J M et al., 2010 *J. Chem. Phys.* **133** 154510
[7] Loa I et al., 2001 *Phys. Rev. Lett.* **87** 125501
[8] Zhou J S et al. 2003 *Phys. Rev. B* **68** 054403
[9] Baldini M et al., 2011 *Phys. Rev. Lett.* **106** 066402
[10] Trimarchi G et al., 2005 *Phys. Rev. B* **75** 035101
[11] Archana Srivastava et al., 2009 *J. Mag. Mag. Mater.,* **321** 3845
[12] Piermarini G J et al., 1975 *J. Appl. Phys.* **49** 3276
[13] Petricek V, Dusek M Jana2000. The crystallographic computing system. Institute of Physics, Praha, Czech Republic
[14] Birch F, 1978 *J. Geophys. Res.* **B83** 1257

Journal of Physics: Conference Series 377 (2012) 012026 doi:10.1088/1742-6596/377/1/012026

# Effect of Dose and Pressure on the Transformation from Neutron-Irradiated Graphite to Amorphous Diamond

**K Niwase[1], T Atou[2], K G Nakamura[2], and T. Iwata[3]**

[1]Hyogo University of Teacher Education, Kato-shi, Hyogo 673-1494, Japan
[2]Materials and Structures Laboratory, Tokyo Institute of Technology, 4259 Nagatsuta, Yokohama, 226-8503, Japan
[3]Nuclear Professional School, School of Engineering, University of Tokyo, Tokai, Ibaraki 319-1188, Japan

E-mail: niwase@hyogo-u.ac.jp

**Abstract.** We previously discovered a novel pathway for the transformation from highly orientated pyrolytic graphite foils irradiated with neutrons into amorphous diamond platelets. Here, we utilized graphite foils irradiated with a neutron dose of about a half of the previous study, and confirmed the pathway for the transformation. Raman spectrum of the irradiated graphite exhibits a broad amorphous one but still possessed weak G and D peaks, appearing in a less disordered state. The irradiated graphite foils were shock compressed at 46, 51 and 54 GPa. At 46 GPa, Raman spectra clearly show the G and D peaks but a weak photoluminescence appears. At 51 GPa, we can find transparent platelets with intense photoluminescence but no diamond Raman peak, implying the transformation to amorphous diamond. At 54 GPa, we obtained transparent platelets with intense photoluminescence and a small and broad diamond peak, suggesting the formation of nano-diamond.

## 1. Introduction

The transformation from graphite to diamond is a long time issue and polycrystalline diamond has been synthesized by direct conversion of graphite under static high pressure and temperature [1]. The ultimate smallest crystalline size of diamond was synthesized from $C_{60}$ fullerene by shock compression and rapid quenching (SCARQ) technique [2]. The obtained material is transparent and comprised wholly of $sp^3$ carbons, suggesting diamond structure, but the diamond crystalline size is within the range of unit cell. Then, it was labeled “amorphous diamond” as it is amorphous in the long-range order and diamond in the short-range order [3]. Graphite, on the other hand, has not been found to transform to amorphous diamond by shock compression but to crystalline diamond by the martensitic mechanism [4], if shock pressure is applied parallel to the c axis.

Recently, we have discovered a novel pathway for the transformation from highly oriented pyrolitic graphite (HOPG) into amorphous diamond [5]. The pathway consists of neutron irradiation, shock compression, and rapid quenching. Defects produced by irradiation (Wigner defects) are considered to work as the nucleation sites for diamond. In the pathway, we can control the introduction of Wigner defects by changing the conditions of irradiation and shock compression.

Here, we aimed to investigate the dose and pressure dependencies on the novel pathway by utilizing HOPG specimen irradiated with neutron to a dose which is about half of the last one and by giving shock compression in three different conditions.

## 2. Experimantal procedure

We utilized highly oriented pyrolytic graphite (HOPG-ZYA) from Union Carbide as a starting material, which was irradiated with fast neutrons at about 333 K to a dose of 1.4 x $10^{24}$ n $m^{-2}$ (E> 1 MeV) in JAERI JRR-2 nuclear reactor. The dose is about half of the previous study, in which the specimen was irradiated to a dose of 2.6x$10^{24}$ n $m^{-2}$ [5]. Thin films of about 8 μm in thickness were prepared by cleavage with a razor. The sample was sandwiched by gold foils, which work for heat sinks. The tungsten flyer was accelerated by a powder gun and was impacted to the front side of the iron capsule. The shock pressures determined by the flyer velocity were 46, 51 and 54 GPa. The samples after the shock compression were observed by using a scanning electron microscope (SEM). Raman measurements were done with a micro-Raman spectroscope (SPEX Raman-500) using an Ar (488 nm) laser.

## 3. Results and discussion

Figure 1 shows Raman spectra for the unirradiated and neutron-irradiated HOPG specimens. The Raman measurements were done in the same condition to compare the relative intensity. The unirradiated HOPG specimen shows the first-order peak near 1580 $cm^{-1}$ (G band) and the second-order Raman band near 2700 $cm^{-1}$. After the neutron irradiation, a broad band ranging from 800 to 1700 $cm^{-1}$ appeared. However, there still remains small G band and a defect peak near 1360 $cm^{-1}$ (D band) appears. This means that the neutron irradiated specimen is in a less disordered state, compared to the specimen used in the previous study [5].

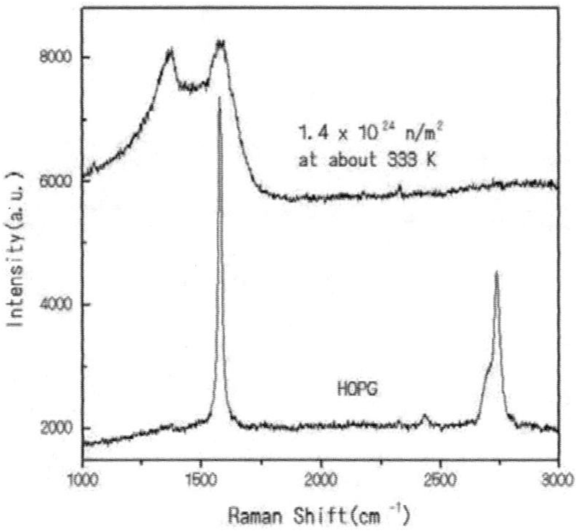


**Figure 1.** Raman spectra of an unirradiated HOPG specimen as a starting material and a HOPG specimen irradiated with fast neutrons at about 333 K to a dose of 1.4 x $10^{24}$ n $m^{-2}$ (E> 1 MeV) in JAERI JRR-2 nuclear reactor. Spectrum of the neutron-irradiated HOPG shows a broad band in a range from 800-1700 $cm^{-1}$. Peaks of the G and D bands still remain in the spectrum.

Figure 2 shows an optical micrograph of neutron-irradiated HOPG recovered after the shock compression. The shock pressure determined by the flyer velocity is 46 GPa. We observed an occurrence of fracture of the graphite foils into platelets, similar to the previous studies [5,6]. A platelet is seen in the center of Fig. 2. A bright square area beside the platelet is gold foil, from which a platelet has fallen off. The colour of the lower part of the platelet is seen to be brighter than that of the upper part.

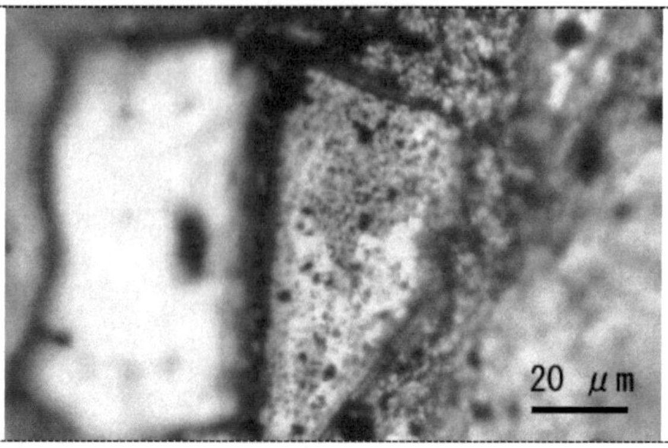


**Figure 2.** Optical micrograph of neutron-irradiated HOPG recovered after the shock compression at 46 GPa. A platelet can be seen in the center of the figure. The colour of the lower part of the platelet is brighter than that of the upper part, suggesting an occurrence of some transparency in the lower area. A bright square area in the left hand side of the platelet corresponds to gold foil, from which a platelet has fallen off.

Figure 3 shows Raman spectra for the neutron-irradiated HOPG specimen and for the platelet shown in Fig. 2, which is obtained after the shock compression at 46 GPa. The Raman measurements were done in the same condition to compare the relative intensity. The Raman intensity is seen to increase after the shock compression due to photoluminescence, of which intensity in the lower part of the platelet is higher than that in the upper part, suggesting some fluctuation on the transformation. The G and D bands are seen to remain after the shock compression.

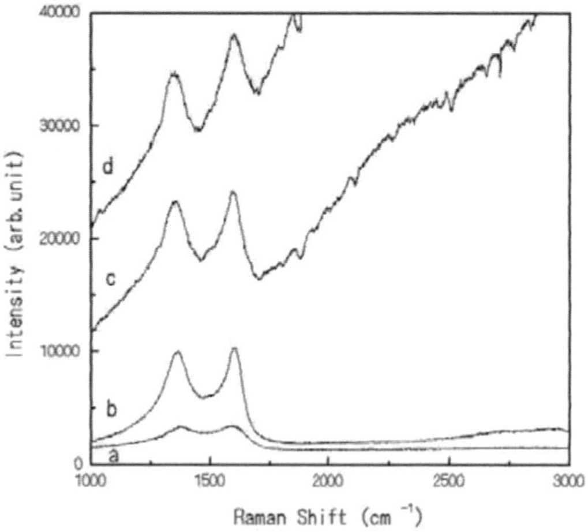


**Figure 3.** (a) Raman spectrum for the neutron-irradiated HOPG specimen. (b) Raman spectrum, obtained after the shock compression at 46 GPa, at upper part of the platelet in Fig. 2. (c), (d) Raman spectra obtained at lower part of the platelet in Fig. 2.

Figure 4 shows optical micrographs of a platelet of 100 μm in length, which was obtained after the shock compression at 54 GPa. Photos (a) and (b) are respectively focused on the surface of gold foil and the recovered specimen. The platelet is seen to be transparent.

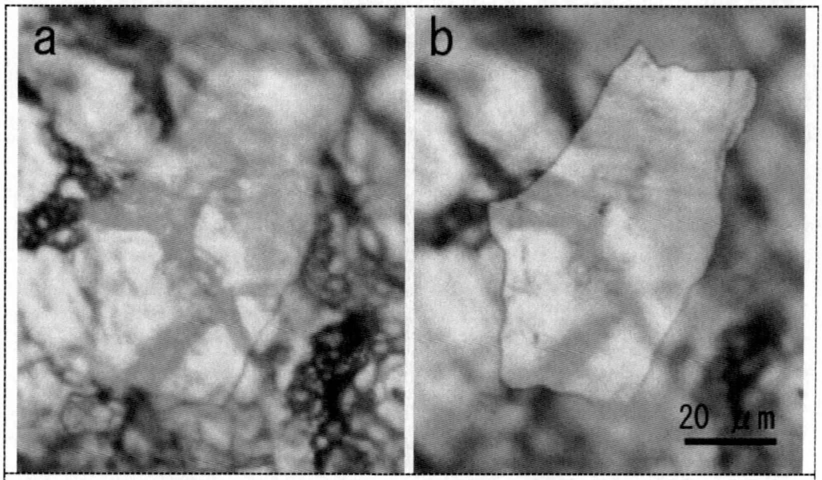


**Figure 4.** Optical micrographs of neutron-irradiated HOPG recovered after the shock compression at 54 GPa. (a) Focused on the gold foil. (b) Focused on a platelet of the recovered specimen.

Figure 5 shows Raman spectra obtained after the shock compression at 54 GPa. Raman spectrum for the neutron-irradiated HOPG specimen is shown at the bottom of the figure for comparison. Intensity of the Raman spectra is seen to remarkably increase after the shock compression. The increase of the background intensity has been seen for the amorphous diamond synthesized from $C_{60}$ fullerene and neutron-irradiated graphite by shock compression, and is due to photoluminescence, probably originating in an existence of defective $sp^3$ bondings [6]. A small and broad peak denoted by arrows can be seen at about 1330 $cm^{-1}$, implying the formation of nano-diamond. We did another shock compression at 51 GPa and found transparent platelets with remarkable photoluminescence. There are slight G and D peaks but no diamond peak, implying the formation of amorphous diamond.

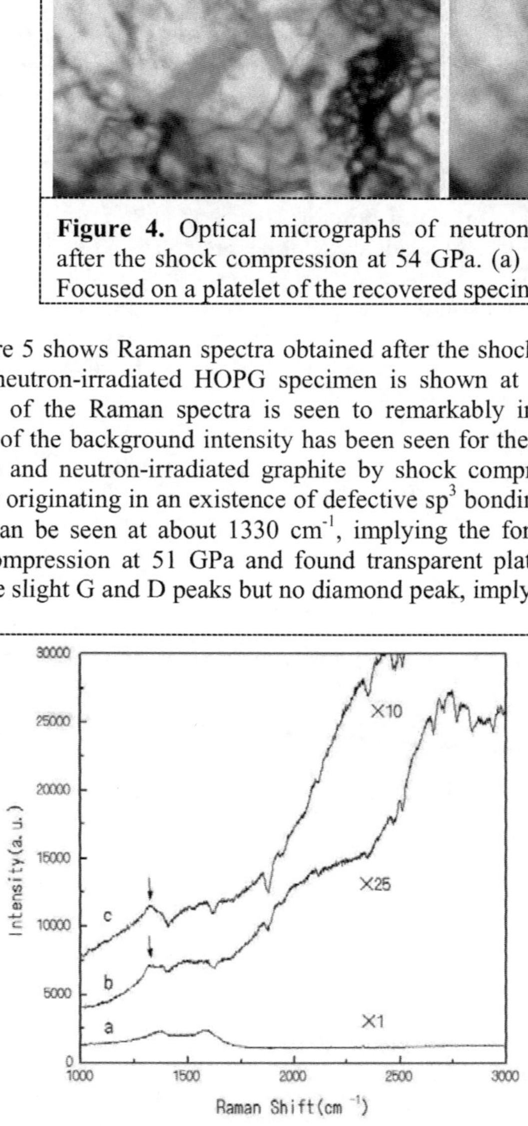


**Figure 5.** (a) Raman spectrum for the neutron-irradiated HOPG specimen. (b), (c) Raman spectra, obtained after the shock compression at 54 GPa, from the platelet shown in Fig. 2. Intensity of the Raman spectra significantly increases after the shock compression due to photoluminescence. A broad peak can be seen at about 1330 $cm^{-1}$, suggesting the formation of nano-diamond.

The neutron irradiated specimen utilized in the present study has been proved to be in a disordered state by means of X-ray diffraction and high resolution transmission electron microscopy [7]. The X-ray profile was highly asymmetrical and the TEM analysis showed breaking and bending of the 0002 lattice fringes which were found in other irradiation studies [8-10]. The result approves the present result of broad Raman spectra shown in fig. 1 and the dislocation accumulation model [11] which attributed the irradiation-induced amorphization to the accumulation of dislocation dipoles, inducing the buckling and the rotation of basal planes. We have proposed a tentative scenario on the synthesis of amorphous diamond platelets from the neutron irradiated HOPG under shock compression [5] as follows. The disordered layered sequences induced by irradiation prevent the martensitic transformation from graphite to diamond under shock compression. Defective sites relating to Wigner defects make a high density of diamond nucleation sites, leading to the synthesis of the amorphous diamond platelets. There remains a question on the kind of defect, relating to the transformation. To clarify the effect of defects on the transformation, it may be interesting to utilize lower damaged graphite specimens before the onset of irradiation-induced amorphization, which possess the layered sequence of the basal planes.

## 4. Conclusions

A novel pathway for the transformation from thin foils of neutron irradiated graphite into amorphous diamond platelet was confirmed with a neutron dose of about a half of the previous study. The Raman spectrum of the neutron irradiated graphite shows a broad amorphous one but is in a less disordered state, compared to the specimen used in the previous study. The graphite foils were shock compressed at 46, 51 and 54 GPa. At 46 GPa, we find the G and D peaks, and additional photoluminescence. At 51 GPa, we obtained transparent platelets with a strong photoluminescence but no diamond peak, suggesting the transformation to amorphous diamond. At 54 GPa, we find transparent platelets with strong photoluminescence. Also, a weak peak at about 1330 $cm^{-1}$ appeared, suggesting the formation of nano-crystallized diamond. Thus, the combined method of irradiation and shock compression should be a promising way to synthesize a various state of diamond such as amorphous diamond and nanodiamond.

## Acknowledgements

This research was partly supported by Collaborative Research Project of Materials and Structures Laboratory, Tokyo Institute of Technology and by JSPS Grant-in Aid (No. 23560787).

## References

[1] Irifune T, Kurio A, Sakamoto S, Inoue T and Sumiya H 2003 *Nature* **421** 599
[2] Hirai H, Kondo K, Yoshizawa N and Shiraishi M 1994 *Appl. Phys. Lett.* **64** 1797
[3] Hirai H, Tabira Y, Kondo K, Oikawa T and Ishizawa N 1995 *Phys. Rev. B* **52** 6162
[4] Erskine D J and Nellis W J 1991 *Nature (London)* **349** 317
[5] Niwase K, Nakamura K G, Yokoo M, Kondo K and Iwata T, 2009 *Phys. Rev. Lett.* **102** 116803
[6] Niwase K, Homae T, Nakamura K G and Kondo K 2006 *Physica (Amsterdam)* **376B–377B** 280
[7] Asthana A, Matsui Y, Yasuda M, Kimoto K, Iwata T and Ohshima K, 2005 *J. Appl. Crystallogr.* **38** 361. The neutron dose for JR and ICM19 samples should be revised to $3\times10^{19}$ n/cm$^2$ and $1.4\times10^{20}$ n/cm$^2$, respectively.
[8] Tanabe T, Muto S and Niwase K 1992 *Appl. Phys. Lett.* **61** 1638
[9] Koike J and Pedraza D F, 1994 *J. Mater. Res.* **9** 1899
[10] Karthik C, Kane J, Butt D P, Windes W E and Ubic R 2011 *J. Nucl. Mater.* **412** 321
[11] Niwase K, *Philos. Mag. Lett.* 2002 **82** 401; *Mater. Sci. Eng. A* 2005 **400–401** 101

Journal of Physics: Conference Series **377** (2012) 012027 doi:10.1088/1742-6596/377/1/012027

# Influence of Cu-site substitution on $La_2Ca_1Ba_2Cu_5O_z$ superconducting system

**J A Bhalodia[1], S R Mankadia[1], S M Dalsaniya[1] and M R Gonal[2]**
[1] Department of Physics, Saurashtra University, Rajkot – 360 005, India
[2] Material Processing Division, BARC, Mumbai – 400 085, India

E-mail: jabrajkot@rediffmail.com

**Abstract.** We have prepared a series of $La_2CaBa_2Cu_{5-x}Co_xO_z$; x = 0.0, 0.1, 0.2, 0.5, 1.0, 2.0, 5.0 (La-2125) compounds by the standard solid state reaction method and characterized for their structural, superconducting, magnetic properties and oxygen content through X-ray diffraction, scanning electron microscopy, d. c. resistivity, magnetic susceptibility and iodometric titration respectively. All the compounds crystallize with the tetragonal $LaBa_2Cu_3O_z$ type structure, space group P4/mmm. Here the effect of higher Co substitution for Cu in the $La_2CaBa_2Cu_{5-x}Co_xO_z$ system has been studied. It is observed that only 2 at. % Co substitution for Cu destroys the superconductivity of the sample. For heavily doped samples (with $x \geq 0.1$) are found non-superconducting presumably because of magnetic pair- breaking effect. These samples do not show superconductivity but are of interest for understanding the interplay between superconductivity and magnetism. Possible reasons for destruction of superconductivity are discussed in this communication.

## 1. Introduction

Since the discovery of high-$T_c$ cuprate superconductors [1], to understand the mechanism of superconductivity in high-$T_c$ oxides has been one of the most interesting research topics. Structurally there are many derivatives of R-123, made up by cation substitutions in the parent compound. Particular among them for the Ca-doped $LaBa_2Cu_3O_{7-\delta}$ compounds, where for a certain doping level of Ca, superconductivity appears along with the structural transition from orthorhombic to tetragonal [2]. The $La_2Ca_1Ba_2Cu_5O_z$ (La-2125) compound is obtained by the simultaneous addition of CaO and CuO to the non-superconducting $La_2Ba_2Cu_4O_z$ which has a tetragonal $LaBa_2Cu_3O_z$ (La-123) structure. The addition of CaO and CuO at the same time results in a superconducting compound with a $T_c$ of 79 K [3], which also has a structure similar to tetragonal La-123.

Substitutions at various crystallographic sites in the triple perovskite Y-123 structure has been a very popular approach of understanding the mechanism of superconductivity in these high temperature superconductors. The effect of substitution by various elements on the physical properties of $RBa_2Cu_3O_{7-\delta}$ (R=Y or rare- earth ion) superconductors have been studied to investigate the origin of superconductivity in such compounds. The $T_c$ of $RBa_2Cu_3O_{7-\delta}$ is not sensitive to rare-earth elements R [4]. Partial rare-earth substitution for La in $La_{2-x}R_xCa_{2x}Ba_2Cu_{4+2x}O_z$ also exhibits almost no effect on superconductivity [5], similar to the observations on rare-earth substituted R-123 systems [4], while the substitution of $Fe^{3+}$, $Co^{3+}$, $Ni^{2+}$, $Zn^{2+}$, $Hf^{4+}$, $Mo^{4+, 6+}$ etc. at the Cu site in R-123 suppresses the $T_c$ [6, 7, 8].

The valence state of the Cu ion is known to be higher than +2. Since a $Co^{3+}$ ion can be in either an S = 0 or 2 state, the substituted $Co^{3+}$ ion will work for removing spins from the two-dimensional $CuO_2$ planes and this may reduce the short-range correlations among $Cu^{2+}$ ions, having $T_c$ decrease as the Co content increases. Keeping mentioned facts in mind, we have initiated a systematic study of such kind of higher doping of Co at Cu site in La-2125 system for verifying the effect of magnetic Co using ρ-T, M(T) and grain growth measurements. Some systems like $Y_1Ba_2(Cu_{1-x}Co_x)_3O_{7-\delta}$ and $Bi_2(Sr_{0.6}Ca_{0.4})_3(Cu_{1-x}Co_x)_2O_z$ are superconducting even when the Co content x is as high as 10 at. %. Therefore, the mechanism for annihilation of superconductivity in $La_2Ca_1Ba_2Cu_{5-x}Co_xO_z$ with x = 0.1 - 5.0 samples should be different from these two systems [7, 9].

## 2. Experimental

All the samples of $La_2Ca_1Ba_2Cu_{5-x}Co_xO_z$ series (x = 0.0, 0.1, 0.2, 0.5, 1.0, 2.0, 5.0) were prepared by standard solid state reaction method. The high purity (>99.9%) starting compounds of $La_2O_3$, $CaCO_3$, $BaCO_3$, CuO, $CoC_2O_4 \bullet 2H_2O$ were taken in the stoichiometric quantities and thoroughly ground under acetone using agate and mortar. The mixture was calcined at 940°C for 24h. The resultant black powder was then palletized and sintered at 950°C for 24h in muffle furnace and slowly cooled to room temperature. The sintered samples were then annealed at 500°C in flowing oxygen gas for 24h and then slowly cooled (1°C/min) to room temperature.

The structure and phase purity of synthesized samples were checked by powder X-ray diffraction (XRD). A JEOL JSM 5600 scanning electron microscope (SEM) was used to study the microstructure of the ceramic samples. The chemical composition was confirmed by energy dispersive analysis of X-ray (EDAX) for all the samples. Electrical resistivity was measured as a function of temperature (5-300 K) with the standard four-probe technique. Electrical contacts were made with silver paste. Magnetic susceptibility was measured as a function of temperature between 5 to100 K. The oxygen content was determined by iodometric double titration method.

## 3. Results and Discussion

Figure 1 shows the Rietveld refined and indexed XRD patterns for the $La_2Ca_1Ba_2Cu_{5-x}Co_xO_z$ (x = 0.0, 0.1) system. The phase purity and structural characteristics of samples were confirmed by Rietveld analysis using the FULLPROF program on XRD data (λ = 1.5412 Å), based on the refined Wyckoff positions at room temperature. The pseudo-voight function was used to analyze the peak profiles. The analysis of XRD pattern shows that x = 0.0 sample exhibit tetragonal structure with the space group P4/mmm. All the samples retain tetragonal structure without transforming to the orthorhombic structure, whereas structural transition from orthorhombic to tetragonal have been observed for the Co-doped Y-123 at a Co concentration of x ~ 0.025 [10]. For x = 0.1 and above samples, small amount of secondary phase is detected and may be identified as La-Cu-Co compound phase. The amount of the secondary phases increased with increasing Co-doping. Tellez at el. found that the Co, Ga, Ni substitution takes place isostructurally at Cu-site in $CaLaBaCu_3O_7$ compound at least up to 8 at. % of the substituent concentration in the material, where as in La-2125 limit is below 2 at.% [11].

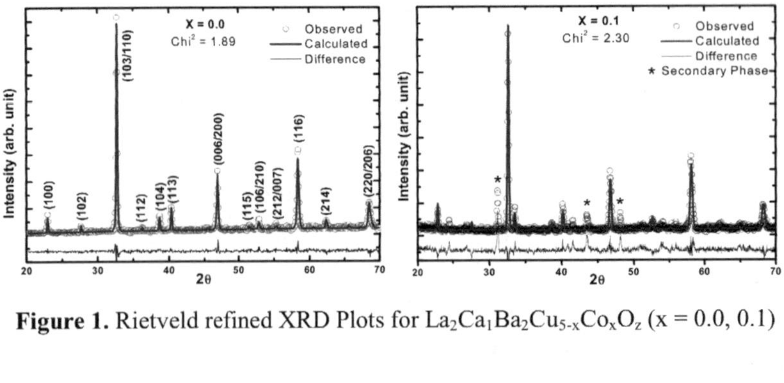


**Figure 1.** Rietveld refined XRD Plots for $La_2Ca_1Ba_2Cu_{5-x}Co_xO_z$ (x = 0.0, 0.1)

Table 1 shows values for the lattice constants – *a, c* and the unit cell volume *V*, with respect to the Co content x. For the lower substitution (x = 0.1 - 2.0), the value of *a*, *c*, and *V* are remains almost level off to constant but increases for the higher substitution (x = 5.0). The increment in *a* and *V* for the higher substitution suggests $Co^{3+}$ ion is in its high-spin configuration state. This is due to the ionic radius of $Co^{3+}$ ion for high-spin state (r = 0.061nm) is larger than that of $Cu^{+2}$ ion(r=0.057nm) [12].

**Table 1.** Values of *a* = *b, c, V*, X-ray density and oxygen content z for $La_2Ca_1Ba_2Cu_{5-x}Co_xO_z$ (x = 0.0 - 5.0).

| Concentration (x) | Unit cell parameters a = b (A°) | c (A°) | (V) (A°)$^3$ | X-Ray Density (g/cm$^3$) | Oxygen Content (z) |
|---|---|---|---|---|---|
| **0.0** | 3.869 | 11.641 | 174.21 | 10.46 | 11.75 |
| **0.1** | 3.931 | 11.635 | 179.79 | 10.09 | 11.45 |
| **0.2** | 3.923 | 11.614 | 178.73 | 10.12 | 11.29 |
| **0.5** | 3.899 | 11.392 | 173.18 | 10.48 | 11.54 |
| **1.0** | 3.882 | 11.359 | 171.17 | 10.54 | 11.32 |
| **2.0** | 3.909 | 11.381 | 173.90 | 10.35 | 11.42 |
| **5.0** | 5.450 | 13.680 | 406.33 | 4.38 | 11.54 |

Figure 2 shows the SEM micrographs for x = 0.1, 1.0, 5.0. The microstructure of the heat-treated pellets shows a uniform grain growth but does not show dense packing of individual particles during heat treatments. The homogeneous packing of tiny crystallites aggregates favours a high sintering activity in the samples. It is also observed that the major phase in the samples is La-2125 with well shaped grains. Microstructure of x = 5.0 shows comparatively smaller grain size than the x = 0.1 sample and segregation is observed. The staff gauge is 1μm.

**Figure 2.** The scanning electron micrograph of the samples $La_2Ca_1Ba_2Ca_{5-x}Co_xO_z$ (x=0.1, 1.0, 5.0).

Figure 3 shows the temperature dependence of resistivity of the $La_2Ca_1Ba_2Cu_{5-x}Co_xO_z$ (x = 0.0, 0.1, 0.5, 5.0) samples from 5 K to 300 K in the warming process. Resistivity for x = 0.0 shows $T_c^{onset}$ at about 79 K. For x = 0.1 – 5.0, resistivity of each of the Co-doped samples increases with decreasing temperature and shows no superconducting transition between 5 and 300 K, even when the Co content x is as low as 0.1. Here, only 2 at. % Co substitution for Cu in La-2125 system destroys superconductivity.

The magnetic property is measured for magnetic moment as a function of temperature. As shown in figure 4(a), for x = 0.0, the magnetic susceptibility measured at field of 500 Oe in the zero field cooled (ZFC) and field cooled (FC) states. As superconducting nature is observed in resistivity measurement for this sample, it exhibits diamagnetic signals near to transition temperature. FC-ZFC fork is also observed near transition temperature. Figure 4(b) shows the substitution of 2 at. % Co at Cu site results in the onset of magnetism in this compound. The plot of M vs T exhibits a broad transition from paramagnetic to antiferromagnetic state around 20 K. This fact indicates that in La-2125 compound, the substitution of a magnetic ion ($Co^{3+}$) at Cu site results in annihilation of superconductivity and at the same time results in the onset of magnetism. Oxygen content was determined through iodometric double titration method and values are listed in table 1.

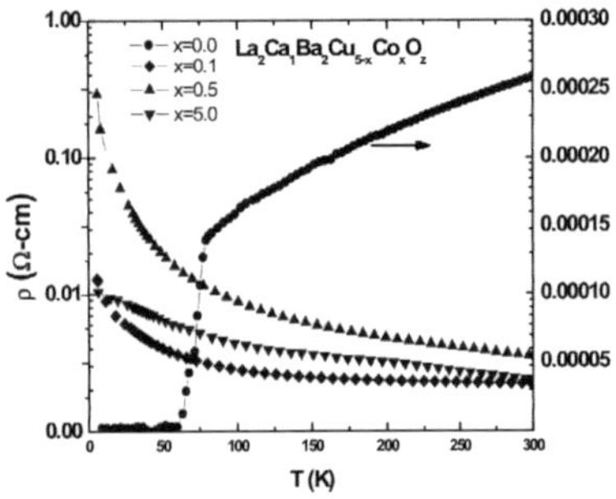


**Figure 3.** Resistivity Vs temperature curves for x = 0.0, 0.1, 0.5 and 5.0 samples.

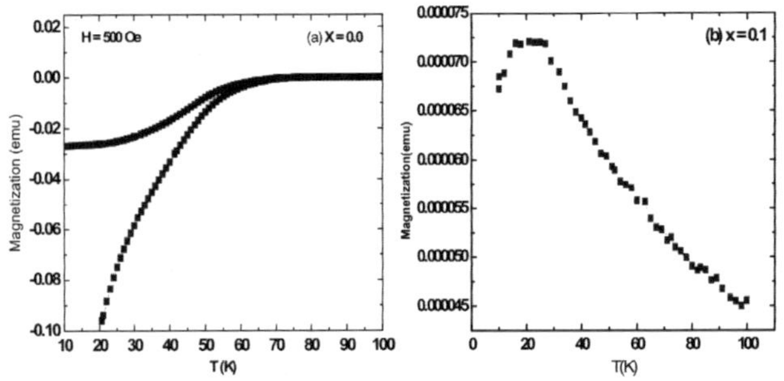


**Figure 4.** Magnetic susceptibility Vs. temperature for the samples $La_2Ca_1Ba_2Ca_{5-x}Co_xO_z$ (x= 0.0, 0.1).

## 4. Conclusions

The effect of substitution of 3d metal Co for Cu in a high temperature superconductor La-2125 was investigated by measuring lattice constant, dc resistivity, magnetic susceptibility, iodometric titration in normal and superconducting states. Superconducting state was observed only in pure copper compound. Only 2 at. % Co substitution at Cu-site in $La_2Ca_1Ba_2Cu_{5-x}Co_xO_z$ samples destroys superconductivity. Such annihilation of superconductivity presumably because of large local paramagnetic moments in the Co element which results into pair braking and high spin state of the Co ions may cause such destruction. However, at the same time the 2 at. % Co at Cu site induces the magnetism in this compound.

## 5. Acknowledgments

Authors from S. U. are thankful to the departmental UGC DRS-SAP & DST-FIST programmes for providing some basic facilities. Part of this work was performed at UGC-DAE Consortium for Scientific Research, Indore. Help from Dr. G. S. Okram, Dr. A. Banerjee and Mr. Ajay Soni from CSR, Indore is thankfully acknowledged.

## 6. References

[1] Bednorz J and Muller K 1986 *Z. Phys. B.* **64** 189
[2] Gu H, Fung K, Wei C, Zhang J, and Yin D 1993 *Physica C* **208** 391
[3] Mankadia S, Dalsaniya S, Okram G, Igalwar P, Gonal M and Bhalodia J 2011 *Proc. Of 55$^{th}$ DAE SSPS 2010 (Manipal)* 1349 (American Institute of Physics) p 893
[4] Hor P, Meng R, Wang Y, Gao L, Huang Z, Bechtold J, Forster K and Chu W 1987 *Phys. Rev. Lett.* **58** 1891
[5] Rayaprol S, Mavami K, Rana D, Thakar C, Thampi R, Kuberkar D, Kulkarni R and Malik S 2002 *J. Supercond.* **15** 211
[6] Kulkarni R, Kuberkar D, Baldha G, Bichile G 1993 *Physica C* **217** 175
[7] Maeda A, Yabe T, Takebayashi S, Hase M and Uchinokura K 1990 *Phys. Rev. B* **41** 4112
[8] Bhalodia J, Shaikh A, Kuberkar D, Baldha G and Kulkarni R 1994 *Mater. Res. Bull.* **29** 89
[9] Sonntag R, Hohlwein D, Hoser A, Prandl W, Schifer W, Kiemel R, Kemmlersake S, Losch S, Schlichenmaier M and Hewat A 1989 *Physica C* **159** 141
[10] Wu X, Pan F, Huang X, Ye C, Jin X and Jiang S 1996 *J. Mater. Sci.* **31** 6113
[11] Tellez D and Aguiar J 2001 *J. Magn. Magn. Mater.* **226** 318
[12] Shannon R 1976 *Acta Crystallogr. Sect. A.* **32** 751

Journal of Physics: Conference Series **377** (2012) 012028 doi:10.1088/1742-6596/377/1/012028

# Bond strengths of New Carbon-nitride-Related material $C_2N_2(CH_2)$

**M Sougawa[1], T Sumiya[1], K Takarabe[1], Y Mori[1], T Okada[2], H Gotou[2], T Yagi[2], D Yamazaki[3], N Tomioka[3], T Katsura[3*], H Kariyazaki[4], K Sueoka[4] and S Kunitsugu[5]**

[1] Okayama University of Science, 1-1 Ridai, Okayama 700-0005, Japan

[2] The Institute for Solid State Physics, University of Tokyo, Kashiwanoha, Chiba 277-8581, Japan

[3] The Institute for Study of the Earth's Interior, University of Okayama, Yamada, Tottori 682-0193, Japan

[4] Okayama Prefectural University, Soja, Okayama 719-1197, Japan

[5] The Industrial Technology Center of Okayama Prefecture, Haga, Okayama 701-1296, Japan

E-mail: s05sm06@std.das.ous.ac.jp

**Abstract.** A new carbon-nitride-related material $C_2N_2(CH_2)$ nanopletelet was synthesized by subjecting a precursor $C_3N_4H_xO_y$+Au in a laser-heating diamond anvil cell (LHDAC) to the pressure of 40 GPa and the temperature of 1200-2000 K. The synthesized $C_2N_2(CH_2)$ was accordingly found to be an orthorhombic unit cell of the space group $Cmc2_1$ with lattice constants $a = 7.625$Å, $b = 4.490$Å, and $c = 4.047$Å. The bulk modulus $B_0$ was determined to be $B_0 = 258 \pm 3.4$ GPa, only the 60 % that of the diamond. $C_2N_2(CH_2)$ consists of the tetrahedrally coordinated C with three C-N single bond and the one C-C single bond, and the bridging carbon with the C-$CH_2$-C bond. The C-N single bond length of the tetrahedron ranges from 1.444 to 1.503 Å. This bond length is close to the C-N single bond of 1.447 to 1.458 Å in the superhard β-$C_3N_4$. The compressibility of the C-N and C-C single bond of $C_2N_2(CH_2)$ ranges from 0.976 to 0.982 with the pressure of 30 GPa. These values are very close to the compressibility of the C-N and C-C single bond of 0.978 to 0.982 in β-$C_3N_4$, cubic-$C_3N_4$, and diamond.

## 1. Introduction

The dense carbon nitride $C_3N_4$ with the C-N single bond was predicted as incompressible as diamond [1], [2]. Its bulk modulus is comparable to or lager than that of diamond. The superhard polymorphism of α-, cubic-, and pseudcubic-$C_3N_4$ were also predicted [3]. This incompressibility is due to the bond strength of the C-N single bond shorter in the bond length than that of the C-C single bond in the diamond. No experimental investigations have been reported on the bond strength of the C-N single bond. Recently, H-Bordon et al. synthesized the orthorhombic $C_2N_2(NH)$ with the space group $Cmc2_1$ at high pressure and high temperature [4]. The chemical bond of $C_2N_2(NH)$ shows that this mainly consists of the tetrahedorally coordinated carbon with four C-N single bond and the C-N(H)-C

[*] Present address: Bayerishes Geoinstitut, University Bayreuth, D-95440 Bayreuth, Germany

bridging bond. $C_2N_2(NH)$ is the first material obtained having the definite C-N single bond expected in the superhard $C_3N_4$. The crystal structure of $C_2N_2(NH)$ is analogous to those of $Si_2N_2(NH)$ and $Si_2N_2O$, which have the Si-N single bond and the Si-N(H)-Si or Si-O-Si bridging bond [5]. The calculated bulk modulus of $C_2N_2(NH)$ is 277 GPa, only 60 % that of diamond. This moderate bulk modulus results from the existence of the C-N(H)-C bridging bond.

A new carbon-nitride-related material $C_2N_2(CH_2)$ was synthesized at high pressure and temperature [6]. This material consists of the tetrahedrally coordinated carbon with three C-N single bond and one C-C bridging bond, and the bridging carbon with the C-$CH_2$-C bond. In this work the bond strength of $C_2N_2(CH_2)$ was conducted by using the high pressure XRD and the first-principles calculation.

## 2. Experimental

$C_2N_2(CH_2)$ was synthesized by subjecting a precursor $C_3N_4H_xO_y$ nanoparticles + Au in a laser-heating diamond anvil cell (LHDAC) to the pressure of 40 GPa and the temperature of 1200-2000 K. A gold (Au) is laser absorbent material. Heating was carried out with the YAG laser for about ten minutes. The temperature was evaluated from the radiation spectrum. The high pressure XRD experiments on synthesized $C_2N_2(CH_2)$ were performed in a diamond anvil cell (DAC) up to the pressure of 30 GPa. The experimental XRD lines of $C_2N_2(CH_2)$ was recorded using a synchrotron radiation source. The $C_2N_2(CH_2)$ was placed in 120 μm diameter holes, which had been drilled in Rhenium gaskets pre-indented to a thickness of about 50 μm. The methanol and ethanol mixture of 4:1 was used as a pressure-transmitting medium. The pressure was determined by the absorbent material (Au) scale put at the same position with the $C_2N_2(CH_2)$.

## 3. Result and discussion

$C_2N_2(CH_2)$ shown in Figure 1 was found to be an orthorhombic unit cell of the space group $Cmc2_1$ with lattice constants $a = 7.625$ Å, $b = 4.490$ Å, and $c = 4.047$ Å. The tetrahedral unit C1-$N_3$C2 and the role of C2 bridging these tetrahedral units along the a-axis are known characteristic in this crystal. The C-N bond length of the C1-$N_3$C2 tetrahedron ranges from 1.444 to 1.503 Å. This bond length is close to the C-N single bond of 1.447 to 1.458 Å in β-$C_3N_4$.

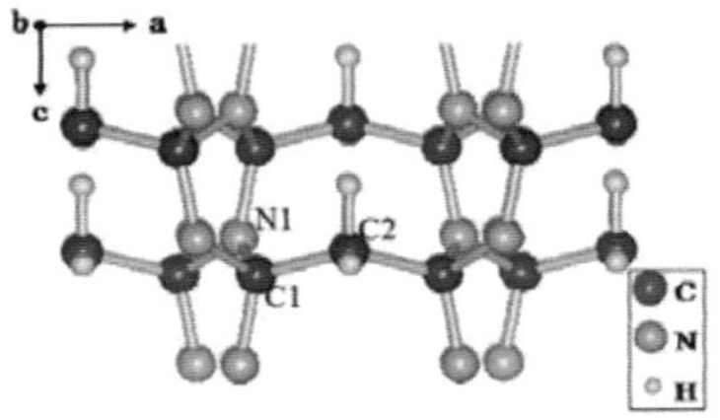


**Figure 1**. Crystal structure of $C_2N_2(CH_2)$.

Figure 2 shows the pressure dependence of the XRD profiles of the synthesized sample. The dot (•) indicates the XRD lines from the orthorhombic phase $C_2N_2(CH_2)$. The square (■) and the star (☆or ★) indicate the XRD lines from the absorbent material (Au) and the sub phases. The XRD profiles of $C_2N_2(CH_2)$ were recorded up to the pressure of 30 GPa and also lower pressure region as well. The relative intensity ratio of each diffraction lines of $C_2N_2(CH_2)$ were not change at high pressure. Therefore, no structural change occurs with the pressure of 30 GPa. The pressure dependence of the lattice constants and relative volume of the orthorhombic $C_2N_2(CH_2)$ was shown in Figure 3. The lattice constants were determined from the diffraction lines of the Miller indices, (110) (200) (111) (020) (310) (021) (311) obtained through measuring the diffraction lines under high pressure. The pressure-volume data of the orthorhombic $C_2N_2(CH_2)$ was fitted to the second-order Birch-Murnaghan equation of state. The bulk modulus of the $C_2N_2(CH_2)$ was determined to be $B_0 = 258 \pm 3.4$ GPa, only

60 % that of diamond. We reported the bulk modulus of $C_2N_2(CH_2)$ is to be tentatively $B_0$ = 374 GPa in the previous conference paper [6]. In the previous estimation, the experiment was done without a pressure transmitting medium and the data at each pressure scattered for each run. The previous value is corrected with this study. The bulk modulus of $C_2N_2(CH_2)$ is close to or slightly smaller than the theoretical value at of $C_2N_2(NH)$.

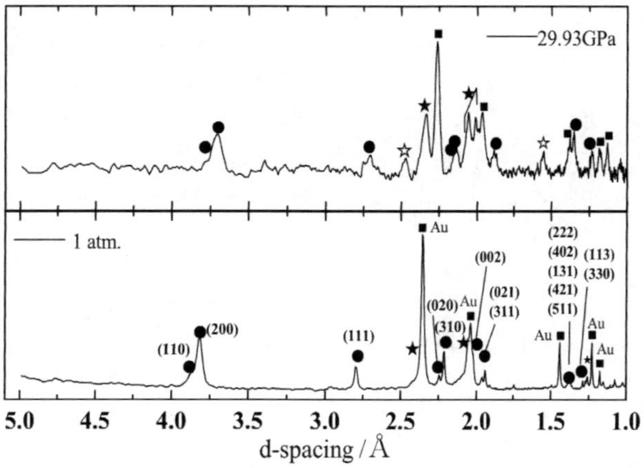


**Figure 2.** Pressure dependence of the XRD profile of the recovered sample. The dots (•) indicates the XRD lines from the $C_2N_2(CH_2)$ and the other XRD lines are sub phase and the absorbent material (Au).

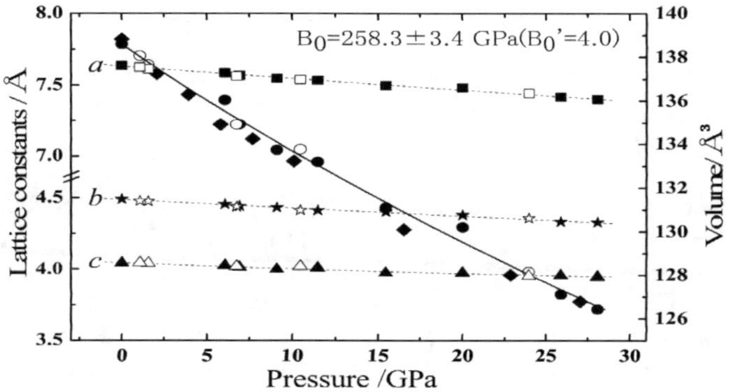


**Figure 3.** Pressure dependence of three lattice constants and the cell volume of $C_2N_2(CH_2)$. The closed symbols indicate the data taken upon compression, while the open symbols are for the data taken upon decompression. The bulk modulus of $C_2N_2(CH_2)$ was determined to be $B_0$ = 258±3.4 GPa with good reproducibility. The pressure was determined by the Au scale in the sample position.

The Rietveld analysis is applied to refine the crystal structural parameters from the diffraction patterns under high pressure but can not be applied to this experiment. This is because the intensity of the XRD profiles was very weak under high pressure. Then, the crystal structural parameters including the atomic positions of $C_2N_2(CH_2)$ were determined by the first-principles calculation, and the pressure dependence of the bond lengths of C-N and C-C single bond in $C_2N_2(CH_2)$ were determined by using the result of the optimized atomic positions. First-principles calculation based on the density-function theory and the local-density approximation (LDA) was performed using the CASTEP code [7]. The lattice constants obtained from the XRD measurement which were not change, and only the atomic positions were geometrically optimized. The compressibility of the C-N single bond of the tetrahedral unit C1N$_3$-C2 of $C_2N_2(CH_2)$ ranges from 0.980 to 0.982. In the case of the C-C single bond of the bridging unit C1-C2H$_2$-C1 of $C_2N_2(CH_2)$, the compressibility of the C-C single bond is 0.976. These values are very close to the compressibility of the C-N and C-C single bond of 0.978 to 0.982 in $\beta$-$C_3N_4$, cubic-$C_3N_4$, and diamond. The bond strength of C-N single bond of the tetrahedron unit C1N$_3$-C2 of $C_2N_2(CH_2)$ is the same as those of $\beta$-$C_3N_4$, cubic-$C_3N_4$, and diamond. The bond strength of the C-C single bond of the bridging unit is also strong. These results are compiled in Table 1.

**Table 1.** Comparison of the compressibility of the C-N and C-C single bond at 1 atm and 30 GPa for $C_2N_2(CH_2)$, hypothetical superhard $\beta$-$C_3N_4$, cubic-$C_3N_4$ and diamond.

| | | $C_2N_2(CH_2)$ | $\beta$-$C_3N_4$ | Cubic-$C_3N_4$ | Diamond |
|---|---|---|---|---|---|
| Bond length at 30 GPa and 1 atm (the ratio of bond length at 30 GPa with that at 1 atm.) | C-N | 1.418/1.445<br>(0.981)<br>1.429/1.455<br>(0.982)<br>1.434/1.464<br>(0.980) | 1.428/1.457<br>(0.980)<br>1.427/1.458<br>(0.979)<br>1.428/1.459<br>(0.979)<br>1.437/1.469<br>(0.978) | 1.444/1.473<br>(0.980) | 1.513/1.540<br>(0.982)<br>(C-C) |
| | C-C | 1.445/1.480<br>(0.976) | - | - | - |

According to the latest study, the bond angles of C1-N1-C1 of the tetrahedral unit C1N$_3$-C2 and C1-C2-C1 of the bridging unit C1-C2H$_2$-C1 decrease by 2 and 3 % with the pressure of 30 GPa. When the bond angle of C1-C2-C1 of the bridging unit of C1-C2(H$_2$)-C1 decrease under high pressure, two tetrahedral units C1N$_3$-C2 connected with the bridging site C2(H$_2$) were slightly rotated in opposite directions. This is the reason why the bulk modulus of $C_2N_2(CH_2)$ is approximately 60 % of diamond.

## 4. Summary

The bond strength of $C_2N_2(CH_2)$ was evaluated by using the high pressure XRD measurement and the first-principles calculation. The compressibility of the C-N single bond of the tetrahedron unit C1N$_3$-C2 of $C_2N_2(CH_2)$ is equivalent to the compressibility of the C-N single bond of superhard $C_3N_4$. However the bulk modulus of $C_2N_2(CH_2)$ was re-determined to be 258 GPa, only 60 % that of the superhard $C_3N_4$. This is because the bond angle of the bridging unit C1-C2H$_2$-C1 is susceptible to the pressure.

## References

[1] Liu A Y and Cohen M L 1989 *Science* **245** 841
[2] Liu A Y and Cohen M L 1990 *Phys. Rev. B* **41** 10727
[3] Teter D M and Hemly R J 1996 *Science* **271** 53
[4] Harvath-Bordon E, Riedel R, McMillan P F, Kroll P, Miehe G, van Aken P A, Zerr A, Hoppe P, Shebanova O, McLaren I, Lauterbach S, Kroke E and Boehler R 2007 *Angew. Chem. Int. Ed.* **46** 1476
[5] Peters D and Jacobs H 1989 *J. Less-Common Met.* **146** 241
[6] Sougawa M, Sumiya T, Takarabe K, Mori Y, Okada T, Gotou H and Yagi T 2010 *J. Phys.: Conf. Ser.* **215** 012137
[7] The CASTEP code is available from Accelrys Software Inc.

# Unusual pressure response of electronic transport properties of a Kondo insulator $CeRu_4Sn_6$

**K Sengupta[1,*], K K Iyer[2], R Ranganathan[1] and E V Sampathkumaran[2]**
[1]Saha Institute of Nuclear Physics, 1/AF Bidhannagar, Kolkata 700064, India
[2]Tata Institute of Fundamental Research, Homi Bhabha Road, Colaba, Mumbai 400005, India

E-mail: ksengupta11@gmail.com

**Abstract.** We report the effect of hydrostatic pressure (up to 20.6 kbar) on the electrical resistivity (1.8 K to 300 K) of a Kondo insulator $CeRu_4Sn_6$, crystallizing in a tetragonal structure with space group *I-4 2m*. The absolute value of resistivity is found to increase monotonically with pressure in the entire temperature range. The hybridization gap, contrary to common observation on other Ce systems, is also found to increase up to 15.1 kbar. Magnetoresistance under pressure appears interesting. For instance, at 1.8 K and at 15.1 kbar, MR is negative till about 100 kOe, showing a minimum around 60 kOe, subsequently showing a sign change; magnetoresistance becomes positive at and above 10 K and at same pressure. These results indicate subtle changes induced by magnetic field on the hybridization gap.

## 1. Introduction

Kondo insulators (KI) have recently attracted much attention because of their unusual physical properties [1]. The electronic structure of KI is thought to derive from the coherent hybridization of 4*f* orbital in a periodic lattice of Kondo ions. In such systems, a small gap/pseudo-gap develops at low temperature in the density of states at the Fermi level.

$CeRu_4Sn_6$ in this regard occupies a special place in this class of system. The Kondo insulating behaviour was first reported by Das and Sampathkumaran [2]. Recent NMR and heat capacity measurements at low temperatures reveal that this system is different from other known KI [3]. Measurements on single crystals showed pronounced magnetic anisotropy and a nodal/line gap has been proposed for this material [4]. It is important to note that while most of KIs form in cubic structure and a few in orthorhombic structure, only $CeRu_4Sn_6$ and $CeCuAs_2$ [5] are reported to form in a tetragonal structure. We considered it worthwhile to investigate how external pressure modifies the electronic transport properties of this compound.

## 2. Experimental details

The sample was prepared by arc melting and followed the same procedure as described in one of earlier works [2]. The electrical resistivity (ρ) measurements, using standard four-probe method, as a function of temperature (*T*) and magnetic-field (*H*) were performed employing a commercial clamp type piston-cylinder cell (easyLab Technologies Ltd, UK) with the help of physical property

* To whom any correspondence should be addressed.

measurements system (PPMS, Quantum Design). The measurements have been taken in a hydrostatic pressure medium using a 1:1 mixture of n-pentane and isopentane.

## 3. Results and discussions

Resistivity of $CeRu_4Sn_6$ at ambient pressure has already been reported in the earlier work [2]. While $LaRu_4Sn_6$ shows normal metallic properties, $CeRu_4Sn_6$ is nonmagnetic and behaves as an insulator from room temperature down to 50 mK [4]. Figure 1 shows resistivity at different applied pressures - 3.2, 6.6, 10.4, 15.1 and 20.6 kbar. It is to be noted that the absolute values are found to increase continuously with increasing pressure in the entire temperature range (see inset of figure 1). For instance, at 20.6 kbar, the ρ(T) at 1.8 and 300 K is enhanced nearly by an order of magnitude compared to that at ambient pressure, and the magnitude of increase is relatively larger at 1.8 K. We attribute it to enhancement of hybridization gap with increasing pressure in this system.

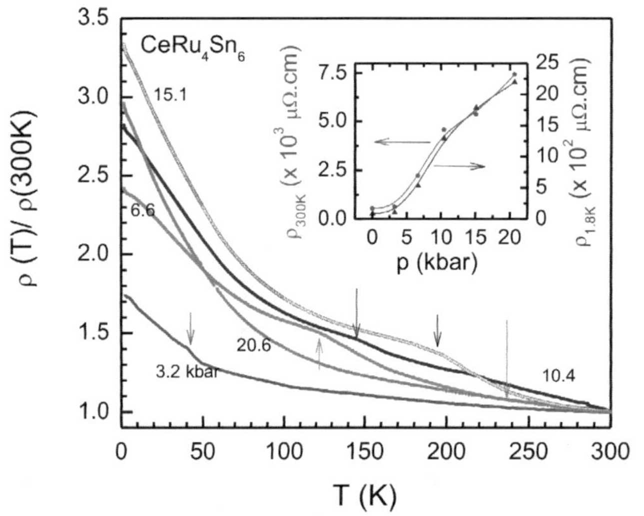


**Figure 1.** Normalized resistivity of $CeRu_4Sn_6$ at different pressures. The arrows in the figure indicate the shoulder. Inset shows variation of resistivity at 1.8 K and 300 K with applied pressure. The continuous lines through data points are guide to eyes.

Figure 2 is derived from figure 1. Figure 2a shows the relative change of ρ (1.8K) with respect to ρ (300K). The ratio increases up to 15.1 kbar, produces a maximum and subsequently starts to decrease. The maximum is marked by an arrow in figure 2a. An increase in the resistivity ratio suggests that the gap increases with pressure up to around the maximum at 15.1 kbar. An initial increase in gap with pressure was also reported for a few Kondo insulators, e.g., CeRhSb [6], but this finding is contrary to that observed in other tetragonal KI $CeCuAs_2$ [7]. The decrease in ρ(1.8K)/ρ(300K) above 15 kbar probably signifies that the gap tends to decrease at higher pressures.

Another novel feature is that ρ(T) develops a shoulder (marked by an arrow in the figure 1) which moves towards a higher temperature with increasing pressure. In figure 2b, the temperature $T_{max}$ where the shoulder of ρ(T) appears in figure 1 is plotted versus pressure. Though the exact origin of this is not clear, we believe that the shoulder arises from the Kondo coherence effect with the characteristic temperature moving up with pressure, as this kind of pressure dependence of Kondo coherence temperature is well-known for the Kondo lattices [6] and the low temperature upturn may be dominated by gap effect. It may be noted that the single-ion Kondo temperature ($T_K$) estimated from the paramagnetic Curie temperature, $\theta_P$, in this system ($T_K \approx |\theta_P|/4$ [8]) at ambient pressure is near 40 K and it is expected to increase with pressure. It is therefore reasonable to assume that the Kondo coherence temperature also increases with pressure.

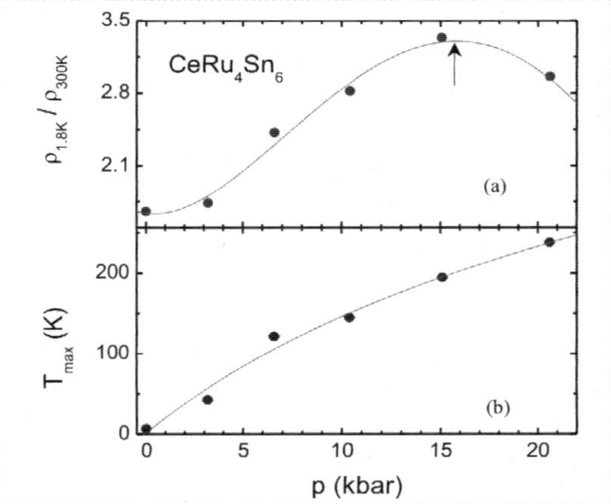


**Figure 2.** (a) The resistivity ratio at 1.8 K to 300 K and (b) variation of $T_{max}$ with pressure for $CeRu_4Sn_6$. The continuous lines through data are guide to eyes. The arrow in figure (a) indicates the maximum in the data.

Magnetoresistance (MR) at an intermediate pressure, 15.1 kbar, is shown in figure 3. The MR is defined as $[\rho(H) - \rho(H = 0)]/\rho(H = 0)$. MR at 10 K and 50 K are positive. At 50 K, it varies quadratically with field over a wide range of field (e.g., till 100 kOe), whereas quadratic variation is restricted to lower field range (till 20 kOe) at 10 K. Dashed lines through data shows $H^2$ fit. Positive MR behaviour is consistent with the Kondo coherence effects, while the deviation from the quadratic behaviour as the temperature is lowered well below 50 K implies interference from some other effects, namely, gap effect. There could be contributions to positive MR from Lorentz force on the conduction electrons. At 1.8 K, a puzzling non-monotonic variation of MR is observed. That is, MR is negative between 0 to around 100 kOe with a minimum around 60 kOe. MR changes sign above 100 kOe. This complicated behaviour at 1.8 K could possibly be related to modification of the hybridization gap and/or interference from Kondo coherence at high magnetic fields and the non-monotonic variation is quite unusual compared to the behaviour of other KIs reported so far [9] at ambient pressure. It is worth noting that MR of $CeRu_4Sn_6$ was also reported [10] to show unusual behaviour at ambient pressure. At 20.6 kbar, the qualitative behaviour of MR measured at 10 K and 50 K remains almost similar to that at 15.1 kbar (not shown here).

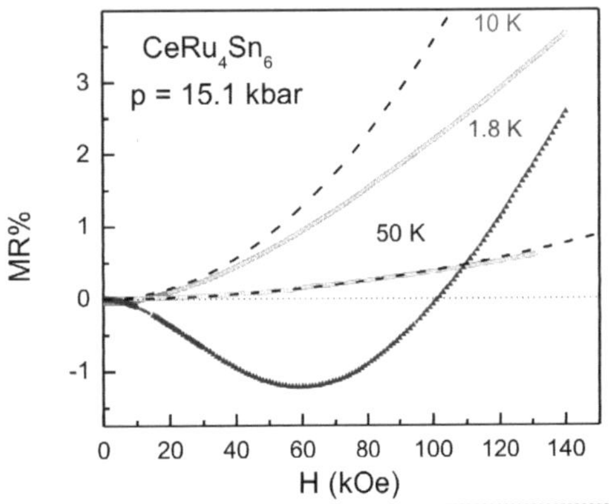


**Figure 3.** Magnetoresistance vs. magnetic fields of $CeRu_4Sn_6$ at different temperature for 15.1 kbar. Dashed lines show region of $H^2$ behaviour of MR. The dotted line indicates MR = 0 axis.

## 4. Summary

We have presented the results of transport and magneto-transport properties of a Kondo insulator $CeRu_4Sn_6$ under hydrostatic pressure. There is an enhancement of electrical resistivity with the application of pressure with the degree of enhancement being larger with decreasing temperature. Features attributable to interplay between the Kondo coherence effect and the gap effect could be seen in the high pressure data. While application of pressure causes a continuous suppression of gap in some compounds, e.g., CeNiSn [11], $SmB_6$ [12] or $CeCuAs_2$ [7] or an increase in gap in some others, e.g., $Ce_3Bi_4Pt_3$ [13]- the behaviour in the present system appears to resemble that in CeRhSb [6], a nonmonotonic response with pressure. It is of interest to study this system at further high pressures to see whether there is any insulator-metal transition in this system.

## 5. References

[1] Riseborough P 2000 *Adv. Phys.* **49** 257
[2] Das I and Sampathkumaran E V 1992 *Phys. Rev.* B **46** 4250
[3] Brüning E M, Brando M, Baenitz M, Bentien A, Strydom A M, Walstedt R E and Steglich F 2010 *Preprint cond-mat/1005*
[4] Paschen S, Winkler H, Nezu T, Kriegisch M, Hilscher G, Custers J and Prokofiev A 2010 *J. Phys.: Conference series* **200** 012156
[5] Sengupta K et al. 2004 *Phys. Rev.* B **70** 064406
[6] Uwatoko Y, Ishii T, Oomi G, Takahasi H, Mori N, Thompson J D, Shero J L, Madru D and Fisk Z 1995 *J. Phys. Soc. Jpn.* **65** 27
[7] Abliz M, Nakano T, Sampathkumaran E V, Jemetio J P F, Doert T, Hedo M and Uwatoko Y 2005 *J. Phys. Soc. Jpn.* **74** 508
[8] Brandt N B and Moshchalkov V V 1984 *Adv. Phys.* **33** 373
[9] Sampathkumaran E V et al. 2005 *Physica* B **359-361** 108
[10] Das I and Sampathkumaran E V 1995 *Phys. Rev.* B **51** 1308
[11] Kurisu M, Takabatake T and Fujiwara H 1988 *Solid State Commun.* **68** 595
[12] Gabáni S, Bauer E, Della Mea M, Flachbart K, Padernoc Y, Pavlíka V and Shitsevalovac N 2004 *J. Mag. Mag. Mater.* **272-276** 397
[13] Cooley J C, Aronson M C and Canfield P C 1997 *J. Appl. Phys.* **55** 7533

23rd International Conference on High Pressure Science and Technology (AIRAPT-23) IOP Publishing
Journal of Physics: Conference Series **377** (2012) 012030 doi:10.1088/1742-6596/377/1/012030

# The Structure of Eu-III

**R J Husband[1], I Loa[1], G W Stinton[1], S R Evans[2], G J Ackland[1] and M I McMahon[1]**

[1]SUPA, School of Physics and Astronomy and Centre for Science at Extreme Conditions, The University of Edinburgh, Edinburgh, EH9 3JZ, UK
[2]European Synchrotron Radiation Facility, 38043 Grenoble, France

E-mail: R.J.Husband@sms.ed.ac.uk

**Abstract.** Previous x-ray diffraction studies have reported Eu to transform from the *hcp* structure to a new phase, Eu-III, at 18 GPa. Using x-ray powder diffraction we have determined that Eu remains *hcp* up to 33 GPa, and that the extra peaks that appear at 18 GPa are from an impurity phase with space group $R\bar{3}c$. Above 33 GPa the diffraction pattern becomes very much more complex, signalling a transition to a phase with a distorted *hcp* structure.

## 1. Introduction

The majority of the lanthanide elements are trivalent due to their $4f^n5d^16s^2$ outer electronic structure. The two exceptions are europium (Eu) and ytterbium (Yb), which are divalent due to their half-filled and filled 4f shells, respectively. As a result, Eu and Yb both exhibit a significantly larger atomic volume than would be expected to be consistent with the general trend observed within the lanthanide series [1]. They also do not follow the general trend of phase transitions under pressure or with increasing atomic number observed in the trivalent lanthanides: *hcp*-(Sm-type)-*dhcp*-*fcc* [2].

Under pressure, the valence of Eu undergoes a continuous transition to a mixed-valence state, and $L_{III}$ x-ray absorption studies indicate that the valence increases to 2.64 at 18 GPa, and is then pressure independent up to 33 GPa [3]. Very recently, Eu has been found to be superconducting at a pressure of approximately 80 GPa, with $T_c = 1.8$ K [4].

Initial high-pressure x-ray diffraction studies by Takemura and Syassen found that Eu exhibits the same body-centred cubic (*bcc*) structure as the divalent alkaline metals at ambient pressure, which transforms to a hexagonal close-packed (*hcp*) structure at 12.5 GPa and then to a new phase (Eu-III) at 18 GPa [1]. Eu-III was initially and tentatively assigned a hexagonal structure closely related to a superstructure of *hcp*.

Krüger *et al.* subsequently performed energy-dispersive x-ray diffraction up to 40 GPa [5]. At pressures exceeding 32 GPa, they found the diffraction pattern to become more complex, with the appearance of new reflections in addition to those observed in the 18–32 GPa region. It was noted that the additional reflections were weak in comparison to the *hcp* peaks, and it was suggested that they could be attributed to superlattice distortions of the *hcp* lattice, or to phase mixing.

Despite this interesting behaviour, very little further attention has been paid to the high-pressure behaviour of Eu until this year, when, prompted by the observation of superconductivity, Bi *et al.* reported results from an x-ray diffraction study up to 92 GPa [6]. They attributed the complex diffraction patterns observed above 18 GPa to arise from a sluggish transition to a mixture of *hcp* and a monoclinic phase with space group *C2/c*. Bi *et al.* did not observe a transition at 33 GPa, but at pressures above 41 GPa, changes in the diffraction pattern were attributed to a mixture of the *C2/c* phase and an orthorhombic phase with the space group *Pnma*, followed by a transition to a pure *Pnma* phase at pressures exceeding 66 GPa. These diffraction results were supported by *ab initio* structure prediction calculations.

In this paper, we present angle-dispersive x-ray diffraction data to illustrate that the Eu-III phase does not consist of pure Eu, but in fact consists of *hcp*-Eu plus a rhombohedral phase, the atomic volume of which suggests it is an impurity.

## 2. Experimental

Angle-dispersive x-ray diffraction was performed at the SRS and ESRF synchrotron sources using monochromatic x-rays of wavelengths 0.44397 Å and 0.4161 Å, respectively. The x-ray beams were collimated to diameters of 50 μm and 15 μm, and the data were collected using MAR345 and MAR555 detectors, respectively.

High-purity Eu samples, supplied by U. Schwarz at the Max-Planck-Institut für Chemische Physik fester Stoffe in Dresden, were loaded into diamond-anvil pressure cells in a dry oxygen-free environment (<1 ppm $O_2$ and <1 ppm $H_2O$) using rhenium and tungsten gaskets. A small piece of ruby was included for pressure calibration. Samples were loaded using helium, mineral oil and no pressure-transmitting media (PTM). The diffraction patterns were integrated using Fit2D and analysed using Rietveld and Le Bail methods with the JANA2006 software [7-9].

## 3. Discussion

In our initial samples, including those loaded using a PTM (helium, mineral oil) and those loaded without a PTM, we observed the appearance of weak non-*hcp* reflections at 18 GPa, in agreement with all previous studies [1,5,6]. In these samples we found that as the pressure is increased further, the intensity ratio between the strong *hcp* peaks and the weak extra peaks remains constant all the way to a phase transition at 33 GPa. This behaviour was also noted by Takemura and Syassen [1]. However, we found that the relative intensity of the *hcp* and extra peaks varies between different samples. Additionally, we left one sample of Eu in mineral oil at 26.1 GPa for one month in order to investigate changes over time. The intensity of the non-*hcp* reflections grew significantly, and many further peaks could be identified. In addition, we found that the Debye-Scherrer rings on the diffraction image corresponding to the non-*hcp* reflections had become spotty, while those from the *hcp* phase remained smooth. This is illustrated in figure 1. All of this behaviour strongly indicated that there were two phases present, and we therefore attempted to index the non-*hcp* peaks as a separate phase.

The extra phase was indexed as having a rhombohedral unit cell with $a$ = 9.293(4) Å, and $c$ = 5.381(4) Å at 26.1 GPa. The lattice parameters of the *hcp* phase at the same pressure are $a$ = 3.159(1) Å and $c$ = 4.869(1) Å. Analysis of the systematic absences of the rhombohedral phase showed the spacegroup to be $R\bar{3}c$ or $R3c$, while density considerations restricted the number of Eu atoms per hexagonal unit cell to 18. Trial refinements showed an excellent fit could be obtained with atoms occupying the 18e ($x$,0,¼) Wyckoff sites of $R\bar{3}c$, with $x$ = 0.800(4). An identical structure was obtained in spacegroup $R3c$. A two-phase *hcp*/ $R\bar{3}c$ Rietveld refinement at 26.1 GPa is shown in figure 2.

The volume per Eu atom of the *hcp* phase at 26.1 GPa is 21.05(1) Å$^3$, while the $R\bar{3}c$ phase at the same pressure is *less* dense, with a volume/atom of 22.36(1) Å$^3$. The decrease in density of 5.86(6) % suggests that the $R\bar{3}c$ phase is not pure Eu, but rather results from a pressure-induced reaction, and due to the small difference in the vol/atom, is perhaps a hydride. The quality of the Rietveld fit shown in figure 2 is excellent, suggesting that whatever the contaminant atoms are, they are weakly-scattering in comparison with europium atoms.

In order to confirm that the $R\bar{3}c$ phase was not pure Eu, we loaded samples with no pressure-transmitting medium and no ruby spheres in a very-high quality glovebox (<0.1 ppm $O_2$ and <0.1 ppm

$H_2O$) in order to obtain contaminant-free samples. In such samples, we observed no additional peaks appearing at 18 GPa, but only a single-phase *hcp* pattern from 12.5 – 33 GPa (see figure 3). However, heating such a sample for 2.5 hours at 100 °C resulted in the reappearance of the spotty Debye-Scherrer rings from the *R–3c* phase.

Peaks corresponding to the Eu-III impurity phase have been observed in all previous x-ray diffraction studies performed to pressures above 18 GPa [1,5,6], and we observed the same peaks in all of our samples loaded with a pressure medium, or pressure calibrant, or loaded in a non-optimum glovebox (>1 ppm $O_2$ and >1 ppm $H_2O$). Indeed, diffraction patterns shown in Ref [5] from the *hcp* phase at 14 GPa show evidence of a contaminant phase at this lower pressure. This illustrates the extreme reactivity of Eu, and the difficulty in loading clean, contaminant-free samples. Our experience suggests that it is more difficult to load contaminant-free Eu samples than it is to load contaminant-free Rb and Cs. The present study also suggests that future studies investigating the pressure-induced superconductivity or valence change of Eu should be combined with x-ray diffraction studies of the same samples to ensure that they are free from contaminant phases.

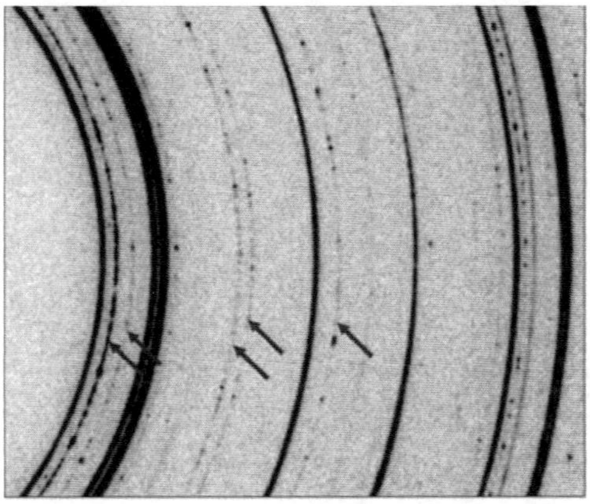

**Figure 1.** A 2D diffraction image from Eu at 26.1 GPa. The Debye-Scherrer (D-S) rings corresponding to the non-*hcp* reflections are spotty, whereas the *hcp* D-S rings are smooth. The arrows mark the five additional reflections observed by Takemura and Syassen [1].

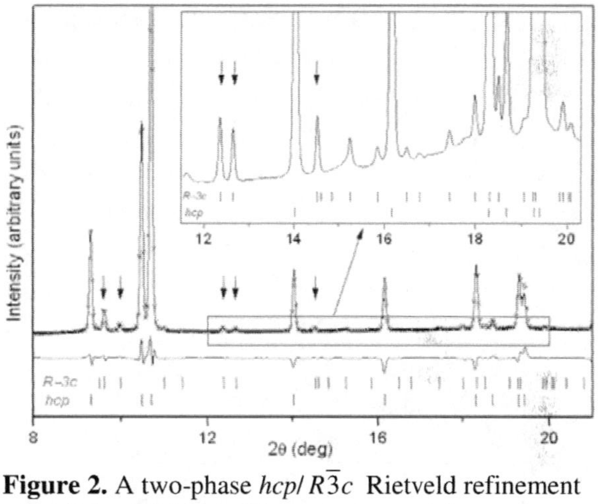


**Figure 2.** A two-phase *hcp*/ $R\overline{3}c$ Rietveld refinement of Eu-III at 26.1 GPa. The experimental data are shown by the crosses and the fit by the solid line. The residuals are shown under the fit and the tick marks show the calculated peak positions. The arrows indicate the Eu-III reflections observed by Takemura and Syassen [1]. The inset shows an enlarged view of the high-angle part of the profile.

Using a contaminant-free sample is also vital for understanding the structural behaviour of europium above 33 GPa, in order to determine which diffraction peaks result from pure-Eu alone. Increasing the pressure of our contaminant-free sample to 33 GPa, we observed the appearance of many additional weak peaks in agreement with Krüger *et al.* [5]. Our data also clearly show splittings of the *hcp* peaks which have not been resolved in previous studies [1,5]. These splittings suggest that the new phase has a distorted-*hcp* structure, and we initially considered the orthorhombic *Pnma* structure proposed by Bi *et al.* for pressures exceeding 41 GPa. A refinement of this structure using the Le Bail method to a profile collected from Eu at 34 GPa is shown in figure 4. The best-fitting lattice parameters are $a$ = 5.291(1) Å, $b$ = 4.713(1) Å, and $c$ = 3.085(1) Å. Although this structure is an orthorhombic distortion of *hcp*, and results in a splitting of the *hcp* diffraction peaks, it cannot account for the observed splittings, and the overall fit is very poor, as illustrated by the inset. There are also a large number of extra peaks not accounted for by this structure. More complex structures must therefore be considered to obtain the correct solution.

In conclusion, we have made high-resolution powder diffraction studies of europium to 37 GPa. These reveal that the long-reported phase transition at 18 GPa to the Eu-III phase is, in fact, a pressure-induced reaction, resulting in the sample becoming a mixture of *hcp*-Eu and a rhombohedral contaminant, perhaps a hydride. Very careful sample preparation and pressure cell loading is required to obtain samples without the contaminant phase. The reasons for the reaction to occur at 18 GPa are unclear, but may be related to the valence of Eu become close to 2.666 at this pressure, perhaps aiding the creation of a stochiometric compound.

**Acknowledgements**
We thank the European Synchrotron Radiation Facility (ESRF) and Daresbury Laboratory (SRS) for provision of synchrotron time and support. We thank M. Hanfland of ID09a (ESRF) and A. Lennie of 9.5HPT (SRS) for their help with the experiments. This work was supported by a research grant and a fellowship (I.L.) from the UK Engineering and Physical Sciences Research Council.

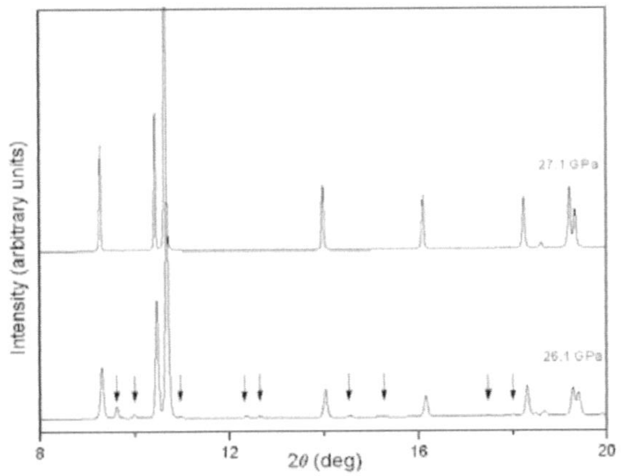


**Figure 3.** Integrated diffraction patterns from a contaminant-free sample of Eu at 27.1 GPa (upper) and a contaminated sample of Eu at 26.1 GPa (lower). Arrows in the lower profile mark the positions of the most intense $R\bar{3}c$ peaks.

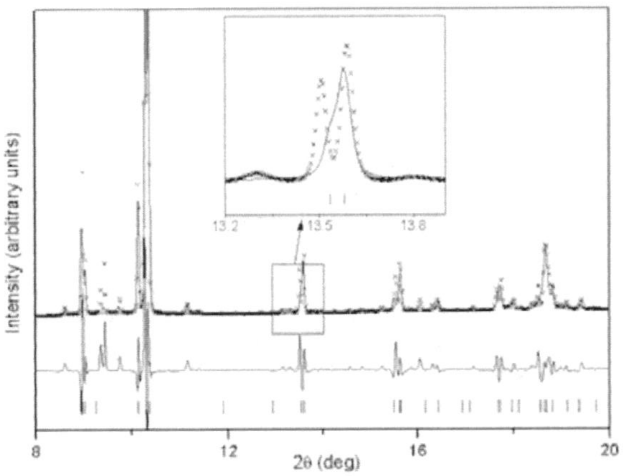


**Figure 4.** Le Bail refinement of a diffraction profile from a contaminant-free sample of Eu at 34 GPa, using the *Pnma* structure proposed by Bi *et al.* [5] for Eu above 41 GPa. The residuals are shown under the fit and the tick marks show the calculated peak positions. The inset shows an enlarged view of the fit to the split (102) peak from the *hcp* phase.

**References**

[1] Takemura K and Syassen K 1985 *J. Phys. F.* **15** 543
[2] Jayaraman A and Sherwood R C 1964 *Phys. Rev.* **134** A691-A692
[3] Rohler J 1986 *Physica* **144B** 27-31
[4] Debessai M, Matsuoka T, Hamlin J J and Schilling J S 2009 *Phys. Rev. Lett.* **102** 197002
[5] Krüger T, Merkau B, Grosshans W A and Holzapfel W B 1990 *High Pressure Research* **2** 193
[6] Bi W, Meng Y, Corneluis A L, Tipton W W, Hennig R G, Zhang Y, Chen C, Schilling J S 2011 *Phys. Rev. B.* **83** 104106
[7] Petrícek V, Dusek M and Palatinus L 2006 *The crystallographic computing system JANA2006*. Institute of Physics, Prague, Czech Republic
[8] Hammersley A P 1997 *ESRF Internal Report* **ESRF97HA02T**
[9] Hammersley A P, Svensson S O, Hanfland M, Fitch A N and Häusermann D 1996 *High Pressure Research* **14** 235-248

# Pressure and Temperature Dependent Structure Of Zircon Type $ThGeO_4$

**S. N. Achary[1][1], S. J. Patwe[1], A. K. Tyagi[1], P. P. Bose[2], R. Mittal[2], A. B. Shinde[2], P. S. R. Krishna[2], S. L. Chaplot[2], C. Narayana[3], G. K. Pradhan[3], S. Banik[4] and S. K. Deb[4]**

[1]Chemistry Division, Bhabha Atomic Research Centre, Mumbai, 400085, India
[2]Solid State Physics Division, Bhabha Atomic Research Centre, Mumbai, 400085, India
[3]Jawaharlal Nehru Centre for Advanced Scientific Research, Bangalore, 560064, India
[4]Indus Synchrotrons Utilization Division, Raja Ramanna Centre for Advanced Technology, Indore, 452013, India

E-mail: sachary@barc.gov.in

**Abstract.** Herein we report the results of high pressure diffraction studies of zircon type $ThGeO_4$. $ThGeO_4$ exhibits anisotropic compressibility with the average compressibility along *a*-axis ($20.8 \times 10^{-4}$/GPa) larger than that along *c*-axis ($9.98 \times 10^{-4}$/GPa). Fitting the pressure dependence unit cell volume to 3rd order Birch-Murnaghan equation of states, the zero pressure bulk modulus ($K_o$) and volume ($V_o$) of 166(5) GPa and 341.6(3) Å$^3$, respectively have been obtained. Preliminary studies on temperature dependent neutron and x-ray diffraction studies on $ThGeO_4$ revealed anisotropic expansion behaviour with larger expansion coefficient along *c*-axis compared to *a*-axis. No structural transition under temperature or pressure is observed in between ambient pressure to 10 GPa and in the temperature range of 25 to 1273K.

## 1. Introduction

Zircon type materials draw a significant research attention for their interesting optical properties and technological applications as solid state laser and scintillator. The zircon type materials often show low thermal expansion coefficients, significantly higher bulk modulus and high radiation stabilities, and these properties make them relevant for several important applications, like waste immobilization, host for minor actinide transmutations in nuclear energy industry [1]. The pressure dependent structural studies of the zircon type materials indicate transformations to denser scheelite and post scheelite type structures at higher pressure [2-4]. Though the nature and magnitude of the thermal expansion or compressibilities of a large number of zircon type materials have been reported in literature, the detailed analyses of the crystal structure with variations in temperature or pressure are relatively under explored in many cases.

A number of silicates of tetravalent cation crystallize in zircon type structure, however the analogous germanates preferentially crystallize in scheelite type structure [5]. Interestingly, the germanates of

[1] To whom any correspondence should be addressed.

Published under licence by IOP Publishing Ltd

thorium exhibit dimorphic behavior depending on preparation conditions [6]. High pressure studies on the scheelite type $ZrGeO_4$ and $HfGeO_4$ show no other polymorphic transformation up to 20 GPa [5]. Higher bulk moduli compared to analogous tungstates and molybdates are observed for scheelite type $ZrGeO_4$ (238 GPa) and $HfGeO_4$ (242 GPa) [5]. Recent high pressure Raman spectroscopic investigations on $ZrGeO_4$ indicate a possible structural phase transition around 12 GPa [7]. High pressure studies revealed a *zircon-scheelite-fergusonite* structural transition sequence for zircon type $ThGeO_4$ and *scheelite-fergusonite* structural transition for scheelite type $ThGeO_4$ [8]. Bulk moduli for both scheelite and zircon type $ThGeO_4$ are about 184 and 186 GPa which are significantly lower than zircon ($ZrSiO_4$) and scheelite type $ZrGeO_4$. In order to correlate the structural parameters with the thermal expansion and compressibilties detailed diffraction studies under non ambient pressure and temperature are carried out and the results are given in this manuscript.

**2. Experimental**

The zircon type $ThGeO_4$ was synthesized by repeated solid state reaction of dry $ThO_2$ and $GeO_2$ around 900°C. The product was characterized by powder X-ray diffraction (XRD) using CuKα radiation and powder neutron diffraction (ND) using neutrons of wavelength 1.249 Å. High pressure X-ray diffraction studies were carried out using DAC (diamond anvil cell) at *Elettra* Synchrotron radiation source. Methanol-Ethanol mixture was used as pressure transmitting medium. Moncohromatic X-ray wavelength of 0.6888 Å was used and the diffraction data were collected using MAR 3450 image plate. Gold was used as pressure marker in the high pressure experiments. The diffraction data were converted to 2D digital data using Fit2D [9]. The observed diffraction data at different temperature or pressure were analyzed by Rietveld method using Fullprof-2K software package [10]. The temperature dependent studies above ambient temperature (in between 300 to 1273 K) were carried out by variable temperature X-ray diffraction method using Philips X-Pert Pro diffractometer coupled with Anton Parr high temperature attachment while below ambient temperature (25-300K) were carried by variable temperature neutron diffraction method, at Dhruva Research Reactors, Trombay.

**3. Results and Discussions**

The formation of zircon type $ThGeO_4$ was confirmed by Rietveld refinement of the observed XRD and ND data using the earlier reported structural data [5,8]. The refined unit cell parameters obtained from the ambient temperature powder ND data are: $a$ = 7.2399(1), $c$ = 6.5425(2) Å and $V$ = 342.93(2) Å$^3$, which are in good agreement with the earlier reported values [5,8]. Final Rietveld refinement plot and other structural parameters for $ThGeO_4$ are shown in figure 1. The crystal structure of zircon type $ThGeO_4$ is made up with the chains of alternate $ThO_8$ polyhedra (bisdisphenoid) and $GeO_4$ (tetrahedra) along $c$-axis. These chains are joined together by sharing other edges of the $ThO_8$ polyhedra along $a$- and $b$-directions. The analyses of structural parameters revealed that the $ThO_8$ bisdisphenoid is made up of two sets of four equivalent Th-O bonds of lengths 2.359(1) and 2.485(1) Å. Similarly, the tetrahedral $GeO_4$ unit has four equivalent Ge-O bonds 1.748(1) Å.

High pressure XRD patterns recorded in the pressurizing conditions were analyzed for their crystal structure. The observed powder XRD patterns at some representative pressures are shown in figure 2a. The diffraction data at ambient pressure indicate characteristic reflections of the zircon type $ThGeO_4$ as well as reflections due to Au (used as pressure marker) and bcc-Fe (gasket). The pressure was determined from the observed unit cell parameters and equation of state *(eos)* of Au. The profile shape of Au indicates the pressure is hydrostatic up to about 10 GPa and beyond it becomes anhydrostatic. The prominent reflections of zircon type $ThGeO_4$ are observed up to 10.3 GPa. The observed XRD pattern at lowest pressure was refined using the position coordinates of the ambient temperature and pressure data. The Rietveld refinement was carried out by considering three phases, pure Au, bcc-Fe and zircon type $ThGeO_4$ together. Unit cell parameters and oxygen position coordinates were refined along with the overall thermal parameters. The XRD data recorded at high pressure were successively

Journal of Physics: Conference Series **377** (2012) 012031 doi:10.1088/1742-6596/377/1/012031

refined using the structural parameters of the previous pressure data. The refined structural parameters for $ThGeO_4$ at different pressures are given in table 1. Rietveld refinement plot for zircon type $ThGeO_4$ at 6.4 GPa is shown in figure 2b.

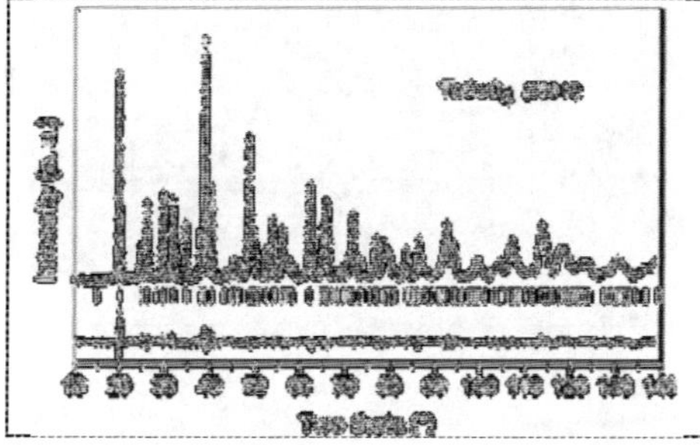

**Figure 1.** Rietveld refinement plot of powder neutron diffraction data of $ThGeO_4$ at 300K (Rp = 3.75, Rwp = 4.97, and $\chi 2$ = 1.83). Tetragonal, Space group: I $4_1$/amd (No. 141); Th: 4a (0,3/4,1/8); Ge: 4b (0,1/4,3/8) and O: 16h (0,y,z), y = 0.4309(2), z = 0.1981(2).

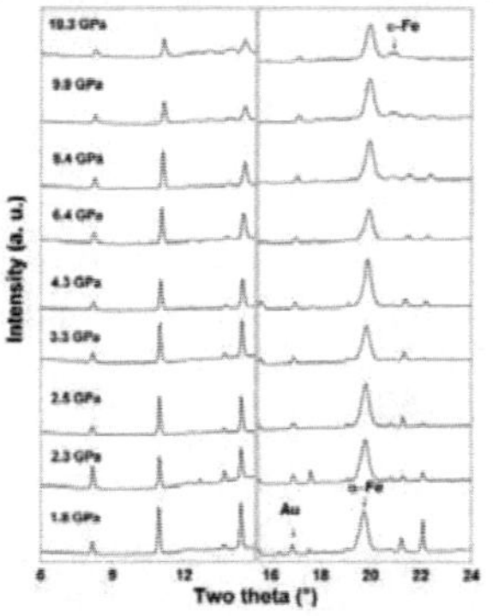


**Figure 2a.** Evolution of XRD patterns with pressure.

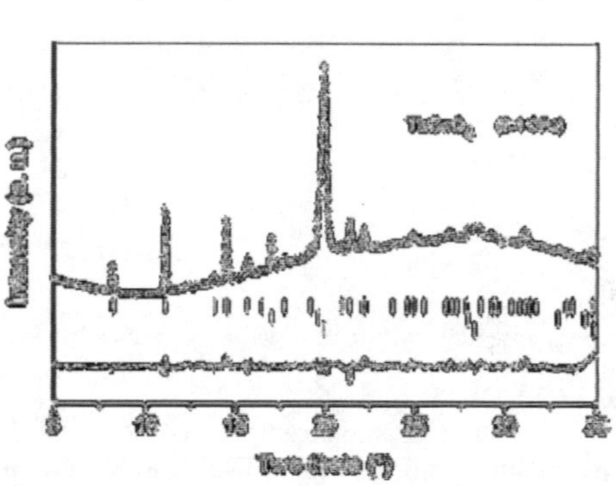

**Figure 2b.** Rietveld refinement plot of XRD data at 6.4 GPa.

The unit cell volume at different pressures were fitted with the 3rd order Birch-Murnaghan *eos* and is depicted in figure 3. The *eos* parameters are also included in figure 3. The zero pressure volume and bulk modulus obtained from the *eos* are 341.6(3) $Å^3$ and 166(5) GPa. The observed bulk modulus for zircon type $ThGeO_4$ (166(5)) GPa is smaller compared to that reported earlier [8]. The average axial compressibility between the 1.8 and 10.3 GPa are: $\beta_a = 20.8 \times 10^{-4}$ /GPa and $\beta c = 9.98 \times 10^{-4}$ /GPa.

Preliminary temperature dependent studies revealed an anisotropic expansion with higher coefficient of expansion along *c*-axis compared to *a*-axis. The observed unit cell parameters at 1273 K are: $a$ = 7.2602(1), $c$ = 6.5808(2) Å and at 25K are: $a$ = 7.2394(2), $c$ = 6.5393(2) Å. The close behavior of expansion and compressibility behavior are reflected from the increase axial ratio (*c/a*) with temperature or pressure. The *c/a* increases from 0.903 to 0.906 as the temperature increase from 25 K to 1273 K, while the *c/a* increases from 0.905 to 0.913 as the pressure increases from 1.8 to 10.3 GPa. The origin of the anisotropy in expansion or compression is related to the crystal structure of $ThGeO_4$. As mentioned earlier, structure of $ThGeO_4$ has chains of edge shared $ThO_8$ and $GeO_4$ units along the *c*-axis. This arrangement causes larger inter-cation repulsion due to the proximity of the highly charged $Th^{4+}$ and $Ge^{4+}$ ions along the *c*-axis. This is reflected in the higher expansion and lower compressibility of *c*-axis compared to *a*-axis. The higher expansion and lower compressibility of *c*- compared to *a*-axis causes an increase in c/a with both temperature as well as pressure.

**Table 1.** Structural parameters of zircon type $ThGeO_4$ at different pressure.

| *P* (GPa) | *a* (Å) | *c* (Å) | *V* (Å)$^3$ | O *y* | O *z* | $B_{ov}$ (Å)$^3$ |
|---|---|---|---|---|---|---|
| 1.79 | 7.205(1) | 6.518(2) | 338.4(1) | 0.437(2) | 0.205(2) | 2.8(5) |
| 2.34 | 7.195(2) | 6.514(3) | 337.2(2) | 0.432(2) | 0.198(2) | 3.0(9) |
| 2.53 | 7.193(2) | 6.506(2) | 336.6(2) | 0.435(2) | 0.203(2) | 2.0(5) |
| 3.35 | 7.177(2) | 6.500(3) | 334.8(2) | 0.433(2) | 0.199(2) | 2.2(6) |
| 4.50 | 7.156(2) | 6.491(3) | 332.4(2) | 0.435(2) | 0.200(2) | 1.3(5) |
| 6.42 | 7.130(3) | 6.475(4) | 329.2(3) | 0.437(3) | 0.201(4) | 3.7(1.0) |
| 8.42 | 7.102(2) | 6.459(3) | 325.7(2) | 0.436(2) | 0.199(3) | 4.4(8) |
| 9.91 | 7.081(4) | 6.462(6) | 324.0(4) | 0.435(4) | 0.197(4) | 6.0(1.4) |
| 10.27 | 7.078(8) | 6.463(12) | 323.7(8) | 0.435(4) | 0.197(5) | 8.02(3.0) |

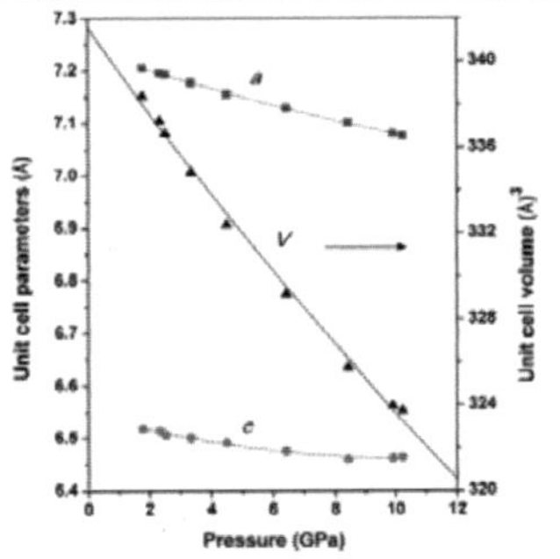


**Figure 3.** Unit cell parameters of $ThGeO_4$ at different pressure (solid lines show *eos* fit). *eos* parameters $V_o$ = 341.6(3) Å$^3$, $K_o$ = 166(5) GPa, K' = 4.0, K'' = -0.02342

## 4. Conclusions

The variation of crystal structure of zircon type $ThGeO_4$ with pressure and equation of state have been investigated by in situ high pressure X-ray diffraction. Anisotropic compressibility and expansion behavior is concluded from the HPXRD and variable temperature structural studies.

## 5. References

[1] Boatner L A 2002 Synthesis, structure, and properties of monazite, pretulite, and xenotime Phosphates*: Geochemical, Geobiological, and Materials Importance, Mineralogy and Geochemistrym vol 48, (Mineralogical Society of America and the Geochemical Society, Chantilly, Virginia)* Eds. M.J. Kohn, J. Rakovan, and J.M. Hughes, 87–123

[2] Mittal R., Garg A B, Vijayakumar V, Achary S N, Tyagi A K, Godwal B K, Busetto E, Lausi A and Chaplot S L 2008 *J. Phys.: Condens. Matter* **20** 075223

[3] Errandonea D and Manjon F J 2008 *Progr. Mater. Sci.* **53** 711

[4] Errandonea D, Lacomba-Perales R, Ruiz-Fuertes J, Segura A, Achary S N and Tyagi A K 2009 *Phys. Rev. B* **79** 184104

[5] Panchal V, Garg N, Achary S N, Tyagi A K and Sharma S M 2006 *J. Phys.: Condens. Matter* **18**, 8241

[6] Achary S N, Patwe S J, Krishna P S R, Sindhe A B and Tyagi A K 2007 *J. Alloys and Compounds* **438** 274

[7] Rao R, Sakuntala T, Achary S N and Tyagi A K 2009 *J. Appl. Phys.* **106** 123517

[8] Errandonea D, Kumar R S, Gracia L, Beltrán A, Achary S N and Tyagi A K 2009 *Phys. Rev. B* **80** 094101

[9] Hammersely A P, Svensson S O, Hanfland M, Fitch A N and Hausermann D 1996 *High Press. Res.* **14** 235

[10] Rodriguez-Caravajal J 1990 "Satellite Meeting on Powder Diffraction" 15th Conf. Inter. Union of Crystallogra. Toulouse, France 127

Journal of Physics: Conference Series **377** (2012) 012032 doi:10.1088/1742-6596/377/1/012032

# Deriving equations of state from non-hydrostatic data

**R Jeanloz[1,2] and B K Godwal[1,*]**
[1]Department of Earth and Planetary Science, University of California, Berkeley, CA 94720, USA
[2]Miller Institute for Basic Research in Science and Department of Astronomy, University of California, Berkeley, CA 94720, USA

*E-mail: godwal@berkeley.edu

**Abstract.** Modeling of non-hydrostatic strains allows extraction of a reliable (quasi-hydrostatic) equation of state from diamond-cell x-ray measurements at high pressures, as illustrated by new data on Os collected to 60 GPa at room temperature: axial- and radial-diffraction measurements are in good agreement with data collected using Ar and He pressure media, as well as with first-principles calculations, in confirming that osmium is the densest but not the most incompressible element. Dynamic-loading methods can generate much higher pressures than static compression, however, shock compression leads to high temperatures there is much interest in compression using ramp waves. This can be accomplished with graded-density mechanical impacts, or with laser-driven pressure waves; other means of maintaining low temperatures include pre-compression and cooling of the sample before it is dynamically compressed by ramp- or multiple-shock waves. Reduced temperatures lead to enhanced strength, which makes it necessary to model both temperature and strength effects in order to extract the equation of state. A unified approach combining analysis of static and dynamic compression measurements offers a means of determining pressure–density equations of state to high compressions.

## 1. Introduction

Static high-pressure experiments use pressure-transmitting media in order to reduce non-hydrostatic (shear) stresses that can have deleterious effects on calibrations and measurements (e.g., of the equation of state) [1-3]. In shock-compression experiments, non-hydrostatic effects exist below the Hugoniot elastic limit [4]. More generally, the degree of non-hydrostaticity depends on a combination of stress, strain rate and temperature [4-5].

Our objective is to identify practical strategies for addressing non-hydrostatic stresses in both static and dynamic high-pressure experiments. These stresses can affect nearly all experiments at pressures above few tens of GPa, a pressure range important for characterizing changes in crystal structures, chemical bonding and a wide range of other properties [6]. The positive aspect of non-hydrostaticity is that it allows determination of practical properties, such as material strength [7]; it can also reduce

[*] To whom any correspondence should be addressed.

kinetic barriers that would otherwise hinder phase transitions. However, shear stresses can significantly affect quantitative measurements of material properties, and the resulting state is not well described by equilibrium thermodynamics (e.g., because of the formation of dislocations, surfaces and other non-equilibrium structures).

We illustrate quantification of non-hydrostatic effects through two examples, static compression of osmium [8] and dynamic compression of diamond [5, 9]: respectively, the stiffest metallic and non-metallic elements known. Although we can vary the state of stress in static experiments, the problem is that we do not yet have reliable means of quantifying the full stress tensor as a function of position across the sample. In contrast, though the large stress gradients and elevated temperatures achieved in dynamic experiments greatly reduce strength effects, it is difficult to disentangle the thermal from shear effects influencing measurements.

## 2. Quasi-Hydrostatic Experiments on Osmium

Osmium powder (99.8 % purity, 10 $\mu$m particle size) was loaded with an argon pressure medium in a 125 $\mu$m diameter sample chamber in a Re gasket that had been indented to a final thickness of 60-70 $\mu$m [10]. The sample was mounted in a symmetric diamond-anvil cell having 300 $\mu$m culets, and probed with an x-ray beam approximately 10 μm wide (beamline 12.2.2 for the Advanced Light Source, ALS). Powder X-ray diffraction patterns were collected in angle-dispersive geometry at room temperature using monochromatic radiation (l = 0.48594 ± 0.00004 Å and 0.61990 ± 0.00004 Å) and a MAR345 image-plate detector [11]. The pressure, determined by ruby fluorescence [12], was taken as an average of several measurements across the sample. The software package FIT2D [13] was used for beam-position and sample-to-detector distance calibrations, and for collapsing the two-dimensional diffraction images to one-dimensional patterns. We minimized non-hydrostatic conditions by thermally annealing the sample at each pressure, with pressurized air being used to heat the entire cell to temperatures of ~40-60° C for 2-5 minutes. We observed significant decreases in pressure variations across the sample after heating, indicating diminishment of shear stresses.

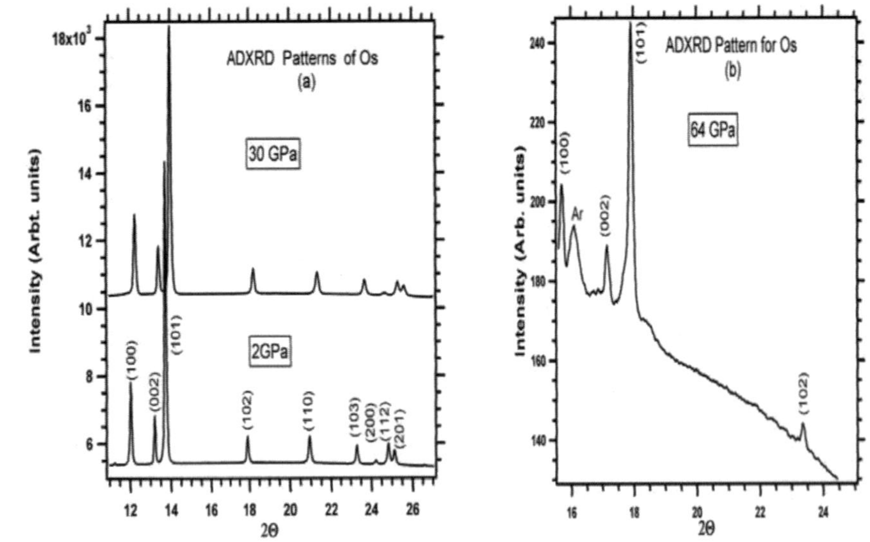


**Figure 1.** Evolution of diffraction patterns of osmium as a function of pressure.

We found Osmium to remain in its ambient hexagonal close-packed structure to the highest pressures of our experiments, 64 GPa (figures 1 (a-b)) and also under non-hydrostatic conditions (figure 4), consistent with results of prior studies. We performed full-spectrum fits of the diffraction profiles with Le Bail and Rietveld methods [14], and were also able to use single-peak fits (PeakFIT V4.06) to determine the lattice parameters and unit-cell volumes based on 9 diffraction peaks up to 30 GPa and 4 peaks beyond this pressure. The pressure at a given volume was determined before and after collecting diffraction patterns, and the pressure–volume (*P–V*) equation of state we find under compression and decompression is in good agreement with prior experimental data, including measurements with a He pressure medium (figure 2) [15-17] and with results of first principles relativistic electronic structure calculations [8].

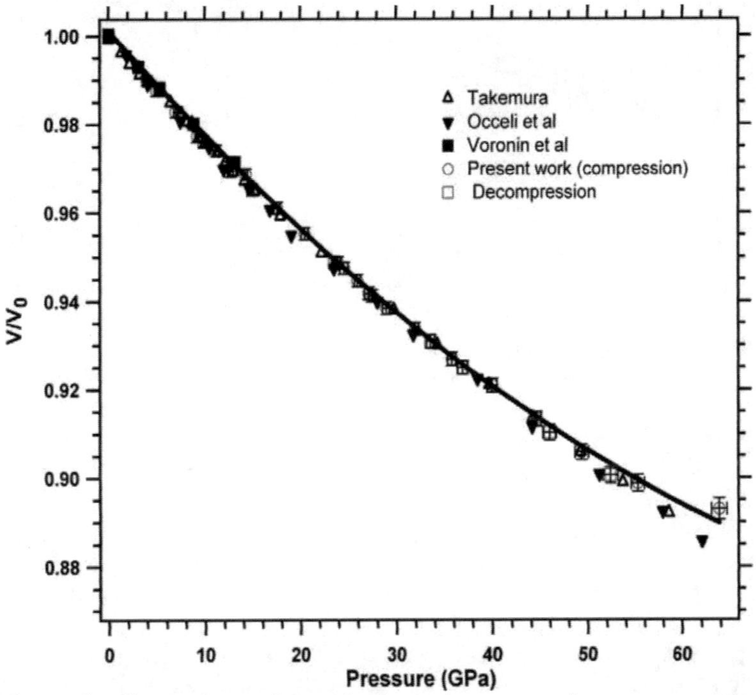


**Figure 2.** Comparison of quasi-hydrostatic compression measurements on Os, with the solid line being a Birch-Murnaghan fit to the present experimental data.

Using the measured zero-pressure volume of $V_0 = 27.977$ ($\pm$ 0.023) Å$^3$, a third-order Birch–Murnaghan [18] fit to the data yields a zero-pressure bulk modulus $K_{0T}$ = 389 ($\pm$ 9) GPa and pressure derivative $K_{0T}'$ = 5.0 ($\pm$ 0.5) (figure 2); here, subscripts *0* and *T* indicate zero pressure and isothermal conditions, respectively. Our values are obtained by a weighted least-squares fit of normalized pressure ($F = P[3f(1+2f)^{5/2}]^{-1}$) as a function of Eulerian strain ($f = 0.5[(V_0V^{-1})^{2/3} - 1]$).

The *c/a* lattice-parameter ratio is also in accord with available experimental values (figure 3) [15-17], though we find no reproducible evidence of subtle shifts that have previously been claimed for *c/a* as a function of pressure [17].

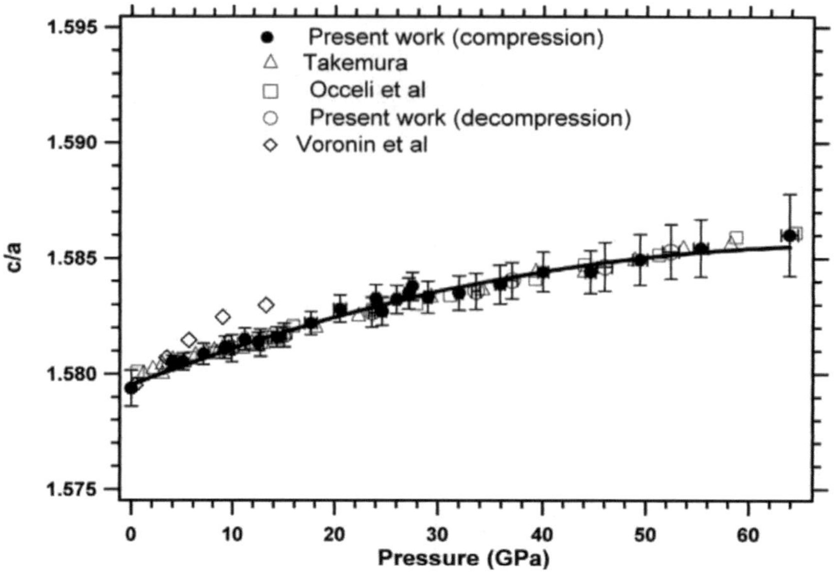


**Figure 3.** Variation in *c/a* ratio of Os with pressure under quasi-hydrostatic conditions (solid line is a guide to the eye).

**3. Non-Hydrostatic Experiments on Osmium**

A modified Mao-Bell type diamond-anvil cell with large openings on two sides [19-20] was used in radial diffraction geometry to collect non-hydrostatic compression data on Os powder (99.8 % purity, 10 $\mu$m particle size) with no pressure medium at beamline 12.2.2 of the ALS. A small platinum flake was embedded in the osmium as an internal pressure standard [21], and a kapton/boron gasket was used to contain the sample. Monochromatic x-rays of 0.49594 ± 0.00004 Å wavelength, and focused to a spot size of 20 x 20 μm, were used to collect diffraction patterns over a 2θ range from $0^0$ to $38^0$ (Mar345 image plate detector placed 287 mm away from the sample). The diffraction images were integrated from 0° to 360° azimuth over 10° sectors for analysis with the MAUD software (figure 4) [22-23].

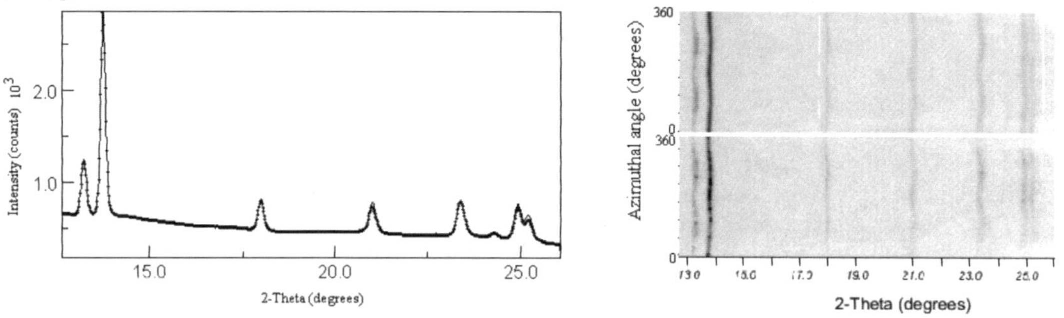


**Figure 4 (a).** Diffraction pattern at 5 GPa.

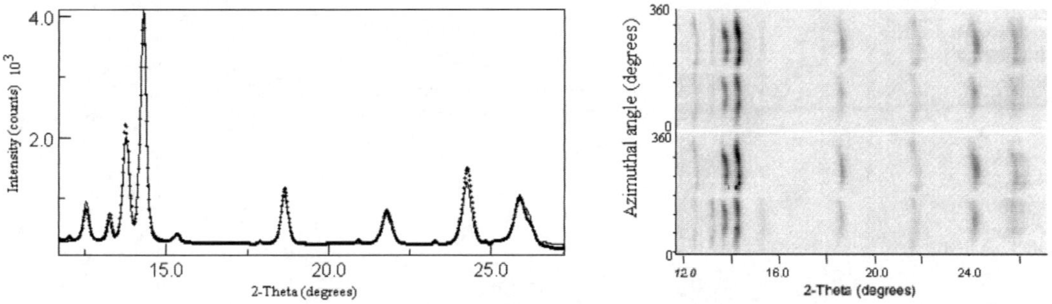


**Figure 4 (b).** Diffraction pattern at 60 GPa

Figure 4. X-ray diffraction patterns of non-hydrostatically compressed Os calculated from MAUD refinements *(left)*, with the 2D experimental patterns *(right)* showing the development of texture at 5 GPa. Osmium remains in the hcp structure under these conditions.

With $\sigma_1$ and $\sigma_3$ being the radial and axial components of stress in the sample, $\sigma_P$ the mean normal stress (defining the hydrostatic pressure), and t = ($\sigma_3$ − $\sigma_1$ ) the deviatoric (uniaxial) stress, the observed *d*–spacings depend on the angle $\eta$ between the diffracting plane normal and the load axis:

$$d_m(hkl) = d_p(hkl)\left[1 + \left(1 - 3\cos^2\eta\right)Q(hkl)\right] \quad (1)$$

based on anisotropic linear elasticity theory [24]. Here $d_P(hkl)$ is the *d* spacing under hydrostatic pressure $\sigma_p$ and $Q(hkl)$ is given by

$$Q(hkl) = \left(\frac{t}{3}\right)\left[\alpha\left\{2G_R(hkl)\right\}^{-1} + (1-\alpha)(2G_v)^{-1}\right] \quad (2)$$

$G_R(hkl)$ is the shear modulus under iso-stress (Reuss) conditions, $G_V$ is the shear modulus under iso-strain (Voigt) conditions, and both $G_R(hkl)$ and $G_V$ are functions of the elastic moduli; α (fraction between 0 and 1) gives the relative weight of iso-stress and iso-strain conditions.

The *d*-spacing varies linearly with (1 – 3 $\cos^2\eta$) and, within the context of the linear-elastic model [24], the intercept gives hydrostatic-pressure values *of d*-spacing, $d_P$ (corresponding to the "magic-angle" $\eta$ = 54.7°). We used the MAUD-refined values of $d_{100}$, $d_{002}$, $d_{101}$, $d_{102}$, $d_{110}$, $d_{103}$, $d_{200}$, and $d_{112}$ for several η in (1) to obtain hydrostatic values and the resulting lattice parameters ('*a*' and '*c*') for different pressures. Similarly, $d_{111}$, $d_{200}$, $d_{220}$, $d_{311}$, $d_{222}$ MAUD-refined values for Pt were used in (1) to obtain hydrostatic values of lattice parameter and hence pressure.

Although with larger error bars and scatter, the compression determined from our non-hydrostatic experiments agrees well with the results obtained from quasi-hydrostatic measurements (figure 5). In the case of Os, therefore, it appears that a reliable *P*–V equation of state is obtained from non-hydrostatic data. We cannot claim that this is a general result based on the present study alone, but the agreement is notable given the stiffness of our sample. Still, we do observe variations in *c*/*a* ratio that correlate with development of lattice-preferred orientation (figure 4), not electronic structure changes [8].

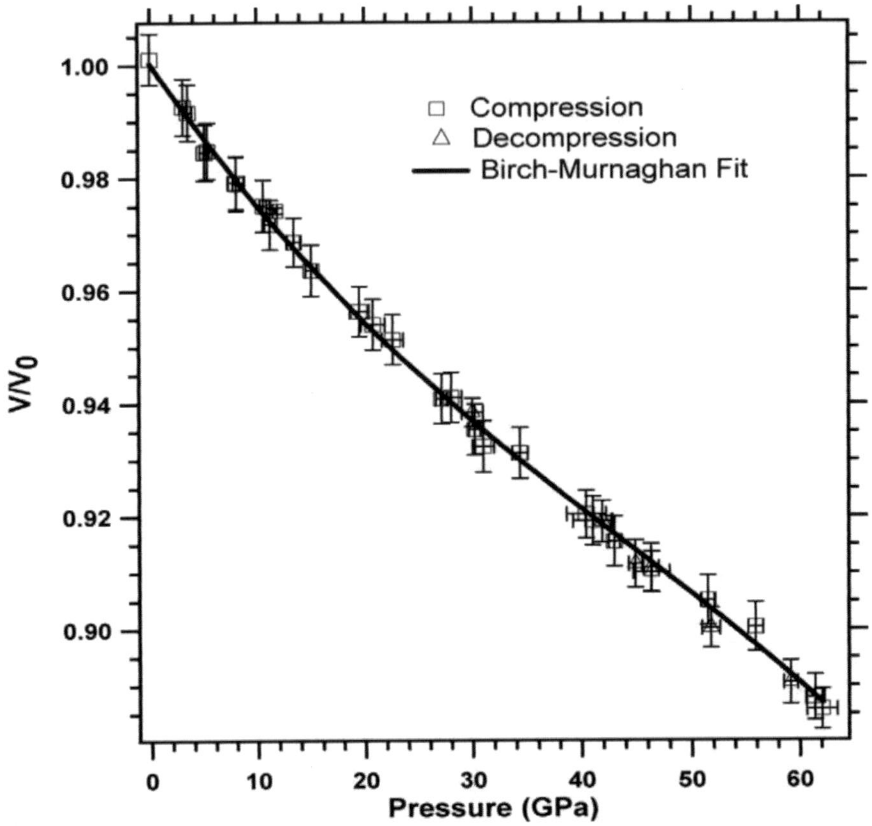


**Figure 5.** Quasi-hydrostatic equation of state derived from non-hydrostatic measurements on Os. The solid line shows the Birch-Murnaghan fit to the present data, yielding $K_{0T}$ = 395 (± 20) GPa and $K_{0T}'$ = 5 (± 1), which compares well with the quasi-hydrostatic results (figure 2) as well as with the ultrasonic value $K_{0T}$ = 406 (±5) GPa for polycrystalline osmium [25].

## 4. Non-Hydrostatic Effects in Dynamic Experiments

Static compression is limited to peak pressures of about 3-5 Mbar, whereas dynamic experiments provide access to the tens of Mbar pressures relevant to the deep interiors of Saturn and Jupiter (~ 30-70 Mbar peak pressures), and of super-Earth and super-giant exoplanets. Ramp compression [26] provides more gradual loading than shock waves [27], and is therefore expected to produce less heating. Ramp compression is thus well suited for reproducing planet-interior conditions, which are largely characterized by isentropic temperature profiles, and also to achieving the highest compressions in samples.

Diamond, formed from the fourth most abundant cosmochemical element, is important both as a constituent of "icy" planets such as Uranus and Neptune and because of its singular properties (e.g., high bulk modulus, thermal conductivity, transparency across a broad spectrum of electromagnetic wavelengths). Thus, it is used as a pressure chamber for dynamic (pre-compression) as well as static high-pressure experiments (e.g., [28]). Diamond exhibits a nearly constant melting temperature at 9-20 Mbar [29-30], and an exceptionally large dynamic strength [5, 9].

Both strength and temperature result in an offset (higher magnitude of stress at a given compression) for the dynamic-compression measurements relative to a fluid response. Therefore, in order to disentangle the combined effects of strength and temperature on dynamic (shock- and ramp-)

compression data, we analyze the derivative of the stress–strain relation that is measured. For an elastic–perfectly plastic material, the shear stress remains constant above the yield stress and the strength-induced offset in stress (relative to the fluid Hugoniot at given strain) remains constant.

Assuming a linear $U_S$-$u_p$ relation for the Hugoniot ($U_S = c + s\,u_p$ for shock velocity $U_S$ and particle velocity $u_p$), a Mie-Grüneisen analysis [28] shows that the Hugoniot and isentrope slopes at a given density ($\varrho = 1/V$) are related by

$$\frac{\left(\frac{\partial \mathrm{P_H}}{\partial \rho} - \frac{\partial \mathrm{P}_S}{\partial \rho}\right)}{c^2} = \frac{\gamma s\left[1 - \frac{\rho_0}{\rho}\right]\left(1 - \left[1 - \frac{\rho_0}{\rho}\right]\right)}{\left(1 - s\left[1 - \frac{\rho_0}{\rho}\right]\right)^3} \qquad (3)$$

where $\gamma$ is the Grüneisen parameter (assumed to depend on density alone). Therefore, if the Hugoniot ($c$ and $s$) is known, the $P$–$\varrho$ slope of ramp-compression measurements can be compared against slopes of both Hugoniot and isentrope as a function of density in order to determine the entropy associated with ramp loading. Once dynamic-compression measurements are related to the Hugoniot and isentrope, and reduced to either with this approach, they can be analyzed jointly with the results of isothermal-compression and wave-velocity (e.g., ultrasonic or Brillouin scattering) experiments through a finite-strain approach involving the normalized pressure as a function of strain, $F(f)$ [31].

Analysis of multi-Mbar data for diamond [5] using (3) suggests that both strength and entropy effects can be important under ramp loading. Our knowledge of material behavior under ramp compression is limited, and we do not have a good microscopic understanding of how samples behave under such conditions: defect production as well as heating may both contribute. Alternative loading paths, including multiple (multi-step) shocks need to be explored in order to better tune and understand strength and entropy effects. Also, experiments achieving much higher compressions by ramp loading at MJ-laser facilities such as NIF can provide testing grounds for microscopic theories of dynamic material response over a wide range of strains and strain rates.

## 5. Conclusions

Real progress is being made in accounting for the effects of non-hydrostaticity in high-pressure measurements, under both static and dynamic conditions. In the case of static measurements, it is important to (i) collect data on decompression as well as compression; (ii) quantify the effects of different states of stress by making measurements with different pressure media and with thermal annealing (modest external heating or more intense laser heating); and (iii) using radial diffraction to complement axial-diffraction measurements. In the case of dynamic loading, ramp, multi-shock and pre-compression experiments are beginning to provide the information needed to distinguish entropy (thermal and defect) from strength effects. Ramp compression may follow a stress–strain path quite distinct from that of an isentrope.

## References

[1] Angel R J, Buzak M., Zhao J, Gatta G D and Jacobsen S D 2007 *J. Appl. Cryst.* **40** 26.
[2] Vohra Y K and Spencer P T 2001 *Phys. Rev. Lett.* **86** 3068
[3] Akahama Y, Kawamura H and Bihan T L 2001 Phys. Rev. Lett. **87** 275503
[4] Rosenberg Z 1993 *J. Appl. Phys.* **74** 752
[5] Bradley D K, Eggert J H, Smith R F, Prisbrey S T, Braun D G, Biener J, Hamza A V, Rudd R E and Collins G W 2009 *Phys. Rev. Lett.* **102** 075503
[6] Ma Y, Eremets M, Oganov A R, Xie Yu, Trojan I, Medvedev S, Lyakhov A O, Valle M, Prakapenka V 2009 *Nature* **458** 0778
[7] Singh A K, Mao H K, Shu J and Hemley R 1998 *Phys. Rev. Lett.* **80** 2157
[8] Godwal B K, Yan J, Clark S M and Jeanloz R 2012 *J. Appl. Phys.* (2012 under publication)

[9] McWilliams R S,Eggert J H, Hicks D G, Bradley D K, Celliers P M, Spaulding D K, Boehly T R, Collins G W and Jaenloz R 2010 *Phys. Rev.* B **81** 014111
[10] Mills R L, Liebenberg D H., Branson D H.and Schmidt L C 1980 *Rev. Sci. Instrum.* **51** 891
[11] Kunz M, MacDowell A A, Caldwell W A, Cambie D, Celestre R S, Domning E E, Duarte R M, Gleason A E, Glossinger J M, Kelez N, Plate D W, Yu T, Zaug, JM, Padmore H A, Jeanloz R, Alivisatos A P and Clark S M 2005 *J. Synch. Rad.* **12(5)** 650-658
[12] Mao H K, Xu J and Bell P M 1986 *J. Geophys. Res.* **B91** 4673
[13] Hammersley A P, Svensson S O, Hanfland M, Fitch A N and Hausermann D *High Press. Res.* 1996 **14** 235
[14] Larson A C and Von Dreele R B 2004 General structure analysis system (GSAS), Los Alamos National Laboratory Report **LAUR** 86-748
[15] Voronin G A, Pantea C, Zerda T W, Wang L and Zhao Y, *J. of Phys. and Chem. Of Sol.* 2005 **66** 706
[16] Takemura K Phys. Rev.B 2004 **70** 012101
[17] Occelli F., Farber D L, Badro J, Aracne C M, Teter D, Hanfland M, Canny B and Couzinet B *Phys. Rev. Lett.* 2004 **93** 095502
[18] F. Birch *Phys. Rev. B* 1947 **71** 809
[19] Liermann H P, Merkel S, Miyagi L, Wenk H R, Shen G, Cynn H, and Evans W J *Rev. Sci. Instrum.* 2009 **80** 104501
[20] Miyagi L, Kunz M, Nasiatka J, Voltolini M, Knight J and Wenk H R, *J. Appl. Phys.* 2008 **104** 103510
[21] Y. Fei, A. Ricolleau, M. Frank, K. Mibe, G. Shen, V. Prakapenka, *PNAS*,104, 9182 (2007)
[22] Lutterotti L, Matthies S, Wenk H R, Schultz A J and Richardson J W 1997 *J. Appl. Phys.* **81** 594
[23] Wenk H R et al. 1998 *J. Appl. Crystallogr.* **31** 262
[24] Singh A K, Balasingh C, Mao H K, Hemley R J and Shu J 1998 *J. Appl. Phys.* **83** 7567
[25] Pantea C, Mihut I, Ledbette H, Betts J B, Zhao Y, Daemen L L, Cynn H and Migliori A *Acta Materialia* 2009 **57** 544
[26] Bastea M, Bastea S, Emig J A, Springer P T and Reisman D B 2005 Phys. *Rev.* B **71** 180101 (R)
[27] Zeldovich Y B and Raizer Y P 1967 Physics of shock waves and high-temperature hydrodynamic phenomena (Academic, New York)
[28] Celliers P M, Loubeyre P, Eggert J H, Brygoo S, McWilliams R S, Hicks D G , Boehley T R, Jeanloz R and Collins G W 2010 *Phys. Rev. Lett.* **104** 184503; Eggert J H, Brygoo S, Loubeyre P, McWilliams R S, Celliers P M, Hicks D G, Boehly T R, Jeanloz R and Collins G W 2008 *Phys. Rev. Lett.* **100** 124503
[29] Eggert J H, Hicks D G, Celliers P M, Bradley D K, McWilliams R S, Jeanloz R, Miller J E, Boehly T R and Collins G W 2010 *Nature Physics* **6** 40
[30] Hicks D G, Boehly T R, Celliers P M, Bradley D K, Eggert J H, McWilliams R S, Jeanloz R and Collins G W 2008 *Phys. Rev.* B **78** 174102
[31] Heinz D L and Jeanloz R 1984 *J. Appl. Phys.* **55** 585

Journal of Physics: Conference Series **377** (2012) 012033 doi:10.1088/1742-6596/377/1/012033

# Electronic phase transitions in cadmium at high pressures

**B K Godwal[1], S V Raju[2], Z Geballe[1] and R Jeanloz[1,3]**
[1]Department of Earth and Planetary Science, University of California, Berkeley, CA 94720, USA
[2]Advanced Light Source, Lawrence Berkeley National Laboratory, Berkeley, CA 94720, USA
[3]Department of Astronomy and Miller Institute for Basic Research in Science, University of California, Berkeley, CA 94720, USA

E-mail: godwal@berkeley.edu

**Abstract.** Elemental solid Cd crystallizes in the hcp structure with a large *c/a* value of 1.89. This leads to anisotropy in the Fermi-surface topology, electron-transport and other physical properties. Application of pressure reduces this anisotropy, with the *c/a* ratio decreasing to the ideal value corresponding to a spherical Fermi surface. There is long standing interest in the detection of departures of *c/a* from a smooth variation with pressure, and in associating such anomalies with electronic topological transitions. Angular x-ray diffraction measurements were carried out on Cd up to 25 GPa at room temperature. Variations of *c/a* with pressure reveal anomalies near 2, 7, 15 and 22 GPa; we find anomalies in the pressure–volume compression curve close to these pressures, which are also associated with electronic topological transitions determined from first-principles calculations. Independent in-situ x-ray powder-diffraction determinations of melting for Cd show departures from Lindemann predictions above 1 GPa, consistent with the occurrence of electronic topological transitions.

## 1. Introduction

Electronic phase transition also known as electronic topological transition (ETT) in materials results whenever the topology of Fermi surface is affected due to temperature, pressure or alloying. The resulting shift in the Fermi level ($E_F$) relative to electronic band extrema leads to subtle electronic phase transitions and causes anomalies in physical properties [1]. Zinc and cadmium crystallize at ambient conditions in the hcp structure with an unusually large *c/a* ratio of lattice parameters with highly anisotropic Fermi surface. Pressure is supposed to alter this anisotropy and hence these materials have been studied intensely under pressure for the observation of electronic phase transitions both experimentally [2-13] and theoretically [14-22]. Inspite of such frantic efforts spanned over two decades the existence of ETT still remains highly controversial. For Zinc which has been studied more

[1] To whom any correspondence should be addressed.

Intensely than cadmium there is broad consensus among the theoretical groups about the existence of ETT, however the experimental groups appear to be divided due to the fact that the resulting compressibility anomaly (or anomaly in *c/a*) is small if it exists at all.

The available limited high pressure studies for Cd suggest that the structural effects of electronic phase transitions are more visible than in Zn and extended over a larger pressure range [22]. Takemura's [6] data on Cd with methanol-ethanol and water mixture as pressure medium although covers the pressure range up to 174 GPa, a close look at the pressure volume curve reveal that there are no data points in the range 5-12 GPa where the strong ETT is predicted [16-17,22]. Pratesi et al's [23] energy dispersive high pressure x-ray diffraction data for Cd up to 17 GPa with mineral oil as pressure transmitting medium indicate anomalies in lattice parameters and *c/a* ratio. Further, the past high pressure x-ray diffraction studies by Speziale et al [24] on $Cd_{0.80}Hg_{0.20}$ alloy, a system with the same structure as Cd but with *c/a* ratio of 1.9 (1.1 % larger than Cd) at ambient conditions showed the presence of anomalies in the pressure dependence of lattice parameters in the extended pressure range which were associated with electronic phase transition according to the electronic structure calculations [24]. However alloying of Cd with Hg could have changed the band filling and hence the number of connectivity of the Fermi Surface sheets leading to the smearing of the electron states and the observable effects [25]. This possibility can not be ruled out because of the lack of our experimental knowledge about electronic structure at high pressures. Hence it is important to study pure Cd for which there is motivation from thermo electric power measurements [9], x-ray diffraction [23] and first principles total energy calculations [22] which indicate possibility of detecting the onset of ETT in Cd by studying the pressure dependence of its elastic properties. We study pure cadmium at room temperature using angle dispersive x-ray diffraction with argon as pressure medium. The primary aim is to improve the accuracy of our measurements compared to Pratesi et al [23] using different pressure medium and to compare it with the available data of Takemua [6] up to the pressure of 25 GPa and correlate the compressibility anomalies with ETTs predicted by theory.

## 2. Experimental details

Rhenium gasket indented to a 50 μm thickness and 100 μm hole served as the sample chamber. The sample material about the size of the sample chamber was cut out from a Cadmium wire. The sample along with few specks of ruby was loaded in a rhenium gasket. For hydrostatic measurements, argon was loaded cryogenically to serve as a pressure-transmitting medium. High-pressure experiments were performed using a symmetric diamond-anvil cell (DAC) beamline 12.2.2 of the Advanced Light Source at Lawrence Berkeley National Laboratory [26]. Powder x-ray diffraction patterns were collected in angle-dispersive geometry using monochromatic radiation ($\lambda = 0.48594 \pm 0.00004$ Å) and a MAR345 image-plate detector at a distance of 295.54 (± 0.01) mm from the sample. The x-ray beam size was approximately 10 μm in the horizontal and vertical directions. Pressure was determined by the ruby-fluorescence method, using the calibration of *Mao et al* [27]. The software package *FIT2D* [28] was used both for beam-position and sample-to-detector distance calibrations, and for collapsing the two-dimensional diffraction images to one-dimensional patterns. As various pressure media used to collect the high pressure data on Cd freeze in the pressure range we have explored [34] we have compared and analyzed all the data sets together.

## 3. Results and discussion

Each of the x-ray diffraction patterns collected exhibits ~9 diffraction lines from the sample. We used all these observed sample lines to determine the unit-cell parameters and the density of Cd at each experimental pressure. The diffraction patterns show that the sample remains in the hcp phase up to 25 GPa (figure 1). Structural parameters as a function of pressure are given in table 1.

**Table 1**. Structural data of Cd for high pressure. The lattice parameters and the axial ratio at ambient pressure at room temperature are a= 2.9788(4) Å, c= 5.6164(6) Å and c/a = 1.8853 (3).

| P(GPa) | a (Å) | c(Å) | c/a | $V/V_0$ |
|---|---|---|---|---|
| 2.0(2) | 2.9673(2) | 5.4867(8) | 1.8491(4) | 0.9694(6) |
| 5.0(3) | 2.9404(4) | 5.3023(13) | 1.8033(6) | 0.9199(8) |
| 6.5(4) | 2.9358(2) | 5.2389(6) | 1.7845(3) | 0.9061(6) |
| 10.5(5) | 2.9089(7) | 5.1345(8) | 1.7651(7) | 0.8718(9) |
| 13.6(6) | 2.9074(1) | 4.963((7) | 1.7070(3) | 0.8418(5) |
| 14.0(6) | 2.9089(7) | 4.9799(8) | 1.7120(7) | 0.8455(9) |
| 17.7(7) | 2.8886(2) | 4.8627(5) | 1.6834(3) | 0.8142(5) |
| 20.3(8) | 2.8731(2) | 4.8091(7) | 1.6738(4) | 0.7966(5) |
| 23.3(8) | 2.8606(3) | 4.7930(9) | 1.6755(5) | 0.7870(6) |

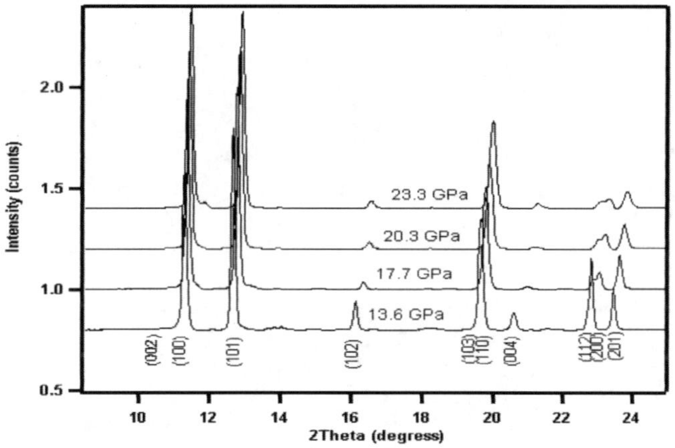


**Figure 1.** Evolution of diffraction patterns with pressure for cadmium.

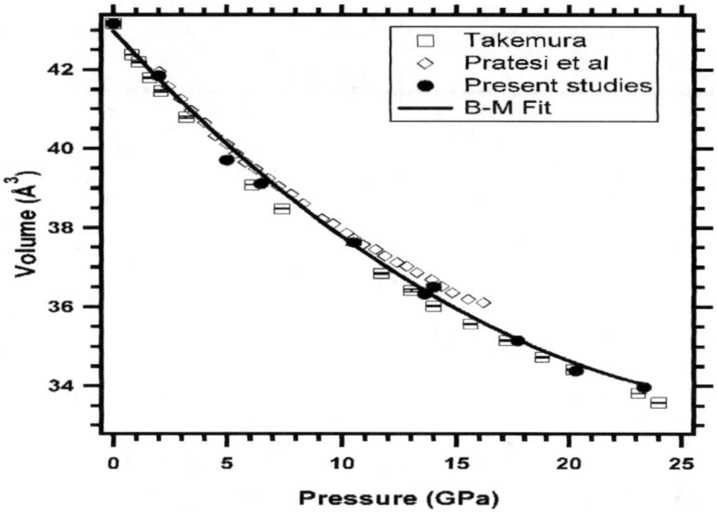


**Figure 2.** Isothermal compression of Cd. The line is the Birch-Murnaghan 300 K isotherm fitted to the data collected in present studies. Error bars are equivalent to the size of the symbols.

A third-order Birch–Murnaghan fit [29] to the pressure dependence of the unit-cell volume fixing the initial volume to the measured value $V_0$ = 43.159 (± 0.02) Å$^3$ (figure 2) yields bulk modulus $K_{T0}$ = 46 (± 6) GPa and pressure derivative $K_{T0}'$ = 6.4 (±1.2), where subscripts 0 and *T* indicate zero pressure and isothermal conditions, respectively. These values are in excellent agreement with values measured by others for Cd, $K_{T0}$ = 46 (± 5) GPa, $K_{T0}'$ = 6 (± 1) [6-8].

It is observed from figure 2 that there are no experimental points in Takemura's [6] measurements in ~5-12 GPa pressure region where the strong anomaly is predicted. Pratesi et al's [23] energy dispersive data with mineral oil as pressure medium lies above Takemura's data. Our data is in between these two data sets with better agreement with Pratesi et al's data upto 10 GPa and close to Takemura's data above 12 GPa. Pratesi et al's data start systematically deviating above 10 GPa from Takemura's and our data. We do not find any observable anomaly in the P-V data. These data sets observed together for the variation of lattice parameters 'a' and 'c' show anomalous behavior near 8-9 GPa and 14-15 GPa (figure 3). The *c*/*a* ratio (figure 4) decreases monotonously from 1.885 to 1.67 through our experimental pressure range of 0-25 GPa. However, the pressure dependence of *c*/*a* shows two clear discontinuities, at 8-10 GPa (*c*/*a* = 1.757 – 1.753) and 14-15 GPa (*c*/*a* = 1.71 – 1.69). The anomaly is clearly seen in Pratesi et al's data of *c*/*a* variation with pressure, which is at close pressure intervals near 8 and 15 GPa. Looking at the variation of *c*/*a* ratio from 18 to 25 GPa we notice an anomaly near 22-23 GPa. If we observe all the data sets together in figures 3 and 4 then we also find anomaly close to 2 GPa.

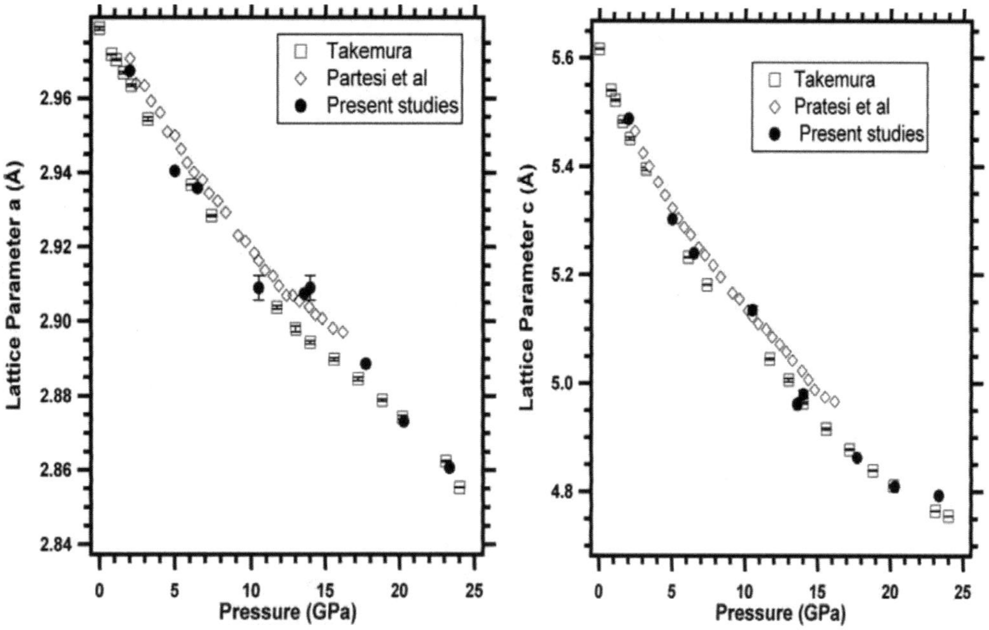


**Figure 3.** Pressure variation of lattice parameters ('a' and 'c') with pressure. Comparison is made with Pratesi et al's [23] and Takemura's data [6]. Pratesi et al's data clearly reveal anomalies. Error bars are equivalent to the size of the symbols.

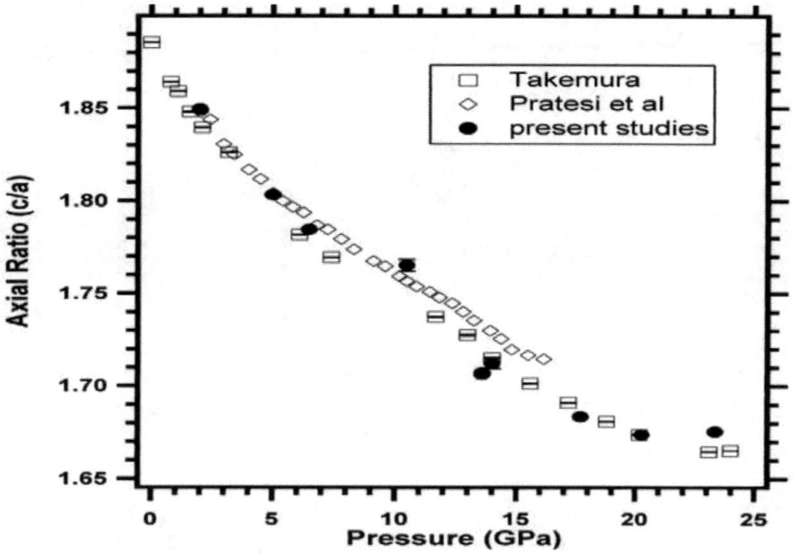


**Figure 4.** Pressure variation of c/a ratio with pressure. We compare our data with Pratesi et al's [23] and Takemura's data [6]. Pratesi et al's data which is in close pressure intervals clearly reveal anomalies in pressure variation of c/a ratio. Error bars are equivalent to the size of the symbols.

In order to better highlight the anomalies in the compression of Cd, we have plotted the data in terms of Eulerian strain ($f = 0.5[(V_0V^{-1})^{2/3} -1]$) and normalized pressure ($F = P[3f(1+2f)^{5/2}]^{-1}$) calculated by fixing the value of $V_0$ to 43.159 (± 0.03) Å$^3$ (figure 5). This representation effectively shows the slope of the $P–V$ data, such that the intercept gives the bulk modulus at zero pressure, $K_{T0}$ [29]. The $F–f$ plot of our data with uncertainties again suggests that there are at least two discontinuities in the compression curve, at 6-8 GPa ($f$ = 0.03-0.040) and at 14-16 GPa ($f$ = 0.06-0.065). It is also observed that these anomalies are also observed in Pratesi et al's [23] and Takemura's data [6] at comparable pressures. Further Takemura's data shown in figure 5 also show anomalies close to 2-3 GPa (f = 0.018-0.02) and 22-23 GPa (f-0.089-0.091).

The two anomalies we find in our data separate regions of the compression curve characterized by similar pressure dependence of the bulk modulus (proportional to the slope of the $F–f$ plot). Fits of the two regions of the $F$-$f$ curve of our data for the pressure region (6-8GPa); $K_{T0}$ = 42.5 (± 3) GPa and $K_{T0}'$ = 4.7 (±1); and $K_{T0}$ = 48 (± 4) GPa and $K_{T0}'$ = 1.7(± 0.8) for the high-pressure region (14-16 GPa). The F-$f$ fit for the anomaly in Takemura's data for the low pressure (2-3 GPa) yield $K_{T0}$ = 40 (± 2) GPa and $K_{T0}'$ = 9 (± 2.5). We note that bulk modulus and its pressure derivative across the two anomalies in our data (figure 5) are different from each other.

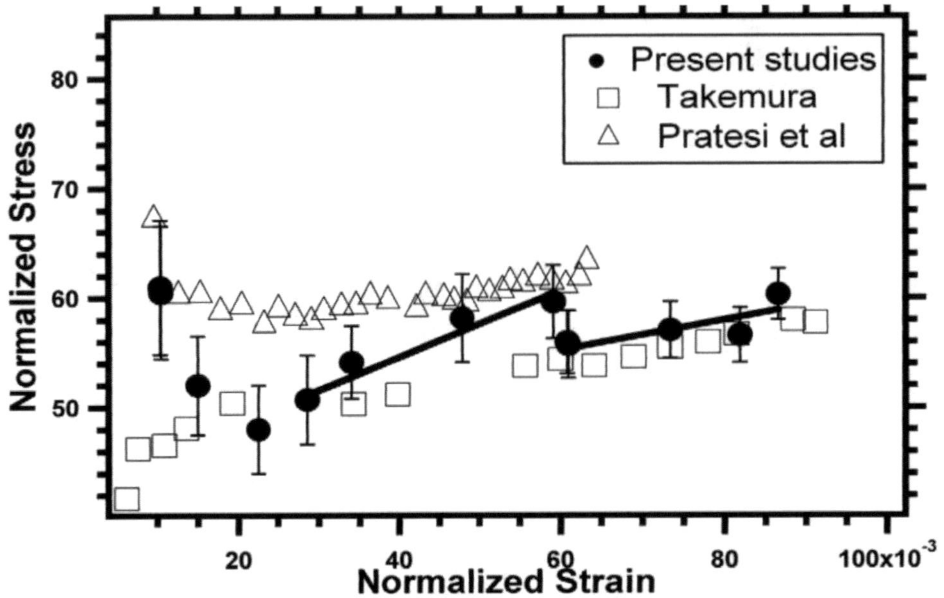


**Figure 5.** Plot of normalized pressure, $F$, versus Eulerian strain, $f$, for Cd ($F = P[3f(1+2f)^{5/2}]^{-1}$, $f = 0.5[(V_0V^{-1})^{2/3} -1]$). The figure also includes extra points corresponding to two measurements at the same pressure (2 and 13.6 GPa) which resulted in slightly different uncertainties in volumes. The lines show linear fits of the two regions of the compression curve (see text).

These anomalies are in excellent agreement with self consistent pseudopotential electronic structure calculations carried out using Quantum Espresso Code [30] which predicts change in the topology of Fermi surface and anomalies in axial ratio, elastic constants and linear compressibility around 2, 7, 15 and 22 GPa [22,30]. However, *Takemura* [6] and *Pratesi et al* [23] interpret their data in terms of a single anomaly at $c/a \sim 1.74$ near 8 GPa.

Independent in-situ x-ray powder-diffraction determinations of melting of Cd up to 10 GPa using resistively heated diamond anvil cell by Raju et al [31-32] and previous melting studies by ex-situ methods [33, 34] show departures from Lindemann predictions [35] above 1 GPa, and are consistent with the occurrence of electronic topological transitions near 2 and 8 GPa in Cd, it is possible that they are responsible for the departure of melting curve from Lindemann law.

Our present studies combined with Takemura's [6] and Pratesi et al's [23] data support four electronic phase transitions up to 25 GPa and differ from the observation of Takemura and Pratesi et al who report only one ETT. The anomalies we have observed are in agreement with the predictions from first principles theories.

## 4. Conclusions

Variation of lattice parameters, axial ratio and equation of state is in agreement with available data to the pressure range of 25 GPa. Our results combined with available data on Cd suggest the occurrence of anomalies in the compressibility at 2, 8, 15 and 22 GPa pressures. Highly accurate first principles electronic structure calculations associate these anomalies with changes in the Fermi surface topology. Thus, our experimental results of high pressure x-ray diffraction by annealing the sample at each pressure step along with our independent melting curve results [32] combined with theoretical calculations support the view that the anomalies of structural parameters and elastic moduli of pure Cd are caused by electronic topological transitions (ETTs).

**References:**

[1] Lifshitz I M 1960 *Sov. Phys.* JETP **11**, 1130
[2] Linch R W, and Drickamer H G 1965 *J. Phys Chem. Sol.* **26**, 63
[3] Potzel W, Steiner M, Karzel H, Schiessl W, Kofferlein M, Kalvius G M, Blaha P 1995 *Phys. Rev. Lett.* **74** 1139
[4] Takemura K 1995 *Phys. Rev. Lett.* **75** 1807
[5] Morgan J G, Von Dreele R B, Wochner P, Shapiro S M 1996 *Phys. Rev.* B **54** 812
[6] Takemura K 1997 *Phys. Rev.* B **56**, 5170
[7] O. Schulte, A. Nikolaenko, and W.B. Holzapfel 1991 *High Pres. Res.* **6**, 169
[8] Schulte O and Holzapfel W B 1996 *Phys. Rev.* B **53**, 569
[9] Godwal B K, Meenakshi S, Rao R S and Vijayakumar V 1998 *J. Phys. Chem. Sol.* **59**, 747
[10] Takemura K 1999 *Phys. Rev.* B **60**, 6171
[11] Klötz S, Braden M, Besson J M 1998 *Phys. Rev. Lett.* **81** 1239
[12] Olijnyk H, Jephcoat A P, Novikov D L, Christensen N E 2000 *Phys. Rev.* B. **62**, 5508
[13] Takemura K , Yamawaki H, Fujuhisa H and Kikegawa T 2002 *Phys. Rev.* B **65**, 132107
[14] Meenakshi S, Vijayakumar V, Godwal B K. and Sikka S K 1992 *Phys. Rev.* **B46**, 14359
[15] Fast L, Ahuja R, Nordström L, Wills J M, Johansson B, Eriksson O 1997 *Phys. Rev. Lett.* **79** 2301
[16] Godwal B K, Meenakshi S, Rao R S 1997 *Phys. Rev.* B **56** 14871
[17] Novikov D L, Katnelson M I, Trefilov A V, Freeman A J, Christensen N E, Svane A, Rodriguez C O 1999 *Phys. Rev.* B **59** 4557
[18] Li Z, Tse J S 2000 *Phys. Rev. Lett.* **85** 5130
[19] Neumann G S, Stixrude L, Cohen R E 2001 *Phys. Rev.* B **63** 54103
[20] Godwal B K, Modak P and Rao R S 2003 *Solid State Comm.* **125**, 401
[21] Qiu S L and Marcus P M J 2003 Phys. Condens. Matter **15**, L755
[22] Qiu S L, Apostol F and Marcus P M 2004 J. Phys.: Condens. Matter **16**, 6405
[23] Pratesi G, DiCicco A, Minicucci M and Itie 2005 J P *J. Phys. Condens. Matter* **17**, 2625
[24] Speziale S, Jeanloz R, Clark SM, Meenakshi S, Vijayakumar V, Verma A K, Rao R S, Godwal B K 2008 *J. Phys. Chem. Solids* **69** 2325
[25] Koudela D, Richter M, Mobius A, Koepernik K and Eschrig H 2006 *Phys. Rev.* B **74** 214103
[26] Kunz M, MacDowell A A, Caldwell W A, Cambie D, Celestre R S, Domning E E, Duarte R M, Gleason A E, Glossinger J M, Kelez N, Plate D W, Yu T, Zaug, JM, Padmore H A, Jeanloz R, Alivisatos A P and Clark S M 2005 *J. Synch. Rad.* **12(5)** 650-658
[27] Mao H K, Xu J and Bell P M 1986 *J. Geophys. Res.*, **B91**, 4673
[28] Hammersley A P et al. 1996 *High Press. Res.* **14** 235
*[29]* Angel R J, Buzak M, Zhao J, Gatta G D and Jacobsen S D 2007 *J. Appl. Cryst.* **40** 26
[30] F. Birch Phys. Rev. B 1947 **71** 809
[31] Srinivasan V, Godwal B K , Grossman J C, and Jeanloz R 2012 (Submitted *Phys. Rev.* B)
[32] Raju S V, Zaug J M, Chen B, Yan J, Knight J W, Jeanloz R, and Clark S M 2011 *J. Appl. Phys.* **110** 023521
[33] Raju S V, Godwal B K, Geballe Z, Williams Q and Jeanloz R, (2012, Submitted to *Phy. Rev.* B)
[34] Errandonea D 2010 *J. Appl. Phys.* **108** 033517
[35] F. A. Lindemann, *Z. Phys.* **11**, 609 (1910)

**Acknowledgements**

We thank the U.S. Department of Energy, Director, Office of Science, and Office of Basic Energy Sciences, under Contract No. DEAC02-05CH11231 and COMPRES for supporting this research at the Advanced Light Source.

# Pressure-driven Phase Transition in CaFeAsF at 40 and 300 K

**S. K. Mishra[1*], R. Mittal[1], S. L. Chaplot[1], S. V. Ovsyannikov[2], D. M. Trots[2], L. Dubrovinsky[2], Y. Su[3], Th. Brueckel[3,4], S. Matsuishi[5], H. Hosono[5], G. Garbarino[6]**

*[1]Solid State Physics Division, Bhabha Atomic Research Centre, Trombay, Mumbai 400 085, India.*
*2Bayerisches Geoinstitut, University Bayreuth, D-95440 Bayreuth, Germany*
*[3]Juelich Centre for Neutron Science, IFF, Forschungszentrum Juelich, FRM II, Lichtenbergstr. 1, D-85747 Garching, Germany.*
*[4]Institut fuer Festkoerperforschung, Forschungszentrum Juelich, D-52425 Juelich, Germany.*
*[5]Frontier Research Center, Tokyo Institute of Technology, 4259 Nagatsuta-cho, Midori-ku, Yokohama 226-8503, Japan*
*[6]European Synchrotron Radiation Facility, BP 220, 38043 Grenoble, France.*

E-mail: skmsspd@barc.gov.in

We carried out systematic investigation of high pressure crystal structure and structural phase transition upto 46 GPa at 40 K and 25 GPa at 300 K in CaFeAsF using powder synchrotron x-ray diffraction experiments. Rietveld analysis of the diffraction data at 40 K reveals a structural phase transition from an orthorhombic to a monoclinic phase at $P_c$ = 13.7 GPa, while increasing pressure. The transition to a lower symmetry phase (orthorhombic to monoclinic) in this 1111 compound under pressure is in contrast with the transition to a high symmetry phase (orthorhombic to tetragonal) as observed in the '122' type compounds. On heating from 40 K at high pressure, CaFeAsF undergoes a monoclinic to tetragonal phase transition around 25 GPa and 200 K. Further, it does not show any post-tetragonal phase transition, as observed in '122' compounds ($CaFe_2As_2$, $BaFe_2As_2$), and remains in the tetragonal phase upto 25 GPa at 300 K. The $dP_c/dT$ slope is found to be positive for CaFeAsF and $CaFe_2As_2$, unlike a negative gradient was found in case of $BaFe_2As_2$.

## 1. Introduction

Iron based superconductor compounds have generated enormous interest because these materials are the first non-copper oxide superconductors with superconducting transition temperature ($T_c$) exceeding 50 K [1-9]. These compounds occur in five different structural classes. Among them the superconductors with the FeAs-1111 structure are known to have the highest superconducting transition temperature. The highest $T_c$ reported so far is about 56 K in Sm-doped SrFeAsF compounds [3]. In SrFeAsF, superconductivity could also emerge by either doping with cobalt directly into the FeAs-layer, or by partial replacement of Sr with rare earth elements (Lanthanum). While, for CaFeAsF, superconductivity was also realized by doping with 3d transition metals (Co and Ni) into the iron sites in the FeAs-layer with the highest $T_c$ of about 22 K in the case of Co-doping [4]. The mechanism of superconductivity in the FeAs-based system remains unclear yet.

At ambient condition, these compounds crystallize in tetragonal symmetry with no magnetic order (i.e paramagnetic in nature). The parent compounds of iron-pnictides undergo a first or second order structural transition below room temperature (typically in the range of 100-210 K), from tetragonal to orthorhombic structure, and magnetic transition from non-magnetic to stripe antiferro magnetic structure. The antiferromagnetic ordering can be suppressed by various ways, namely, by changing the carrier concentration, by charge neutral doping [3-5], and by applying external pressure [6-7]. High pressure experiments play an important role in the field of superconductivity by three primary ways: (i) to test theories of superconductivity and reveal important systematic, (ii) to improve the properties of known superconductors and (iii) to create new superconductors. In this paper, we report systematic investigation of high pressure crystal structures and a structural phase transition in CaFeAsF at ambient and low temperature using powder synchrotron x-ray diffraction experiments and Rietveld analysis of the diffraction data. A detailed manuscript including this work will be published elsewhere [10].

## 2. Experiment

The x-ray diffraction studies were performed on the CaFeAsF powder samples at the ID27 high-pressure beam line of the European Synchrotron Radiation Facility using a monochromatic beam ($\lambda$ = 0.3738 Å ) focused to 3×2 $\mu m^2$. The diamond anvil cells with 150 $\mu$m cullet diamonds with stainless steel gasket and helium as pressure transmitting media were used, which provides the best hydrostatic conditions. During the measurements, we determined the pressure using two ruby balls. The pressure difference determined from ruby balls was always below 0.1–0.2 GPa. The diffraction patterns were collected with a CCD camera and the intensity vs $2\theta$ patterns were obtained using the FIT2D software. The structural refinements were performed using the Rietveld refinement program FULLPROF [11]. During the measurement, the CaFeAsF sample was first cooled to 40 K, and then pressure was increased. The diffraction patterns indicate preferred orientation of the samples along [211] direction.

## 3. Results and Discussion

### 3.1 Structural response of CaFeAsF to Pressure at Low Temperature (40 K)

Inspection of powder synchrotron x-ray diffraction profile of the CaFeAsF at ambient condition confirmed the primitive tetragonal symmetry with space group *P4/nmm* which is consistent with the literature. On cooling, splitting of a (220) peak of the tetragonal phase at 300 K unambiguously confirms the orthorhombic structure at 40 K, as shown in the inset of Figure 1. The pressure-dependent powder x-ray diffraction measurements were carried out at 40 K. Detailed analysis of diffraction patterns reveal that these have dissimilar broadening in various peaks with pressure. The most pronounce changes occur in peak around 2θ=11 degree. Observations of dissimilar broadening in diffraction profiles suggest a possibility of a structural phase transition with pressure. To explore the possibility, we refined powder diffraction data using orthorhombic structure with space group *Cmma*. We found that all the diffraction patterns could be successfully accounted only for up to 12 GPa. A progressive worsening of the quality of the Rietveld fits with increasing pressure was found for pressure above 12 GPa. The most apparent signature of the subtle structural transformation that occurs at above 12 GPa is an inability of the orthorhombic structure (space group *Cmma*) to account satisfactory for the peaks around 11 degree, shown in Fig. 2 (a). Extra broadening (splitting) of the peaks suggests either lowering of the symmetry or coexistence of another high symmetry phase.

Recently, we found that in $BaFe_2As_2$ orthorhombic and tetragonal phases coexist over a large pressure range above 29 GPa at 33 K [5]. In view of this, we made an attempt to refine diffraction data using phase coexistence model. We found that if the profile around 11.5 degree was fitted (forcefully), the calculated peaks around 2.5, 7.9, 9.8 degree arising due to the tetragonal phase are not observed in the experiment data (see Fig. 2(b)). This unsatisfactory quality of the Rietveld fit clearly ruled out the possibility to the phase coexistence.

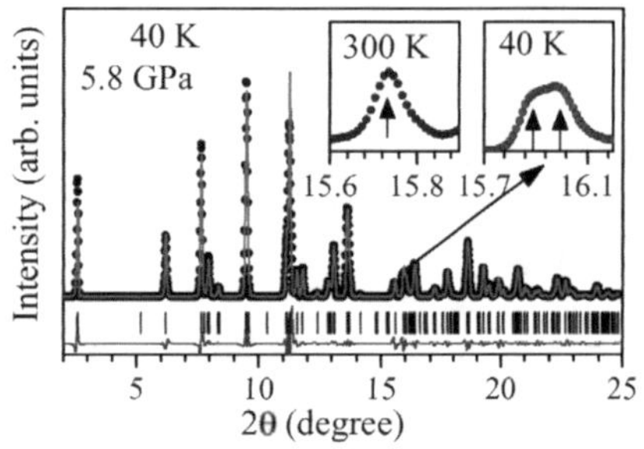


**Fig 1.** (Color online) Observed (black solid circle), calculated (continuous red line), and difference (bottom blue line) profiles obtained after the Rietveld refinement of CaFeAsF at 5.8 GPa & 40 K, in orthorhombic phase (space group *Cmma*). Insets show the (220) reflection of the tetragonal phase and its spitting/broadening at 40K. This is an unambiguous signature for the orthorhombic structure at low temperature.

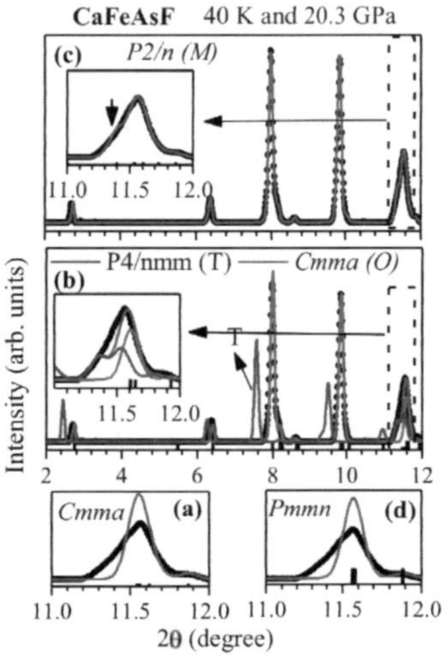


**Fig. 2.** (Color online) Observed (solid black circle) and calculated (continuous red line profiles obtained after the Rietveld refinement of CaFeAsF at 40 K & 20.3 GPa using different models namely: (*a*) an orthorhombic (*Cmma*), (*b*) a combination of tetragonal (*P4/nmm*) and an orthorhombic (*Cmma*); (*c*) monoclinic (*P2/n*) and (*d*) an orthorhombic (*Pmmn*) phases, respectively.

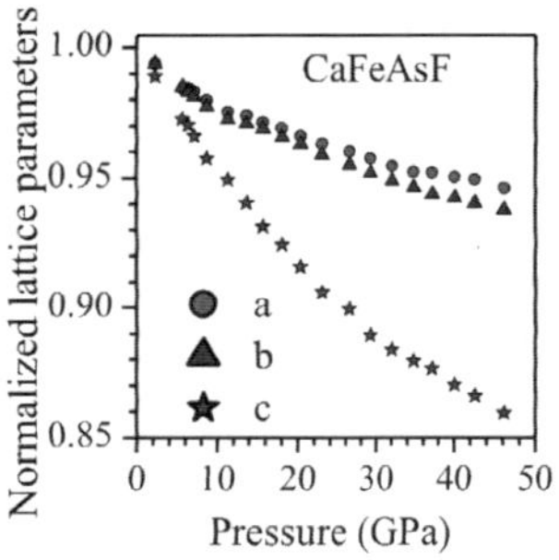


**Fig. 3.** (Color online) Pressure evolution of the normalized lattice parameters of CaFeAsF at 40 K obtained from Rietveld analysis.

Critical inspection of diffraction data only show the splitting in some peaks but not any additional superlattice peak in diffraction patterns. Absence of superlattice peak clearly ruled out the possibilities of integral multiplication of lattice parameters with respect to orthorhombic cell. Thus, it suggests the lowering of symmetry from the orthorhombic structure with space group *Cmma*. We have further explored various possibilities, namely, orthorhombic symmetry with space group *Pbcm, Pmmn;* monoclinic structure with space group *P12/n1* etc. (see Fig. 2 (c) and (d)) to identify the correct space group. We found that orthorhombic space groups could not fit the structural data very well, but the monoclinic structure with space group *P12/n1* (in *cab* setting *P112/n*) could successfully index all the peaks (see Fig. 2(c)). It is well documented in the literature that many iso-structural Fe-based materials undergo a structural phase transition to the monoclinic phase with temperature and pressure [8]. Rietveld refinements employing this structural model are satisfactory for all the diffraction patterns up to the highest pressure measured by us. A careful inspection and analysis of diffraction data reveal that CaFeAsF transforms to the monoclinic structure at $P_c$= 13.7 GPa even though the monoclinic distortion is quite small. The monoclinic angle (β) as a function of pressure shows a sharp discontinuity at the transition pressure at 13.7 GPa and very small further change from $90.5^{o}$ to $91.2^{o}$ on increase in pressure up to 46.2 GPa. Discontinuity in the pressure evolution of the monoclinic angle clearly suggests a first order phase transition in CaFeAsF. Figure 3 depict the pressure evolutions of the normalized structural parameters of CaFeAsF at 40 K. It is evident from this figure that the response of the lattice parameters to pressure is strongly anisotropic with the interlayer spacing (along <001>) showing a significantly larger contraction than the basal plane dimensions (along <100> and <010>).

### 3.2 Structural response of CaFeAsF to Pressure at High Temperature (300 K)

Due to experimental limitations, we could not increase pressure on CaFeAsF beyond 46 GPa at 40 K. However, to see the effect of temperature on the newly stabilized monoclinic phase in CaFeAsF (at 40 K and around 40 GPa), we carried out measurements at different conditions and details are given elsewhere [10]. Careful analysis of the diffraction data show that there is an abrupt change in the diffraction pattern at 25 GPa and 200 K, at which the monoclinic splitting/broadening in Bragg reflections around at $11.5^{0}$ and $16.5^{0}$ disappears. Detailed Rietveld analyses of the diffraction data reveals that the sample undergoes a structural phase transition from the monoclinic to the tetragonal phase at around 25 GPa and 200 K. It retains in the tetragonal phase at 25 GPa and 300 K. We note that while the phase transition is observed with pressure at 40 K, no transition is found at 300 K. In case of $SmFeAsO_{0.81}F_{0.19}$ and $LaFeAsO_{0.9}F_{0.1}$ [9], no phase transition is found also at 300 K up to 20 and 32 GPa respectively.

## 4. Conclusions

In summary, we have carried out systematic investigation of high-pressure crystal structures and structural phase transition upto 46 GPa at 40 K and 25 GPa at 300 K in CaFeAsF using powder synchrotron x-ray diffraction experiments and Rietveld analysis of the diffraction data. It is interesting to note that at 40 K, CaFeAsF shows the orthorhombic to monoclinic phase transition at $P_c$ =13.7 GPa, whereas at room temperature the tetragonal phase is stable upto 25 GPa. The observation of $dPc/dT>0$ is similar to that observed in $CaFe_2As_2$ in which a collapsed tetragonal phase transition occurs at a lower pressure (0.3 GPa) at a lower temperature (50 K), in comparison with 1.7 GPa at 300 K. However, this behaviour of dPc/dT may not hold for Ba/Sr compounds. In addition to this, the transition to a lower symmetry phase (orthorhombic to monoclinic) in the '1111' compounds (CaFeAsF/SrFeAsF) under pressure is in contrast with the transition to a high symmetry phase (orthorhombic to tetragonal) in the '122' ($BaFe_2As_2$/ $CaFe_2As_2$)-type compounds.

## Acknowledgments

S. K. Mishra and R. Mittal thank Department of Science and Technology (DST), India for providing financial support to carry out synchrotron x-ray diffraction at European Synchrotron Radiation Facility, Grenoble, France.

[1] Paglion J. and Greene R L 2010 *Nat. Phys. **6**, 645*, Mazin I. I. and Johannes M. D 2009 *Nat. Phys. **5**, 141*; Wang Fa and Lee Dung-Hai, 2011 *Science **332**, 200.*
[2] Imai T., *et al* 2009 *Phys. Rev. Lett.* **102**, 177005;
[3] Wu G, Xie *et al.* 2009 *Phys.: Condens. Matter* 21 142203.
[4] Xiao Y., *et al.* 2009 *Phys. Rev. B*, **79** 060504, Ishida, K., *et al.* 2009 *J. Phys. Soc. Jpn* 78, 062001.
[5] Mittal R., *et al.*, 2011 *Phys. Rev B*, **83**, 054503.
[6] Wang C, *et al.* 2009, Phys. Rev B 79, 054521.
[7] Okada H., *et al.*, 2010 Phys. Rev. B, **81**, 054507; *ibid*, 2010 *J. of Phys.: Confer. Series* 200 012151.
[8] Bao W., *et al.* 2009 *Phys. Rev. Lett.* **102**, 247001.
[9] Garbarino G., *et al.*, 2011 *Phys. Rev. B* **84**, 024510; *ibid* 2008 *Phys. Rev. B **78**, 100507.*
[10] Mishra S K *et al.*, 2011 *Phys. Rev. B* (to be published).
[11] Rodriguez-Carvajal J., 1992 *Physica B* **192**, 55.

# Melting of Sn to 1 Mbar

**R Briggs[1], D Daisenberger[1], A Salamat[2], G Garbarino[2], M Mezouar[2], M Wilson[3] and P F McMillan[1,4]**

[1] Department of Chemistry and Institute of Shock Physics, Christopher Ingold Laboratories, University College London, 20 Gordon Street, London, WC1H 0AJ, UK

[2] European Synchrotron Radiation Facility, BP 220, F-38043, Grenoble, France

[3] Department of Chemistry, Physical and Theoretical Chemistry Laboratory, University of Oxford, South Parks Road, Oxford, OX1 3Q, UK

E-mail: p.f.mcmillan@ucl.ac.uk

**Abstract.** The melting point of Sn was determined between 20-105 GPa using laser-heated diamond anvil cell experiments, coupled with *in situ* synchrotron X-ray diffraction studies. In agreement with previous LH-DAC speckle experiments, we observe a flattening of the melting slope between P = 40–60 GPa. However, we also observe that this plateau is followed by a further increase in the melting slope above P ~ 70 GPa, leading to a remarkably high melting point of $T_m$ = 5500 K by P = 105 GPa.

## 1. Introduction

The determination of melting points of elements and compounds to pressures in the megabar range is of interest for planetary sciences, as well as fundamental and applied materials physics studies. Typically such studies have been carried out using shock methods accompanied by empirical or *ab initio* simulations [1-3]. Recent advances in laser-heated diamond anvil cell (LH-DAC) techniques have extended static determinations into the megabar pressure range, with melting temperatures recorded up to between 6000-7000 K [4]. Several techniques have been explored to provide diagnostics of the melting event during LH-DAC experiments. Two main methods have emerged as providing the most understandable results. The first is a laboratory-based technique that relies on observation of the onset of changes in the laser 'speckle' derived from reflections and interference of coherent light reflected from the sample surface [5]. The observation of rapid fluctuations in the speckle pattern during laser heating is presumed to signal the melting event, although it is also possible that surface reconstruction phenomena or chemical reaction events may also contribute. A second technique has emerged recently based on synchrotron X-ray scattering (SXS) that relies on the observation of a liquid S(Q) pattern produced within the bulk sample [6].

In a first set of LH-DAC + SXS experiments performed on elemental Pb, Dewaele *et al.* studied the melting relation to below megabar pressures and determined $dT_m/dP$ values in close correspondence with both *ab initio* theory and shock wave studies [6]. Unexpected divergences had been noted previously between melting curves for elements determined by the speckle technique and the results of *ab initio* calculations, SXS melt results and shock wave experiments [4,7]. The advent of the SXS method has now begun to resolve some of these discrepancies by identification of chemical

[4] To whom any correspondence should be addressed.

Published under licence by IOP Publishing Ltd

reactions occurring during the LH-DAC experiments, with C derived from the diamond surfaces reacting to form metal carbides [4].

Sn is a well-known element that occupies an intermediate position in the periodic table below the semiconductors Si and Ge and above metallic Pb. Its melting relations have been studied using explosive gas gun techniques and it is established that the Hugoniot crosses the melting curve near 50 GPa [8]. The LH-DAC speckle method was used to study Sn melting to P ~ 68 GPa and T ~ 2300 K. Those results found that flattening of the melt curve occurred above 38 GPa to achieve a constant value near $T_m$ ~ 2300 K [9]. Here we obtained new data on the melting of Sn to P > 1 Mbar using the LH-DAC heating and the SXS technique. The data below 60 GPa are in agreement with the speckle results . However we find an sharp rise occurring in the melting temperature above P ~70 GPa, leading to the remarkably high melting value $T_m$ = 5500 ± 500 K at P ~ 105 GPa.

## 2. Experimental Methods

Experiments were conducted at beamline ID27 of the European Synchrotron Radiation Facility (λ = 0.3738 Å). Samples of pure Sn powder (99.99% Aldrich, packed under Ar) were pre-compressed into discs (~ 5μm thick) and loaded into membrane driven diamond anvil cells under an inert atmosphere. 200 μm or 150/300 μm bevelled diamond anvils were used. NaCl or KBr were used as pressure transmitting media and to thermally insulate the diamonds. Re was used as a gasket material and was pre-indented to ~30 μm thickness. Ruby grains were added to the sample chamber as an internal pressure gauge: pressures were also established during the X-ray diffraction experiments using the NaCl equation of state [10]. The samples were compressed to the target pressure and then heated simultaneously from both sides using Nd:YAG IR lasers (λ = 1064 nm). To reduce thermal gradients, the lasers were slightly defocused on each side. The X-ray beam was focused to a 2 x 2 $\mu m^2$ area at the center of the hot spot. Thermal emission spectra were recorded from the same 2 x 2 $\mu m^2$ area and fitted to Planck and Wien functions to determine T. X-ray diffraction patterns were recorded at 1–2 s time resolution to monitor the state of the sample continuously. Raw XRD patterns were integrated as a function of 2θ using FIT2D [11]. XRD refinements were carried out using POWDERCELL [12] and GSAS [13-14]. No oxides or carbides were observed to form during any of the laser heating experiments, with no differences in results from loadings conducted under inert atmosphere or ambient. The appearance of a liquid S(Q) signature in the XRD patterns was taken to indicate the onset of melting (Fig. 1) [15-16].

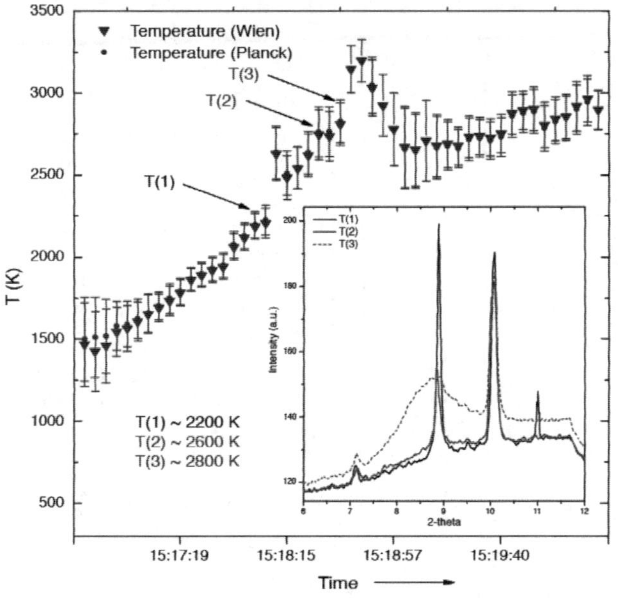


Figure 1: Temperature profiles of LH-DAC samples determined by fitting Planck and Wien functions to the thermal emission spectra. This diagram illustrates typical results from a heating series at P~49 GPa. The inset shows integrated XRD patterns (λ=0.3738 Å) for 3 different T values. The liquid S(Q) scattering is clearly observed to appear in pattern 3, corresponding to the arrowed position in the heating ramp.

### 3. Results and Discussion

Our LH-DAC melting data determined between 20-105 GPa are shown in Figure 2. The results obtained below P < 40-60 GPa agree well with previous shock data and LH-DAC melting points determined using the laser speckle method [8-9]. The melt slope increases smoothly above the ß-Sn / bct-Sn / liquid triple point until P ~ 40 GPa, above which the melting relation flattens to attain a nearly constant $T_m$ value. A similar result has been found for a wide range of elements studied using the LH-DAC speckle technique and has been used to support new models for the melting event and melt structures [5], though they are not always supported by *ab initio* calculations or shock studies [7]. In the case of Sn, the shock Hugoniot crosses the melting curve at at P ~ 50 GPa and T ~ 2300 K [8]. Isentropic shock compression experiments have been carried out using Sn samples preheated to T ~ 600 K [17], but these provide no further information on the higher P melting behavior. Recently Hu *et al.* studied shock properties of Sn using sound velocity measurements as a diagnostic technique to indicate onset of melting or other phase transformation phenomena [18]: however, the incipient melting temperature suggested by these authors was significantly lower than that suggested by previous shock studies [8] or the static melting results.

In our present study using LH-DAC + SXS techniques, we observe a sharp rise occurring in the melting slope at above P > 70 GPa, leading to a remarkably high melting value of $T_m$ ~ 5500 ± 500 K at P ~ 105 GPa that could not have been predicted from the previous experiments. The usual explanation for such a change in melting slope would be occurrence of an underlying solid-solid transition, with an associated triple point. Sn is reported to undergo a *bct* to *bcc* phase transition in this region. However, the two polymorphs coexist over a wide pressure range with an upper limit of coexistence reported at P ~ 56 GPa, that does not correspond to the pressure where a triple point would be expected from our melt data (P ~ 70 GPa) [19,20]. Even ignoring the absolute transition pressure values that might be affected by the use of different pressure-transmitting media and the presence of non-hydrostatic environments, the volume change associated with the *bct- bcc* transition is minimal (~ 0.5 %) and could not account for the dramatic rise in the melting temperature. Likewise, if entropy effects were responsible, this would require the entropy of *bcc*-Sn to be significantly higher than that of *bct*-Sn. This is unlikely. We are now examining alternative models to explain the unusual melting properties of this apparently simple element with unexpectedly complex behavior.

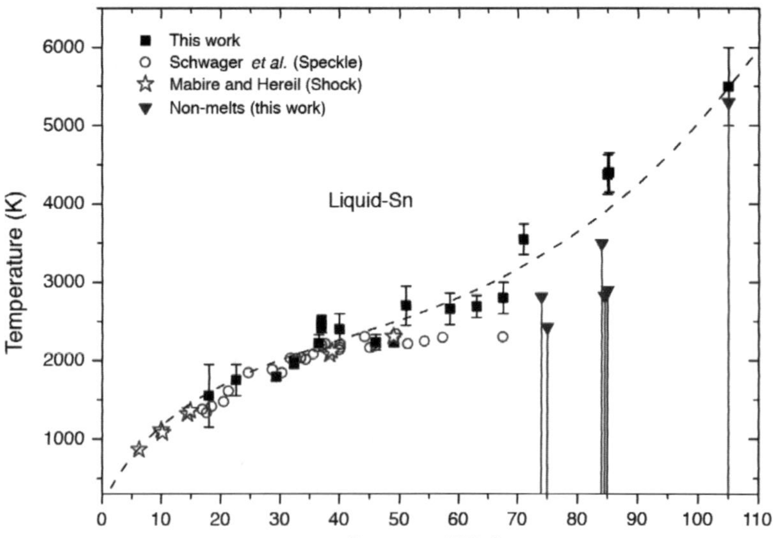


Figure 2: Melting curve of Sn to P > 1 Mbar. Our data are shown as filled squares. Open circles represent LH-DAC speckle data [9] and open stars are shock data [8]. Triangles represent experimental runs where only solid-Sn was observed in the LH-DAC X-ray diffraction patterns.

**Acknowledgements**
This project is supported by funding to PFM from the Institute of Shock Physics (ISP) as a subcontract from Imperial College London. RB and DD are supported directly by this contract. We acknowledge the European Synchrotron Radiation Facility and the ID27 beamline team for access to and support with the synchrotron radiation experiments.

**References**

[1] Brown J and McQueen R G 1986 *J. Geophys. Res.* **91-B7** 7485
[2] Streitz F H, Glosli J N and Patel M V 2006 *Phys. Rev. Lett.* **96** 225701
[3] Wu C J, Söderland P, Glosli J N and Klepeis 2009 *Nature Materials* **8-3** 223
[4] Dewaele A, Mezouar M, Guignot N and Loubeyre P 2010 *Phys. Rev. Lett.* **104** 255701
[5] Errandonea D, Schwager B, Ditz R, Gessmann C, Boehler R and Ross M 2001 *Phys. Rev. B.* **63** 132104
[6] Dewaele A, Mezouar, Guignot N and Loubeyre P 2007 *Phys. Rev. B.* **76** 144106
[7] Taioli S, Cazorla C, Gillan M J and Alfè D 2007 *Phys. Rev. B.* **75** 214103
[8] Hereil P-L and Mabire C 2000 *J. Phys. IV. France* **10** 799
[9] Schwager B, Ross M, Japel S and Boehler R 2010 *J. Chem. Phys.* **133** 084501
[10] Decker D L 1971 *J. Appl. Phys.* **42-8** 3239
[11] Hammersley A P, Svensson S O, Hanfland M, Fitch A N and Häusermann D 1996 *High Pressure Research* **14** 235
[12] Kraus W and Nolze G 1996 *J. Appl. Cryst.* **29** 301
[13] Larson A C and Von Dreele R B, *Los Alamos National Lab Report* (1994)
[14] Toby B H 2001 *J. Appl. Cryst.* **34** 210
[15] Santamaría-Pérez D, Ross M, Errandonea D, Mukherjee G D, Mezouar M and Boehler R 2009 *J. Chem. Phys.* **130** 124509
[16] Errandonea D, Somayazulu M, Häusermann D and Mao H K 2003 *J. Phys. Cond. Mat.* **15** 7635
[17] Davies J-P and D. B. Hayes 2004 *AIP Conference Proceedings*
[18] Hu J, Zhou X, Dai C, Tan H and Li J 2008 *J. Appl. Phys.* **104** 083520
[19] Olijnyk H and Holzapfel W B 1984 *Journal De Physique* **C8-11** 153
[20] Desgreniers S, Vohra Y K and Ruoff A L 1989 *Phys. Rev. B.* **39-14** 10-359

Journal of Physics: Conference Series **377** (2012) 012036 doi:10.1088/1742-6596/377/1/012036

# High Pressure Phase Transitions in Yttria, $Y_2O_3$

**Preyoshi P. Bose[1], M. K. Gupta[1*], R. Mittal[1], S. Rols[2], S. N. Achary[3], A. K. Tyagi[3] and S. L. Chaplot[1]**

[1]*Solid State Physics Division, Bhabha Atomic Research Centre, Trombay, Mumbai, 400085, India*
[2]*Institut Laue-Langevin, BP 156, 38042 Grenoble Cedex 9, France*
[3]*Chemistry Division, Bhabha Atomic Research Centre, Trombay, Mumbai, 400085, India*

E-mail: mayankg@barc.gov.in

We have carried out neutron inelastic scattering measurement of the phonon density of states and lattice dynamic calculations for $Y_2O_3$ using ab-initio density functional perturbation theory and interatomic potential model. The calculations are found to be in good agreement with experimental data, indicating that the potential model can be used for the calculation of phase diagram of $Y_2O_3$. The model is then used for free energy calculation to understand the stability of various phases as a function of pressure and temperature.

## 1. Introduction

Yttria ($Y_2O_3$) is an industrially and technologically very useful ceramic material used to increase the ductility of aluminum oxide (alumina ceramics) and to stabilize the cubic and tetragonal phases of zirconia, which is particularly important in the design of solid oxide fuel cells. Dense polycrystalline form of yttria is used in nuclear applications. Raman spectroscopy measurements [1] on yttria single crystals reported two phase transitions, namely, cubic to monoclinic and monoclinic to hexagonal at 12 and 19 GPa respectively. Recent high pressure X-ray diffraction experiments [2] on cubic $Y_2O_3$ at room temperature showed coexistence of the cubic, monoclinic and hexagonal phases in a pressure range from 15 GPa to 25.6 GPa. Kinetics seems to play a major role in determining the sequence of phase transitions.

We have carried out extensive lattice dynamical studies to provide insight into the phase transition mechanism of yttria. The phonon spectrum has been calculated using the shell model as well as *ab-initio* approach. We have also performed inelastic neutron scattering investigations to derive the phonon density of states (PDOS) of $Y_2O_3$ at ambient pressure and temperature. The shell model calculations results are in good agreement with the inelastic neutron data and *ab-initio* calculations. This allowed testing of our model, which is used to explore the high pressure and high temperature regime of the phase diagram. The model has been used for calculation of free energy in various phases of $Y_2O_3$. Our approach allows us to calculate the pressure dependence of the vibrational energy contribution to the Gibbs free energy, which is computationally expensive to calculate using *ab-initio* approach as cubic and monoclinic phase of $Y_2O_3$ has 40 and 30 atoms/primitive cell. The details of work have been published elsewhere [3].

## 2. Experiment

The polycrystalline sample in the cubic phase of $Y_2O_3$ was characterized using X-ray diffraction. The analysis of powder diffraction data indicated that sample is in single phase. The inelastic neutron scattering experiments were performed using the thermal time-of-flight neutron spectrometer IN4 at the Institute Laue Langevin (ILL) located in Grenoble (France). A quantity of 25 grams of polycrystalline $Y_2O_3$ sample was placed inside a thin aluminum sample holder. The incident neutron wavelength of 2.4 Å (14.2 meV) was chosen which allowed the spectra to be obtained in the neutron energy gain mode at 300 K. We used a Fermi chopper speed of 28000 RPM to allow for a time focusing condition in the inelastic regime (Inelastic time focusing). The detector bank of IN4 covers scattering angles up to 120°. After usual correction, the spectra at all detector angles were summed up to increase the statistics and to average out the coherent effect which is necessary to allow for a correct estimation of the phonon density of states (the so called "incoherent approximation").

## 3. Computational Details

We have performed calculations using potential model and first principles. The pseudopotential generated by PBE exchange correlation functional [4] under generalized gradient approximation (PBE-GGA) has been used for *ab-initio* calculations in cubic phase of $Y_2O_3$. The calculations have been carried out using plane wave basis sets and Quantum Espresso software package [5]. We have chosen 4×4×4 k mesh for self-consistent field calculations with energy cutoff of 45 Rydberg, which is found to be sufficient for convergence of the order of meV. The k point mesh has been generated using Monkhorst-pack method [6] Dynamical matrices were calculated on a 3×3×3 mesh in the irreducible Brillouin zone.

The shell model is based on a transferable potential which consists of long-range Coulmbic interaction, short range Born-Mayer type repulsive terms, and weakly attractive van der Waals terms. The form of the interatomic potential used in our model is given by the following expression:

$$V(r) = \frac{e^2}{4\pi\varepsilon_0}\frac{Z(k)Z(k')}{r} + a\ \exp\left(\frac{-br}{R(k)+R(k')}\right) - \frac{C}{r^6}$$

Where, a=1822eV and b=12.364 are empirical constants. We have successfully used [25, 26] this set of parameters in the lattice dynamical calculations of several complex solids. The term $C_{ij}$ =100 eVÅ$^6$accounts for the vander Waals interaction between O-O pairs. The effective charge Z(k) and radii R(k) parameters used in our calculations are Z(Y)= 2.4 , Z(O)= -1.6 , R(Y)= 1.931Å and R(O)= 1.89 Å. The polarizability of the oxygen atoms is introduced in the framework of the shell model [27, 28]. The shell charge and shell core force constants for oxygen atoms are −3.75 and 170 eV/Å$^2$ respectively. The calculations are carried out using the software DISPR [7] developed at Trombay.

## 4. Results and Discussions

### *4.1 Phonon density of states*

The comparison between the experimental and calculated neutron-weighted phonon density of states of yttria in the cubic phase using the potential model as well as ab-inito approach is shown in figure 1a. The experimental spectrum can be divided into two parts: the low frequency part (below 40 meV) consists of a series of weak features (14, 17 and 20 meV) superimposed to a general curve having a maximum at 20 meV. The high frequency part features intense contributions at 45 meV (a multi-component peak), 60 meV and 72 meV. No modes have been observed for frequencies higher than 80 meV. The general characteristics of the experimental features are well reproduced by the calculations. Both the calculations agree very well with the experimental data. In particular, the calculations reproduce very well the low frequency part of the spectrum where the modes involving mainly Y vibrations dominate the dynamics, as indicated by the partial density of states (see Fig. 1b). The comparison of calculated partial density of states (Fig. 1b) of various atoms in cubic $Y_2O_3$ from both the potential as well as *ab-initio* calculations indicates slight difference in the phonon spectra for oxygen atoms above 40 meV.

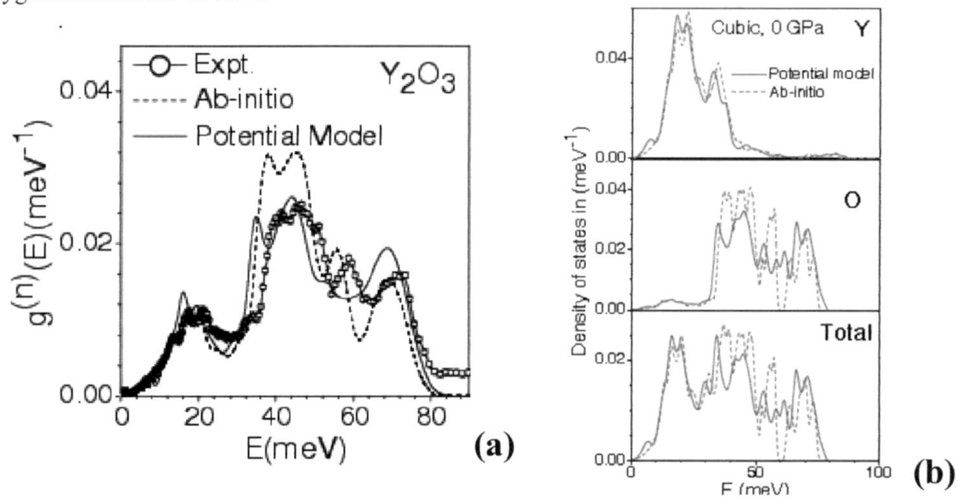


**Figure 1 (a)** The comparison of the calculated and experimental neutron weighted phonon density of states for cubic phase of yttria at ambient pressure [3]. The *ab-initio* as well as potential model calculations in the cubic phase are carried out at 0 K, while the measurements were performed at 300 K. **(b)** The calculated partial densities of states in various phases of $Y_2O_3$. The calculations in the cubic phase at 0 GPa are carried out using both the *ab-initio* and potential model.

### *4.2 Free energy and phase diagram*

The calculated phonon spectra from the model as well as *ab-initio* calculations agree very well with the experimental data, indicating that potential model can be used to obtain the phase diagram of yttria. The Gibbs free energy was calculated using the model potential in the various phases of $Y_2O_3$. The Gibbs free energy of the $n^{th}$ phase is given by

$$G_n = \phi_n + PV_n - TS_n$$

Where, $\phi_n$, $V_n$ and $S_n$ respectively relate to internal energy, lattice volume and vibrational entropy of the $n^{th}$ phase. The Gibbs free energy of the various phases of $Y_2O_3$ are compared at different pressures for a fixed temperature, which allows deriving the phase diagram presented in figure 2(a). We find that above 15 GPa at 300 K the monoclinic phase has lower free energy as compared to the cubic phase, and the phonon modes are stable. The calculations therefore predict the transition from cubic to monoclinic phase at this critical pressure, while experimental data[1] report this transition at 12 GPa at room temperature.

The difference between the free energies of the competing monoclinic and hexagonal phases of $Y_2O_3$ as a function of pressure (at 300 K) is reported in figure 3a. The calculations show that the transition from the monoclinic to the hexagonal phase at 48 GPa (Fig. 2(b)) is of first order in nature, as can be seen from the non zero slope of the free energy difference (Fig. 2(b)) between the two phases at the transition pressure. The experimental data [12] show that the hexagonal phase is stable at pressures above a pressure of 26 GPa. The Gibbs free energy calculation qualitatively reproduces the

Journal of Physics: Conference Series **377** (2012) 012036 doi:10.1088/1742-6596/377/1/012036

phase diagram of yttria. This is satisfactorily since it is very difficult to reproduce free energy differences with requisite high accuracy for phase diagram calculation. It is also difficult task to identify equilibrium phases from experiments at high pressure due to large hysteresis. The transition pressure is also affected by the kinetics and also depends on the doping [8] of rare earth ions in $Y_2O_3$. Such effects are not accounted for in our investigations.

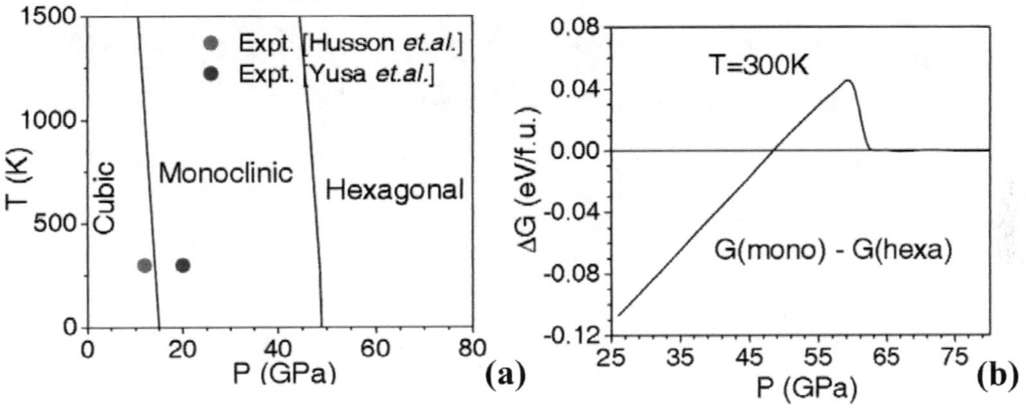


**Figure 2. (a)** The calculated phase diagram obtained from free energy calculations [3]. The closed and open circles are the experimental data [1,2] for the cubic to monoclinic and monoclinic to hexagonal phase transitions respectively. **(b)** The difference in the free energies of hexagonal and monoclinic phases of $Y_2O_3$ as a function of pressure at 300 K.

As shown in Fig. 2 (b) the difference becomes zero above 62 GPa suggesting the presence of a second order transition between the two phases. The small difference in free energy can be tuned by the incorporation of a small amount of impurity if the first order transition does not already occur at lower pressure. This might be the origin of the coexistence of the monoclinic and hexagonal phases at high pressure [2] or for the two different sequences of phase transition [8]for pure $Y_2O_3$ (cubic-hexagonal) and Eu-doped $Y_2O_3$ (cubic-monoclinic-hexagonal) with temperature.

## 5. Conclusions

We have reported inelastic neutron scattering measurements on polycrystalline samples of $Y_2O_3$ in the cubic phase. The measurements are in agreement with *ab-initio* as well as shell model calculations. The free energy calculations in various phases of $Y_2O_3$ help us to study the relative stability of high-pressure phases.

## References

[1] Husson E., Proust C., Gillet P. and Itie J. P. 1999 Mat. Res. Bull. **34** 2085.
[2] Yusa H., Tsuchiya T., Sata N. and Ohishiz Y. 2010 Inorg. Chem. **49** 4478.
[3] Preyoshi P. Bose *et al* 2011 Phys. Rev. B **84**, 094301.
[4] Perdew J. P., Burke K. and Ernzerhof M. 1996 Phys. Rev. Lett. **77** 3865.
[5] Monkhorst H. J. and Pack J. D., 1976 Phys. Rev. B **13** 5188.
[6] Giannozzi P., *et al*., 2009 J. Phys. Condens. Matter **21** 395502.
[7] Chaplot S. L. Unpublished.
[8] Wang L., Pan Y., Ding Y., Yang W., Mao W. L., Sinogeikin S. V., Meng Y., Shen G. and Mao H. 2009 Appl. Phy. Lett. **94** 061921.

# Raman study of Si clathrates $Sr_8Al_xGa_{16-x}Si_{30}$ under High Pressure

**Y Imaeda[1], T Kume[1†], S Sasaki[1], H Shimizu[1], K Kishimoto[2], N Ikeda[2] and T Koyanagi[2]**
[1]Department of Materials Science and Technology, Gifu University, 1-1 Yanagido, Gifu 501-1193, Japan
[2]Graduate School of Science and Engineering, Yamaguchi University, Ube, Yamaguchi 755-8611, Japan

E-mail: kume@gifu-u.ac.jp

**Abstract.** High-pressure Raman scattering experiments were carried out for Si clathrates ($Sr_8Al_xGa_{16-x}Si_{30}$) with structures of type I and type VIII. The type I clathrate ($Sr_8Al_7Ga_9Si_{30}$) indicated spectral changes at 13 GPa and 19 GPa. The spectral shape was characteristic of amorphous above 19 GPa. For type VIII ($Sr_8Al_8Ga_8Si_{30}$), the spectral shape changed to amorphous like one at 20 GPa. After releasing the pressure after the amorphizations, the measured Raman spectra were the same as that of type I, for both the cases of type I and VIII, which suggests that type VIII structure irreversibly transforms to type I via an amorphous phase.

## 1. Introduction

Semiconductor clathrates consist of polyhedron cages formed by the host atoms (E) of group 14 elements. Each cage accommodates alkali, alkali-earth, or halogen atom as the guest atom (G)[1-3]. There are several types of structures in the clathrates, such as type I, II and so on. In most clathrates, there are two or more kinds of cages in the structure. As shown in Fig. 1(a), type I structure (space group Pm-3n, No. 223) has two kinds of cages: $E_{20}$ pentagonal dodecahedra centred by G1 and $E_{24}$ tetrakaidecahedra centred by G2. There are two G1 (2a site) and six G2 (6d site) guest atoms per $G_8E_{46}$ unit cell [1-3]. However, type VIII structure (space group I-43m, No. 217) is known to consist of only one kind of cage (Fig. 1(b)). This is unique in a sense of one kind of cage. Eight guest atoms at 8c site are enclosed in the identical eight cages per unit cell, resulting in the same compositional formula as that of Type I ($G_8E_{46}$)[4-7].

The clathrates of $Sr_8Al_xGa_{16-x}Si_{30}$ are able to take both of type-I and VIII structures by the composition of Al and Ga; the Ga-rich ($0 < x < 7$) and Al-rich ($8 < x < 13$) conditions lead to type I and type VIII, respectively[7]. According to detailed structural study of $Sr_8Al_xGa_{16-x}Si_{30}$, the unit cell volume of type VIII is smaller than that of type I[7], implying the type VIII structure more favourable than type I under high pressure. For type I clathrates, high pressure studies have been extensively performed [8-11], but the study of type-VIII clathrate under high pressure has not been reported yet.

† To whom any correspondence should be addressed.

In this works, the high pressure Raman measurements were carried out for type VIII ($Sr_8Al_7Ga_9Si_{30}$) and type I ($Sr_8Al_8Ga_8Si_{30}$) clathrates in order to clarify the structural stability and the phase transition of type I and VIII clathrates under high pressure. For type I clathrate, spectral changes were observed at 13 GPa and 19 GPa. Above 19 GPa, amorphous-like spectra were obtained. Also for type VIII, a change of the spectral shape to amorphous-like one was observed at 20 GPa. For both cases of type I and VIII, the Raman spectra characteristics of type I were observed after releasing the pressure, suggesting that type VIII structure irreversibly transforms to type I through an amorphous state.

## 2. Experimental

The quaternary samples of $Sr_8Al_xGa_{16-x}Si_{30}$ were synthesized by using Sr (99.9%), Al (99.999%), Ga (99.9999%), and Si (99.999%) as starting materials [7]; they were weighted to the composition ratio of $Sr_8Al_xGa_{16-x}Si_{30}$ ($0 \leq x \leq 16$). $Sr_8Al_xGa_{16-x}Si_{30}$ ingots were prepared by the arc melting method. The obtained ingots were ground to fine powder and then sintered by spark plasma sintering (SPS) with graphite dies. Sintering was carried out under the following conditions: sintering temperature, 1023 K; sintering time, 60 min; and pressing pressure, 40 MPa. The structural details obtained by Rietveld analysis [12] for XRD data have been already described elsewhere [7].

High pressure experiments have been carried out using a diamond anvil cell (DAC) with a culet of 0.3 mm in diameter. A hole of 0.15 mm in diameter drilled on a pre-indented rhenium gasket of about 0.04 mm in thickness served as the sample chamber. Samples used in this study were $Sr_8Al_7Ga_9Si_{30}$ (type I) and $Sr_8Al_8Ga_8Si_{30}$ (type VIII). Each sample was cut and polished into a plate of ~50 μm in size with ~20 μm in thickness. The sample was loaded into DAC with Ar pressure medium. Raman spectra were measured using a spectrometer equipped with a triple polychromator and a charge coupled device detector. A solid state laser with a wavelength of 532 nm was used for excitation. Pressure calibration was carried out by the ruby method[13].

## 3. Results and discussion

Figure 2 (a) and (b) show Raman spectrum of $Sr_8Al_7Ga_9Si_{30}$ (Type I) and $Sr_8Al_8Ga_8Si_{30}$ (Type VIII) at 1 atm, respectively. Each figure shows the Raman spectrum before and after high pressure experiments. Let us discuss the spectra before high pressure experiments. The spectra were consistent with those reported in the previous study[14]. For the clathrate type-I structure (Pm-3n) without any

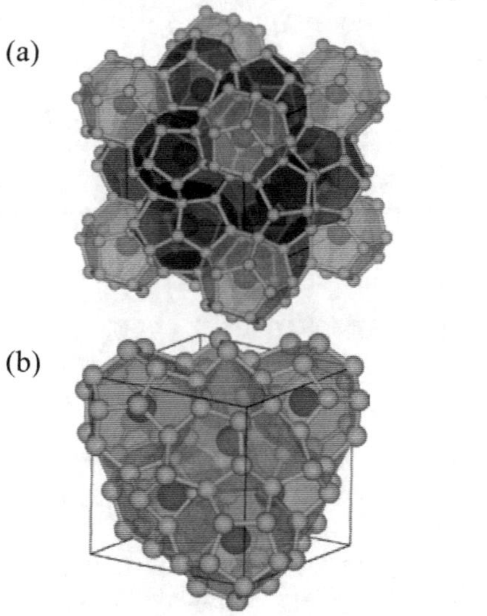


**Figure 1.** Crystal structures of (a) Type I, (b) Type VIII.

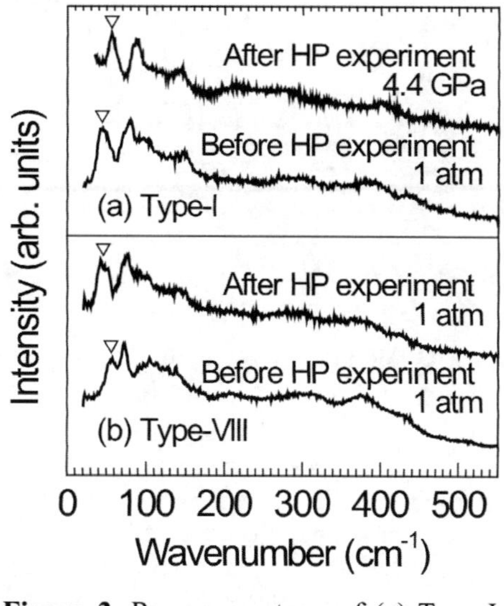


**Figure 2.** Raman spectrum of (a) Type I, (b) Type VIII before and after high pressure experiment.

defects and disorders, the first order Raman-active modes are predicted by the group theory to have $3A_{1g} + 8E_g + 9T_{2g}$ symmetries (20 peaks, 46 modes) under the assumption of the guest being at the on-centre position (6d site) [15], where two Raman signals of symmetry $E_g$ and $T_{2g}$ are expected to appear in a low-frequency region as the vibrational modes of the guest atoms in the large $E_{24}$ polyhedrons (6d sites). The vibrations associated with guest atoms in the small $E_{20}$ polyhedrons are not Raman active. For the type-VIII structure (I -43m), we can derive the first order Raman-active modes to be $4A_1 + 6E + 13T_2$ symmetries (23 peaks, 55 modes), where four Raman signals of symmetry $A_1$, E, and $2T_2$ are expected to appear in a low frequency region as the vibrations of the guest atom in the cage (8c site). For the clathrate compounds, the vibrations of guests and hosts are separated because the interaction of guest and host is much weaker than covalent bonds between the host atoms. Furthermore, the atomic mass of Sr (87.62) is quite larger than those of the host atoms. Therefore, the guest vibrations are observed in the lowest frequency region. For the present samples, the peaks of 30–65 $cm^{-1}$ (open triangles) correspond to the guest Sr vibrations, whereas the peaks above 65 $cm^{-1}$ are due to the host cages. The guest vibration of type VIII has a higher frequency than that of type I. This is because the guest of type VIII is more tightly enclosed in the cage than that of type I [14].

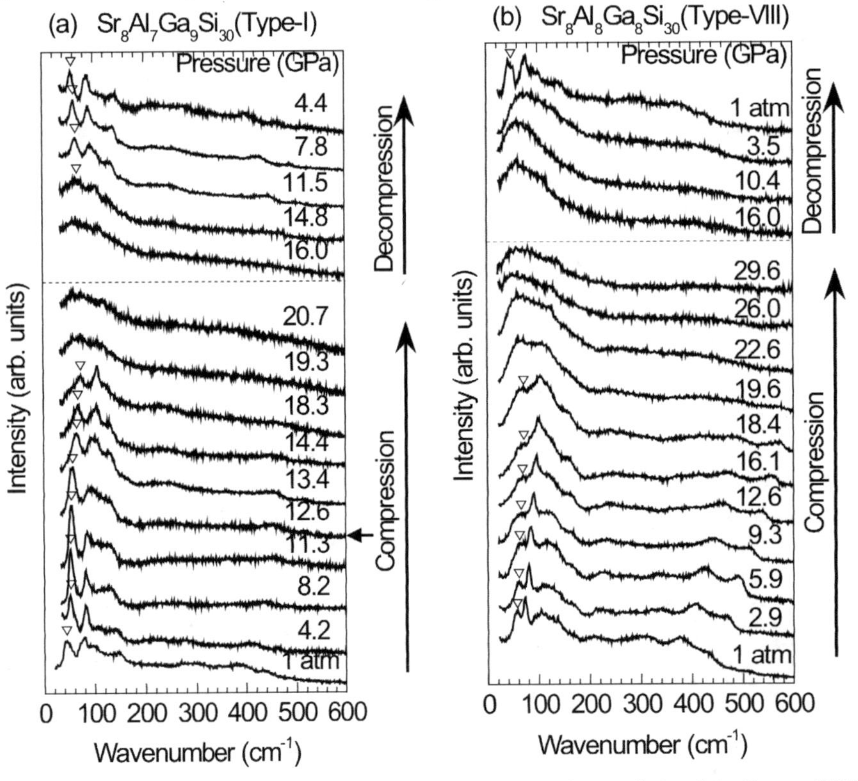


**Figure 3.** Raman spectra of (a) Type I ($Sr_8Al_7Ga_9Si_{30}$), (b) Type VIII ($Sr_8Al_8Ga_8Si_{30}$) measured under high pressures.

Figures 3 (a) and (b) show Raman spectra of $Sr_8Al_7Ga_9Si_{30}$ (type I) and $Sr_8Al_8Ga_8Si_{30}$ (type VIII) obtained under high pressures, respectively. From Fig. 3 (a), the type I clathrate shows spectral changes around 13 and 19 GPa, suggesting structural phase transitions. At 19 GPa where all the peaks disappear and only a broad band is observed, the occurrence of amorphization is strongly suggested. When the pressure is released (upper panel of Fig. 3(a), the Raman peaks were completely recovered. Thus, this amorphization is reversible phenomena. The reversible amorphization for the covalent bonding materials have been reported on the clathrasils (clathrate of $SiO_2$) by Tse et al.[16]. In Fig. 3 (b), we note that the Raman spectrum of type VIII also changed at ~20 GPa, and amorphous like spectra were observed. When the pressure was released, the Raman peaks appeared again. It should be noted, however, that the obtained spectrum was not of type VIII, but of type I as shown in the upper panel of Fig. 3(b). As mentioned earlier, in the case of type I, the spectrum after releasing the pressure was of type I. The facts in the above suggest that type I structure is more favourable than type VIII under high pressure.

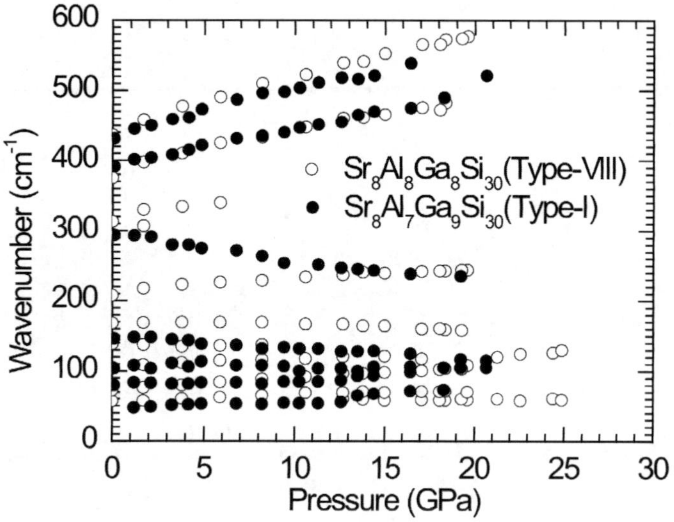


**Figure 4.** Pressure dependence of Raman peaks of Type I (closed circles) and Type VIII (open circles).

Figure 4 shows Rama frequencies of type I and VIII as a function of pressure. Let us discuss the highest frequency mode in each type clathrate, which corresponds to stretching vibration between host atoms. The higher frequency of this mode roughly means the shorter covalent bond between hosts, i.e., the smaller unit cell. Comparing the highest frequency peaks between type I and VIII at high pressures e.g. at 15 GPa, we may consider that the unit cell of type VIII is smaller than that of type I. This implies that type VIII structure is more favourable than type I. This is not compatible with the fact that the transformation from type VIII to type I. In order to discuss in more detail about the structural stability of type I and VIII, it is highly required to perform high-pressure XRD experiments. Such the experiment will be made in near future, and the results will be published elsewhere.

**Acknowledgments**

This work has been supported by KAKENHI (Grant-in-Aid for Scientific Research) on the Priority Areas "New Materials Science Using Regulated Nano Spaces-Strategy in Ubiquitous Elements" from the Ministry of Education, Culture, Sports, Science and Technology of Japan.

**References**

[1] Kasper J S, Hagenmuller P, Pouchard M, Cros C 1965 *Science* **150** 1713
[2] Bobev S, Sevo S C 2000 *J. Solid State Chem.* **153** 92
[3] Kovnir K A, Shevelkov A V 2004 *Rus. Chem. Rev.* **73** 923
[4] Paschen S, Carrillo-Cabrera W, Bentien A, Tran V H, Baenitz M, Grin Yu and Steglich F 2001 *Phys Rev.* B **64** 214404
[5] Leoni S, Carrillo-Cabrera W, Grin Y 2003 *J. Alloys Comp.* **350** 113
[6] Sasaki Y, Kishimoto K, Koyanagi T, Asada H and Akai K 2009 *J. Appl. Phys.* **105** 073702
[7] Kishimoto K, Ikeda N, Akai K and Koyanagi T 2008 *Appl. Phys. Expr.* **1** 031201
[8] Kume T, Fukuoka H, Koda T, Sasaki S, Shimizu H and Yamanaka S 2003 *Phys. Rev. Lett.* **90** 155503

[9] Kume T, Koda T, Sasaki S, Shimizu H and Tse J S 2004 *Phys. Rev.* B **70** 052101
[10] Shimizu H, Imai T, Kume T, Sasaki S, Kaltzoglou A and Fässler T F 2008 *Chem. Phys. Lett.* **464** 54
[11] Kume T, Ohno S, Sasaki S, Shimizu H, Ohishi Y, Okamoto N L, Kishida K, Tanaka K and Inui H 2010 *J. Appl. Phys.* **107** 013517
[12] Izumi F, Momma K 2007 *Solid State Phenom.* **130** 15
[13] Mao H K, Bell P M, Shaner J W, Steinberg D J 1978 *J. Appl. Phys.* **49** 3276
[14] Shimizu H, Takeuchi Y, Kume T, Sasaki S, Kishimoto K, Ikeda N, Koyanagi T 2009 *J. Alloys Compd.* **487** 47
[15] Takasu Y, Hasegawa T, Ogita N, Udagawa M, Avila M A, Suekuni K, Ishii I, Suzuki T, Takabatake T 2006 *Phys. Rev.* B **74** 174303
[16] Tse J S, Klug D D, Ripmeester J R, Desgreniers S, Lagarec K 1994 *Nature* **369** 724

# Observation of Rattling Vibrations in Clathrate under High Pressure and Low Temperature

**K.Funahashi[1], I. Yajima[1], T. Kume[1†], S. Sasaki[1], H. Shimizu[1], T.Takabatake[2]**
[1]Department of Materials Science and Technology, Gifu University, 1-1 Yanagido, Gifu 501-1193, Japan
[2]Graduate School of Sci. and Eng., Hiroshima Univ., Higashi Hiroshima, Hiroshima 739-0046, Japan

E-mail: kume@gifu-u.ac.jp

**Abstract.** An experimental system for low-frequency Raman measurements at low temperatures and high pressures was constructed in order to research low frequency vibrations of guest atoms in guest-host materials such as semiconductor clathrates. Raman measurements in the range to 10 $cm^{-1}$ were attained under low temperature and high pressure by arranging a diamond anvil cell fixed on a cryostat in a quasi-back-scattering geometry. Raman spectra of a clathrate compound $Eu_8Ga_{16}Ge_{30}$ were measured using this experimental system. The low frequency Eu vibration so called the rattling vibration located at ~20 $cm^{-1}$ was clearly observed under high pressures and low temperatures.

## 1. Introduction

Group-14 elements are known to form clathrate compounds which consist of polyhedron cages encapsulating guest atoms (M) such as Na, Ba [1-3]. These clathrates are expected to be potential materials for thermoelectric devices because these have low thermal conductivity despite high electrical conductivity. The low thermal conductivity has been considered to result from rattling vibrations of the guest atoms, which are like an Einstein mode with large amplitude. Since the frequencies of the guest rattling are significantly low, these phonon dispersions *anti-cross* the acoustic phonon ones, leading to retardation of the thermal propagation [4]. Such the properties of the rattling modes arise from the weak guest-host interaction. The rattling vibration frequencies depend on the size of the cage; the larger cage induces a shift of the guest toward off-centre position and causes the more distinct rattling nature. In addition, the rattling frequency shows an anomalous change with temperature decreasing, which was interpreted with high unharmonicity. Thus, it is important to measure the rattling vibrations as a function of temperature for various case sizes controlled with pressure.

Previously, we investigated the guest vibrations of various types of clathrate compounds by Raman experiments under high pressure [5-9]. However, the experimental conditions have been restricted to room temperature and a frequency region higher than ~25 $cm^{-1}$. The rattling vibrational frequencies of our interesting sample, i.e. $Eu_8Ga_{16}Ge_{30}$ are at about 20 $cm^{-1}$[10,11], and thus, in order to investigate

† To whom any correspondence should be addressed.

these rattling vibrations, it is required to extend the experimental conditions to lower temperatures and lower frequencies.

In this work, we constructed an experimental system for low-frequency (LF) Raman measurements at low temperatures (LT) and high pressures(HP). Using the constructed LF-LT-HP Raman system, we successfully observed the rattling vibrations of Eu in a clathrate ($Eu_8Ga_{16}Ge_{30}$) at about 20 cm$^{-1}$ under low temperatures and high pressures.

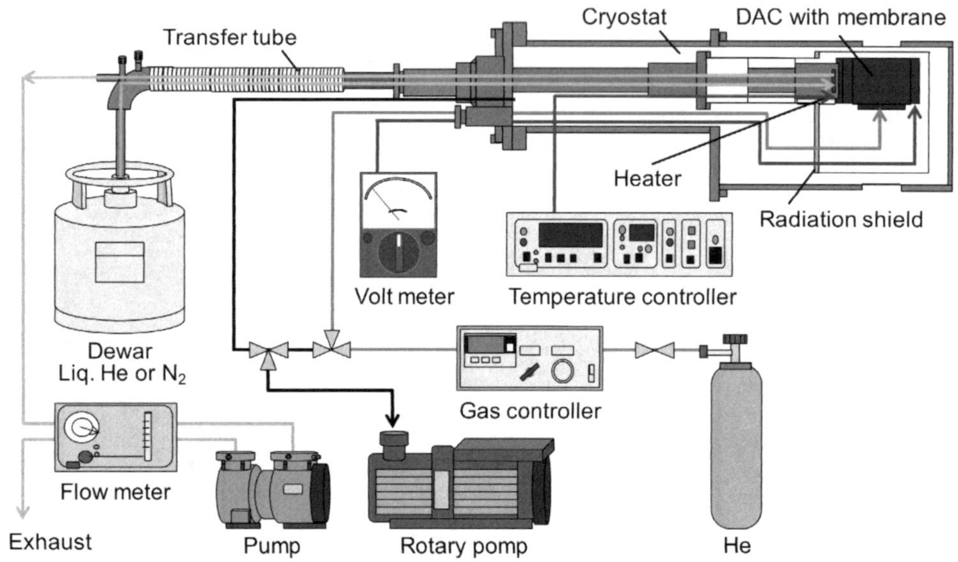


**Figure 1.** LT-HP-LF Raman scattering measuring device.

## 2. Experimental

Figure 1 schematically shows an experimental system which was constructed for LF Raman measurements under LT and HP. A diamond anvil cell (DAC) and a helium flow cryostat were used for HP and LT experiments. Because our Raman measurements are carried by using a microscopic optical system, the stability of the sample position is very important. Therefore, we adopted a gas flow type cryostat which is free from vibration. The pressure of the sample was controlled under low temperature by a gas driven membrane. In Fig. 2, we indicated details around the DAC. The DAC was mounted on a cold head of the cryostat. The temperatures were measured by thermocouples which were fixed on the cold head and on a DAC surface which is the opposite side of the cold head. We estimate approximately the temperature of the sample by taking the average of the temperatures measured at these two places. Pressure is determined by a ruby fluorescence method, taking into account the shift of the fluorescence wavelength with temperature [12,13].

For low frequency Raman measurements, we adopt a quasi-back scattering geometry as shown in Fig. 2. It should be noted that the excitation light is incident on the sample in DAC with an oblique direction. The light scattered in direction vertical to the culet is collected by a microscopic objective lens, and then introduced to a spectrometer equipped with a CCD detector. Radiation of 532 nm from a solid-state laser was incident with a power of 5 mW. The resolution of the spectra is about 1.5 cm$^{-1}$. The use of the pseudo-back scattering geometry and a polished plate-like sample effectively decreases in the Rayleigh signal. As a result, it became possible to detect the Raman signals in frequency range as low as about 10 cm$^{-1}$ under low temperature and high pressure.

Using the constructed system for the LT-HP-LF Raman measurement, we attempted to make experiments on a single crystal of $Eu_8Ga_{16}Ge_{30}$. Details of synthesis of the sample were described

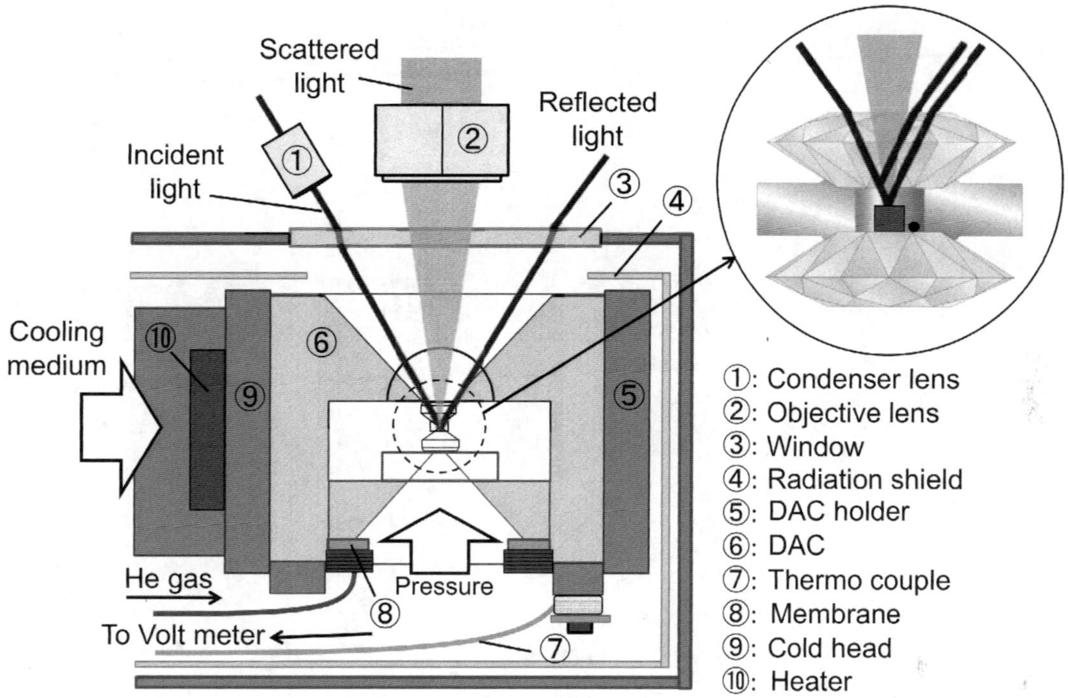


**Figure 2.** Details around DAC and pseudo-back scattering geometry.

elsewhere [4]. The single crystal $Eu_8Ga_{16}Ge_{30}$ was cut and polished into ~50 μm in size with ~20 μm in thickness. The small plate sample was placed into the sample chamber made on a rhenium gasket, and Ar pressure medium filled the sample chamber.

### 3. Results and discussion

Figure 3 shows Raman spectra of $Eu_8Ga_{16}Ge_{30}$ which were measured at high pressures (0.7, 4.1, 6.8 GPa) and room temperature (RT) and at high pressures and low temperatures (4.1 GPa and 250 K, and 6.7 GPa and 202 K).We successfully observed clear signals of the rattling vibration of Eu guest atoms occurring around 20-30cm$^{-1}$ under HP-LT conditions. The obtained signals were well consistent with the previous Raman date measured at ambient pressure [11]. As pressure increases to 4 GPa at RT, the guest rattling frequency shifted to higher frequencies. However, this shift is not observed in 4-7 GPa. On the other hand, for the host vibration at 175 cm$^{-1}$ (which was observed at 174 cm$^{-1}$ under 1 atm and assigned as $A_{1g}$ symmetry [10]), the high-frequency shift was observed up to 7 GPa. Changing of the temperature did not largely affect the Raman spectra, and the frequency of each peak position remained almost the same. This means that the volume contraction on cooling is very small under high pressures above 4 GPa.

According to the previous result reported under ambient pressure [11], the line width of the guest atom located at off-centre position shows broadening as temperature decreases, because the split of the guest peak becomes more obvious. The present spectrum of guest vibration at 4 GPa seems to become slightly broader at 250 K than RT. On the other hand, the width of the spectrum measured around 7 GPa was less sensitive to the temperature, implying that the guest atoms are located near the centre of the cage at 7GPa.

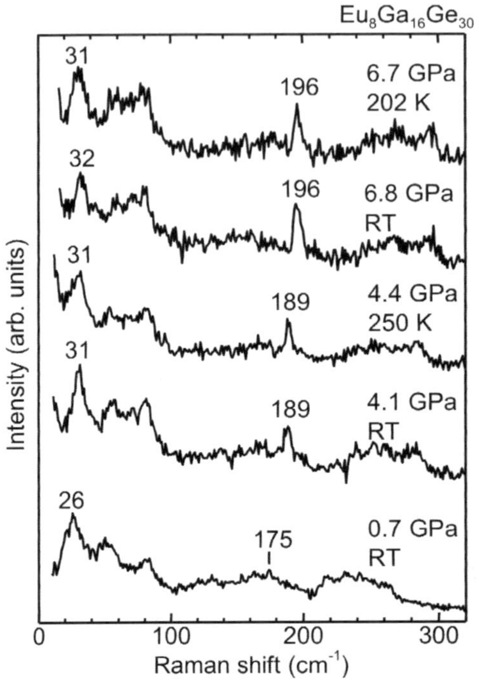


**Figure 3.** Raman spectra of $Eu_8Ga_{16}Ge_{30}$ measured at various temperature and pressure conditions. The lowest frequency peak corresponds to Eu rattling vibration. The host vibration of 175 cm$^{-1}$ was assigned as $A_{1g}$ mode, according to the spectra obtained at 1 atm [10].

If we perform the LF Raman measurements for various pressures under low temperature, we will clarify the properties of the rattling vibration for various conditions of guest off centre or on centre. Moreover, LT-HP XRD measurements of $Eu_8Ga_{16}Ge_{30}$ are planed to determine a guest position as a function of pressure. Such the Raman and XRD experiments will be much informative for understanding of the rattling vibration of the clathrate compounds.

## 4. Summary

We developed, in the present work, an experimental system for low-frequency (LF) Raman measurements at low temperatures (LT) and high pressures (HP). The use of the quasi-back scattering geometry allowed us to make the Raman measurements to 10 cm$^{-1}$ under high pressure. Temperature and pressure are controlled using the gas-driven membrane system and the He cryostat. Using the developed LT-HP-LF Raman system, we successfully observed the rattling vibrations of Eu in a clathrate ($Eu_8Ga_{16}Ge_{30}$) at 20-30 cm$^{-1}$ under low temperatures and high pressures. In near future, the rattling vibration will be clarified under low temperature as a function of pressure (cage size), and the results will be published elsewhere.

## Acknowledgments

This work has been supported by KAKENHI (Grant-in-Aid for Scientific Research) on the Priority Areas "New Materials Science Using Regulated Nano Spaces-Strategy in Ubiquitous Elements" from the Ministry of Education, Culture, Sports, Science and Technology of Japan.

## References


[1] Kasper J S, Hagenmuller P, Pouchard M, Cros C 1965 *Science* **150** 1713

[2] Bobev S, Sevo S C 2000 *J. Solid State Chem.* **153** 92

[3] Kovnir K A, Shevelkov A V 2004 *Rus. Chem. Rev.* **73** 923

[4] Christensen M, Abrahamsen A B, Christensen N B, Juranyi F, Andersen N H, Lefmann K, Andersen J, Bahl C R H and Ifersen B B 2008 *Nature Materials* 7 811

[5] Kume T, Fukuoka H, Koda T, Sasaki S, Shimizu H and Yamanaka S 2003 *Phys. Rev. Lett.* **90** 155503

[6] Kume T, Koda T, Sasaki S, Shimizu H and Tse J S 2004 *Phys. Rev. B* **70** 052101

[7] Tse J S, Iitaka T, T. Kume, H. Shimizu, K. Parlinski, H. Fukuoka, and S. Yamanaka 2005 *Phys. Rev. B* **72** 155441

[8] H. Shimizu, T. Imai, Kume T, Sasaki S, Kaltzoglou A and Fässler T F 2008 *Chem. Phys. Lett.* **464** 54

[9] Kume T, Ohno S, Sasaki S, Shimizu H, Ohishi Y, Okamoto N L, Kishida K, Tanaka K, and Inui H 2010 *J. Appl. Phys.* **107** 013517

[10] Takasu Y, Hasegywa T, Ogita N, Udagawa M, Avila M A, Suekuni K, Ishii I, Suzuki T and Takabatake T 2006 *Phys. Rev. B* **74** 174303

[11] Takasu Y, Hasegawa T, Ogita N, Udagawa M, Avila M A, Suekuni K and Takabatake T 2008 *Phys. Rev. Lett.* **100** 165503

[12] Mao H K, Bell P M, Shaner J W and Steinberg D J 1978 *J. Appl. Phys.* **49** 3276

[13] Ragan D D, Gustavsen R and Schiferl D,1992 *J. Appl. Phys.* **72** 5539

# Magnetic EXAFS study of Fe-Ni invar alloy under high pressure using nano-polycrystalline diamond anvils

**Ken Matsumoto[1], Hiroshi Maruyama[1], Naoki Ishimatsu[1], Naomi Kawamura[2], Masaichiro Mizumaki[2], Tetsuo Irifune[3], and Hitoshi Sumiya[4]**

[1] Graduate School of Science, Hiroshima University, 1-3-1, Kagamiyama, Higashi-Hiroshima, Hiroshima 739-8526, Japan.
[2] Japan Synchrotron Radiation Research Institute, SPring-8, 1-1-1 Kouto, Sayo, Hyogo 679-5198, Japan.
[3] Geodynamics Research Center, Ehime University, 2-5 Bunkyo, Matsuyama, Ehime 790-8577, Japan.
[4] Electronics and Materials R&D Laboratories, Sumitomo Electric Industries, 1-1-1 Koyakita, Itami, Hyogo 664-0016, Japan.

**Abstract.** The magnetic EXAFS technique was applied to study the local magnetic structure in Fe-Ni Invar alloy under high pressures. Nano-polycrystalline diamond was used to apply pressure in a diamond anvil cell. As a result, we succeeded in recording a glitch-free spectrum with a high accuracy by preventing Bragg diffraction from the diamonds. As the pressure was increased, the ferromagnetic correlation was strongly reduced, which was more remarkable for Fe atoms than for Ni atoms. The canting effect of the magnetic moment may be crucial for a rapid decrease in the ferromagnetic correlation. This local magnetic structure in the Invar alloy is consistent with a model based on the non-collinear spin structure.

## 1. Introduction

Extended X-ray absorption fine structure (EXAFS) is a well-established technique used to provide information about the structural and electronic properties of various materials. This technique has been also applied to study the local structure under extreme conditions. However, it is impossible to directly probe local magnetic structure in magnetic materials by means of EXAFS. The physical origin of magnetic EXAFS, defined as X-ray magnetic circular dichroism in the EXAFS region, has been considered to be the exchange contribution to the backscattering amplitude caused by neighboring magnetic atoms [1]. Thus, magnetic EXAFS is a powerful technique for examining the ferromagnetic correlation in the local environment.

In order to study pressure-induced magnetic phase transition, we apply magnetic EXAFS to a high pressure experiment using a diamond-anvil cell (DAC). Although DAC has been widely used as a device for generating high pressures, Bragg reflections from the single-crystal diamond are in general strong and superimpose a spikelike profile on the spectrum. Therefore, the magnetic EXAFS spectrum is subject to distortions in the oscillatory profile and to reduction in the data accuracy. To prevent the influence of Bragg diffraction, we introduced nano-polycrystalline diamond (NPD) anvils [2, 3]. Since NPD is harder and more mechanically isotropic than the single-crystal diamonds, it is suitable for applying higher pressure by DAC.

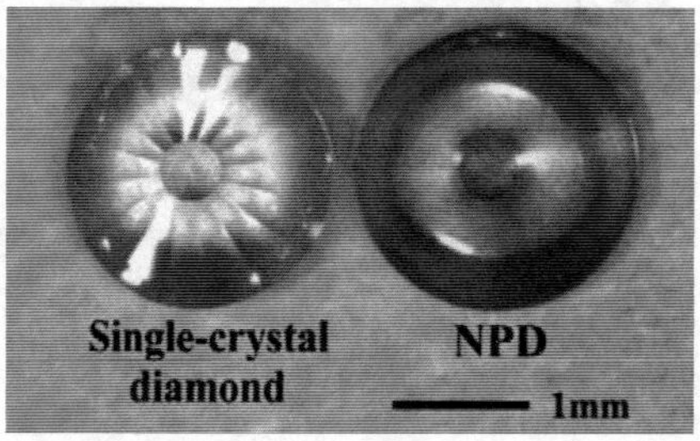


**Figure 1.** Photograph of diamond anvils.

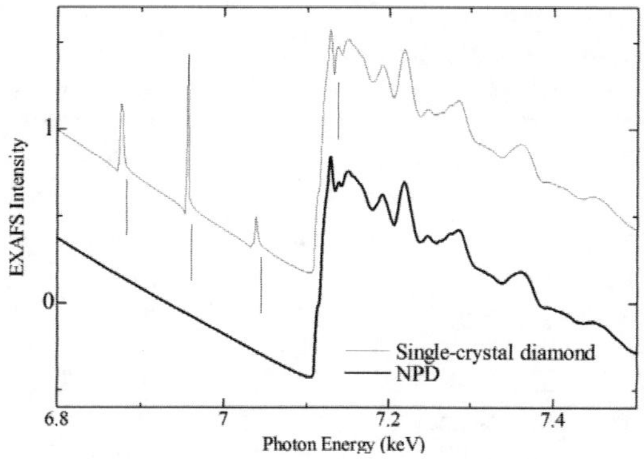


**Figure 2.** EXAFS spectra obtained with single-crystal diamond anvil and NPD anvil.

In this paper, magnetic EXAFS at high pressures is presented for fcc $Fe_{65}Ni_{35}$ Invar alloy, in which the DAC with NPD anvils was employed for the first time. Fe and Ni metals were also measured at ambient pressure as reference materials in bcc and fcc structures. Fe-Ni Invar alloy has been regarded as a prototypical material that exhibits anomalously low thermal expansion [4]. Because the magnetic EXAFS is element-specific and electronic-shell selective, it is appropriate for comparing Fe and Ni atoms in the local environment.

## 2. Experiment

X-ray absorption and dichroic spectra were recorded at the Fe and Ni *K*-edges using the helicity-modulation method on beamline 39XU at SPring-8 [5], which was equipped with a diamond 111 double-crystal monochromator, Kirkpatrick-Baez focusing mirrors [6], and a diamond $\lambda/4$-plate to produce a circularly polarized beam. A magnetic field of 0.6 T was sufficient to magnetically saturate the sample. The generated pressure was estimated by the ruby fluorescence method. Two kinds of diamond anvil, single-crystal and nano-polycrystalline diamonds are compared in the photograph in figure 1. Although the NPD exhibits a dark yellow due to nitrogen impurities and lattice defects [7], it has a high transparency for X-rays. Figure 2 shows the raw EXAFS spectra of Fe metal recorded using the DAC with single-crystal diamond and NPD. The latter gives a much better spectrum without any spikelike noise. We succeeded in recording a glitch-free spectrum with a high accuracy over a wide range of photon energy (6 - 10 keV). The experiment was carried out at room temperature.

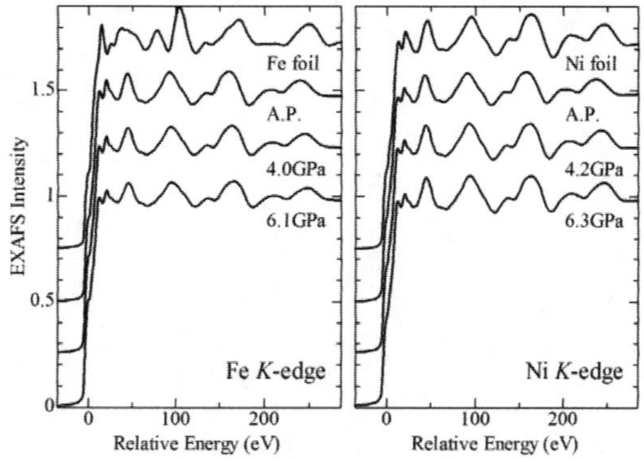


**Figure 3.** EXAFS spectra at the Fe *K*-edge (left panel) and Ni *K*-edge (right panel) in Fe-Ni invar alloy, Fe and Ni foils.

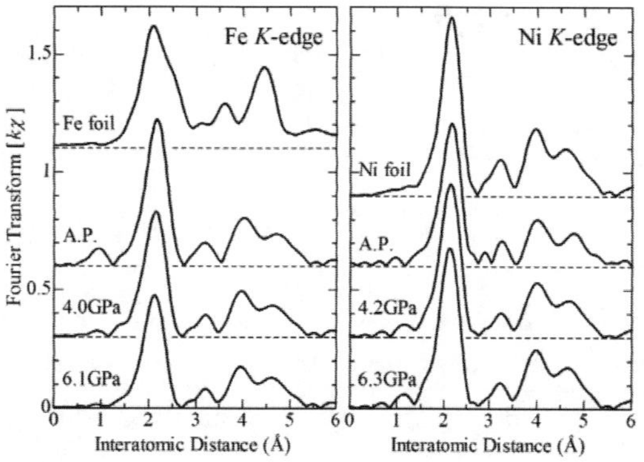


**Figure 4.** Pressure variation of RDF for Fe and Ni absorbing atoms.

Journal of Physics: Conference Series **377** (2012) 012039 doi:10.1088/1742-6596/377/1/012039

### 3. Results and Discussion

Figure 3 shows the pressure variation of the normalized EXAFS spectrum in $Fe_{65}Ni_{35}$ alloy at the Fe and Ni *K*-edges in comparison with those of pure Fe and pure Ni metals. Difference in the spectral profile between Fe and Ni metals is ascribed to the difference in crystal structure, fcc and bcc. It is clearly seen that the EXAFS oscillation in the Fe-Ni alloy closely resembles that of fcc Ni metal. As the pressure increases, the spectral shape and oscillatory intensity show no drastic changes at either edge, but the oscillation period becomes slightly longer because of the contraction of lattice spacing. These trends indicate that the $Fe_{65}Ni_{35}$ alloy maintains the fcc structure even at 6 GPa and does not undergo any structural transition, and the local environment is notably similar around Fe and Ni absorbing atoms.

For the EXAFS analysis, we used the following conventional procedure [8]: background removal, base-lining of oscillation, determination of absorption edge energy, normalization of spectral intensity by edge jump, and then Fourier transform to obtain the radial distribution function (RDF). As shown in figure 4, the RDF has an intense peak at an interatomic distance of about 2.2 Å, which corresponds to the nearest neighbor (NN) interatomic pair around Fe and Ni atoms; the subsidiary peaks correspond to 2nd, 3rd, and 4th NNs. The RDF profile is well reproduced by the theoretical simulation using FEFF code [8]. It should be noted that the chemical structure of the alloy is quite stable in the local environment even at high pressures.

Figure 5 shows the magnetic EXAFS spectrum at the Fe and Ni *K*-edges, compared with those of Fe and Ni metals. As the pressure is increased up to 6 GPa, the magnetic dichroic amplitude is significantly reduced, and the magnetic EXAFS oscillation is also drastically suppressed by the applied pressure. However, the magnetic EXAFS pattern is very similar for both the edges, which indicates that the Fe and Ni atoms are in similar magnetic states in the local environment. On the other hand, the magnetic EXAFS shows a different pattern between Fe and Ni metals.

The pressure variation of the magnetic RDF is shown in figure 6, together with those of Fe and Ni metals. The magnetic RDF has an intense peak at about 2.2 Å, which corresponds to the ferromagnetic NN spin pairs around the Fe and Ni atoms. The subsidiary peaks are associated with the 2nd (3.2 Å), 3rd (3.8 Å) and 4th (4.5 Å) NN pairs, which are also ferromagnetically coupled with the absorbing atom [9]. For bcc Fe metal, the two prominent peaks are located at about 2.2 Å and 4.5 Å, as a result of the bcc structure. The latter peak is enhanced by the focusing effect of multiple scattering paths, in addition to contribution of the single scattering path [10]. For fcc Ni metal and Fe-Ni alloy, the main peak is located at about 2.2 Å, and the intensity is significantly reduced as the pressure is increased up to 6 GPa.

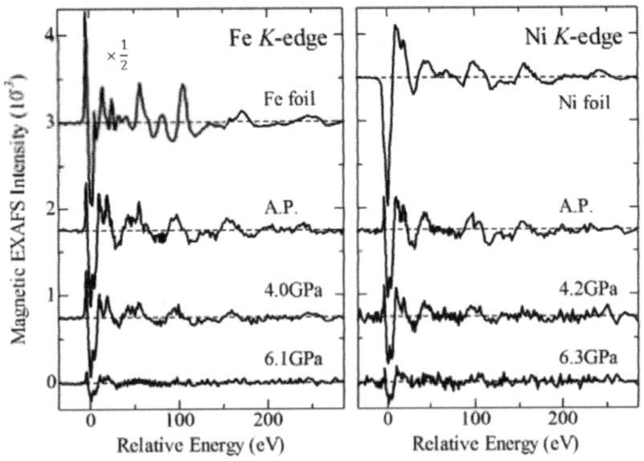


**Figure 5.** Magnetic EXAFS spectra at the Fe *K*-edge (left panel) and Ni *K*-edge (right panel) in Fe-Ni invar alloy, Fe and Ni foils.

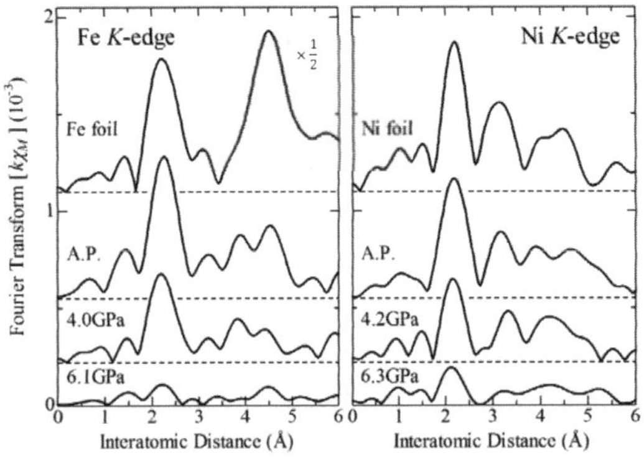


**Figure 6.** Pressure variation of magnetic RDF for Fe and Ni absorbing atoms.

Magnetic RDF intensity is a measure of strength in ferromagnetic correlation. From the pressure variation in figure 6, a drastic reduction is observed in the NN peak intensity for both the Fe and Ni *K*-edges. This trend contrasts with that of the RDF for the chemical structure (cf. figure 4). This fact demonstrates that the ferromagnetic correlation is strongly reduced by the applied pressure. This feature is common to both Fe and Ni absorbing atoms. However, Fe and Ni atoms show different tolerance for the pressure; that is, the ferromagnetic correlation around the Fe atom is subject to a larger reduction than that around the Ni atoms. There is a considerable decline in the ferromagnetic correlation for the Fe NN pairs. It is possibly caused by a reduction in the magnitude of magnetic polarization or by an increase in canting angle of magnetic moments. It should be emphasized that the canting effect of the magnetic moment is crucial for a rapid decrease in the ferromagnetic correlation for the Fe NN pairs. This local magnetic structure in the Invar alloy is consistent with the theoretical prediction based on the non-collinear spin structure [11].

## 4. Conclusion

The magnetic EXAFS study of Fe-Ni Invar alloy under high pressure has been made using a DAC with NPD anvils for the first time. We demonstrate that the NPD is very effective at avoiding Bragg diffraction from the diamond anvils. It was revealed that the ferromagnetic correlation is drastically suppressed by the applied pressure, and that the suppression is more remarkable for Fe atoms than for Ni atoms. This trend probably involves the canting effect of the magnetic moment, so that non-collinear spin configuration increasing at high pressures should be taken into account from the viewpoint of local magnetic structure [12]. Magnetic EXAFS is an effective probe for investigating local magnetic structure under extreme conditions.

## Acknowledgement

This study was performed at SPring-8 with the approval of PRC-JASRI (Nos. 2009A1302 and 2009B1495).

## References

[1] Schütz G, Knülle M and Ebert H 1993 *Phys. Scr.* **T49A** 302
[2] Irifune T, Kurio A, Sakamoto S, Inoue T and Sumiya H 2003 *Nature* **421** 599–600
[3] Sumiya H, Irifune T, Kurio A, Sakamoto S and Inoue T 2004 *Mater. Sci.* **39** 445–50
[4] Guillaume C E 1897 *C. R. Acad. Sci.* **125** 235
[5] Kawamura N, Ishimatsu N and Maruyama H 2009 *J. Synchrotron Radiat.* **16** 730–6
[6] Yumoto H, Hirata K, Nisawa A, Ueno G, Sato M, Son J-Y, Koganezawa T, Machida M, Muro T, Hirosawa I, Suzuki M, Kawamura N, Mizumaki M, Ohashi H, Yamamoto M, Watanabe Y and Goto S 2009 *Proc. SPIE* **7448** 74480Z
[7] Sumiya H, Harano K, Kagi H and Irifune T 2009 (in press) *Jpn. J. Appl. Phys.* Optical characteristics of nano-polycrystalline diamond synthesized by direct conversion from graphite
[8] Ravel B and Newville M 2005 *J. Synchrotron Radiat.* **12** 537–41
[9] Popescu V, Güßmann M, Fähnle M and Schütz G 2009 *Phys. Rev. B* **79** 014440
[10] Kurde J, Ponpandian N, Luo J, Weis C and Baberschke K 2007 *Phys. Rev. B* **76** 224418
[11] Schilfgaarde M V, Abrikosov I A and Johansson B 1999 *Nature* **400** 46–9
[12] Matsumoto K, Maruyama H, Ishimatsu N, Kawamura N, Mizumaki M, Irifune T and Sumiya H 2011 *J. Phys. Soc. Jpn.* **80** 023709

# Melting of potassium to 22 GPa

**E E McBride, O Narygina, G W Stinton and M I McMahon**
SUPA, School of Physics and Astronomy, and Centre for Science at Extreme Conditions, The University of Edinburgh, Edinburgh EH9 3JZ, UK

E-mail: e.mcbride@ed.ac.uk

Using *in-situ* x-ray diffraction, the melting curve of potassium was determined to 22 GPa and was found to be remarkably similar to that of sodium, and strikingly different to that reported previously. The existence of a maximum in the *bcc* phase was determined at 5.8(5) GPa and 530(10) K; the melting temperature was then observed to decrease over several GPa, flattening out at the *bcc-fcc*-liquid triple point at 13.6(3) GPa and 466(10) K, before further decreasing from 15.6(3) GPa to a minimum at 19(1) GPa and 390(10) K. It then regained a positive slope and increased rapidly at a rate of 65(5) K/GPa In the *tI*19 phase it was observed that the guest chains "melted" before the host structure at 20.3(3) GPa and 420(10) K, but were solid at 22.5(3) GPa and 350(10) K.

## 1. Introduction

The alkali metals at ambient conditions are well described by the nearly-free electron (NFE) model but show a significant departure from this behaviour with increasing pressure. Anomalous melting, where the gradient of the melting curve becomes negative over large pressure regions, has been observed in rubidium (Rb) and caesium (Cs) [1], sodium (Na) [2] and lithium (Li) [3]. To-date, potassium (K) appears to be an exception to this trend, with no observed melting maximum, at least not to 14.5 GPa [4].

Further evidence of a departure from the NFE model is seen in the emergence of structural complexity with pressure. At atmospheric pressure, the alkali elements are all body-centred cubic (*bcc*), and with increasing pressure they all transform to the closer-packed face-centred cubic (*fcc*) structure. Further application of pressure, however, sees them depart from this "simple" behaviour, with the emergence of low-symmetry open-packed structures [5-11]. The high-pressure phases of K have much in common with its group-I neighbours, with the incommensurate host-guest structure *tI*19.2 (hereafter referred to as *tI*19) being observed in Na, K and Rb [6-8], and the *oP*8 and *hP*4 forms observed in both Na and K [6, 9-11].

Given the similar structural transitions of K and the other alkali metals, and given that anomalous melting has been observed in all others, it is remarkable that K does not follow this trend of a melting maximum. We have therefore reinvestigated its melting temperature using diamond anvil cell (DAC) techniques and *in-situ* x-ray diffraction.

## 2. Experiment

All experiments were conducted using high-purity (99.95+%) samples of K (Aldrich Chemical Company) which were loaded under argon in a dry, oxygen-free atmosphere (< 1 ppm $O_2$, < 1 ppm $H_2O$), without a pressure transmitting medium to prevent contamination. Samples were loaded into

Published under licence by IOP Publishing Ltd

membrane DACs with tungsten or composite rhenium-iridium gaskets. $SrB_4O_7:Sm^{2+}$ was used as a pressure calibrant.

The sample temperature was controlled using external resistive heaters and measured using a thermocouple on the back of one of the diamonds. *In-situ* x-ray diffraction measurements were performed at beamlines ID09a and I15 at the European Synchrotron Radiation Facility and Diamond Light Source, respectively, using MAR 555 and MAR 345 area detectors. For each x-ray exposure, the samples were rocked ±7° to ensure good powder diffraction images, and to ensure that very few crystalline Bragg reflections were missed if the sample had annealed into a single crystal.

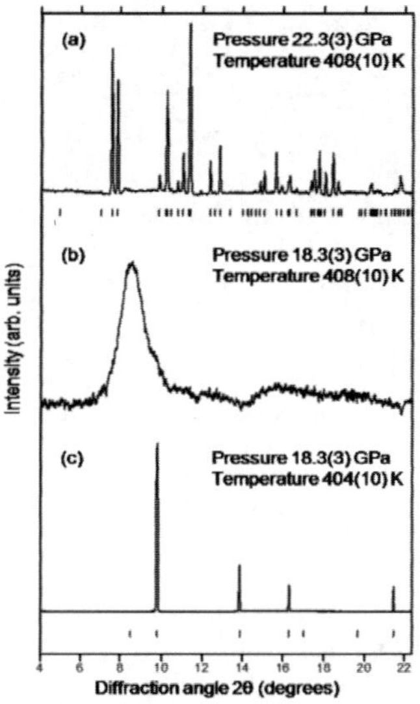


Figure 1. Representative integrated profiles from the crystalline (a) *tI*19 and (c) *fcc* phases, and the (b) liquid phase. The background in the liquid profile has been subtracted, using the method of Sanloup *et al.* [12]. Tick marks below profiles (a) and (c) show the calculated peak positions for the *tI*19 and *fcc* phases, respectively. The lowest angle (111) peak from the *fcc* phase is absent, as the sample was a single crystal and this reflection was not observed during the ±7° oscillation during the x-ray exposure.

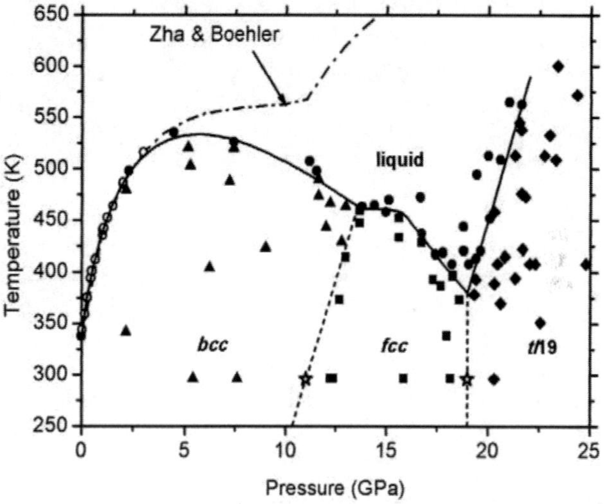


Figure 2: The melting curve of K to 22 GPa. Unfilled circles are data to 3 GPa from refs. [4] and [13], and the solid black line below 13.6 GPa is a fit to the Kechin equation [14] using this data combined with our own. The continuation of the solid line above 13.6 GPa is a guide to the eye. Stars indicate room temperature phase transition pressures [7], and the dashed lines are proposed phase boundaries based on this study. The dot-dashed line is the previously determined melting temperature [4].

## 3. Results

Representative integrated x-ray diffraction profiles in figure 1 show evidence of melting. It is clear that there is a sharp transition from the *fcc* structure at 18.3(3) GPa and 404(10) K to the liquid phase at 18.3(3) GPa and 408(10) K. By maintaining the temperature at 408(10) K and increasing pressure by 4 GPa potassium then recrystallises in the *tI*19 phase. At high temperatures, K anneals easily into single crystals. As a result, integrated profiles have missing reflections, and inaccurate peak intensities. Integrated profiles were thus analysed using Le Bail rather than Rietveld methods.

The melting curve of K, as determined by our measurements, is shown in figure 2, along with the previously-determined melting curve of Zha & Boehler [4]. Our results indicate the clear presence of a melting maximum in the *bcc* phase. Melting of the *bcc* phase was observed at ambient pressure,

2.2(3) GPa, 4.5(3) GPa, 7.4(3) GPa, and 11.5(3) GPa at 338(10) K, 498(10) K, 536(10) K, 526(10) K and 498(10) K, respectively. There is clearly a maximum in the melting temperature between 4.5(3) GPa and 7.4(3) GPa.

Up to 3 GPa, our melting results are in good agreement with previous studies (see refs. [4] and [13]); these results are shown by the unfilled circles in figure 2. Combining these previous results with our own, and using our *bcc*-*fcc*-liquid triple point at 13.6(3) GPa and 466(10) K, we performed a fit using the Kechin equation [14] to determine the location of the melting maximum in the *bcc* phase. This fit locates the maximum at 5.8(5) GPa and 530(10) K.

Above the triple point the melting temperature remains (almost) flat for several GPa. It should be noted that the presence of a second maximum, like that observed in Cs [1], cannot be ruled out, but it was too slight to be observed in this study. This region of nearly constant melting temperature ends as the melting curve decreases from 15.6(3) GPa and 462(10) K to a minimum at 19.0(3) GPa and 390(10) K. The location of this minimum coincides with the transformation from *fcc*-K to *tI*19-K; the phase boundary can be seen to be almost vertical. In the *tI*19-K phase the melting temperature rises very rapidly, increasing at a rate of 65(5) K/GPa.

Upon heating the *tI*19 phase, it was observed that the guest chains "melted" before the sample, consistent with a weak host-guest interaction. This is illustrated in figure 3, which shows 2D diffraction images collected from the *tI*19 phase both below and above the chain-melting transition. At 22.5(3) GPa and 350(10) K (figure 3(a)) Bragg reflections from the guest phase are clearly visible, although elongated perpendicular to the chain direction, indicating only short-range inter-chain correlations. The (101) guest reflection is circled. Figure 3(b) shows the 2D image from the same sample at 20.3(3) GPa and 420(10) K. The Bragg reflections from the guest chains have disappeared, as a result of the loss of long range order, and only some diffuse scattering remains. The calculated position of the (101) reflection is circled. The change in the guest reflections is quantified in figure 4, which shows 1D integrations of the 2D images in figure 3, performed over the same small regions encompassing the (101) reflection. At 22.5(3) GPa and 350(10) K there is a distinct Bragg peak and this has clearly disappeared at 20.3(3) GPa and 420(10) K. Initial observations at higher pressures indicate the chain melting temperature is almost independent of pressure.

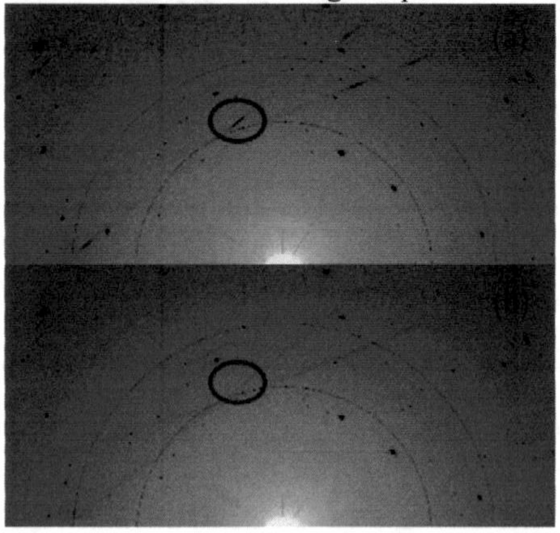

Figure 3: 2D diffraction images of the *tI*19 phase collected at (a) 22.5 GPa and 350 K, and at (b) 20.3 GPa and 420 K. The position of the (101) guest reflection is circled in both images.

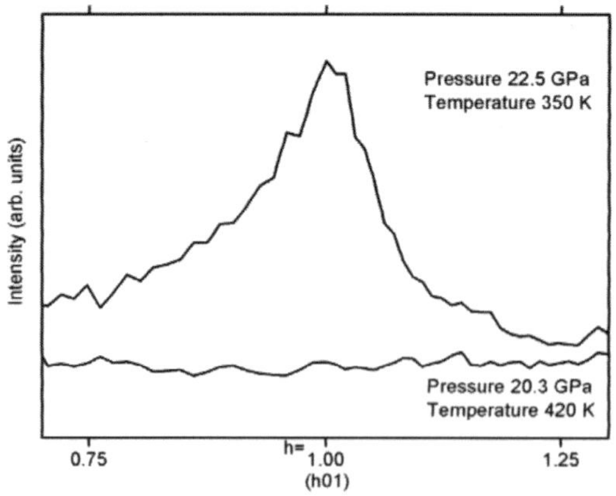


Figure 4: 1D profiles obtained by integrating the 2D diffraction images over the same small region encompassing the (101) reflection. There is a distinct Bragg peak at 22.5(3) GPa and 350(10) K which is absent at 20.3(3) GPa and 420(10) K.

## 4. Discussion

The melting behaviour of K is remarkably similar to that of Na, where there is also a melting maximum in the *bcc* phase followed by a decrease in the melting temperature to a dramatic minimum of 300 K at 118 GPa. After this, the gradient of the melting curve regains a positive slope and increases rapidly [2], although the increase does not appear to be as rapid as in K. From the Clausius-Clapeyron relation, a change in the melting curve gradient from positive to negative is an indication of densification of the liquid phase relative to the underlying solid. A change in the sign of the gradient might thus imply a structural change in either the liquid or the solid phase. In Na, as in K, the maximum in the melting temperature occurs in the *bcc* phase, several GPa (approximately 6 GPa in K and 30 GPa in Na) before the transition in the solid from the *bcc* to the *fcc* phase.

To further understand this anomalous melting behaviour, Raty *et al.* performed *ab initio* calculations on the evolution of the melting temperature of Na [15]. They showed that the anomalous melting could be explained by a sequence of structural and electronic transitions in the liquid phase. They found that at the maximum in the melting temperature, *bcc*-like to *fcc*-like structural transitions were occurring in the liquid. Furthermore, their study attributed the negative slope of the melting temperature above the crystalline *bcc* to *fcc* transition to electronic and structural transitions in the liquid. Not only did they report a drop in calculated co-ordination number above 60 GPa, to that of the solid above 103 GPa, but they also detailed a three-fold drop in conductivity, indicating electronic changes occurring in the liquid.

Due to the remarkable similarity of the melting curves of Na and K, we suggest that similar mechanisms to those detailed above for Na could explain the melting behaviour of K. The advantage of K is that the anomalous melting behaviour occurs at much lower, and hence experimentally accessible, pressures than Na, and so additional studies are planned to investigate the co-ordination and electronic properties of liquid-K.

**Acknowledgements**
We thank S Evans and M Hanfland of ID09a at the European Synchrotron Radiation Facility, and A K Kleppe, H Wilhelm, and A P Jephcoat of I15 at the Diamond Light Source, for provision of synchrotron time and support. We also thank E Gregoryanz for discussions. This work was supported by a research grant from the UK Engineering and Physical Sciences Research Council.

**References**
[1] Boehler R and Zha C-S 1986 *Physica B+C* **139-140** 233
[2] Gregoryanz E, Degtyareva O, Somayazulu M, Hemley R J and Mao H-K 2005 *Phys. Rev. Lett.* **94** 185502
[3] Guillaume C L, Gregoryanz E, Degtyareva O, McMahon M I, Hanfland M, Evans S, Guthrie M, Sinogeikin S V and Mao H-K 2011 *Nature Physics* **7** 211
[4] Zha C-S and Boehler R 1985 *Phys. Rev. B* **31** 3199
[5] Marqués M, McMahon M I, Gregoryanz E, Hanfland M, Guillaume C L, Pickard C J, Ackland G J and Nelmes R J 2011 *Phys. Rev. Lett.* **106** 095502
[6] Gregoryanz E, Lundegaard L F, McMahon M I, Guillaume C L, Nelmes R J and Mezouar M 2008 *Science* **320** 1054
[7] McMahon M I, Nelmes R J, Schwarz U and Syassen K 2006 *Phys. Rev. B* **74** 140102
[8] McMahon M I and Nelmes R J 2001 *Phys. Rev. Lett.* **87** 055501
[9] Lundegaard L F, Marqués M, Stinton G, Nelmes R J and McMahon M I 2009 *Phys. Rev. B* **80** 020101
[10] Marqués M, Ackland G J, Lundegaard L F, Stinton G, Nelmes R J and McMahon M I 2009 *Phys. Rev. Lett.* **103** 115501
[11] Ma Y, Erements M, Oganov A Rm Xie Y, Trojan I, Medvedev S, Lyakhov A O, Valle M and Prakapenka V 2009 *Nature* **458** 07786
[12] Sanloup C, Gregoryanz E, Degtyareva O and Hanfland M 2008 *Phys. Rev. Lett.* **100** 075701
[13] Luedemann H D and Kennedy G C 1968 *Journal of Geophysical Research* **73** 2795
[14] Kechin V V 2001 *Phys. Rev. B* **65** 052102
[15] Raty J-Y, Schwegler E and Bonev S A 2007 *Nature* **449** 448

# Magnetic States in Fe, Co, Ni hydrides under High Pressure Probed by X-ray Magnetic circular dichroism

**N Ishimatsu[1], T Shichijo[1], Y. Matsushima[1], H Maruyama[1], N Kawamura[2], M Mizumaki[2], T Matsuoka[2], K Takemura[3]**

[1] Graduate School of Science, Hiroshima University, 1-3-1 Kagamiyama, Higashihiroshima 739-8526, Japan

[2] JASRI/SPring-8, 1-1-1, Kouto, Sayo, Hyogo 679-5198, Japan

[3] NIMS, 1-1 Namiki, Tsukuba Ibaraki 305-0044, Japan

naoki@sci.hiroshima-u.ac.jp

**Abstract.** Magnetic states of Fe, Co, and Ni hydrides under pressure have been investigated by X-ray magnetic circular dichroism (XMCD) at the *K*-edge. At room temperature, Fe, Co, and Ni metals are hydrogenated to ferromagnetic FeH, ferromagnetic CoH, and paramagnetic NiH, respectively. A sharp XMCD profile is observed after hydrogenation, which is attributed to the hydrogen-induced reconstruction of the electronic structure. The pressure dependence of XMCD demonstrates that the ferromagnetic state of FeH is less stable than that of CoH under pressure.

## 1. Introduction

Metal hydride systems have recently attracted considerable interest because of their potential applications as hydrogen storage materials and of hydrogen-induced modification to crystal- and electronic structures. 3*d* transition metal (TM) monohydrides can be synthesized by a direct reaction with $H_2$ fluid under gigapascal pressures [1]; hydrogen causes drastic changes in their magnetic properties. When ferromagnetic 3*d* TMs such as Fe, Co, and Ni undergo hydrogenation, they exhibit appearance or disappearance of the ferromagnetic states accompanied with large volume expansion and structural transition [2].

An investigation into the magnetic states of the TM hydrides and the pressure dependence of these states enables a fundamental understanding of the influence of hydrogen on the electronic structure of TM. For this purpose, we synthesized Fe, Co, and Ni monohydrides and measured the X-ray magnetic circular dichroism (XMCD) spectra at the *K*-edge. XMCD is a spectroscopic technique that probes the magnetic state with element- and orbital selectivities. Although the $1s \rightarrow 4p$ dipole transition is dominant at the *K*-edge, the XMCD spectrum is sensitive to the 3*d* magnetic states because of TM 4*p*-3*d* hybridization with neighbouring atoms [5,6]. In this paper, the magnetic states of these hydrides are discussed [3,4], and the stability of the ferromagnetic phase under pressure is also investigated.

## 2. Experimental Procedure

Polycrystalline metal foils 3–5 μm in thickness were prepared as the starting material. A tiny foil was set into a diamond anvil cell (DAC) with $H_2$ fluid that was initially pressurized to 0.18 GPa. We employed a versatile gas-loading system to load $H_2$ fluid to the DAC [7]. The $H_2$ fluid functioned as a pressure-transmitting medium and as the source of hydrogen. X-ray absorption near edge structure (XANES) and XMCD were measured on the beamline 39XU at the SPring-8 facility [8]. The XMCD spectra were obtained using the helicity-modulation method. All the measurements were carried out at room temperature. A magnetic field of 0.6 T was applied perpendicular to the sample surface. The pressure applied to the sample was measured through the conventional ruby fluorescence method.

## 3. Results and Discussion

Figure 1 shows the XANES and XMCD profiles of the TMs and the TM hydrides. The XANES and XMCD spectra show significant change caused by hydrogenation. XANES spectra of the TMs are found to be different from each other, while those of the TM hydrides are almost the same, consisting of a suppressed shoulder **a** and a crest **b**. The characteristics of the XANES spectra mean that TM 4*p* unoccupied density of states (DOS) is very similar among TM hydrides.

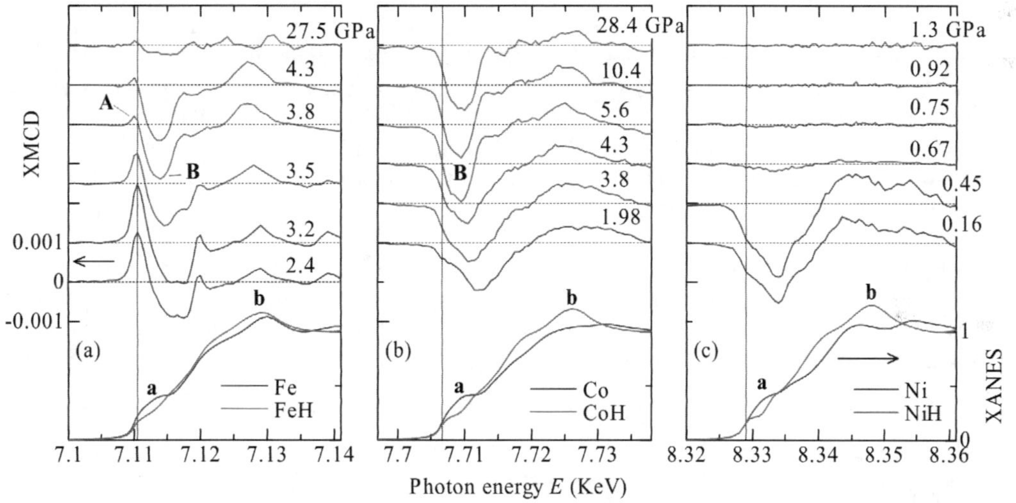


**Figure 1.** XANES (lower part of the graph) and XMCD (upper part) spectra at selected pressures. The blue and red lines correspond to the spectra of TMs and TM monohydrides, respectively. The green lines represent the spectra of intermediate phases that correspond to the coexistence of TM and TM monohydride for Fe and Ni, and correspond to TM hydride with a small hydrogen content H/TM<1 for Co. The vertical lines indicate the absorption edge $E_0$.

The pressure ranges required to synthesize TM monohydrides were determined from the XMCD and X-ray diffraction measurements [4]. Hydrogenation of Fe to FeH was observed between 3.2 and 3.8 GPa. The hydrogenation of Co to CoH is gradual, in a wide pressure range of 2.7–5 GPa, while that of Ni was observed over a narrow and low pressure range of 0.67–0.84 GPa.

The XMCD spectrum of NiH completely vanishes after hydrogenation, revealing that NiH is paramagnetic [2]. In contrast to that, XMCD spectra of FeH and CoH do not disappear. The remaining XMCD shows that both hydrides are ferromagnetic.

In the XMCD spectra of FeH and CoH, the negative peak **B** shifts toward the absorption edge $E_0$. In addition, the peak **B** exhibits sharp profiles with large amplitude as compared to those of Fe and Co. The XMCD profile near $E_0$ is interpreted as influence from the "orbital moment density" of the 3*d* state via the 4*p*-3*d* hybridization [5,6]. Therefore, the sharp profile indicates that the orbital moment density of the $3d$ state is concentrated just above $E_F$. According to the DOS calculated previously [9], this feature is probably attributed to the electronic structure of the unoccupied minority $3d$ DOS and the antibonding state between H and TM. We note that both the states are formed within an energy range of $\sim E_F + 6$ eV thorough the hydrogenation.

In the hydrogenation of Fe to FeH, it is observed that the positive peak **A** is significantly suppressed. The experimental result demonstrates that the electronic structure of FeH approaches to strong ferromagnetism from the weak ferromagnetism [5,6], as predicted from the theoretical calculation [9]. On the other hand, the absence of the positive peak **A** suggests that CoH maintains the electronic structure classified as strong ferromagnetism; strong ferromagnetism is similar to the electronic structure of Co. The present results show that the XMCD spectrum at the $K$-edge probes the reconstruction of the electronic structure, in particular the changes that occur in the $3d$ unoccupied DOS as a result of hydrogenation. The XMCD profile of FeH is similar to that of CoH except for the small positive peak **A** even though these hydrides have different crystal structure, i.e. dhcp for FeH, fcc for CoH [2]. This result indicates that the crystal structure has little effect on the XMCD profile.

Stability of the ferromagnetic state is investigated by the pressure dependence of XMCD. At the highest pressure, the XMCD amplitude of FeH becomes very small at 27.5 GPa, while the XMCD profile of CoH is almost unchanged at 28.4 GPa (figure 1). To clarify this trend, the pressure dependence of XMCD is summarized in figure 2, where the integrated intensity of XMCD is plotted. Because the XMCD profile changes largely by hydrogenation, marked variations occur at the corresponding pressure ranges. The intensity of FeH decreases linearly up to 20 GPa, after which the slope becomes steep with further increase in pressure. Extrapolating this decrease, we evaluate the critical pressure $p_c$ to be 29.5 GPa; $p_c$ is the pressure at which the XMCD spectrum vanishes. FeH is paramagnetic above $p_c$, and Curie temperature would be reduced below room temperature. In the case of CoH, the intensity of CoH linearly decreases up to 28.4 GPa; a sudden drop is absent. The slope of the CoH is two times smaller than that of the FeH below 20 GPa. We also note that the weak pressure dependence is comparable to the reported behavior of Co and Ni metals; the ferromagnetic states of Co and Ni are known to remain above 100 GPa [10,11]. The present result indicates the high stability of the ferromagnetic state in CoH under pressure.

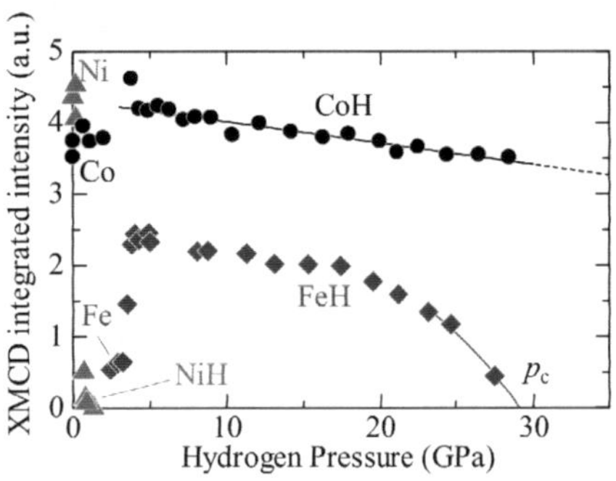


**Figure 2.** Integrated intensity of CoH (closed circles), FeH (closed diamonds), and NiH (closed triangles) as a function of pressure. $P_c$ denotes the estimated critical pressure of FeH.

Appearance or disappearance of the ferromagnetic state can be explained by the $3d$ DOS around $E_F$. In the case of TM hydrides, hydrogenation causes remarkable volume expansion ($\Delta V = \sim 2$ Å$^3$ per formula unit) [4,12,13], which leads to higher $3d$ DOS with a narrow width. The reduced overlap between TM $3d$ orbitals plays an important role in the electronic structure. Taking Stoner criterion into account [14], the electronic structure of FeH and CoH with higher densities around $E_F$ is suitable to stabilize ferromagnetic states. In contrast, most of the $3d$ DOS of NiH is occupied, and $E_F$ is, consequently, located above the upper threshold of the $3d$ DOS [15]. The electronic structure with low densities around $E_F$ results in the paramagnetic state of NiH.

Under higher pressure, broadening of the 3*d* DOS is introduced by the enhanced overlap between TM 3*d* orbitals due to the lattice contraction. This reduces the DOS around $E_F$, resulting in the instability of the ferromagnetic state. The high stability of the ferromagnetic state of CoH implies that its electronic structure is less influenced by the applied pressure, whereas the electronic structure of FeH is strongly modified. The marked contrast in the stability is probably attributed to the different profiles of the 3*d* DOS around $E_F$.

## 4. Conclusion

The magnetic states of Fe, Co, and Ni hydrides under pressure have been investigated by XANES and XMCD. The XMCD spectra reveal that Fe, Co, Ni metals are hydrogenated to ferromagnetic FeH, ferromagnetic CoH, and paramagnetic NiH, respectively. The sharp XMCD profiles near $E_0$ indicate that the orbital moment density of 3*d* states is concentrated just above $E_F$, which is probably attributed to the hydrogen-induced electronic structure around $E_F$, *i.e.* antibonding states, and the higher 3*d* DOS with a narrow width. The pressure dependence of XMCD shows that the ferromagnetic state of FeH is less stable than that of CoH under pressure. Our study demonstrates that the XMCD spectra at the *K*-edge sensitively probe the changes that occur in the electronic structure as a result of hydrogenation.

## Acknowledgments

We appreciate T. Oguchi (Osaka University), T. Tsumuraya (AIST), Y. Matsuura, T. Shishido (Hiroshima University), for their beneficial discussions with regard to the first-principles calculation. This work was supported by the Grants of the NEDO project "Advanced Fundamental Research on Hydrogen Storage Materials." The experiments were performed at SPring-8 with the approval of PRC-JASRI (Nos. 2008B1284, 2009B1011, 2009B1011, 2010A1005, and 2010B1520). This study has also been approved by the Photon Factory Program Advisory Committee (Proposal No. 2009G659).

## References

[1] Fukai Y 2005 *The Metal-Hydrogen system, 2nd ed.* (Springer,Berlin)
[2] Antonov V 2002 *J. Alloys and Compd.* **330-332**, 110.
[3] Ishimatsu N, Matsushima Y, Maruyama H, Tsumuraya T, Oguchi T, Kawamura N, Mizumaki M, Matsuoka M and Takemura K 2010 *Mater. Res. Soc. Symp. Proc.* **1262** W04-02
[4] Ishimatsu N *et al.*, in preparation.
[5] Igarashi J and Hirai K 1994 *Phys. Rev. B* **50** 17820
[6] Igarashi J and Hirai K 1996 *Phys. Rev. B* **53** 6442
[7] Takemura K, Sahu P C, Kunii Y and Toma Y 2001 *Rev. Sci. Instrum.* **72** 3873
[8] Kawamura N, Ishimatsu N and Maruyama H 2009 *J. Synchrotron Rad.* **16** 730
[9] Elsässer C, Zhu J, Louie S. G, Fähnle M and Chan C T 1998 *J. Phys.: Condens. Matter* **10** 5081
[10] Ishimatsu N, Kawamura N, Maruyama H, Mizumaki M, Matsuoka T, Yumoto H, Ohashi H and Suzuki M 2011 *Phys. Rev. B* **83** 180409
[11] Iota V, Klepeis J H P, Yoo C S, Lang J, Haskel D and Srajer J 2007 *Appl. Phys. Lett.* **90** 042505
[12] Badding J V, Hemley R J and Mao H K 1991 *Science* **253** 421
[13] Fedotov V K, Antonov V E, Antonova T E, Bokhenkov E L, Dorner B, Grosse G and Wagner F E 1999 *J. Alloys Compd.* **291** 1
[14] Mohn P 2006 *Magnetism in the solid state* (Springer, Berlin) p. 79
[15] Vargas P and Christensen N E 1987 *Phys. Rev. B* **35** 1993

# Shock pressure measurements in Polyvinyl alcohol (PVA) films using multi-frame optical shadowgraphy

S **Chaurasia**[1], **S Tripathi**[1], **P Leshma**[1], **J Pasley**[2] **and M Kumar**[3]

[1]High Pressure and Synchrotron Radiation Physics Division, Bhabha Atomic Research Centre, Mumbai, 400085, India
[2]York Plasma Institute, Department of Physics, University of York, York, YO10 5DQ, UK
[3]Radiation and Photochemistry Division, Bhabha Atomic Research Centre, Mumbai, 400085, India

E-mail: pgshivanand@gmail.com

**Abstract.** The knowledge of the equation of state (EOS) of materials at high pressures in excess of 10 Mbar is important in several branches of physics including astrophysics and inertial confinement fusion. It is possible to access this high pressure regime in the laboratory using shock waves launched by the interaction of a high power laser with a solid target. To study laser driven shock waves in plastic (Polyvinyl alcohol) $(C_2H_4O)_n$ targets, a multiframe optical shadowgraphy technique has been developed, with spatial and temporal resolution of 12 μm and 500 ps respectively. The experiments were performed using the 1064 nm 20 J /500 ps Nd: Glass laser at BARC. The focused laser intensity on target was varied between 6 x $10^{13}$ W/cm$^2$ and 2.7 x $10^{14}$ W/cm$^2$. The experimental data have been compared with the results of previous experimental and theoretical studies. The results are also found to be in agreement with SESAME data. The maximum pressure attained in the experiments was 30 Mbar, achieved with a laser intensity of 2.7 x $10^{14}$ W/cm$^2$.

## Introduction

The knowledge of the equation of state (EOS) of materials at high pressures in excess of 10 Mbar is important in several branches of physics including astrophysics [1] and inertial confinement fusion [2]. It is possible to access this high pressure regime in the laboratory using shock waves launched by the interaction of a high power laser with a solid target. The use of lasers in this role enables the control of shock pressure profile, and the use of a relatively simple experimental geometry. On the other hand, laser-plasma interactions tend to generate suprathermal electrons, which can preheat the material ahead of the shock front. It is well known that the use of shorter laser wavelength is an effective approach to controlling suprathermal electron production and the resulting preheating of the target. Understanding the compressibility of ignition-relevant ablator materials is crucial for inertial confinement fusion (ICF) target designs [2, 3].In these experiments, a precise knowledge of the equation-of-state is necessary in order to

Published under licence by IOP Publishing Ltd

accurately predict the behaviorof the imploding fuel capsule. In the field of planetary astrophysics, it is important to understand the equation of state of relevant materials at multi-Mbar pressures [4]. In laser driven equation of state experiments, it is advantageous to incorporate an low atomic number ablator, so as to restrict keV ablation temperatures to elements of low atomic number irrespective of the sample material, reducing the risk of generating hard x rays; and the ability to use a single ablator of known properties for experiments on less well-understood sample materials. By appropriate choice of material, an ablator may also increase the efficiency of conversion of laser energy to pressure in the sample [5, 6] by using the impedance mismatch technique. Plastics are attractive ablator materials as they are stable to handle and readily available with compositions of low atomic number, such as polystyrene $(C_8H_8)_n$, polyparaxylylene-N (parylene-N)$(C_8H_8)_n$, and polyethene $(C_2H_4)_n$. Previous studies have investigated the properties of the ablation plasma, including two-dimensional effects in planar targets [7], and have investigated the generation and mitigation of preheat in the sample [8, 9]. Damian C. Swift et al., have done extensive studies on the properties of plastic ablators in laser-driven material dynamics [10]. S. X. hu et al, has measured plastic ablator compressibility and shock pressure at the laser intensities from 0.5 to 1.5 x $10^{15}$W/cm$^2$ at the Omega laser [11] using an x-ray shadowgraphy technique. Here we describe the development of a novel complimentary diagnostic technique for EOS studies, based upon multi-frame optical shadowgraphy.

**Experimental Set-up:** The experiment was performed with a 20J/ 500 ps Nd: Glass MOPA laser system. The details of the laser system are described elsewhere. The laser beam was focused at normal incidence on to the target, which is held under vacuum at the focal point of a f/10 lens. The focal spot was ~ 100μm in diameter and the achieved intensity was up to 2.7x$10^{14}$W/cm$^2$. The dynamic motion of the rear surface of the foils was studied using transverse two frames optical shadowgraphy [12]. For the shadowgraphy a1% leak of the main laser beam was taken and split into two with the help of polarizing beam splitter. One of the two probe beams was delayed by a desired amount and was made collinear with the other using polarizing optics. These two beams of wavelength 1064 nm were converted to 532 nm using a SHG crystal (KD*P) to backlight the target at two different time delays after irradiation of target with the main beam. The two time delays were 2.92ns, and 6.393ns with respect to arrival of the peak of the main beam at target. The delay between the probe beams and the main beam was measured with vacuum bi-planar photodiodes (rise time 80 ps) and a 500 MHz oscilloscope and was verified with the optical streak camera with temporal resolution of 20 ps. The beams are propagated along the same optical path and illuminate the target edge such that the target foil is aligned at the centre of the field of view as shown in the shadowgrams in figure 1(a)-1(c). The magnification at the focal plane of the imaging lens was 3.457; the spatial and temporal resolutions of the set up were 12μm and 500ps respectively. The targets used were PVA (ρ=1.26 g/cc) foils of thickness 15μm. The width of foils was 1.5-2 mm which is much greater than the focal spot. The target films were prepared by the solution casting method.

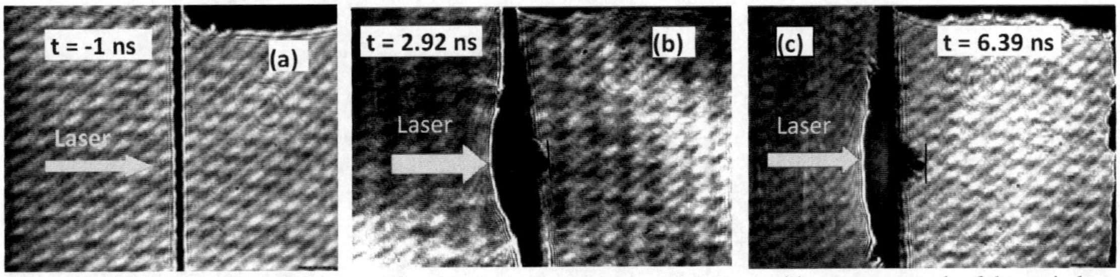


**Figure 1.**Shadowgraph recorded at a delay of (a) -1 ns (b) 2.92 ns (c) 6.93 ns with respect to peak of the main laser pulse on target.

**Result and discussion:** When a laser pulse is focused on to the target foil, the surface gets ablated and a blow off plasma is created which expands away from the target surface and exerts a pressure on foil. Due to this pressure, the foil is accelerated. The free surface displacement of the foil was measured with multiframe optical shadowgraphy at different time delays as shown in figure 1(a) - (c). The particle velocity ($u_p$) is taken as half of the free surface velocity with a correction factor expressed by Benuzzi-Mounaix et al [13]. The foil acceleration was measured at various intensities ranging from 6 x $10^{13}$ W/cm$^2$ to 2.7 x $10^{14}$ W/cm$^2$. The particle velocity of the foil scales with laser intensity as $u_p = I^{\alpha}$, Where $\alpha = 0.85$ as shown in figure 2. The shock velocity and shock pressure were calculated using standard equation of state relations

$$u_s = a + bu_p \tag{1}$$

$$\mathrm{P} - \mathrm{P}_0 = \rho_0 u_s u_p \tag{2}$$

Where $u_s$, $u_p$, $\rho_0$ and P are the shock velocity, particle velocity, density of the material and pressure respectively, a and b are constants. The values of a, b and $\rho_0$ for the PVA are 0.246, 1.565, and 1.26 g /cc respectively. The shock pressures are calculated using equation 2. The scaling of shock pressure with laser intensities is shown in figure 3. The published work of S. X. Hu et al [11] and Eliezer et al [5] are also presented along with our result in figure 3. Our results are in good agreement with the other results. The Hugoniot curve (shock pressure with particle velocity) constructed with our experimental data have been compared with the results of previous experimental and theoretical studies by Koenig et al [14] at LULI, data from HELEN, AWE, UK and Eliezer et al shown in figure 4.The results are also found to be in agreement with the theoretical data available in the LASL data book and SESAME data. The maximum pressure attained in the experiments was 30Mbar, attained with a laser intensity of 2.7 x $10^{14}$ W/cm$^2$.

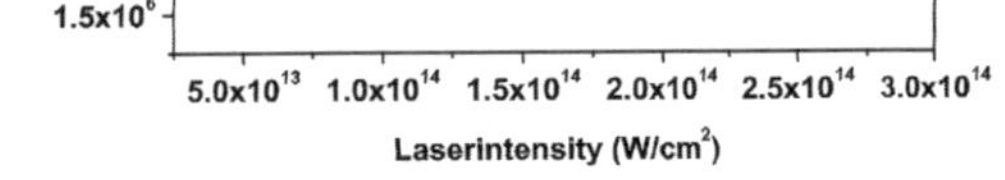


**Figure 2.**Scaling of particle velocity with laser intensity.

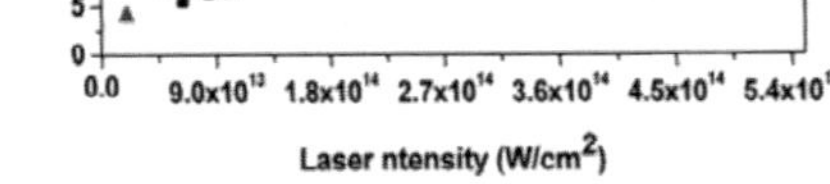


**Figure 3.** Scaling of shock pressure with incident laser intensity.

**Conclusion:** The laser driven shock studies in PVA plastic have been performed using a novel multi-frame optical shadowgraph with a spatial and temporal resolution of 12 μm and 500 ps respectively. The laser intensity on target was varied between 6 x $10^{13}$ W/cm$^2$ to 2.7 x $10^{14}$ W/cm$^2$. The experimental data have been compared with the results of previous experimental and theoretical studies. The results are also found to be in agreement with SESAME data. The maximum pressure attained in the experiments was 30Mbar, achieved with a laser intensity of 2.7 x $10^{14}$ W/cm$^2$

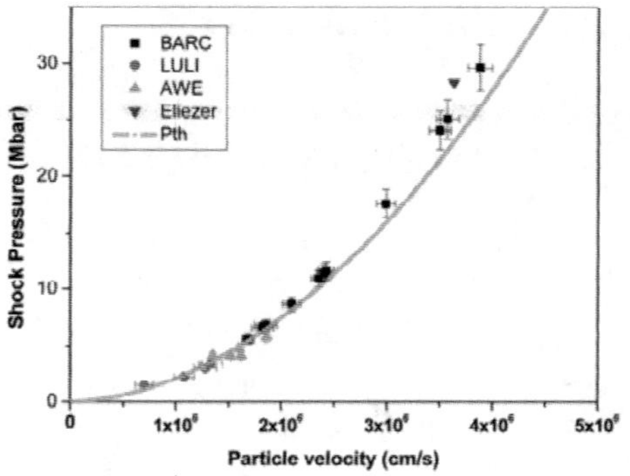


**Figure 4.**The experimental Hugoniot EOS data points for PVA (■) in our experiments, Koenig et al, LULI (●), HELEN, AWE (▲) and Eliezer et al (▼) show good agreement with SESAME curve (—).

**Acknowledgement:** The authors wish to acknowledge the support and continuous encouragement received from Dr. S. M. Sharma, Head of the HP&SRPD and Dr. S. Kailas, Director of the Physics Group at BARC. Also, we acknowledge the excellent support given by C. G. Murali and Diwakar S. Munda in design and implementation of laser electronics and smooth operation of laser system. Author M. Kumar acknowledges the support received from Dr. S. K. Sarkar, Head of the RPCD, and Dr. T. Mukherjee, Director of the Chemistry Group.

## References

[1] Remington B A, Drake R P and Arnett D 2000 *Phys. Plasmas* **7** 164.
[2] Lindl J 1995, *Phys. Plasmas* **2** 3933.
[3] Bodner S E *et al.*2000 *Phys. Plasmas* **7** 2298.
[4] Lee K K M et al.2006, *J. Chem. Phys.* **125** 014701.
[5] Eliezer L, Gitter A D, Krumbein D and Salzmann 1981 *Phys. Lett* **86A** 464.
[6] Chaurasia S, Leshma P, Pasley J, Tripathi S and Kumar M (to be published).
[7] Grun J, Decoste R, Ripin B H, and Gardner V 1981 *Appl. Phys. Lett.* **39** 545.
[8] Delettrez J, Bradley D K, Jaanimagi P A, and Verdon C P 1990 *Phys. Rev. A* **41** 5583.
[9] Colvin J D and Kalantar D H, in *Shock Compression of Condensed Matter*, edited by M. D. Furnish, et al, 2006 AIP Conf. Proc. No. 845
[10] Swift D C and Kraus R G 2008 *Phys. Rev. E* **77** 066402.
[11] Hu S X et al. 2008 *Phys. Rev. Lett.***100** 185003.
[12] Chaurasia S, Leshma P, Tripathi S, Murali C G, Munda D S, Sharma S M, Kailas S 2010 *BARC News Letter* **317** 13.
[13] Benuzzi-Mounaix et al 2002 *Phys. Plasmas* **9** 2466.
[14] Koenig M *et al.*1998 *Appl. Phys. Lett* **72** 1033.

Journal of Physics: Conference Series **377** (2012) 012043 doi:10.1088/1742-6596/377/1/012043

# Shock Deformation of Coarse Grain Alumina above HEL

**A K. Mukhopadhyay[1]*, K D Joshi[2], A Dey[1], R Chakraborty[1], A Rav[2], J Ghosh[1], S Bysakh[1], S K Biswas[1] and S C Gupta[2]**
[1]Central Glass and Ceramic Research Institute, Kolkata-700032, CSIR, India,
[2]Applied Physics Division, Bhabha Atomic Research Centre, Mumbai-400085, India
E-mail: *anoopmukherjee@cgcri.res.in

**Abstract.** Asymmetric shock recovery experiments with a two stage gas gun were conducted on a 10 μm grain size alumina at 6.5 and 12 GPa shock pressures levels which were more than three to six times as high as the Hugoniot Elastic Limit (HEL) of the same alumina and the shock recovered alumina fragments were characterized by X-ray diffraction (XRD), nanoindentation, scanning electron microscopy (SEM), field emission scanning electron microscopy (FE-SEM) and transmission electron microscopy (TEM). Based on these results a new qualitative damage model was developed to explain the deformation mechanisms of shock loaded alumina.

## 1. Introduction
HEL data on Alumina is available only for fine grain sizes [1-2] and rarely for coarse grain size [3]. This is what forms the basic scope and objective of the present work.

## 2. Experimental Materials and Methods
Symmetric shock experiments were conducted on a 10 μm grain size coarse alumina ceramic (density 3.98 $gm.cc^{-1}$) with a gas gun to identify its Hugoniot Elastic Limit (HEL) which was found from experimental measurement to be about ~2 GPa [3] which was a bit too much on the lower side than expected for alumina [1,2] presumably due to the large grain size, residual porosity and / or error due to the uncertainties of measurement. To understand the damage initiation and their subsequent growth mechanisms in coarse grain alumina subjected to shock impact at levels much above the HEL, additional asymmetric shock recovery experiments with a stainless steel flyer plate were deliberately conducted at 6.5 and 12 GPa with the same gas gun on the same alumina [4]. Detailed characterization of the shock recovered alumina fragments were done by XRD, nanoindentation, SEM, FE-SEM and TEM. Further experimental details are given elsewhere [3-6].

## 3. Results and Discussions
XRD of as received alumina (ARA) and shocked alumina (SA) from 6.5 and 12 GPa shock recovery experiments showed evidences of peak shift (figure 1a,b) to lower angles confirming the presence of a strained lattice [3,4]. The typical nanoindents in ARA (figure 1c) and the corresponding load (P) – depth (h) plots for both ARA and SA samples (figure 1d) showed elastic-plastic deformation behaviour. SEM images of the SA samples from 6.5 GPa shock recovery experiments confirmed the characteristic presence of intergranular macrocracking (figure 2a) and intragranular microcleavages (figure 2b) with grain localized micro / nano-scale deformations, while the corresponding FE-SEM images show presence of both in-plane and out of plane shear deformation band of about 50-100nm

width (figure 2c,d) in single grains suggesting thereby the important role played by shear induced deformation and fracture processes and that the deformation and fracture processes indeed originate at the sub-microstructural or nanostructural length scale regime of the microstructure [5,6].

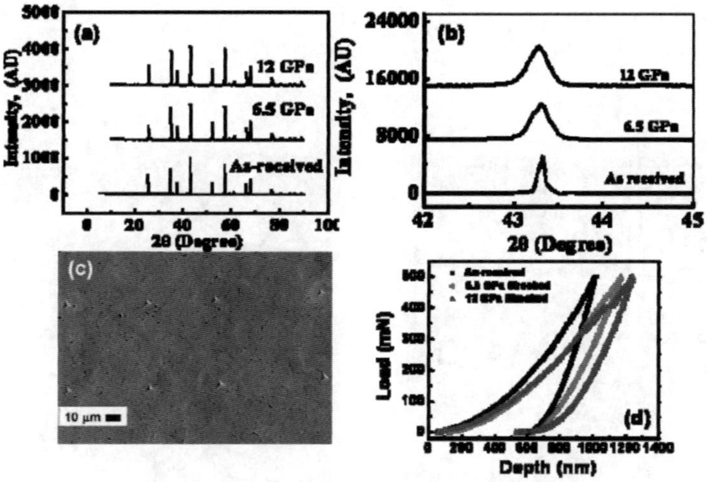


**Figure 1.** XRD (a,b), microstructure (c) and P-h plots (d) of ARA and SA.

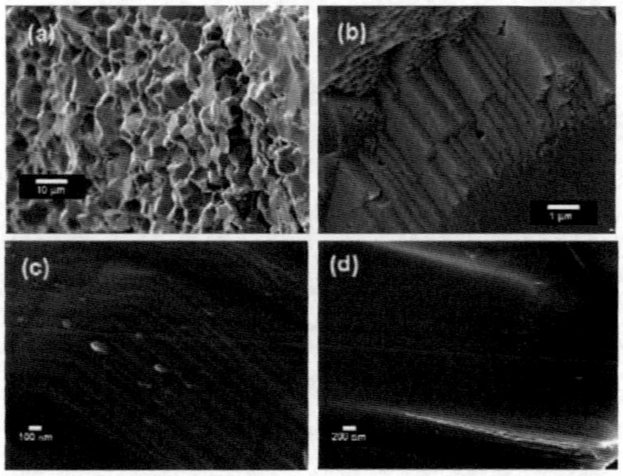


**Figure 2.** SEM (a,b) and FESEM (c,d) images of SA at 6.5 GPa peak stress.

The SEM image of SA samples from the 12 GPa shock recovery experiments showed extensive grain boundary crack formation (figure 2a) and grain localized microcleavages (figure 2b) while the FE-SEM images showed high spatial density of 40-60 nm thick shear deformation bands in adjacent grains (figure 2c) as well as 20-40 nm thick shear deformation bands in mutually perpendicular directions (figure 2d) in single grains. Further FE SEM images of the SA 12 GPa sample showed microwing crcak formation (figure 4a) and parallel array of grain localized shear deformation bands oriented at about $70^0$ angle to the horizontal direction (figure 4b). The presence of such high shear stress component lead to TEM images which showed extensive presence of dislocations impeded at grain boundary (figure 4c, 6.5 GPa) and dislocation entanglement within single grains (figure 4d, 6.5 GPa). Thus, the shock recovered alumina samples form 6.5 and 12 GPa shock experiments showed a multitude of microdamages. These included micro-cleavages, grain-boundary microcracks, micro-wing crack formation, and extensive shear induced deformations. In addition fractures localized at grains, grain boundaries and triple grain junctions, grain localized dislocations and their pile up impeded at grain boundaries were observed (figures 2-4). The shock at higher pressure created more extensive surface and sub-surface damages. These microdamages lead to the presence of a stronger indentation size effect (ISE) in the nanoindentation experiments in the SA (6.5 and 12 GPa) samples compared to that of the ARA sample (figure 5) wherein the solid lines show that the presence of ISE in all the data could be nicely described by the Nix and Gao model [7].

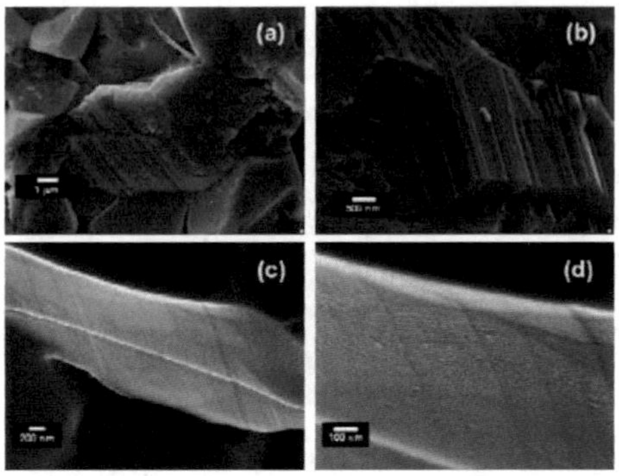


**Figure 3.** SEM (a,b) and FESEM (c,d) images of SA at 12 GPa peak stress.

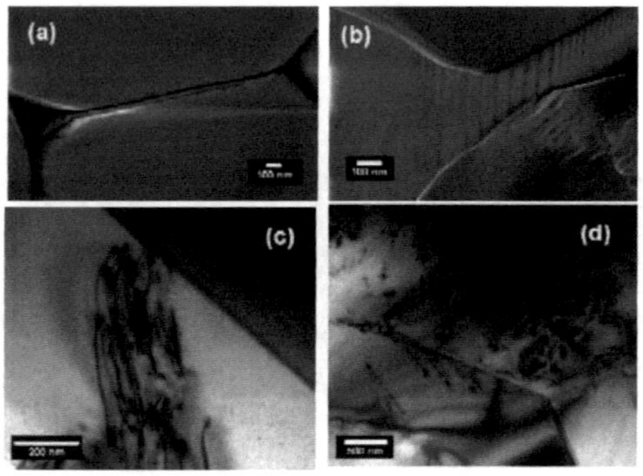


**Figure 4.** SEM (a,b) at 12 GPa and TEM (c,d) of SA at 6.5 GPa peak stresses.

Numerous efforts have been made earlier to model the high strain rate shock induced impact failure of brittle ceramics but no particular importance was given to coarse grain alumina ceramics [7-13]. These models either considered strength of the intact and the comminuted material [7], or assumed [8] an existing distribution of micro-cracks that grow according to dynamic fracture mechanics. Others [9] considered nucleation and growth of penny-shaped microcracks in multiple-planes and a balance between time dependent fracture and rate-dependent plasticity that allowed yield surfaces to vary with strain rate of the shock pulse delivered [10]. Recent attempts [1] tried to establish the role of grain localized plasticity in determining HEL. Efforts were made to depict [11] that surface of a crack slide to form wing cracks under multiaxial compression when the resolved applied shear stress exceeds frictional stress due to normal forces on the crack surfaces. For the present alumina ceramics the critical resolved shear stress for shear fault formation was estimated [3] to be 2.4-4 GPa which was much lower than the applied shock pressure of 6.5 GPa. These estimates matched also with those reported by others [2] for alumina. Therefore, shear stress induced deformation was quite possible for the present alumina ceramics. Further, it has been argued [3] that the dislocation pile up length of the present alumina sample would be on the higher side as it had a comparatively larger grain size e.g. ~ 10 μm, and consequently, a lower magnitude of critical shear stress would be necessary to nucleate a microcrack (figure 3a) at the head of the dislocation pileup (figure 4c) from a favourably oriented slip plane. Recent micromechanical finite element modeling [12] indeed have showed that the onset of plastic deformation in alumina results from grain boundary shear (that must first occur in order for wing cracks to form) that creates a tensile stress acting on two neighbouring suitably oriented grain boundaries (figure 4b) leading to mode I fracture thus forming a "micro-wing crack'' (figure 4a). On continued loading, these tensile microcracks began to join hands with each other eventually to form cracked microstructural scale slabs in the case with no lateral confinement, i.e. similar to the case of the present experimental conditions. The molecular dynamic simulation work [13] on dynamic failure of glass also supports this picture. However, from the results of the shock recovery experiments (figs.

1-4) we can possibly not guarantee whether the cracks and / or micro scale defects formed exactly in compression during loading or under tension during unloading. Ceramics being characteristically brittle most of the defects supposedly generate due to the compressive as well as shear stress during loading at such high impact stress [1, 2, 8-9] and the additional sub-major defects might grow and / or generate under tension during unloading [10-13].

Based on the experimental data and evidences from SEM, FE-SEM and TEM observations of the present work a new, simplistic, qualitative damage model (figure 6) was developed to explain the deformation mechanisms of coarse grain alumina ceramics shock loaded to stress levels much above the HEL. As indicated in the model (figure 6), apart from the characteristic brittle fracture; inelastic damage growth, coalescence and fragmentation process in shocked alumina can happen by shear fracture initiation in a favourably oriented single grain at a single plane (i,i'), or at multiple planes (ii,ii'), and / or by twist fracture initiation in a favourably oriented grain at a single plane (iii,iii'), or at multiple planes (iv,iv') and the process repeats itself within a few μs at multiple grains (v). It is suggested that when these cracks grew, coalesced and propagated very fast the fragmentation of shocked alumina happened. However, future efforts shall be directed to verify this picture by computer simulation.

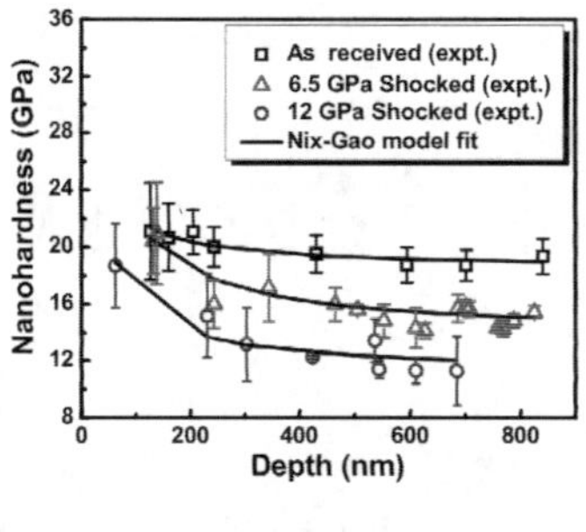


**Figure 5.** ISE in ARA and SA samples.

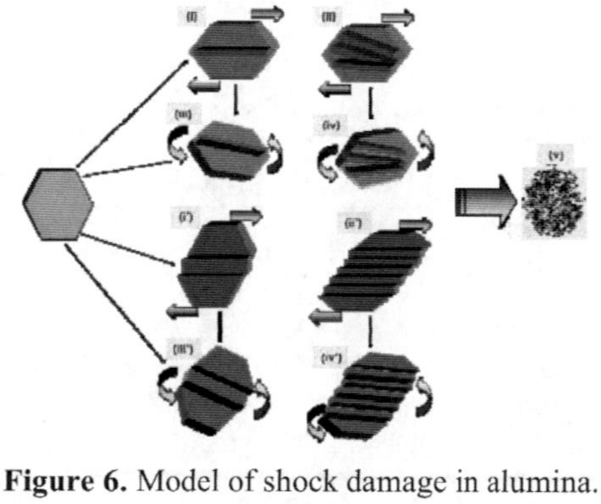

**Figure 6.** Model of shock damage in alumina.

## 4. Summary

Based on the experimental results a new qualitative damage model was developed to explain the deformation mechanisms of shock loaded coarse grain alumina.

**Acknowledgments**

The authors appreciate Director's kind encouragements for and financial support of BRNS (Project No. GAP 0219) and CSIR (NWP 0027, NWP 00299, SRFship of A Dey) to this work.

**References**

[1] Bourne N K 2006 *Proc. R. Soc. A* **462** 3213
[2] Chen M W, McCauley J W, Dandekar D P and Bourne N K 2006 *Nat. Mater.* **5** 614
[3] Mukhopadhyay A K, Joshi K D, Dey A, Chakraborty R, Rav A, Biswas S K and Gupta S C 2010 *J. Mater. Sci.* **45** 3635
[4] Mukhopadhyay A K, Joshi K D, Dey A, Chakraborty R, Rav A, Mandal A K, Ghosh J, Bysakh S, Biswas S K and Gupta S C 2010 *Mater. Sci. Eng. A*, **527** 6478
[5] Mukhopadhyay A K, Joshi K D, Dey A, Chakraborty R, Rav A, Mandal A K, Ghosh J, Bysakh S, Biswas S K and Gupta S C 2011 *Ceram. Int.* **37** 2365

Journal of Physics: Conference Series **377** (2012) 012043 doi:10.1088/1742-6596/377/1/012043

[6] Chakraborty R, Dey A, Mukhopadhyay A K, K. D. Joshi, Rav A, Mandal A K, Bysakh S, Biswas S K and Gupta S C 2011 *Mett. Mater. Trans. A* in press
[7] Johnson G R. and Holmquist T J 1992 A computational constitutive model for brittle materials subjected to large strains Shock-wave and High Strain-rate Phenomena in Materials, ed. M. A. Meyers, L. E. Murr and K. P. Staudhammer (Marcel Dekker Inc. : New York) pp 1075-1081
[8] Rajendran A M 1994 *Int. J. Impact Eng.* **15** 749
[9] Espinosa H D, Zavattieri P D, Dwivedi S K 1998 *J. Mech. Phys. Solids.* **46** 1909
[10] Grady D E 1998 *Mech. Mater* **29** 181
[11] Horii M, Nemat-Nasser S 1986 *Phil. Trans. R. Soc. Lond. A* **319** 337
[12] Warner D H, Molinari J F 2006 *Acta Mater.* **54** 5135
[13] Lu Z, Nomura K I, Sharma A, Wang W, Zhang C, Nakano A, Kalia R, Vashishta P 2005 *Phys Rev. Lett.* **95** 135501

Journal of Physics: Conference Series **377** (2012) 012044 doi:10.1088/1742-6596/377/1/012044

# A New Mechanism for ISE in Alumina Shock Deformed at 6.5 GPa

**R. Chakraborty[1]*, A. Dey[1], A. K. Mukhopadhyay[1], K. D. Joshi[2], , A. Rav[2], J. Ghosh[1], A. K. Mandal[1], S. Bysakh[1], S. K. Biswas[1] and S. C. Gupta[2]**
1Central Glass and Ceramic Research Institute, Kolkata-700032, CSIR, India,
2Applied Physics Division, Bhabha Atomic Research Centre, Mumbai-400085, India

E-mail: *chak.riya2@gmail.com

**Abstract.** With a view to understand how high strain rate flyer plate impact affects the nanohardness of a coarse (10 μm) grain high density (3.978 $gm.cc^{-1}$) alumina, load controlled nanoindentation experiments were conducted with a Berkovich indenter on as received alumina (ARA) and shocked alumina fragments (SA) of obtained from a flyer plate shock impact study at 6.5 GPa. The results showed that the nanohardness of the shocked alumina (SA) samples was much lower than that of the as received alumina (ARA) samples and that the indentation size effect (ISE) was mild in the ARA but quite severe in the SA samples. Extensive additional characterization by field emission scanning electron microscopy (FE-SEM), transmission electron microscopy (TEM) and a physical model based analysis of the experimental load depth data were utilized to provide a new explanation for the presence of strong indentation size effect in the shock recovered alumina.

## 1.Introduction

In spite of the wealth of literature on hardness of alumina ceramics which primarily determines there resistance against compressive failure [1,2] there effect of shock damage on nanohardness of alumina ceramics has not been thoroughly investigated [3]. This is what forms the basic objective of the present work.

## 2. Experimental Materials and Methods

The samples used in the present work were alumina discs prepared by pressureless sintering at 1310°C in air (dia~48 mm and thickness~2.5 mm, density~3.978 $gm.cc^{-1}$ and grain size ~ 10 μm). The alumina samples were deliberately shocked at > $10^4 s^{-1}$ strain rate by an asymmetric flyer plate impact with a SS304 plate at a pressure of 6.5 GPa in a two stage gas gun [3-6]. The shock recovered alumina fragments had a large statistical variety of shapes and sizes with a nominal average size of about 0.04-0.05 by 0.03-0.04 by 0.02-0.03 mm for the tiniest ones. Their tiny size left no other option than to use the nanoindentation technique [6-7] to evaluate their nanohardness. Thus, the nanohardness of the shocked alumina (SA) and as received alumina (ARA) samples were measured at various loads in the range of 10-1000 mN utilizing a Berkovich indenter of 150 nm tip radius and 65.3° semi-apex angle [3-6] using the Oliver-Pharr Model [7] by the load controlled nanoindentation technique (Fischerscope H100-XYp; Fischer, Switzerland; Resolutions- depth ± 1nm, load ± 0.2 μN, DIN 50359-1 standard). Microstructural characterizations were done by SEM (s430i, Leo, UK), FESEM (Supra VP35 Carl Zeiss, Germany) and TEM (Tecnai, G2 30, S-Twin, 300 KV, FEI, The Netherlands).

## 3. Results and Discussions

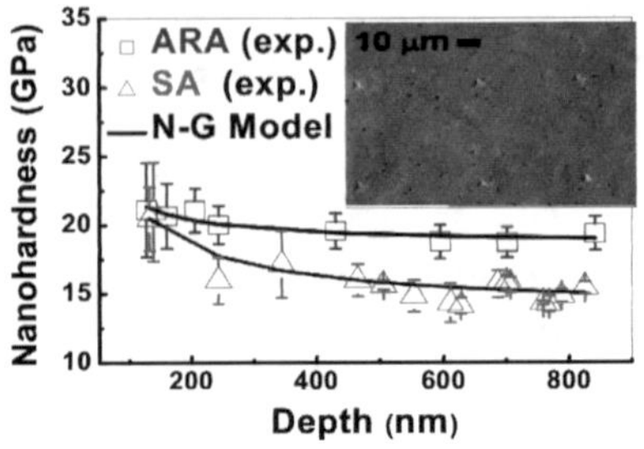


**Figure 1.** ISE in ARA and SA samples.

The indentation size effect (ISE) was mild in as received alumina (ARA) but quite strong (figure 1) in the shocked alumina (SA) samples where nanohardness (H) degraded. The data could be nicely explained by the well established Nix and Gao (N-G) model [8] in both the cases. The inset of figure 1 shows a typical array of nanoindents in an FE-SEM image. The empirical power law dependence of depth (h) on load (P), $P=ch^n$ [9] are shown for ARA and SA samples in correspondence in figures 2(a,b). The exponent n was « 2 for SA samples but ~2 for the ARA samples. So, ISE was strong in the SA but quite mild in the ARA samples (figure1). As n was ≤ 2, $d^2P/dh^2$ decreased with h, figures 2(c,d). The maximum shear stress ($\tau_{max}$) below the tip of the nanoindenter [6,10,11] at 100 and 1000 mN were ~36 and 20 GPa, respectively, for the ARA and SA samples based on the appropriate critical load values for ARA (figures 3a,b) and SA (figures 3c,d) samples. As $\tau_{max}$ is » $\tau_{theor}$ ~ 3 GPa [1], it justified the prolific presence of grain localized shear deformation band induced microcracks (figure 4a), intragranular shear bands without microcrack formation (figure 4b), shear deformation at triple grain junction (figure 4c) and possibly shear induced intraganular microcleavage (figure 4d).

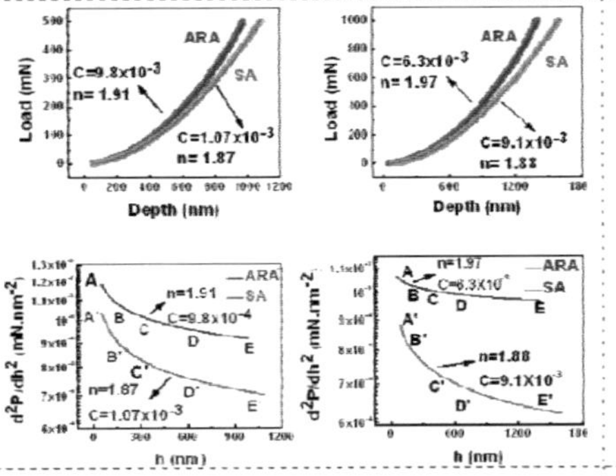


**Figure 2.** Power law fit of P-h plots (a) P=500 and (b) P=1000 mN (c,d) corresponding $d^2P/dh^2$ vs. h plots of ARA and SA samples.

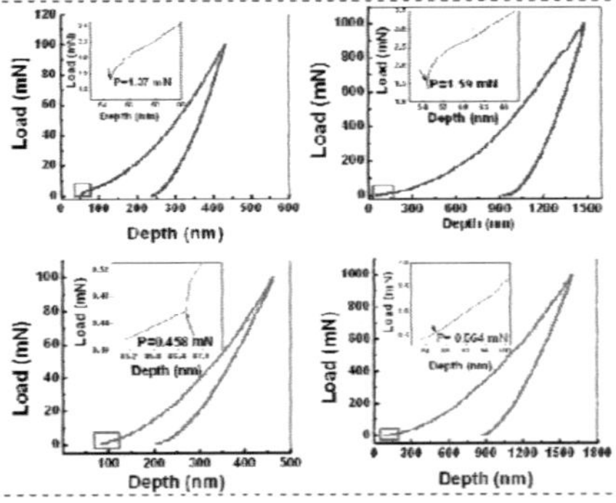


**Figure 3.** Critical loads from P-h plots of ARA (a) P=100 and (b) P=1000 mN and SA (c) P=100 and (d) P=1000 mN samples.Insets show the critical load determinations.

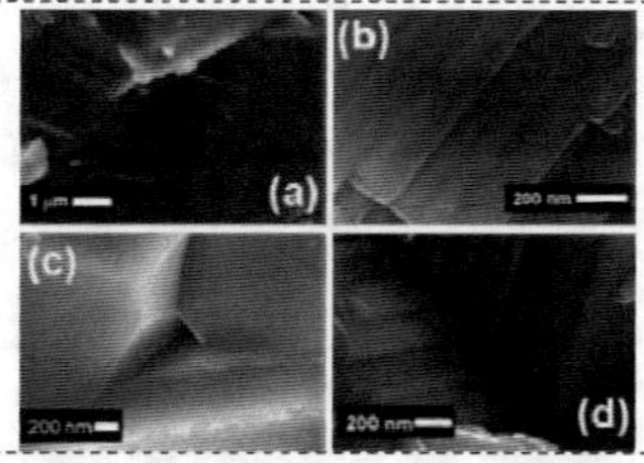


**Figure 4.** FE-SEM images of SA sample (a) grain localized shear deformation band induced microcracks (b) intragranular shear bands without microcrack formation (c) shear deformation at triple grain junction (d) extensive intragranular microcleavage.

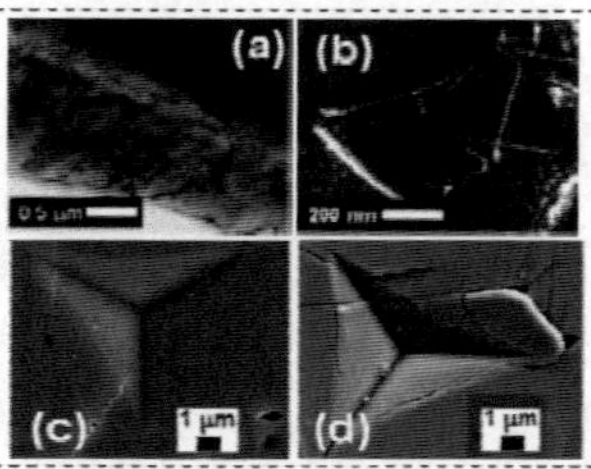


**Figure 5.** TEM images of SA samples (a) dislocation and deformation band at a grain edge (b) dislocation entanglement in a grain. FE-SEM images of nanoindent in (c) ARA (d) SA samples.

As $\tau_{max}$ is » $\tau_{theor}$ the corroborative TEM images showed huge amount of dislocations (figures 5a,b). Similarly, the FE-SEM images (figure 5d) of the nanoindentation cavities for the SA samples showed the relatively predominant presence of slip lines, deformation bands and microcracks and their comparative absence (figure 5c) in the ARA samples. All these evidences (figures 2-4) prove that the shock induced nano and microscale defects were much more in the SA than in the ARA samples. As a consequence, the *'multiple micro pop-in'* [12] in loading cycles occurred much more for the SA than in the ARA samples (figures 6a,b) which happened due to the nanoscale displacement jumps that could continuously and frequently occur [13], especially more acutely in the SA samples than in the ARA samples. Hence, as shown in the qualitative model, the ARA sample contains relatively lesser number of microstructural defects (figure 6c) as compared to their relatively much larger population in the SA (figure 6d) sample. Therefore, the extent of interaction between the penetrating nanoindenter and the microstructural damages is increased at a much faster rate thereby giving rise to the experimental observation (figure 1) of a strong ISE present in the case of SA samples.

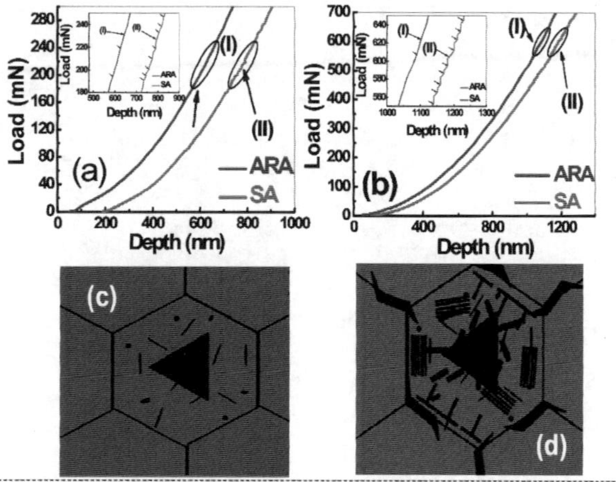


**Figure 6.** Serrations signifying 'multiple micro pop-in' events in P-h plots of ARA and SA samples (a)P=300 and (b) 700 mN; model of ISE in ARA (c) and SA samples (d).

**Summary**

The nanohardness of the shock alumina (SA) was much lower than that of the as received alumina (ARA) samples. ISE was mild in ARA but strong in SA samples. The depth dependencies of nanohardness in both the cases could be successfully explained by the Nix and Gao model. It is proposed that the strong ISE in SA samples happened due to the highly localized micro-tensile stresses in the suitably oriented grains of the damaged microstructure. It is proposed further that the relative frequency of the multiple 'micro pop-in' events would actually govern the relative rate of penetration by the nanoindenter into a given alumina ceramic microstructure and thereby the relative extent of the indentation size effect. Theoretical estimate of shear stress magnitude close to the nanoindentation tip was much higher than the theoretical shear strength of alumina and indicated that the first displacement bursts observed in the experimentally measured load versus depth plots could well be linked to the initiation of plastic deformation through the nucleation of shear bands highly localized in the material. Finally a new, qualitative model was proposed to explain the whole scenario of nanoindentation responses in the ARA and SA samples.

**Acknowledgments**

The authors appreciate Director's kind encouragements for and financial support of BRNS (Project No. GAP 0219) and CSIR (NWP 0027, NWP 00299, SRFship of A Dey) to this work.

**References**

[1] Rosenberg Z, Yaziv D, Yeshurun Y and Bless S J 1987 *J. Appl. Phys.* **62** 1120
[2] Mashimo T, Hanaoka Y and Nagayama K 1988 *J. Appl. Phys.* **63** 327
[3] Mukhopadhyay A K, Joshi K D, Dey A, Chakraborty R, Rav A, Biswas S K and Gupta S C 2010 *J. Mater. Sci.* **45** 3635
[4] Mukhopadhyay A K, Joshi K D, Dey A, Chakraborty R, Rav A, Mandal A K, Ghosh J, Bysakh S, Biswas S K and Gupta S C 2010 *Mater. Sci. Eng.* A, **527** 6478
[5] Mukhopadhyay A K, Joshi K D, Dey A, Chakraborty R, Rav A, Mandal A K, Ghosh J, Bysakh S, Biswas S K and Gupta S C 2011 *Ceram. Int.* **37** 2365
[6] Chakraborty R, Dey A, Mukhopadhyay A K, K. D. Joshi, Rav A, Mandal A K, Bysakh S, Biswas S K and Gupta S C 2011 *Mett. Mater. Trans.* A, in press
[7] Oliver W C and Pharr G M 1992 *J. Mater. Res.* **7** 1564
[8] Nix W D and Gao H 1998 *J. Mech. Phys. Solids* **46** 411
[9] Ebisu T and Horibe S 2010 *J. Eur. Ceram. Soc.* **30** 2419
[10] Packard C E and Schuh C A 2007 *Acta Mater.* **55** 5348
[11] Shang H, Rouxel T, Buckley M and Bernard C 2006 *J. Mater. Res.* **21** 632
[12] Subhash G, Maiti S, Geubelle P H and Ghosh D 2008 *J. Am. Ceram. Soc.* **91** 2777
[13] Nowak R, Sekino T and Niihara K 1996 *Phil. Mag. A* **74** 171

Journal of Physics: Conference Series **377** (2012) 012045 doi:10.1088/1742-6596/377/1/012045

# Mechanical Properties of Shock Treated Aluminium Alloy Al 2024-T4

**K.D. Joshi[a], A.K. Mukhopadhyay[b], A. Dey[b], Amit S. Rav[a], S. Biswas[b] and Satish C. Gupta[a]**

[a]Applied Physics Division,, Bhabha Atomic Research Centre, Mumbai, India, 400085

[b]Mechanical Test Section, Central Glass and Ceramic Research Institute, Kolkata, India, 700032

E-mail: kdj@barc.gov.in

**Abstract.** Plate impact experiment has been carried out on Al 2024-T4 alloy using single stage gas gun. The dynamic yield strength and spall strength of Al 2024-T4 sample has been determined to be 0.35 GPa and 1.43 GPa, respectively, from free surface velocity history measured using VISAR. The sample recovered after unloading from peak shock pressure of 4.4 GPa along with an unshocked sample is analyzed for mechanical properties using nano-indentation and scanning electron microscopy (SEM). The nano-indentation measurements reveal that the hardness and Young's modulus for unshocked sample remains unchanged as a function of load (equivalently depth), however, the same for shocked sample decreases monotonically with increase of load up to ~40 mN and on further increase of load it remains unchanged, suggesting the (i) increase in hardness of shock loaded sample; (ii) the increase in hardness is limited to certain depth, which in our case is 845.12 ± 43.16 nm.

## 1. Introduction

Aluminium alloy, Al 2024-T4 exhibits good machinability, high ultimate tensile strength (UTS), high yield strength and low density. These mechanical properties of this alloy make it a potential candidate for applications in amours, automobile and aerospace industries where the strength to weight ratio becomes an important parameter. For such applications it is not only the mechanical properties under quasistatic conditions but also under high strain rate conditions become crucial. Thus, it is important to study the mechanical properties at high strain rates. The shock loading experiments are best suited for such studies, as, strain rates ranging from $10^3$/s - $10^9$/s have been achieved in such experiments [1, 2]. Further, the shock compression experiments designed properly can generate not only the high compressive stresses but also high tensile stresses in the materials [1-3] enabling to investigate material behaviour *e.g.* dynamic yield strength, spall fracture etc., in negative pressure regime also. The effect of yielding and spall fracture is manifested in the free surface velocity ($U_{fs}$) history of the impacted sample, which, can be measured using interferometry techniques such as velocity interferometer system for any reflector (VISAR) [4] and optically recorded velocity interferometer system (ORVIS) [5].

In the present work we have determined the dynamic yield strength and spall strength of Al2024-T4 alloy from measured free surface velocity history. Further, a comparative study of mechanical properties of this alloy has been carried out before and after shock treatment.

## 2. Experimental Method

The Al2024 –T4 alloy having 92.56 % of Al, 4.4 % of Cu, 1.4 % of Mg, 0.6 % of Mn, 0.5 % of each Fe and Si and remaining Cr, Zn and Ti is procured from Indian Aluminium Company Limited (INDAL). The initial density ($\rho_0$), static yield strength and ultimate tensile strength of the alloy obtained from static test is 2.78 g/cc, 0.324 GPa and 0.469 GPa, respectively. The target is prepared in the form of a disk of diameter 57 mm and thickness 10 mm and subjected to shock pressure of 4.4 GPa by impacting it with a flyer plate of same material with thickness 5 mm, accelerated to a velocity Vp = 0.56 km/s in a 63 mm bore single stage light gas gun [6]. The interaction of tensile waves resulted from the reflection of shock wave from the target free surface and from the flyer free surface generated large tensile stress in the target which exceeded a limit, termed as spall strength ($\sigma_s$), causing spall fracture. The dynamic yield strength (Y) and spall strength have been determined from the velocity history of free surface of target plate recorded using VISAR.

In order to understand the effect of shock loading on mechanical properties of Al2024-T4, the nano-indentation and scanning electron microscopy (SEM) measurements have been carried out on shock treated and unshocked samples of this material. Micromechanical characterization of polished samples was done using a nanoindentation machine (Fischerscope, H100XYp, Switzerland) in a load range between 0.4 and 100 mN, with force and depth sensing resolutions of 0.2 μN and 0.1nm.The machine was calibrated with a BK7 reference glass block (Schott, German) equipped with a standard Berkovich indenter. Experiments were conducted using a load-time sequence. For each loading-unloading cycle, loading and unloading lasted 30s, respectively. During each test run, a personal computer collected and stored data for the load and displacement as the indenter was driven into the sample (loading segment) and then withdrawn from it (unloading segment). The raw data were then used to construct the load-displacement plot. For each load a 5x5 square array of 25 indentations were made. These positions were randomly selected without any particular bias. The well-known Oliver Pharr model [7] has been utilized to calculate the micromechanical properties of Al2024-T4 in as-received condition and after shock treatment. The samples from fracture surface of the shocked Al2024-T4 sample were carbon coated prior to insertion in the SEM.

## 3. Results and Discussions

In figure 1 we display the recorded fringe pattern and free surface velocity profile derived from this for Al2024-T4 target plate during unloading from peak shock pressure of 4.4 GPa. This peak shock pressure is determined using expression: $P = \rho_0 D U_p$. Where, $U_p$ is the particle velocity at the flyer-target impact interface which, in our experiment, carried out in symmetric impact configuration, is half (0.28 km/s) of the measured impact velocity (0.56 km/s). The shock velocity $D$ in the target corresponding to this particle velocity is obtained to be 5.73 km/s from the known Hugoniot of Al2024 alloy [10]. The Hugoniot elastic limit ($\sigma_{HEL}$) and spall strength ($\sigma_s$) are determined from the free surface velocity profile using the following expressions [1, 8, 9]:

$$\sigma_{HEL} = \frac{1}{2} U_H \rho_0 c_l ; \sigma_s = \frac{1}{2} \Delta U_{fs} \rho_0 c_b$$

Where, $U_H$ is free surface velocity at $\sigma_{HEL}$. The $\Delta U_{fs}, c_b, c_l$ are pull back velocity of free surface, bulk sound speed and longitudinal sound speed, respectively. The pull back velocity is defined as $U_f - U_m$ where, $U_f$ is the peak free surface velocity and $U_m$ is free surface velocity just ahead of spall pulse. The values of $C_l$ and $C_b$ have been taken to be 6.37 km/s and 5.37 km/s from literature [10]. The values of $\sigma_{HEL}$ and $\sigma_s$ are determined to be 0.70 GPa and 1.43 GPa, respectively, with corresponding strain rates of ~$10^4$/s. The value of Y deduced from $\sigma_{HEL}$ is 0.355 GPa. These values of $\sigma_s$ and Y are higher than the ultimate tensile strength of 0.469 GPa and yield strength of 0.324 GPa, respectively, reported from quasistatic tests.

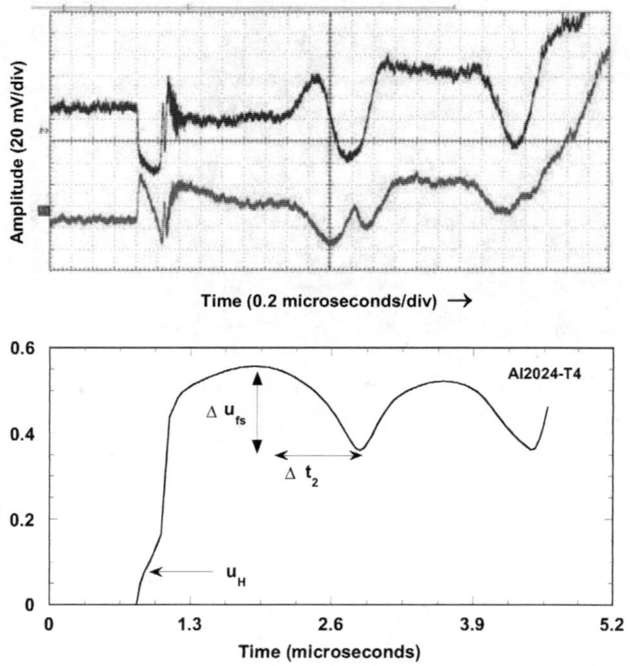


**Figure 1** The fringe shift recorded for S and P waves (top) and corresponding free surface velocity profile (bottom) determined from the fringe shift for Al2024-T4 alloy

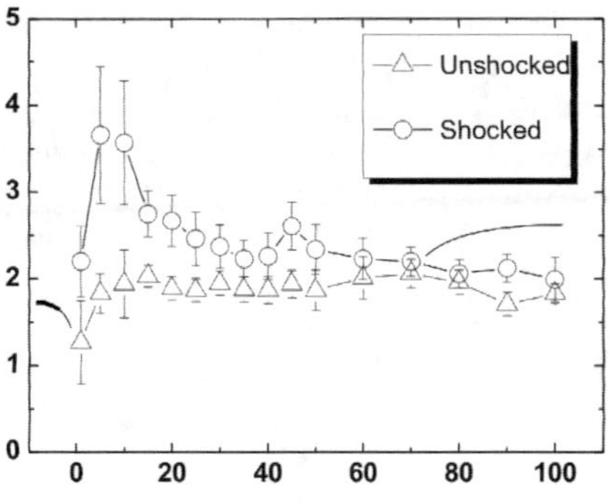


**Figure.2** The measured hardness for unshocked and shock treated Al2024-T4 alloy

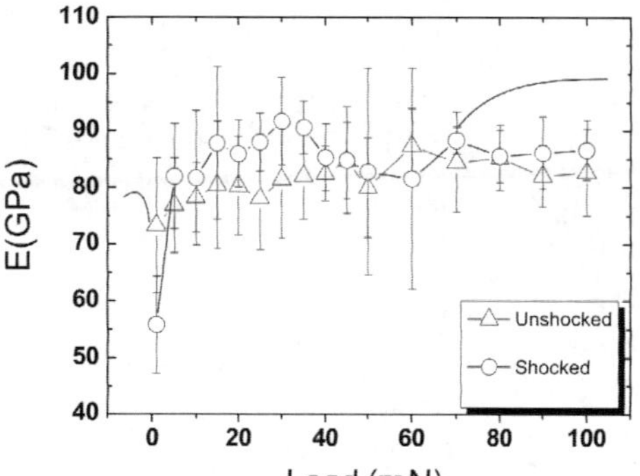


**Figure.3** The measured Young's modulus for unshocked and shock treated Al2024-T4 alloy

The nano-indentation analysis (Figure. 2) reveals that the hardness (H) for unshocked sample remains unchanged as a function of load, however, the same for shocked sample decreases monotonically with increase of load up to ~40 mN and on further increase of load it remains unchanged. Also, the hardness of shock treated sample is higher than that of the unshocked sample and

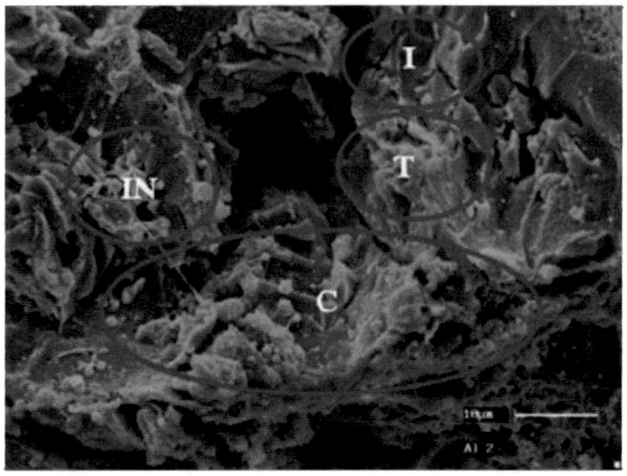


**Figure 4** SEM picture of shock treated Al2024-T4 sample showing different types of fractures (IN, C, T and I explained in text).

this difference in hardness decreases with increasing load and for loads ≥ ~50 mN the hardness of both the samples become almost equal, indicating that the shock treatment has increased the hardness of the Al2024-T4 alloy and the increase in the hardness is limited up to a depth of about 845.12 ± 43.16 nm only. This was measured by a capacitive transducer attached to the machine. It had a resolution of ±1 nm. The Young's modulus (E) was relatively unchanged within errors except for the point at the minimum applied load of 0.4 nm (Figure 3). In figure 4 we display SEM image of fractured surface of shock treated Al2024-T4. As marked in figure, the SEM image reveal a very highly deformed fracture surface comprising of large areas of trans-granular (T) fracture with some localized zones of both intra (IN)- and inter- granular (I) crack propagation and a large amount of comminuted smaller chunks (C) or pieces which might have resulted as debris from the shock experiment. The transgranularly fractured portion shows fracture through the grain itself. The intragranularly fractured portion indicates failure occurring within the grain. The intergranularly fractured portion exhibits failure occurring in between the grains. The agglomerate of shock fractured debris appeared as the comminuted chunks.

**References**

[1] G.I. Kanel, S.V. Razorenov, A.V. Utkin, V.E. Fortov, K. Baumung, H. U. Karow, D. Rusch and V. Licht 1993 *J. Appl. Phys* **74(12)** 7162.
[2] E. Moshe, S. Eliezer, Z. Henis, M. Werdiger, E. Dekel, Y. Horovitz, and S. Maman, I. B. Goldberg D. Eliezer 2000 *Appl. Phys. Lett.* **76** 1555.
[3] Joshi K.D. and Gupta Satish C. 2007 *High Press. Res.* **27** 259.
[4] Barker L.M. and Hollenbach R.E. 1972 *J. Appl. Phys.* **43** 4669.
[5] Bloomquist D. and Sheffield S.A. 1983 *J. Appl. Phys.* **54** 1717.
[6] Satish C. Gupta, R.G. Agrawal, J.S. Gyanchandani, S. Roy, N. Suresh, S. K. Sikka, A. Kakodkar and R. Chidambaram Shock Compression of Condensed Matter 1991 (Eds. S.C. Schmidt, R.D. Dick, J.W. Forbes and D.G. Taksar, Elsevir Science,1992) p 839.
[7] Oliver W C, Pharr G M. 1992 *J. Matter Res* **7** 1564.
[8] E.B. Zaretsky, G.I. Kanel, S.V. Razorenov and K. Baumung 2005 *Int. J. Impact Engg.* **31** 41.
[9] K.D. Joshi, Amit. S. Rav, Satish C. Gupta and S. Banerjee 2010 *J. Phys.Conf. Series* **215** 012149.
[10] Marsh S.P.1980 *LASL Shock Hugoniot data* University of California Press 1980 p.166 and 213

Journal of Physics: Conference Series **377** (2012) 012046 doi:10.1088/1742-6596/377/1/012046

# Damage Evolution Mechanisms in Plate Impact of Alumina

**A K. Mukhopadhyay[1]*, K D Joshi[2], A Dey[1], R Chakraborty[1], A Rav[2], J Ghosh[1], S Bysakh[1], and S C Gupta[2]**
[1]Central Glass and Ceramic Research Institute, Kolkata-700032, CSIR, India,
[2]Applied Physics Division, Bhabha Atomic Research Centre, Mumbai-400085, India

E mail: [1]**anoopmukherjee@cgcri.res.in*

**Abstract.** Asymmetric shock recovery experiments with a two stage gas gun were conducted on a 10 μm grain size alumina at 6.5 GPa shock pressures levels which was more than three times as high as the Hugoniot Elastic Limit (HEL) of the same alumina and the shock recovered alumina fragments were characterized by field emission scanning electron microscopy (FE-SEM) and transmission electron microscopy (TEM). Based on these results a new qualitative damage model was developed to explain the deformation mechanisms of shock loaded alumina.

## 1. Introduction

Numerous damage models have been developed earlier to describe the high strain rate shock induced impact failure of mainly fine grained brittle ceramics [1-9] but significant attention had not yet been given to coarse grain alumina ceramics [10]. The Bourne model [1,2] emphasized the role of grain localized plasticity in determining HEL of fine grained alumina while the Johnson-Holmquist model considered strength of the intact and the comminuted ceramics [3]. The Rajendran model assumed [4] that a distribution of micro-cracks exist in a fine ceramic microstructure and their growth is dictated by the principles of dynamic fracture mechanics. The model proposed by Espinosa et al. [5] considered nucleation and growth of penny-shaped microcracks in multiple-planes of fine grained ceramics. The Grady model proposes that a fine competitive balance between time dependent fracture and rate-dependent plasticity linked processes dominates the shock induced local yielding in brittle ceramics as a function of strain rate [6]. The Horii and Neemat-Nasser model [7] proposed that surface of a crack present in a fine ceramic microstructure can slide to form wing cracks under multiaxial compression when the resolved applied shear stress exceeds frictional stress due to normal forces on the crack surfaces. Micromechanical finite element modeling [8] have proved that the onset of plastic deformation in alumina results from grain boundary shear. The molecular dynamic simulation of the dynamic failure of glass also [9] supports this picture. Therefore, the objective of the present work was to propose a new damage evolution mechanism in plate impact induced high strain rate deformation and fracture of a dense, coarse grain e.g. 10 μm alumina ceramic.

## 2. Experimental Materials and Methods

A pressureless sintered alumina of density higher than 99.9% and 10 μm grain size was used in the present work [10]. At first the symmetric shock experiments were conducted on the alumina ceramic with a two stage gas gun to identify its Hugoniot Elastic Limit (HEL). The HEL was measured to be about 2 GPa commensurate with its large grain size [10]. To understand the damage initiation and their subsequent growth mechanisms in coarse grain alumina subjected to shock impact at levels more than three times the HEL, additional asymmetric shock recovery experiments were deliberately conducted at 6.5 GPa with a stainless steel flyer plate with the same gas gun on the similar 10 μm grain size alumina [11-13] processed in an identical manner to result in a similar microstructure. Detailed characterization of the shock recovered alumina fragments were done by FE-SEM and TEM. Further experimental details are given elsewhere [10-13].

### 3. Results and Discussions

The experimental results are shown in figures 1-3 and the model in figures 4a-d. Based on the experimental data and evidences from the FE-SEM and TEM observations of the present work (figures1-3) the new damage model (figure 4) was developed. The FE-SEM images of the shocked alumina (SA) samples confirmed the characteristic presence of intergranular microcracking, grain boundary cracks (figure 1a) and grain localized micro / nano-scale deformations (figure 1b) in suitably oriented grains, the presence of both in-plane and out of plane shear deformation band of about 50-100nm width (figure 1c) in single grains and a characteristic presence of intraganular shear induced microcleavages where the individual shear fractured plates of a single grain (figure 1d) were only 40-50 nm thick. This observations suggest (i) the important role played by shear induced deformation and fracture processes (ii) the deformation and fracture processes indeed originate at the sub-microstructural or nanostructural length scale regime of the microstructure [12,13]. The FE-SEM image of Fig. 2a showed further the most important role played by grain orientation where a micro-wing crack had formed only in between the large and suitably oriented grains at the centre of the photomicrograph while the other grains were relatively unaffected. As indicated in the model (Figure 4a), apart from the characteristic brittle fracture; inelastic damage growth, coalescence and fragmentation process in shocked alumina can happen by shear fracture initiation in a favourably oriented single grain when its boundary with the adjoining grain shears out [7,8] to form a micro-wing crack (figures 4a and 2a). The formation of the micro-wing crack creates a tensile stress [7,8] acting on two neighboring suitably oriented grain boundaries leading to mode I fracture (figure 4b). On continued loading, these tensile microcracks began to join hands with each other to form eventually a cracked microstructural scale slab in the case with no lateral confinement which was similar to the case of the present experimental conditions. In addition, as a result of the propagation of the shock wave through the present alumina sample; a significant amount of shear stress becomes operative [3-13] both in-plane and out of plane (figure 1b,c) in the shock loaded microstructure. In-fact the FE-SEM image of figure 2b showed micro-shear deformation bands only on suitably oriented grain facets and grain faces at the triple grain junction while other grain faces were less affected. For the present alumina ceramics the critical resolved shear stress was estimated [10] to be 2.4-4 GPa which was much lower than the applied shock pressure of 6.5 GPa. These estimates (2.4-4 GPa) matched also with those reported by others [2] for alumina. Therefore, shear stress induced deformation and / or micro-fracture was quite possible to happen in the present alumina ceramics. Due to such high shear stress the single and/ or multiple deformation bands can form in a suitably oriented grain and / or grain assembly in the microstructure (figures 1b,c and 4b). Also, in FE-SEM images of figures 2c and d very severe intragranular microcleavage had occurred in a suitably oriented single large grain while the grains around this large grain did not suffer so much of microcleavages. Such microclevage can only occur due to micro-shear (figure 4c) and / or micro-twist fracture (figure 4d). The micro-shear induced failure (SF) can occur due to in-plane shear stresses either at a single plane ($SF_s$) or at multiple planes ($SF_m$) of a suitably oriented grain [5] and / or grain assembly as indicated in figure 4c. Similarly, the micro-twist induced failure (TF) can occur due to out-of-plane shear stresses either at a single plane ($TF_s$) or at multiple planes ($TF_m$) of a suitably oriented grain and / or grain assembly as indicated in figure 4d. Both of these processes can also lead to grain boundary micro-crack formation as indicated in figures 4c,d and was indeed experimentally observed (figures 1a,b). These two processes can work independent of each other or may be simultaneously operative depending on the actual energetics of the process which physically

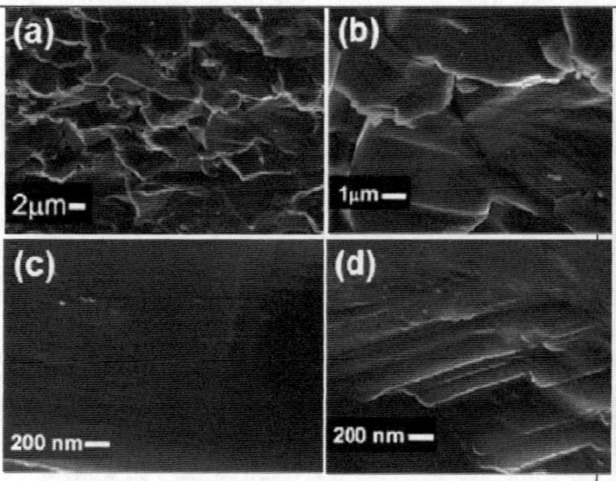


**Figure 1** FE-SEM images of SA (a) microfracture (b) shear (c) mutually perpendicular deformation bands (d) microcleavage.

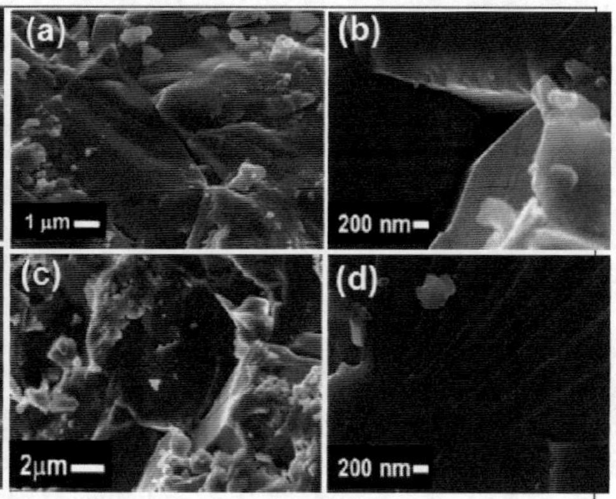


**Figure 2** FE-SEM images of SA (a) micro-wing crack (b) micro-shear at grain boundary (c) microcleavage (d) details of (c).

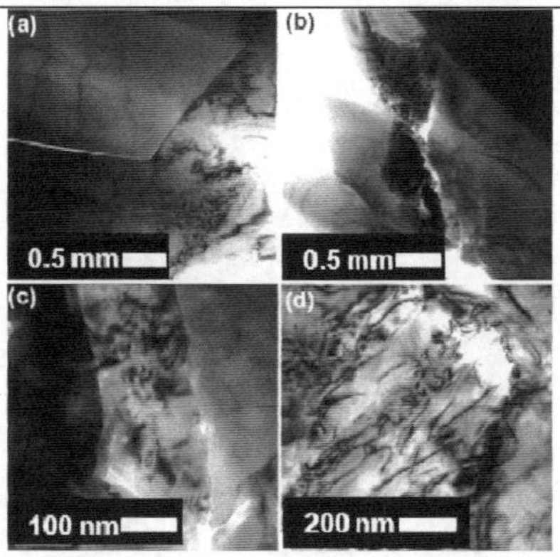


**Figure 3** TEM images of SA - dislocations at (a) triple grain junction (c) grain edge (d) in a single grain (b) crack at grain boundary.

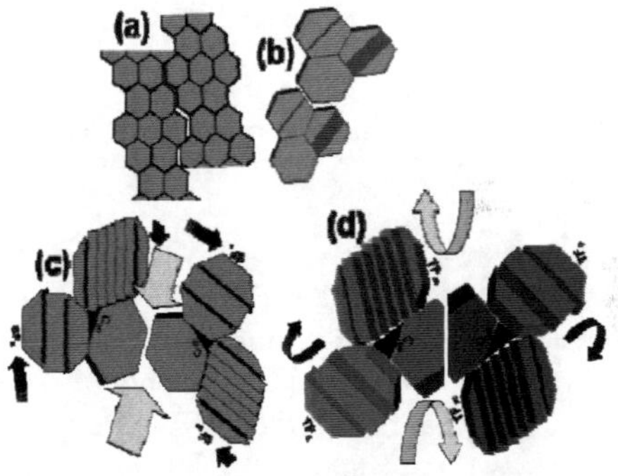


**Figure 4** Model of shock damage (a) micro-wing crack (b) shear band (c) micro-shear (d) micro-twist induced fracture.

depends on the locally active magnitude of the shock wave stress component and the relative orientation of the grain and / or grain assembly with respect to the direction of propagation of the shock wave and it's own immediate microstructural surroundings.

The TEM images showed dislocation entanglement at triple point along with deformation band in adjacent grains (figure 3a), grain boundary micro-crack formation, and a region of very fine (5-50 nm) agglomeration of adhered debris smeared on the fractured grain (figure 3b), (c) micro-cracks and dislocations around highly plastically deformed grain (figure 3c), dislocations and deformation band formation in a single grain (figure 3d); thus confirming a major role of high shear stresses. Further, it has been argued [10] that the dislocation pile up length of the present alumina sample would be on the higher side as it had a comparatively larger grain size e.g. ~ 10 μm, and consequently, a lower magnitude of critical shear stress would be necessary to nucleate a micro-crack (figure 3b) at the tip of the dislocation pileup from a favourably oriented slip plane (figures 3c,d). Such a micro-crack can then quite easily act as a failure initiation point. As the shock wave actually propagates in three dimensions, these micro-cracks grow very fast within a few μs at the nano and / or micro-structural length scale and link up by repetitive process through a multituude of grains with each other to coalesce at the microstructural length scale first. Eventually when many such events happen within a very small (few μs) time scale, the fracture propagates to cover the macrostructural length scale. The quick repetition of such a generic process is proposed to lead to the fragmentation of the shock loaded coarse grained high density alumina as observed the present work.

## 4. Summary

The shock recovered alumina fragments from a plate impact test at 6.5 GPa with a two stage gas gun were characterized by FE-SEM and TEM. In addition to the characteristic brittle compressive fracture, the shock fragments showed characteristic damage evolution features of grain localized micro / nano-scale deformations, micro-cleavages, grain boundary micro-cracks, micro-wing crack formation, extensive shear induced deformations and fractures localized at grains and grain boundaries as well as grain localized dislocations and damaged grain boundary smeared with nanoscale debris. Based on these results a new damage model based mainly on micro-shear and micro-twist fracture initiation, coalescence and propagation was developed to explain the failure mechanisms of shock loaded dense, coarse grain alumina in plate impact experiments.

## Acknowledgements

The authors appreciate Director's kind encouragements for and financial support of BRNS (Project No. GAP 0219) and CSIR (NWP 0027, NWP 0029, SRFship of Dey A) to this work.

## References

[1] Bourne N K 2006 *Proc. R. Soc.* **A 462** 3213
[2] Chen M W, McCauley J W, Dandekar D P and Bourne N K 2006 *Nat. Mater.* **5** 614
[3] Johnson G R. and Holmquist T J 1992 A computational constitutive model for brittle materials subjected to large strains *Shock-wave and High Strain-rate Phenomena in Materials*, ed. M. A. Meyers, L. E. Murr and K. P. Staudhammer (Marcel Dekker Inc. : New York) pp 1075-1081
[4] Rajendran A M 1994 *Int. J. Impact Eng.* **15** 749
[5] Espinosa H D, Zavattieri P D, Dwivedi S K 1998 *J. Mech. Phys. Solids.* **46** 1909
[6] Grady D E 1998 *Mech. Mater* **29** 181
[7] Horii M, Nemat-Nasser S 1986 *Phil. Trans. R. Soc. Lond.* **A 319** 337
[8] Warner D H, Molinari J F 2006 *Acta Mater.* **54** 5135
[9] Lu Z, Nomura K I, Sharma A, Wang W, Zhang C, Nakano A, Kalia R, Vashishta P 2005 *Phys Rev. Lett.* **95** 135501
[10] Mukhopadhyay A K, Joshi K D, Dey A, Chakraborty R, Rav A, Biswas S K and Gupta S C 2010 J. Mater. Sci. 45 3635
[11] Mukhopadhyay A K, Joshi K D, Dey A, Chakraborty R, Rav A, Mandal A K, Ghosh J, Bysakh S, Biswas S K and Gupta S C 2010 *Mater. Sci. Eng.* A, **527** 6478
[12] Mukhopadhyay A K, Joshi K D, Dey A, Chakraborty R, Rav A, Mandal A K, Ghosh J, Bysakh S, Biswas S K and Gupta S C 2011 *Ceram. Int.* **37** 2365
[13] Chakraborty R, Dey A, Mukhopadhyay A K, K. D. Joshi, Rav A, Mandal A K, Bysakh S, Biswas S K and Gupta S C 2011 *Mett. Mater. Trans.* A, in press

# Energetic materials under high pressures and temperatures: stability, polymorphism and decomposition of RDX

**Z A Dreger**
Institute for Shock Physics, Washington State University, Pullman, WA 99164, USA

E-mail: dreger@wsu.edu

**Abstract.** A recent progress in understanding the response of energetic crystal of cyclotrimethylene trinitramine (RDX) to high pressures and temperatures is summarized. The optical spectroscopy and imaging studies under static compression and high temperatures provided new insight into phase diagram, polymorphism and decomposition mechanisms at pressures and temperatures relevant to those under shock compression. These results have been used to aid the understanding of processes under shock compression, including the shock-induced phase transition and identification of the crystal phase at decomposition. This work demonstrates that studies under static compression and high temperatures provide important complementary route for elucidating the physical and chemical processes in shocked energetic crystals.

## 1. Introduction

Energetic materials (EMs) are a class of materials with a large amount of stored energy in chemical bonding which can potentially be released. Generally stable at ambient conditions, these materials decompose (deflagrate, detonate) with proper stimulus. The practical application of energetic materials often involves exposure to shock waves, therefore of particular interest are their properties and performance at high pressures and temperatures. The fast time scale of processes occurring during shock compression poses significant experimental challenges to fully quantify the molecular mechanisms involved in the energetic materials at these conditions. On the other hand, static compression experiments, despite of differences in rate of loading, states of stress, and thermal conditions can provide critical knowledge to gain insight into processes in shocked energetic materials. In particular, detailed information on polymorphism, chemical stability, and thermal decomposition obtained from static experiments can be useful to understanding the reactive behaviour of energetic crystals over the range of pressures and temperatures relevant to those under shock initiation.

Here, we present recent progress in understanding the response of energetic crystal of cyclotrimethylene trinitramine (RDX) to high pressures and high temperatures [1-3]. RDX is an important EM used extensively in explosives and monopropellants formulations. Despite of these significant applications the fundamental knowledge on this crystal at high pressures and temperatures has been both limited and inconsistent. Two RDX phases, $\alpha$ and $\gamma$ have been known for long time [4, 5], but even though the detailed structure of the $\gamma$-RDX has been determined recently [6]. The transition between these two phases takes place at ~ 4 GPa. Later on it was demonstrated that both phases transform to the other phase at high temperatures [7, 8]. It was proposed that this new, a HP-HT phase is actually the $\beta$-RDX, which was previously observed only at ambient pressure. Furthermore, the same studies provided the phase diagram for three RDX phases: $\alpha$, $\gamma$ and HP-HT. However, the

Published under licence by IOP Publishing Ltd

results from the two studies were largely inconsistent regarding the phase boundaries and the triple point. For example, the triple point locus differed by more than 100 K in those two studies. Also, it was reported that the HP-HT phase is primarily responsible for RDX decomposition at high pressures and temperatures. Given that the behaviour of RDX at high pressures and temperatures is of primary interest for understanding the shock initiation, the proper characterization of the HP-HT phase is essential. Therefore, in this work we present recent studies on RDX at high pressures and temperature to address the previous inconsistencies and to complement understanding of processes under shock compression [1, 2, 9-12]. In particular, we focus on: (i) determining details of the vibrational structure of the HP-HT phase, (ii) assessing the claimed association of this phase with β-RDX, (iii) providing the refined P-T diagram of RDX, (iv) examining the stability and decomposition of the HP-HT phase, and (v) relating the static compression results to those under shock compression.

## 2. Experimental Approach

All experiments were performed with single crystals. A large oriented crystals used in shock experiments were provided by Dr. D. E. Hooks from Los Alamos National Laboratory. Small crystals for static experiments were grown from acetone solvent. Static high pressure experiments were performed in a heated diamond anvil cell (DAC). Cryogenically loaded argon was used as a pressure transmitting medium. Ruby fluorescence was used to monitor pressure. Temperature of the DAC was controlled using a resistive heater wrapped around the cell. The sample temperature was monitored with iron-constantan thermocouples. A micro-Raman system (T64000, JY-Horiba) equipped with a microscope (Olympus BX-40), utilizing a 532 line from a cw diode-pumped solid-state laser, was used for Raman spectra measurements. Experimental details for static compression measurements can be found elsewhere [1].

For shock experiments, oriented, single crystals of RDX were prepared with dimensions of 7 x 7 x 0.4 $mm^3$ and confined between two optical windows. Shock waves were generated by impacting the target assembly with quartz or sapphire crystals launched by a single stage light-gas gun. Because of the impedance mismatch between the optical windows and RDX crystals, the sample was compressed under stepwise loading. Three experimental techniques were used to obtain the Raman and emission spectra. High spectral resolution Raman measurements were performed with a single 20 ns pulse from a Nd:YAG (532 nm) laser. Time-resolved Raman measurements were obtained with a 2 μs pulse from a tunable dye (514.5 nm) laser. Raman experiments were performed to stresses up to 10 GPa. In the emission experiments, light produced by the shocked crystal was detected by the spectroscopic system to provide time-resolved emission spectra. Emission experiments were performed for stresses between 7 and 20 GPa. Details of shock experiments and detection techniques can be found elsewhere [10, 11, 13, and 14].

## 3. Results and Discussion

### *3.1. Static compression*

#### *3.1.1. HP-HT phase*

First we addressed issue concerning the proposed association of the HP-HT phase with the β-RDX. Because the HP-HT phase has limited chemical stability under the pressure-temperature conditions where it is produced, the detailed characterization of its structure is rather challenging. However, it turned that this phase can be quenched to room temperature. In Figure 1, we demonstrate that the quenching can preserve both the shape of vibrational spectra and the quality of single crystal of the HP-HT phase. Note the spectrum at high temperature has broader peaks due to the thermal effects. Having the HP-HT phase at room temperature, we found that this phase is stable over the wide range of pressures; however, it transforms to α-RDX at ~ 0.6 GPa.

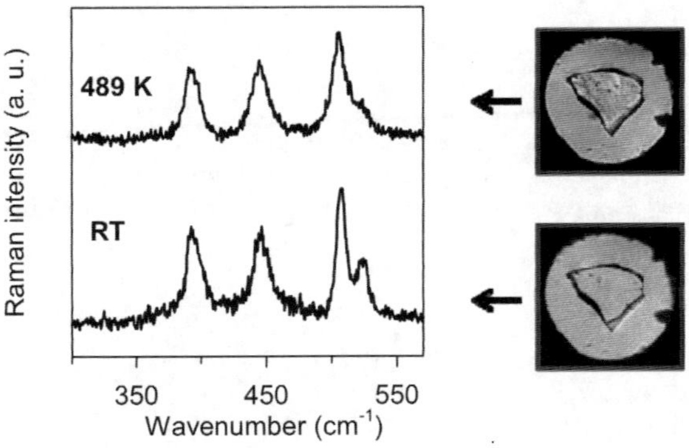


**Figure 1.** An example of Raman spectra of the HP-HT phase at high temperature and at room temperature (RT) with corresponding images of RDX crystal in a diamond anvil cell. Spectra were acquired for 10 s. Both spectra and images were obtained at 4.2 GPa.

To examine the association of the HP-HT phase with the β-RDX, we compared the Raman spectra of the stable form of the HP-HT phase with the Raman spectra of β-RDX grown from nitrobenzene solution. The spectra in the low frequency and the CH stretching range are shown in Figure 2. A clear difference can be noticed between two sets of spectra, demonstrating that the HP-HT phase is not the β-RDX as was previously suggested, but it is a new phase [1, 3]. This phase was labeled as ε-RDX, and its crystal structure was determined as having the same space group as γ-RDX [3].

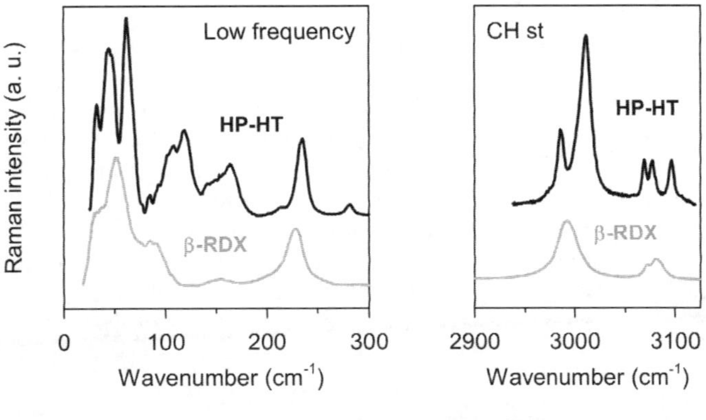


**Figure 2.** Comparison between Raman spectra of the HP-HT RDX (black lines) and β-RDX (grey lines). The HP-HT spectrum was obtained at 0.8 GPa and the β-RDX at ambient pressure.

*3.1.2. New phase diagram and decomposition of ε-RDX*

To determine the pressure and temperature locus for the RDX phases, the Raman and optical imaging measurements were performed at a specific pressure and with gradually increasing temperature. All experiments were carried out at pressures ranging from 0 to ~7 GPa, and at temperatures ranging from ~295 K to 550 K. Identification of characteristic features in Raman spectra permitted determination of the boundaries between three phases. In Figure 3, we show new phase diagram of RDX [2], along with the previously published [7, 8]. The Figure 3a is from this work and the lines represent the best fit to the experimental data; while the Figure 3b is from previous studies. This experiment allowed us to locate the triple point between RDX phases, as being approximately at 3.7 GPa and 466 K. Also, we found that: (i) both the α-ε and the γ-ε transitions are irreversible at the phase boundaries and (ii) the γ-phase can decompose at high temperature without involving the ε-RDX. The results presented here

Journal of Physics: Conference Series **377** (2012) 012047 doi:10.1088/1742-6596/377/1/012047

clearly differ from previously reported results [7, 8]. Although it is difficult to determine the source of these discrepancies, we believe that previous results could be affected by: (i) sample morphology; polycrystals or powders, (ii) nonhydrostaticity and pressure gradients, and (iii) interactions between the sample and pressure transmitting media.

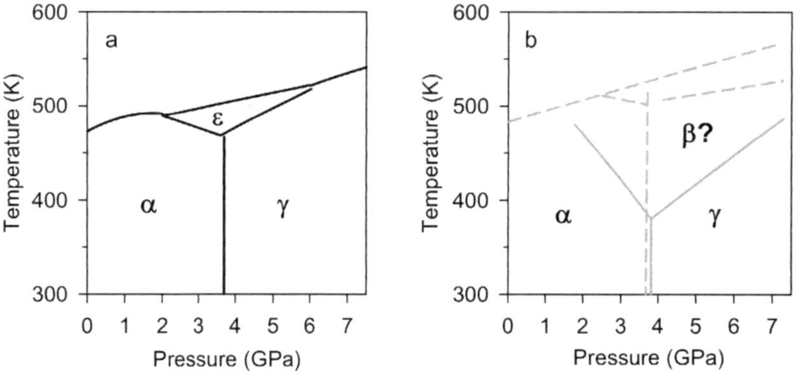


**Figure 3.** Phase diagram of RDX: a) This work, b) Previous works; solid lines [7], dashed lines [8].

As can be noticed from the new phase diagram, the ε-RDX appears to exist in a limited range of pressure and temperature from 2.8 to 6.0 GPa, and above ~ 465 K. Since this phase has also limited chemical stability we looked into its thermal decomposition kinetics. By relating a decrease in Raman intensity with time to the fractional extent of decomposition (FED), we found that the FED obeys autocatalytic-type reaction. Fitting the data to the Prout-Tompkins model [15] provided the global decomposition rates at several pressures and temperatures. The obtained rates increased with temperature and decreased with pressure. Therefore, the activation volume for ε-RDX decomposition was found to be positive, in contrast to α-RDX for which the negative values were reported [8].

### *3.2. Relevance of static results for shock compression*

#### *3.2.1. The α-γ phase transition under shock compression*

The α-γ phase transition is well established under static compression and recognized by characteristic changes in the Raman spectra [9]. However, it was unclear if this phase transition can occur under shock compression conditions. To address this issue we performed shock experiments using the high spectral resolution Raman system for detecting characteristic features in the spectra. As shown in Figure 4a, our experiments to different peak stresses on crystals impacted at (111) plane revealed that changes in Raman spectra occur ~ 4 GPa [10]. Having data under static compression, see Figure 4b, we found that the changes in Raman spectra obtained under shock compression are similar to those under static compression. Therefore, this result demonstrates that the same α-γ phase transition, as observed under static compression, takes place under shock compression on a nanosecond time scale. Furthermore, the shock studies on crystals with different orientation indicated that the α-γ transition occurs regardless of crystal orientation [10].

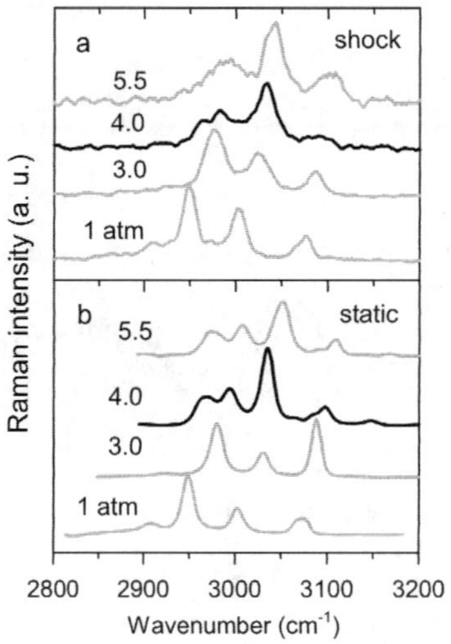


**Figure 4.** Raman spectra of RDX crystal at several pressures: a) shock compression results for (111) oriented crystal, b) static compression results. Note similar spectra at 4 GPa under shock and static compression. Pressures are given in GPa.

*3.2.2. RDX phase at the shock-induced decomposition*

The experiments at low stresses (< 5.5 GPa) did not reveal signatures of decomposition. Therefore, to gain insight into the shock-induced decomposition the experiments were performed at higher stresses [11]. At stresses higher than7 GPa, we observed, in addition to changes in the Raman spectra, a gradual raise of light emission from the crystal with time. This shock-induced emission dominated the Raman spectrum at latter times. Because, the emitted light is associated with the chemical decomposition, the presence of Raman and emission signals at the same time can be used to identifying the phase of decomposing crystal under shock compression. In Figure 5, we present an attempt to establish the phase of the shocked-decomposing crystal.

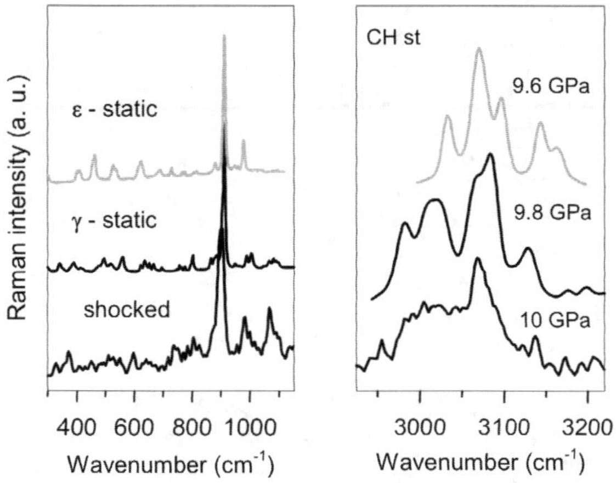


**Figure 5.** Comparison of Raman spectra of the shocked-decomposing crystal with Raman spectra of the γ- and ε-RDX measured under static compression.

For this purpose we compared the Raman spectra obtained on the shocked-decomposing crystal with the Raman spectra of the ε- and γ- phase obtained at static compression. When comparing these spectra one should keep in mind that the spectrum of shocked crystal, in contrast to other spectra, was obtained with single pulse laser and, therefore has much lower signal-to-noise ratio. Nevertheless, we find several instances: region of 3000 $cm^{-1}$, below 400 $cm^{-1}$, and above 1000 $cm^{-1}$ indicating that the spectrum of shocked-decomposing crystal is similar rather to the γ-phase than to ε-phase. Therefore, this result implies that the shock-induced decomposition in RDX takes place from the γ-phase. This finding has important implication for the modeling of the shock-induced decomposition of RDX.

## 4. Final Remarks

We summarized recent work on energetic crystal of RDX under high pressures and temperatures.
Using Raman spectroscopy and optical imaging we provided new phase diagram for RDX crystal. In particular, we provided evidence that: (i) the HP-HT phase is a new RDX phase; ε-RDX, (ii) the ε-RDX exists in a narrow P-T domain and decomposes autocatalytically with the rates decreasing with pressure, and (iii) the γ-RDX decomposes without involving the ε-RDX. Finally, we demonstrated the use of static compression results in understanding the phase transition and decomposition under shock compression.

**Acknowledgments**
Prof. Y. M. Gupta, and Drs. J. E. Patterson and D. E. Hooks are sincerely thanked for contributing to this project. This work was supported by the founding from ONR/MURI and DOE/NNSA.

## References

[1] Dreger Z A and Gupta Y M 2010 *J. Phys. Chem.* A **114** 7038
[2] Dreger Z A and Gupta Y M 2010 *J. Phys. Chem.* A **114** 8099
[3] Millar D I, Oswald I D H, Barry C, Francis D J, Marshall W G, Pulham C R and Cumming A S 2010 *Chem. Commun.* **46** 5662
[4] Choi C S and Prince E 1972 Acta Cryst. B **28** 2857
[5] Olinger B, Roof B and Cady H 1978 *Proc. Int. Symp. on High Dynamics Pressures*, Paris, France, p 3
[6] Davidson A J, Oswald I D H, Francis D J, Lennie A R, Marshall W G, Millar D I, Pulham C R, Warren J E and Cumming A S 2008 *Cryst. Eng. Comm.* **10** 162
[7] Baer B J, Oxley J and Nicol M 1990 *High Pressure Res.* **2** 99
[8] Miller P J, Block S and Piermarini G J 1991 *Combus. Flame* **83** 174
[9] Dreger Z A and Gupta Y M 2007 *J. Phys. Chem.* B **111** 3893
[10] Patterson J E, Dreger Z A, Gupta Y M 2007 *J. Phys. Chem.* B **111** 10897
[11] Patterson J E, Dreger Z A, Miao M S, Gupta Y M 2008 *J. Phys. Chem.* A **112** 7374
[12] Dang N C, Dreger Z A, Gupta Y M and Hooks D E 2010 *J. Phys. Chem.* A **114** 11560
[13] Hemmi N, Dreger Z A, Gruzdkov Y A, Winey J M and Gupta Y M 2006 *J. Phys. Chem.* B **110** 20948
[14] Hemmi N, Zimmerman K A, Dreger Z A and Gupta Y M 2011 *Rev. Sci. Instrum.* **82** 083109
[15] Brown W E, Dollimore D and Galwey A 1980 *Reactions in the Solid State, Comprehensive Chemical Kinetics*, eds Bamford C H and Tipper C F H (Amsterdam: Elsevier) p 41

Journal of Physics: Conference Series **377** (2012) 012048 doi:10.1088/1742-6596/377/1/012048

# Hypervelocity impact of tungsten cubes on spaced armour

**Pradeep S Chandel, Dharmanshu Sood, Rajeev Kumar, Prince Sharma, Bhupinder Sewak, Vikas Bhardwaj, Manoj Athwal, Vikas Mangla, Ipsita Biswas and Manjit Singh**
*Terminal Ballistics Research Laboratory, DRDO, Min. of Defence, Sector 30, Chandigarh -160 030, INDIA*

E-mail: pradeep137@yahoo.co.in; pradeep.tbrl@gmail.com

**Abstract:** The paper summarizes the experimental observations and simulation studies of damage potential of tungsten alloy cubes on relatively thin mild steel spaced armour target plates in the velocity regime 1300 – 4000 $ms^{-1}$ using Two Stage Light Gas Gun technique. The cubes of size 9.5 mm and 12 mm having mass 15 g and 30 g respectively were made to impact normally on three target plates of size 300 mm × 300 mm of thickness 4, 4 and 10 mm at 100 mm distance apart. Flash radiography has been used to image the projectile-target interaction in the nitrogen environment at 300 mbar vacuum at room temperature. The results reveal clear perforation by 9.5 mm cube in all the three target plates up to impact velocity of about 2000 m/s. While 12 mm cube can perforate the spaced armour upto impact velocity of 4000 m/s. This shows that 9.5mm tungsten alloy cube is not effective beyond 2000 m/s while 12 mm tungsten alloy cube can defeat the spaced armour upto 4000 m/s. The simulation studies have been carried out using Autodyn 3D nonlinear code using Lagrange solver at velocities 1200 – 4000 m/s. The simulation results are in good agreement with the experimental findings.

## 1. Introduction

The penetration capabilities of fragments cannot be estimated using available empirical relationships which are not valid for hypervelocity impacts. The basic theories have focused principally to predict the mean fragment size through energy and momentum balance principles [1-4] and on statistical issues of mean fragment size distributions [5-8]. However, there are number of unresolved issues within the development of statistical energy-balance theories of fragmentation which led to a pressing need for hypervelocity experimental data in order to optimize the fragment parameters to cause the desired damage to the target.

The implementation and validation of these basic theories in numerical wave propagation codes is needed to investigate dynamic fragmentation in large scale complex problems. Dynamic failure and fragmentation data of solids is required to characterize and understand the material behavior under intense stress-wave loading conditions. Although the static data exist for these materials, but this does not suffice to in-depth theoretical understanding of solids under highly dynamic regime.

The present study addresses the threshold breakup speed and damage potential of the tungsten alloy cube fragments of size 9.5mm and 12mm made to impact normally upon relatively thin plates (Plate Thickness/Fragment Diameter, $T/d < 1$) in a vacuum 300 mbar with plate spacings greater than several fragment diameters. The experiments in the velocity regime 1300 - 4000 ms-1 have been carried out using Two Stage Light Gas Gun facility at room temperature. At lower impact velocity, the fragments perforate thin plates without much deformation or mass loss. Above threshold impact speed, the projectile starts to disintegrate into smaller fragments. As the impact speed increases, the high speed larger particles from 12mm cube fragment debris cloud perforated second and third target plates but relatively smaller debris cloud particles from 9.5mm cube fragment could not perforate all the three plates. The axis of the cloud is aligned with the line of fire. Dispersion of fragment debris particles also increases with increase in impact speed.

The simulation studies have been carried out using Autodyn 3D nonlinear code using Lagrange solver at velocities 1200 – 4000 m/s.

## 2. Experimental Details

The experiments were conducted using Two Stage Light Gas Gun Facility at TBRL, Chandigarh using Polycarbonate Sabots, custom designed in-house to impact the target plate edge on. The Target assembly is rigidly fixed within the Chamber-3 (see Fig 1) along the axis line. The projectile assembly (as shown in Fig 2) is launched by 29 mm and 40 mm calibers into the chambers, which are first evacuated to 30mBar and then filled with nitrogen gas to 300 mBar.

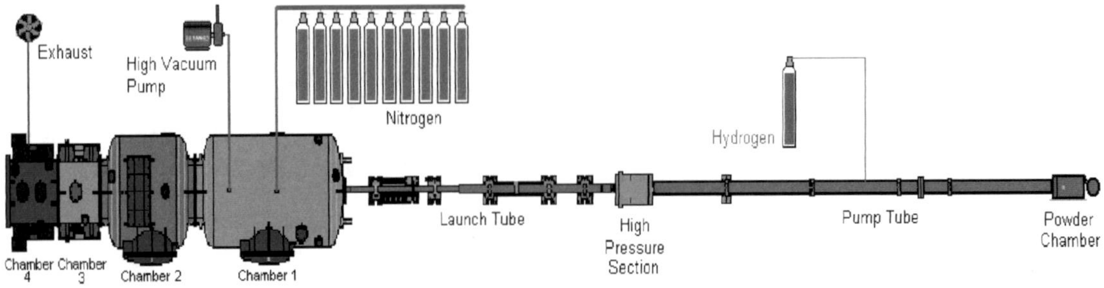


Figure 1: Layout of Two Stage Light-Gas Gun Facility for Cube Impact Trials

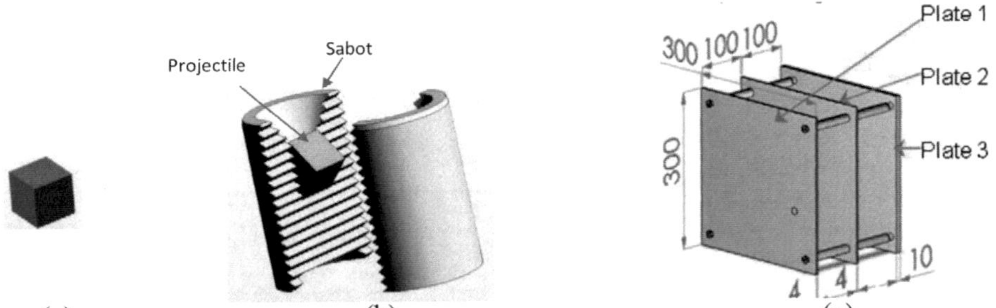


(a) (b) (c)

Figure 2: Schematic diagram of (a) Tungsten Alloy Cube (b) Projectile Assembly (c) Target Plates

The tungsten alloy cubes mounted in polycarbonate sabots were launched at impact velocities from about 1200 to 4000 m/s and made to impact the target array edge-on. The target array consists of three plates made up of mild steel of size 300 mm × 300 mm of thickness 4, 4 and 10 mm at 100 mm distances apart. Polycarbonate sabots were separated through forces produced by inert nitrogen gas atmosphere filled at 300 mbar in chamber 2. Sabot fragments were arrested upstream by sabot catcher plates and did not reach the target plates.

Laser beam sensors and Aluminium foil sensors were deployed along the flight path in Chamber-1 and 3 to record the velocity of the impactor (shown in figure 3). The cube impacting edge-on and fragment – plate particle clouds were radiographed at two stations by two 150 keV x-ray tubes (Make: Scandiflash) placed orthogonal to the line of fire as shown schematically in figure 4. The delay times were calculated from the predicted impact velocity and the x-ray tubes were triggered independently by orthogonally placed laser barriers during flight and by PCB sensors on the target plates. The impact velocity was predicted by optimizing various ballistic parameters through CESAR simulation software. PCB sensors are affixed on the target plates to trigger orthogonal flash X-rays as well as to measure arrival time of impactor.

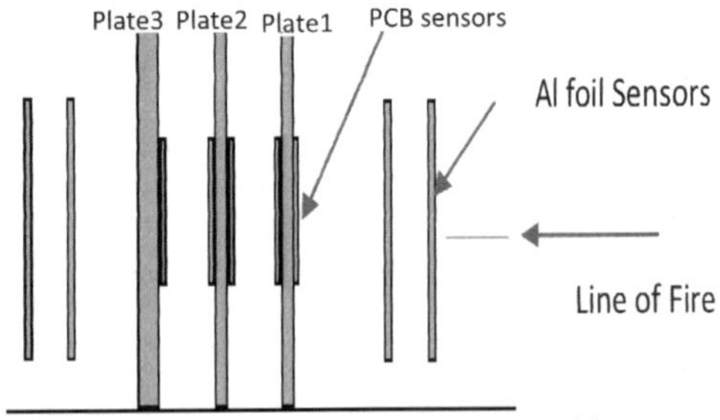


Figure 3: Schematic of Target Assembly

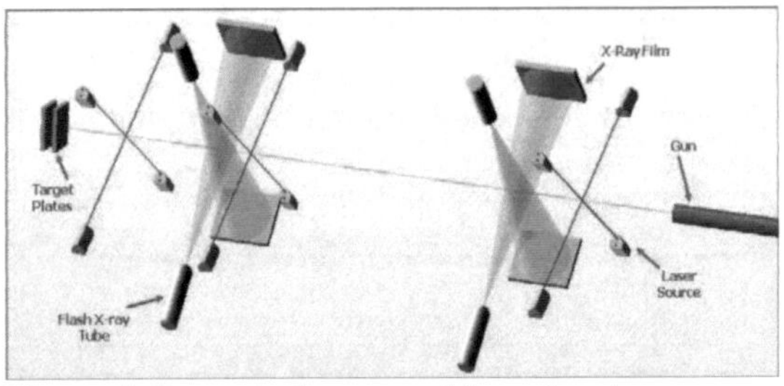


Figure 4: Layout of x-ray and laser barriers

## 3. Results and Discussion

An extensive experimental and numerical study was carried out to analyze the penetration effectiveness of 9.5mm and 12mm cube fragments. Experiments were conducted using two-stage light gas gun. About 8 to 10 experiments were conducted for each cube size in the velocity range from about 1300 – 4000 m/s against identical spaced armour plate assemblies. Numerical simulations were carried out using explicit finite element code ANSYS AUTODYN to verify the observations made from experimental results. Since the problem involved large deformations and fragmentation of the projectile as well as target, smooth particle hydrodynamics (SPH) solver has been used in the analysis. The target and projectile material was modeled using Steel 1006 and Tung. Alloy properties available in AUTODYN material library. The summary of material properties is given in Table1.

Figure 5(a) and 5(b) shows the experimental and simulation results for impact of 9.5mm cube. The radiograph and simulation shows the cube impacting edge-on and dispersions of the fractured cubes behind the second target plate at impact velocity of about 4000 m/s. It has been observed that 9.5 mm cube is effective in the velocity window 1300-2000 ms-1 only. Above this threshold impact speed, the projectile starts to disintegrate into smaller particles. With the increasing impact velocity, the fragment breakup become more severe; the sizes of debris particles decrease and the number of particles increase. It can also be seen that the dispersion of fragment debris also increases with increasing impact velocity. The perforating debris particles loose momentum and may perforate or produce cratering effect on the next plate. As the impact velocity increases, the density of fragments increases. While 12 mm cube can perforate all the three target plates even up to 4000 m/s. Figure 6 shows that as the cube size is increased, the size of fragment debris particles and dispersion is relatively lower that 9.5 mm cube fragment. The larger fragment particles possess sufficient momentum to perforate subsequent plates.

Impact pressure and relief wave mechanisms are predominant in the generation of fragment and plate particle population clouds. Impact pressure at the interface can be relieved laterally as well as in the direction normal to the interface. For small T/d (plate thickness to fragment diameter) ratio, radial velocities in the plate are relatively small but as the T/d ratio increases, radial components become greater [9-10]. This was revealed by flash radiographs as shown in figures 5 and 6. Thus the impact velocity which governs stress intensity and normalized plate thickness T/d which governs the relative magnitude of radial stress are the predominant factors in fragment and plate fragmentation.

Table1: Material properties used for numerical simulation.

| Projectile – Tung. Alloy | | Target – Steel 1006 | |
|---|---|---|---|
| Equation of State - Shock | | Equation of State - Shock | |
| Reference density | 17.0 g/cc | Reference density | 7.896 g/cc |
| Gruneisen coefficient | 1.54 | Gruneisen coefficient | 2.17 |
| Parameter C1 | 4.029e+003 m/s | Parameter C1 | 4.569e+003 m/s |
| Parameter S1 | 1.237 | Parameter S1 | 1.49 |
| Specific Heat | 134.000015 J/Kg K | Specific Heat | 451.999969 J/Kg K |
| Strength Model – Johnson Cook | | Strength Model – Johnson Cook | |
| Shear Modulus | 1.60e+008 kPa | Shear Modulus | 8.180001e+007 |
| Yield Stress | 1.506e+006 kPa | Yield Stress | 3.500000e+005 |
| Hardening Constant | 1.77e+005 kPa | Hardening Constant | 2.750000e+005 |
| Hardening Exponent | 0.12 | Hardening Exponent | 0.36 |
| Strain Rate Constant | 0.016 | Strain Rate Constant | 0.022 |
| Thermal Softening Exponent | 1.00 | Thermal Softening Exponent | 1.00 |
| Melting Temperature | 1.723e+003K | Melting Temperature | 1.811e+003K |

| Reference strain rate | 1.00 $s^{-1}$ | Reference strain rate | 1.00 |
|---|---|---|---|

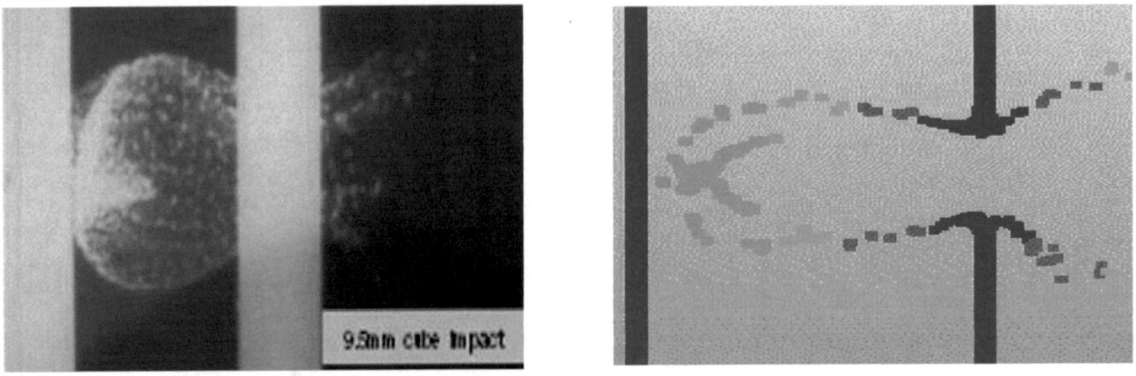


Figure 5: Impact of 9.5mm cube at velocity 4000 m/s (a) Experimental (b) Numerical Simulation.

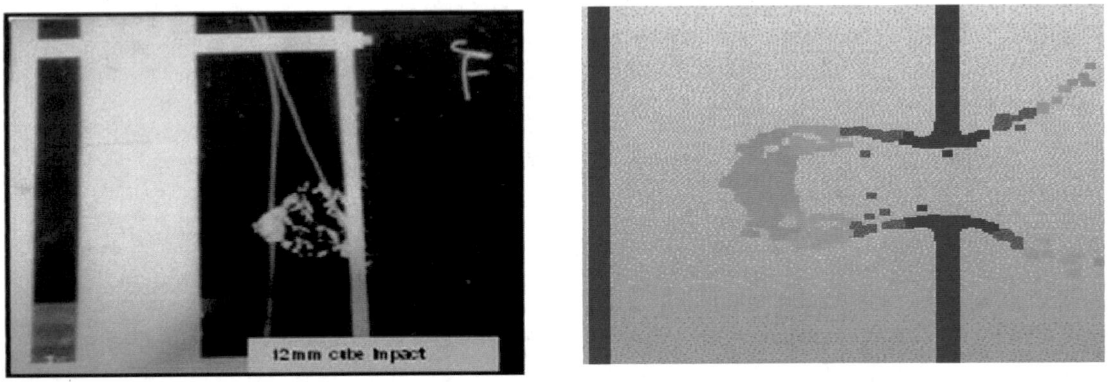


Figure 6: Impact of 12mm cube at velocity 4000 m/s (a) Experimental (b) Numerical Simulation.

## 4. Conclusions

It is concluded that small size 9.5 mm tungsten alloy cube is not effective beyond 2000 m/s while 12 mm tungsten alloy cube can defeat the spaced armour upto 4000 m/s. The impact velocity and normalized plate thickness T/d are the predominant factors responsible for impactor and plate fragmentation. The simulation results carried out using Autodyn 3D explicit finite element code using SPH solver at velocities 1200 – 4000 m/s support the experimental findings.

5. **Acknowledgements:** The authors are thankful to Mr Sohan Lal, Tech 'A' and Mr GN Jha Tech 'A' for their assistance in conducting experimental trials.

## 6. References

[1] Kipp ME and Grady DE 1985 *J. Mech. Phys. Solids* **33** 399-415
[2] Grady DE and Kipp ME 1987 *Int. J. Impact Engg* **20** 293-308
[3] Dickinson DL, Yatteau JD and RF Recht 1987 *Int. J. Impact Engg.* **5** 249-260
[4] Walker JD 2001 *Int. J. Impact Engg.* **26** 809-822
[5] Englman R, Jaeger Z and Levi A 1984 *Phil. Mag.* B **50** 307-315
[6] Grady DE and Kipp ME 1985 *J. Appl. Phys.* **58 (3)** 1210-1222
[7] Brown WK 1989 *J. Astrophys. Astr.* **10** 89-112
[8] Grady DE 1990 J. Appl. Phys. **68** 6099-6105
[9] Zukas JA 1990 *High Velocity Impact Dynamics, John Wiley & Sons Inc.*
[10] TBRL Report 2008 No. TBRL/TECH/GG/324/08

Journal of Physics: Conference Series **377** (2012) 012049 doi:10.1088/1742-6596/377/1/012049

# Equation of state measurements using single Fabry-Perot velocimeter

**A K Saxena, T C Kaushik, A M Rawool and Satish C Gupta**
Applied Physics Division, Bhabha Atomic Research Centre, Trombay, Mumbai - 400085, India

E-mail: a_saxena@barc.gov.in

**Abstract.** Accurate measurement of projectile and free surface velocity of target is a crucial part of any high pressure physics experiments involving hyper velocity impact. Fabry Perot velocimeter (FPV) is a widely accepted technique for single surface velocity measurements. In the present work a possibility of using single FPV for two surface velocity measurements is experimentally explored. The laser light is launched in such a way that it illuminates both the surfaces under study and reflected light coming from the flyer as well as from the target is used to generate interference fringes. Therefore by recording fringe displacements due to Doppler shift in reflected light corresponding to movement of both the surfaces at different time instants, velocity profile of both target and projectile can be computed. This technique has been utilized for equation of state measurements of polyurethane based retro-reflective tape at 12 GPa and aluminum at 15 GPa. Measured peak particle velocities at the target-glass interface have been found in good agreement with the reported equation of state data.

## 1. Introduction

High pressure equations of state (EOS) studies by projectile impact often require the velocity measurement of minimum two surfaces. These measurements are commonly the velocity of projectile and particle velocity in target. Optical velocimeters [1, 2], which are based on measuring the Doppler shift in the light reflected from the moving surface are widely used for such measurements. One such technique viz. Fabry-Perot velocimeter (FPV) utilizes the property of Fabry-Perot interferometer that its fringe diameters are related to the wavelength of the incident light. Therefore by recording the time evolution of fringe diameters on a streak camera, velocity profile of the moving surface can be computed by knowing the equations of Fabry-Perot interferometer and Doppler shift. Since two streak cameras and two interferometers are required for measuring velocity of two surfaces, such measurements become complex and expensive. In the present work a possibility of using single FPV for two surface velocity measurements is demonstrated. The laser light is launched in such a way that it illuminates both the surfaces under study and reflected light coming from the flyer as well as from the target is used to generate interference fringes. In this arrangement, during the initial flyer movement, expanding fringes are expected to be seen in the streak record along with unchanged fringes produced by the fraction of light reflected from stationary target. After impact, these unchanged fringes would also show a jump according to the movement of target surface. Therefore by recording both fringe movements on same record, velocity profile of both the surfaces can be computed.

Here we report the experimental viability of this concept being investigated on an electrically exploding foil accelerator [3] (also known as electric gun) for EOS studies of polyurethane based retro-reflective tape and aluminum target.

## 2. Experimental Setup

To carry out high pressure impact experiments an 8 kJ portable electrically exploding foil accelerator has been utilized. The principle of electrically exploding foil accelerator (EEFA) is to use the energy initially stored in a fast capacitor bank to explode a thin metallic bridge foil sandwiched in between two dielectric sheets backed by a heavy tamper on one side and an appropriate size barrel on the other. Due to fast explosion, foil material expands and punches out a section of the dielectric sheet (flyer) positioned adjacent to it and drives it up in the barrel to suitably high velocities. The present system is capable in accelerating 10 mm diameter thin (~125-150 μm) dielectric or metal laminated dielectric flyers to a velocity of 2-4 km/s. The accelerated flyers made an impact to the target placed at the other side of the barrel covering only half the barrel area and leaving other half for the probing of flyer. Schematic of the experimental setup is shown in fig.1. The thin target sample is supported by a 1 mm thick glass plate to maintain its planarity till the time of impact. The experiments have been performed under symmetric impact condition by using the flyers made of 125 μm thick Kapton sheet with a layer of target material bonded over it.

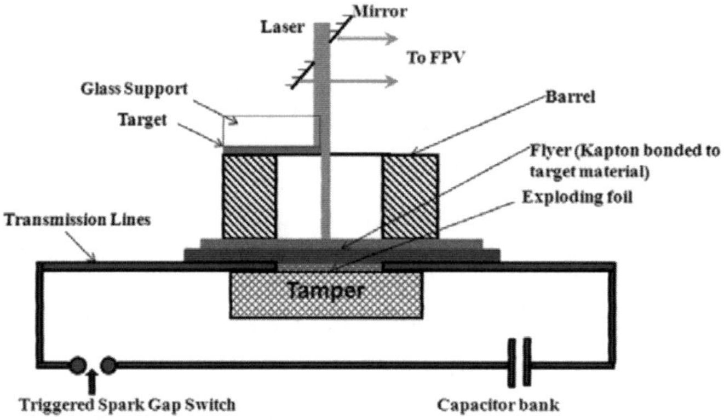


**Figure 1.** Schematic of experimental setup.

A single frequency (532 nm) laser with beam diameter of 2.5 mm is launched partially over the target and partially over the flyer through a hole created in the mirror, which also direct the scattered light to a collimating lens (not shown in Fig.1). The collimated light is then used to generate the interference pattern using a Fabry-Perot etalon. The diameter of interference fringe produced by Fabry-Perot interferometer depends upon the wavelength of the incident light. Hence by considering the equation for relativistic Doppler shift and the equations of Fabry-Perot interferometer, velocity of moving projectile may be written in terms of the instantaneous diameter of Fabry-Perot fringes [4].

$$v = \frac{c\lambda_L}{4d\mu}\left(i + \frac{D_{R_0}^2 - D_{L_0}^2}{D_{L_1}^2 - D_{L_0}^2}\right) \tag{1}$$

Where μ is the refractive index of the material present between the two mirrors of Fabry-Perot interferometer, which are separated by a distance 'd', and 'i' represents the total number of new fringes moved out from the center of the fringe pattern. $D_{L0}$ and $D_{L1}$ are the diameters of the innermost and next to innermost bright fringes produced by the Fabry-Perot interferometer when flyer/ target are at rest. $D_{R0}$ is the instantaneous diameter of the innermost bright fringe while flyer/ target are in motion. Hence if fringe diameters are known at different time instances velocity profile can be obtained by this relation. Time evolution of interference fringe diameters is recorded on a streak camera.

### 3. Experimental Results

Initial experiments have been carried out on a 30 μm thick retro-reflective tape, which is used because of its good reflective properties maintained even at the arrival of shock. The projectiles made of retro-reflective tape (30 μm) bonded over the Kapton sheet (125 μm) are accelerated in a barrel of length 2.0 mm. These projectiles impact on the target made of 30 μm thick retro-reflecting tape supported by 1 mm thick glass plate. Time evolution of fringe diameters with computed time resolved velocity profiles of both target and flyer are shown in Fig.2 (a). The initial part of the streak record shows the fringe movements according to flyer movement along with stationary fringes generated from the stationary target, at a later stage on impact these stationary fringes also show a jump corresponding to

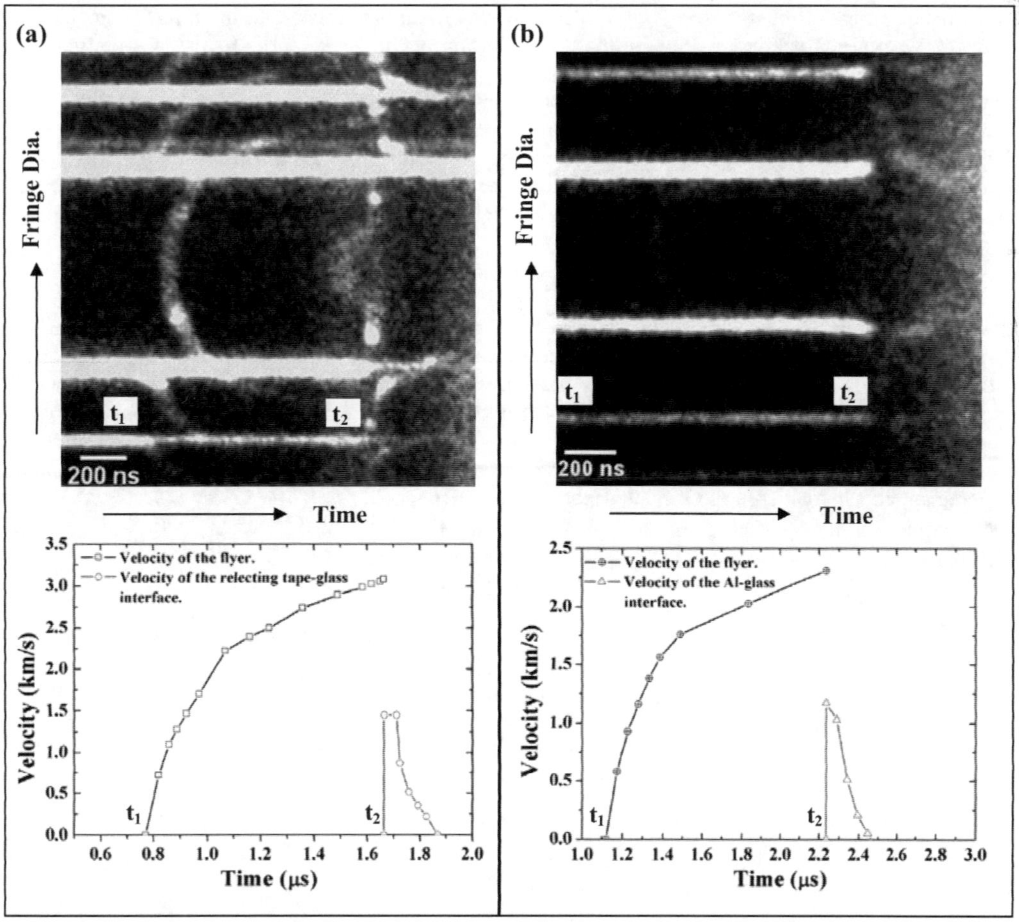


**Figure 2.** Streak record of FPV (top) and inferred velocity profiles (bottom) for (a) retro-reflective tape at 12 GPa (b) aluminum at 15 GPa. (**$t_1$**= Time at which flyer movement starts, **$t_2$**= Time at which target-glass boundary moves.)

the movement of tape- glass interface. The analysis of obtained streak record shows a flyer velocity of 3.1 km/s and tape-glass interface peak velocity of 1.6 km/s. As retro-reflective tape is mostly made of polyurethane with a thin reflecting metal layer, the EOS of polyurethane as reported in reference-[5] is used for the theoretical prediction of interface velocity. On comparison measured peak interface velocity has been found to be 30 % higher than this theoretical prediction which might be due to the presence of thin metallic layer embedded in the tape.

The results of experiments carried out in a similar way on 30 μm thick aluminum target are shown in Fig.2 (b). In these experiments though the major portion of the laser spot is focussed on target but due to loss in reflectivity on shock arrival the fringe intensities are low. The analysis of obtained streak record shows a flyer velocity of 2.3 km/s and aluminum-glass interface peak velocity of 1.2 km/s, which has been found to be matching well with the reported EOS data within experimental accuracies of 8%, which is higher in present case due to finite absolute error (±0.2 km/s) involved with the present FPV measurements, but at higher pressures or at higher velocities contribution of this error will become less significant.

## 4. Conclusions

In the present work experimental feasibility of using single FPV for two surface velocity measurements has been successfully demonstrated on retro-reflective tape and aluminum targets at 12 GPa and 15 GPa respectively. Single streak records showing the details of flyer and target velocity profiles have been obtained by illuminating the Fabry-Perot interferometer by laser light reflected from flyer and target both. In these experiments complete velocity profiles of both surfaces have been recorded by operating streak camera at slower speeds, but with some prior information about flyer velocity, streak camera may be operated at faster speeds to record only final flyer velocity and complete target velocity profile to achieve finer temporal details.

## 5. References

[1] Barker L M and Hollenbach R E 1972 *J. Appl. Phys.* **43** 4669.

[2] Durand M, Laharrague P, Lalle P, Le Bihan A, Morvan J, and Pujols H 1977 *Rev. Sci. Instrum.* **48**, 275.

[3] Saxena A K, Kaushik T C, and Gupta Satish C 2010 *Rev. Sci. Instrum.* **81**, 033508.

[4] McMillan C F, Goosman D R, Parker N L, Steinmetz L L, Chau H H, Huen T, Whipkey R K, and Perry S J 1988 *Rev. Sci. Instrum.* **59**, 1.

[5] Marsh S P 1980 *LASL Shock Hugoniot Data* (London: University of California Press).

# Suppression of instability by double ablation in tungsten doped polyvinyl alcohol foils

**Leshma Peedikakkandy and S Chaurasia**

High Pressure & Synchrotron Radiation Physics Division, Bhabha Atomic Research Centre, Mumbai 400085, India

leshma@barc.gov.in

**Abstract**. In Inertial fusion Energy (IFE) research stable acceleration of fusion targets is a significant problem due to hydrodynamic instabilities. This paper presents the results of the experiments done to investigate the effects of doping 20% of Tungsten (W) (by weight) in Polyvinyl Alcohol (PVA) polymer foils for suppression of instability during laser ablative acceleration. A 20J, 1.060μm, 900ps, Nd: Glass laser system with a focusable intensity of 3 to $9.6 \times 10^{13} W/cm^2$ was used in the experiment. It is observed that the doped PVA targets yielded stable and enhanced foil acceleration as compared to the undoped PVA foils.

## Introduction

Stable laser driven ablative acceleration is a crucial problem in direct drive Inertial Confinement Fusion (ICF) scheme using high gain targets [1]. In the fast ignition scheme energy requirement is lower than the central ignition scheme for fast heating and high density compression. High density compression of more than thousand times the liquid density has, therefore, become a critical subject not only in direct drive but also in fast ignition concept. Recently Murakami et al. has suggested a new technique called as Impact fast ignition scheme. In case of Impact Fast Ignition scheme macro particles of sufficiently high density and mass are accelerated by laser induced plasma ablation to velocities above $10^8$cm/s [2]. The main obstacle in achieving such macroparticle velocities is the Rayleigh-Taylor (RT) instability, which prevents the macro particles from stable acceleration and destroys them before colliding with the precompressed DT fuel. The Laser driven ablative acceleration of thin foil targets and suppression of RT instablity [3, 4] has been a subject of considerable work for many years in ICF scheme. The RT instability growth rate can be expressed using Takabe-Bodner like formula [5]

$$\gamma = \sqrt{\frac{kg}{1+kL}} - \beta k \upsilon_a \tag{1}$$

where k is perturbation wave number, g is foil acceleration, L is plasma density scale length at the ablation surface, β is numerical coefficient determining the effectiveness of the ablation effect, $v_a$ is ablation velocity. RT instability makes any small perturbation in the ablation surface to grow exponentially with time 't' ($e^{\gamma t}$) causing density variations in the foil leading to its fragmentation. Thus stable acceleration of foils to high velocities requires mitigation of RT instability. Several concepts to suppress the ablative RT instability have been proposed, including impulsive acceleration [6], control of isentropic distribution of picket pulse [7], by radiation emitted from a thin metallic overcoat layer [8]. More recently Fujioka et al. proposed a new RT suppression scheme named as SMART

(Stabilization of the RT instability by multiablation in self-radiation target) or Double Plasma Ablation (DA) scheme to suppress RT instability which uses targets doped with high Z elements [9]. In this scheme, a new ablation surface due to X rays from the high Z ions is produced in addition to the electron ablation surface due to electron energy transport. This helps in reducing the instability by creating a longer density gradient scale length. Similar work has been done by our group long back to measure the enhanced shock pressure using 22% W doped Zapon plastic foil [10] but the reason for this was not clear at that time.

**Experiment**

The experiment was performed with the 20 J, 1.06 μm, 900 ps Nd: Glass laser system at Bhabha Atomic Research Centre, Mumbai, India. The laser pulse was focused normally on the target foils to a spot diameter of 100μm with intensity of 3– $9.6x10^{13}$ W/cm$^2$. The energy output from laser is stable with shot to shot energy variation not more than ± 5%. The angular expansion of plasma was determined by ion collectors FC2 and FC1 placed at a distance of 53cm from the target and at an angle of $46^0$ and $55^0$ respectively from the target normal as shown in figure 1. The X-ray emission from the ablating plasma was measured using semiconductor diodes with 5μm Aluminium filter (0.7-1.56 keV) and B10 filter (>0.9 keV) for soft X-rays and 12μm Titanium filter (3-5 keV) for hard X-rays. The accelerated foil movement was determined using two frame optical shadowgraphy with a magnification of 3.45 and spatial and temporal resolutions of 12 μm and 900 ps respectively [11]. For each laser energy, the accelerated foil movements were recorded at four optical delays such as 2.92, 4.82, 6.39 and 8.29 ns with respect to the arrival of main laser pulse at target as shown in figure 2. The targets used were 15μm plain Polyvinyl Alcohol (PVA) foils and 20% by weight Tungsten (W) doped PVA foils.

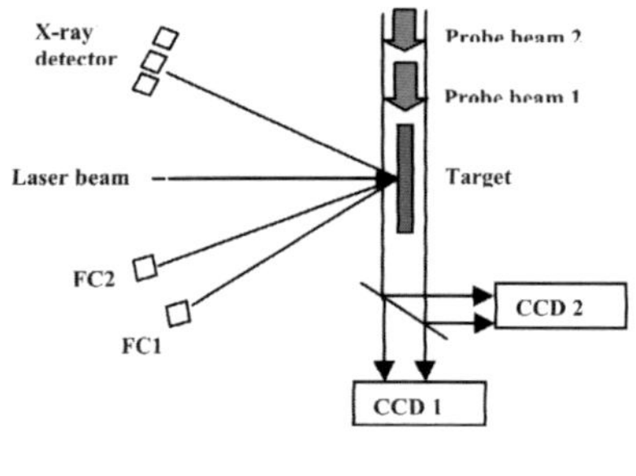


**Figure 1.** Experimental set up

**Results and discussion:** Laser focused on to the target foil results in the instantaneous formation of plasma which expands in the direction opposite to the laser with a high velocity. This expansion of low density coronal plasma exerts a backward pressure on the remaining high density portion of the foil thereby accelerating it in a direction opposite to plasma expansion [4]. The shadowgrams of PVA and PVA (W) are given in figure 2. From figure 2a-2h it is clear that doped foils are having stable accelerations almost up to 8ns. The flyer velocity of doped foils is also observed to have increased to more than twice as compared to plain PVA foils for laser intensity in the range of 3- $9.6x10^{13}$W/cm$^2$ figure 3. The reason for stability and enhanced foil motion can be explained using the equation 1. From equation 1 it is clear that the instability growth rate is inversely proportional to the ablation velocity and ablation density scale length at the ablation surface. Fujioka et al has shown using 1D hydrodynamic simulation code (ILESTA-1D) that in the high Z doped plastics foils, the radiation driven (x-ray radiation produced by high Z doped material) ablation front (RA) formed ahead of the electron-driven ablation (EA) front forms a density plateau region (i.e. enhanced density scale length).

These two ablation fronts propagate with different velocities, the plateau region extended with time. The radiation generated near the EA front propagates diffusively through the LTE plasma and non-uniformity of the laser irradiation is expected to be thermally smoothed out before it is transmit to the RA front. Also, the ablation velocity is mass ablation rate per unit surface area. The x-ray driven ablation rate is much higher than the direct drive ablation rate. This indicates that in the doped foil ablation velocity is also higher than in undoped foil. Thus, in high Z doped PVA target, the instability suppression occur because of larger ablation velocity and longer density scale length.

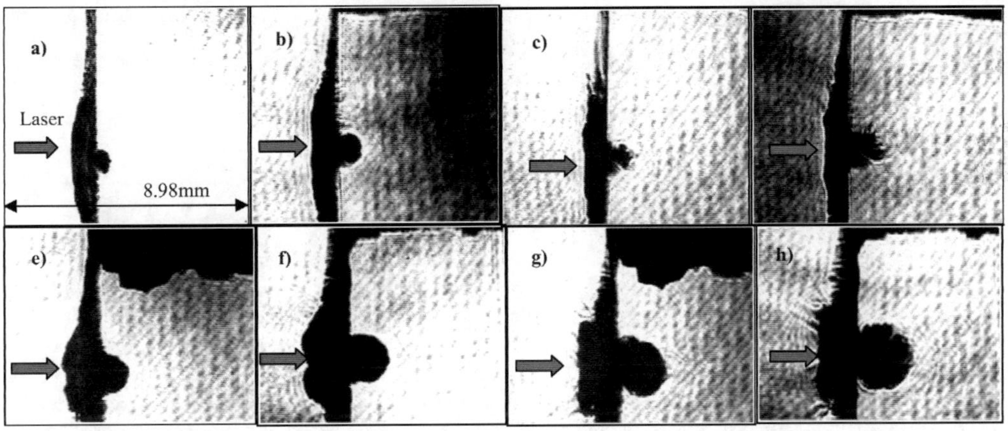


**Figure 2.** Shadowgraph records of foil free surface (flyer) movement at time delays of 2.92, 4.82, 6.39 and 8.29 ns after laser irradiation at laser intensity of 8x $10^{13}$ W/cm$^2$. 2(a)-(d) shows the 15 μm PVA free surface movement. 2(e) - (f) shows 15 μm Tungsten doped PVA free surface movement. Red arrow in the figure denotes the direction of laser beam.

The soft and hard X-ray emissions from PVA and PVA (W) are shown in figure 4. There is an obvious enhancement of X-ray yield from doped foils which is responsible for the radiation driven ablation ahead of electron driven ablation. The soft X-ray yield (>0.9keV) from doped targets were on an average higher than 3.7 times of that from the undoped targets, while the hard X-ray (3-5keV) enhancement was higher than 6.7 times.

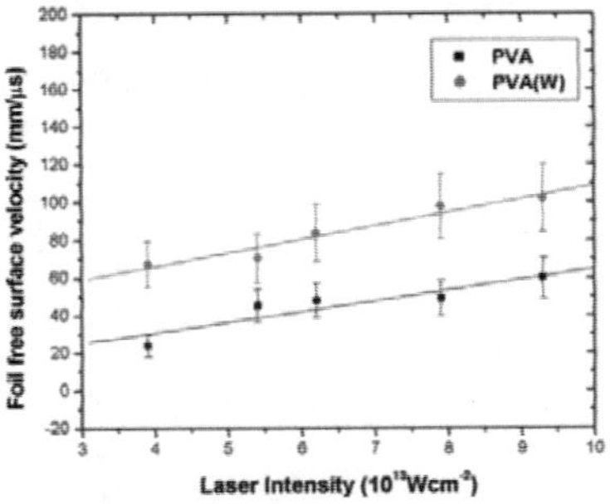


**Figure 3.** Foil free surface velocity variation with laser intensity plot at 2.92ns for PVA and PVA(W)

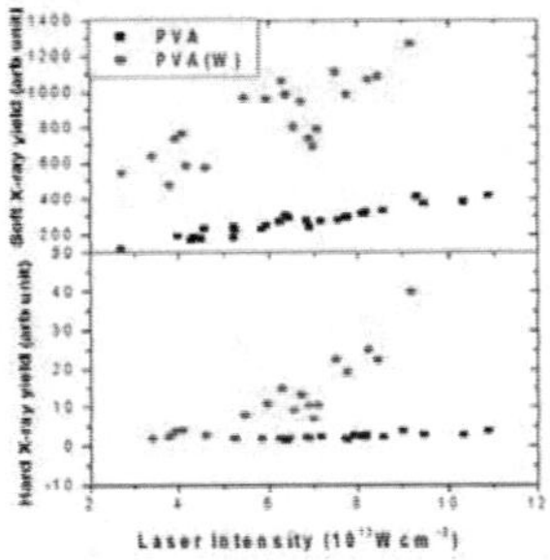


**Figure 4.** Variation of soft and hard X-ray yields for PVA and PVA(W) with laser intensity

The angular distribution of ions having velocities ~ $0.3\text{x}10^8$cm/s from PVA and PVA (W) measured using ion collectors is shown in figure 5. The figure shows that the ion emission from PVA and W doped PVA has a double peak slightly away from the centre. This dip in ion amplitude at target normal is due to the spatial laser intensity distribution of laser beam, which shows double peak in spatial intensity distribution along the radius. But this effect is more pronounced for plain PVA target while in the case of doped PVA the ion angular distribution was almost smoothened out. Also, from the shadowgraphy it is seen that the plasma expansion of doped PVA is narrower and directional to the target normal than the undoped PVA foil similar to the result recorded by ion collectors.

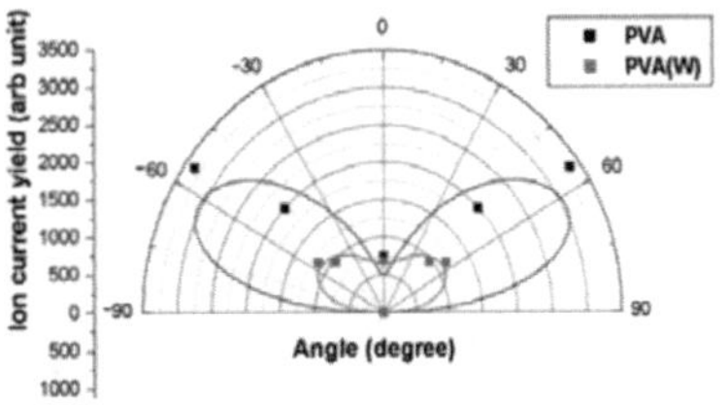


**Figure 5.** Angular distribution of ions for PVA and PVA(W)

**Conclusion:** It is found that a few percent doping of high Z dopant in polymer targets considerably improved various important parameters of the ablation plasma as well as the flyer foil. The presence of high Z ions lead to enhanced absorption of laser light by the plasma; this is evident from the increase in X-ray yield from the ablation plasma of PVA (W). The flyer foil velocity is 2 times higher for doped foils than undoped foil and they accelerated stably for more than 8ns after laser irradiation. The RT instability is therefore significantly suppressed by doping of high Z elements into low Z materials. The suppression of RT leads to the flyer foil being denser in case of PVA (W). Such dense flyers find use in impact studies, various ICF schemes and equation of state studies.

**Acknowledgment:** The authors wish to acknowledge the support and continuous encouragement received from Dr. S. M. Sharma (Head, HP&SRPD, BARC) and Dr. S. Kailas (Director, Physics Group, BARC). Also, we acknowledge the excellent support given by C. G. Murali and Diwakar S. Munda in smooth operation of laser system.

## References

[1] Bodner S E 1981, *Fusion Energy* **1**, 221
[2] Murakami M, Nagatomo H, Azechi H, Ogando F, Perlado M and Eliezer S 2006, *Nucl. Fusion* **46** 99
[3] Ripin B H et al. 1980, *Phys. Fluids* **23** 1012
[4] Decoste R, Bodner S E, Ripin B H, McLean E A, Obenschain S P and Armstrong C M 1979, *Phys. Rev. Lett.* **42** 1673
[5] Takabe H, Mima K, Montierth L and Morse R L 1985, *Phys. Plasmas* **28**, 3676
[6] Lindl J D and Mead W C 1975, *Phys. Rev Lett.* **34** 1273
[7] McKenty P W, Goncharov V N, Town R P J, Skupsky S, Betti R and McCrory R L 2011, *Phys. Plasmas* **8** 2315
[8] Bodner S E, Colombant D G, Schmitt A J and Klapisch M 2000, *Phys. Plasmas* **7** 2298
[9] Fujioka S et al. 2004, *Phys. Rev. Lett.* **92** 195001
[10] Dhareshwar L J, Naik P A and Pant H C 1986, *Pramana* **27** 435
[11] Chaurasia S, Leshma P, Tripathi S, Murali C G, Munda D S, Gupta N K, Dhareshwar L J, Sharma S M and Kailas S 2010, *Bhabha Atomic Research Centre Newsletter* **317** 13

Journal of Physics: Conference Series **377** (2012) 012051 doi:10.1088/1742-6596/377/1/012051

# Studies on Shock Attenuation in Plastic Materials and Applications in Detonation Wave Shaping

**Ritu Khurana, P C Gautam, Rajwant Rai, Anil Kumar, A C Sharma & Dr. Manjit Singh**
Terminal Ballistics Research Laboratory, DRDO, Ministry of Defence, Sector 30, Chandigarh -160 030, INDIA

E-mail: ritu1976@ymail.com

Abstract. Pressure in plastic materials attenuates due to change of impedance, phase change in the medium and plastic deformation. A lot of theoretical and experimental efforts have been devoted to the attenuation of shock wave produced by the impact of explosive driven flyer plate. However comparatively less work has been done on the attenuation of shock waves due to contact explosive detonation. Present studies deal with the attenuation of explosive driven shock waves in various plastic materials and its applications in design of Hybrid Detonation Wave Generator In present work shock attenuating properties of different polymers such as Perspex, Teflon, nylon, polypropylene and viton has been studied experimentally using rotating mirror streak camera and electrical position pins. High explosive RDX/TNT and OCTOL of diameter 75-100mm and thickness 20 to 50mm were detonated to induce shock wave in the test specimens. From experimental determined shock velocity at different locations the attenuation in shock pressure was calculated. The attenuation of shock velocity with thickness in the material indicates exponential decay according to relation US = UOexp(-αx). In few of the experiments manganin gauge of resistance 50 ohms was used to record stress time profile across shock wave. The shock attenuation data of Viton has successfully been used in the design of hybrid detonation wave generator using Octol as high explosive. While selecting a material it was ensured that the attenuated shock remains strong enough to initiate an acceptor explosive. Theoretical calculation were supported by Autodyne 2D hydro-code simulation which were validated with the experiments conducted using high speed streak photography and electrical shock arrival pins. Shock attenuation data of Perspex was used to establishing card gap test and wedge test in which test items is subjected to known pressure pulse by selecting the thickness of the plastic material.

## 1. Introduction

The detonation of high explosives in contact with the solid materials produces interface pressure of the order of hundreds of Kbar depending upon the shock impedances of the two materials. The high dynamic pressure exerted by the detonating explosives results in the formation of a shock wave which travels with a velocity more than sound speed in that material. Because of the Taylor wave following the detonation front in plane unsupported detonation, the pressure time profile across the shock wave in solid is triangular in shape. Very high shock pressure excites different attenuation mechanisms such as irreversible compression & heating, plastic deformation and phase changes. As a result of the interaction of Taylor wave and other attenuation mechanisms the shock wave generated in solids has a decreasing velocity till the shock wave becomes elastic wave.

Ideally a flat topped shock wave is necessary for the shock velocity to stay constant over the measuring distance. A flat topped shock wave is produced by the collision of two flat metal plates [1,2]. In this method an explosively accelerated thin plate impacts on a target of like material. Two shock waves are generated at the interface of impact, one travelling into the target and other travelling back into thin flyer plate. When the latter reaches the rear free surface of the flyer plate a rarefaction fan is generated which eventually overtakes and attenuates the shock wave in the target material. A square shock pulse is thus produced whose amplitude depends upon the flyer velocity and time duration depends upon the thickness of the flyer. A lot of theoretical and experimental efforts have been devoted to the attenuation of shock wave produced by the impact of explosive driven flyer plate. However, comparatively less work has been done on the attenuation of shock waves due to contact explosive detonation. Drummond [3] studied the attenuation of explosive driven shock waves in metals. He analyzed the propagation of plane shock waves in solids and developed a procedure for calculating the attenuation of the shock wave. He determined the equation of state of the burned explosive gas from shock attenuation characteristics of the metals. He also obtained, theoretically, the approximate solution for attenuation of plane shock wave in metals and compared the results with experimental values by adjusting the value of specific heat ratio for detonation products. Present studies deal with the attenuation of shock waves in various plastic material and applications in detonation wave shaping. There were two motives for the present studies. First; to select the materials with energy absorption capacity for protecting the equipments and materials in shock environment and second; to find out the shock attenuating characteristics of plastic materials for their use in design of hybrid detonation wave generators.

## 2. Shock Propagation in Solid Materials

Consider an explosively generated shock wave in an inert specimen. Since the mass of the explosive is finite, the pressure behind the shock front falls off rapidly due to rarefaction waves. The peak pressure decays as the wave propagate outward because the energy distributed over the constantly increasing area and also is due to finite dissipation of energy in the transition through shock front. Figure1(a) shows a test set up for loading the shock waves in solids by contact explosive detonation and Figure1(b) shows progressive attenuation of the peak shock pressure with distance travelled in the solid specimen. The peak pressure and the shock velocity decreases till the shock wave becomes an elastic wave. Since the sound velocity is increasing function of pressure the shock pulse becomes broader with distance traveled in solid specimen.

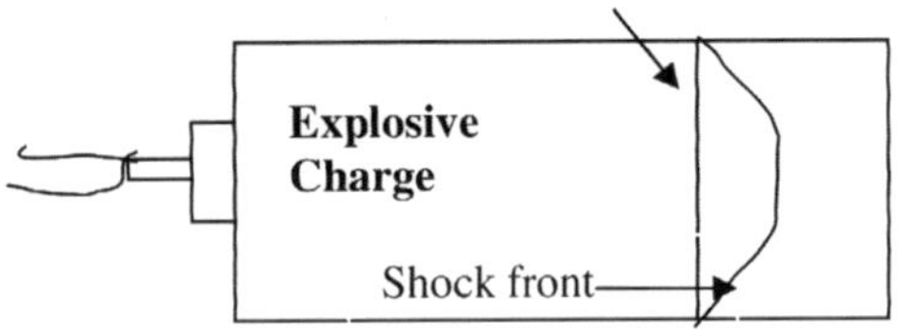


Figure 1(a). Test set up for shock loading of solids by contact explosive shock front

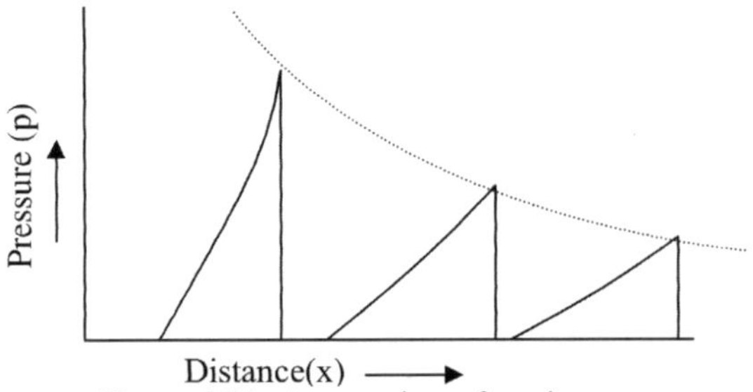


Figure 1(b). Attenuation of peak pressure t with the distance traveled in the solid specimen

The decrease of pressure of shock wave in travelling a unit distance in the solids is dependent on the pressure gradient in non uniform wave profile behind the shock front and the ratio of velocities of rarefaction wave and the shock wave in the specimen. If Us is the velocity of the shock front then pressure can be calculated from shock jump condition

$P = \rho_o Usup$ (1)

Where up is the particle velocity. Shock velocity and particle velocity are related by a linear equation

Us = co + bup (2)

Where co is the bulk sound speed in the media. b constant

The pressure attenuation with distance is written as

$$\frac{dP}{dx} = \left( \frac{u_p + c}{U_s} - 1 \right) \frac{dP}{ds} \quad (3)$$

where c is sound velocity in compressed solid, dP is change in pressure over a distance dx in solids and dP/ds is pressure gradient behind the shock front. The pressure gradient dP/ds is a function of the length of explosive.

In the near region of shock loading the shock velocity attenuates according to an exponential law

$$U_s = U_{so} \exp(-\alpha x) \quad (4)$$

where Uso is the amplitude of transmitted shock velocity at the explosive solid interface and □ is attenuation constant. A more general law of attenuation in form of mach number is described by equation

$$M_S = (M_{so} - 1)\exp(-\alpha x) + 1 \quad (5)$$

where Ms = Us/co, is the ratio of shock velocity to the sound velocity at normal pressure. At explosive solid interface, x = 0 and Ms = Mso and at x = □, Ms = 1, which means the shock wave degenerate into an elastic wave at very large distance. The attenuation coefficient $\alpha$ in equation (2) depends on the material, incident shock velocity and shape of the shock front.

## 3. Experimental procedure

Cylindrical test specimen of nylon, Perspex, teflon & polypropylene of dia 75-100mm were machined out of the rod. High explosive RDX/TNT (60:40) of dia 75-100mm and thickness 20 to 50mm were detonated to induce shock wave in the test specimen shown in figure 2. High explosives were center initiated to induce a spherically diverging shock wave in the test specimen. A flat bottomed hole of diameter 8mm were drilled in the test specimen terminating at a depth of 5mm from explosive solid interface. Perspex pallets of thickness 5-6mm were inserted into the hole leaving 0.25mm air gap in between the two pallets. The sequential light flashes generated by the shock ionization of successive air gap were recorded on a streak camera through an optical fibre. Rotating mirror streak camera with maximum writing rate of 10mm/microsecond was used to record the shock arrival at different locations inside the test specimen. Cylindrical test specimen of nylon, teflon & polypropylene of dia 75-100mm were machined out of the rod. High explosive RDX/TNT (60:40) of dia 75-100mm and thickness 20 to 50mm were detonated to induce shock wave in the test specimen shown in figure 2(a). High explosives were center initiated to induce a spherically diverging shock wave in the test specimen. A flat bottomed hole of diameter 8mm were drilled in the test specimen terminating at a depth of 5mm from explosive solid interface. Perspex pallets of thickness 5-6mm were inserted into the hole leaving 0.25mm air gap in between the two pallets. The sequential light flashes generated by the shock ionization of successive air gap were recorded on a streak camera through an optical fibre as shown in Figure2(b) Rotating mirror streak camera with maximum writing rate of 10mm/microsecond was used to record the shock arrival at different locations inside the test specimen.

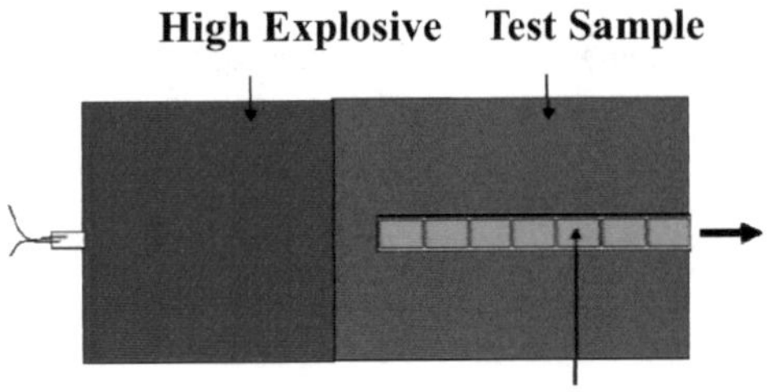


Figure2(a). Sketch of experimental assembly used for shock attenuation

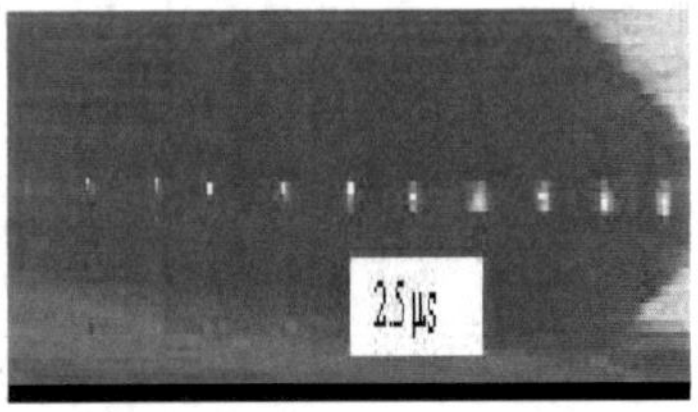


Figure2(b) Streak camera record of shock attenuation in plastic materials

The shock velocity has been found to attenuate exponentially following the relation

$$U_S = U_O \exp(-\alpha x) \qquad (1)$$

where Uo is the initial shock velocity at explosive plastic interface, $\alpha$ is attenuation constant and x is thickness in plastic material.

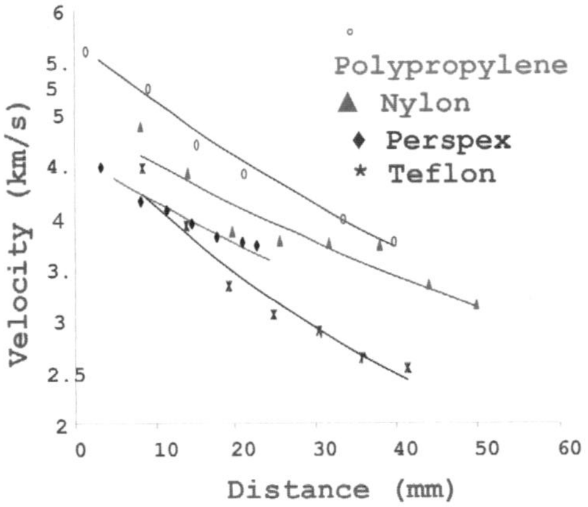


Shock Attenuation relations
Polypropylene($\rho$ = 0.93g/cc)
Us = 5.70 exp (-0.01x)
Nylon($\rho$ = 1.14g/cc)
Us = 5.42 exp (-0.0008x)
Perspex ($\rho$ = 1.18g/cc)
Us = 4.59 exp (-0.0102x)
Teflon($\rho$ = 2.14g/cc)
Us = 5.06 exp(-.0177x)

Figure3. Experimental data on shock attenuation in different plastic materials

Experimentally found out shock attenuation data of different materials is plotted in Fig 3.

**4. Shock Attenuation of Viton**

Shock attenuation characteristics of Viton has been found out experimentally where Viton in the form of discs of diameter 100mm and 10mm thickness were processed. High speed streak photography and electrical shock arrival pins were deployed to monitor shock velocity at different locations inside the sample material. High explosive Octol (70:30) of diameter 100mm and thickness 30 to 40mm were detonated to induce shock wave in the test specimen. High explosives were center initiated to induce a spherically diverging shock wave in the test specimen. Half cut view of explosive assembly and layout of pins inside the test specimen is shown in Fig 4(a). Electrical shock arrival pins were used to record shock velocity at each interface of Viton discs. All the pins were used in a circular layout of radius 20mm as shown in Fig 4(b). A typical oscilloscope record is shown in Fig 5(a). A number of experiments were conducted and the experimental data has been found to fit into an exponential relation

$$U_S = 5.788 \exp(-0.01x)$$

where 5.788 is the initial shock velocity in Km/s, 0.01 is attenuation constant and x is the distance travelled in Viton.

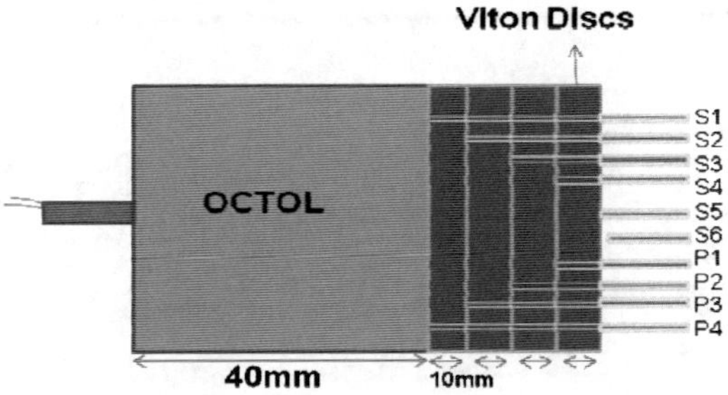


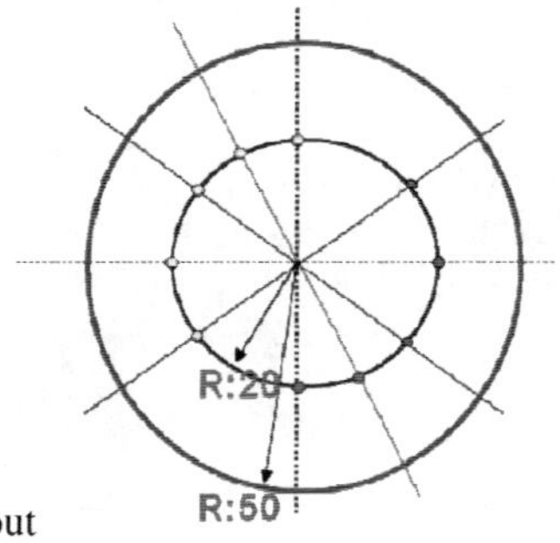


Fig4(a).Experimental Setup (b) Pins Layout

The experimental data is plotted in Figure 5(b).

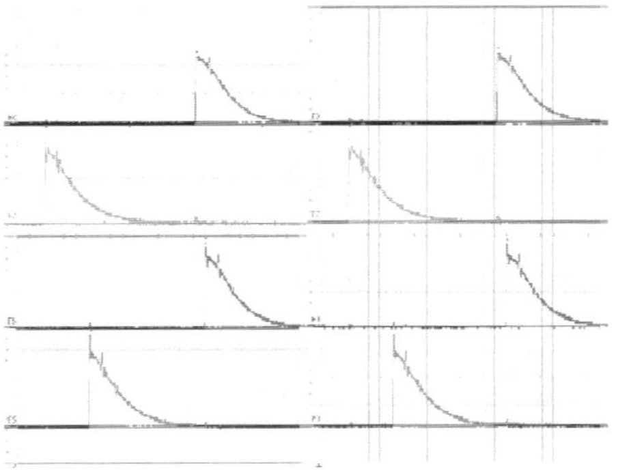

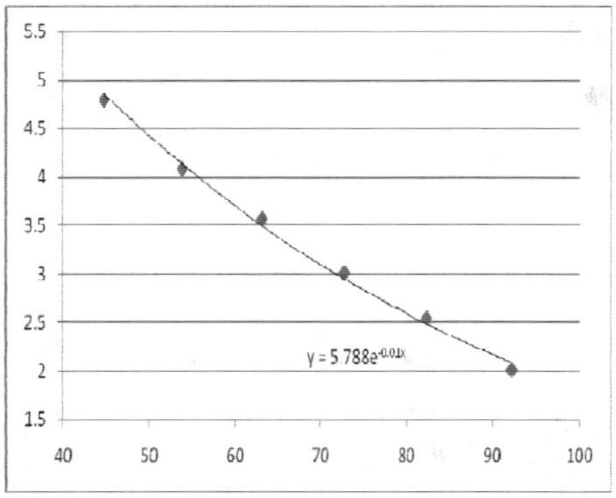


Figure5(a) Oscilloscope Record of shock arrival pins

Figure 5(b) Experimental data of attenuation in Viton

**5. Equation of State**

Equation of state is important thermodynamic relation between P,V and T or E. Many EOS are available for solids but for the application under consideration shock EOS is more appropriate. Shock equation of state is a linear relation between shock velocity and particle velocity as given by US = a+bUp. Shock velocity was found out experimentally at different distances from shock attenuation relations. The particle velocity was found from free surface velocity measurement. A total of four experiments were conducted with Viton material of varying thickness of 20, 30, 40, 50 & 60mm. Electrical shock arrival pins were deployed to monitor free surface velocity, one pin in contact with Viton disc and other 3mm raised. The results of four experiments has been plotted in the graph between shock velocity and particle velocity in Fig6.

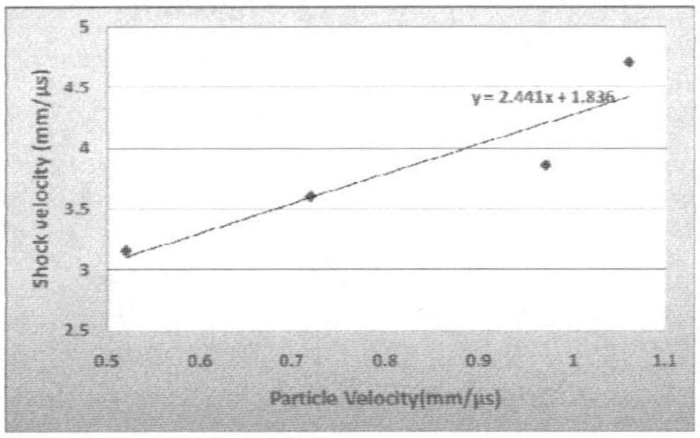


Figure6. Experimental data on particle velocity

So experimental data has been found to fit in the following linear relation Us = 1.836+2.441Up

### 6. Applications in detonation wave shaping

#### Design of Octol- Viton Spherical Explosive Lens

Based on shock attenuation data and equation of state found for viton a spherical wave lens has been designed by using combination of high VOD explosive Octol and viton an inert material[4]. Based on the calculations a preliminary design was made as shown in Fig 7.

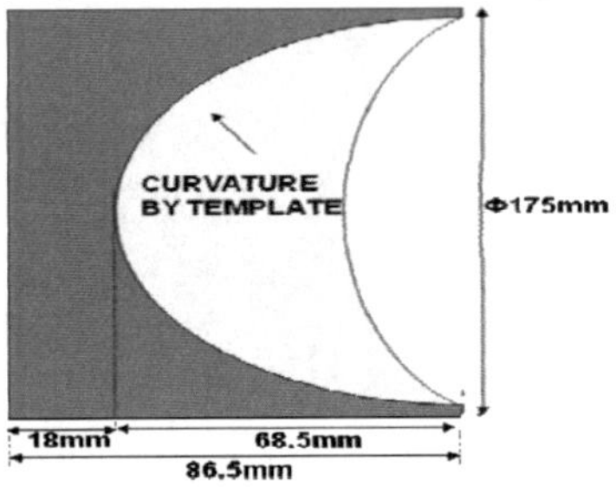


Figure7. Preliminary Design

### 7. Simulation Setup

Preliminary design was verified through numerical simulation using Autodyne2D hydrocode.

Simulation was done in 2D axial symmetry about x-axis. Eulerian space was created with uniform mesh density of 0.2mm square cell. JWL equation of state was used to represent explosive Octol having density of 1.82g/cc. Shock equation of state for Viton was used of density 1.91g/cc. Material models are shown in Appendix 'A'. Cycle 0 of simulation setup is shown in Fig8. Pressure time profile at exit surface of Viton vs time is shown in Fig 9(a) and snapshots of detonation/shock propagation in the designed lens are shown in Fig 9(b).

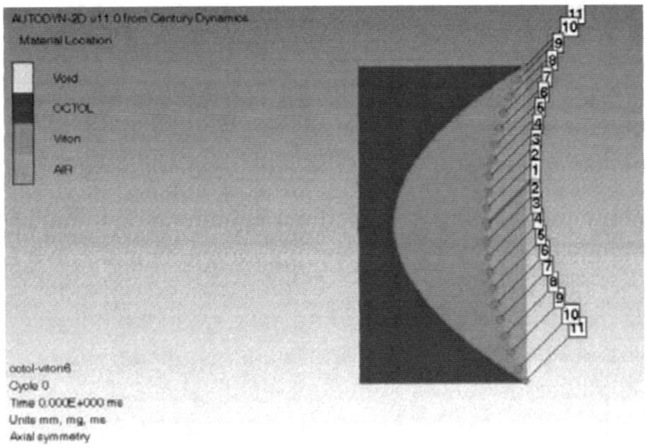


Figure8. Simulation setup for Octol/Viton

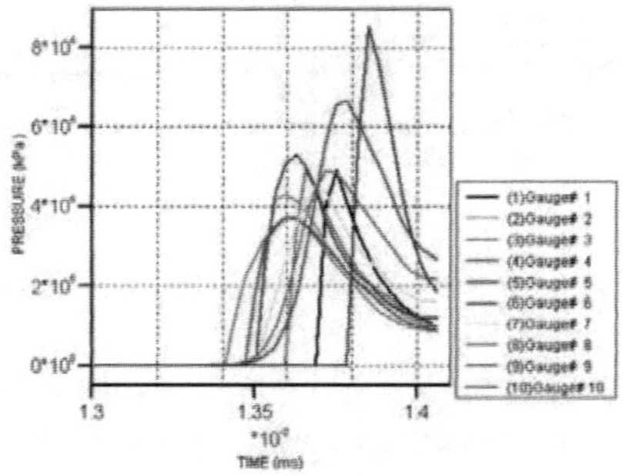


Figure9(a). Pressure history at exit surface of Viton vs time

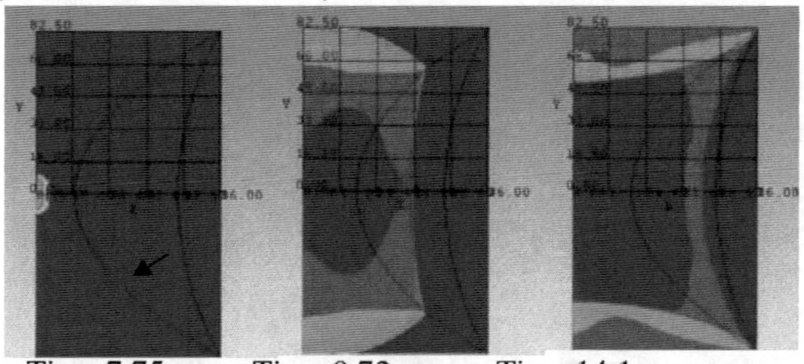

Time:7.75μs Time:9.72 μs Time:14.1 μs
Figure9(b). Simulation result showing shock wave arrival at front face of Viton

Simulation was used to verify the contour between the explosive and inert material so that the shock wave reaches the Viton surface simultaneously. On viton surface insitu casting of explosive was done. The design was then validated through experiments

## 8. Validation of Design through Experimental Evaluation

High speed streak photography and electrical ionization pins were deployed to monitor planarity of shock arrival at the front surface of Viton. Fig 10 shows the layout of electrical and piezoelectric pins to monitor simultaneity in shock wave arrival at front surface of Viton.

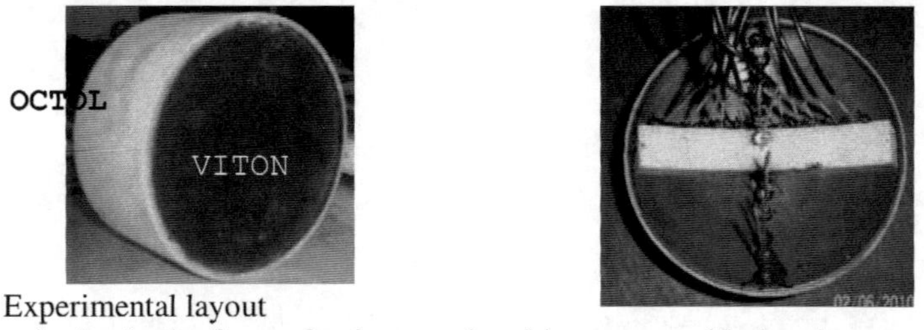


Figure 10. Experimental layout

Acceptor explosive in the form of strip was placed in contact with the viton to study the initiation of detonation. The simultaneity in detonation wave reaching the explosive free surface was recorded by a streak camera. The final photograph of the spherical explosive lens at the time of trial is shown in Fig 11.

Figure 11. Trial Photographs

## 9. Results

Fig 12(a) shows the shock wave arrival at Viton free surface as recorded by the shock arrival pins and Fig 12(b) shows the shock arrival at Viton and detonation wave arrival at free surface of acceptor explosive.

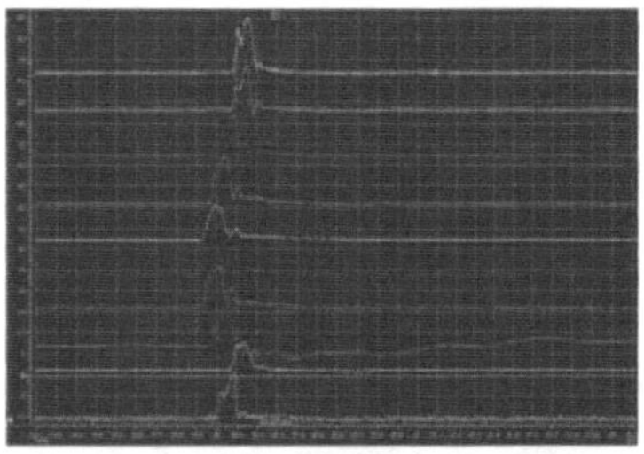

Figure 12(a) Oscilloscope record

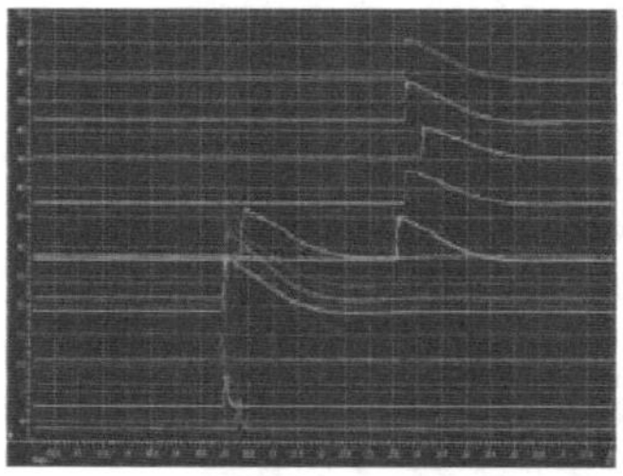

Figure 12(b) Oscilloscope record

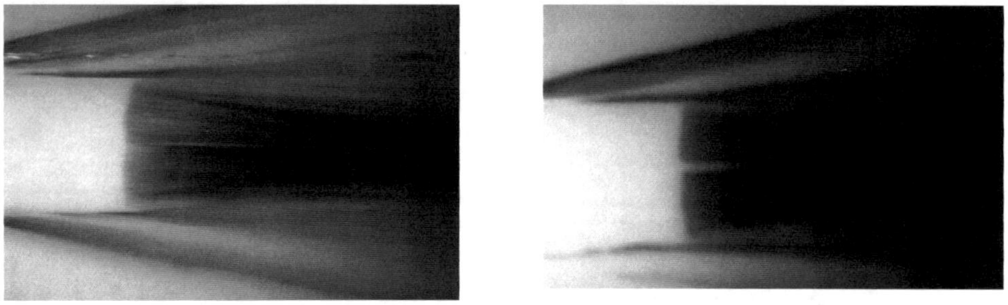

Figure 13(c) Streak Camera Record

The measured simultaneity in shock wave arrival at Viton surface is 0.18μsec which is acceptable. However the simultaneity in detonation wave arrival at explosive free surface was recorded by streak camera as shown in Fig 13(c) and the measured simultaneity is between 0.13μs to 0.18μs which is within the acceptable limit..

The shock attenuation data of Perspex was used in card gap test experiments. Perspex of density 1.186 g/cc was used whose linear Hugoniot relation is given by

$Us = a + b\ Up$

where a = 2.561 mm/μs and b = 1.595.

In order to standardize the geometry, trials were conducted with the same experimental set up where Tetryl was used as the donor. Two pellets each of diameter 2 inch and length 1 inch were used followed by Perspex as the attenuator material[5]. The experiments were conducted using rotating mirror streak camera which is driven by compressed air and has a maximum writing rate of 10 mm/μs.

Records were obtained by viewing the Perspex in velocity measurement mode where the event direction is parallel to the slit but perpendicular to the optical axis as shown in figure 14

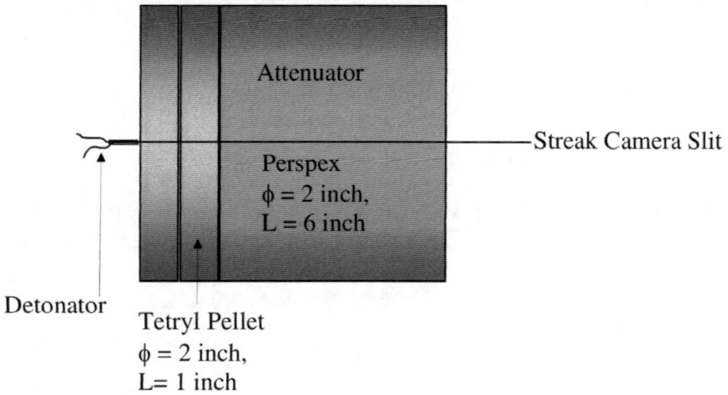


Figure 14. Trial Set-Up for Recording Shock Attenuation in Perspex

The results were analyzed on image analyzer and it was found that the shock wave attenuates with distance traveled in the Perspex. The shock pressure has been found to attenuate exponentially with the distance traveled in Perspex. The following relation was fitted after experiments

$P= P0\ e\text{-}\alpha\ x$

$P= 136\ e\text{-}0.0358x$

where P is the pressure at a distance x in the Perspex and Po is the initial pressure at interface, $\alpha$ is the attenuation constant whose value in this case is 0.0358.

## 10. CONCLUSION

Shock attenuation relations for different polymers has been found out experimentally. Based on shock attenuation properties of polymers spherical wave lens of diameter 175mm have been designed. The developed DWG has been test fired and simultaneity of the order of 0.14μsec has been achieved. The shock attenuation of data of Perspex has been used to establish card gap test.

## 11. Acknowledgements

The authors express their sincere thanks to Dr. Manjit Singh, Director, TBRL, for his encouragement and kind permission to publish the paper. The support provided by Anil Kumar and Gaurav Goel is also acknowledged.

## 12.References

[1] Al, Donald R. Cuvan, Non hydrodynamics attenuation of shock waves in, Journal of applied physics, Vol 34, No. 9, Pg 2677-2685

[2] W.E.Drummond, Explosive induced shock waves, Journal of Applied Physics, Vol 28, 1437.

[3] Kenneth J. Graham and Peter J. Cahill, Super Large-Scale Gap Tests on Energetic Formulations', 42nd Joint Propulsion Conference & Exhibit

[4] Ritu khurana,Bhupinder Sewak, Rajwant Rai & Dr. Manjit Singh, Hybrid detonation wave generator, 5th International High Energy Materials Conference & Exhibit (HEMCE -2005), 23-25 Nov. 2005, DRDL Hyderabad

[5] Super Large-Scale Gap Tests on Energetic Formulations', Kenneth J. Graham and Peter J. Cahill, 42nd Joint Propulsion Conference & Exhibit

# Laser induced shock studies at RRCAT, Indore

**P.A. Naik[1], V. Arora, S. Bagchi, Y.B.S.R. Prasad, S. Barnwal, and P.D. Gupta**
Laser Plasma Division, Raja Ramanna Centre for Advanced Technology,
Indore 452 013, M.P., India

E-mail:panaik@rrcat.gov.in

**Abstract.** The knowledge of material response under shock compression is important in terms of proper understanding of the process and in view of the wide range of technological applications associated with it. At the Laser Plasma Division, RRCAT, Indore, among other things, we are also involved in studying the material response from two different perspectives. On one front, we have developed capability for acquiring Equation of State (EOS) of materials using laser generated shock waves with long pulse lasers (>200 ps). On the other front, we have started shock experiments on single crystals using time resolved x-ray diffraction technique, using ultra-short pulse lasers (< 1 ps). In these initial experiments, characteristic sub-picosecond, high brightness $K_{\alpha}$ x-ray pulses generated by Ti: sapphire femtosecond laser produced plasmas of different target materials (Ti, Fe, and Cu) were used to probe the effect of compression generated by moderately intense (GW/cm2) sub-nanosecond laser (200 ps) pulse irradiation on the crystal surface. The dynamics of the shock propagation is manifested in terms of the temporal evolution of the rocking curve of shocked sample. The shock velocity deduced from these measurements is 12 km/s, consistent with the predicted velocities and probe depth. The observed maximum compression is 0.4 % which corresponds to a pressure of 0.8 GPa.

## 1. Introduction

The study of material response under extreme conditions is of great importance in several fields, in particular material science, astrophysics, geophysics, and research of target design in the context of inertial confinement fusion (ICF) studies [1,2,3]. The equation of state (EOS) data is necessary for hydrodynamic simulations dealing with nuclear fusion [4,5]. In the recent years, shock compression of condensed matter has emerged as a subject of considerable importance for a variety of research investigations such as phase transitions [6] (both structural and magnetic), strain propagation [7], etc. The simple technique of x-ray diffraction offers great potential in terms of direct observation of lattice distortion because of the inherently large penetration depth of x-rays in matter. Time resolved x-ray diffraction thus enables real time visualization of atomic arrangement in the sub ps time scale [8]. The x-ray bursts from sub-picosecond laser produced plasmas have been utilized for time-resolved study of processes like ultrafast excitation of the crystal lattice [9], non-thermal melting of semiconductors [10], acoustical phonon excitations [11]. In particular, the characteristic K-shell line

[1] To whom any correspondence should be addressed.

radiation ($K_\alpha$) has generated much interest to be used as a narrow bandwidth and ultrafast probe to study fast processes in matter. Depending on the interaction parameters, the hot electrons are generated through collective mechanisms such as resonance absorption [12], vacuum heating, j x B heating etc. These electrons penetrate into the cold target material to generate continuum hard x-ray bremsstrahlung and characteristic $K_\alpha$ line radiation. The photon flux time, time duration, and monochromaticity of the $K_\alpha$ emission is primarily determined by the hot electrons parameters such as the incident number, energy and transport into the bulk solid [13].

Laser Plasma Division at RRCAT, Indore, is engaged in research related to shock physics in two areas. One area is EOS studies at megabar pressures using high power lasers and the other area is time resolved x-ray diffraction using ultra-short pulse lasers. The latter studies are at GPa (tens of kilobar) pressures. The article is organised as follows. Section 2 gives the experimental description and the results of the EOS studies. Section 3 gives the experimental description and the results of the time resolved x-ray diffraction studies. Our future plans in both these areas are presented in section 4.

## 2a. EOS experiments with high power lasers

EOS at pressures up to a few Mbar can be measured in static experiments using diamond anvil /gas gun. For higher pressures (>10 Mbar) the laser-induced shock wave technique is useful. Laser produced plasmas had been used to generate pressures up to 100 Mbar in different materials to construct their EOS [14,15]. The shock pressure can be varied in a controllable way from less than 10 kbar up to ~800 Mbar, by using lasers with different energy, time duration, and focal spot. The strain rates achievable in laser-produced shock waves vary from about $10^6$ to $10^9$ $s^{-1}$. An additional important advantage of laser-induced shock wave is the possibility to synchronize the shock with a variety of optical, x-ray, and electronic diagnostics. EOS of copper up to 20 Mbar, gold and lead up to 10 Mbar, Parylene-C and brominated CH at pressures up to 8 Mbar, have been made using the HELEN laser at AWE, UK [16]. Absolute measurements of the EOS of iron at pressures in the range of 1–8 Mbar have been made using the LULI Laser at Ecole Polytechnique, France [17]. EOS of dielectric materials like sapphire ($Al_2O_3$) and lithium fluoride (LiF) up to 20 Mbar was measured using the Omega laser at Univ. Rochester, USA [18]. The Hugoniot of tantalum up to pressures of 40 Mbar was measured with the Gekko/Hyper laser at Institute of Laser Engineering, Japan [19]. EOS of low density foams in the range of 60–130 mg/cm$^3$ up to pressures of 3.6 Mbar was measured with the PALS laser at Prague [20]. In general, determination of EOS involves measurement of the shock and particle velocities in the medium. In these experiments, an intense laser pulse is focused on a foil (whose EOS is to be determined) kept in vacuum. The ablating plasma launches a strong shock wave in the foil material due to momentum recoil of the ablating material. During the duration of the laser pulse, a continuous shock wave is launched in the inward direction from the ablating target surface. If the thickness of the foil is such that the shock transit time is less than the laser pulse duration, there is no shock rarefaction. The shock wave travels with a velocity $u_s$ and particles behind the shock wave attain a velocity $u_p$. If the laser spot size on the foil is much more than the foil thickness, one can assume 1-D shock propagation. The various material parameters in the shocked region like density ($\rho_1$), shock pressure ($P_1$), and kinetic energy of the compressed material ($E_1$) are related to the initial material properties, the shock velocity ($u_s$), and the particle velocity ($u_p$) by the standard Hugoniot relations $\rho_o u_s=\rho_1(u_s-u_p)$, $\rho_o u_s u_p=P_1-P_o$, $\rho_o u_s\{E_1-E_o+½u_p^2\}=P_1 u_p$. The shock velocity is experimentally measured from the time it takes for the shock wave to reach the rear side of the foil. This, in turn, is measured from the shock luminosity due to compression. These propagation times are usually on ps time scale and hence fast optical streak cameras are required to record the luminosity variation. The particle velocity is determined experimentally using optical techniques like Velocity Interferometer System for Any Reflector (VISAR) or Optically Recording Velocity Interferometer System (ORVIS). These instruments give the free surface velocity of the rear-side of the foil. The particle velocity is generally equal to or slightly less than half the free surface velocity. In absence of instruments like VISAR, it is still possible to obtain the particle velocity independently by using optical shadowgraphy techniques where two pulses with known delay are used for probing the plasma from the tangential direction and the particle velocity can be measured [21,22]. In addition, it is possible to use chirped pulse diagnostics like chirped pulse reflection [23], frequency domain interferometry [24] etc., to obtain the shock breakout time and particle velocity more accurately. When it is not possible to determine the particle velocity experimentally, one uses the standard relation $u_s = u_o + \alpha\, u_p$ to get the particle velocity from the shock velocity, when $u_o$ and $\alpha$ are known for the material [25].

## 2b. Experimental Setup and Results

The initial experiments at RRCAT were conducted on a 200 ps Nd:YAG commercial laser system at laser energy of 2 J (max) to obtain the EOS of gold by using Al+Au and Al+Cu layered targets using the impedance matching method [26]. At laser intensities ~3-5x$10^{13}$ W/cm$^2$, shock pressures in the range 4.5 to 6.5 Mbar for Al, 9-13 Mbar for Au, and 9-10.5 Mbar Cu were obtained [27]. In order to extend these studies to higher pressures, experiments were then carried out using an Nd: glass laser system at laser energy 10 J with pulse duration of 600 ps, with Al as reference material. The laser output was converted to second harmonic (532 nm) to enhance the temporal contrast. The laser energy was focused on 8μm Al foils at an intensity of $10^{13}$ W/cm$^2$. An (S-20) optical streak camera with a time resolution of 8 ps was used for these measurements. In order to have a time reference or *fiducial*, a portion of the laser light was collected at an earlier stage, converted into second harmonic using a KDP crystal, and then passed through an optical fiber up to the entrance slit of the streak camera. The length of the fibre was adjusted such that the time of laser light reaching the end of the fibre is exactly equal to that of the laser light reaching the streak camera (in absence of the foil). The shock velocity was estimated by measuring the shock propagation time defined by the time delay between *fiducial* and the shock arrival at the rear surface of the foil (figure 1).The streak camera was triggered in advance by the laser pulse using a photodiode to compensate for the inherent time delay of the streak camera circuitry. A shock velocity of 21 km/s was calculated from the measured transit time (figure 2). The corresponding particle velocities are 12 km/s, as estimated from the LASL data-book [25] which correlates the shock velocity and particle velocity by the relation $u_s = 5.38 + 1.34u_p$. The shock pressure was calculated to be 7 Mbar from the standard relationship $P = \rho_o u_s u_p$, where $\rho_o$ is the solid density of aluminium (2.7 gm/cm$^3$).

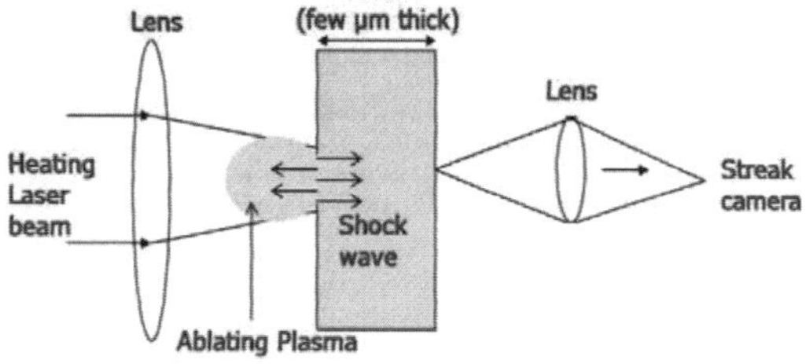


Figure 1: Schematic experimental setup for shock speed measurement

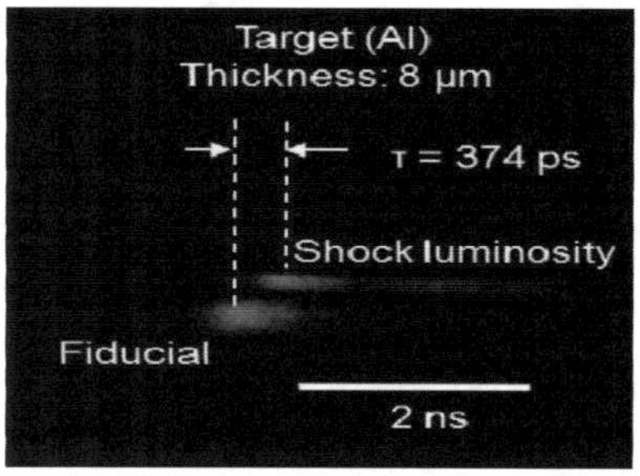


Figure 2: CCD image of the streaked image showing the shock luminosity signal along with the fiducial marker.

## 3a. Time Resolved X-ray Diffraction Experiments with Ultra-short Pulse Lasers

EOS studies require long pulse (>200 ps) lasers whereas time resolved x-ray diffraction (TXRD) needs ultra-short lasers (< 1 ps). TXRD is used to probe the lattice strains on sub-picosecond time-scales under extreme conditions of temperature and pressure. The diffraction pattern shifts have been used to determine the average lattice compression following elastic–plastic deformation. The appearance of new diffraction lines has been used to examine solid–solid phase transitions. Vanishing of the reflected Bragg peak can give information about solid-liquid transition. Although 3rd and 4th generation synchrotron sources can be used for this purpose, characteristic x-rays produced with ultrashort lasers offers a cheaper and compact alternative. The TXRD technique was first demonstrated with a shocked silicon (111) crystal by Kalantar *et. al.* [28] using Nova laser. bcc to hcp phase transition in Fe at 13 GPa shock produced by laser irradiation at $2\times10^{10}$ to $1 \times 10^{12}$ W/cm$^2$ (2 – 6 ns) was probed with 1.85 Å iron K-shell x-rays by Hawreliak *et. al.* [8]. X-ray diffraction was used with VULCAN laser to measure shear strains in single crystals of copper shocked to pressures in excess of 100 GPa (1 Mbar) by Murphy *et al* [29].

In a typical TXRD setup, a major part of the laser beam energy is focused on to a solid target to generate an efficient high brightness, sub-picosecond x-ray probe beam. The line radiation emitted is used as monochromatic x-ray source for probing the transient structural modifications in the crystalline sample induced by the remaining fraction of the laser beam referred as pump laser. The rocking curves of sample are recorded as a function of diffraction angles for different time delays between pump and probe pulse. Commonly, the experiment is performed in Bragg geometry. It is well known that in the case of Bragg reflection, a large number of layers participate in the process and the diffraction pattern is cumulative contribution of all the participating layers. From Bragg's law, the angle of diffraction is related to the change in inter-planar spacing. The lattice strain can be estimated using $\Delta d/d = -\cot\theta_B$ where d is inter-planar spacing, $\theta_B$ is Bragg angle. In such experiments, dynamics of the lattice deformation is studied from the shift / broadening of the diffracted x-ray peak and the change in reflectivity. In addition to this, exploiting the fact that different x-ray photon energies have different penetration depth (defined as 1/e attenuation length) in the material, keeping the external excitation source parameters invariant, it is possible to infer about the characteristics of shock wave propagation inside the optically opaque samples.

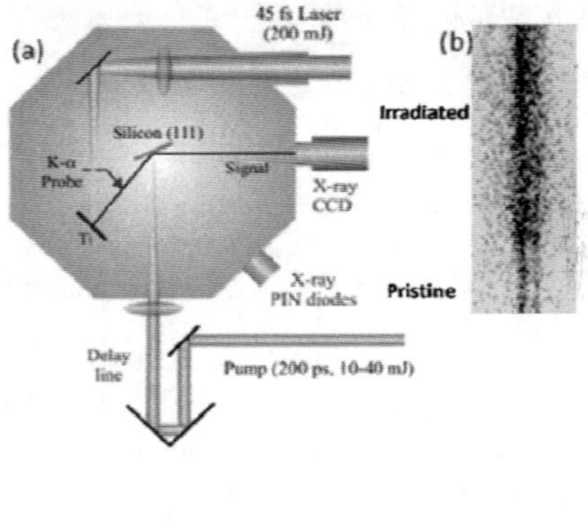


Figure 3: (a) Schematic experimental setup (b) CCD image of the diffracted x-rays showing irradiated and pristine parts of the crystal.

### 3b. Experimental Setup and Results

Experiments were conducted using the 10 TW Ti: Sapphire laser system. The schematic of experimental setup is shown in figure 3 (a). The $K_\alpha$ x-ray probe was generated by focussing the 45 fs laser pulses onto solid (titanium, iron and copper) targets. A part of the uncompressed (200 ps) pulse was used to irradiate a 500 μm thick flat Si (111) crystal (d = 6.271 Å) at an intensity of 6 GW /cm$^2$. The time delay between the 200 ps pump laser pulse and the probing x-ray pulse was adjusted by an optical delay line. A positive delay here means the pump laser pulse is leading the probe x-ray pulse. The diffracted x-ray spectrum was recorded on an x-ray CCD camera. The point x-ray source allowed a direct imaging of a part of the crystal surface on to the x-ray CCD camera. This enabled simultaneous recording of the diffracted x-ray spectrum from the laser irradiated as well as pristine area of the crystal. Figure 3 (b) shows the space resolved CCD image of the diffracted iron $K_\alpha$ x-rays from laser irradiated Si (111) surface at a delay of +600 ps. The lower part of the same picture shows the pristine sample where the $K_\alpha$ lines ($K_{\alpha1}$ and $K_{\alpha2}$) are clearly identified. The upper part of the picture shows blurring of the $K_\alpha$ lines because of the non-uniform lattice compression attributed to cumulative effects of the laser induced compression wave and the associated thermal broadening of the lattice.

The measured rocking curves of Si (111) irradiated by 200 ps laser pulse at a fluence of 2.3 J-cm$^{-2}$ for various delay times between -300 and +1800 ps are shown in figure 4. The x-ray spectrum shows that the diffraction profiles of Fe $K_{\alpha 1}$ (6403.8 eV) and $K_{\alpha 2}$ (6390.8 eV) are well resolved for zero and negative delays. It is observed that the diffraction pattern broadens with increasing time delay up to +1200 ps. After that, the broadening reduces and finally comes back to original state for delays larger than +1500 ps. The x-ray beam is diffracted from the sample at a particular angle within a thin layer of the crystal where the lattice spacing is such that the Bragg condition is satisfied. Different layers strained to different extents under the influence of the compression wave, will diffract at other angles leading to the broadening. Furthermore, at early times, the diffraction will occur from both shocked material and the underlying unperturbed crystal. It may be noted that the broadening of the diffracted signals towards higher angles implies lattice compression induced by pump laser beam. The compression of lattice is either due to the laser ablation of very thin surface of the silicon or stress caused at the front of thermal expansion due to the surface energy deposition [30]. On the other hand, the spread towards lower angles reveals the signature of thermal disordering effect indicating the role of the thermal wave. Nevertheless, the time evolution of the rocking curve signifies the propagation of laser induced shock waves inside the crystal. At later times, the separation of $K_\alpha$ peak signifies the passing of shock waves beyond the maximum probe depth inside the crystal.

The rocking curve of silicon irradiated at same laser parameters (200 ps, 2.3 J-cm$^{-2}$) was also measured with Ti $K_\alpha$ (4.5 keV, Bragg angle: 26°) and Cu $K_\alpha$ (8.05 keV, Bragg angle: 14.2°).The profiles of the rocking curves are similar to that measured with Fe $K_\alpha$ (4.5 keV, Bragg angle: 18°). Figure 5 shows the FWHM of $K_{\alpha 1}$ line radiation as function of delay between pump and the probe pulse for Ti, Fe and Cu. It is observed that the diffraction pattern broadens with increasing delay to reach a maximum and thereafter the $K_{\alpha 1}$width decreases and come close to the pristine. The time for the maximum broadening is 650 ps, 1160 ps, and 1870 ps for Ti, Fe, and Cu respectively. FWHM of the rocking curve of irradiated sample shows broadening of 3.2 ± 0.3 times compared to the rocking curves of pristine sample for all the three probe x-ray lines. The high energy x-rays have longer penetration depth and are therefore expected to yield information about the strain over a greater crystal depth.

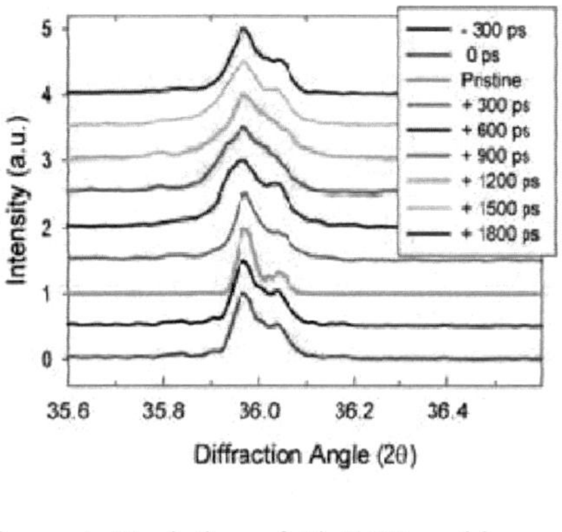


Figure 4: Evolution of Si (111) rocking curve for various time delays, at a fluence of 2.3 J-cm$^{-2}$.

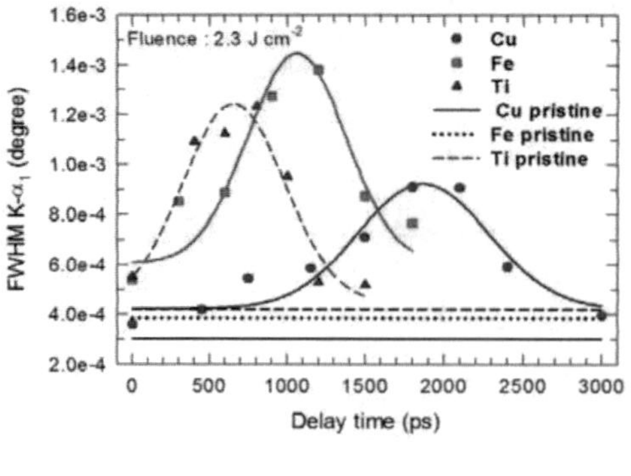


Figure 5: FWHM of $K_{\alpha 1}$ line radiation as function of delay between pump and the probe pulse for Ti, Fe, and Cu.

A quantitative knowledge of the velocity of propagating compression front (shock wave) requires code based on the dynamical diffraction theory applied to the strained and laser-shocked crystal.

Nevertheless, some important features can be realized by approximating the delay time for maximum broadening occurs when shock arrived at depth equals to penetration depth of x-ray probe. The penetration depths for Ti, Fe, and Cu $K_\alpha$ for their respective incident angle are 2.7, 2.9, and 8.3 μm respectively. The mean propagation speed of the laser induce compression wave, deduced from the maximum broadening time of the three different x-ray probe is 12 km/s which is very close to the sound velocity of 9.4 km/s in Si. However, the reduction in intensity of x-ray beam in passage through sample has to be taken into account. Nevertheless, the derived shock velocity values are in broad agreement with those obtained under similar irradiation conditions. Maximum lattice strain was estimated from the shift (up to which diffraction signal appears) of 0.074° which corresponds to a compression of 0.4 %. Hironaka *et al.* [31] observed a maximum lattice strain of 1.05 % corresponding to the maximum pressure of 2.18 GPa, whereas the maximum lattice strain of 0.4 % corresponds to 0.83 GPa (8.3 kbar).

## 4. Future Plans

On the EOS front, the particle velocity is being estimated at present indirectly from the relation $u_s=u_o+\alpha u_p$ by assuming coefficients $u_o$ and $\alpha$ obtained from literature. At present, the laser system is being upgraded to deliver laser pulses with energies up to 200 J in 400 ps at second harmonic (0.5 TW). After a year or so, we will be in position to carry out EOS studies at high energies and intensities. In the time resolved diffraction studies, in future, we have plans to study time resolved x-ray diffraction on shorter time scales (tens of fs). We are involving more and more material science groups (UGC-DAE-CSR, Indore, Materials and Advanced Accelerator Sciences Division of RRCAT, Laser Materials and Device Development Division, RRCAT) for experiments where the time resolved x-ray diffraction technique could be used for carrying out interesting investigations. Isochoric heating, i.e. heating the matter at solid density ($10^{22}$ $cm^{-3}$) to a very high temperature (1 eV to 1keV), in a very short time (< 1 ps) so that the heating is *without change in volume,* is one of the least explored areas relevant to structures of planetary and stellar interiors as well as inertial confinement fusion (also called "Warm Dense Matter"). The fast heating can be achieved using 1) x-rays, 2) hot electrons, or 3) protons generated by ultrashort, ultrahigh intensity lasers. With the 150 TW laser to be installed in our laboratory by Feb. 2012, and with our indigenous 50 TW (25 J in 500 fs) under construction, we should soon be in position to start experiments in this new exciting area of research.

## References

[1] Celliers P M , Collins G W, Da Silva L B, Gold D M, Cauble R, Wallace R J, Foord M E and Hammel B A, 2000 *Phys. Rev. Lett.* **84** 5564
[2] Remington B A, Drake R P, Takabe H and Arnett D, 2000 *Phys. Plasmas.* **7** 1641
[3] Benuzzi A *et. al.,* 2006 *Plasma Phys. & Control. Fusion* **48** B347
[4] Godwal B K, Sikka S K, and Chidambaram R, 1981 *Phys. Rev. Lett.* **47**, 1144
[5] Cauble R *et al.*, 1998 *Phys. Rev. Lett.* **80** 1248
[6] Lübcke A *et. al.*, 2010 *New J. Phys.* **12** 83043
[7] Nicoul M, Shymanovich U, Tarasevitch A, Linde D.V, and Sokolowski-Tinten K 2011 *Appl. Phys. Lett.* **98** 191902
[8] Hawreliak J, Kalantar D H , Stölken J S, Remington B A and Lorenzana H E 2008 *Phys. Rev. B* **78** 220101R
[9] Sokolowski-Tinten K, Blome Ch, Dietrich C, Tarasevitch A, von der Linde D, Horn-von Hoegen M, Cavalleri A, Squier J A and Kammler M 2001 *Phys. Rev. Lett.* **87** 225701
[10] Rousse A *et. al.,* Nature 410, 65 (2001)
[11] Uschmann I, Kämpfer T, Zamponi F, Lübcke A. Zastrau U, Loetzsch R, Höfer S, Morak A, and Förster E 2009 *Appl. Phys. A* **96**, 91
[12] Kruer W L 2003 *The physics of laser–plasma interactions* (Westview, New York)
[13] Reich Ch, Gibbon P, Uschmann I and Förster E 2000 *Phys. Rev. Lett.* **84** 4846
[14] Batani D *et. al.*, 2004 *Phys. Rev. Lett.* **92** 065503.
[15] Batani D *et. al.*, 2002 *Phys. Rev. Lett.* **88** 235502.
[16] Rothman S D, Evans A M, Horsfield C J, Graham P and Thomas B R, 2002 *Phys. Plasmas* **9** 1721.

[17] Benuzzi-Mounaix A, 2002 *Phys. Plasmas* **9** 2466.
[18] Hicks D G, Celliers P M, Collins G W, Eggert J H and Moon S J, 2003 *Phys. Rev. Lett.* **91** 035502
[19] Ozaki N *et. al.,* 2004 *Phys. Plasmas* **11** 1600
[20] Dezulian R *et. al*., 2006 *Phys. Rev. E* **73** 047401.
[21] Chourasia S , Leshma P, Tripathi S, Murali C G, Munda D S, Sharma S M, Kailas S, Gupta N K and Dhareshwar L J 2010 *BARC Newsletter* **317** 13
[22] Peter G and Janez M, 2011 *Optics Letters* **36** 2782
[23] Chen J P, Li R X, Zeng Z N, Wang X T, Wang W Y, Jiang Y H, Cheng C F and Xu Z Z, 2003 *J. App. Phys.* **94** 858
[24] Benuzzi-Mounax A, Koenig M, Boudenne J M, Hall T A, Batani D, Scianitti F, Masini A and Di Santo *D* 1999 *Phys. Rev. E* **60** R2488
[25] Marsh S P 1980 *LASL Shock Hugoniot Data* (Berkeley, CA, Univ. of California Press)
[26] Zeldovich Ya B and Raizer Yu P, 1967 *Physics of Shock Waves and High Temperature Hydrodynamic Phenomena* (Academic Press, New York)
[27] Shukla M, Upadhyay A, Senecha V K, Khare P, Bandyaopadhyay S, Rai V N, Navathe C P, Pant H C, Khan M and Godwal B K 2003 *Laser & Part. Beams* **21** 615
[28] Kalantar D H *et. al*., 1999 *Rev. Sci. Instrum.* **70** 629
[29] Murphy W J *et. al.,* 2010 *J. Phys.: Cond. Matter* 22, 065404
[30] Kishimura H, Morishita H, Okano Y H, Hironaka Y, Kondo K, Nakamura K G and Atou T 2006 *Phys. Rev. B* **74** 224301
[31] Hironaka Y, Yazaki A, Saito F, Nakamura K.G, Kondo K, Takenaka H and Yoshida M 2000 *Appl. Phys. Lett.* **77** 1967

Journal of Physics: Conference Series **377** (2012) 012053 doi:10.1088/1742-6596/377/1/012053

# Effect of very high pressure on life of plants and animals

**F Ono[1,8], Y Mori[1], M Sougawa[1], K Takarabe[1], Y Hada[2], N Nishihira[3], H Motose[4], M Saigusa[4], Y Matsushima[5], D Yamazaki[6], E Ito[6], and N L Saini[7]**
[1]Department of Applied Science, Okayama University of Science, 1-1 Ridaicho, Kitaku, Okayama 700-0005, Japan
[2]Department of Biosphere-Geosphere System Science, Okayama University of Science, 1-1 Ridaicho, Kitaku, Okayama 700-0005, Japan
[3]Okayama Ichinomiya High School, 221 Narazu, Kitaku, Okayama 701-1202, Japan
[4]Department of Biology, Okayama University, 1-3 Tsushima-Naka, Kitaku, Okayama 700-8539, Japan
[5]Department of Physics, Okayama University, 1-3 Tsushima-Naka, Kitaku, Okayama 700-8539, Japan
[6]Institute for Study of Earth's Interior, Okayama University, Misasa, Tottori, 682-0193, Japan
[7]Dipartimento di Fisica, Universita di Roma "La Sapienza", Piazzale Aldo Moro 2, 00185 Rome, Italy

E-mail: fumihisa@das.ous.ac.jp

**Abstract.** We studied the tolerance of living organisms, such as a small animal (Milnesium tardigradum), a small crustacean (*Artemia*), non-vascular plants or moss (*Ptichomitrium* and *Venturiella*), and a vascular plant (*Trifolium*) to the extremely high hydrostatic pressure of 7.5 GPa. It turned out that most of the high pressure exposed seeds of white clover were alive. Those exposed to 7.5 GPa for up to 1 day and seeded on agar germinated roots. Those exposed for up to 1 hour and seeded on soil germinated stems and leaves. Considering the fact that proteins begins to unfold around 0.3 GPa, it seems difficult to understand that all the living samples which have been investigated can survive after exposure to 7.5 GPa.

## 1. Introduction

Effects of hydrostatic pressure on biological samples have mostly been investigated to the order of several hundred MPa. It was reported that proteins began to unfold around this pressure, where most bacteria and many other organisms die [1]. However, under the pressure of 0.6 GPa, which was two times higher than the pressure mentioned above, tardigrades (*Milnesium tardigradum*) were shown to survive [2]. Further extensions of the pressure to a few GPa order were made by Sharma et al. [3] and Margosch et al. [4] on some microbial activities.

On the other hand, using a technique of shock compression, survival of some microorganisms was reported after exposure to very high sock pressure up to several ten GPa for a very short time [5]. This result shows that the mechanism of unfolding of proteins is different from that under static high pressure, or the unfolding does not take place up to such a high pressure when applied for a very short time of the order of $10^{-6}$ s.

[8] To whom any correspondence should be addressed.

Attempts to extend the investigations on mm-size animals and seeds of higher plants under static compression to an extremely high pressure range up to several GPa may produce a new research field, and introduce us some new or unknown phenomena and information. The present authors group have been extending the static pressure range to 7.5 GPa using a cubic anvil press.

## 2. Experimental Procedure

Specimens of living animal, tardigrades and spore cases of mosses, *Ptychonitrium sinense* and *Venturiella sinensis* have been collected from the field close to our university. Some other specimens, eggs of *Artemia* and seeds of *Trifolium lepens L* (white clover) have been purchased from companies. Before high pressure exposure, these living specimens were dehydrated by keeping them in a sealed desiccator of the size of 0.5x0.4x0.7 $m^3$ with silicagel of about 0.5 kg in weight for more than one week. These specimens were put in a small teflon capsule with an inner diameter and length of 1.6 and 1.8 mm, respectively. As the liquid pressure medium, Fluorinate PC77 (perfluorocarbon, $C_8F_{18}$, Sumitomo 3M) was adopted and put into the teflon capsule together with the living specimens. It was proven that soaking in 100% fluorinate has no practical effect on the life of living specimens. The capsule was placed in the center of a cube which was made of pyrophyllite with an edge length of 6.0 mm.

To generate hydrostatic pressure up to 7.5 GPa a cubic anvil press was adopted. A pyrophyllite cube was placed at the centre of six tungsten-carbide anvils with a front edge length of 4.0 mm. These anvils were compressed by a 250-ton press. The pressure was determined by using a calibrated curve [6] of the relation between the press load and the actual pressure established before the experiment. The error of the intensity of pressure was smaller than 0.3 GPa. In the present apparatus, the press load was controlled automatically to keep the intensity of the pressure constant during the operation at the maximum pressure.

The pressure was increased from ambient to the maximum pressure of 7.5 GPa at a rate of 0.3 $GPamin^{-1}$. From our former experiments observing temperature dependence of magnetic susceptibility on increasing and decreasing pressure using the same high pressure equipment [6], it was proven that the temperature of the sample had been kept constant within the relative error of about 1 K. The pressure was kept constant at the maximum pressure for various duration of time from 15 minutes to 6 days and then decreased down to the ambient pressure with the same rate as on increasing. Those high pressure experiments were made at room temperature. Immediately after the pressure was released, the specimens were brought out from the teflon capsule and they were cultivated or seeded on a humid media. Then, their survival, germination and growth rates were investigated using a microscope.

## 3. Results and Discussion

### *3.1 Tardigrades*

A terrestrial tardigrade (*Milnesium tardigradum*) is a small animal with 0.6-0.7 mm in length. This animal is known to show very strong resistivity against extreme environmental conditions at cryptobiotic state, i.e., dehydrated state called "tun". The first experimental search for life under extremely high hydrostatic pressures of several GPa order was made by the present authors' group for this animal [7].

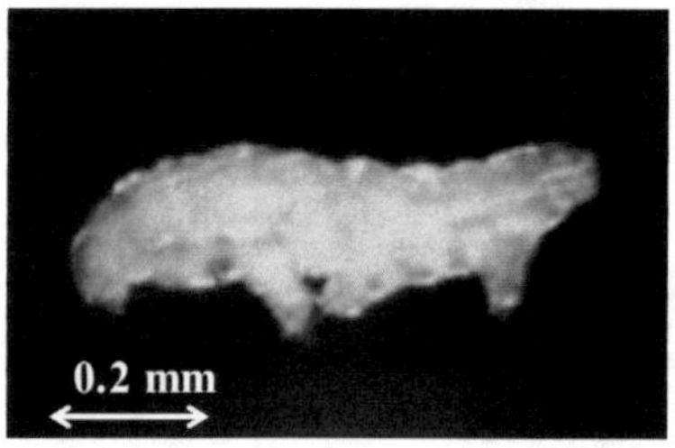


**Figure 1.** One of the 20 tardigrades exposed to 7.5 GPa for 3 hours and brought back to ambient pressure. All of them were found alive and moving around actively.

It was found that tardigrades could survive under 7.5 GPa for up to 13 hours. Figure 1 shows a tardigrade which was exposed to 7.5 GPa for 3 hours and brought back to ambient pressure. This tardigrade was moving around actively. Those tardigrades could live as long as control. Judging from their movements and the length of their life span after the exposure to high pressure, they had no remarkable damages in the body and the organs. It is a very striking result that an animal is still alive after being exposed to such a very high hydrostatic pressure.

*3.2 Artemia* eggs

*Artemia*, a kind of planktons and called brine shrimps, are found worldwide in inland saltwater lakes, but not in oceans. They have a biological life cycle of one year. Their eggs in a dried state, called cryptobiosis with a moisture level of below 10 % are metabolically inactive and can remain alive for so many years as to keep the species without evolution since the Triassic period. They hatch into *Nauplii*, *Artemia* larvae several hours after soaked in seawater. The hatching rate of dried eggs of *Artemia* after exposed to 7.5 GPa for various durations up to 3 days was investigated by the present authors' group [8, 9].

The *Artemia* eggs exposed to 7.5 GPa for up to 48 hours began to hatch around 15-18 hours after soaked in seawater. The hatching rate reached maximum around 40-70 hours after soaked into seawater. Observed hatching ratio (hatched/total number of eggs) of 80-90 % was as high as for control. Hatching to *Nauplius* was observed under a microscope, and no difference was confirmed between the high-pressure exposed group and control. After hatched into *Nauplius*, the high pressure exposed *Artemia* moved around as seen in figure 2 with no practical differences in speed and frequency with control.

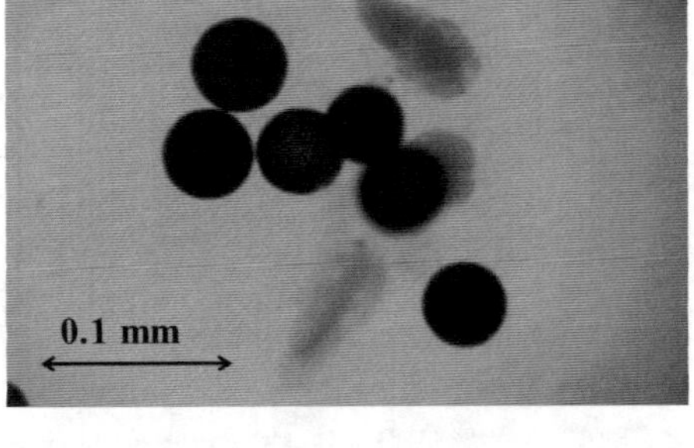


**Figure 2.** *Artemia* exposed to 7.5 GPa for 2 days. Some of the eggs hatched into *Nauplii*. The photograph was taken 24 hours after soaked in sea water.

*3.3 Ptychomitrium*

*Ptychonitrium sinense* belongs to *Grimmiaceae*, and are mostly found on rocks at sunshiny places. In a dried atmosphere, the leaves of *Ptychomitrium* frizzle and their colour turns brown. On the other hand, when they get wet, they become straight and green, again. The spores of *Ptychomitrium* are spherical in shape with a diameter of about 20 μm. A spore case of *Ptychomitrium* contains tens of thousands of spores. Once spores are dried, they become resistant against extreme environmental conditions and remain alive for a long period. When they are seeded on a humid media, they begin to germinate within one week.

The investigation was extended to *Ptychomitrium* [10], and the germination ratio (the ratio of the number of germinated spores against the total number of the spores seeded on an agar) was investigated after exposing its spores to 7.5 GPa for various durations up to 6 days. Figure 3 shows *Ptychomitrium* spores exposed to 7.5 GPa for 6 days, and seeded on agar. In this photograph, germination of the spore is evident. However, no frizz in germinated protonema was seen in this figure. It is the characteristics of *Ptychomitrium* that their protonema frizzle when they are dried to the same extent as in the present case of being incubated on an agar for three weeks. After being exposed to 7.5 GPa for such a long time as 6 days, the mechanism of the characteristics of frizz has been distracted.

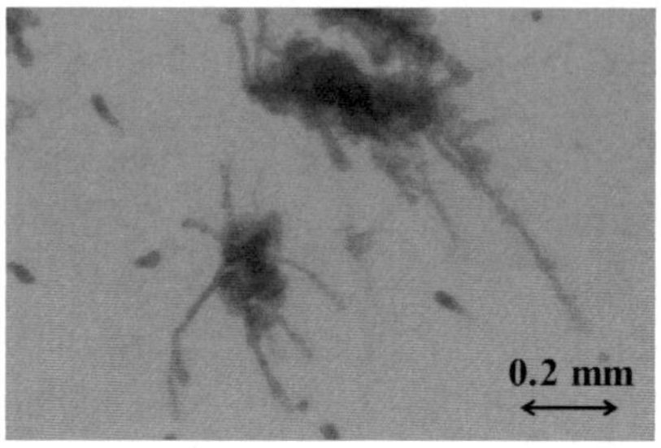


**Figure 3.** *Ptychomitrium* exposed to 7.5 GPa for 6 days. The photograph was taken 3 weeks after seeded on agar.

*3.4 Venturiella*

*Venturiella sinensis* belongs to *Erpodiaceae*, being one of the popular mosses seen in moderate climate areas such as Japan, Korea and China. The spores of *Venturiella* are spherical in shape with a diameter of about 20 μm. A spore case, being in the size of 0.5 mm in diameter and 1 mm in length contains tens of thousands of spores. Once spores are put in a dried atmosphere, they become resistant against extreme environmental conditions and remain alive for long years. When they are seeded on a humid media, they begin to germinate within one week.

The high pressure experiment was further extended to this moss, and germination ratio was investigated after exposing its spores to 7.5 GPa for various durations up to 6 days. One week after seeded, both the spores of high pressure exposed group up to three days and control began to germinate. The germination ratio of these high-pressure exposed groups were 70-90%, which is as high as control [11].

After being incubated for five months, those spores exposed to 7.5 GPa for up to 3 days grew up to have green leaves. On the other hand, those spores exposed to 7.5GPa for 6 days, only 4 individuals in a hundred were germinated. To count the numbers of the germinated spores, we had seeded only about a hundred spores in tens of thousands in the spore case exposed to 7.5 GPa on an agar. One of the germinated spores is shown in figure 4. Considering that tardigrades can stand against 7.5 GPa for up to 13 hours, the pressure tolerance of moss is about one order stronger than that of tardigrades [7].

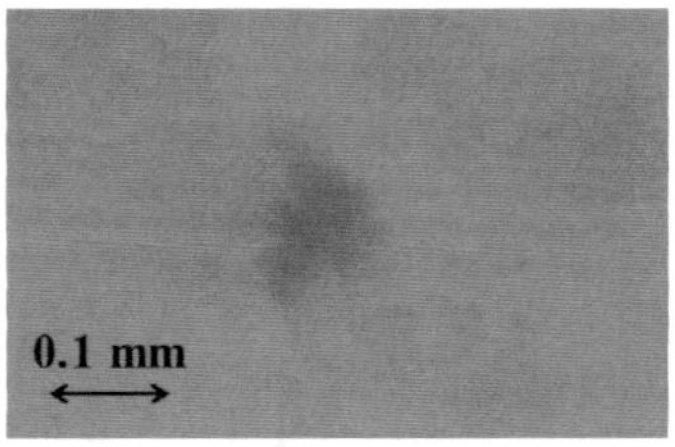


**Figure 4.** One of the four germinated *Venturiella* in about a hundred spores exposed to 7.5 GPa for 6 days. The photograph was taken four weeks after seeded. This agar was wrapped by a transparent plastic sheet so as to avoid contamination during the observation and incubation.

*3.5 Trifolium lepens L*

The experiment has been extended, further, to a vascular plant, white clover. White clover, *Trifolium lepens L* is distributed worldwide and often introduced as a pasture crop. Since white clovers contain proteins and are so abundant worldwide, they are considered to be a survival food. The high pressure study has been extended to this vascular plant.

A few seeds of white clover were exposed to 7.5 GPa for various durations from 30 minutes to 6 days, and then, seeded on agar or directly on sowing soil in flower pods. It was found that they were tolerable against 7.5 GPa for only one hour to germinate stems with trifoliate leaves [12]. However, they were tolerable for 1 day to germinate roots. Figure 5 (a) shows white clovers exposed to 7.5GPa for 30 minutes and (b) for 1 hour. These photographs were taken six and a half months after seeded. Eight months after seeded, a flower came out from the group exposed to 7.5 GPa for 30 minutes as seen in figure 6.

(a) (b)

**Figure 5.** (a) White clovers exposed to 7.5GPa for 30 minutes, and (b) for 1 hour and seeded on sowing soil. The photographs were taken six and a half months after seeded.

**Figure 6.** Eight months after seeded, a flower came out from the white clover seed exposed to 7.5 GPa for 30 minutes.

A close investigation was made under a microscope to examin damages in the cells of roots, stems and leafs after the exposure to the very high pressure. As seen in figure 7, no differences were observed between the high pressure exposed group and control. However, the average length of the roots of the clover exposed to 7.5 GPa for 1 hour was about a half of those exposed for 30 minutes.

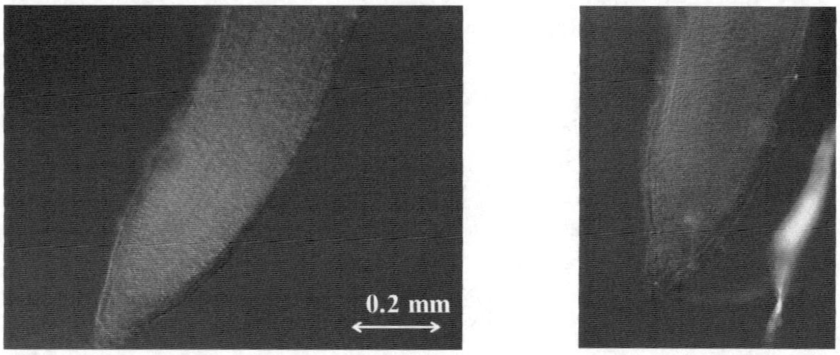


**Figure 7.** Light-microscopy images of roots of white clover exposed to 7.5 GPa for 30 minutes (left) and for 1 hour (right) after treated to be transparent. No differences of the cells can be seen, except for the larger diameter and the shorter length for the high pressure exposed group for longer time (right).

The present investigation may give the possibility that very small creatures can travel through the space in a large meteorite, and reach the earth from other planet or galaxy. It may also give a support to the recent report on findings of some DNA building blocks in a meteorite by Callahan et al. [13].

## 4. Conclusion

Judging from the fact that all the living creatures we have examined, both animals (tardigrades and *Artemia*) and plants (*Ptychomitrium*, *Venturiella* and white clover) were alive after exposure to 7.5 GPa, it was suggested that most of the proteins of those creatures which unfolded at the early stage of the compression remain principally unchanged after exposure to the very high pressure of 7.5 GPa. It was also suggested that the unfolding of proteins was completely reversible up to 7.5 GPa.

The maximum hydrostatic pressure applied to all the living specimens investigated in the present experiments corresponds to that in the upper mantle, at the depth of 180 km from the surface of the earth. Considering only the magnitude of the pressure and imagining if there would be some way to escape from being heated up to such a high temperature as in the earth's interior, it seems possible for a very small creature to travel through space in a large meteorite made of some sedimentary rock.

## 5. Acknowledgment

This paper presents a result of a joint research program carried out at the Institute for Study of the Earth's Interior, Okayama University.

## 6. References

[1] Silva J L, Luan P, Glaser M, Voss E W and Weber G 1992 *J. Virol.* **66** 2111
[2] Seki K and Toyoshima M 1998 *Nature* **395** 853
[3] Sharma A, Scott J H, Cody G D, Fogel M L, Hazen R M, Hemley R J and Huntress W T 2006 *Nature* **295** 1514
[4] Margosch D, Ehrmann M A, Buckow R, Heinz V, Vogel R F and Gänzle M G 2006 *Appl. Environ. Microbiol.* **72**, 3476
[5] Horneck G, Stoeffler D, Ott S, Hrnemann U, Cockell C S, Moeller R, Meyer C, De Vera J P, Fritz J, Schade S, Artemieva N A 2008 *Astrobiology* **8** 17
[6] Matsushita M, Miyoshi M, Endo S and Ono F 2005 *Phys. Rev.* **B 72** 214404
[7] Ono F, Saigusa M, Uozumi T, Matsushima Y, Ikeda H, Saini N L and Yamashita M 2008 *J. Phys. Chem. Sol.* **69** 2297
[8] Ono F, Minami K, Saigusa M, Matsushima Y, Mori Y, Takarabe K, Saini N L and Yamashita M 2010 *J. Phys. Chem. Solids* **71**, 1127
[9] Minami K, Ono F, Mori Y, Takarabe K, Saigusa M, Matsushima Y, Saini N L and Yamashita M 2010 *J. Phys. Conf. Series* **215** 012164
[10] Nishihira N, Shindou A, Saigusa M, Ono F, Matsushima Y, Mori Y, Takarabe K, Saini N L and Yamashita M 2010 *J. Phys. Chem. Solids* **71** 1123
[11] Ono F, Mori Y, Takarabe K, Nishihira N, Shindo A, Saigusa M, Matsushima Y, Saini N L and Yamashita M 2010 *J. Phys. Conf. Series* **215** 012165
[12] Nishihira N, Iwasaki T, Shinpou R, Hara A, Ono F, Hada Y, Mori Y, Takarabe K, Saigusa M, Matsushima Y, Saini N L and Yamashita M 2011 *SMEC2011*, presented and to be published in *J. Appl. Phys.*
[13] Callahan M P, Smith K E, James Cleaves II H, Ruzicka J, Stern J C, Glavin D P, House C H and Dworkin J P 2011 *Proc. National Acad. Sci. USA* Aug. 11 Early Edition 13995

Journal of Physics: Conference Series **377** (2012) 012054 doi:10.1088/1742-6596/377/1/012054

# Microbial activity in deep marine sediments: does pressure make the difference?

**Aude Picard and Timothy G. Ferdelman**

Max Planck Institute for Marine Microbiology, Biogeochemistry Department, Celsiusstr.1, 28359 Bremen, Germany

MARUM, Center for Marine Environmental Sciences, Bremen, Germany

E-mail: aude.picard@uni-tuebingen.de

**Abstract.** We attempted to evaluate the effects of high hydrostatic pressure on microbial heterotrophic activity in deep marine sediments from the Atlantic Ocean. We investigated the potential respiration rates (acetate/glucose oxidation to $CO_2$) in oxic sediments recovered from up to ~4500 m water depth. Incubations were performed at ambient pressure and at near in situ pressure (~40-45 MPa) with sediments stored at ambient pressure and at in situ pressure. Potential respiration rates in sediments stored at ambient pressure were lower when measured at in situ pressure than when measured at ambient pressure, independently of the substrate used. It appears that the pressure of storage is critical since potential respiration rates of sediments stored at in situ pressure were higher than in the counterpart sediments stored at ambient pressure.

## 1. Introduction

Oceans cover 2/3 of the Earth's surface; marine environments including the deep-sea, the sub-seafloor sediments and the basaltic crust account for the largest high-pressure (HP) habitat for microorganisms. Microbiologists began to consider the effects of HP in sediments as early as the end of the 19$^{th}$ century [1]. More than half a century later, Zobell and Johnson characterized microorganisms that grow and metabolize optimally at HP as "barophilic" [2]. Yayanos reported in 1979 the isolation of the first bacterium in pure culture growing better at HP [3], and nowadays coined "piezophilic" [4]. Today, it is clear that piezosensitive and piezophilic microorganisms share the same HP habitats, however the relative proportion and activity of each group are still unknown [5]. In the last decades it was shown that microbial activity rates in deep-sea water samples are usually higher when measured at in situ pressure conditions [6 and references therein]. Some similar observations have been made in sediments [e.g. 7], however the effects of HP on microbial activity in sediments, and more generally on microbial activity important for biogeochemical cycles, have not been extensively studied.

In this communication, we present new data on the effects of the incubation and storage pressure on microbial heterotrophic activity in deep-sea oxic sub-seafloor sediments recovered in the Atlantic Ocean.

## 2. Material and methods

### 2.1. Site description and sediment collection

We visited the North Pond area in the Atlantic Ocean (23°N) during cruise MSM 11/1 on R/V Maria S. Merian in 2009 [8]. Sediment for microbial activity measurement was sampled aerobically and stored at 4°C. Samples used in this study and their incubation/storage conditions are listed in Table 1.

**Table 1**. List of samples used in this study and their pressure conditions of storage and incubation.

| Station (water depth) | Sampling depth (cm below seafloor) | Storage pressure at 4°C (MPa) | Incubation pressure at 4°C (MPa) |
|---|---|---|---|
| GeoB 13501 (4480 m) | 42.5, 105.5, 201.5, 301.5, 401.5, 501.5, 601.5, 701.5, 801.5 | 0.1 | 0.1/45 |
| GeoB 13506 (4143 m) | 57.5, 131.5, 231.5, 331.5, 431.5, 531.5 | 0.1 | 0.1/45 |
| GeoB 13502 | 157 | 40 | 0.1/40 |
| (4250 m) | 204.5 | 0.1 | 0.1/40 |

### 2.2. High-pressure incubation

Sediment samples were stored and incubated at 4°C in steel pressure vessels similar to those described by Zobell and Openheimer [9]. Sediment slurries were contained in plastic syringes or in small hungate tubes without headspace and pressurized in water using a multifluid hand pump (Enerpac MP-1000).

### 2.3. Microbial heterotrophic activity measurements

Microbial heterotrophic activity was defined here as the oxidation of $^{14}C$-acetate or $^{14}C$-glucose to $^{14}CO_2$ in oxic conditions. Briefly, $^{14}C$-acetate or $^{14}C$-glucose was added to the sediment slurries. Incubation was performed at 4°C at ambient pressure or close to the in situ pressure (40-45 MPa). At 3 different time points, the amount of $^{14}CO_2$ was measured and potential respiration rates were calculated. The complete procedure was described elsewhere [8].

## 3. Results

### 3.1. Effects of high pressure on microbial heterotrophic activity profiles in oxic sediments

Potential respiration rates were evaluated in two sediment cores, GeoB 13501 and GeoB13506, recovered at water depths of 4480 and 4143 m, respectively. Sediment samples from GeoB 13501 were amended with $^{14}C$-glucose and $^{14}C$-acetate and incubated at 4°C at ambient pressure and at near in situ pressure (45 MPa). Potential respiration rates, shown in Fig. 1, are comparable with both substrates (<0.1 nmol organic compound respired $cm^{-3}$ $d^{-1}$), although a bit lower using glucose. With glucose, rates measured at high pressure are all lower than rates measured at 0.1 MPa (Fig. 1A and 1B). Using acetate, rates measured at high pressure are in the same range as those measured at atmospheric pressure, although rates appeared to be slightly enhanced during the 8-day incubation under pressure (Fig. 1C and 1D).

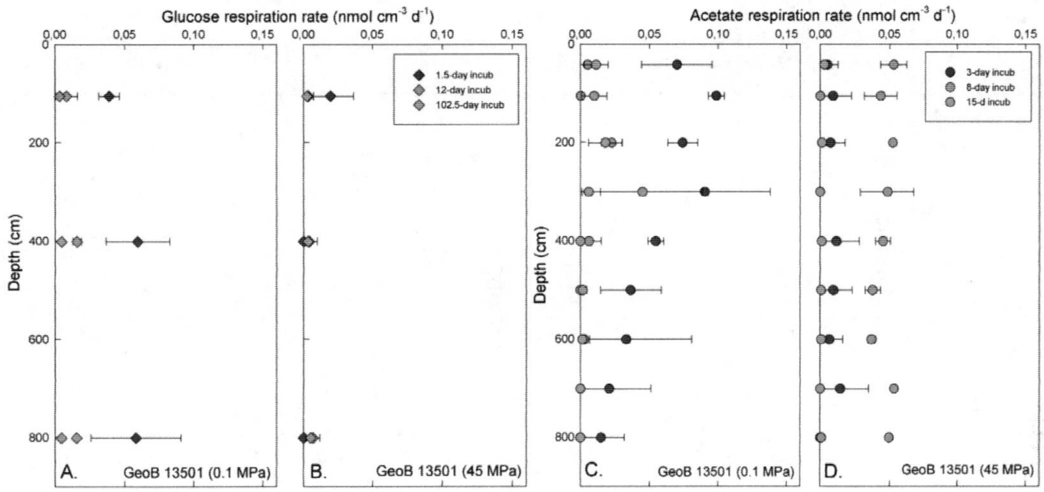


**Figure 1.** Microbial activity profiles at Station GeoB 13501. Glucose respiration rates measured at 0.1 MPa (A) and 45 MPa (B). Acetate respiration rates measured at 0.1 MPa (C) and 45 MPa (D).

At Station GeoB13506, rates are clearly lower at high pressure at all incubation times (Fig. 2).

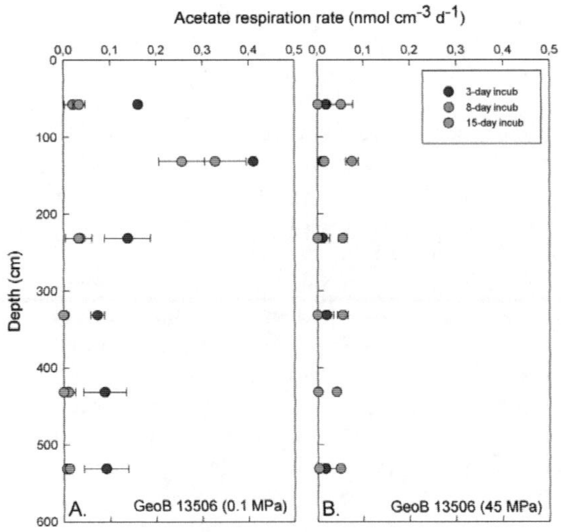


**Figure 2.** Microbial activity profiles measured at 0.1 MPa (A) and 45 MPa (B) at Station GeoB 13506.

### 3.2. Effects of storage pressure conditions on microbial activity in sediments

One sediment sample was selected in Core GeoB 13502 (157 cm below seafloor) and was pressurized at 40 MPa shortly after sampling. The pressure vessel was stored at 4°C and pressure was checked regularly for ~8 months until the experiment was performed. Since it was not possible to obtain two samples from the same depth, we used the sample taken at 204.5 cm in a similar lithological horizon as a reference stored at 0.1 MPa. The activity rates were evaluated with acetate in the two samples and incubations were performed both at 0.1 and 40 MPa (Fig. 3). Activity was apparently increased in the sample that has been stored under pressure, independently of the pressure of incubation.

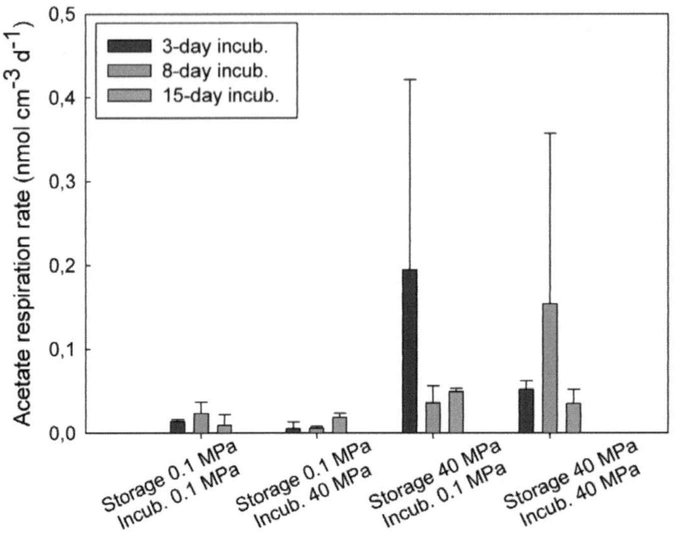


**Figure 3.** Acetate respiration rates measured in samples from GeoB 13502. The sample stored at 0.1 MPa was taken at 204.5 cm depth below seafloor, while the sample stored at 40 MPa was taken at 157 cm below seafloor. Both samples are of similar lithology.

## 4. Discussion

We show here that pressure inhibits microbial activity in deep-sea sediments recovered and stored at ambient pressure. Our results suggest also that microbial activity is preserved in sediments that have been stored at in situ pressure. How pressure affects microbial activity in sediments is not clearly established yet. Ideally, microbial activity has to be measured as soon as possible after the recovery of the sediment samples. However, in practice, an important amount of sediment samples are stored for several weeks to several months (especially when the samples have to be shipped to the laboratory via container) until performing the experiments. We suggest here that quickly restoring the in situ pressure after the recovery of deep-sea sediments could maintain the initial metabolic potential of the microbial community. Further technical developments would be necessary to sample, store and incubate deep-sea sub-seafloor sediment samples without any decompression. In conclusion, more studies on deep environmental samples are needed to draw a general conclusion. Moreover studies on the effects of high pressure on "simple" systems, such as microbial isolates from deep environments, would also greatly improve the understanding on the effects of high pressure on more complex systems, such as sub-seafloor sediments. The field of "high-pressure geomicrobiology" is obviously still in its infancy.

## Acknowledgments

The authors would like to thank the captain Klaus Bergmann and the crew of the R/V Maria S. Merian; chief scientist Heiner Villinger, and all participants of the MSM 11/1 cruise. A.P. was funded through DFG-Research Center/Excellence Cluster "The Ocean in the Earth System" and the Max Planck Society. Cruise MSM11/1 was funded by the German Science Foundation (DFG).

**References**

[1] Certes A 1884 *C.R. Hebd. Acad. Sci.* **99** 385
[2] Zobell CE and Johnson FH 1949 *J. Bact.* **57** 179
[3] Yayanos AA, Dietz AS and Van Boxtel R 1979 *Science* **205** 808
[4] Yayanos AA 1995 *Annu. Rev. Microbiol.* **49** 777
[5] Fang J, Zhang L and Bazylinski DA 2010 *Trends Microb.* **18** 413
[6] Nagata T, Tamburini C, Aristegui J, Baltar F, Bochdansky AB, Fonda-Umani S, Fukuda H, Gogou A, Hansell DA, Hansman RL, Herndl GJ, Panagiotopoulos C, Reinthaler T, Sohrin R, Verdugo P, Yamada N, Yamashita Y, Yokokawa T, Bartlett DH 2010 *Deep-sea Res. II* **57** 1519
[7] Kallmeyer J, Ferdelman TG, Jansen KH and Jorgensen BB 2003 *J. Microbiol. Methods* **55** 165
[8] Picard A and Ferdelman TG 2011 *Front. Microbio.* **2** 263
[9] Zobell CE and Oppenheimer CH 1950 *J. Bacteriol.* **60** 771

# Germination of vegetable seeds exposed to very high pressure

**Y Mori**[1], **S Yokota**[1] **and F Ono**[1,2]
[1]Department of Applied Science, Okayama University of Science, 1-1 Ridaicho, Kitaku, Okayama 700-0005, Japan

E-mail: fumihisa@das.ous.ac.jp

**Abstract.** Effects of high hydrostatic pressure were investigated on vegetable seeds in the GPa range to examine the potentialities of breed improvement by high-pressure processing. Specimens of several seeds of broccoli (*Brassica oleracea* var. *italica*), Turnip leaf (*Brassica rapa* var. *perviridis*) and Potherb Mustard (*Brassica rapa var. nipposinica*) were put in a teflon capsule with liquid high pressure medium, fluorinate, and inserted into a pyrophillite cube. By using a cubic anvil press a hydrostatic pressure of 5.5 GPa was applied to these seeds for 15 minutes. After being brought back to ambient pressure, they were seeded on humid soil in a plant pot. Many of these vegetable seeds began to germinate within 6 days after seeded.

## 1. Introduction

Concerning the biological studies under high pressures, the techniques for generation of high pressure have firstly been applied industrially to the field of food science [1, 2]. In these studies the pressure was extended to the order of several hundred MPa, where most bacteria and many other organisms die. The high pressure denaturation techniques have been utilized to preserve foods for longer times. The effects of high pressure on living things have also been intensively investigated in the field of deep sea biology [3], where many new species of bacteria and planktons are being found.

On the other hand, there are rather few biological studies under high pressures to the order of several GPa. A several experiments have been conducted by Sharma et al. [4], Margosch et al. [5], and Vanlint et al. [6], where the pressures were extended to a few GPa. The present author's group have been extended the presdsure further to the order of several GPa [7-12]. These experiments have been conducted from bacteria to small animals and plants. It was shown that both small animals (such as tardigrades [7] and *Artemia* [8, 9]), and mosses (*Ptychomitrium* [10] and *Venturiella* [11]) in their cryptobiotic state, can survive under high hydrostatic pressure of 7.5 GPa. Apart from elevated temperature, this pressure corresponds to that in the upper mantle, at the depth of 180 km below the earth's surface. From these series of studies, it becomes clear that the smaller the size of animals and plants is, the stronger the tolerance would be against high pressure.

The study of biological samples under very high pressures could contribute to the development of agricultural applications for the selection of very strong species of vegetables and fruits which can tolerate severe environmental conditions. For the first step to turn the research to such direction, we selected white clover, *Trifolium lepens L* [12], and found that such a vascular plant could tolerate against 7.5 GPa. Furthermore, it was also shown that though high pressure treated white clovers

[2] To whom any correspondence should be addressed.

germinated and grew up more slowly than control (not exposed to high pressure), they have stronger resistance and living longer than control.

Considering a possibility for application of our techniques, these experiments have been further extended to some popular winter vegetables to look for improvement of breed.

## 2. Experimental Procedure

Specimens of seeds of broccoli (*Brassica oleracea* var. *italica*), Turnip leaf (*Brassica rapa* var. *perviridis* (Komatsuna)) and Potherb Mustard (*Brassica rapa var. nipposinica* (Mizuna)) were purchased from Takii Seed co., Japan. They were dehydrated by keeping them in a low humidity container of about 20 % for a week. Then, they were put in a teflon capsule with an inner diameter and length of 5.0 and 5.0 mm, respectively. As the liquid pressure medium, Fluorinate PC77 (perfluorocarbon, $C_8F_{18}$, Sumitomo 3M) was adopted and put into the teflon capsule together with 10-20 seeds of the three kinds of vegetables. It was proven that soaking in 100% fluorinate has no practical effect on the life of tardigrades [7], *Artemia* [8, 9], mosses [10, 11] and white clover [12]. The capsule was put in the center of a cube which was made of pyrophyllite with an edge length of 19.0 mm.

To generate hydrostatic pressure up to 5.5 GPa a cubic anvil press was adopted. The pyrophyllite cube was compressed by six tungsten-carbide anvils with a front edge length of 15.0 mm. These anvils were compressed by a 1500-ton press. The pressure was determined by using a calibrated curve of the relation between the press load and the actual pressure established before the experiment. The error of the intensity of pressure was smaller than 0.3 GPa. In this apparatus, the press load was controlled automatically to keep the intensity of the pressure constant during the operation at the maximum pressure.

The pressure was increased from ambient to the maximum pressure of 5.5 GPa at a rate of 0.14 GPa$min^{-1}$. The pressure was kept constant at the maximum pressure for 15 minutes, and decreased down to the ambient pressure with the same rate as on increasing. Those high pressure experiments were made at room temperature. Figure 1 shows the pyrophyllite cube and the teflon capsule after exposed to the maximum pressure. Immediately after the pressure was released, the seeds were brought out from the teflon capsule and seeded on a humid sowing soil on a tray. The tray was put on the interior of a room by the side of a bright window, and sometimes brought outside on a fine day in winter.

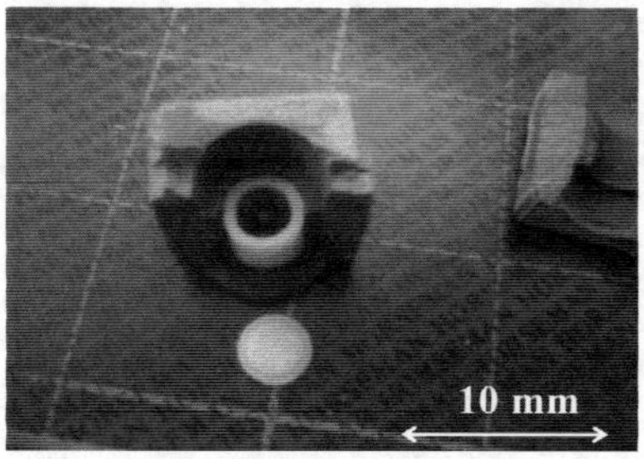


**Figure 1.** A photograph of the pyrophyllite cube, the Teflon capsule with the seeds of Turnip leaf after exposed to 5.5 GPa for 15 minutes.

## 3. Results and Discussion

A photograph of the tray is shown in figure 2. This photograph was taken two weeks after seeded. It is evident that most of the seeds exposed to high pressure were alive and germinated. Furthermore, the germination rate of the three kinds of vegetables exposed to high pressure was only a little smaller than the controls.

Observed percentages of the germinated seeds for high pressure exposed broccoli, Turnip leaf and Potherb Mustard were 40%, 58% and 83%, respectively, while 58%, 83% and 100% for control, respectively. These germination ratios were almost as large as 80 % as indicated by Takii Seed co., except for broccoli. Though the germination ratio of the high pressure exposed seeds were about 20-

30% smaller than control, it was proven that all the three vegetables can tolerate the high pressure of 5.5 GPa.

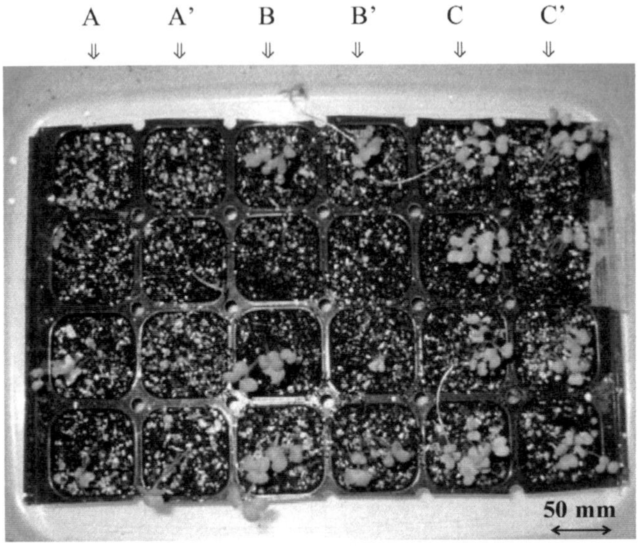


**Figure 2.** A photograph of sowing soil tray, two weeks after seeded. (A) broccoli and (A') control, (B) Turnip leaf and (B') control, (C) Potherb Mustard and (C') control.

**Figure 3.** Broccolis after transferred onto bigger size pots; (a) high pressure exposed group and (b) control.

The germinated vegetables were transferred from the sowing soil tray onto bigger size pots on the day two weeks after seeded. Broccolis just after the transfer were shown in figure 3. A photograph of high pressure exposed Potherb Mustard incubated for two months is shown in figure 4, where some grown up leaves are clearly seen.

The average height observed for germinated Potherb Mustard for both high pressure exposed and control was plotted in figure 5 as a function of elapsed days after seeded. In this figure it is seen that the leaves germinated from the high pressure exposed seeds became a little taller than control. The number of the samples shown in figure 5 was 20, and this number is the statistical basis of the present investigation.

Considering the fact that diamond powders industrially used are synthesized under a pressure around 5.5 GPa and at high temperature [13], where graphite changes into diamond structure, it seems difficult, even only counting the pressure, to understand that all of these three winter vegetables survive after being exposed to such a high pressure of 5.5 GPa.

**Figure 4.** Germinated leaves of Potherb Mustard exposed to 5.5 GPa for 15 minutes and incubated for two months.

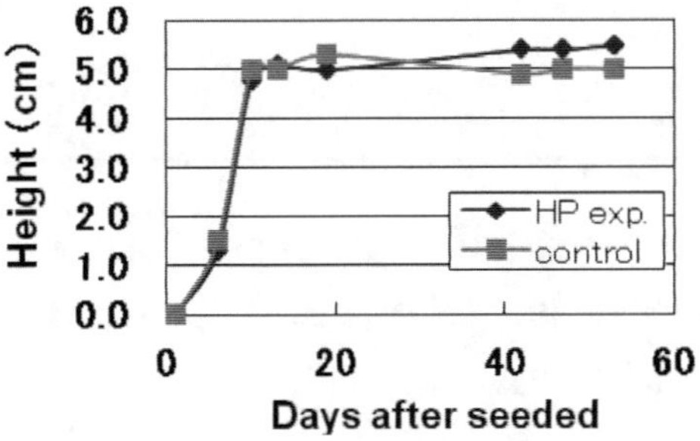


**Figure 5.** Average height of high pressure exposed Potherb Mustard as a function of elapsed days after seeded.

## 4. Conclusions

It was found from the present study that all the three kinds of seeds of typical winter vegetables, broccoli (*Brassica oleracea* var. *italica*), Turnip leaf (*Brassica rapa* var. *perviridis*) and Potherb Mustard (*Brassica rapa var. nipposinica*) germinated after exposed to 5.5 GPa for 15 minutes. Observed percentage of the germinated seeds was 20-30% smaller than control, but they grew up as rapidly and tall as control. The results of the present study may have a possibility to look for a strong species that stands against severe environmental conditions.

## 5. References

[1] Hatakeyama I, Kanno T, Yoshiyama and K, Fujio Y 1994 *J. Food Sci.* **59** 164
[2] Reddy N R, Solomon H M, Tetzloff R C, Rhodehamel E J, Balasubramanian V M and Palaniappan S 2003 *J. Food Prot.* **66** 1402
[3] Kato C, Li L, Nogi Y, Nakamura Y, Tamaoka J and Horikoshi K 1998 *Appl. Environ. Microbiol.* **64** 1510
[4] Sharma A, Scott J H, Cody G D, Fogel M L, Hazen R M, Hemley R J and Huntress W T 2006 *Nature* **295** 1514
[5] Margosch D, Ehrmann M A, Buckow R, Heinz V, Vogel R F and Gänzle M G 2006 *Appl. Environ. Microbiol.* **72**, 3476
[6] Vanlint D, Mitchell R, Bailey E, Meersman F, McMillan P F, Michiels C W and Aertsen A 2011 *mBio* **2** e00130
[7] Ono F, Saigusa M, Uozumi T, Matsushima Y, Ikeda H, Saini N L and Yamashita M 2008 *J. Phys. Chem. Sol.* **69** 2297
[8] Ono F, Minami K, Saigusa M, Matsushima Y, Mori Y, Takarabe K, Saini N L and Yamashita M 2010 *J. Phys. Chem. Solids* **71**, 1127
[9] Minami K, Ono F, Mori Y, Takarabe K, Saigusa M, Matsushima Y, Saini N L and Yamashita M 2010 *J. Phys. Conf. Series* **215** 012164
[10] Nishihira N, Shindou A, Saigusa M, Ono F, Matsushima Y, Mori Y, Takarabe K, Saini N L and Yamashita M 2010 *J. Phys. Chem. Solids* **71** 1123
[11] Ono F, Mori Y, Takarabe K, Nishihira N, Shindo A, Saigusa M, Matsushima Y, Saini N L and Yamashita M 2010 *J. Phys. Conf. Series* **215** 012165
[12] Nishihira N, Iwasaki T, Shinpou R, Hara A, Ono F, Hada Y, Mori Y, Takarabe K, Saigusa M, Matsushima Y, Saini N L and Yamashita M 2011 *SMEC2011*, presented and to be published in *J. Appl. Phys.*
[13] Bundy F P, Hall H T, Strong H M and Wentorf R H 1955 *Nature* **176** 51

# Electrical Properties of amphiboles from the Kola super deep borehole, Russia, at mantle pressure and temperature conditions

**G.Parthasarathy[1] and Felix Gorbatsevich[2]**
[1] CSIR-National Geophysical Research Institute, Uppal Road , Hyderabad, AP.500007, India.
[2] Geological Institute, Kola Science Centre, Russian Academy of Science, Apatity, Murmansk Region, 184209, Russia

E-mail: drg.parthasarathy@gmail.com

**Abstract.** We present here for the first time, the temperature and pressure dependence of the electrical resistivity of ferro-actinolite amphibole from the World's deepest borehole, the Kola super-deep borehole (KSDB) up to the depth of 12.84 km. High-pressure and high-temperature measurements were carried out by using an opposed anvil system up to 700 K and 4 GPa. Ten samples from different depths (from the surface to the deepest 12,890 m ) were investigated for their compositional, structural and electrical behaviour. Ferro-actinolite sample exhibits a thermally induced phase transformation to cummingtonite-clinopyroxene—quartz assemblages at 780 K and 0.5 GPa. The transformation temperature is found to be decreasing with the increase of pressure and reaches a value of 680 K at 6.0 GPa. The conductivity activation energy, (determined from the temperature dependence of conductivity) of the ferroactinolite is found to be decreasing from 0.8 eV at 0.5 GPa to 0.30 eV at 6 GPa. The present study indicates that the amphiboles at deeper crustal level is more conducting and bears significance in understanding the physical properties of rocks at the deep continental crust.

## 1. Introduction

High-pressure studies on the amphiboles attracted much attention among the earth and planetary scientists because of the role of dehydration of amphiboles in the volcanic activities. [1] There are few volcanic activities such as in Muriah, Indonesia, may be related to the pressure induced dehydration of K-amphibole [1] The Kola Super deep bore hole (KSDB) the world's deepest bore hole is of 12.899 km depth.) is located in the north-eastern part of the Kola Peninsula (north-western Russia) where crystalline basement of Precambrian Baltic shield is exposed (Figure.1) . There are two main geological units have been revealed by drilling. Alternating sequence of Proterozoic metasedimentary and metamagmatic rocks composes the upper part of the section (0-6840 m). Mainly biotite-plagioclase gneisses intercalated with amphibolites represent the lower (Archean) part [2]. Core samples from the depth of 560 m, 4673 m, 9190 m, 12148 m were collected and characterised by using X-ray fluorescence , electron microprobe , and X- ray diffraction techniques. The detailed

[1] To whom any correspondence should be addressed.

electrical properties of these amphiboles from the deep borehole (KSDB-3) are of great importance in understanding the continental crustal evolution , and P-T-conditions of crystallization of the complex formation. To the best of our knowledge there are no previous reports on the high-pressure electrical properties of the amphiboles from this borehole.

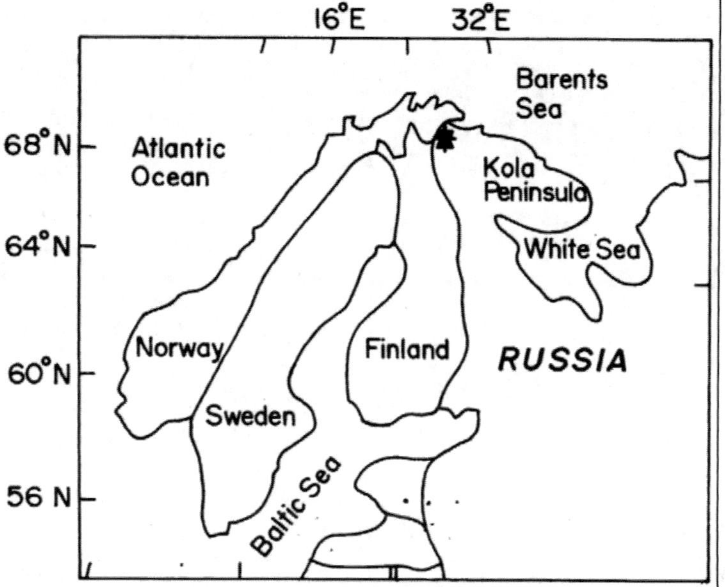


Figure.1 Location Map of the Kola super deep borehole, the world's deepest borehole is located in the north-eastern part of the Baltic shield, and adjacent area in the Kola peninsula, Russia. The symbol* indicates the exact location of the borehole.

## 2. Experimental Methods :

High-pressure and high-temperature electrical resistivity measurements were carried out on well characterized amphiboles samples, by using opposed anvil cell technique and solid pressure transmitting medium. We used internal pressure standard of high-purity bismuth, which exhibit three pressure induced solid-solid phase transitions at 2.5, 2.7 and 7.4 G Pa respectively [3]. In the pressure range lower than 2.5 GPa, manganin was used as a calibrant. In addition to bismuth, pressure-induced solid-solid phase transitions in thallium and tellurium were also used as a fixed points [4]. The typical uncertainty involved in the pressure measurements is 5 %. Chromel wires used as heating elements. The maximum temperature can be reached is typically 900 K. Temperature measurements were made by K-type thermocouples, and typical uncertainty involved is ± 5 K in the temperature range of 500-800 K [4]. While calculating the electrical resistivity of sample at high-pressures, the effect of pressure on the dimensions of the sample was ignored as it is estimated to be with in the experimental uncertainties [5,6].

## 3. Results and Discussions

Geological settings, mineralogical and geochemical data on the selected samples are discussed elsewhere[ 7,8]. The electrical resistivity of actinolite samples from the KSDB obeys the Arrehenius behaviour for the entire temperature range of the present investigations , which is expressed by the Arrehenius equation $\rho=\rho_0 \exp(\Delta E/kT)$, where $\rho$ is the electrical resistivity in ohm* m ; $\rho_0$ = pre exponential factor; $\Delta E$= Activation energy in eV; k= Boltzmann's constant and T= temperature in K. The activation energy for electronic conduction $\Delta E$ has been found to be decreasing from 0.8 eV at room pressure to 0.30 eV at 6 GPa. At room pressure the actinolite undergoes a thermally induced phase transformation from ferro-actinolite to cummingtonite-clinopyroxene-quartz and water assemblages at 780 ± 4 K.. The high temperature product of actinolite consisting of cummingtonite has been characterised by the Fourier transform infrared spectroscopy, in the frequency range 3600 to 3700 cm $^{-1}$. and the powder XRD methods.

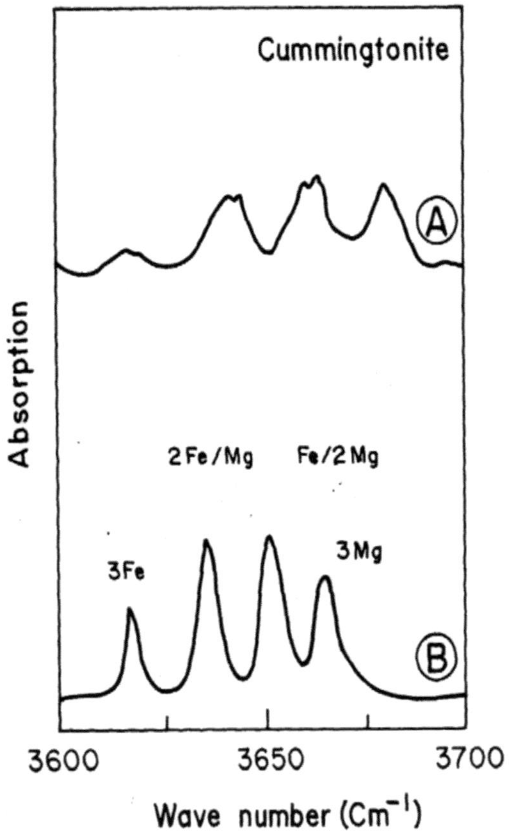


**Figure2**. Fourier transform infrared spectra of the Kola superdeep sample (A) and the surface analogue of cummingtonite sample [ref 3] confirming the actinolite sample transforms to cummingtonite phase at high-temperature. The vibrational mode assignment of OH stretching vibration of cummingtonite at 3620, 3640, 3655, and 3670 cm$^{-1}$ were attributed to vibration of cation configurations of 3Fe; 2Fe/Mg; Fe/2Mg; and Mg sites in amphiboles. The full width at half maximum of these bands is typically 3 to 6 cm$^{-1}$ for well crystallised cummingtonite. The width is almost double for the high-temperature product of the KSDB ferroactinolite (A), indicating the disorder in the cummingtonite phase.

KSDB amphiboles and their high-pressure and high temperature phases were characterised by powder X-ray diffraction and Fourier transform infrared (FTIR ) vibrational spectroscopic methods. The vibrational assignments were made based on the structural model proposed by Hawthrone [9] . The presence of strong XRD peaks of (-1 6 3), (-3 0 2), (1 0 4 ), (-3 6 5) indicate the high pressure product is cummingtonite with P21/m symmetry. Earlier studies of thermal analyses showed that actinolite transforms irreversibly to cummingtonite-clinopyroxene—quartz assemblages at 780 K.

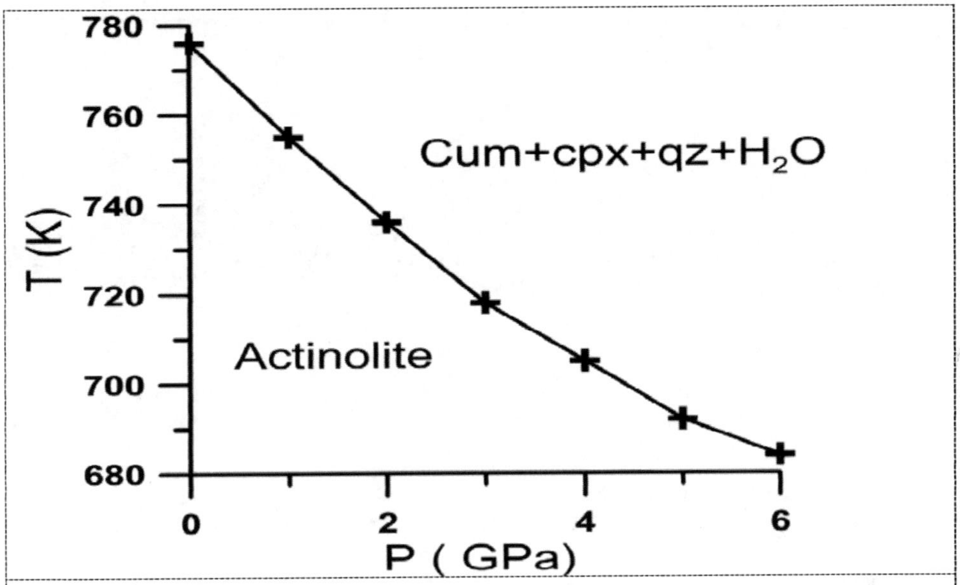


**Figure 3.** Phase diagram of actinolite derived from the electrical resistivity and FTIR methods. The uncertainties in the values of temperature and pressure are smaller than the sizes of the symbols. The high temperature mineral assemblages are cummingtonite (Cum)-clinopyroxene (cpx) —quartz (qz) . The transformation temperature at different pressures was determined by measuring the temperature dependence of the electrical resistivity of actinolite . ( Details are explained in the text)

The transition temperature is found to be decreasing with the increase of pressure.(Figure.3). The electrical resistivity of actinolite amphiboles from the the KSDB section showed a distinct slope change in the electrical resistivity –reciprocal temperature plot at about 780 K. at 0.5 GPa, indicating a possibility of a phase transition of actinolite. The powder XRD studies on the heat treated samples show 20 distinct XRD lines corresponding to cummingtonite phase with cell parametr a= 0.9526 nm, b=1.8190, c= 0.5320nm and β= 102.4°, revealing that the observed phase transition is a thermally induced phase transformation and decomposition of actinolite into cummingtonite-clinopyroxene-quartz and water assemblages. This transition temperature decreases from 780 K at 0.5 GPa to 680 K at 6.0 GPa. ( Figure 3). It is well known that amphibole lattice preferred orientations play a major role for crustal seismic anisotropy . This behaviour is consistent with the decrease of observed deep continental crustal velocity, where the amphibolites is one of the major mineral assemblages [10]. The present study of high pressure and high temperature phase stability of amphibole is relevant to the understanding of the mantle petrology under hydrous conditions.

## 4. References

[1] Inoue T, Irifune T, Yurimoto H, Miyagi I 1998, *Physic. Earth. Planet. Interior* **107,** 221
[2] Kern H, Popp T, Zharikov A, Labanov KV, and Smornov Yu P 2001 *Tectonophysics* **338** 113
[3] Parthasarathy G. 2006 *J.Applied Geophysics* **58** 321
[4] Parthasarathy G 2011 *American Mineralogist.* **96** 860
[5] Chandra U, Sharma P, Parthasarathy G 2010 *American Mineralogist* **95** 870

[6] Chandra U, Parthasarathy G, Sharma P 2010 *Canadian Mineralogist* **48** 1137
[7] Gorbatsevich F, Mitrofanov F, Gillen C, and Guberman D 2005 *Geophys. Research Abstracts,* **7** 154-1- 154-3.
[8] Mitronov FP, and Gorbatsevich FF 2000 *The Results of the Study of the Deep Substance and Physical Processes in the Kola superdeep Borehole section down to a depth of 12, 261meters* (Geological Institute, Kola Science Centre, Russian Academy of Science).
[9] Hawthrone FC 1983 *Canadian Mineralogist* **21** 173
[10] Tatham DJ, Lloyd GE, Butler WH, Casey M, 2008 *Earth and Planetary Science Letters* **267** 118

## Acknowledgments

The authors thank the anonymous reviewers for their constructive comments and suggestions. One of us (GP) is grateful to PLANEX , Department of space, Government of India for their support . He is thankful to the Director, CSIR- National Geophysical Research Institute for the permission to publish this work. This paper is dedicated to Dr. R. Chidambaram , Principal Scientific Adviser to the Government of India, on the occasion of his 75$^{th}$ Birthday.

Journal of Physics: Conference Series **377** (2012) 012057 doi:10.1088/1742-6596/377/1/012057

# Impact metamorphism at the Cretaceous-Paleogene Boundary and High-pressure Mineral Physics

**G. Parthasarathy**[1]
Council of Scientific and Industrial Research (CSIR), National Geophysical Research Institute, Uppal Road , Hyderabad, 500007, India.

E-mail: drg.parthasarathy@gmail.com

**Abstract.** We discuss the occurrences of Iridium, high-pressure phase of fullerene and other carbonaceous matter in the Earth close to the Cretaceous-Paleogene boundary (KPB) in India along with other mineralogical data. Several sources of fullerenes at the KPB have been proposed, including an impact origin . The presence of toluene-insoluble high-pressure phase of fullerene $C_{60}$ in the acid resistant carbonaceous matter extracted from the iridium-rich layers of the Intertrappean sediments of Anjar, India, has been characterized by ultra violet-visible and Fourier transform infrared (FT-IR) spectroscopic techniques. The FT-IR spectroscopic measurements on the insoluble fullerenes show strong absorption peaks at 1381,741, 608, and 542 $cm^{-1}$ wavenumbers, that are characteristic of a non-hydrostatic high pressure and temperature phase of $C_{60}$ fullerene. The presence of Ir-rich layers, associated with high-pressure phase of fullerene, at Anjar strongly support impact- metamorphism at the KPB.

## 1. Introduction

Alverez et al. [1] have hypothesised that a single large asteroid killed the dinosaurs and caused the mass extinction of many other organism at the Cretaceous-Tertiary (K/T) boundary. The K–T boundary is a geological signature, usually a thin band, dated to 65.5 ± 0.3 Ma (million years) ago. *K* is the traditional abbreviation for the Cretaceous period, and *T* is the abbreviation for the Tertiary period. The boundary marks the end of the Mesozoic era and the beginning of the Cenozoic era, and is associated with the K-T- extinction event, a mass extinction. With "Tertiary" being discouraged as a formal time or rock unit by the International Commission on Stratigraphy, the K–T extinction event is now called the Cretaceous–Palaeogene (or K–Pg) event by many researchers [2] . Hence in this article we would like to refer this by " Cretaceous-Paleogene boundary " (KPB) instead of K-T boundary, 65 Ma (million years ago) in the Earth's geological history. This meteorite impact created a crater at Yukutan, Gulf of Mexico, called the Chicxulub crater, which is about 180 to 300 km diameter in size.

[1] To whom any correspondence should be addressed.

The KPB is marked in many places by a distinctive layer of clay often highly enriched in elements such as iridium and osmium relative to the layers above and below [3]. The unusual iridium enrichment was consistent with the hypothesis of Alverez et al [1], that this elements having an extraterrestrial origin. However, the KPB layer is now generally believed to have been precipitated by the impact of a 10-kilometer-wide asteroid which created the Chicxulub Crater. Observation micro - spherules deposit and Iridium anomalies at the KPB layer indicated a possible origin of a shock metamorphic event at KPB [2], and made a convincing case for the evidence for the mass extinction at the end of the Cretaceous. In India there are few KPB sections were reported at Anjar and Meghalaya [4]. Bhandari [3] has identified the KPB layer within the Deccan intertrappean sediments at Anjar, about half way between Flow III and Flow IV. ( Figure.1). Mineralogical and geochemical analyses of the KPB section of Anjar were discussed elsewhere [5-7]. Here we report the first occurrence of natural fullerenes from the Indian subcontinent, in the intertrappean sediments of Anjar (Kutch), which are believed to encompass the KPB . We have also determined the spectroscopic properties of $C_{60}$ fullerene that demonstrate the presence of its high-pressure-temperature phase. The co-occurrence of fullerene and other impact signatures at Anjar as well as at other KPB sites indicates that it can be used as a definitive geochemical marker for identifying impact horizons in terrestrial sediments. The occurrence of high-pressure phase of fullerene and irdium in Anjar KPB section have provided evidence for the impact metamorphism during the K-P- boundary.

## 2. Experimental :

The Anjar volcano-sedimentary sequence is located at the western periphery of the Deccan flood basalt province and consists of nine lava flows and at least four intertrappean beds [3,4]. The third intertrappean bed is well developed (about 6 m thick) and consists of cherty limestone, shale and mudstone (Fig. 1). Three thin limonitic layers are present in the lower 1.5 m of the intertrappean bed which was exposed in pit BG-1 [3,4]. The limonitic layers have high concentration of iridium (650 to1333 pg/g) and osmium (650 to 2230 pg/g) compared to less than 100 pg/g measured in other horizons of this pit. Based on the geochronological, geochemical, palaeomagnetic and palaeontological data [3,4], it has been shown that the limonitic layers were deposited during the KPB event. The deposition rate of sediments in the intertrappean bed is much higher than in the contemporaneous marine sections, and therefore they offer a better time resolution, allowing a high-resolution sequence stratigraphy of K/T transition events and their relation to Deccan volcanism to be constituted. Clay minerals from the Flow III-intertrappean bed were treated with acid digestion technique, for the removal of carbonates and silicates. The acid resistant residue were analyzed with Fourier transform infrared and UV-Visible spectroscopic methods. The acid resistant residue were further treated with toluene for the extraction of toluene soluble fullerene.

The clay mineral fraction was removed by washing with double distilled water after dispersal with a peptizing agent and ultrasonic treatment. The details of sample preparations and extraction procedures are discussed elsewhere [8,9]. The carbonaceous matter was extracted from the powdered samples using a standard acid-digestion method . All the samples were treated with 12N HCl for 24 hours followed by 60% HF for 18 hours. The samples were then heated in 60% HF at $70^0$C for one hour. The resulting solid residue was washed with about one liter of double distilled water and dried overnight at $80^0$C. A fraction of the carbon-rich residues was treated with toluene for the extraction of fullerenes, which were then characterized by mass spectroscopy and $^{13}$C-NMR spectroscopy.The electron impact ionization mass spectra were obtained with a VG AUTOSPEC-mass spectrometer and a OPUS.V3.IX data system. The conditions of measurements were as follows: source temperature: 250 $^o$C ; electron energy: 70 eV; trap current: 200 mA; resolution: 1000; mass range scanned: 300-900 atomic mass unit (amu); scan time : 5 seconds per decade of mass; interscan delay: 0.5 s. Samples were loaded in a quartz cup, introduced through direct inlet probe and instantly heated to 250 $^o$C . The mass spectrometer was calibrated up to 900 amu with a reference sample of high-purity perfluorokerosene. The samples were scanned between 300 and 900 amu to improve the sensitivity of detection of $C_{60}$ and $C_{70}$.Fourier transform-infrared spectroscopic (FT-IR) studies on the carbon-rich residue extracted from the Ir-rich layers of Anjar samples have been carried out in ambient conditions, using BIORAD 175C FT-IR spectrometer adopting potassium bromide pellet method. Samples were scanned in the frequency range of 4000 $cm^{-1}$ to 400 $cm^{-1}$. Each sample has been scanned three times. Typical uncertainty in the wave numbers is $\pm$ 3 $cm^{-1}$ in the frequency region 2000 to 4000 $cm^{-1}$.

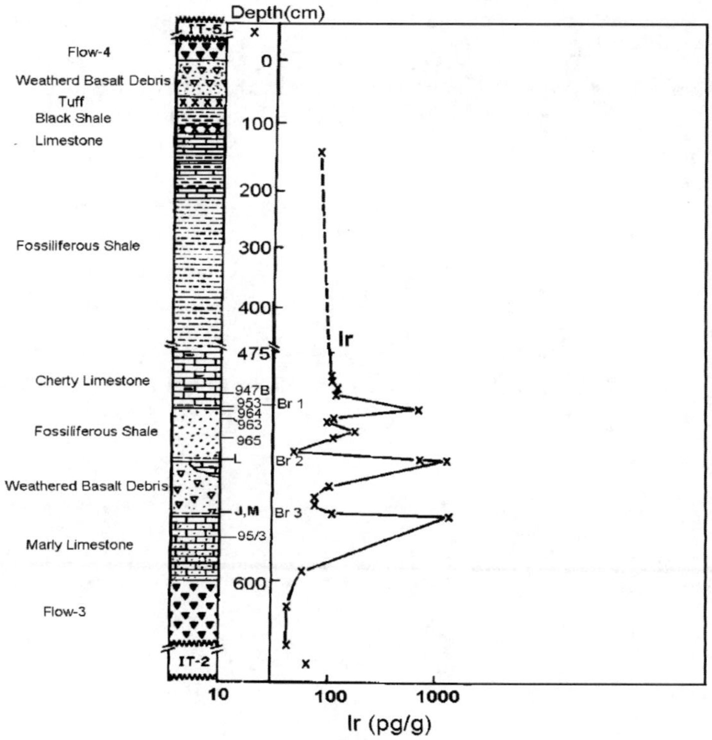


**Fig. 1**

- **Figure.**1 Location of the samples in the third Intertrappean section, between Flow III and Flow IV. Three different Ir bearing samples are clearly indicated as Br-1 , Br-2 and Br-3. The depths were measured from the base of Flow IV. Intertrappean IT2 and IT 5 are also indicated to mark the stratigraphic locations of the samples[ Ref 8].

Journal of Physics: Conference Series **377** (2012) 012057 doi:10.1088/1742-6596/377/1/012057

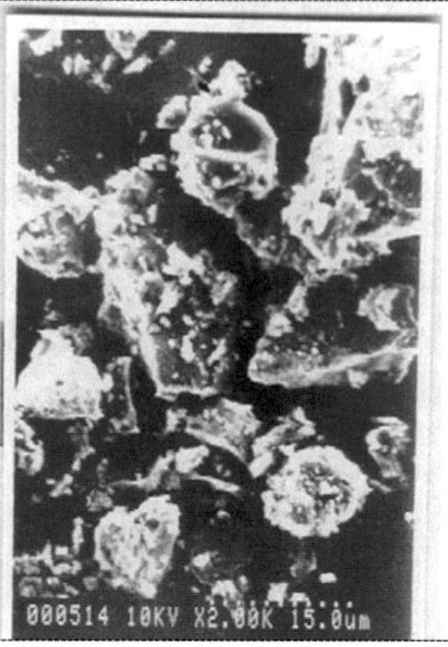


**Figure 2.** Scanning electron microscopic (SEM) images of the spherules collected from the Anjar intertrappean Iridium-rich layers. The spherules provide additional evidence for the shock metamorphism at the Cretaceous-Palaeogene boundary. The clay mineral in which the sphrules are embedded is identified as sepiolite-palygorskite group smectite. We also observed that smectites are least in the iridium poor section.

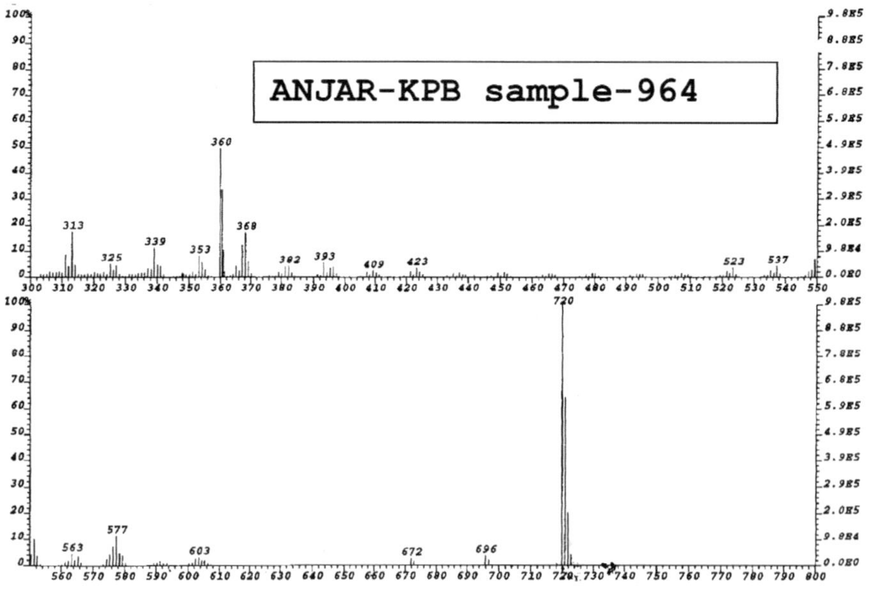


**Figure .3.** The electron-impact ionization mass spectrum of toluene soluble fullerene sample extracted from the iridium rich layer of the sample-964. The X-axis is mass (m/z) in atomic mass unit and y-axis is intensity in arbitrary unit. The peaks at 720 atomic mass unit indicate the presence of $C_{60}$ in the Anjar KPB samples. The fullerene amount is estimated by using a standard $C_{60}$ and found to be about 800 ppb (parts per billion). Please note that we did not observe any Helium in our samples.

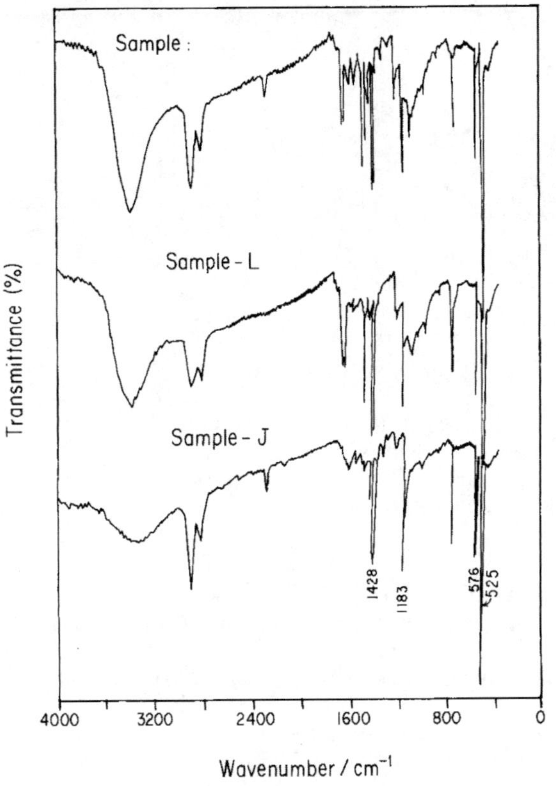


**Figure-4** Fourier transform infrared (FTIR) spectra of fullerene and iridium rich samples 964, L and J. The absorption bands at 1429, 1183, 577, and 528 are assigned to $C_{60}$ and the strong peaks at 740, and 509 cm $^{-1}$ are characteristic of high-pressure temperature (HPT) phase of $C_{60}$. The broad peaks near 3400 cm $^{-1}$ are due to the hydrous and polyaromatic hydrocarbons. The additional vibrational bands at 2928, 2850 and 1380 cm$^{-1}$, which are characteristic of aliphatic CH, $CH_2$ and $CH_3$ groups probably occurring as impurities, have also been observed The presence of HPT- phase of $C_{60}$ provides an unambiguous evidence for the shock metamorphism at the KPB

The presence of polycyclic hydrocarbons, mono- and bicyclic alkanes, adamantine, diamantine, and methyldiamantanes in the K-P- boundary sediments at Kawaruppu has been reported [10]. Such poly hydrocarbons are also seen in the Anjar samples (figure.4)., which could be a precursor materials for the formation of fullerenes at the KPB. It is possible to suggest that the occurrence of fullerenes and an increase in carbon soot content are considered to be important geochemical markers of the KPB. The other carbonaceous matter, which are observed at the KPB sediments include polycyclic hydrocarbons , mono- and bicyclic alkanes and diamondiod hydrocarbons [10]. About 26 polycyclic aromatic hydrocarbons from naphthalene to coronene were reported from the KPB sediments at Kawaruppu, Hokkaido, Japan, indicating a possible combustion origin of these hydrocarbons within the KPB clays. The nanoparticles of primary iron minerals like hematite, goethite or limonite observed in KPB layer can also form in an impact event [6] . This fact, combined with the otherwise rare occurrence of these carbonaceous materials in nature, makes fullerenes potential geochemical markers for impact boundaries .In a nearby Meghalaya site, however Ni rich spinel has been found and therefore it is clear that impact metamorphism during KPB event has delivered its high-pressure products in the Indian subcontinent [3] . In conclusion, the systematic mineralogical investigations on the KPB section of the Deccan Trap revealed presence of the HPT phase of $C_{60}$ . The coexistence of high-pressure temperature (HPT) phase of $C_{60}$ with high iridium concentrations at Anjar KPB makes it a good geochemical marker for the impact by the extraterrestrial material [11].

### 3. References

[1] Alvarez L W, Alvarez W, Asaro F, and Michel HV.1980 *Science* **208** 1095
[2] Reimold WU 2007 *South African J Geology* **110** 1
[3] Bhandari N, 1998, *Proc. Indian Academy of Sciences*, **107**, 251
[4] Hildebrand AR, Penfield GT, Kring DA, Pilkington M, Camargo ZA, Jacobson S, and Boynton WV. 1991 *Geology* **19** 867
[5] Bhandari N, Gupta M, Pandy J, and Shukla PN. 1994.*Chemical Geology***113**, 45
[6] Bhandari N, Verma HC, Upadhyay C, Tripathi A, and Tripathi RP 2002 *Catastrophic Events and Mass Extinctions: Impact and Beyond* Eds C. Koeberl and K.G. MacLeod, Geological Society of America Special Paper No 356 pp 201-211
[7] Parthasarathy G, Bhandari N, Vairamani M, Kunwar AC, and Narasaiah B 2002 *Catastrophic Events and Mass Extinctions: Impact and Beyond* Eds C. Koeberl and K.G. MacLeod, Geological Society of America Special Paper No.356 . Pp 345-350
[8] Parthasarathy G, Srinivasan R, Vairamani M, Ravikumar K, and Kunwar AC 1998 *Geochim.Cosmochim.Acta* **62** 3541
[9] Parthasarathy G, Bhandari N, Vairamani M, and Kunwar AC 2008 *Geochim.et Cosmochim.Acta* **72** 978
[10] Shimoyama A and Yabuta H 2002 *Geochemical Journal* **36** 173
[11] French BM and Koeberl C 2010 *Earth Science Reviews* **98** 123

### Acknowledgments

This paper is dedicated to Dr. R. Chidambaram , Principal Scientific Adviser to the Government of India, on the occasion of his $75^{th}$ Birthday. Author is grateful to R. Chidambaram, N.Bhandari (Physical Research Laboratory, Ahmedabad), A.C. Kunwar, M. Vairamani (CSIR-Indian Institute of Chemical Technology , Hyderabad ) and W.U. Reimold ( Humboldt-University, Berlin) for their very kind encouragements and support. Financial support from Space Application Centre, Physical Research Laboratory-Indian Space Research Organization, Government of India under PLANEX program is gratefully acknowledged. I thank the anonymous reviewers for their critical comments and suggestions, which improved the manuscript.

# Pressure dependent electronic properties of $\alpha$-$Be_3P_2$

**K B Joshi**[1] **and U Paliwal**

Department of Physics, University College of Science, M.L. Sukhadia University, Udaipur- 313001, India

E-mail: cmsmlsu@gmail.com

**Abstract.** The first-principles linear combination of atomic orbitals method implemented in the CRYSTAL code has been applied to compute equilibrium lattice constant and bulk modulus of $\alpha$-$Be_3P_2$. The electronic band structure calculations are also performed to report bandgap. The properties are computed at the level of density functional theory and hybrid calculation. The effect of pressure on the bandgap is studied and the pressure coefficient and volume deformation potential are computed. The computed lattice parameters are well in agreement with the experimentally reported data. Band structure calculation at the level of density functional theory reveals that $\alpha$-$Be_3P_2$ is a direct bandgap semiconductor with $E_g$=2.39 eV. The pressure dependent calculations show that the bandgap decreases almost linearly with pressure giving negative value of pressure coefficient.

## 1. Introduction

The phosphides and nitrides of II group elements are mostly found in the $II_3$-$V_2$ unit. In the bulk form, these have technological applications in optoelectronics, detectors, IR sensors, solar cells ultrasonic multipliers and Hall generators. The larger excitonic radii and narrow bandgap in these materials offer promising applications at nanoscale. Compared to II-VI and III-V semiconductors, the $II_3$-$V_2$ type phosphides and nitrides are studied less both in the bulk and nanophase [1-4]. It has mostly been attributed to the complex crystal structures and large number of atoms. Major studies are directed to the $Cd_3P_2$, $Zn_3P_2$, $Mg_3N_2$, $Mg_3P_2$, $Be_3N_2$ etc. [1-7]. In this work we compute lattice constant and bulk modulus of $\alpha$-$Be_3P_2$. The pressure dependent isothermal band structure calculations are performed to report volume deformation potentials and the pressure coefficient of bandgap. The calculations are performed using the first-principles linear combination of atomic orbitals methods (LCAO) at the level of density functional theory (DFT) and hybrid scheme implemented in the CRYSTAL code [8].

It is known that DFT under local density approximation gives incorrect estimate of fundamental bandgap in semiconductors [9,10]. The GGA shows only some improvement in predicting bandgap. HF theory, gives good results in systems where screening is unimportant but overestimates bandgap in general. Quasi particle (QP) shift and GW calculations performed on ZnO, CdO and InN revealed improvement in the computed bandgaps. These calculations suggest a correction term to be included in the bandgap computed form the DFT. According to the ¼ recipe of Fuchs *et al* [11] the correction in Kohn-Sham eigenvalues due to QP shift is negligibly small when hybrid functional mixing 25% of

[1] To whom correspondence should be addressed.

Published under licence by IOP Publishing Ltd

Journal of Physics: Conference Series **377** (2012) 012058 doi:10.1088/1742-6596/377/1/012058

HF exchange with the 75% DFT exchange is taken to compute band structure [9,10]. Therefore, in this work we report electronic band structure calculations using hybrid scheme. The calculations at the level of DFT taking Perdew-Burke-Ernzerhof (PBE) exchange and correlation functionals are also reported.

## 2. Computational detail

Beryllium phosphide ($Be_3P_2$) may be found in the *bcc* anti-bixbyite and tetragonal structures. The α-$Be_3P_2$ crystallises in cubic anti-bixbyite type structure belonging to the space group 206. The lattice constant of α-$Be_3P_2$ is $a$=10.15 Å [12,13] and the unit cell contains 80 atoms, 48 (Be) and 32(P). Each beryllium atom is tetrahedrally coordinated with phosphorous and each phosphorous atom is coordinated by six beryllium atoms [12,13]. The beryllium atom situated at ($x_1$,$y_1$,$z_1$) occupies *48e* Wyckoff position while crystallographically different phosphorous atoms P1 at ($x_2$,0.0,0.25) and P2 at (0.25,0.25,0.25) occupy *24d* and the *8b* Wyckoff positions, respectively [13,14].

The first-principles periodic LCAO method implemented in the CRYSTAL code [8,15] is applied considering the density functional theory (DFT) and the hybrid scheme. In the current non-cellular method, each crystalline orbital $\psi(\mathbf{r},\mathbf{k})$ is a linear combination of Bloch functions $\varphi(\mathbf{r},\mathbf{k})$ defined in terms of local functions $\varphi(\mathbf{r})$, normally referred as atomic orbitals. The local functions are expressed as linear combination of certain number of individually normalized Gaussian type functions. The 5-111G [16] and 85-21d1G [17] Gaussian basis sets were taken for Be and P, respectively. The gradient corrected PBE exchange and correlation functionals [18] were taken. Additionally, hybrid calculation adopting hybrid functional which is very similar to the B3LYP scheme is also performed. However, total exchange is taken care of by mixing 75% PBE exchange with the 25% Fock exchange. The LYP correlation functional is replaced by the PBE correlation functional and all other non-local parameters were same as in the B3LYP scheme. The self-consistent calculations were performed considering Monkhorst and Pack [8] net of 11×11×11 points. The self consistency was achieved within 14 cycles.

To perform geometry optimization, force calculations were performed which enables to find the atomic positions of Be and P atoms. To begin with experimental atomic positions available for α-$Be_3N_2$ [7] were taken for geometrical optimization. The positions were optimized by performing force calculations (OPTGEOM option).The total energy curve i.e. E(V) was generated to obtain the lattice constant (say, $a_1$). The atomic coordinates were optimized at $a_1$. Considering these optimized coordinates, the E(V) curve is generated to obtain refined lattice constant (say, $a_2$). Thereafter, the atomic coordinates were optimized at $a_2$. Finally the total energy curve for α-$Be_3P_2$ was obtained and coupled with the Murnaghan equation of state [19] to propose equilibrium lattice constant and the bulk modulus for α-$Be_3P_2$ computationally.

## 3. Results and discussion

The total energy curves, E(V) generated considering DFT and the hybrid scheme, show that energy is lower in the entire range of volume for the hybrid scheme. The equilibrium lattice constants, bulk moduli and pressure derivative of bulk moduli deduced by coupling E(V) curves with Murnaghan equation of state for the two calculations are given in Table 1. The optimized atomic positions are also listed in the table. It is obvious that the exchange and correlation schemes affect atomic positions upto 1 part in $10^3$. Also, the current calculations and published results on atomic positions agree upto 1 part in $10^2$. The lattice constant obtained from the hybrid scheme is closer to the experiment than the PBE-GGA. The lattice constant predicted by PBE-GGA scheme is merely 0.36% higher than the experiment compared to the 0.42% predicted by the FP-LAPW method. For hybrid scheme difference is smaller. Naturally, the bulk modulus proposed by PBE-GGA is larger than the FP-LAPW method. The difference is, however, very small (~14 GPa). Moreover, the density 2.24 gm/cm$^3$ given by current calculations is in very good agreement with 2.25 gm/cm$^3$ reported experimentally.

**Table 1.** Optimized atomic positions, lattice constants (*a*), bulk modulus ($B_0$) and its pressure derivative ($B^{/}$) and density (ρ) of $Be_3P_2$.

| | Atomic positions | | | | *a* (Å) | $B_0$ (GPa) | $B^{/}$ | ρ (g/cm$^3$) |
|---|---|---|---|---|---|---|---|---|
| | $x_1$ | $y_1$ | $z_1$ | $x_2$ | | | | |
| PBE-GGA | 0.3824 | 0.1333 | 0.3735 | 0.9926 | 10.1862 | 114.23 | 2.716 | 2.24 |
| Hybrid | 0.3837 | 0.1344 | 0.3732 | 0.9920 | 10.1741 | 115.97 | 2.742 | 2.24 |
| FP-LAPW[a] | 0.3790 | 0.1300 | 0.3740 | 0.9940 | 10.1929 | 99.087 | 3.600 | -- |
| Expt. | --- | | | | 10.1500[b] | -- | -- | 2.25[c] |

[a]Ref. [14], [b]Ref [13], [c]Ref [20]

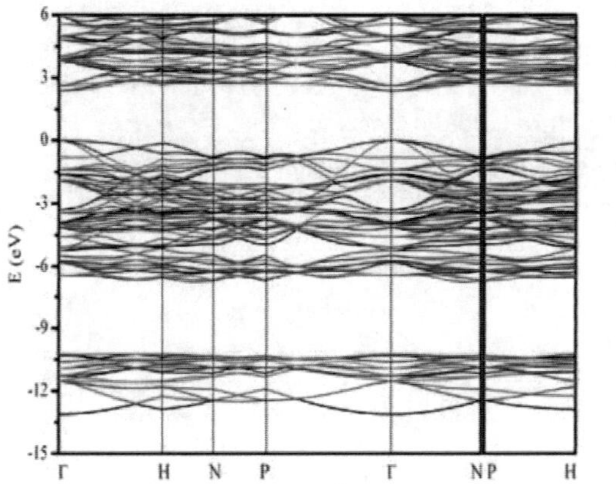


**Figure 1.** Band structure of α-$Be_3P_2$ calculated using PBE-GGA functional at zero pressure.

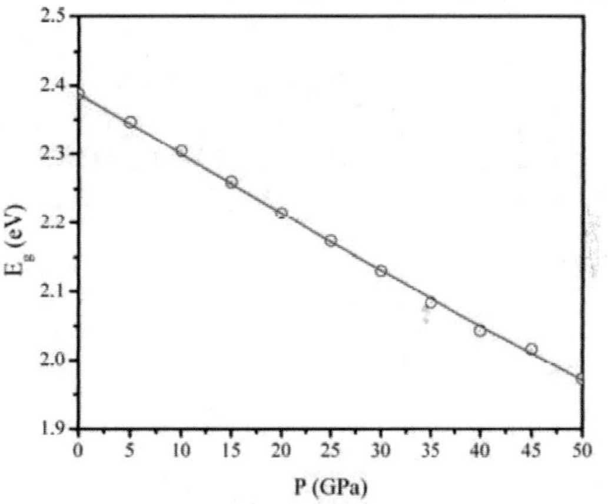


**Figure 2.** Pressure dependent bandgap of α-$Be_3P_2$ calculated using PBE-GGA functional.

The electronic band structure computed from density functional PBE-GGA is shown in figure 1. The upper valence bands are predominantly of the phosphorous 3*p* type states. The lower valence bands are the phosphorous 3*s* states with a minor contribution from Be 2*s* states. The lower conduction bands are dominated by the beryllium 2*s* and 2*p* states. Other bands arising from the 1*s*, 2*s*, and 2*p* states of phosphorous and 1*s* of beryllium are deep below the lowest band visible in figure 1. The band structure exhibits direct bandgap of 2.39 eV. Bandwidth of the upper valence bands is 6.79 eV and lower bands are extended within 2.86 eV. As discussed earlier, hybrid scheme has also been adopted to compute band structure. The overall topology of the band structure obtained from hybrid functional, not shown here, is similar to the band structure shown in figure 1. However, the direct bandgap predicted by the hybrid calculation is 3.75 eV. Bandwidth of the upper and valence bands is 7.61 eV and 3.82 eV, respectively. Notably, the bandgap 1.6 eV, proposed by FP-LAPW method is lower than the current results.

The application of hydrostatic pressure causes shift of conduction band edge relative to the valence band edge owing to change in volume. The change allows direct estimation of the effective hydrostatic deformation potentials for the bandgap. As the pressure derivatives of band energy differences tend to be correctly calculated without resorting to GW or QP calculations, band structure calculations combined with structural study from first-principles methods enable computation of deformation potentials and pressure dependent bandgaps [9,21]. In order to examine pressure dependence of the bandgap, calculations were performed upto 50 GPa and no change in the nature of the bandgap was seen. To know the pressure dependence a quadratic equation was fitted to the computed bandgaps [22]:

$$E_g(P) = E_g(0) + \alpha_p P + \beta_p P^2 , \quad (1)$$

where $E_g$ is the fundamental gap (eV), P is pressure (GPa) while $\alpha_p$ and $\beta_p$ are the pressure coefficients. The $\alpha_p$ obtained after fitting Eq. 1 to the fundamental gap is -0.89 (meV kbar$^{-1}$) while $\beta_p$ is ~$10^{-4}$ (meV kbar$^{-2}$). It suggests almost linear change in bandgap as shown in figure 2. The $\alpha_p$ helps to obtain volume deformation potentials ($a_v$=-$\alpha_p B_0$) useful in thin films applications of semiconductors [22]. The volume deformation potential $a_v$ suggested for $\alpha$-$Be_3P_2$ by the current study is 1.016 eV.

## 4. Conclusions

Structural and electronic properties of $\alpha$-$Be_3P_2$ are studied applying the first-principles method considering the PBE-GGA and hybrid functional. The optimized atomic positions, lattice constant and the bulk modulus are presented. The proposed lattice constant and bulk modulus follow general trend and in very good agreement with experiment. The band structure calculations show that the $\alpha$-$Be_3P_2$ is a direct bandgap semiconductors. The PBE-GGA calculations predict bandgap of 2.39 eV while hybrid calculation gives 3.75 eV. The band structure calculations performed at various pressures, considering the low pressure limit, show that the bandgap decreases linearly with pressure. The computed pressure coefficient $\alpha_p$ is -0.89 (meV kbar$^{-1}$) while the volume deformation potential $a_v$ suggested by the current study is 1.016 eV.

**Acknowledgement:**
UP is grateful to the CSIR, New Delhi for awarding Senior Research Fellowship.

**References**

[1] Zdanowicz W and Zdanowicz L 1975 *Annu. Rev. Mater. Sci.* **5** 301
[2] Madelung O 2005 *Semiconductor Data Handbook* III edition (Springer-Verlag, Berlin)
[3] Shen G, Chen P, Bando Y, Golberg D and Zhou C 2008 *Chem. Mater.* **20** 7319
[4] Gregory D H 2001 *Coord. Chem. Rev.* **215** 301
[5] Imai Y and Watanabe A 2006 *J. Mater. Sci.* **41** 2435
[6] Armenta M G M, Reyes-Serrato A and Borja M A 2000 *Phys. Rev. B* **62** 4890
[7] Reckeweg O, Lind C, Simon A and DiSalvo F J 2003 *Z. Naturf.* **58b** 159
[8] Dovesi R, Saunders V R, Roetti C, Orlando R, Zicovich-Wilson C M, Pascale F, Civalleri B, Doll K, Harrison N M, Bush I J, D'Arco Ph and Llunell M 2006 *CRYSTAL06 User's Manual* (University of Torino, Torino)
[9] Bechstedt F, Fuchs F and Kresse G 2009 *Phys. Status Solidi* b **246** 1877
[10] Onida G, Reining L and Rubio A 2002 *Rev. Mod. Phys.* **74** 601
[11] Fuchs F, Furthmuller J, Bechstedt F, Shishkin M and Kresse G 2007 *Phys. Rev. B* **76** 115109
[12] Stackleberg M V and Paulus R 1935 *Z. Phys. Chem. Abt. B* **22** 305
[13] Wyckoff R W G 1986 *Crystal Structures* (Krieger, Malabar, FL)
[14] Mokhtari A 2008 *J. Phys.: Condens. Matter* **20** 135224
[15] Evarestov R A 2007 *Quantum Chemistry of Solids: The LCAO First principles Treatment of Crystals* Springer Series in Solid State Science, Vol. 153 (Springer-Verlag, Heidelberg)
[16] Dovesi R, Pisani C, Ricca F and Roetti C 1982 *Phys. Rev. B* **25** 3731
[17] Zicovich-Wilson C M, Bert A, Roetti C, Dovesi R and Saunders V R 2002 *J. Chem. Phys.* **116** 1120.
[18] Perdew J P, Burke K and Ernzerhof M 1997 *Phys. Rev. Lett.* **77** 3865; **78** 1396(E)
[19] Murnaghan F D 1944 *Proc. Natl. Acad. Sci. USA* **30** 244
[20] El-Maslout A, Motte J –P, Courtois A, Protas J and Gleitzer C 1975 *J. Sol. Stat. Chem.* **15** 223
[21] Zhu Y Z, Chen G D, Ye H, Walsh A, Moon C Y and Wei S –H 1999 *Phys. Rev. B* **77** 245209
[22] Wei S -H and Zunger A 1999 *Phys. Rev. B* **60** 5404

Journal of Physics: Conference Series **377** (2012) 012059 doi:10.1088/1742-6596/377/1/012059

# Thermal properties of a novel compound $PbVO_3$

**Atahar Parveen [1], Archana Srivastava and N. K. Gaur**
Department of Physics, Barkatullah University, Bhopal 462026, India

E-mail: ataharparveen@gmail.com

**Abstract.** Effect of pressure on thermal properties of a multiferroic (MF) novel compound $PbVO_3$ is studied using an atomistic approach. The temperature dependence of the specific heat ($C_P$) (20K $\leq T \leq$ 300 K) of $PbVO_3$ has been studied and compared with the earlier experimental work. Also, this work reports, probably for the first time the relation of pressure to the Debye temperature ($\theta_D$) along with other properties like bulk modulus (B), molecular force constant (*f*), Reststrahlen frequency ($\upsilon_0$), thermal expansion ($\alpha$) and Grüneisen parameter of $PbVO_3$ with the Modified Rigid Ion Model (MRIM).

## 1. Introduction

The $AVO_3$ (A = Sr, Ba, and Pb) type perovskite-based oxide compounds are very interesting. Among which the prototype $PbVO_3$ shows unusual behaviours at high pressure. $PbVO_3$ adopts a layered structure [1-5] with isolated layers of corner-shared $VO_5$ pyramids due to the strong tetragonal distortion (space group *P4mm*); Tetragonal $PbVO_3$ has attracted considerable attention as a promising candidate of MF material [1, 6-11]. Except for potential application as MF material, tetragonal $PbVO_3$ also exhibits intriguing and fantastic fundamental physics. The $ABO_3$-type FE perovskite materials often show a temperature- and/or pressure-induced tetragonal to cubic (T-to-C) structural transition, corresponding to ferroelectric to paraelectric (FE-to-PE) phase transition [4] which is an unusual metal to insulator transformation [1]. Quite recently, with the application of high-temperature and high-pressure techniques, $PbVO_3$ was successfully synthesized for the first time. The most pure $PbVO_3$ samples were obtained in 40-60 kbar pressure range and at 700-750°C [2]. This compound was found to possess a very large tetragonal distortion (*c/a*, 1.23) [5, 11]. Only at high pressure (about 2 GPa), can the T-to-C phase transition be realized at room temperature [5]. The well depth of V atoms in $PbVO_3$ decreases gradually when the lattice constant is suppressed. As a result, the ferroelectric instability of $PbVO_3$ is weakened and the tetragonal to cubic transition becomes available under pressure, being consistent with the experimental observations that the transition appears at about 2 GPa at room temperature [5].

The $AVO_3$ compounds are considered to possess many potential applications in fields such as multiferroic devices, high-temperature solid oxide fuel cells, high *Tc* superconductivity, and so on [3]. Despite their potential importance, relevant investigations on the $AVO_3$ compounds are relatively scarce. The magnetic properties have been well studied; but a systematic study on the thermal properties on the compound is lacking. We present the cohesive and thermal properties of a novel compound $PbVO_3$ [1] in distorted tetragonal and cubic phase. Earlier we have successfully used MRIM to study the thermal properties of some perovskite manganites and other inorganic compounds [12-15] which motivated us to compute the pressure / temperature dependent thermal properties of

$PbVO_3$. The essentials of the MRIM formalism and the results obtained from its application are presented in subsequent sections.

## 2. Formalism of RIM

We have formulated the modified rigid ion model (MRIM) by keeping the effect of long-range (LR) Coulomb attractions, the short-range (SR) Hafemeister–Flygare-type (HF) [16] overlap repulsion effective up to the next nearest neighbour atoms and the van der Waals attraction due to dipole–dipole (d–d)interactions. The potential describing the formalism of MRIM is expressed as

$$\phi = \frac{-e^2}{2}\sum_{kk'} Z_k Z_{k'} r_{kk'}^{-1} - \sum_{kk'} C_{kk'} r_{kk'}^{-6} + \sum_i \left[ \begin{array}{l} n_i b_i \beta_i^{kk'} \exp\left\{\left(\frac{r_k + r_{k'} - r_{kk'}}{\rho_i}\right)\right\} + b_i n_i' \beta_i^{kk} \exp\left\{\left(\frac{2r_k - r_{kk}}{\rho_i}\right)\right\} \\ + b_i \frac{n_i'}{2} \beta_i^{k'k'} \exp\left\{\left(\frac{2r_{k'} - r_{k'k'}}{\rho_i}\right)\right\} \end{array} \right] \quad (1)$$

The contribution of van der Waal's (vdW) attraction for the dipole-dipole interaction is determined by using the Slater- Kirkwood Variational (SKV) method [17]. The various terms in Eq (1) have their usual meaning as defined in our earlier papers [12-14].

The model parameters, hardness ($b$) and range ($\rho$) are determined from the equilibrium condition.

$$\left[d\phi/dr\right]_{r=r_0} = 0 \text{ and } B = 1/9Kr_0\left[d^2\phi/dr^2\right]_{r=r_0} \quad (2)$$

**Table 1.** The average cation radius at A-site, Tolerance factor (t) and model parameters of $PbVO_3$ (Cubic-C, Tetragonal-T)

| $r_k$ (Å) (A-site) | Temp. (K) | Pressure (GPa)/ Structure | t | $\rho_1$(Å) (V-O) | $b_1 \times 10^{-19}$(J) (V-O) | $\rho_2$(Å) (Pb -O) | $b_2 \times 10^{-19}$ (J) ( Pb -O) |
|---|---|---|---|---|---|---|---|
| 1.49 | 295 | 4-6 /C | 0.999 | 0.374 | 1.877 | 0.644 | 0.235 |
| 1.29 | 90 | 0 /T | 0.930 | 0.195 | 0.145 | 0.490 | 0.022 |
| 1.29 | 530 | 2 /T | 0.930 | 0.183 | 0.107 | 0.465 | 0.016 |

## 3. Results and discussions

The input data like the lattice parameters and interatomic distances at various pressures and temperatures are taken from [2, 4 and 5]. The ionic radii for distorted tetragonal and cubic structure of $PbVO_3$ with coordination no. 8 and 12 respectively were taken from [18] (Table1). Using the input data and the vanderwaals coefficients (using SKV method [17]) we computed the model parameters, hardness ($b_1$, $b_2$) and range ( $\rho_1$, $\rho_2$) corresponding to the ionic pairs $V^{4+}$-$O^{2-}$ and $Pb^{2+}$ - $O^{2-}$ . The bulk modulus calculated on the basis of AIM theory [19] similar to our previous work [12-14] could not be compared due to lack of experimental data. But the calculated bulk modulus, 91.5GPa for cubic $PbVO_3$ is in fair agreement with the bulk modulus, 97.3GPa for multiferroic $BeFeO_3$ [20] (Table 2). The cohesive energy of $PbVO_3$ is computed using eqn. (1) and is reported in table 2. The negative value of the cohesive energy indicates that this compound is stable at ambient temperature and pressure. Calculated cohesive energy of $PbVO_3$, is quite comparable with the reported value of the lattice energy for $LaVO_3$, -140.07eV [21]. Further, to ascertain the validity of the MRIM we have used the generalized Kapustinskii equation [22] on the lines of our earlier work [12-14]. The lattice energy obtained using the Kapustinskii equation is close to the MRIM results (average deviation being less than 3%). The cohesive energy results conclude that tetragonal $PbVO_3$ is more stable than cubic $PbVO_3$. An empirical prediction for the size of the distortion is given by the tolerance factor, $t$ ($t = (r_{Pb}+ r_O) / 2(r_V + r_O)$.The tolerance factor $t$ (Table 1) satisfies the condition that it lies between 0.8 and 1 for orthorhombic and is almost 1 for cubic structure [23].

Our results on Debye temperatures correlate well with the earlier experimental results on $CaVO_3$ [24]. Also it can be clearly seen in Table 2 that the Debye temperature ($\theta_D$) of the tetragonal $PbVO_3$ is

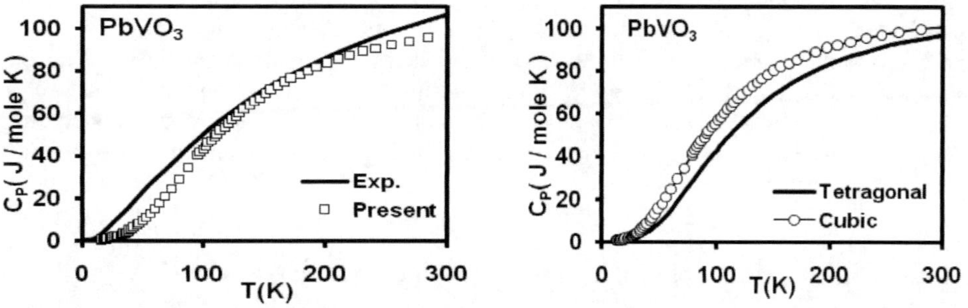


**Figure. 1** Variation of calculated heat capacity (Cp) with temperature compared with the experimental data of Tsirlin *et al.* [1] (left) and Comparison of specific heat of $PbVO_3$ in tetragonal and cubic phases (right).

higher than that of cubic and $\theta_D$ increases slightly with increase in temperature and pressure for tetragonal $PbVO_3$. The value of Gruneisen parameter (Table 2) seems to be reasonable since its value ranges from 2 - 3 for perovskite type compounds as reported earlier [25].

**Table 2.** Bulk modulus, Cohesive and Thermal properties of $PbVO_3$ (Cubic-C, Tetragonal-T)

| Temp. (K) | Pressure(GPa)/ Structure | $B_T$ *(Gpa) (AIM)* | $\phi$ *(eV)* | $\phi$ *(eV) (kapustinskii equation)* | $f$ *(Nm^-1)* | $\nu$ *(THz)* | $\theta_D$ *(K)* | $\gamma$ |
|---|---|---|---|---|---|---|---|---|
| 295 | 4-6 /C | 91.5<br>97.3[a] | -131.19 | -132.93 | 18.08 | 7.74 | 393.27<br>368[c] | 1.15 |
| 90 | 0 /T | 116.7 | -138.62 | -140.57 | 27.31 | 9.52 | 483.38 | 2.10 |
| 530 | 2 /T | 119.77 | -138.81<br>-140.07[b] | -140.89 | 28.69 | 9.75 | 495.38 | 2.18<br>[2-3][d] |

[a]ref [20], [b]ref [21], [c] ref. [24], [d] ref [25]

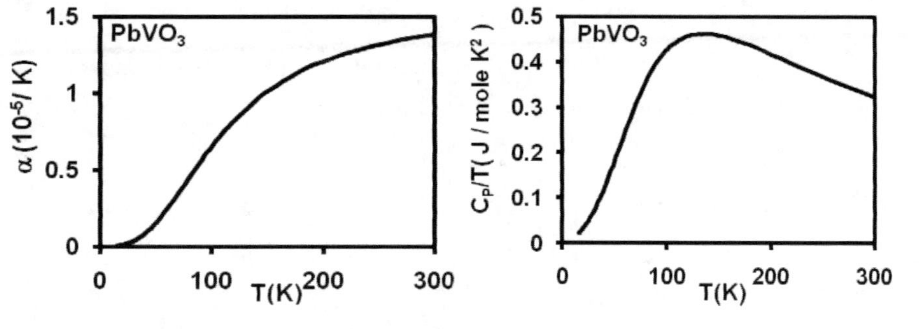


**Figure. 2** Temperature dependence of thermal expansion coefficient (left) and Variation of $C_P$ /T as a function of temperature (right).

The heat capacity ($C_P$) (20K $\leq T \leq$ 300 K) for this multiferroic compound is shown in Figure1. Also we have compared the specific heat for the two pressure-induced phases (tetragonal and cubic), Figure 1concludes that higher the pressure, higher is the specific heat of $PbVO_3$. The specific heat curve is smooth between 1.8 and 270 K and suggests the absence of phase transitions in $PbVO_3$ in this temperature range. The conclusion is supported by thermal expansion curve depicted in Figure 2. Our results on specific heat have quite good match with the experimental work of Tsirlin *et al* [1]. Besides we have computed the $C_P/T$ curve at temperature 20K $\leq T \leq$ 300 K and displayed in Figure 2.

Thus, a satisfactory prediction of the cohesive and thermal properties for the multiferroic compound attained by us is remarkable in view of inherent simplicity and less parametric nature of Modified Rigid Ion Model (MRIM). Our results on specific heat can be further improved by including the magnetic contribution. The theoretical values of thermal properties presented in this paper can serve as a guide to the experimental workers in the future.

**Acknowledgements**
Atahar Parveen is thankful to DST for providing financial support under WOS-A.

**References**
[1] Tsirlin A A, Belik A A, Roman V S, Antipov E V, Muromachi E T and Rosner H 2008 *Phys Rev B* **77** 092402
[2] Roman V S, Chernaya V V, Tsirlin A A, Chizhov P S, Sklovsky D E, and Antipov E V 2004 *Chem Mater* **16** 3267-3273
[3] Yan S Y, Xie Y, Liu T and Yu H 2010 *J Phys Condens Matter* **22** 125501
[4] Ming X, Yin J W, Wang X L, Wang C Z , Huang Z F , Chen G 2010 *Solid State Sciences* **12** 938–945
[5] Belik A A, Azuma M, Saito T, Shimakawa Y and Mikio Takano 2005 *Chem Mater* **17** 269-273
[6] Uratani Y , Shishidou T, Ishii F, Oguchi T 2005 *Jpn J Appl Phys* Part 1 **44** 7130
[7] Singh D J 2006 *Phys. Rev. B* **73** 094102
[8] Martin L W, Zhan Q, Suzuki Y, Ramesh R, Chi M, Browning N, Mizoguchi T and Kreisel J 2007 *Appl Phys Lett* **90** 062903
[9] Kumar A, Martin L W, Denev S, Kortright J B, Suzuki Y, Ramesh R and Gopalan V 2007 *Phys Rev B* **75** 060101(R)
[10] Kumar A, Podraza N J, Denev S, Li J, Martin L W, Chu Y H, Ramesh R, Collins R W and Gopalan V 2008 *Appl Phys Lett* **92** 231915
[11] Shpanchenko R V, Chernaya V V, Tsirlin A A, Chizhov P S, Sklovsky D E, Antipov E V, Khlybov E P, Pomjakushin V, Balagurov A M, Medvedeva J E, Kaul E E and Geibel C 2004 *Chem. Mater* **16** 3267
[12] Srivastava Archana and Gaur N K 2009 *J Phys Condens Matter* **21** 096001-14
[13] Srivastava A and Gaur N K 2009 *J Magn. & Magn. Mater* **321** 3854
[14] Parveen Atahar, Srivastava A, Gaur N K 2010 *Int. Ferroelectrics* **121** 129
[15] Parveen Atahar, Srivastava A, Gaur N K 2011 *AIP conf series* **1349** 975; Gaur N K, Parveen Atahar, Srivastava A and Thakur Rasna 2011 *AIP conf series* **1349** 973
[16] Hafemiester D W and Flygare W H 1965 *J Chem Phys* **43** 795-800
[17] Slater J C and Kirkwood J G 1931 *Phys Rev* **37** 682-97
[18] Shannon R D1976 *Acta Crystallogr A* **32** 751
[19] Pendas A M, Costales A, Blanco M A, Recio J M and Luana V 2000 *Phys Rev* B **62** 13970-8
[20] Zhua J L, Fenga S M, Wanga L J, Jina C Q, Wangb X H, Lib L T , Lic Y C, Lic X D and Liuc J 2010 *High Pressure Research* **30** 265–272
[21] Khan R T A and Akhtar M J 2006 *Solid State Commun* **137** 110
[22] Glasser Leslie 1995 *Inorg. Chem* **34** 4935-6
[23] Fang C M and Ahuja R 2006 *Phys Earth Planet Inter* **157** 1
[24] Belik A A and Muromachi E T 2006 *J Solid State Chem* **179** 1650–1658
[25] Radaelli P G and Cheong S W 2002 *Phys. Rev.* B **66** 094408

Journal of Physics: Conference Series **377** (2012) 012060 doi:10.1088/1742-6596/377/1/012060

# Enhanced Anharmonicity Under Pressure

**Ion Errea**[1,2,3], **Bruno Rousseau**[2,3], **Aitor Bergara**[1,2,3]

[1]Materia Kondentsatuaren Fisika Saila, Zientzia eta Teknologia Fakultatea, Euskal Herriko Unibertsitatea, 644 Postakutxatila, 48080 Bilbao, Basque Country, Spain
[2]Donostia International Physics Center (DIPC), Paseo de Manuel Lardizabal 4, 20018, Donostia, Basque Country, Spain
[3]Centro de Física de Materiales CFM - Materials Physics Center MPC, Centro Mixto CSIC-UPV/EHU, Paseo Manuel de Lardizabal 5, 20018 Donostia, Basque Country, Spain

E-mail: ion.errea@ehu.es

**Abstract.** Contradicting common sense, pressure does not monotonically harden the phonons in many systems but makes some specific modes soften at given points of the first Brilloiun zone, even inducing dynamical instabilities that drive structural phase transitions. As the harmonic part of the ionic potential becomes smaller, higher order terms turn out to be more and more important. In $AlH_3$, for instance, anharmonicity suppresses the predicted high superconducting transition temperature at 110 GPa in agreement with experiments. Furthermore, anharmonicity stabilizes the high-pressure simple cubic phase of calcium even at zero temperature, explaining its mechanical stability. We will review the calculations performed in these two systems and show that anharmonicity can be tackled making use of perturbation theory or the so called self-consistent harmonic approximation.

## 1. Introduction

The discovery that lithium was a good superconductor reaching the outstanding 20 K value for $T_c$ at 48 GPa [1, 2] clearly addressed that Mathias's rule [3] is obsolete when determining whether an element might be a good superconductor or not under pressure. Furthermore, it is now well established experimentally that many elements that do not superconduct at ambient pressure become superconductors when compressed regardless of their average number of valence electrons [4]. For instance, although Mathias's rule forbids superconductors with one and six valence electrons, some alkaline metals, like the above mentioned lithium and cesium, and several group VI elements, such as oxygen, sulfur, selenium and tellurium, have been found to become superconductors [1, 2, 5, 6, 7, 8]. Moreover, calcium has recently set a new record for the superconducting transition temperature of an element, reaching 29 K at 216 GPa [9]. Remarkably, pressure not only enhances superconductivity in the elements, but several binary alloys also have been reported to dramatically increase their $T_c$ [10, 11].

The enhancement of the superconducting transition temperature is often related to the counterintuitive softening of some specific phonons in a given region of the first Brillouin zone (1BZ). This is counterintuitive because when atomic distances are reduced, interatomic forces are expected to become larger, thus leading to a phonon frequency increase. When phonons frequencies are lowered, the electron-phonon coupling is enhanced. This is rather straightforward

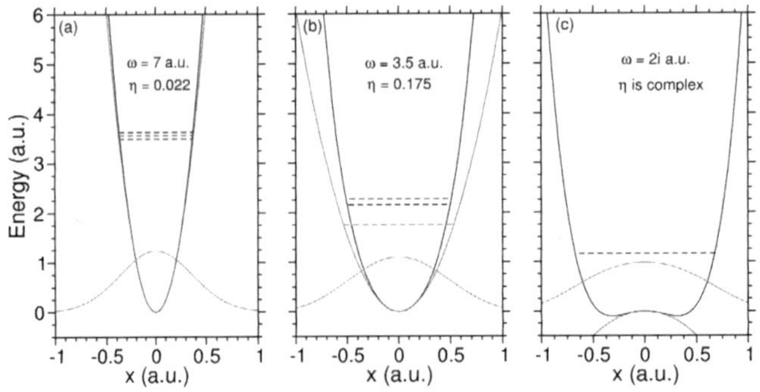


**Figure 1.** The $\frac{1}{2}\omega^2 x^2 + \alpha x^4$ potential of Eq. (2) (black curve), its harmonic part (red curve) and the ground state eigenfunction of $\hat{H}$ (green curve) for different values of $\omega$ keeping the same $\alpha$. The harmonic ground state energy, $\varepsilon_h = \omega/2$ (red dashed line), its first order correction given by perturbation theory, $\varepsilon_p = \omega/2 + 3\eta$ (blue dashed line), and the ground state numerical energy, $\varepsilon_n$ (black dashed line), are depicted as well.

considering that the single mode electron-phonon coupling parameter is given by

$$\lambda_\nu(\mathbf{q}) = \frac{1}{\pi N(\varepsilon_F)} \frac{\gamma_\nu(\mathbf{q})}{\omega_\nu^2(\mathbf{q})}, \tag{1}$$

where $N(\varepsilon_F)$ is the density of states at the Fermi level, $\gamma_\nu(\mathbf{q})$ is the phonon linewidth associated to the electron-phonon interaction of mode $\nu$ at momentum $\mathbf{q}$ and $\omega_\nu(\mathbf{q})$ is its frequency. Therefore, when a specific mode softens in a given region of the 1BZ, its $\lambda_\nu(\mathbf{q})$ is considerably enhanced. The emergence of large values of $T_c$ under pressure in lithium [12, 13, 14], sulfur [15], yttrium [16] and tellurium [17] perfectly fits into this picture since in all these cases a given phonon softens at a specific point in the 1BZ. Indeed, it seems that the lower the frequency of the softened mode, the larger $T_c$ [12, 16, 17]. The theoretical predictions recently performed in the group III hydrides also follow this trend since the predicted $T_c$ is larger the lower the hydrogen optical modes [18, 19].

The origin of the phonon softening has often been attributed to Fermi surface nesting, which is related to the electronic complexity and departure from free electron-like behavior induced by pressure [20]. This seems to be the case in the above mentioned Li [12, 21] and S [15]. However, strong Kohn anomalies can appear even in systems with spherical Fermi surfaces provided the value of the pseudopotential is large enough at $2k_F$ [22], where $k_F$ is the Fermi momentum. Interestingly, when increasing the electron density, the value of the pseudopotential at $2k_F$ is normally enhanced and, therefore, phonon anomalies are expected to appear under pressure even within the free electron-like model.

Regardless of its origin, softening leads to an important consequence: the increasing relevance of anharmonic effects. The reason is rather straightforward: if the harmonic part of the ionic potential is small, the anharmonic terms become qualitatively important and should not be neglected. This is exemplified by the one-dimensional anharmonic potentials of Fig. 1. In the figure, the eigenvalues of the Hamiltonian

$$\hat{H} = \frac{1}{2}\hat{p}^2 + \frac{1}{2}\omega^2\hat{x}^2 + \alpha\hat{x}^4 = \omega\left[\hat{a}^\dagger\hat{a} + \frac{1}{2} + \eta(\hat{a} + \hat{a}^\dagger)^4\right] \tag{2}$$

are plotted for different values of $\omega$ keeping $\alpha$ fixed. In Eq. (2), $\eta = \alpha/4\omega^3$, $\hat{x}$ and $\hat{p}$ are position and momentum operators, and $\hat{a}^\dagger$ and $\hat{a}$ creation and annihilation operators. In the figure, the harmonic ground state eigenvalue, $\varepsilon_h = \omega/2$, its first order correction given by perturbation theory, $\varepsilon_p = \omega/2+3\eta$, and the ground state eigenvalue obtained solving numerically the associated Schrödinger equation, $\varepsilon_n$, are compared for three different values of the harmonic frequency, $\omega$. As shown in Fig. 1(a), when the harmonic frequency is large and $\eta$ small, the harmonic part of the potential approximates very closely the total potential, and $\varepsilon_h$, $\varepsilon_p$ and $\varepsilon_n$ are very similar. Thus, the harmonic approximation is extremely accurate. However, when $\omega$ is reduced as in Fig. 1(b), the harmonic energy starts to differ from the exact value, but first-order perturbation theory still yields a good approximation to the energy. Finally, as depicted in Fig. 1(c), if the harmonic frequency is slightly imaginary, then, the harmonic approximation completely breaks down and it is impossible to apply perturbation theory directly as the energy of the unperturbed system has no lower bound. On the other hand, the exact solution still yields a positive energy and, despite being wider, the wave function is similar to those obtained previously and peaks at $x = 0$.

The main message to be drawn from this is that the harmonic approximation can often lead to a wrong interpretation of a system's behavior. For instance, we might think that a system is completely unstable while it is observed experimentally and that eigenvalues are too low with respect to the real excitations of the system. Undoubtedly, taking into account anharmonicity can correct all these issues. Considering that under pressure phonon softening becomes very important, anharmonicity becomes crucial as well. In order to stress this idea, here we will briefly review the strong effect that anharmonicity has in the superconducting properties of $AlH_3$ at 110 GPa [23] and of simple cubic (sc) calcium at 50 GPa [24].

## 2. Suppression of $T_c$ in $AlH_3$

It has been suggested that metal hydrides could be good superconductors at experimentally attainable pressures as the metal ions act to precompress hydrogen [25], which is itself expected to be a large $T_c$ superconductor at very high pressures. In this context, alane ($AlH_3$) has received considerable attention: a metallic phase of cubic symmetry has been found to be stable above 70 GPa [19, 26]. Computations based on the harmonic approximations suggested that this phase could have a $T_c$ as large as 24 K at 110 GPa, but measurements could not detect any superconducting transition down to 4 K.

As it turns out, most of the electron-phonon coupling in this system comes from a small region of the first Brillouin zone. In order to identify the origin of the discrepancy between computed and measured $T_c$, selected representative phonon modes at the X point were studied further [23]. Frozen phonon calculations suggested that, although the system is stable, anharmonic terms significantly renormalize the shape of the effective ionic potential, akin to the situation presented in Fig. 1(b). Although indicative, the potential obtained from frozen ionic displacements commensurate with the unit cell is insufficient to describe quantitatively the anharmonic renormalization of the phonon frequencies; a method was devised to extract all the necessary anharmonic coefficients from *ab initio* calculations in order to compute the frequency renormalization to lowest order in perturbation theory [23].

It was found that the harmonic phonon frequencies of the modes in question were renormalized by anharmonicity from 19 to 32 meV for one mode and from 87 to 109 meV for the other. Given the relationship between electron-phonon coupling and frequency, as given by Eq. (1), it is clear that an increase in frequency suppresses the coupling; in the case of $AlH_3$, it was argued that this suppression could be sufficient to reduce the predicted $T_c$ to below 5 K, consistent with the observation that no transition appears to occur down to 4 K.

## 3. Stabilization and superconductivity of simple cubic calcium

The situation in sc Ca is even worse considering that, as it has been noted already [27, 28, 29, 30], it presents imaginary phonon frequencies throughout the whole 1BZ. In this sense many

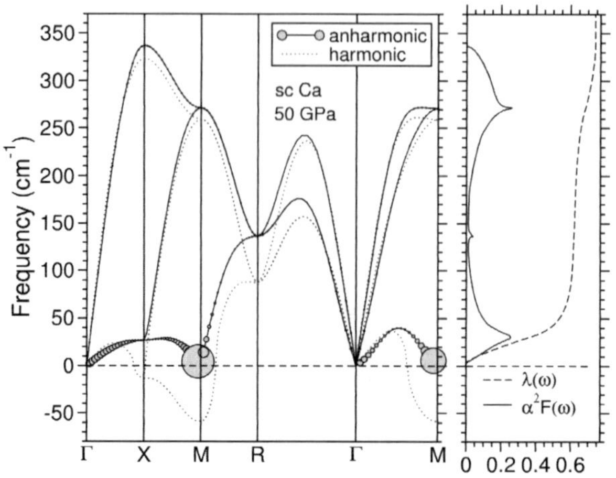


**Figure 2.** (Left panel) Harmonic phonon spectra and renormalized anharmonic phonon spectra at 0 K of sc Ca at 50 GPa. For the 0 K anharmonic branches the value of $\lambda_\nu(\mathbf{q})$ is proportional to the area of each filled circle. (Right panel) At zero temperature, the anharmonic results for the integrated electron-phonon coupling parameter, $\lambda(\omega)$, and the Eliashberg function, $\alpha^2F(\omega)$.

authors have doubted whether the sc is the real ground state structure of Ca in the 32-119 GPa pressure range where it is experimentally found [31, 32, 33] and alternative phases have been proposed [34]. Therefore, we are in the situation sketched in Fig. 1(c) and the method based on perturbation theory used in Section 2 cannot be employed. However, as we have recently demonstrated [24], the self-consistent harmonic approximation (SCHA) [35, 36] can be employed to build a variational nonperturbative approach that allows to tackle the problem of imaginary phonons efficiently. According to our results the sc phase is dynamically stable when anharmonicity is taken into account [24].

The basic idea behind the SCHA is that the most adequate phonon frequencies of the system are not those given by the harmonic approximation, but those that minimize the free energy. The ionic Hamiltonian $\hat{H} = \hat{T} + \hat{U}$, with $\hat{T}$ the ionic kinetic energy and $\hat{U}$ the potential, is redefined as $\hat{H}_0 + \hat{H}_1$, with $\hat{H}_0 = \hat{T} + \hat{U}_2^0$ and $\hat{H}_1 = \hat{U} - \hat{U}_2^0$. $\hat{U}_2^0$ is an arbitrary harmonic term that yields real frequencies. In this case, the Gibbs-Bogoliubov inequality states that the exact free energy of the system is always lower than the free energy calculated with $\hat{H}_0$ plus the statistical quantum average of $\hat{H}_1$,

$$F \leq F_0 + \langle \hat{H}_1 \rangle_0. \tag{3}$$

Hence, if we find the $\hat{U}_2^0$ that minimizes Eq. (3), the frequencies that diagonalize $\hat{U}_2^0$ will be the renormalized anharmonic frequencies of the system. In the calculation, the $\hat{U}$ potential was expanded up to fourth order and anharmonic coefficients were obtained in the whole 1BZ. As in the case of $AlH_3$, these coefficients were estimated taking numerical second derivatives of dynamical matrices calculated in supercells [24].

As shown in Fig. 2, the imaginary branches of sc Ca at 50 GPa become positive when anharmonicity is treated according to the SCHA. Thus, quantum anharmonic effects fully stabilize the high pressure sc phase of Ca in agreement with experiments. Moreover, the renormalized anharmonic phonon spectra allowed us to estimate the renormalized Eliashberg function. This is simply done substituting in Eq. (1) the harmonic frequencies by the anharmonic frequencies. The Eliashberg function is shown in Fig. 2 and shows a clear peak related to the low energy mode. Integrating it we obtain the value of $\lambda = 0.74$ for the electron-phonon coupling parameter and $\omega_{log} = 53\,\mathrm{K}$ for the logarithmic frequency average. With these values and using the standard $\mu^* = 0.1$ in the Allen-Dynes modified McMillan equation [37] we obtain $T_c = 2.1\,\mathrm{K}$. We should stress that this value is in close agreement with experimental results [38, 39] and that could not have been estimated earlier as a consequence of the imaginary frequencies that the harmonic approximation yields.

## 4. Conclusions

In this work we have argued that due to the rather common softening of phonon modes under pressure, anharmonicity becomes crucial in many cases and cannot be neglected in the calculations. We have briefly reviewed the effect that anharmonicity has in $AlH_3$ at 110 GPa and sc Ca. In the former, the frequencies renormalized by anharmonicity reduce the value of the electron-phonon coupling parameter and, as a consequence, suppress $T_c$ as hinted by experiments. In the latter, we have seen that anharmonicity stabilizes the structure and that superconductivity can only be understood including anharmonic corrections in the calculations. We also have presented two different methods to treat anharmonicity: perturbation theory and the SCHA. As a final remark, we should stress that whenever soft phonons appear under pressure anharmonicity might have a crucial effect in the predicted value of $T_c$.

## Acknowledgments

The authors are grateful the Department of Education, Universities and Research of the Basque Government, UPV/EHU (Grant No. IT-366-07) and the Spanish Ministry of Science and Innovation (Grant No. FIS2010-19609-C02-00) for financial support. Computer facilities were provided by the DIPC.

## References

[1] Shimizu K, Ishikawa H, Takao D, Yagi T and Amaya K 2002 *Nature* **419** 597
[2] Struzhkin V V, Eremets M I, Gan W, Mao H k and Hemley R J 2002 *Science* **298** 1213
[3] Matthias B T 1955 *Phys. Rev.* **97** 74–76
[4] Ashcroft N W 2002 *Nature* **419** 569
[5] Wittig J 1970 *Phys. Rev. Lett.* **24** 812–815
[6] Shimizu K, Suhara K, Ikumo M, Eremets M I and Amaya K 1998 *Nature* **393** 767
[7] Struzhkin V V, Hemley R J, Mao H k and Timofeev Y A 1997 *Nature* **390** 382
[8] Akahama Y, Kobayashi M and Kawamura H 1992 *Solid State Communications* **84** 803 – 806
[9] Sakata M, Nakamoto Y, Shimizu K, Matsuoka T and Ohishi Y 2011 *Phys. Rev. B* **83** 220512
[10] Sanfilippo S, Elsinger H, Núñez Regueiro M, Laborde O, LeFloch S, Affronte M, Olcese G L and Palenzona A 2000 *Phys. Rev. B* **61** R3800–R3803
[11] Debessai M, Matsuoka T, Hamlin J J, Gangopadhyay A K, Schilling J S, Shimizu K and Ohishi Y 2008 *Phys. Rev. B* **78** 214517
[12] Kasinathan D, Kuneš J, Lazicki A, Rosner H, Yoo C S, Scalettar R T and Pickett W E 2006 *Phys. Rev. Lett.* **96** 047004
[13] Profeta G, Franchini C, Lathiotakis N N, Floris A, Sanna A, Marques M A L, Lüders M, Massidda S, Gross E K U and Continenza A 2006 *Phys. Rev. Lett.* **96** 047003
[14] Bazhirov T, Noffsinger J and Cohen M L 2010 *Phys. Rev. B* **82** 184509
[15] Degtyareva O, Magnitskaya M V, Kohanoff J, Profeta G, Scandolo S, Hanfland M, McMahon M I and Gregoryanz E 2007 *Phys. Rev. Lett.* **99** 155505
[16] Yin Z P, Savrasov S Y and Pickett W E 2006 *Phys. Rev. B* **74** 094519
[17] Mauri F, Zakharov O, de Gironcoli S, Louie S G and Cohen M L 1996 *Phys. Rev. Lett.* **77** 1151–1154
[18] Kim D Y, Scheicher R H, Mao H k, Kang T W and Ahuja R 2010 *Proc. Natl. Acad. Sci. USA* **107** 2793–2796
[19] Goncharenko I, Eremets M I, Hanfland M, Tse J S, Amboage M, Yao Y and Trojan I A 2008 *Phys. Rev. Lett.* **100** 045504
[20] Rousseau B, Xie Y, Ma Y and Bergara A 2011 *Eur. Phys. J. B* 1–14
[21] Rodriguez-Prieto A, Bergara A, Silkin V M and Echenique P M 2006 *Phys. Rev. B* **74** 172104
[22] Rousseau B, Errea I and Bergara A 2010 *Journal of Physics and Chemistry of Solids* **71** 1159 – 1164
[23] Rousseau B and Bergara A 2010 *Phys. Rev. B* **82** 104504

[24] Errea I, Rousseau B and Bergara A 2011 *Phys. Rev. Lett.* **106** 165501
[25] Ashcroft N W 1968 *Phys. Rev. Lett.* **21** 1748–1749
[26] Pickard C J and Needs R J 2007 *Phys. Rev. B* **76** 144114
[27] Errea I, Martinez-Canales M, Oganov A R and Bergara A 2008 *High Press. Res.* **28** 443–448
[28] Gao G, Xie Y, Cui T, Ma Y, Zhang L and Zou G 2008 *Solid State Commun.* **146** 181 – 185
[29] Teweldeberhan A M and Bonev S A 2008 *Phys. Rev. B* **78** 140101
[30] Teweldeberhan A M, Dubois J L and Bonev S A 2010 *Phys. Rev. Lett.* **105** 235503
[31] Yabuuchi T, Nakamoto Y, Shimizu K and Kikegawa T 2005 *J. Phys. Soc. Jpn.* **74** 2391–2392
[32] Gu Q F, Krauss G, Grin Y and Steurer W 2009 *Phys. Rev. B* **79** 134121
[33] Mao W L, Wang L, Ding Y, Yang W, Liu W, Kim D Y, Luo W, Ahuja R, Meng Y, Sinogeikin S, Shu J and Mao H k 2010 *Proc. Natl. Acad. Sci. USA* **107** 9965–9968
[34] Oganov A R, Ma Y, Xu Y, Errea I, Bergara A and Lyakhov A O 2010 *Proc. Natl. Acad. Sci. USA* **107** 7646–7651
[35] Hooton D J 1955 *Philosophical Magazine Series 7* **46** 422–432
[36] Koehler T R 1966 *Phys. Rev. Lett.* **17** 89–91
[37] Allen P B and Dynes R C 1975 *Phys. Rev. B* **11** 1895–1905
[38] Okada S, Shimizu K, Kobayashi T C, Amaya K and Endo S 1996 *J. Phys. Soc. Jpn.* **65** 1924–1926
[39] Yabuuchi T, Matsuoka T, Nakamoto Y and Shimizu K 2006 *J. Phys. Soc. Jpn.* **75** 083703

# High Pressure Phase Transition Phenomena in CeS and LaS

**Madhu Singh[1], Sadhna Singh[2*], Faisal Shareef M.[2] and Purvee Bhardwaj[2]**
*[1]Govt. SGS PG College, Ganjbasoda- 464221, India*
*[2] High Pressure Research Lab, Department of Physics, Barkatullah University, Bhopal -462026, India*
**E-mail:drsadhna100@gmail.com*

**Abstract.** In present study we have investigated the high pressure phase transition pressures, volume collapses, elastic and thermo-physical properties of cerium and lanthanum sulphides using the three body interaction potential (TBIP) approach. The CeS and LaS belong to the wide class of binary rare-earth monochalcogenides which crystallize in NaCl-type structure and undergo a transformation into CsCl-type structure under pressure. The values of phase transition pressure and associated volume collapse estimated by us are found to be well suited with experimental values.

## 1. Introduction:

When pressure is increased CeS and LaS undergo first order NaCl (B1) to CsCl (B2) structural phase transition which has been shown by high pressure x-ray diffraction experiments. The phase transition from B1 to B2 in CeS and LaS would likely to occur in the range of 30-35 GPa and 27-28 GPa respectively [1, 2]. Theoretically, the modified inter-ionic potential theory [3] found the transition pressure $P_T$ equal to 32.9 GPa and the tight binding linearized-MT orbital (TB-LMTO) calculations [1] found it in the pressure range between 30 and 35GPa for CeS while the full-potential linear muffin-tin orbital study on LaS found it at 29 GPa[2]. The present study aims to shed more light on various high pressure properties of CeS and LaS using an effective potential model (TBIP). A brief account of the theoretical approach is presented in section 2 and the results are discussed in section 3.

## 2. Potential model and Method of computation:

To study these compounds under pressure, we have developed and employed three body interaction potential model (TBIP) [4]. At room temperature, the atoms get sufficient thermal energy to vibrate at its mean position. This vibration results in an overlap of the ions. Hence, the electron shell associated with each ion is subjected to deformation as a consequence of which there occurs a small amount of charge transfer between the overlapping ions. These transferred charges interact with other charges of the lattice and give rise to many body interactions. The three body interaction is dominant among the many body interactions. These TBI effects have been incorporated in the Gibb's Free Energy (G = U + P V – TS) as a function of pressure.

The Lattice energies for B1 and B2 phases are as follows:

$$U_{B1}(r) = -\alpha_m z^2 e^2 / r - (12\alpha_m z e^2 f(r))/r + 6b\beta_{ij}\exp[(r_i + r_j - r)/\rho] + 6b\beta_{ii}\exp[(2r_i - 1.414r)/\rho] + 6b\beta_{jj}\exp[(2r_j - 1.414r)/\rho] \quad (1)$$

$$U_{B2}(r') = -\alpha'_m z^2 e^2 / r' - (16\alpha'_m z e^2 f(r'))/r + 8b'\beta_{ij}\exp[(r_i + r_j - r')/\rho] + 3b'\beta_{ii}\exp[(2r_i - 1.154r')/\rho] + 3b'\beta_{jj}\exp[(2r_j - 1.154r')/\rho] \quad (2)$$

Published under licence by IOP Publishing Ltd

These lattice energies consists of long–range coulomb energy modified with three body interactions corresponding to the nearest neighbour separation r(r') [first and second terms in equation (1) and (2)] and energy due to the overlap repulsion extended up to the second neighbour ions (rest of the terms). There are three model parameters (b, ρ, f(r)) in the present TBIP, namely hardness, range and three-body force parameter respectively. The $\alpha_m$ ($\alpha'_m$) are the Madelung constants, $\beta_{ij}$ is the Pauling coefficient, ze is the ionic charge and $r_i$ ($r_j$) are the cation (anion) radii. At absolute zero of temperature, the Gibbs free energy (G) is the enthalpy (H) itself. For any two structures real (B1) and hypothetical (B2) at phase transition pressure (P), we have $G_{B1}(r) = G_{B2}(r')$. At the phase transition pressure, the difference of free energies (ΔG) approaches zero .The pressure at which $\Delta G \rightarrow 0$ is called the phase transition pressure. The associated volume with this pressure shows a sudden collapse in volume indicating that the phase transition is of first order.

**Table 1.** Input crystal data and Model parameters

| | Input Parameters | | | Model parameters | | |
|---|---|---|---|---|---|---|
| Compound | $r_i$(Å) | $r_j$(Å) | r(Å) | b($10^{-19}$J) | ρ(Å) | f(r) |
| CeS | 0.91 | 1.24 | 2.883 [a] | 4.5213 | 0.3225 | -0.0090 |
| LaS | 1.04 | 1.24 | 2.925 [b] | 3.5592 | 0.3098 | -0.0085 |

[a] Ref [4] [b] Ref[9]

To understand the elastic properties of the present compounds we have calculated second order elastic constants ($C_{11}$, $C_{12}$ and $C_{44}$) their pressure derivatives, combination of the SOECs, bulk modulus, third order elastic constants etc. following Shanker et al. [5]. In order to understand the thermophysical properties we have calculated the molecular force constant (f), infrared absorption frequency ($\nu_0$), Debye temperature($\Theta_D$), Grunneisen parameter (γ) and ratio of volume expansion coefficient to specific heat at constant volume which are directly derived from the cohesive energy expression[6].

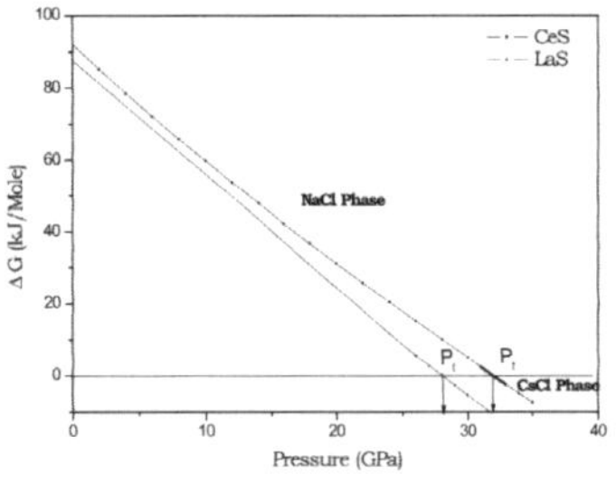


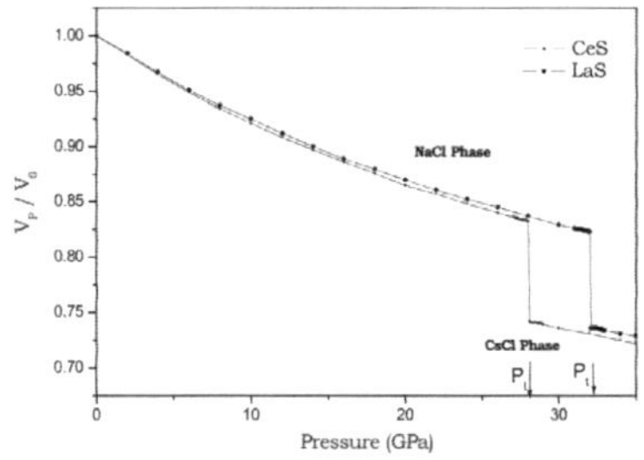


**Figure 1.** Variation of ΔG with Pressure P (GPa) **Figure 2.** Variation of $V/V_0$ with Pressure

**3. Results and Discussion:**

The input data and calculated model parameters are shown in table 1. The structural phase transition pressures are calculated by minimizing the Gibb's free energies. We have plotted the Gibb's free energy differences ΔG against pressure (P) for Ces and LaS which is shown in the figure 1. This figure shows that on increasing pressure ΔG decreases and we reach a point at which ΔG approaches zero, that pressure is the phase transition pressure (Pt) which indicated by arrows. The phase transition pressures thus obtained are listed in table 2 along with the available experimental and other theoretical data.

**Table 2.** Calculated transition pressure and volume collapse

| | Transition Pressure(GPa) | | | Volume collapse (%) | | |
|---|---|---|---|---|---|---|
| Compound | Present | Expt. | Others | Present | Expt. | Others |
| CeS | 32 | - | 30-35 [a], 32.9[b], 33.7[c] | 8.5 | - | 7.0[b], 7.5[c] |
| LaS | 28 | 27-28[d] | 29.3[d], 24.9 [e] | 9.0 | 9.5[d] | 10.3[d], 8.40[e] |

[a] Ref [1] [b] Ref [3] [c] Ref [11] [d] Ref [2] [e] Ref [10]

The values of difference in relative volume changes [$\Delta V(P)/V(0)$] associated with compression have been obtained and plotted in figure 2. The magnitude of relative volume changes at the transition pressure for CeS and LaS are 8.5% and 9% respectively and they are also compared with the available data in table 2.

**Table 3.** Second order elastic constants ($\times 10^{12}$dyne/cm$^2$).

| Compound | $C_{11}$ | | $C_{12}$ | | $C_{44}$ | | $C_{44}/B_T$ |
|---|---|---|---|---|---|---|---|
| | Present | Exp./Others | Present | Exp./Others | Present | Exp./Others | |
| CeS | 1.71 | 2.3497[a]/1.5628[b] | 0.64 | 0.130[a]/0.400[b] | 0.52 | 1.1105[a]/0.4014[b] | 0.12 |
| LaS | 1.66 | 2.279[c]/2.34[d] | 0.60 | 0.180[c]/0.23[d] | 0.49 | 0.222[c]/0.25[d] | 0.13 |

[a] Ref [12] [b] Ref [3] [c] Ref [2] [d] Ref [11]

The calculated values of the second order elastic constants ($C_{11}$, $C_{12}$ and $C_{44}$) are given in table 3. The study of elastic behaviour under pressure is well known to supply useful information about change in the character of covalent and ionic forces induced in the crystal as it is subjected to the phase transformation. We have plotted the variations of $C_{11}$, $C_{12}$ and $C_{44}$ under pressure in figure 3. It is clear that from these figures that $C_{11}$ varies largely under the effect of pressure as compared to the variations in the $C_{12}$ and $C_{44}$. Since the elastic constant $C_{11}$ represents elasticity in length, a longitudinal strain produces a change in $C_{11}$. The $C_{12}$ and $C_{44}$ are related to the elasticity in shape, which is a shear constant. A transverse strain causes a change in shape without any change in volume. Therefore $C_{12}$ and $C_{44}$ are less sensitive to pressure as compared to $C_{11}$. According to the Vukcevich [7] for a crystal to be in the stable phase shear elastic constant $C_{44}$ is non-zero. We obtained non zero value of $C_{44}$ as tabulated in table 3 which satisfy condition given by Vukcevich. Demarest et.al [8] proposed a slight modification of the Born stability criterion that a phase transition takes place when the ratio $C_{44}/B_T$ reaches a critical value in the neighbourhood of 0.14. On the same guidelines we have obtained exactly the same trends from TBIP as earlier reported by Singh [9] which is also tabulated in table 3 and is shown in figure 4.

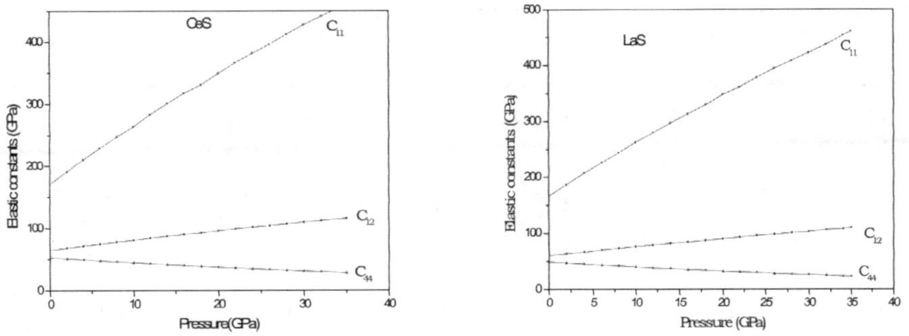


**Figure 3.** Variation of $C_{11}$, $C_{12}$ and $C_{44}$ with Pressure P (GPa) for CeS and LaS

Further the pressure derivatives of elastic moduli $dB_T/dP$, $dC_S/dP$ and $dC_{44}/dP$ along with combination of second order elastic constants $C_S$ and $C_L$ ($B_T$, $C_S$ and $C_L$ are the bulk modulus, shear modulus and the tetragonal modulus of the crystal respectively.) are calculated to investigate anharmonic properties of proposed compounds and are tabulated in table 4.

**Table 4.** Pressure derivatives of elastic moduli and their combinations ($\times 10^{12}$dyne/cm$^2$).

| Compound | $dB_T/dP$ | $dCs/dp$ | $dC_{44}/dP$ | $C_S$ | $C_L$ |
|---|---|---|---|---|---|
| CeS | 5.037 | 3.883 | -0.112 | 0.531 | 1.70 |
| LaS | 5.091 | 3.883 | -0.112 | 0.532 | 1.62 |

The TOEC's at zero pressure are given in the table 5 for CeS and LaS which shows that the $C_{111}$, $C_{112}$,

$C_{166}$ are negative and $C_{123}$, $C_{144}$, $C_{456}$ are positive in magnitude. The investigated values of thermo physical properties are well tabulated in table 6. The comparison of the present values with that of experimental and other theoretical results is not possible because of lack of availability but their values are similar and close to those obtained for the other monochalcogenides having same NaCl type structure.

**Table 5.** Third order elastic constants ($x10^{12}$dyne/cm$^2$).

| Compound | $C_{111}$ | $C_{112}$ | $C_{123}$ | $C_{144}$ | $C_{166}$ | $C_{456}$ |
|---|---|---|---|---|---|---|
| CeS | -29.1495 | -2.9602 | 1.0937 | 0.8917 | -2.1273 | 0.8674 |
| LaS | -28.6522 | -2.7699 | 1.0344 | 0.8432 | -1.9493 | 0.8203 |

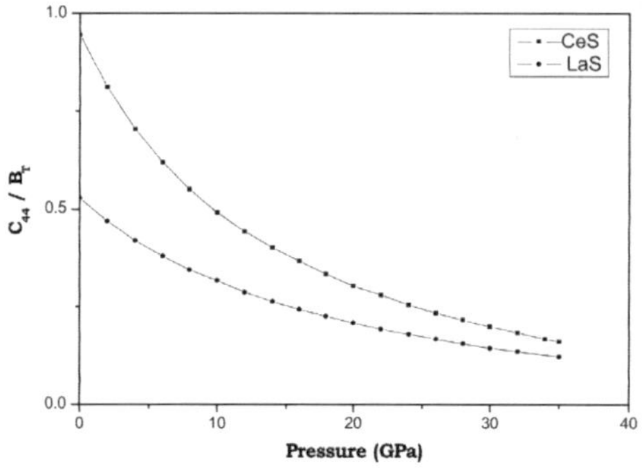


**Figure 4.** Variation of $C_{44}$ / $B_T$ with Pressure P (GPa)

## 4. Conclusion:

We may conclude that the TBIP model is well suited for analyzing the structural, elastic and thermophysical properties of CeS and LaS showing B1 to B2 structural phase transition.

**Table 6.** Calculated thermo physical properties

| Compound | f ($10^4$ dynes/cm) | $\nu_0$ ($10^4$dynes/cm) | $\Theta_D$(K) | $\gamma$ | $\alpha_v/C_v$ ($10^3$ J) |
|---|---|---|---|---|---|
| CeS | 7.6930 | 6.704 | 320. 6 7 | 1.544 | 7.55 |
| LaS | 7.8551 | 6.732 | 324.33 | 1.616 | 7.960 |

**Acknowledgement:**

Two of our authors (SS and FS) are thankful to BRNS Mumbai for financial support of this work.

## References

[1] Svane A, Temmerman W, Szotek Z 1999 *Phys. Rev. B* **59** 7888-92
[2] Vaitheeswaran G, Kanchana V, Heathman S, Idiri M, Le Bihan T, Delin A and Johansan B 2007 *Phys. Rev. B* **75** 184108-114
[3] Srivastava V, Bandyopadhyay A K, Jha P K and Sanyal S P 2003 *J. Phys. Chem. Solids* **64** 907
[4] Singh R K 1982 *Phys. Rep. (Netherlands)* **85**, 259
[5] Shanker J, Jain V C, Singh J P 1980 *Phys Rev. B* **22** 1083
[6] Thakur K P 1978 *J. Chem. Phys.* **75** 300
[7] Vukcevich M R 1972 *Phys. Status solidi (b)* **54** 435
[8] Demarest H H, Otta R, Anderson A L 1977 *High Pressure Research Applications in Geophysics* (Academic Press, New York) pp. 281
[9] Singh R K, Singh S 1992 *Phy. Rev. B* **45** 1019
[10] Salghetti-Drioli F, Wachter P, Degiorgi L 1999 *Solid State Commun.* **109** 773
[11] Bouhemadou A, Khenata R, Maamache M 2006 *J. Mol. Struc.:Theochem* **777** 5-10
[12] Fulde P, Keller J, Zwicknagl G 1988 *Solid State Physics* (Academic Press, New York vol 41) pp1

Journal of Physics: Conference Series **377** (2012) 012062 doi:10.1088/1742-6596/377/1/012062

# Pressure induced phase transition of XN (X=B, Ga) compounds

**M Sarwan, S Singh and R Dubey**
High Pressure Research Lab, Department of Physics, Barkatullah University, Bhopal - 462026, India

Email: madhusarwan@gmail.com

**Abstract.** The pressure induced phase transition, volume collapse and elastic properties have been investigated by means of two different models (1) modified interaction potential model (MIPM) which includes the three body interaction, covalency effect, overlap repulsive interaction up to second neighbour ions and van der Waal interaction. (2) Interaction potential model (IPM) which exclude the covalency effect in previous (IPM) model. The BN and GaN transform from B3 to B1 structure under high pressure ($p_t$= 555 GPa and 37GPa) and show the volume collapse of 3.6% and 16.53%. The elastic constants have also been obtained. The results are in good agreement with the experimental and theoretical data where available.

## 1. Introducton

Group III- nitrides generally crystallize in wurtzite (W) structure, but zinc-blend (ZB) structure have been synthesized by epitaxial growth on suitable substrates [1]. The ZB phase has several advantages for device applications including easy cleavage, smaller band gap, and higher carrier mobilities [1]. Cubic BN exhibits three different structures hexagonal, cubic zinc-blend structure and wurtzite structure. But in zinc-blend structure the properties of BN such as high melting point, high thermal conductivity and extreme hardness make it useful for protective coatings which are related to their mixed ionic and covalent bonding [2]. The experimental studies on III-nitrides are less reported. Among these the elastic constants at zero pressure have been performed experimentally for BN only [3]. By using FP-LAPW method Zaoui et al.[4] found that the transition pressure from ZB (zinc-blend) →RS (rock-salt) is 550 GPa. Furthermore Christensen et al [5] concluded that the pressure was 850 GPa by using linear muffin-tin orbital (LMTO) calculation. The transformations from ZB→RS have also been reported by Serrano at pressure 42.1GPa for GaN [6]. N.E. Christensen et. al have been studied the optical and structural properties of III-V nitrides by using band structure and total energy calculations. These calculations were made by means of the linear muffin tin orbital (LMTO) method. They conclude that all nitrides transform from W →RS phase and BN transform from ZB→RS at pressure 850 GPa. First-principle pseudo potential method shows that BN transform from ZB→RS phase at pressure 1088 GPa with a volume reduction of 3.1% [7]. M.B. Kanoun et.al [8] was studied the pressure dependence of elastic constants of BN, AlN, GaN and InN by using first-principle full potential augmented-plane wave plus local orbital (FP-APW+lo) calculations with density functional theory in local density approximation.

The structural, mechanical and electronic properties of III- nitrides compounds have been studied by various first-principle calculations. Inspite of the above electronic, structure calculations, there is no study available using phenomenological model. The nature of interatomic forces is not well understood in these compounds. So phenomenological models based on various interatomic interaction energies to determine stable structures become important. First-principle calculations have their own disadvantages as generally van der Waals interactions are ignored in them. In the present paper, we report the structural and elastic properties of XN (X= B, Ga) compound by using modified interaction potential model (MIPM) and interaction potential model (IPM). Since the compounds are partially covalent in nature so the IPM is modified by taking covalency effect.

## 2 Present potential model and method of computation

To study the phase transition, elastic and thermo physical properties of XN (X=B, Ga) we have developed the (1) MIPM and (2) IPM. The MIPM consist of long-range coulomb attraction, three-body interaction modified by covalency effect, short-range overlap repulsive interaction and van der Waal interaction. A complete study has been made with IPM, which is similar to MIPM except that covalency effects have been excluded from it. The application of pressure on the crystal causes the decrease in their volume, which in turn leads to an increased charge transfer (or exchange) between the overlapping electron shells of the adjacent ions [9].

These TBI effects have been incorporated in the Gibbs free energy (G=U+PV-TS). Here, U is the lattice energy which at T=0K is equivalent to the lattice energy, S is the vibrational entropy at absolute temperature T. At T=0K and pressure P, the Gibbs free energy for the zinc-blend (B3, real) and NaCl (B1, hypothetical) structures are given by

$$G_{B3}(r) = U_{B3}(r) + PV_{B3} \tag{1}$$

$$G_{B1}(r') = U_{B1}(r') + PV_{B1} \tag{2}$$

With $V_{B3}$ ($3.08\,r^3$) and $V_{B1}$ ($2.00\,r'^3$) as unit cell volumes for B3 and B1 phases, respectively. The first term in (1) and (2) are lattice energies for B3 and B1 structures and they are expressed as

$$U_{B3}(r) = \frac{-\alpha_M e^2 Z^2}{r} - \frac{4\alpha_M e^2 Z f_m(r)}{r} - \left[\frac{C}{r^6} + \frac{D}{r^8}\right] + 4b\beta_{ij}\exp(r_i + r_j - r_{ij}/\rho)$$
$$+ 6b\beta_{ii}\exp(2r_i - 1.63r_{ij}/\rho) + 6b\beta_{jj}\exp(2r_j - 1.63r_{ij}/\rho) \tag{3}$$

$$U_{B1}(r) = \frac{-\alpha'_M e^2 Z^2}{r'} - \frac{6\alpha'_M e^2 Z f_m'(r)}{r'} - \left[\frac{C}{r'^6} + \frac{D}{r'^8}\right] + 6b\beta_{ij}\exp(r_i + r_j - r'_{ij}/\rho)$$
$$+ 6b\beta_{ii}\exp(2r_i - 1.414r'_{ij}/\rho) + 6b\beta_{jj}\exp(2r_j - 1.414r'_{ij}/\rho) \tag{4}$$

With $\alpha_M$ and $\alpha'_M$ as the Madelung constants for ZnS and NaCl structures, respectively. C (C') and D (D') are overall van der Waal coefficients of $B_3$ ($B_1$) phases, $\beta_{ij}$ ( i,j = 1,2 ) are the Pauling coefficients defined as $\beta_{ij} = 1 + (Z_i/n_i) + (Z_j/n_j)$ with $Z_i$ ($Z_j$ ) and $n_i$ ($n_j$ ) are the valence and the number of electrons of the i (j)th ion. Ze is the ionic charge and b (ρ) are the hardness (range) parameters, r(r') are the nearest neighbour separation for ZnS (NaCl) structure. $f_m(r)$ is the modified three-body force parameter which includes the covalency effect with three-body interaction $r_i$ ($r_j$) are the ionic radii of ions i (j). These lattice energies consist of long range Coulomb energy (first term ), modified three-body interactions corresponding to the nearest neighbour separation r (r') (second term ), van der Waal interaction (third term ), energy due to the overlap repulsion represented by Hafemeister and Flygare ( HF) type potential and extended up to the second neighbour ions (forth, fifth and sixth terms ).

We have included the effect of covalent forces in three-body force parameter f(r) on the lines of Motida [10]. In the present model $f_m(r)$ is assumed to be constructed by two parts, i.e.

1. due to covelancy $f_{cov}$ (r)
2. due to overlap distorted effect (three-body interaction) $f_{TBI}(r)$

Now modified three-body parameter $f_m(r)$ becomes

$$f_m(r) = f_{TBI}(r) + f_{cov}(r) \tag{5}$$

## 3. Results and discussion

The present interaction potential model (IPM) contains three model parameters b, ρ and f(r), namely hardness, range and three-body force parameter. We have modified the three-body force parameter $f_m(r)$ by including covalency effect in the modified interaction potential model (MIPM). These parameters are calculated by using the first derivative of lattice energy (U) and equilibrium condition expressed as:

$$\left|\frac{dU}{dr}\right|_{r=r0} = 0 \quad \text{and} \quad B1 + B2 = -1.261\,Z[Z + 8f(r)] \tag{6}$$

The input data and model parameters are given in table 1. As the stable phase is always associated with minimum of energy, we have minimised $G_{B3}$ (r) and $G_{B1}$ (r') at different pressures in order to obtain inter ionic separations r and r' for B3 and B1 phases respectively associated with minimum energies.

**Table 1** Input and model parameters of XN (X=B, Ga) compounds.

| | Input parameters | | | Model parameters | | | |
|---|---|---|---|---|---|---|---|
| | $r_1$ (Å) | $r_2$(Å) | a(Å) | b($10^{-12}$ ergs) | ρ (Å) | f(r) | $f_m$(r) |
| BN | 1.2 | 1.71 | 3.615 | 0.076 | 0.314 | -0.0647 | -0.0053 |
| GaN | 0.62 | 1.71 | 4.50 | 0.1557 | 0.160 | -0.0234 | -0.0364 |

We have evaluated the corresponding $G_{B3}$ (r) and $G_{B1}$ (r') and their respective differences $\Delta G=G_{B3}(r)-G_{B1}(r')$. A positive value of ΔG at zero pressure justifies the stability of B3 phase. As the pressure is increased these values of ΔG decrease and approach to zero at the transition pressure ($P_t$). Beyond this pressure ΔG becomes negative as the phase B1 become stable. It is found that the phase transition pressures from MIPM (IPM) is at 555 (556) GPa and 37 (38) GPa for BN and GaN are presented in table 2 and compared with the available results. The first order phase transition involving discontinuity in volume take place at the phase transition pressure. At elevated pressures, the materials undergo structural phase transition associated with a sudden change in the arrangement of the atoms. The atoms are rearranged in new position leading to a new structure. Experimentally one usually studies the relative volume change associated with the compressions. The discontinuity in volume at the transition pressure is obtained from the phase diagram. The compression curves are plotted in figure 1 to show equation of state (EOS). The values of the volume collapses (%) are depicted in table 2. It is clear from figure 1 that the volume of BN decreases up to 555 GPa, an abrupt decline in volume is observed at this pressure which is observed in first-order transformation associated with structural change from B3→ B1 phase. Similar trend is observed in GaN too. The EOS obtained from MIPM show good agreement with available theoretical results [7, 11] than IPM, which show a small deviation from MIPM and others [7, 11]. This difference is due to the exclusion of covalency effect in IPM.

**Table 2** Input and model parameters of XN (X=B, Ga) compounds.

| | Phase transition pressure (GPa) | Volume collapse (%) | Elastic constants (GPa) | | |
|---|---|---|---|---|---|
| | | | $C_{11}$ | $C_{12}$ | $C_{44}$ |
| **BN** | | | | | |
| MIPM | 555 | 3.6 | 819 | 189.6 | 475 |
| IPM | 556 | 3.7 | 817 | 188 | 474 |
| Exp. [3] | - | - | 820 | 190 | 480 |
| Other [7, 4] | 1088, 555 | 3.1 | 801.26 | 159 | 456.9 |
| **GaN** | | | | | |
| MIPM | 37 | 16.53 | 278 | 177 | 186.9 |
| IPM | 38 | 16.0 | 277 | 176 | 185 |
| Exp. | - | - | - | - | - |
| Other [11, 8] | 37 | 17 | 274.2 | 166.1 | 199 |

and give important information about nature of forces operating in solids. The study of elastic properties is also essential to understand the chemical bonds and cohesion of material. In order to obtain the values of elastic constants we have used the method reported in our earlier work [9]. The second order elastic constants (SOECs) have been obtained and they are given in table 2. The calculated SOECs are compared with experimental data and others [3, 7-8]. It is clear from table 2 that values of SOECs obtained from MIPM are good agreement with experimental data than IPM and others [7-8] where available.

## 4. Conclusion

We have applied Modified interaction potential model(MIPM) and interaction potential model (IPM) to investigate the structural phase transition and elastic properties of XN (X=B, Ga) compounds. An overall assessment shows that, our values of MIPM near to available experimental data and they are in general better matching than other theoretical results and IPM. Finally it may be concluded that the present MIPM has successfully predicted the compression curves giving the phase transition pressures, volume collapse and elastic constants for XN compounds. Our results where no experimental results are available may be tested with different theoretical and experimental methods in future.

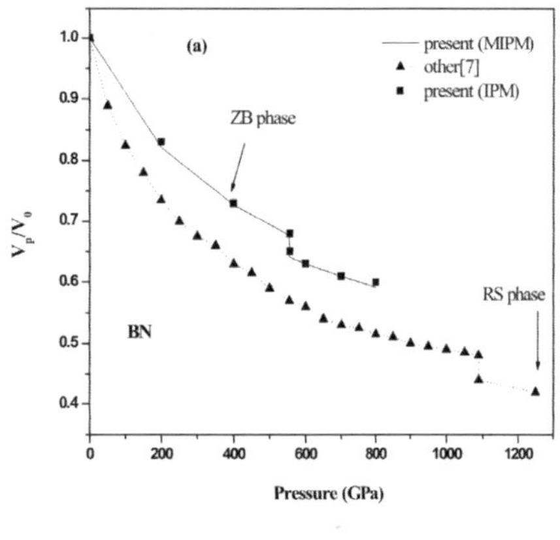


**Figure 1.** The Variation of relative volume with pressure for BN.

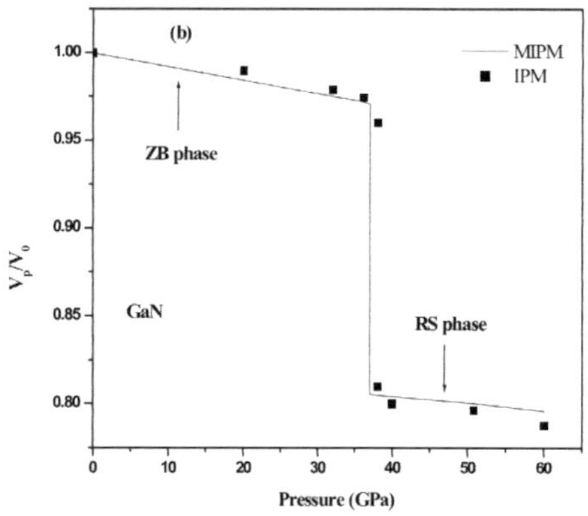


**Figure 2** The Variation of relative volume with pressure for GaN.

**Acknowledgement**

The authors are thankful to the U.G.C. New Delhi for financial support.

**Refrences**

[1] Ploog K H, Brandt O, Yang H, Menniger J and Klann R 1997 Solid State Electron. **41** 235
[2] Kanoun M B, Merad A E, Merad G, Cibert J and Aourag H 2004 Solid State Electron. **48** 1601
[3] Grimsditch M, Zouboulis E S and Polian A 1994 J. Appl. Phys. **76** 832
[4] Zaoui A and Haj Hassan F El 2001 J. Phys. Cond. Mat. **13** 253
[5] Christensen N E and Gorczyca I 1994 Phys. Rev. B **50** 4397
[6] Serrano J, Rubio A, Hernandez E, Munoz A and Mujica A 2000 Phys. Rev. B **62** 16612
[7] Cui S, Feng W, Hu H and Feng Z 2010 Cent. Eur. J. Phys. **8** 628
[8] Kanoun M B, Merad A E, Merad G, Cibert J and Aourag H 2004 Solid State Electron. **48** 1601
[9] Singh R K 1982 Phys Rep. **85** 259; Singh R K and Singh S 1992 Phys. Rev. B **45** 1019.
[10] Motida K 1986 J. Phys. Soc. Jpn. **55** 1636
[11] Xia H, Xia Q and Ruoff A L 1993 Phys. Rev. B **47** 12925

# Thermophysical properties of B1-LiF

**S G Khambholja[1], B Y Thakore[1*], N K Bhatt[1], P.N. Gajjar[2] and A R Jani[1]**
[1]Department of Physics, Sardar Patel University, Vallabh Vidyanagar, 388 120, Gujarat, India
[2]Department of Physics, University School of Science, Gujarat University, Ahmedabad, 380 009, Gujarat, India

E-mail: *brijmohanthakore@rediffmail.com

**Abstract.** Lithium fluoride (LiF) is an important technological material. It is chemically stable, thermal insulator, having found application in the field of high pressure as a pressure transmitting medium. In the present paper, we report various thermophysical properties of rock salt structured LiF using plane wave pseudopotential density functional theory within local density approximation in conjunction with quasi harmonic Debye model. Ground state properties are studied using density functional theory. To include effect to ionic motion at finite temperatures, quasi harmonic Debye model is used. Calculated ground state and finite temperature/pressure thermophysical properties are found to be in good agreement with experimental and other theoretical results.

## 1. Introduction

Among all alkali halides and ionic crystals, LiF has found much attraction due to its interesting physical properties. It is a material with low absorption, low scattering factor for sample measurement through Diamond Anvil Cell (DAC). As a result, it has found application as pressure transmitting medium as well as pressure sensor. Structural stability of LiF up to about one mega bar pressure also makes it an important material [1]. Previously, experimental and theoretical work on LiF is performed mostly at ambient conditions [2]. Studies on thermophysical properties of LiF are reported in limited range of temperature and pressure [3]. Liu et al [1] have studied thermophysical properties up to 37 GPa pressure and ~1000 K temperature. Most recently, Smirnov [4] has studied various ground state properties of LiF using full potential linear muffin tin orbital method as implemented in LMTART code with few modifications. Further, he has studied pressure dependence of elastic constants and structural stability under pressure. In his work, thermal contribution at finite temperature was taken into account using modified Debye model. In the present work, we describe ground state and finite temperature/pressure thermophysical properties of rock salt structured (B1) LiF using plane wave pseudopotential density functional theory as implemented in Quantum ESPRESSO package [5] in conjunction with quasi harmonic Debye model [6]. The paper is prepared as follows. The next section 2 describes the theoretical method used in the present study. The section 3 describes the results and discussion. The last section 4 describes the conclusions.

## 2. Method of Calculation

In the present work, ground state properties of LiF are calculated using plane wave pseudopotential density functional theory within local density approximation (LDA) as implemented

Published under licence by IOP Publishing Ltd

in Quantum ESPRESSO code [5]. It is known that LiF crystallizes in rock salt (B1) structure and remain stable up to very high pressure. Total energy calculation of B1-LiF was performed at various values of lattice constants in the step of 0.1 au. Before that, a convergence test was performed and it gave an 8x8x8 Monkhrost-Pack [7] **k**-point grid for Brillouin zone integration along with an energy cut-off of 100 Ry. Norm conserving pseudopotentials were used for the calculation of total energy. The calculated total energies were fitted to Murnaghan equation of state [8], in order to estimate equilibrium lattice constant, isothermal bulk modulus and its pressure derivative. Full phonon dispersion along major symmetry directions in B1-LiF are calculated using density functional perturbation theory [5]. A 2x2x2 grid of **q** point was used for the computation of force constants, which were further interpolated to calculate phonon frequencies. Further, elastic stiffness constants are calculated from the long wavelength limit of phonon frequencies along [100] and [110] planes [9]. Once the ground state properties are obtained, free energy at elevated temperature can be obtained using quasi harmonic Debye model [6]. In this approach, non equilibrium GIBBS function is given by,

$$G^{*}(V;p,T) = E(V) + VP(V) + A_{vib}(\Theta(V),T) \quad (1)$$

Here, the first term on right side is the optimized energy. Second represents the corresponding pressure. Third term corresponds to the thermal ionic contribution at finite temperature. Thermal electronic contribution is not included, as LiF is a wide band gap material. More details about quasi harmonic Debye model can be found out in Ref. [6]. The important input is the Poisson's ratio of LiF, which in the present work is calculated from the presently calculated elastic constants.

## 3. Results and Discussion

In the present work, we have calculated various ground state and finite temperature thermophysical properties of B1-LiF. Calculated cohesive energy is shown in Figure 1. The calculated energy is fitted to Murnaghan equation of state in order to predict structural properties viz; equilibrium lattice constants, isothermal bulk modulus and its pressure derivative. The calculated data is shown in Table 1 along with corresponding experimental and theoretical findings. A good agreement is observed. Equilibrium lattice constant is slightly underestimated compared to experimental results, which may be attributed to LDA as exchange-correlation functional. Further, full phonon dispersion is calculated using density functional perturbation theory. Calculated phonon frequencies along major symmetry directions show good agreement with experimental values [2]. We have also calculated phonon frequencies at high pressure (P = 85 GPa). We observe that at such high pressure, phonons along $\Gamma \rightarrow X$ and $\Gamma \rightarrow L$ directions are greatly affected. Also, longitudinal phonon frequencies along [100] plane shows maximum near 0.7 $Å^{-1}$ and drops at zone boundary. This feature is generally observed in the heavy elements like lead [10], but not in the light compounds like LiF. Such feature in lead was explained in terms of spin-orbit coupling by Verstraete et al [10]. Thus we predict that spin-orbit coupling might be playing important role at high pressures. Also, LO-TO splitting is found to decrease with pressure along $\Gamma \rightarrow X$ direction. From the long wavelength limit of phonon frequencies along [100] and [110] direction, we have calculated elastic stiffness constants and the results are shown in Table 1. Again a good agreement is observed, except that $C'$ is slightly overestimated.

We have calculated various thermodynamic properties using GIBBS program. Calculated room temperature thermal equation of state is shown in Figure 3 along with experimental data [1]. Here also, a good agreement is observed. Isothermal bulk modulus and Debye temperature are calculated at different temperatures and the results are shown in Figure 4 and Figure 5, respectively along with experimental results [12]. The trend of variation of bulk modulus and Debye temperature with temperature matches with the experimental results. The presently calculated temperature coefficient of bulk modulus is -0.033 GPa/K, while the experimental value is -0.0218 GPa/K. Similarly, the temperature coefficient of Debye temperature with temperature is -0.133, while the experimental value is -0.15. We have also calculated temperature variation of specific heat. Figure 6 shows the calculated data with the experimental value [13]. The value of room temperature specific heat agrees well with the experimental values. No other values of specific heats at constant pressure or at constant volume were found for comparison.

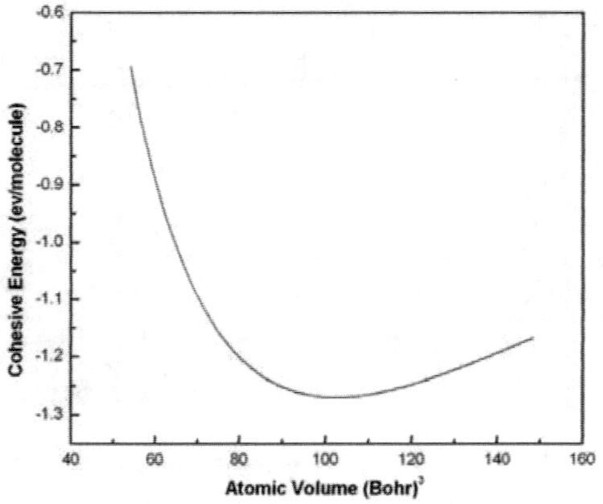


**Figure 1.** Cohesive energy of LiF as a function of atomic volume.

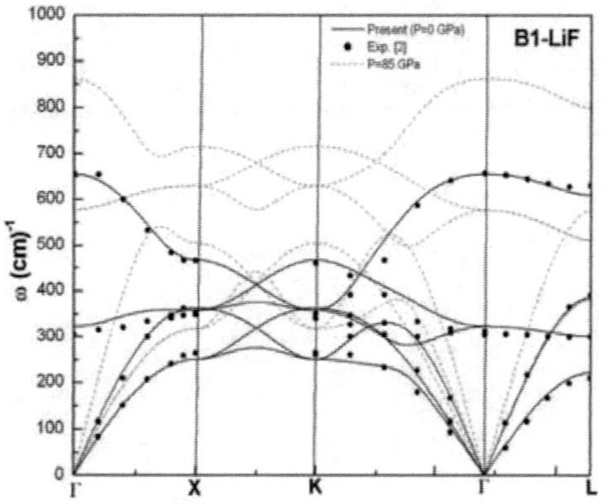


**Figure 2.** Phonon dispersion in LiF in ground state and at high pressure, along with experimental data.

**Table 1.** Calculated structural and elastic properties of B1-LiF.

| Properties | Present | Exp. | Others [4] |
|---|---|---|---|
| $a_o$(Å) | 3.944 | 4.026 [11] | 4.022 |
| $B_o$ (GPa) | 76.4 | 73.0 [1] | 73.8 |
| $B_o'$ | 3.36 | 3.90 [1] | 3.76 |
| $C'$ (GPa) | 53.775 | 41.1 [12] | 41.3 |
| $C_{44}$ (GPa) | 68.36 | 64.9 [12] | 66.0 |
| $\sigma$ | 0.216 | --- | --- |

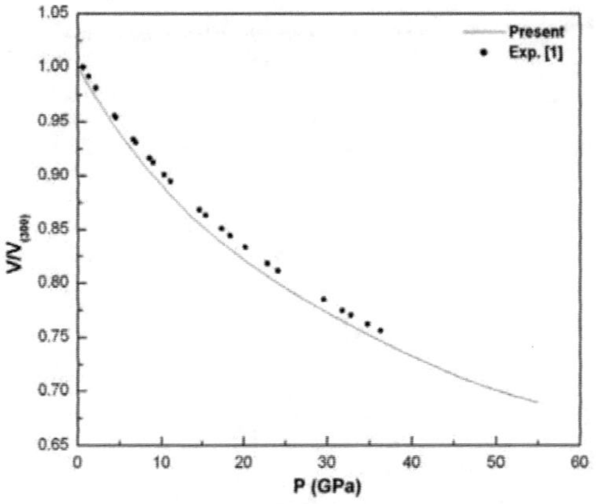


**Figure 3.** Room temperature thermal equation of state of LiF.

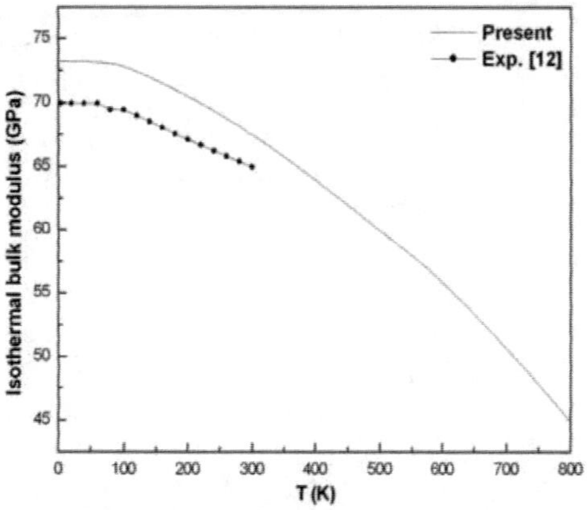


**Figure 4.** Isothermal bulk modulus at different temperatures along with experimental results.

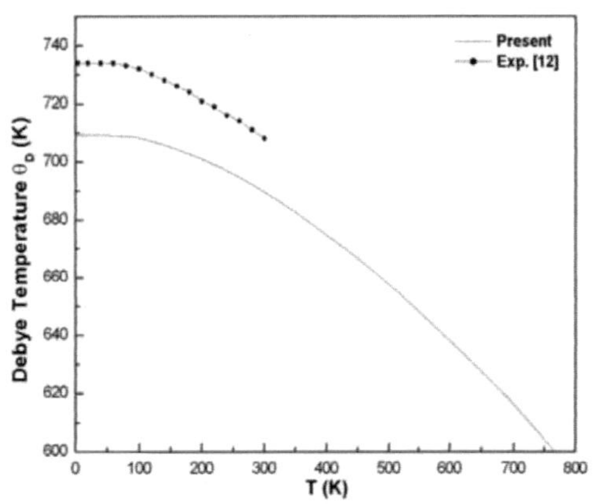


**Figure 5.** Calculated temperature dependent Debye temperature along with the experimental results.

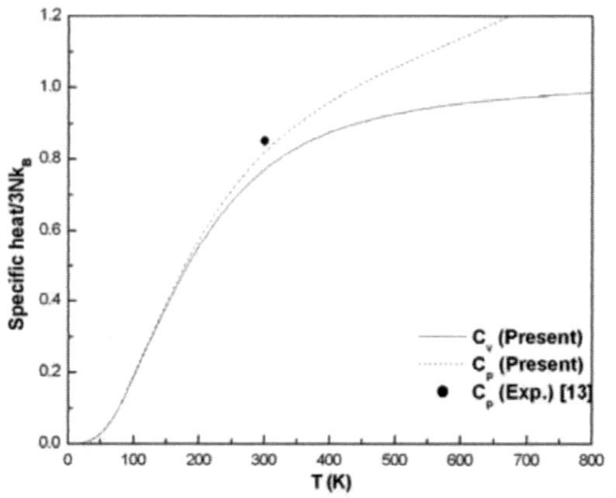


**Figure 6.** Specific heats as a function of temperature along with the experimental results.

In conclusion, we conclude that present work reports various ground state and finite temperature thermophysical properties of rock salt structured LiF. Our presently calculated structural and lattice mechanical properties agree well with experimental values. Phonon frequencies at high pressures show features, generally not observed in light elements or compounds. To the best of our knowledge, phonons in LiF are reported first time using density functional perturbation theory. Calculated thermodynamic properties namely room temperature equation of state, temperature variation of bulk modulus, Debye temperature and specific heat show good agreement with corresponding experimental values, confirming present approach in determining finite temperature thermophysical properties of LiF. Present results form a useful set of data of finite temperature/pressure thermophysical properties of LiF, which may be useful for further research in this field.

## 4. References

[1] Liu J, Dubrovinsky L, Ballaran T and Crichton W 2007 *High Pressure Research* **27** 483
[2] Dolling G, Smith H G, Nicklow R M, Vijyaraghvan P R and Wilikinson M K 1968 *Phy. Rev.* **168** 970
[3] Yagi T 1978 *Phys Chem Solids* **39** 563
[4] Smirnov N A 2011 *Phys Rev B* **83** 014109
[5] www.pwscf.org
[6] Blanco M A, Francisco E and Luana V 2004 *Comp. Phys. Commu.* **158** 57
[7] Monkhrost H J and Pack J D 1976 *Phys. Rev. B* **13** 5188
[8] Murnaghan F D 1944 *Proc. Natl. Acad. Sci.* USA **30** 244
[9] Karch K, Pavone P, Windl W, Schutt O and Strauch D 1994 *Phys. Rev. B* **50** 17054
[10] Verstraete M J, Torrent M, Jollet F, Zerah G and Gonze X 2008 *Phys. Rev. B* **78** 045119
[11] Ullrich H J, Uhlig A, Geise G, Horn H and Waltinger H 1992 *Mikrochim Acta* **107** 283
[12] Briscoe C V and Squire C F 1957 *Phys. Rev.* **106** 1175
[13] Landolt B 1961 *Zahlenwerte und Funktionen* (Springer-Verlag, Berlin vol **2**) pt 4

Journal of Physics: Conference Series **377** (2012) 012064 doi:10.1088/1742-6596/377/1/012064

# Pressure induced Structural Phase Transition in SrS

**B Y Thakore*, A.Y.Vahora, S G Khambholja, N K Bhatt and A R Jani**
Department of Physics, Sardar Patel University, Vallabh Vidyanagar, 388 120, Gujarat, India

*E-mail: brijmohanthakore@rediffmail.com

**Abstract.** Pressure induced structural phase transition in SrS from NaCl (B1) structure to CsCl (B2) structure has been studied using plane wave pseudopotential density functional theory as implemented in Quantum Espresso code. Ultrasoft pseudopotential along with generalised gradient approximation (GGA) has been used for total energy calculation. The electronic band structure has been calculated for B1 phase of SrS. We also report the vibrational properties such as phonon frequencies along major symmetry directions using density functional perturbation theory. The phase transition pressure is calculated by comparing enthalpies of both phases and its presently calculated value is 17.95 GPa, in very good agreement with the experimental results.

## 1. Introduction

The alkaline earth chalcogenides are technologically important material as they have found application ranging from catalysis to microelectronics [1-3]. They are important closed-shell ionic system crystallizing in the NaCl-type (B1) structure at ambient condition except for the MgTe and the beryllium chalcogenides. Among the alkaline earth chalcogenides, very little information is available for SrS. By X-ray diffraction experiment [4], SrS was found to undergo the structural phase transition from B1 to B2 at 18 GPa pressure. Khenata et al. [5] found the structural phase transformation occurring at 18 GPa, and reported the elastic constants in B1 phase at ambient condition. The non-metallic SrS after B1 to B2 phase transition was predicted to become metallic through band gap overlap mechanism [6-7]. Uğur et al. [8] have studied phonon properties of SrS under high pressure. Yan et al. [9] studied the elastic properties of SrS. The electronic band structure and optical properties of SrS are mainly discussed in other literature [10-12]. In present paper, we report the results of our theoretical study of structural and vibrational properties of SrS using plane wave pseudopotential density functional theory as implemented in Quantum Espresso code [13]. We have also calculated the phase transition pressure of SrS from B1 to B2 phase. In this section, we give the basic introduction. In the second section we give the information about methodology of the calculation and third section us discussed results and finally we conclude in the fourth section.

## 2. Calculation

The calculation of total energy of B1 and B2 phases of SrS has been carried out using ultrasoft pseudopotential within the generalised gradient approximation (GGA) using the scheme of Perdew et al. [14]. Convergence test gave a K.E. cut off of 55 Ry and 8x8x8 Monkhorst-Pack special **k**-point grid for Brillouin zone sampling. The calculated total energies were fitted to Murnaghan equation of state [15] in order to estimate structural parameters.

Published under licence by IOP Publishing Ltd

Journal of Physics: Conference Series **377** (2012) 012064 doi:10.1088/1742-6596/377/1/012064

We have calculated the phonon dispersion curve and elastic constants of SrS. In phonon calculation, Brillouin zone sampling was performed with a line width of $10^{-3}$ Ry. To obtain the full phonon spectrum, dynamical matrices were evaluated on a 4x4x4 grid in q space for the B1 structure. These matrices were then Fourier-interpolated to obtain the phonon dispersion curve. From the phonon frequencies, we have calculated the phonon density of state (p-dos). Further, electronic band structure is also calculated.

The phase transition pressure from B1 to B2 phase is calculated by calculating enthalpies of both the phases. At T=0 K, enthalpy is given by,

$$H(P) = E_{tot}(V) + V\,P(V).$$

At a given pressure, the stable structure is the one whose enthalpy has the lowest value. The phase transition pressure is the pressure at which, the enthalpies of the two phases are equal.

## 3. Results and Discussions

In order to calculate the ground state properties of SrS, the total energies are calculated in both phases for different volume around the equilibrium cell volume $V_0$. The plot of total energy versus volume for SrS in both the phases are shown in Figure 1.It can be seen from E-V curves that the NaCl-type structure is the most stable at ambient conditions. The calculated total energies are fitted to Murnaghan equation of state [15] to determine the ground state properties namely lattice parameter ($a_0$), bulk modulus ($B_0$), first order pressure derivative of bulk modulus ($B'_0$). The calculated equilibrium parameters in both structures are given in Table 1, which also contains the results of previous first principal calculation and experimental data. Our present lattice constant is slightly larger than experimental work. This is essentially due to the GGA.

**Table 1.** The calculated ground state properties of B1 and B2 phases of SrS along with experimental and theoretical data.

| Properties | B1 phase | | | B2 phase | | |
|---|---|---|---|---|---|---|
| | **Present** | **Exp.** | **Others** | **Present** | **Exp.** | **Others** |
| $a_0$ (Å) | 6.040 | 6.024[a] | 6.035[b], 6.076[c] | 3.655 | 3.61[a] | 3.655[b], 3.680[c] |
| $B_0$ (GPa) | 53.6 | 58.0[a] | 48.0[b], 53.9[c] | 49.1 | ... | 51.0[b], 50.6[c] |
| $B'_0$ | 3.55 | ... | 3.96[b], 4.66[c] | 4.35 | ... | 4.38[b], 4.50[c] |

[a] Ref.[5], [b] Ref.[8], [c] Ref.[16].

**Table 2.** Elastic constant (in GPa) of SrS in B1 phase.

| | **Present** | **Others** |
|---|---|---|
| $C_{11}$ | 129.1 | 141.0 [4], 113.9 [22] |
| $C_{12}$ | 15.85 | 17.2 [4], 19.4 [22] |
| $C_{44}$ | 74.53 | 62.5 [4], 30.3 [22] |

**Table 3.** Transition pressures and Volume collapse for the B1→B2 transition in SrS

| | **Present** | **Exp.** | **Others** |
|---|---|---|---|
| Phase transition pressure | 17.95 | 18[a] | 18.3[b], 17.1[c],19.23[d] |
| Volume collapse (ΔV/V (%)) | 11.34 | 11.4[e] | 11.04[f], 10.8[g] |

[a] Ref.[5], [b] Ref.[17], [c] Ref.[18] [d] Ref.[19] [e] Ref.[20] [f] Ref.[4] [g] Ref.[21].

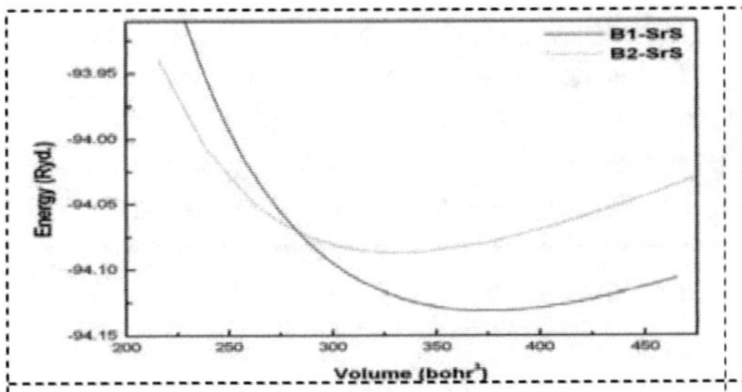


**Figure 1.** Energy-Volume relation for B1 and B2 phases of SrS.

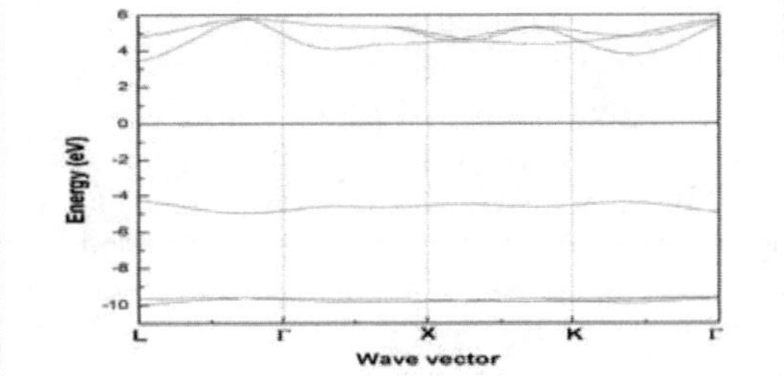


**Figure 2.** Calculated electronic band structure for B1 phase of SrS.

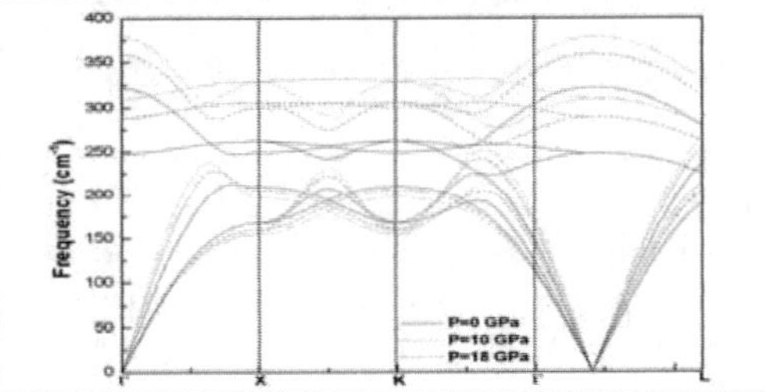


**Figure 3.** Calculated Phonon dispersion curve of SrS along high symmetry directions.

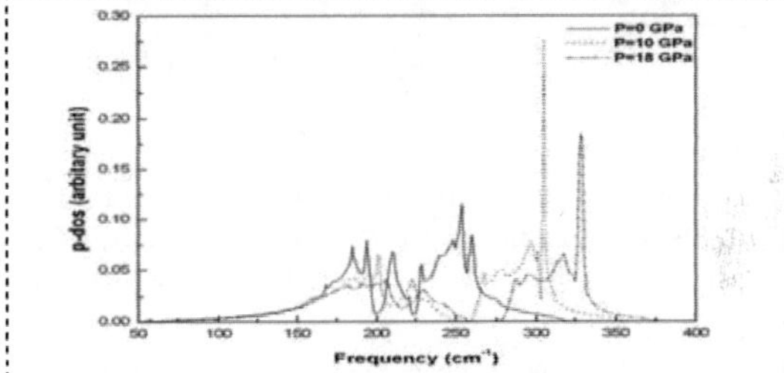


**Figure 4.** Calculated Phonon density of state (p-dos) of SrS.

The calculated electronic-band structure for the B1 phase along the various symmetry directions is shown in Figure 2. From the plot of band structure, we predict that SrS in B1 phase is insulator, which is in agreement with the experimental data.

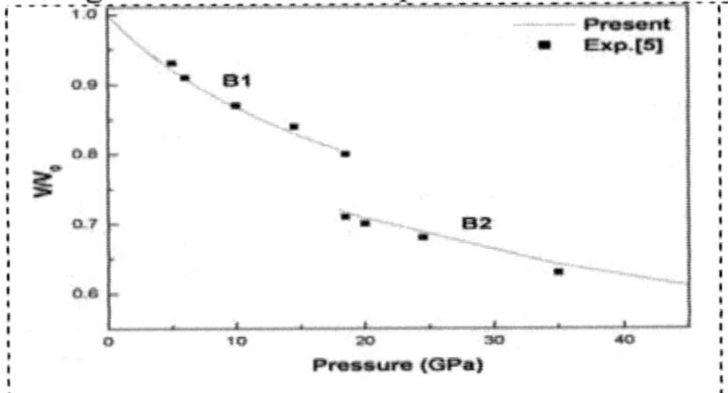


**Figure 5.** Calculated EOS for both phases of SrS.

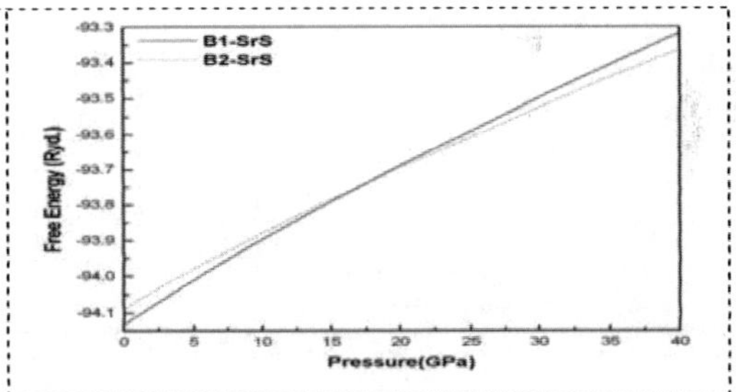


**Figure 6.** The calculated Gibbs free energy versus pressure for both phases of SrS.

Figure 3 shows the phonon dispersion curve along the high-symmetry direction for SrS in B1 phases at various pressures. We see that all the phonon frequencies have positive value at ambient condition; no modes appear to be soft, indicating that this structure is mechanically stable. The experimental results of phonon dispersion curve for SrS is not available yet. Phonon dispersion relation is also calculated at 10 GPa and 18 GPa pressure. The longitudinal phonon frequencies along [100] plane become hard with pressure. On the other hand, transverse phonon frequencies become soft. Thus, we belive that softness of phonon frequencies along [100] and [110] plane might be one of the reasons behind phase transition under pressure. From the phonon frequencies, we have also calculated the pressure dependent phonon density of state (p-dos), which is having the sharp peak is at 250 $cm^{-1}$ at ambient condition, as shown in Figure 4. The computed p-dos at high pressure show that peak shifts towards heavy frequency side. We have also calculated the elastic constants ($C_{11}$, $C_{12}$ and $C_{44}$) in B1 phase of SrS using long wavelength limit of phonon frequencies. The experimental values of elastic constants are not available yet. On the other hand, clear difference is observed in the previously reported theoretical elastic constants in Ref. [4] and Ref. [22]. Thus, in the absence of any experimental results, presently computed values form a new set of reliable data of elastic constants of B1-SrS at ambient condition. Figure 5 shows the pressure-volume relation for SrS along with the experimental results. It is evident from the figure that presently calculated results agree well with available experimental values. Figure 6 represents the pressure variation of enthalpy for both phases. It is evident from the graph that upon compression SrS undergoes a first order structural phase transition at about 17.95 GPa pressure, which is in excellent agreement with the experimental results of 18 GPa. The volume collapse due to phase transition is calculated. Table 2 summarized the transition pressures $p_t$ and the volume collapse of SrS, along with the other theoretical and experimental data. The presently calculated value of volume collapse is 11.34 %, which is in very good agreement with the experimental value o 11.4 %.

## 4. Conclusion

The structural stabilities under pressure, elastic and vibrational properties of SrS are investigated using first principles calculations based on density functional theory. The presently calculated cohesive properties of B1 and B2 phase of SrS namely lattice constant, bulk modulus, first order pressure derivative of bulk modulus are found to be in good agreement with experimental and other theoretical results. The electronic band structure of B1-SrS shows the insulating nature, in agreement with the experimental and other theoretical findings. The calculated values of phonon frequencies along major symmetry directions and elastic constants confirm that SrS is mechanically stable in B1 structure. Further, it is observed that *softening* of phonon frequencies with pressure may be one of the reasons behind the structural phase transition from B1 to B2 phase. The phase transition pressure and volume collapse from B1 to B2 phase is also in excellent agreement with experimental results.

## 5. Acknowledgements

The authors SGK and BYT acknowledge the financial support of UGC, New Delhi, India.

## 6. References

[1] Asano S, Yamashita N, Nakao Y 1978 *Phys.Stat.Sol.B* **89** 663
[2] Pandey R, Sivaraman 1991 *Phys.Chem.Solids* **52** 211
[3] Nakanishi Y, Hatanaka Y, Shimapka G 1992 *Appl.Surf.Sci.* **66** 515
[4] Khenta R, Baltache H, Rerat M 2003 *Physica B* **339** 208
[5] Syassen K 1985 *Phys.Stat.Sol.* **91** 11
[6] Syaseen K, 1986 *Physica B* **139-140** 277
[7] Bukowinski M.S.T 1982 *J.Geophys.Res.***87** 303
[8] Sule Uğur 2009 *Material science and engineering B* **162** 116
[9] Cheng Yan, Lu Lai-Yu, JIA Ou-He, GOU Qing-Quan 2008 *Commun.theor.Phys.* **49** 1611
[10] Pandey R, Jaffe J.E and Barry Kunz 1991 *Phys.Rev.B* **43** 9228
[11] Stepanyuk V. S, 1989 *Phys.Stat. Sol.***155** 215
[12] Ching W.Y, Gan F and Huang M.Z 1995 *Phys.Rev.B* **52** 1595
[13] Baroni S, Corso A, Gironcoli S.de, Giannozzi P, Cavazzoni C, Ballabio G, Scandolo S, Chiarotti G, Focher P, Pasquarello A, Lassonen K, Trave A, Marzari N, Kokalj, http:// www.pwscf.org.
[14] Perdew J.P, Wang Y 1992 *Phys.Rev.B* **45** 13244
[15] Murnaghan F.D 1944 *Proc.Natl,Acad.Sci.U.S.A.* **50** 697
[16] Lu.L.Y, Tan J.J, Jia O.U, Chen X.R, 2007 *Physica B* **399** 66
[17] Kholiya K and Gupta B.R.K 2008 *Phase Transitions* **81** 403
[18] Shameen Banu I.B, Rajagopalan M, Palanivel B, Kalpana G, Shenbagaraman P 1998 *J.Low Temp. Phys.* **112** 211
[19] Camp P.E.V, Doren V.E.V and Martins J.L 1995 *Phys. Stat. Sol. (b)* **190** 193
[20] Ross M and McMahan A.K 1982 *Phys. Rev. B* **26** 4088
[21] Xiao-Cui Y, Ai-Min H, Jie Y, Yong-Hao H, Gang P, Chun-Xiao G, Guang-Tian Z 2008 *Chin.Phys.Lett* **25** 1807
[22] Chang Yan, lU Lai-Yu, JIA Ou-He and GOU Qing-Quan, 2008 *Commun Theor. Phys.* **49** 1611

# High Pressure Behavior of PuBi and NpBi Compounds

**Ashok K Ahirwar[1,2], Mahendra Aynyas[2] and Sankar P Sanyal[3]**

[1]Department of Physics, Sadhu Vaswani College, Bhopal- 462030, India.
[2]Department of Physics, C.S.A. Govt.P.G. College, Sehore-466001, India.
[3] Department of Physics, Barkatullah University, Bhopal - 462026, India.
E-mail: ashokahirwar6@gmail.com

**Abstract-**The structural phase transition properties of two mono-bismuthides (NpBi, PuBi) are investigated by using a modified inter-ionic potential theory (MIPT), employing the minimization technique of Gibbs free energy expressed as a function of pressure. We have also presented the equation of state, which shows an appreciable volume collapse at the phase transition pressure and associated volume collapse are generally in good agreement with their experimental findings. We have also reported the elastic and thermal properties for these compounds, for the first time.

## 1. Introduction

In recent years, there has been considerable interest in theoretical and experimental studies of actinide compounds (AnX) with NaCl (B1) structure. These compounds exhibit a rich variety of electronic and magnetic properties. One of the common problem observed in these compounds, the degree of localization and itinerancy of the 5*f* electrons in An-An atom [1,2,4]. This property depends on the bonding distance between An atoms and consequently on the lattice parameter of the compounds. It is therefore, interesting to note that, by applying pressure the lattice parameter can be changed appreciably, and thereby reveling the effect of electron localization on structural, mechanical, thermal and elastic properties of actinides compounds [3,5]. The structural at high pressure properties for NpBi have been investigated by Genesini et. al. [6] using high pressure X-ray diffraction method. It has been reported that it undergoes first order structural phase transitions from NaCl structure ($B_1$) to body centered tetragonal structure (BCT) around 8.5 GPa with volume collapse 12% and isothermal bulk moduli ($B_T$) also observed to be 145 GPa at ambient pressure. The lattice parameter ($a_o$) of NpBi was determined to be 6.3732 Å by Benedict et.al. [7]. The high pressure behavior in PuBi has been studied by Meresse et. al. [8] using X-ray diffraction method. It was found that PuBi undergoes a NaCl type structure to body centered tetragonal type structure (BCT) at 10 GPa with volume change 12%. Petit et.al. [10] have also reported first principles calculations on plutonium compounds using SIC-LSD scheme technique.

## 2. Methodology

In the present analysis, we have considered three types of interionic forces that are responsible for cohesion in crystals with NaCl structure, viz. long range Coulomb, Flygare form of short range repulsive and van der Wall multipole interaction. Accordingly, the crystal energy can be expressed as

$$U(r) = \sum \frac{Z_m^2 e^2}{r_{ij}} + \sum_{ij} b\beta_{ij} \exp\left[\frac{\left(r_i + r_j - r_{ij}\right)}{\rho_{ij}}\right] + \sum_{ij} C_{ij} r_{ij}^{-6} + \sum_{ij} D_{ij} r_{ij}^{-8} \quad (1)$$

Where the first term represents the usual Coulomb energy with $r_{ij}$ as the interionic separation. The second term corresponds to Flygare form of short range repulsive while third and fourth terms in equation (1) represent the van der Wall's dispersive forces due to dipole-dipole and dipole-quadrupole interactions, with $C_{ij}$ and $D_{ij}$ as coefficients. $Z_m$ is the modified ionic charge due to Coulomb screening effect, $\beta ij$ are the Pauling coefficients, b and ρ are the short range parameters, which can be determined from the equilibrium condition and bulk modulus ($B_T$). Details of the method of calculation of elastic and thermal properties are given in our earlier paper [3].

**3.Result and Discussion**

In the present paper, we have investigated the structural phase transition, elastic and thermal properties of NpBi and PuBi compounds using the methodology outlined in section 2. The input data and output parameters are given in Table 1 and the cohesive and structural properties at high pressure are presented in Table 2. An inspection of the Table 2 reveals that the present MIPT has satisfactory explained the structural stability, cohesive and phase transition properties of these compounds. It is also seen from Table 2 that ΔG is positive which implies that the zero pressure NaCl phase is stable. The table also contains the calculate values of structural phase transition pressure for NpBi and PuBi from NaCl (B1) phase to body centered tetragonal (BCT) phase which are 9 Gpa and 11 GPa with 12% and 10% volume collapse respectively. These results are good agreement with experimental and other theoretical results [6-8].

**Table 1. input and output model parameters for NpBi and PuBi compounds.**

| | Input Parameters | | | | | Output Parameters | |
|---|---|---|---|---|---|---|---|
| **Solid** | **$r_+$ (Å)** | **$r_-$ (Å)** | **$r_0$ (Å)** | **$B_T$ (GPa)** | **$Z_m^2$** | **b ($10^{-19}$J)** | **ρ (Å)** |
| **NpBi** | 0.80 | 1.03 | 3.186[a] | 145[b] | 3.90 | 425.11 | 0.185 |
| **PuBi** | 0.89 | 1.03 | 3.179[c] | 61[c] | 2.755 | 50.24 | 0.290 |

[a]Ref. [7]. [b]Ref.[6]. [c]Ref.[8].

**Table 2. Cohesive energy, Phase transition pressure and Volume collapse for NpBi and PuBi.**

| **Solid** | | **Equilibrium constant(R) (Å)** | | **Cohesive energy (KJ/mole)** | | **ΔU (KJ/mole)** | **$P_T$ (GPa)** | **$V/V_0$ (%)** |
|---|---|---|---|---|---|---|---|---|
| | | **$R_{B1}$** | **RBCT** | **$U_{B1}$** | **$U_{BCT}$** | | | |
| **NpBi** | Pre. | 3.199 | 3.334 | -2855.15 | -2803.04 | 52.1 | 9 | 12 |
| | Exp. | 3.186[a] | - | - | | - | 8.5[b] | 12[b] |
| **PuBi** | Pre | 3.055 | 3.231 | -1953.03 | -1904.01 | 49.02 | 11 | 10 |
| | Exp | 3.179[c] | - | - | | - | 10[c] | 12[c] |

[a]Ref. [7]. [b]Ref.[6]. [c]Ref.[8].

It is obvious that the energy differences (ΔG) have attained their values to be zero at the phase transition pressure ($P_t$ = 9 &11 GPa) for NpBi and PuBi respectively. These values are in better agreement with their measured data [6-8]. Also, there is an abrupt change in Figures (1) – (2) of volume discontinuity at the phase transition pressures showing the compression of the lattice due to which the NaCl structure becomes unstable at high pressure and transforms to body centered tetragonal (BCT) phase.

At this point it can be emphasized that at high pressure the short range forces for these compounds increase significantly, which are responsible for change in the coordination number and phase transformation.

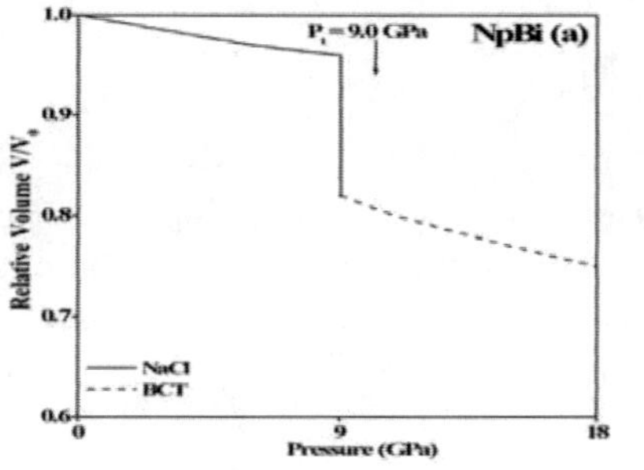


**Figure 1.** Equation of state for NpBi

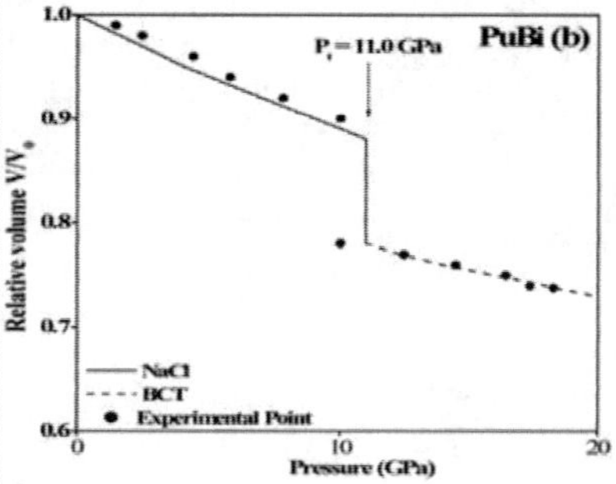


**Figure 2.** Equation of state for PuBi

The elastic properties are important parameters for crystals. They provide valuable information about the nature of the forces and bonding characteristics of solids. The calculated values of second order elastic constant (SOEC) for NpBi and PuBi at ambient as well as high pressure, by using present approach are obtained by using the methodology as described in section 2.To the best of our study no experimental and theoretical measurement on elastic constant have been reported so for, NpBi. There are some useful information about the second order elastic constants of PuBi observed by Srivastava et. al [9]. A look at from Figure-3, that the $C_{11}$ increase linearly with pressure and $C_{44}$ decreases with pressure up to 18 GPa. This similarity has also been observed in other rare earth compounds [3,5,13]. Our results for second order elastic constants (SOEC) satisfy the stability condition criteria for cubic crystal, $C_{11}$-$C_{12}$> 0, $C_{44}$ > 0, $B_0$ >0, by using elastic constants namely $C_{11,}$ $C_{12}$ and $C_{44}$. Due to absence of any earlier work, we could not compare our calculated values.

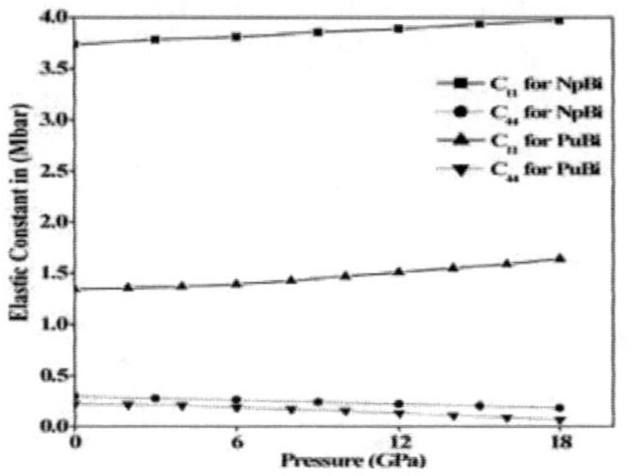


**Figure 3** variation of elastic constants with pressure.

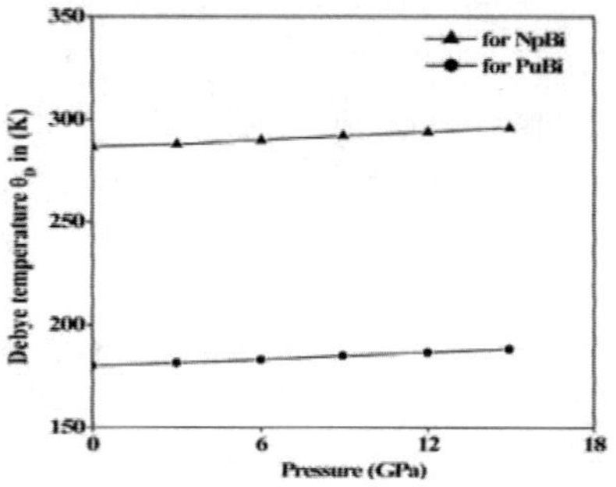


**Figure 4** variation of Debye temperature with pressure.

The Debye temperature is an important fundamental parameter of solid. The calculated values of debye temperature for these compounds are 281 K and 179 K at ambient pressure.

We have also calculated the variation of $\theta_D$ with respect to pressure for these compounds and predicted them in Figure 4. It is clear from the figure that $\theta_D$ increases as pressure increases for both the compounds, which show quite normal behavior in this type of compounds.

## 4.Conclusions

We present the results of our investigation, employing the modified inter-ionic potential theory (MIPT). Our calculation on structural phase transition pressure, bulk modulus and volume collapse agree well with the available experimental data for NpBi and PuBi compounds. We have also analyzed the elastic, mechanical, thermal properties for both the compounds for first time. Our theory is adequately suitable for explaining the structure stability and nature of inter- atomic forces for both the compounds.

## 5 References

[1] Aynyas M, Kaurav N, Sanyal S P 2002 J. *Phys. Chem. Solids* **63** 821-826.
[2] Leger J M 1993 *Physica B* **190** 84-91.
[3] Singh A, Aynyas M, Srivastava V, Sanyal S P 2009 *Physica B* **404** 1852-1857.
[4] Jha P K, Sanyal S P 1997 *Phys. Stat. Solidi* B **200** 13-18.
[5] Ahirwar Ashok K, Aynyas M, Arya B S, Sanyal S P 2011 *J. Nuclear Mat.* **413** 30-35.
[6] Benedict U, Holzapfel W B 1993 Handbook of Phy. and Chem. of Rare-earths, Lanth./Acts.Phy.-1,Vol.17 Edited by K.A.Gschneidner, Jr. L.Eyring, G.H. Lander and G.R. Choppin P 277.
[7] Chemical Thermodynamics of Neptunium and Plutonium 1994 (OECD Nuclear energy, Agency ed.) Vol.4 Chapter 11 P 187.
[8] Meresse Y, Heathman S, Rijkeboer C, Rebizent J 1999 *J. Alloys and Comp.* **284** 65-69.
[9] Srivastava V, Sanyal S P 2004 *J. Alloy and Comp.* **366** 15-20.
[10] Petit L, Savne A, Temmerman W M, Szotek Z 2002 *Eur. Phys. J.* B **25** 139-146.
[11] Hafemeiester D W, Flygare W H 1965 *Journal of Phy. & Chem.* **43** 795-800.
[12] Srivastava V, Bandyopadhyay A K, Jha P K, Sanyal S P 2003 *J.Phys.Chem.Solids* **64** 907-912.
[13] Rukmangad A, Aynyas M, Sanyal S P 2009 *Ind. Journal of Pure & Appl. Phys.* **47** 114-118.
[14] Sun Z M, Li S, Ahuja R, Schneider J M 2004 *Solid State Commun.* **129** 589-592.
[15] Wachter P, Filzmoser M, Rebizant J 2001 *Physica B* **293** 199-223.
[16] Jansiukiewiz C, Karpus V 2003 *Solid State Commun.* **128** 167-169.

## Acknowledgements

The authors are thankful to Madhya Pradesh Council of Science and Technology (MPCST), Bhopal and University Grants Commission (UGC), New Delhi for the financial assistance.

# Pressure induced zincblende to rocksalt phase transition in AlN nanocrystal

**Anurag Srivastava**[1] **and Neha Tyagi**
Advance Material Research Lab, ABV - Indian Institute of Information Technology and Management, Gwalior (M.P.) 474010, India

E-mail: profanurag@gmail.com

**Abstract.** Structural phase transition in Aluminium nitride (AlN) nanocrystal has been studied within the framework of density-functional theory, using both the local-density as well as generalized gradient approximation as exchange correlation functionals. The study observes that under the application of pressure AlN nanocrystal transforms from its original zincblende (B3) type phase to hypothetical rocksalt (B1) type phase within the pressure range of 46GPa to 56GPa, which is comparatively larger than its bulk counterpart. The lattice parameter, bulk modulus and pressure derivatives of AlN nanocrystal in its original B3 type phase as well as hypothetical B1 type phase have also been computed as ground state properties. The mechanical strength of the AlN nanocrystal has been analysed in terms of volume collapse at transition pressure and bulk modulus.

## Introduction

In the last few years, nanostructures of semiconducting materials have attracted a great attention of scientific community as they are expected to play a significant role in the development of future nanoscale technology, particularly in the optoelectronic and field-emission devices. Aluminium nitride (AlN), with its wide band gap, high thermal conductivity and high melting point can be used in various electronic applications. Lu et al.[1] have synthesized the hexagonal and cubic AlN nanocrystals with average grain size of about 2 nm and 4nm. Hexagonal AlN nanocrystals with average diameters in the 10–20 nm range have been prepared by the reaction of hexamethyldisilazane with $AlCl_3$ under solvothermal conditions and a characteristic photoluminescence band is observed around 420nm [3]. Tolbert and Alivisatos [4, 5] are the pioneer workers in the field of high pressure structural transformation in nanocrystals, discussed the stability of CdSe nanocrystals in wurtzite and rocksalt phases and observed structural phase transformation under the application of pressure and also reported the effect of crystal size on the transition pressure. The hexagonal AlN nanocrystals have a particle size of 10 nm shows wurtzite to rocksalt phase transition at around 14.5 GPa, which is significantly lower than the transition pressure of 22.9 GPa observed for the bulk AlN by using the same technique with bulk modulus ($B_0$) of 321 ± 19 GPa in wurtzite type phase of AlN nanocrystal, which is larger than that of the bulk AlN as $B_0$ of 208 GPa[6]. Recently, Costales et al.[7] have investigated a periodic cluster approach to simulate AlN nanocrystal and predicted that the small-size

[1] profanurag@gmail.com

nanocrystal display graphitic-like layers and nonbuckled wurtzite structure. Looking to the bulk aspects of AlN, the wurtzite to rocksalt structural phase transformation occur at around 20.91GPa and zincblende to rocksalt at around 19.69GPa [8] by using LDA, whereas in recent study of structural phase transformation in AlN nanocrystal with an average size of 10nm and 45nm by Lei et al.[9] the wurtzite to rocksalt transformation have been observed at 14.5GPa and 21.5GPa respectively. In view of above facts and researches on nanomaterials, we thought it pertinent to explore the possibility of carrying out an investigation structural transformation in AlN nanocrystal from its B3 to B1 type phase.

## Computational Method

First principle calculations have been performed using the Atomistix Tool Kit (ATK) based on density functional theory. In the calculation, local density approximation with Perdew Zunger [10] type parameterization and the generalized gradient approximation with revised Perdew Burke Ezeroff parameterization [11,12 ] has been used. The plane wave functions are expressed by plane waves with cutoff energy 100Ryd. The Brilloun-zone (BZ) integration is performed with a Monkhorst-Pack scheme using 1x1x50 k points. The cutoff energy and the number of k points are varied to test the convergence and are found to converge within the force tolerance of 0.05eV/Å for the reported value. We have assumed roughly spherical nanocrystals and their diameter can be calculated by using the relationship of lattice parameter and number of atoms present in the nanocrystal, given elsewhere [13]. The transition pressure is evaluated by calculating ground state energy versus cell volume (E-V) curves. First, the ground state energy of the fully relaxed unit cell is calculated. Then, a series of ground state energies are calculated for unit cells scaled by a set of scaled factors. The calculated total energy for the considered structures at different volumes around equilibrium has been fitted to the Murnaghan's equation of state [14]. In the present calculation we have assumed the roughly spherical nanocrystal whose diameter *(d)* can be calculated by using the following relationship:

$$d = a\left[\frac{3N}{4\pi}\right]^{\frac{1}{3}}$$

Where, *'a'* is the lattice parameter and *'N'* number of atoms present in the nanocrystal.

## Results and Discussions

The present paper discusses the ab-initio study of structural phase transformation in AlN nanocrystal from its zincblende (B3) type phase to hypothetical type rocksalt (B1). The computation of total energy has been performed within the density-functional theory framework, using both the local-density and generalized gradient approximation as exchange correlation functional. To understand the ground state properties of AlN nanocrystal in its B3 and B1 type phase the equilibrium lattice parameter, bulk modulus and pressure derivatives have been computed and summarized in Table-1 along with their bulk counterpart. The computed ground state parameters for bulk AlN in its B3 and B1 type phases are in close match with the other reported values [8,15-19], whereas the calculated ground state parameters for AlN nanocrystal in its two phases could not get compared due to absence of any experimental and theoretical data. Our calculated bulk modulus and pressure derivatives for AlN nanocrystal in its B3 and B1 type phases are comparatively smaller than the bulk counterpart. The calculated total energies are plotted as a function of volume for AlN nanocrystal in B3 and B1 type phases using LDA and GGA potentials in Figure 1(a) and (b). The equilibrium volume of B1 type phase of AlN nanocrystal is lower than that of B3 type phase, as in the case of bulks. The transitions have been calculated through the two pairs of E –V curves, and the E and V values at each tangent point are listed in table 2. From the obtained energy and volume data at the points of structural transition, the volume collapse ($-\Delta V_t/V_0$) at transition pressure ($P_t$) has been calculated. The observed structural phase transitions from B3 to B1 in AlN nanocrystal are 46.16GPa and 55.73GPa using LDA and GGA potentials respectively, which are comparatively larger than their bulk counterpart. One of the reasons behind this increase in transition pressure is probably the decrease in the volume collapse at the transition pressure. In the present theoretical approach we have not considered the surface energy differences and internal energy differences; otherwise these two factors also play a vital role in explaining the elevation of transition pressure. Figure 2 (a) and (b) shows the plots of relative volume as a function of pressure for AlN nanocrystal using LDA and GGA potentials, where the volume collapse at transition pressure are calculated and summarized in table 2. The calculated transition pressure and volume collapse for bulk AlN are in good agreement with the other reported data [8,15].

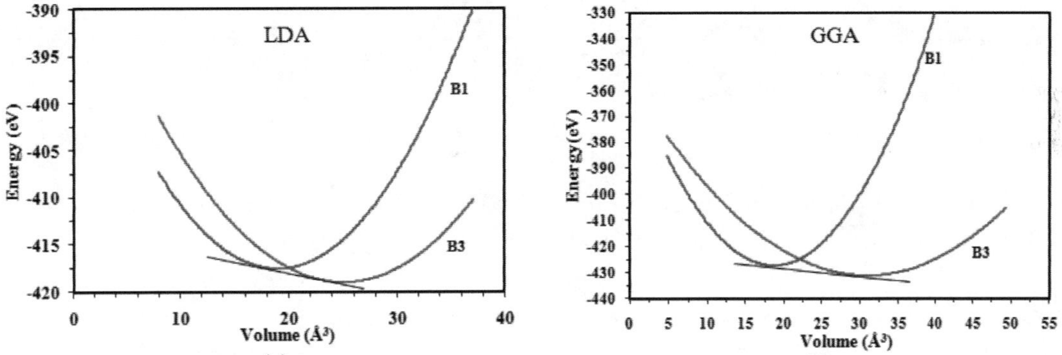


(a) (b)

Figure 1(a-b). Energy vs. volume curves for AlN nanocrystal.

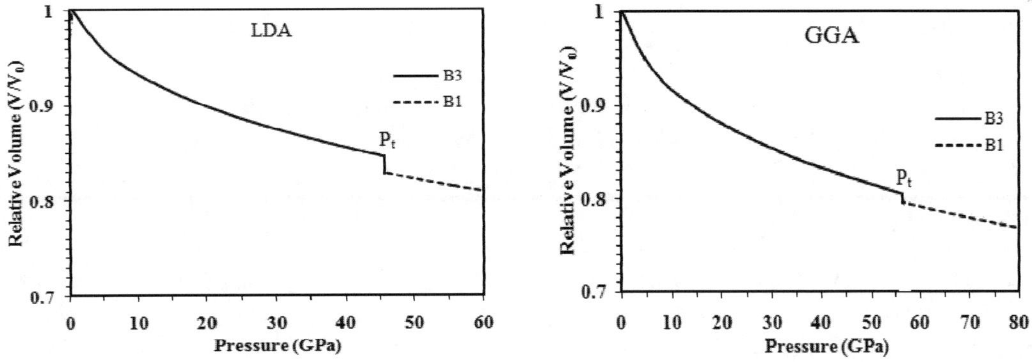


(a) (b)

Figure 2. Relative volume vs. pressure curves for AlN nanocrystal.

**Conclusion**

The present study computes the structural phase transition in AlN nanocrystal from zincblenbe (B3) to rocksalt (B1) structure at high pressure and calculation found that the transition pressure of AlN nanocrystal is higher than that of the bulk AlN. The calculated lattice parameter, bulk modulus and pressure derivatives of bulk AlN in its B3 and B1 type phases are in good agreement with the other reported data.

**Table 1.** The calculated lattice parameter (a), bulk modulus ($B_0$) and pressure derivatives ($B_0'$) for AlN nanocrystal.

| Structure | Lattice Parameter (a) | | Bulk Modulus ($B_0$) | | $B_0'$ | |
|---|---|---|---|---|---|---|
| | B3-AlN | B1-AlN | B3-AlN | B1-AlN | B3-AlN | B1-AlN |
| Nanocrystal | 4.348[a],<br>4.515[b] | 4.017[a],<br>4.113[b] | 186.150[a],<br>134.940[b] | 261.341[a],<br>226.364[b] | 3.684[a],<br>3.197[b] | 4.155[a],<br>3.210[b] |
| Bulk | 4.328[a],<br>4.407[b],<br>4.36[8],<br>4.38[8],<br>4.302[15],<br>4.37[16] | 3.993[a],<br>4.104[b],<br>4.031[8],<br>4.052[8],<br>3.978[15],<br>4.045[17] | 200.673[a],<br>181.016[b],<br>208.27[8],<br>197.92[8],<br>210[15] | 276.958[a],<br>230.469[b],<br>274.87[8],<br>260.60[8],<br>272[15],295<br>±17[18],221<br>±5.0[19] | 3.715[a],<br>3.404[b],<br>3.95[8],<br>3.91[8],<br>3.7[15] | 4.887[a],<br>3.544[b],<br>4.02[8],<br>3.99[8],<br>3.8[15],3.5±0.<br>4[18],<br>4.8±1.0[19] |

[a]LDA, [b]GGA, [8]Ref, [15]Ref, [16]Ref [17]Ref, [18]Ref, [19]Ref

**Table 2.** The calculated results of structural phase transition for AlN nanocrystal.

| Structure | Volume ($Å^3$) | | $-\Delta V_t/V_0$ (%) | Energy (eV) | | $P_t$ (GPa) |
|---|---|---|---|---|---|---|
| | B3-AlN | B1-AlN | | B3-AlN | B1-AlN | |
| Nanocrystal | 73.421[a],<br>82.166[b] | 57.882[a],<br>62.109[b] | 16.94[a],<br>18.28[b] | -417.062[a],<br>-422.728[b] | -416.610[a],<br>-422.034[b] | 46.16[a],<br>55.728[b] |
| Bulk | 72.372[a],<br>76.409[b] | 56.991[a],<br>61.041[b],<br>55.29[8],<br>56.37[8] | 19.74[a],<br>19.85[b],<br>20.99[8],<br>20.56[8],<br>20.4[15] | -417.767[a],<br>-423.528[b] | -417.680[a],<br>-422.973[b] | 6.324[a],<br>22.89[b],<br>19.69[8],<br>25.75[8],<br>7.1[15] |

[a]LDA, [b]GGA,[8]Ref,[15]Ref

**Acknowledgement**
Authors gratefully acknowledge ABV-Indian Institute of Information Technology and Management, Gwalior for providing the infrastructural support. One of us Neha Tyagi is thankful to IIITM for the award of Ph.D. scholarship.

**References**

[1] Lu S X, Tong Y H, Liu Y C , Xu C S , Lu Y M, Zhang J Y, Shen D Z and Fan X W 2005 *J. Phys. and Chem. of Solids* **66** 1609
[2] Zhang Yingjiu, Liu Jun, He Rongrui, Zhang Qi, Zhang Xiaozhong, and Zhu Jing 2001 *Chem. Mater.* **13** (11) 3899
[3] Sardar Kripasindhu and Rao C N R 2005 *Solid State Sciences* **7** 217
[4] Tolbert S H and Alivisatos A P 1994 *Science* **265** 373
[5] Tolbert S H and Alivisatos A P 1995 *J. Chem. Phys.* **102** 4642
[6] Wang Zhongwu, Tait Kimberley, Zhao Yusheng, Schiferl David, Zha Changsheng, Uchida Hinako and Downs Robert T 2004 *J. Phys. Chem. B* 2 **108** 11506
[7] Costales Aurora, Blanco M A, Francisco E, Solano C J F, Pendas A Martín 2008 *J. Phys. Chem. C* **112** (17) 6667
[8] Saib S and Bouarissa N 2005 *Eur. Phys. J. B* **47** 379
[9] Lei W W, Liu D, Zhang J, Cui Q L and Zou G T 2008 *J. Phys.: Conf. Ser.* **121** 162006
[10] Perdew J P and Zunger A 1981 *Phys. Rev. B* **23** 5048

[11] Zhang Y and Yang W 1998 *Phys. Rev. Lett.* **80** 890
[12] Hammer B , Hansen L B and Norskov J K 1998 *Phys. Rev. B* **59** 7413
[13] Sapra Sameer and Sarma D D 2004 *Phys. Rev. B* **69** 125304
[14] Murnaghan F D 1944 *Proc. Nat. Acad. Sci. USA* **30** 244
[15] Serrano J, Rubio A, Hernandez E, Munoz A, Mujica A 2000 *Phys. Rev. B* **62** 16612
[16] *Properties of Group-III Nitrides*, edited by J. H. Edgar, EMIS Datareviews Series (IEE, London, 1994)
[17] Vollstadt H, Ito E, Akaishi M, Akimoto S and Fukunaga O 1990 *Proc. Jpn. Acad., Ser. B: Phys. Biol. Sci.* **66** 7
[18] Uehara S, Masamoto T, Onodera A, Ueno M, Shimomura O, Takemura K 1997 *J. Phys. Chem. Solids* **58** 2093
[19] Xia Q, Xia H, Ruoff, A L 1993 *J. Appl. Phys.* **73** 8198

Journal of Physics: Conference Series **377** (2012) 012067 doi:10.1088/1742-6596/377/1/012067

# Study of phase transition and cohesive energy in MgO

N Munjal[1], P Bhambhani[1], G Sharma[1,*], V Vyas[1], B K Sharma[2]

[1] Department of Physics, Banasthali University, Banasthali -304022, India
[2]Department of Physics, University of Rajasthan, Jaipur -302005, India

E-mail: gsphysics@gmail.com

**Abstract.** In this paper, we present first-principles study of phase transition and cohesive energy of MgO. The calculations are performed on the basis of density functional theory (DFT) based on linear combination of atomic orbitals (LCAO) method. In the present calculations, the generalized gradient approximation (GGA) proposed by Perdew-Burke- Ernzerhof (PBE) and Becke's scheme are considered to treat the correlation and exchange effects respectively. The study reveals that the stable phase of MgO is B1. The study also suggests B1 to B2 phase transition at 260.75 GPa and 10.98 eV cohesive energy of MgO in the stable phase. The calculated data are found in good agreement with the previous theoretical and experimental investigations.

## 1. Introduction

Magnesium oxide (MgO) is a white solid mineral present in earth's lower mantle. MgO is a II–VI compound formed by an ionic bond between one magnesium and one oxygen atom. It is one of the most fundamental material for industrial sciences due to its wide band gap (~7.833eV) [1], high dielectric constant, and ability to form ternary crystalline phase. It is used in numerous applications such as an insulator, in medicines, cement, desiccant etc. At ambient conditions, MgO crystallizes in rocksalt (B1) structure with space group $(Fm\bar{3}m)$. Under pressure, MgO undergoes a structural phase transition from B1 to B2 structure. However, no experimental evidence is found on such a transition.

Several theoretical and experimental studies on electronic and structural properties of MgO have been reported earlier [2 -12]. The structural and electronic properties of MgO in B1 and B2 structures have been investigated by Chang and Cohen [2] using pseudopotential method within the local density approximation (LDA). Causa et al. [3] investigated the electronic structure and stability of different crystalline phases of MgO using Hartree-Fock (HF) theory. Jaffe et al. [4] has used LDA and GGA calculations for high pressure phase transition in ZnO and MgO. The structural properties of MgO have been investigated by Baranov et al. [5] using the LDA and GGA schemes. Alfe et al. [6] have performed the quantum Monte-Carlo calculations of the structural properties and B1 to B2 phase transition in MgO. The ground and excited state properties of MgO, ZnO, and CdO polymorphs have been studied by the Schleife et al. [7] using Vienna *ab-initio* simulation package (VASP). Zhu et al. [8] reported the properties of MgO in B1 phase using full potential linearized augmented plane wave (FP-LAPW) method. The total energy of MgO in B1, B2 phases have been calculated as a function of unit cell volume using FP-LAPW method within the framework of the DFT by Gueddmin et al. [9]. Using the plane wave pseudo potential density functional method within the GGA, the structural, electronic and optical properties of MgO have been reported by Liu et al. [10]. Anderson and Andreatch [11] and Y. Fei [12] performed experimental studies to calculate the several structural properties of MgO in B1 phase. Since MgO is a constituent of earth's mantle, its electronic structure inside the earth will considerably changed compared to that at normal pressure. Therefore, it is valuable to study the effect of pressure on the structural properties of MgO.

A few theoretical and experimental studies on cohesive energy of MgO have been reported earlier. In these studies, different methods have been used to calculate the cohesive energy of MgO. For instance, cohesive energy and elastic constant of MgO have been calculated by Calois et al. [13] using partially scaled wave functions. Chang and Cohen [2] have calculated the cohesive energy both theoretically and experimentally. Causa et al. [3] have computed the cohesive energy of MgO for B1, B2, $B8_1$ structures using the HF theory. Jaffe et al. [4] also calculated the cohesive energy of ZnO and MgO in B4, B3, B2, and B1 phases using LDA and GGA approximations. Verma and Bhardwaj [14] have calculated the cohesive energy for alkali halides and alkaline earth chalcogenides using an empirical relation. The main aim of the present work is to study the phase transition and cohesive energy in MgO

## 2. Method of calculation

The *ab-initio* periodic LCAO calculations are performed using quantum mechanical CRYSTAL06 [15] program package. In this approach crystalline orbitals are expanded over the basis set of atomic orbitals. The calculations are performed by solving Kohn-Sham equations self-consistently under DFT. This code includes various schemes like HF, DFT and the hybrid. These schemes differ from each other in the form of Hamiltonian operators. In the present calculations, the all electron basis set for Mg and O have been taken from http:// www.crystal.unito.it. The Kohn-Sham Hamiltonian is constructed by considering the exchange and correlation scheme of Becke [16] and PBE [17] respectively. The self-consistent calculations are performed considering 165 **k** points in the irreducible Brillouin Zone with sufficient tolerances. To achieve self-consistency, 45% mixing of successive cycles is considered and self consistency is achieved within 10 cycles. Using this method, the total energy as a function of primitive cell volume for the B1 and B2 structure and cohesive energy of MgO are computed.

## 3. Results and Discussion

The total energy as a function of cell volume for both phases of MgO are plotted in figure 1. The energy states of B1 phase is below the B2 phase, indicating the stability of B1 phase. In figure 1 the curves represent our computed values and the symbols represent fitting results to the Birch-Murnaghan equation with the following formula [18, 19].

$$E(V) = E_0 + \frac{9V_0B_0}{16}\left\{\left[\left(\frac{V_0}{V}\right)^{2/3} - 1\right]^3 B_0' + \left[\left(\frac{V_0}{V}\right)^{2/3} - 1\right]^2 \left[6 - 4\left(\frac{V_0}{V}\right)^{2/3}\right]\right\} \tag{3.1}$$

where $E_0$ is minimum energy, $V_0$ is corresponding volume, $B_0$ is the bulk modulus at zero pressure and $B_0' = dB_0/dP$. The lattice constant $a_0$, bulk modulus $B_0$ and its pressure derivative at zero pressure for both phases are calculated by fitting the above equation. These results are summarized in table 1.

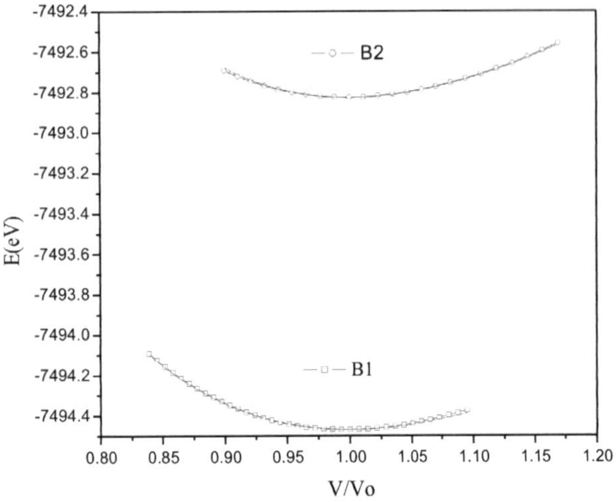


**Figure 1.** First-principles energy volume curves of B1 and B2 structures of MgO calculated using DFT-LCAO method.

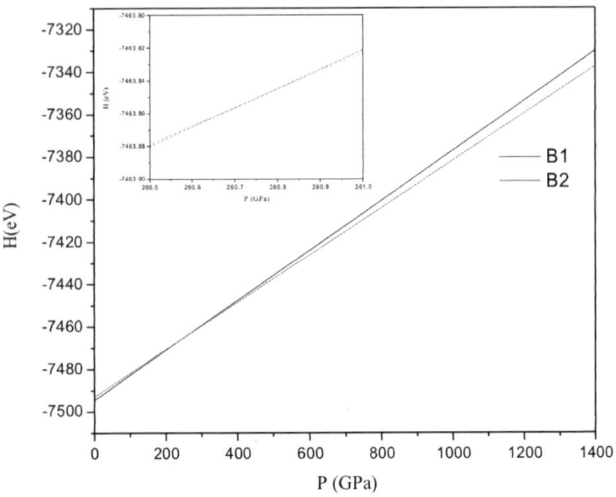


**Figure 2.** Enthalpy calculated from first-principles method for the B1 and B2 structures of MgO using PBE correlation functional. At the transition pressure ($P_t$) 260.75 GPa curves cross each other.

Table 1. Structural properties of MgO

| | present | Experimental | Other calculations |
|---|---|---|---|
| **B1** | | | |
| a (Å) | 4.220 | 4.213[12] | 4.098[6], 4.254[7],4.185[8], 4.259[9], 4.215[10] |
| $B_o$ (GPa) | 178.9 | 178.0[20], 160.0[12] | 183.0[6], 148.6[7], 174.0[8], 145.68[9], 157.40[10] |
| B' | 4.455 | 4.15[12] | 4.30[7], 4.24[8], 4.23[9], 4.16[10] |
| $P_t$ (GPa) | 260.75 | >227[21] | 1050[2], 220[3], 515[5], 611.12[11] |
| $E_{coh}$ (eV) | 10.98 | 10.33[2] | 11.817[5], 10.045[5], 9.96[2], 7.27[3] |
| **B2** | | | |
| a (Å) | 2.610 | - | 2.661[4], 2.656[7], 4.227[9] |
| $B_0$ (GPa) | 205.0 | - | 152.60[4], 140.30[7], 134.33[9] |
| B' | 4.05 | - | 3.39[4], 4.10[7], 4.24[9] |

Here, we have considered zero temperature in our calculations. At this temperature the thermodynamically stable phase is that one with lowest enthalpy, H = E + PV at a given pressure (figure. 2). The transition occurs when the enthalpy of the lower pressure phase coincides with some other phase at high pressure. After the point of intersection, the enthalpy of B2 becomes less than B1 phase. Hence, at low pressure the stable phase is B1 and at high pressure the stable phase is B2. The transition pressure is obtained from crossing of the enthalpy (H) versus pressure (P) curves. We obtain that MgO transform from B1 to B2 phase at about 260.75 GPa. This is consistent with experimental result reported by Duffy et al. [21] and is within the theoretical range of 198-1050 GPa. Theoretical and experimental transition pressures are listed in table 1.

Journal of Physics: Conference Series **377** (2012) 012067 doi:10.1088/1742-6596/377/1/012067

Cohesive energy is one of the important parameter, used to understand the nature of chemical bonding in solids. Its magnitude explains stability and chemical reactivity of solids. It also helps to determine the structure, since different possible structures have different cohesive energies. The cohesive energy from the Compton profile data can be derived as follows [22]

$$E_{coh} = \int_0^\infty p_z^2 \left[ J_s(p_z) - J_{fa}(p_z) \right] dp_z \tag{3.2}$$

where $J_s$ and $J_{fa}$ represents Compton profiles of solid and free atoms respectively. The present DFT-PBE values ($J_s$) of B1 phase are used to calculate the cohesive energy of MgO. The free-atom Compton profile ($J_{fa}$) values are taken from Biggs et al. [23]. The calculated cohesive energy along with available data are given in table 1. From this table, it is evident that our calculated value of $E_{coh}$ shows fairly good agreement with earlier calculations.

## 4. Conclusion

In summary, first-principles study of phase transition and cohesive energy in MgO using CRYSTAL06 program package are performed. The calculated structural properties are found in good agreement with the previous results. We conclude that at ambient conditions the stable phase is B1 and the phase transition from B1 to B2 occurs at 260.75 GPa. The cohesive energy is also found to be in good agreement with the earlier calculations.

## 5. References

[1] Whited R C, Flaten C J and Walker W C 1973 *Solid State Commun.* **13** 1903
[2] Chang K J and Cohen M L 1984 *Phys. Rev. B* **30** 4774
[3] Causa M, Dovesi R, Pisani C and Roetti C 1986 *Phy. Rev. B* **33** 1308
[4] Jaffe J E, Snyder J A, Lin Z and Hess A C 2000 *Phys. Rev. B* **62** 1660
[5] Baranov A N, Stepanyuk V S, Hergert W, Katsnelson A A, Settels A, Zeller R and Dederichs P H 2002 *Phys. Rev. B* **66** 155117
[6] Alfe D, Alfredsson M, Brodholt J, Gillan M J, Towler M D and Needs R J 2005 *Phys. Rev. B* **72** 014114
[7] Schleife A, Fuchs F, Furthmuller J and Bechstedt F 2006 *Phys. Rev. B* **73** 245212
[8] Zhu Y Z, Chen G D, Ye H, Walsh A, Moon C Y and Wei S H 2008 *Phys. Rev. B* **77** 245209
[9] Gueddim A, Bouarissa N and Villesuzanne A 2009 *Phys. Scr.* **80** 055702
[10] Liu Z J, Du Y X, Zhang X L, Qi J H, Tian L N and Guo Y 2010 *Phys. Status Solidi B* **247** 157.
[11] Andreson O L and Andreatch P J 1966 *J. Am. Ceram. Soc.* **49** 404
[12] Fei Y 1999 Am. Miner. **84** 272
[13] Calais J L, Makila K, Mansikka K, Pettersson G and Vallin J 1971 *Phys. Scr.* **3** 39
[14] Verma A S and Bhardwaj S R 2006 *J. Phys: Condens. Matter* **18** 8603
[15] Dovesi R et al. 2006 *CRYSTAL06 User's manual* University of Torino, Torino
[16] Becke A D 1993 *J.Chem. Phys.* **98** 5648
[17] Perdew J P, Burke K and Ernzerhof M 1996 *Phys. Rev. Lett.* **77** 3865
[18] Birch F 1978 *J. Geophys. Res.* **83** 1257
[19] Poirier J P 2000 *Introduction to the Physics of the Earth's Interior*, Cambridge University Press, Cambridge
[20] Drickarmer H G, Lynch R W, Clendenen R L and Perez-Albuerne E A 1966 *Solid State Phys.* **19** 135
[21] Duffy T S, Hemley R J and Mao H K 1995 *Phys. Rev. Lett.* **74** 1371
[22] Holt R S and Cooper M J 1980 *Philos. Mag. B* **41** 117
[23] Biggs F, Mandelsohn L B and Mann J B 1975 *Atomic Data Nucl. Data Tables* **16** 201

Journal of Physics: Conference Series **377** (2012) 012068 doi:10.1088/1742-6596/377/1/012068

# First-principles study of B1 to B2 phase transition in PbS

P Bhambhani[1], N Munjal[1], G Sharma[1*], V Vyas[1], B K Sharma[2]

[1] Department of Physics, Banasthali University, Banasthali -304022, India
[2]Department of Physics, University of Rajasthan, Jaipur -302005, India

E-mail: gsphysics@gmail.com

**Abstract.** The high pressure structural phase transition in PbS has been studied by means of first-principles total energy calculations which are based on linear combination of atomic orbitals (LCAO) method within local density approximation (LDA). In the present study, the exchange scheme of Becke and correlation functional of von-Barth-Hedin (VBH) are employed. It is observed that more stable phase for PbS is NaCl type (B1) and PbS transforms to the CsCl type (B2) structure under high pressure (22.8 GPa). The calculated value of transition pressure ($P_t$) from B1 to B2 structure is found in good agreement with the earlier experimental and theoretical investigations.

## 1. Introduction

Lead sulfide (PbS) which is known as lead salt has attracted considerable attention because of its direct narrow band gap (0.41 eV), high carrier mobility, high dielectric constant and positive (negative) temperature (pressure) coefficient [1,2]. PbS is very promising material for technological applications in the field of infrared (IR) devices, diode lasers and thermo-photovoltaic energy converters [3].

Knowledge of the electronic and structural properties of PbS under pressure is important in the semiconductor device industry. In the last few decades, structural phase transition in PbS under high pressure attracts new interest. Under normal conditions, PbS crystallizes in the rocksalt (RS) type (B1) structure. Structural phase transformations in PbS compound are reported by Bridgman [4] in 1940. Chattapadhyay et al. [5] have also made high pressure X-ray diffraction study on the structure phase transitions in PbS with synchrotron radiation and found that PbS undergoes structural phase transition from the NaCl type phase to the CsCl type phase at about 21.5 GPa. Knorr et al. [6] have performed an experimental study of the high pressure phase transition in PbS. High pressure structural phase transition in PbS has been reported by R. Ahuja [7] by using full potential linear muffin-tin-orbital (FP-LMTO) method within the generalized gradient approximation (GGA). Recently, using the full potential linearized augmented plane wave (FP-LAPW) method based on density functional theory (DFT) within GGA, high pressure phase transition of PbS has been investigated by Bencherif et al. [8].

## 2. Method of calculation

The present calculations are performed in the framework of density functional theory (DFT). We have employed the linear combination of atomic orbitals (LCAO) method as implemented in the CRYSTAL06 code [9]. The exchange scheme of Becke [10] and correlation functional of von Barth-Hedin (VBH) [11] are considered in the computation. In the present calculations, the basis sets for Pb and S have been taken from http:// www.crystal.unito.it. Following the standard truncation criteria of the code, the calculations are carried out considering 29 $\vec{k}$ points in the irreducible wedge of the first

Brillouin zone, taking advantage of the symmetry. To achieve self-consistency 55% mixing of successive cycles was considered and the self-consistency was achieved within 8 cycles. Using this method, the total energy as a function of cell volume for the B1 and B2 structure of PbS are computed.

## 3. Results and Discussion

We have calculated the total energy of PbS in NaCl and CsCl phases as a function of cell volume (figure 1). The energy states of B1 phase is below the B2 phase indicating the stability of B1 phase. In figure 1, the curves represent our computed values and the symbols represent fitting results to the Birch-Murnaghan equation of states [12,13]. The Birch-Murnaghan equation of states is given as follows

$$E(V)=E_0+\frac{9V_0B_0}{16}\left\{\left[\left(\frac{V_0}{V}\right)^{2/3}-1\right]^3B_0'+\left[\left(\frac{V_0}{V}\right)^{2/3}-1\right]^2\left[6-4\left(\frac{V_0}{V}\right)^{2/3}\right]\right\} \quad (3.1)$$

where $E_0$ is minimum energy, $V_0$ is corresponding volume, $B_0$ is the bulk modulus at zero pressure and $B_0' = dB_0/dP$. The equilibrium lattice constant $a_0$, bulk modulus $B_0$ and pressure derivative of the bulk modulus $B_0'$ are computed for both phases at zero pressure by fitting the above equation. These results are summarized in table 1.

In figure 2, enthalpy of PbS in B1 and B2 phases are plotted as a function of pressure. The transition from B1 phase to B2 phase occurs at about 22.8 GPa. The transition pressure is obtained by plotting the difference curve of enthalpy ($\Delta H$) in terms of pressure (P). At transition pressure (22.8 GPa) the difference becomes zero. Theoretical and experimental transition pressures are listed in table 2.

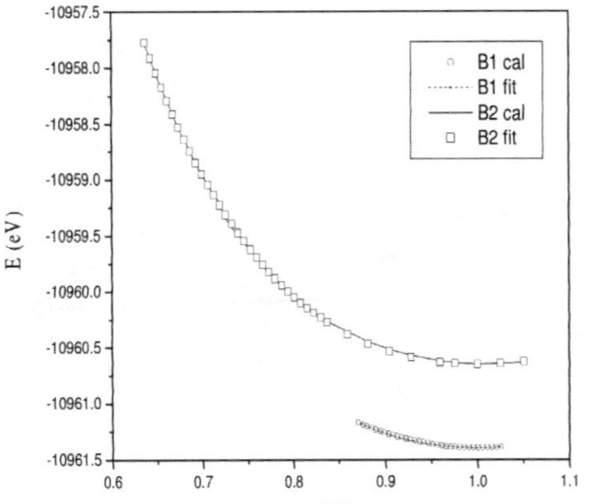


**Figure 1.** The total energy of PbS in NaCl and CsCl phases are plotted as a function of cell volume.

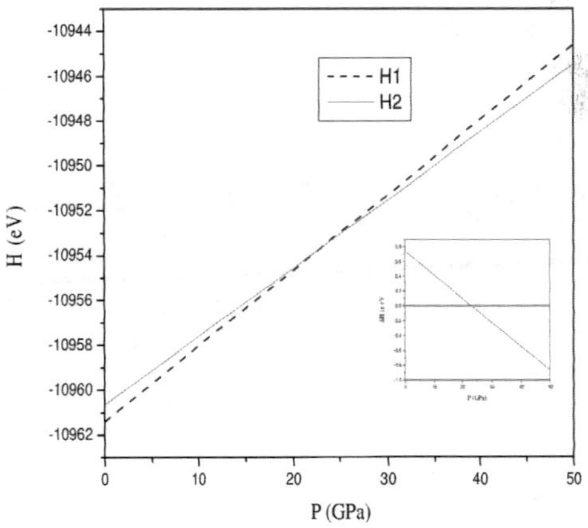


**Figure 2.** Enthalpy of PbS in NaCl and CsCl phases are plotted as a function of pressure.

**Table 1.** Structural properties of PbS.

| structures | | a(Å) | $B_0$ | $B_0'$ |
|---|---|---|---|---|
| B1 | Present | 5.99 | 64.063 | 3.8 |
| | Experimental | 5.929[a] | 52.9[a] | - |
| | | 5.936[b] | 62.8[b] | |
| | | 5.940[g] | | |
| | Other calculations | 6.000[c] | 54.63[c] | 4.10[c] |
| | | 5.906[d] | 66.3[d] | 4.38[d] |
| | | 6.012[e] | 52.0[e] | |
| | | 6.006[f] | 56.0[f] | 3.6[f] |
| B2 | Present | 3.649 | 70.714 | 3.8 |
| | Experimental | 3.289[g] | | |
| | Other calculations | | 59.0[f] | 3.6[f] |

[a] Reference [14]
[b] Reference [17]
[c] Reference [8]
[d] Reference [16]
[e] Reference [15]
[f] Reference [7]
[g] Reference [5]

**Table 2.** Phase transition pressure $P_t$ (GPa) for PbS

| | $P_t$ (GPa) |
|---|---|
| Present | 22.8 |
| Experimental | 21.5[a] |
| | 21.5[b] |
| Other calculations | 25.30[c] |
| | 25.0[d] |

[a] Reference [5]
[b] Reference [14]
[c] Reference [8]
[d] Reference [7]

Journal of Physics: Conference Series **377** (2012) 012068 doi:10.1088/1742-6596/377/1/012068

## 4. Conclusions

We have performed the study of structural properties (transition pressure, equilibrium lattice constants and bulk modulus) of PbS using LCAO method within LDA. The total energy of PbS in NaCl and CsCl phases as a function of cell volume are used to estimate the structural parameters. The computed values are found in good agreement with the earlier experimental and theoretical data.

## 5. References


[1] Dalven R 1973 *in: Solid State Physics, edited by Ehrenreich H,Seitz F and Turnbull D (Academic Press, New York,)* **28** 179

[2] Suzuki N, Sawai K and Adachi S 1995 *J. Appl. Phys.* **77** 1249

[3] Zogg H, Fach A, John J, Masek J, Muller P,Paglino C and Buttler W 1994 *Opt. Eng.* **33** 1440

[4] Bridgman P W1940 *Proc. Am. Acad. Sci.U.S.A* **74** 21

[5] Chattopadhyay T, Schnering H G von, Grosshans W A and Holzapfel W B 1986 *Physica* **139/140B** 356 and references therein.

[6] Knorr K, Ehm, Hytha M, Winkler B and Depmeier W 2003 *Eur. Phys. J.* **B 31** 297-303

[7] Ahuja R 2003 Phys. *Stat. Sol. (b)* **235** 341

[8] Bencherif Y, Boukra A, Zaoui A and Ferhat M 2010 *Mat. Chem. and Phys.* **126** 707

[9] Dovesi R, Saunders V R, Roetti C, Orlando R, Zicovich-Wilson C M, Pascale F, Civalleri B, Doll K, Harrison N M, Bush I J, D'Arco Ph and Liunell M 2006 *CRYSTAL06 User's manual,* University of Torino, Torino

[10] Becke A D 2003 *J Chem. Phys.* **98** 5648

[11] Barth U von and Hedin L 1972 *J. Phys. C: Solid State Phys.* **5** 1629

[12] Birch F 1978 *J. Geophys* Res. **83** 1257

[13] Poirier J P 2000 *Introduction to the Physics of the Earth's Interior*, Cambridge University Press, Cambridge

[14] Madelung O, Schulz M and Weiss (Eds.) H 1983 *Numerical Data and Functional Relationships in Science and Technology, Landolt-Bornsteein,New Series*, **17** (Springer , Berlin)

[15] Hummer K, Gruneis A and Kresse G 2007 *Phys. Rev.* **B 75** 195211

[16] Wei S H and Zunger A 1997 *Phys. Rev.* **B 55** 13605

[17] Madelung O, Rossler U and Schulz (Eds.) M 2005 *Semiconductors: Group IV Elements, IV-IV and III-IV compounds, Landolt- Bornstein,New Series,Group III* **vol. 41**,Pt. A, Springer-Verlag, Berlin

# Modifying the Electronic Properties of Nano-Structures Using Strain

**V K Lamba**[1] **and D Engles**[2]
[1]Electronics & Communication Engineering, Haryana College of Technology & Management Kaithal, Haryana-136027, INDIA
[2]Electronics Technology, Guru Nanak Dev University, Amritsar, Punjab, INDIA

E-mail: lamba_vj@hotmail.com

**Abstract.** We used density-functional theory based Non equilibrium green function simulations to study the effects of strain and quantum confinement on the electronic properties of Germanium & Silicon NWs along the [110] direction, such as the energy gap and the effective masses of the electron and hole. The diameters of the NWs being studied in a range of 3-20 Å. On basis of our calculation we conclude that the Ge [110] NWs possess a direct band gap, while Si [110] NWs possess indirect band gap at nanoscale. The band gap is almost a linear function of strain when the diameter of Ge NWs D < 10 Å while shows parabolic behaviour for, D > 15 Å; & for Si it is linear in behaviour. On doping silicon wire we found that the bandgap shows parabolic behaviour for change in strain. We also concluded that the band gap and the effective masses of charge carries (i.e. electron & hole) changes by applying the strain to the NWs. Our results suggested that strain can be used to tune the band structures of NWs, which may help in design of future nanoelectronic devices.

## 1. Introduction

One-dimensional nanostructures, such as NWs, and nanotubes have attracted extensive research [1-4] efforts since early 90s. They are expected to play important roles as a functional component in nano & flexible electronics [3-5]. Therefore, it is of great importance to study their electronic properties. One of the key for such an application is the electrical response, which monitors the change of environment caused by external factors [5-8]. It is thus important to understanding the electron transport and the effect of external properties on nano structures. Several work [4-9] have actually been carried out to study the electron transport in NWs. Various aspects such as *I-V* curve, conductance, nanocontact, surface defect, disorder, doping, band gap, density of state, effective mass of charge carriers, and structural effects have been explored [6-12]. Researchers [1-8] found that the band gap, conductance, etc of NWs depends on several factors such as, size, crystalline orientation, surface chemistry, doping, etc. It is well known that strain is not avoidable during epitaxial growth if there is a lattice mismatch, in many applications NWs are usually embedded in some materials within the coatings brings strains to the wires and in the field of microelectronics, strain has become a routine factor to engineer band gaps of semiconductors [9-12]. Thus we studied the effect of strain on band structure of Ge and Si NWs of different diameter having [110] orientation.

[1] To whom any correspondence should be addressed.

Published under licence by IOP Publishing Ltd

## 2. Modelling & Simulations

We are using a two-probe model, and the approach we used is employed successfully in many theoretical studies for calculating the electron transport properties in NWs, or nanotubes [12-17]. We focus on the effects of length, and strain on the electronic properties of the NWs. The largest system studied contains 512 (NW) + 200 (electrodes) atoms. Geometry optimization is performed using Gaussian until the absolute value of force acting on each atom is less than 0.04 eV/Å. The transport calculations are performed using NEGF [12-14], combined with DFT. The Landauer [18] formalism is implemented in ATOMISTIX TOOLKIT [18-19] and *I-V* characteristics are calculated by,

$$I = \frac{2e^2}{h}\int T(E,V_b)[f_l(E) - f_r(E)]dE \qquad (1)$$

where $e$, $h$, and $f_{l(r)}$ are electron charge, Planck's constant, and the Fermi distribution functions at left(right) electrode, respectively. $T(E,V_b)$ is the transmission coefficient at energy $E$ and bias voltage $V_b$. The core electrons are described using ultrasoft Vanderbilt pseudo-potentials. The dangling bonds in the NWs surface are saturated by hydrogen atoms. Valence electrons are expanded in double $\zeta$-plus polarization (DZP) and single-$\zeta$ plus polarization (SZP) basis sets for the NWs (Si and H/Ge and H atoms) and metal electrodes (Au atom), respectively. The band gap of wires is defined by the energy difference between the bottom of the conduction band (conduction-band edge (CBE) and the top of the valence band (valance-band edge (VBE). The effective masses of the electron and hole can be readily calculated using equation 2 from the band structure of the wire.

$$m^* = \hbar^2(d^2E/dk^2) \qquad (2)$$

## 3. Results & Conclusions

In the present work we have used Ge and Si whose electrical properties as a bulk at room temperature is given in table 1

**Table 1.** Electronic properties of Germanium & Silicon at 300° K.

| Properties | Germanium | Silicon |
|---|---|---|
| Band Gap (eV) | 0.66 | 1.12 |
| Hole Mobility ($Cm^2V^{-1}S^{-1}$) | 1800 | 450 |
| Electron Mobility($Cm^2V^{-1}S^{-1}$) | 3800 | 1500 |
| Intrinsic resistivity (Ω.cm) | 46 | $3.25 \times 10^5$ |

Ge NWs show a direct band gap at Γ, in Fig. 1, we present the band gap of Ge and Si NWs with varied diameters i.e. 4 to 20 Å, as demonstrated in literature.[19] The band gap of the Ge NWs in fig 1 is decreased when the diameter of wire is increased.

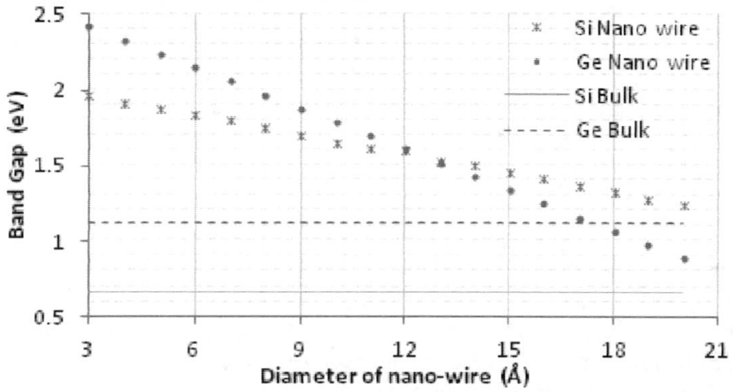


**Figure 1:** Bandgap of Ge, and Silicon NWs of different dimensions at room temperature

This effect is primarily due to quantum confinement. Our predicted size dependence of the band gap in Ge NWs is in a good agreement with the reference [21-22]. In case of Si NWs the band gap also decreases with increase in diameter as shown in fig 1. The variation in band gaps as a function of uniaxial strain for several different sized Ge NWs is plotted in Fig. 2(a), for the wire with a diameter of 5 & 10 Å, the band gap variation with strain is almost linear.

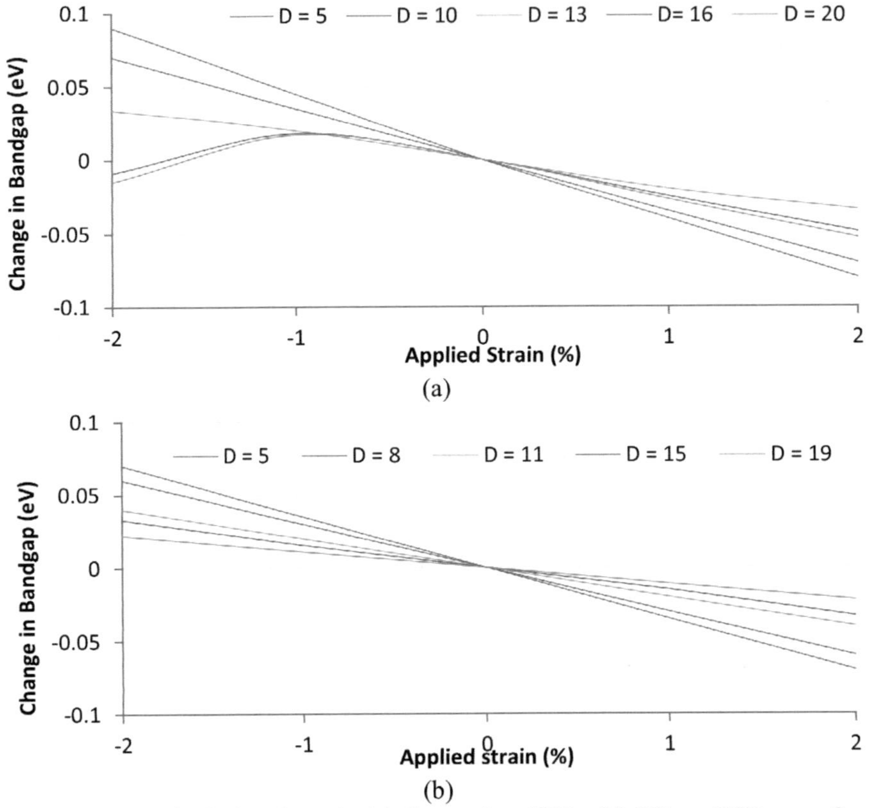


**Figure 2:** The change in the band gap in (a) Germanium NWs, (b) Silicon NWs, as a function of uniaxial strain, positive strain refers to expansion while negative strain corresponds to compression.

The gap decreases with expansion and increases with compression. The gap variation with strain in the wire with diameter of 13 Å has a more modest change in the gap for a given strain, compared with the 10 Å wire. However, for the wire with diameter of 16 & 20 Å, the gap variation with strain, exhibits a nearly parabolic behaviour, the gap drops not only under expansion, but also under compression beyond 2%. Thus we conclude that the strain effect on the band gap in Ge wires is strongly dependent on its size. On, examining the variation in energies, of VBE, and CBE with strain, shown in fig 3 for Ge NWs having diameter 10 and 16 Å. The energies of the CBE and VBE decrease with expansion while increasing with compression. Since the band gap is given by the energy difference between the CBE and VBE, it is also a nearly linear function of strain for 10 Å curve.
However, for the 16 Å wire, the energies of the CBE and VBE, are not linear functions with strain. Generally, both energies of the CBE and VBE are reduced under expansion and increased with compression. It is because once the axial strain is applied, the bonds in the $x$ and $y$ directions will change due to the Poisson effect. The reduction in Ge-Ge bond lengths makes the electron cloud of the VBE and CBE orbital's more efficiently shared by Ge atoms. This effect results in an increased electron-nucleus Coulomb attraction, thus an appreciable decrease in energies of both the VBE and CBE. In addition, we found that the orbital of the CBE is more delocalized than that of the VBE.

Thus, the electron cloud of the CBE is more effectively shared by Ge atoms in compared to that of the VBE. As a result, the energy of the CBE is more sensitive to strain than that of the VBE and, similarly for case of Si nano wireas shown in fig 2(b).

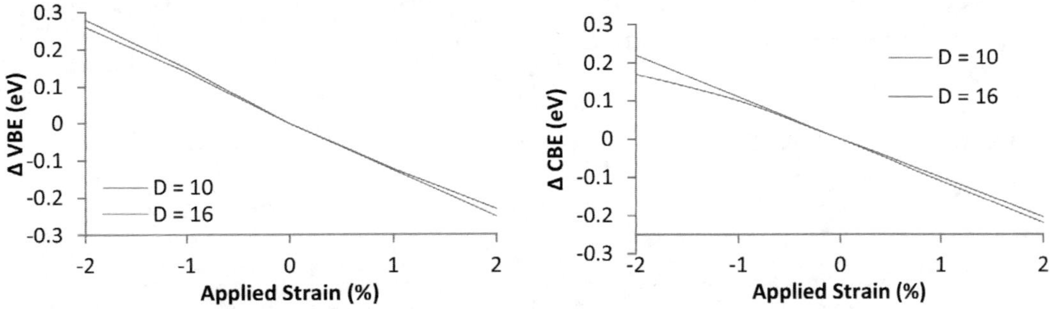


**Figure 3:** The changes of CBE and VBE energies in Ge NWs are plotted as a function of strain.

## 4. Summary

We conclude that the Ge [110] NWs possess a direct band gap, while Si [110] NWs possess indirect band gap at nanoscale. The band gap is almost a linear function of strain when the diameter of Ge NWs D < 10 Å while shows parabolic behaviour for, D > 15 Å; & for Si it is linear in behaviour. On doping Si wire we found that the bandgap shows parabolic behaviour for change in strain. We also concluded that the band gap and the effective masses of charge carries (i.e. electron & hole) changes by applying the strain to the NWs. Our results suggested that strain can be used to tune the band structures of NWs, which may help in design of future nanoelectronic devices.

## 5. Acknowledgement

We acknowledge the support by C.S.I.R. and AICTE under Grant No. 22(0519)10/EMR-II, and 8024/RIFD/MOD-145/2010-11respectively.

## 6. References

[1] Cui Y, Wei Q, Park H and Lieber C M 2001 *Science* **293** 1289.
[2] Hahm J and Lieber C M 2004 *Nano Lett.* **4** 51.
[3] Cui Y and Lieber C M 2001 *Science* **291** 851.
[4] Tian B, Zheng X, Kempa T J, Fang Y, Huang J and Lieber C M 2007 *Nature* **449** 885.
[5] Ma D, Lee C S, Au F C K, Tong S Y and Lee S T 2003 *Science* **299** 1874.
[6] Chan T L, Ciobanu C V, Chuang F C, Lu N, Wang C, and Ho K M 2006 *Nano Lett.* **6** 277.
[7] Svizhenko A, Leu P W and Cho K 2007 *Phys. ReV. B* **75** 125417.
[8] Fernandez-Serra M V, Adessi C, and Blase X 2006 *Nano Lett.* **6** 2674.
[9] Fernandez-Serra M V, Adessi C, and Blase X 2006 *Phys. ReV. Lett.* **96** 166805.
[10] Ponomareva I, Menon M, Richter E, and Andriotis A N 2006 *Phys. ReV. B* **74** 125311.
[11] Datta S 1995 *Electronic Transport in Mesoscopic Systems*; Cambridge Univ. Press, Cambridge.
[12] Brandbyge M, Mozos J L, Ordejon P, Taylor J and Stokbro K 2002 *Phys.ReV. B* **65** 165401.
[13] Taylor J, Guo H, and Wang J 2001 *Phys. ReV. B* **63** 121104.
[14] Crljen Z, Grigoriev A, Wendin G and Stokbro K 2005 *Phys. ReV. B* **71** 165316.
[15] Cheng D, Kim W Y, Min S K, Nautiyal T and Kim K S 2006 *Phys. ReV. Lett.* **96** 096104.
[16] Odbadrakh K, Pomorski P, and Roland C 2006 *Phys. ReV. B* **73** 233402.
[17] Buttiker M, Imry Y, Landauer R, and Pinhas S 1985 *Phys. ReV. B* **31** 6207.
[18] Atomistix ToolKit version 2.0, Atomistix A/S (www.atomistix.com).
[19] Bruno M, Palummo M, Marini A, and Del Sole R 2007 Surf. Sci. **601** 2707.
[20] Arantes J T, and Fazzio A 2007 Nanotechnology **18** 295706.
[21] Medaboina D, Gade V, Patil S K R and Khare S V 2007 Phys. Rev. B **76** 205327.
[22] Kagimura R, Nunes R W and Chacham H 2007 *Phys. Rev. Lett.* **98** 026801

# Vibrational properties of BaClF, BaBrF and BaIF under high pressure

**N. Yedukondalu and G. Vaitheeswaran***
Advanced Centre of Research in High Energy Materials, University of Hyderabad, Hyderabad, - 500046, Andhra Pradesh, India.

*E-mail: gvaithee@gmail.com

**Abstract.** We have studied the Raman active optical phonons of layered BaClF, BaBrF and BaIF scintillators at ambient as well as at high pressures. The calculations were carried out using density functional perturbation theory within the local density approximation. We observed the softening of low and high frequency modes $E_g(1)$ and $E_g(3)$ under pressure. The calculated optical dielectric and Born effective charge tensors are found to be isotropic even though the crystal structures are anisotropic.

## 1. Introduction

Mixed barium halofluorides BaXF (X = Cl, Br, I) crystallize in the primitive tetragonal PbClF-type (matlockite) structure. The materials can act as host lattices for rare earth ions[1,2]. Europium activated BaClF and BaBrF are used for X-ray storage phosphors for medical imaging via photo stimulated luminescence[3-5]. Our recent investigation also revealed that these materials can act as storage phosphors[6]. BaClF : $Sm^{+2}$ can be used as a luminescence sensor in the pressure range of 5 GPa[7]. The electronic structure determines the physical and chemical properties of materials. Extensive theoretical studies have been devoted to understand the structural, electronic and optical properties of MXF (M = Ca, Sr, Ba, Pb and X = Cl, Br, I) compounds [8-14]. In addition to that high pressure X-ray diffraction study has been performed by several groups [15-17], in which they observed structural phase transitions in BaClF and BaBrF at 21 and 27 GPa respectively. High pressure phase transition in BaClF is also studied by using molecular dynamics[18]. The thermal expansion coefficients of BaClF and BaBrF were determined by Kurobori *et al* [19]. Ultrasonic pulse echo and Brillouin scattering techniques have been used for the measurement of elastic constants for BaClF[20].

The reported lattice dynamical calculations using shell model[21], provide a theoretical understanding of elastic constants, equation of state, phonons, and thermo dynamic properties of MXF (M = Sr, Ba, Pb and X = Cl, Br, I) compounds. Raman spectra analysis of BaClF, BaBrF and BaIF single crystals were carried out at ambient conditions[22,23], in which the polarizations and frequencies of the vibrations are interpreted in terms of symmetry co-ordinates. High pressure Raman spectroscopic measurements on BaClF[24] and BaIF[25] revealed that ambient tetragonal structure of BaClF transforms to a mixture of tetragonal and orthorhombic structures at 10.8 GPa and both of these phases coexist upto 21.1 GPa, at which a new monoclinic phase is observed and all three phases coexist upto 25.6 GPa. While the ambient phase of BaIF found to be extremely stable and it undergoes a gradual layer to non-layer transition under high pressure. To the best of our knowledge there are no theoretical calculations exploring the vibrational properties of the above mentioned compounds under pressure. So, it is worthwhile to perform the phonon calculations under pressure in order to understand the structural phase transitions in these series of compounds. In the present study, we report the Raman active optical phonons of layered BaClF, BaBrF and BaIF at ambient as well as at high pressures.

## 2.Method of calculation

First principles calculations were performed using the CAmbridge Series of Total Energy Package[26] based on the density functional theory (DFT). We used norm conserving pseudo potentials [27] for the electron-ion interactions to calculate the total energies as they are well suited for the calculation of phonons. The exchange correlation potential of Ceperley and Alder[28] parameterized by Perdew and Zunger[29] in the local density approximation for electron-electron interactions are used. The following plane wave basis orbitals were used in the calculations, Ba:$5s^2$ $5p^6$ $6s^2$, F:$2s^2$ $2p^5$, Cl:$3s^2$ $3p^5$, Br:$4s^2$ $4p^5$, I:$5s^2$ $5p^5$. To conform the convergence of the calculations for the Brillouin-zone sampling, we tested the plane wave cut off energy and K-mesh according to the Monkhorst-pack grid scheme[30]. From which, we fix the cut off energy as 450 eV and K-mesh are 8x8x6, 8x8x6 and 6x6x4 for BaClF, BaBrF and BaIF respectively. We used a criterion for self-consistency that the energy is converged to $5x10^{-7}$ eV/atom and the force to $10^{-4}$ eV/Å.

## 3.Results and Discussion

### *3.1.Raman active modes*

In our recent investigation[6], we studied the structural (under pressure), electronic, bonding and optical properties of BaClF, BaBrF and BaIF. In the present work, we have calculated the zone centre phonon frequencies of BaClF, BaBrF upto 30 GPa with a step size of 5GPa, and for BaIF upto 60 GPa in steps of 10 GPa. The BaXF(X = Cl, Br and I) crystallizes in the primitive tetragonal symmetry $D_{4h}^7$ or *P4/nmm* PbClF type structure with two formula units (six atoms) per unit cell. The barium and halogen (Cl, Br and I) atoms are located at $C_{4v}$ (or 4mm) whereas the fluorine atoms are at $D_{2d}$ (or $\bar{4}$2m) symmetry site. According to group theory, the symmetry decomposition of modes is as follows: $\Gamma_{18} = 2A_{1g}+B_{1g} +3E_g +3E_u+3A_{2u}$, in which six are Raman active ($2A_{1g}$ , $B_{1g}$ , $3E_g$) and four are IR active ($2A_{2u}$ , $2E_u$) modes. $E_g$(1,2,3) modes correspond to atomic motion of Ba, Cl and F along **b** -axis respectively. The anti symmetric vibration of Ba and Cl belongs to the modes of $A_{1g}$(1,2) along **c**-axis. The $B_{1g}$ mode corresponds to the anti symmetric vibration of F-atoms along **c**-axis. The calculated Raman active optical modes are given in Table.1 along with experimental values, they agree quite well with the experiments for BaClF[22,24], BaBrF[22], BaIF[23,25].

The pressure dependence of Raman active optical modes of BaClF, BaBrF and BaIF are shown in Fig.1(a), Fig.1(b) and Fig.1(c) respectively. It can be clearly seen that the vibrational frequencies of $A_{1g}$(1), $E_g$(2), $A_{1g}$(2) and $B_{1g}$ increases with pressure, which implies that the hardening of the modes in these three compounds. In BaClF, $E_g$(1) and $E_g$(3) modes have been soften upto 20 and 10 GPa respectively. This might be an indication for the structural phase transition that occurs in BaClF, which agrees with the tetragonal phase co-existing with orthorhombic phase at 10.8 GPa and further these two phases co-existing with monoclinic phase at 21.1GPa[17,24].

**Table.1:** The calculated and experimental (in parenthesis) Raman active optical modes of BaClF, BaBrF and BaIF at ambient pressure.

| Mode | BaClF[22,24] | BaBrF[22] | BaIF[23,25] |
|---|---|---|---|
| $E_g$(1) | 81.34(89,82) | 74.46(76) | 66.29(70,40) |
| $A_{1g}$(1) | 128.36(125,132) | 99.85(105) | 76.34(79,81) |
| $E_g$(2) | 135.54(142,145) | 104.71(109) | 99.17(105,74) |
| $A_{1g}$(2) | 161.64(162,165) | 122.94(119) | 107.46(113,113) |
| $B_{1g}$ | 216.73(212,215) | 208.85(215) | 210.11(203,205) |
| $E_g$(3) | 308.72(247,255) | 315.90(240) | 217.82(218,221) |

The softening of $E_g(1)$ mode is also observed in BaBrF and BaIF upto 30 and 60 GPa respectively, this could be an evidence for the reported structural phase transition in BaBrF at 27 GPa[16] and in BaIF around 55 GPa[25]. $E_g(3)$ mode decreases within pressure range 10-20 GPa for BaBrF and 10-40 GPa for BaIF, this is due to enhancement of intramolecular interaction under pressure. $A_{1g}(1)$ and $E_g(3)$ modes of BaIF are present upto 60 GPa in our calculation, but these modes are absent in experiment[25] from 30 and 55 GPa respectively.

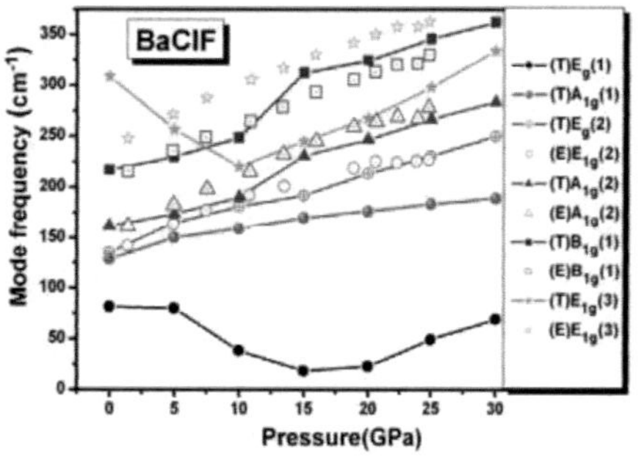


(a)

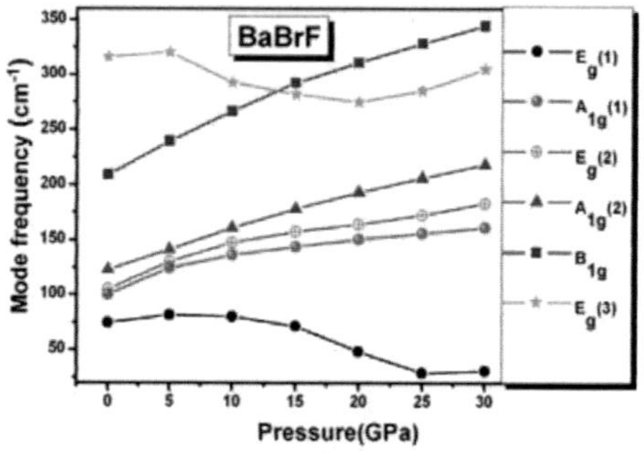


(b)

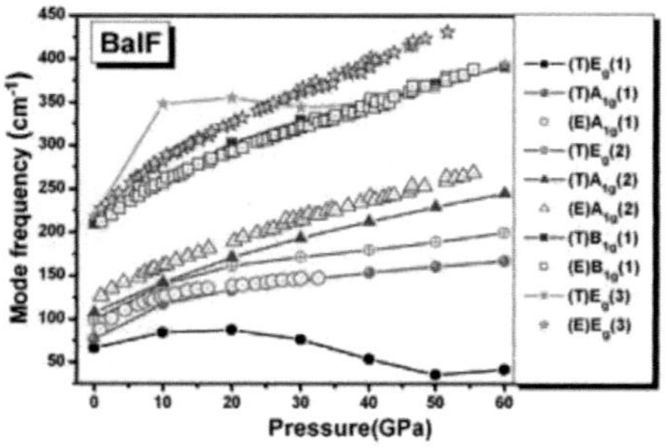


(c)

**Fig.1:** The calculated pressure dependence of Raman active optical modes for (a) BaClF[24], (b) BaBrF and (c) BaIF[25] along with the experiment. (E) stands for experiment and (T) for theory.

*3.2. Born effective charge tensors*

Born effective charge($Z^*$) and optical dielectric($\varepsilon^\infty$) tensors are fundamental quantities for understanding the lattice dynamics in polar insulators. The form of $Z^*$ directly results from site symmetry of crystal. The Born effective charge tensor $Z^*_{ij,k}$ is defined as the linear change in the polarization per unit cell created along the direction *i*, when the atom *k* is displaced along the direction *j*

$$Z^*_{ij,\,k} = V\frac{\partial p_i}{\partial r_j^k}$$

where, V is volume of the unit cell, $p_i$ , macroscopic polarization along the *i*–axis, and $r_j^k$ is the displacement of the kth atom along *j*-axis. The effective charge tensors of Ba, X (X = Cl, Br and I) and F atoms and the dielectric constants of BaXF compounds are reported in Table.2. The effective charge tensors vanish, when they summed over all the atoms in the unit cell *i.e* $\sum_k Z^*_{ij,\,k} = 0$. The neutrality guarantees that the polarization remains invariant under rigid translation of the crystal and also suggests that the calculations are well converged. Due to tetragonal symmetry of crystals the effective charge tensors for each atom, which are diagonal and obey $Z^*(xx) = Z^*(yy)$. The $Z^*(xx)$, $Z^*(zz)$ components of each atom in all these three compounds are very similar, indicating that the effective charge tensor is nearly isotropic. This is even more true from dielectric tensor, where the differences between $\varepsilon^{\infty}(xx)$ and $\varepsilon^{\infty}(zz)$ are very small, therefore the BaXF compounds are isotropic in view point of dielectric and effective charge tensors. Our recent investigation also revealed that these are optically isotropic materials even though strong anisotropy from structural and elastic properties[6]. We observed that the accumulation of the effective charges for Ba and halogen ions along ***a*** or ***b***-axis is more compared to ***c***-axis. This is in good agreement with the experiment for BaClF[31]. It is also clear from the Born effective charges, the fluoride ion seems to be rigid since its dynamical charge is less deviation from its formal charge (-1). The accumulation of effective charges on Ba and F (except for BaBrF along **a**-axis) increases whereas it is decreases on halogen (X = Cl, Br, I) from BaClF to BaBrF then to BaIF.

**Table.2**: The calculated effective dynamical charge tensors perpendicular $Z^*(xx)$ and parallel $Z^*(zz)$ to the **c**-axis and high frequency dielectric constants of BaCF, BaBrF and BaIF at ambient pressure.

| Parameter | BaClF | | | BaBrF | | | BaIF | | |
|---|---|---|---|---|---|---|---|---|---|
| | Ba | Cl | F | Ba | Br | F | Ba | I | F |
| $Z^*(xx)$ | 2.58 | -1.46 | -1.22 | 2.62 | -1.43 | -1.19 | 2.69 | -1.39 | -1.30 |
| $Z^*(zz)$ | 2.57 | -1.35 | -1.22 | 2.59 | -1.31 | -1.28 | 2.70 | -1.30 | -1.40 |
| $\varepsilon^{\infty}(xx)$ | | 3.15 | | | 3.43 | | | 3.96 | |
| $\varepsilon^{\infty}(zz)$ | | 3.12 | | | 3.38 | | | 3.86 | |

## 4. Conclusions

The vibrational properties of BaClF, BaBrF upto 30 GPa and BaIF upto 60 GPa have been studied using density functional perturbation theory within the local density approximation. The calculated Raman active optical modes at ambient pressure are consistent with the experimental results. The softening in the $E_g(1)$ and $E_g(3)$ modes under pressure might be responsible for pressure-induced structural phase transitions in BaClF, BaBrF and BaIF. The calculated optical dielectric tensor and effective charge tensors show the isotropic nature inspite of strong anisotropy from their crystal structures.

**Acknowledgements**

N.Y.K thanks DRDO through ACRHEM for financial support, and the CMSD, university of Hyderabad for providing computational facilities.

**References**

[1] Kiss Z J and Weakliem H A 1964 Bull. Am. Phys. Soc. **9** 89
[2] Kiss Z J and Weakliem H A 1965 Phys. Rev. Lett. **15** 457
[3] Baetzold R C 1987 Phys. Rev. B **36** 9182
[4] Takahashi K, Kohda K, Miyahara J, Kanemitsu Y, Amitani K and Shionoya S 1984 J. Lumin. **31** 266
[5] Takahashi K, Miyahara J 1984 J. Electrochem.Soc. **132** 2492
[6] Yedukondalu N, Ramesh Babu K, Bheema Lingam Ch, Singh D J, Vaitheeswaran G and Kanchana V 2011 Phys. Rev. B **83** 165117
[7] Comodi P, Zanazzi P F 1993 J. Appl. Crystall. **26** 843
[8] Kalpana G, Palanivel B, Shameem Banu I B and Rajagopalan M 1997 Phys. Rev. B **56** 3532
[9] Kanchana V, Vaitheeswaran G and Rajagopalan M 2003 J. Phys.:Condens. Matter **15** 1677
[10] El haj Hassan F, Akbarzadeh H, Hashemifar S J and Mokhtari A 2004 J. Phys. Chem. Solids **65** 1871

[11] El haj Hassan F, Akbarzadeh H and Hashemifar S J 2004 J. Phys.: Condens. Matter **16** 3329
[12] Ali Hussain Reshak, Charifi Z and Baaziz H 2008 Physica B **403** 711
[13] Ali Hussain Reshak, Charifi Z and Baaziz H 2007 J. Solid State electron. **51** 1133
[14] Ali Hussain Reshak, Charifi Z and Baaziz H 2007 Eur. Phys. J. B **60** 463
[15] Shen Y R, Englisch Chudinovskih L, Porsch F, Haberkorn R, Beck H P and Holzapfel W B 1994 J. Phys.: Condens. Matter **6** 1994
[16] Decremps F, Fischer M, Polian A, Itie J P and Sieskind M 1999 Phys. Rev. B **59** 4011
[17] Subramanian N, Chandra Shekhar V, Sahu P Ch, Mohammad Yousuf and Govinda Rajan K 1998 Phys. Rev. B **58** R555
[18] Liu M, Kurobori T and Hirose Y 2001 Phys. Stat. Sol. (b) **225** R20
[19] Kurobori T, Hirose Y and Takeuchi M 2000 Phys. Stat. Sol. (b) **220** R11
[20] Decremps F, Fischer M, Polian A, Sieskind M 1998 Eur. Phys. J. B **5** 7
[21] Mittal R, Chaplot S L, Sen A, Achary S N and Tyagi A K 2003 Phys. Rev. B **67** 134303
[22] James Scott F 1968 J. Chem. Phys. **49** 2766
[23] Sieskind M, Ayachour D, Merle J C, and Boulou J C 1990 Phys. Stat. Sol. (b) **158** 103
[24] Sundarakannan B, Ravindran T R, kesavamoorthy R and Satyanarayana S V M 2002 Solid state Communications **124** 385
[25] Decremps F, Gauthier M, Chervin J C, Fischer M and Polian A 2002 Phys. Rev B **66** 024115
[26] Segall M D, Lidan P L, Probert M J, Picard C J, Hasnip P J, Clark S J and Payne M C 2002 J.Phys:Condens. Matter **14** 2717
[27] Hamann D R, Schluter M and Chiang C 1979 Phys. Rev. Lett. **43** 1494
[28] Ceperley D M and Alder B J 1980 Phys. Rev. Lett. **45** 566
[29] Perdew J P and Zunger A 1981 Phys. Rev. B **23** 5048
[30] Monkhorst H J and Pack J B 1976 Phys. Rev. B **13** 5188
[31] Sieskind M, Ayadi M and Zachmann G 1986 Phys. Stat. Sol. (b) **136** 489

Journal of Physics: Conference Series **377** (2012) 012071 doi:10.1088/1742-6596/377/1/012071

# Order-disorder transition in $ZnGeP_2$ and $ZnSnP_2$

**Tista Basak**[1], **Mala N Rao**[1] **and S L Chaplot**[1]
[1]Solid State Physics Division, Bhabha Atomic Research Centre, Mumbai-400 085.

E-mail: mala@barc.gov.in

**Abstract.** Lattice dynamical calculations of $ZnGeP_2$ and $ZnSnP_2$ have been performed to investigate the possibility of the existence of the ordered chalcopyrite-to-disordered zincblende phase transition exhibited by some similar compounds and to understand the role played by phonons. The calculated values of the lattice parameters, bulk modulii and vibrational frequencies are in good agreement with the available experimental data. We observe that with increase in pressure the transverse acoustic branches and the lowest lying optic mode show a softening, which is eventually followed by elastic instability (at 3 GPa and 6.5 GPa respectively for $ZnGeP_2$ and $ZnSnP_2$) of the chalcopyrite phase. However, the disordered cubic phase is found to be dynamically and mechanically stable at these pressures.

## 1. Introduction

Ternary $A^{II}B^{IV}C^{V}_2$ semiconducting crystals having chalcopyrite structure (space group I-42d) are ternary analogues of the binary $A^{III}B^{V}_2$ zinc-blende semiconductors (space group F-43m). The $ZnGeP_2$ (ZGP) crystal, in particular, has been used extensively for the manufacture of nonlinear optical devices such as the optical parametric oscillator in the near infra-red region [1]. High pressure studies on ternary semiconductors have shown [2] that the chalcopyrite structure (body-centred tetragonal, Z=2, Zn, Ge, P at Wyckoff positions 4a [(000), (½0¾)], 4b[(00½)(½0¼)], 8d[($x$¼⅛)(-$x$¾⅛)(¼,-$x$,⅞)(¾$x$⅞)] respectively) becomes unstable under pressure and transforms either to the NaCl-type structure (if ratio of averaged cationic radius to the anionic radius is smaller than 0.414) or to a chemically disordered zinc-blende phase (face-centred cubic, Z=4). According to this criterion, for $ZnGeP_2$ and $ZnSnP_2$, a transition to the rock-salt phase is not expected.

The isothermal phase behaviour studies of ZGP as a function of pressure by using optical polarizing microscopy revealed that this compound transformed rapidly at (20.0+0.5) GPa to small segments of black opaque phases indicating a phase transition [3]. Further, even at ambient pressure, x-ray diffraction patterns of these compounds have shown [4, 5] the presence of both chalcopyrite and zincblende phases. The calculation of defect energies by Zapol et al. [6] have predicted that cation sublattice disorder in the form of antisite defect pairs (Zn in Ge site, and Ge in Zn site, respectively) may play a role behind the existence of the disordered cubic zinc-blende phase. Lattice dynamical calculations within the framework of the rigid-ion model [7] and shell model [6] have earlier been used to study the vibrational properties of chalcopyrites. Experimental studies of vibrational properties of ZGP and $ZnSnP_2$ have been limited to polarized Raman and infrared measurements [7, 8]. However, to the best of our knowledge, the role played by phonon/elastic instabilities in the possible order-disorder transition and/or coexistence of phases exhibited by these compounds has not yet been studied. In this paper, we have adopted a transferable potential model to (i) to study the vibrational

Journal of Physics: Conference Series **377** (2012) 012071 doi:10.1088/1742-6596/377/1/012071

properties of $ZnGeP_2$ and $ZnSnP_2$ in both phases at ambient and high pressure; (ii) to investigate the contribution of elastic instabilities to the phase transition.

## 2. The Potential Model and Group Theory

The calculations have been carried out using the current version of the program DISPR [9], developed at BARC. The interactions among the atomic constituents of the lattice are described by using a two-body interatomic potential function consisting of Coulomb interactions, short range interactions of the Born-Mayer type and van der Waals interaction. The atoms are modelled by using the shell model. The potential parameters have been optimized such that the condition of dynamical equilibrium is satisfied. In the computations, the ordered chalcopyrite structure is as given in Refs. [4, 10], whereas the disordered zincblende structure [4] is generated both as a supercell and in the virtual crystal approximation (VCA). The supercell was generated by constructing a x b x 2c supercell of 16 atomic sites. Eight atomic sites are occupied by the atoms Zn and Ge, randomly but with equal probability, and the remaining eight sites are occupied by P atoms. In the VCA, one atomic site in the primitive unit cell is assumed to be occupied by an ion of mass and ionic radius equal to an average of the masses and ionic radii of Zn and Ge, whereas the other site is occupied by P ion. In the present case, both these methods produce similar results for the vibrational frequencies because the difference in masses and ionic radii of Zn and Ge is small. The phonon modes at the zone centre for the chalcopyrite structure are classified as

$$\Gamma = 1A_1+2A_2+3B_1+4B_2+7E$$

The E modes are doubly degenerate. The $A_1$, $B_1$, $B_2$ and E modes are Raman active; modes $B_2$ and E are both Raman and infrared active and one each of the $B_2$ and E modes are acoustic. The $A_2$ modes, on the other hand, are inactive in both Raman and infrared scattering experiments.

## 3. Results and discussion

### *3.1. Structure and elastic properties*

The calculated structure parameters as well as bulk modulii are in good agreement with the experimental data obtained at ambient temperature and pressure. The calculated elastic constants also do not deviate greatly from the values obtained from previous theoretical studies.

**Table 1.** Calculated and experimentally determined values for lattice constants and bulk modulii

| Compound | a (Å) | | c (Å) | | V (Å$^3$) | | B (GPa) | |
|---|---|---|---|---|---|---|---|---|
| | Expt | Calc | Expt | Calc | Expt | Calc | Expt | Calc |
| chalcopyrite | | | | | | | | |
| $ZnGeP_2$ | 5.46 [4] | 5.45 | 10.71 [4] | 10.30 | 159.6 [4] | 152.8 | 76.6-79.6[3] | 74.7 |
| $ZnSnP_2$ | 5.65 [10] | 5.66 | 11.30[10] | 10.83 | 180.5 [10] | 173.1 | 69.4 [11] | 63.5 |
| zincblende | | | | | | | | |
| $ZnGeP_2$ | 5.44[4] | 5.35 | | | 160.9 [4] | 153.1 | | 81.8 |
| $ZnSnP_2$ | | 5.47 | | | | 163.7 | | 69.4 |

### *3.2. Raman-active phonons at ambient and high pressures*

The computed phonon frequencies at the zone centre for $ZnGeP_2$ and $ZnSnP_2$ are, on an average, within 10% of the measured Raman and infrared values. The theoretically calculated phonon frequencies at the Brillouin zone centre for the chalcopyrite phase of $ZnSnP_2$ along with available Raman and their computed Gruneisen parameters are shown in Table 2. For ZGP, the computed variation of the Raman-active phonon frequencies with pressure is compared with the experimental data [8] in Fig. 1.

**Table 2.** Calculated and measured [7] Raman-active frequencies of chalcopyrite $ZnSnP_2$ and the corresponding Gruneisen parameters $\gamma$

| Symmetry of vibrational mode | Calculated ($cm^{-1}$) | Experimental ($cm^{-1}$) | Calculated $\gamma$ |
|---|---|---|---|
| $A_1$ | 297 | 309 | 1.19 |
| $B_1$ | 297 | 353 | 1.63 |
| $B_2$ | 316 | 347 | 1.74 |
| $B_2$ | 359 | 365 | 0.85 |
| E | 77 | | -0.62 |
| E | 232 | 317 | 1.71 |
| E | 298 | 328 | 1.84 |
| E | 353 | 342 | 0.75 |

We observe that all the modes (except the lowest-lying E mode) exhibit a hardening with pressure (i.e. positive mode Gruneisen parameter) for both $ZnGeP_2$ and $ZnSnP_2$. The lowest-lying E mode corresponds to the transverse acoustic mode in the zincblende structure and its softening (i.e. negative Gruneisen parameter) is related to the phase transition to the rock salt phase.

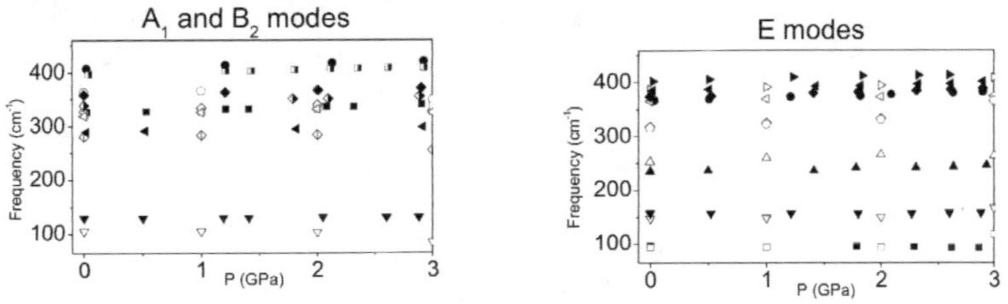


**Figure 1.** Comparison of the computed (open symbol) and experimental [8] (solid symbol) Raman-active phonon frequencies for the chalcopyrite phase of $ZnGeP_2$ with pressure.

*3.3 Mode Grüneisen parameters of transverse acoustic modes and high-pressure elastic constants*

Figure 2 represents the variation of the Grüneisen parameters of the transverse acoustic phonons along the [001] and [100] directions for chalcopyrite $ZnGeP_2$ and $ZnSnP_2$ respectively. In case of both $ZnGeP_2$ and $ZnSnP_2$, the mode Grüneisen parameters are negative along both the [001] and [100] direction. In addition, the variation of mode Grüneisen parameter over the entire Brillouin zone along the [001] direction is almost linear while it shows maxima close to the zone centre and the zone boundary along the [100] direction, for both $ZnGeP_2$ and $ZnSnP_2$. The negative mode Gruneisen parameters of the transverse acoustic modes are related to the mechanical instability of the chalcopyrite structure (as shown later in this section), this behaviour is similar to that observed in neutron scattering experiments on $AgGaSe_2$ [12].

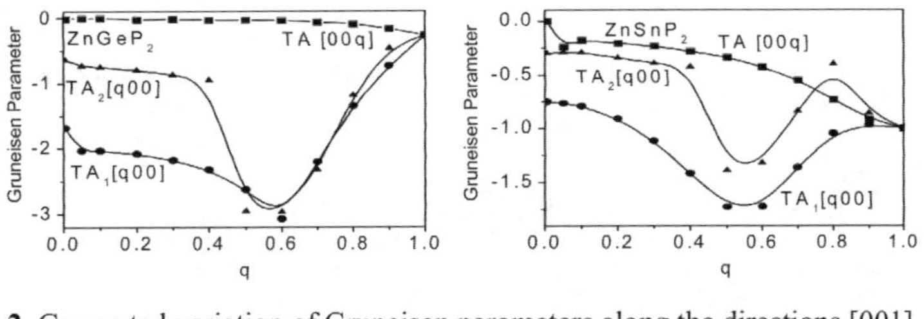


**Figure 2.** Computed variation of Gruneisen parameters along the directions [001] and [100] for chalcopyrite $ZnGeP_2$ and $ZnSnP_2$ respectively.

The computed variation of the elastic constant $C_{66}$ with pressure for both $ZnGeP_2$ and $ZnSnP_2$ is shown in Fig. 3. It is observed that with pressure there is a softening of this elastic constant which implies that the chalcopyrite phase becomes mechanically unstable with increase in pressure (at 3 GPa and 6.5 GPa respectively for $ZnGeP_2$ and $ZnSnP_2$). The stability range of $ZnSnP_2$ in the chalcopyrite phase is thus greater than ZGP.

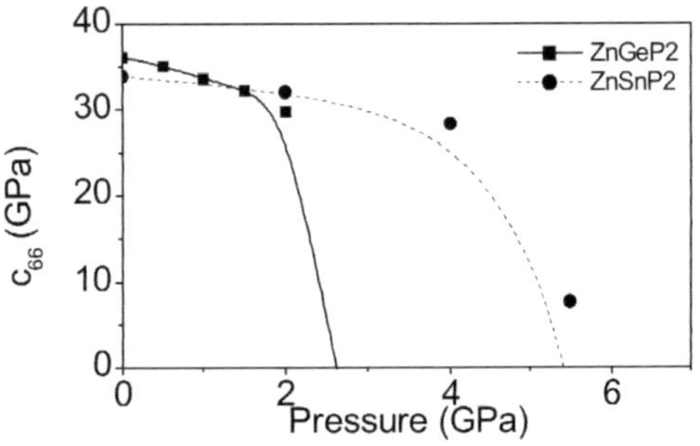


**Figure 3.** Computed variation of elastic constant $C_{66}$ with pressure for chalcopyrite $ZnGeP_2$ and $ZnSnP_2$.

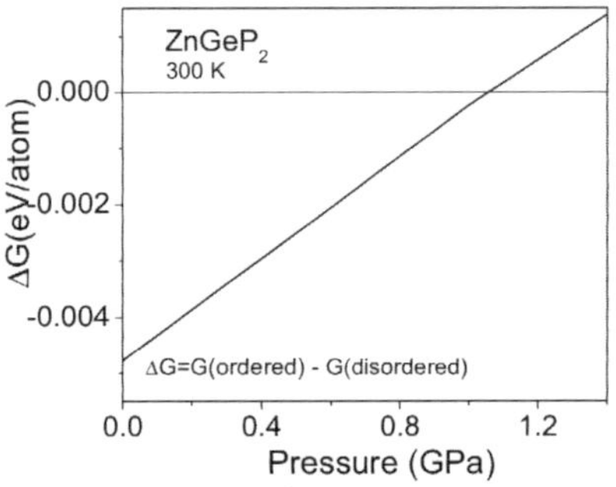


**Figure 4.** Computed variation of the Gibbs' free energy with pressure for the ordered chalcopyrite and disordered zincblende phase

However, in these computations, the disordered zincblende phase is found to be dynamically and mechanically stable at pressures upto 10 GPa. This indicates that a high pressure phase transition from ordered chalcopyrite to disordered zincblende is probable, in qualitative agreement with pressure-driven order-disorder transition observed in experiments [3]. The calculated Gibbs' free energy difference (Fig. 4) between the ordered and disordered phases, in the pressure range of mechanical stability, is very small (~0.005 eV/ atom); this is consistent with experimental [4, 5] findings of coexistence of both chalcopyrite as well as zincblende phases at ambient pressure.

## 4. References

[1] Okunev A O, Verozubova G A, Trukhanov E M, Dzjuba I V, Galtier P R J and Said Hassani S A 2009 *J. Appl. Cryst.* **42** 994; Mason P D, Jackson D J and Gorton E K 1994 *Opt. Commun*, **110** 163.

[2] Carlone C, Olego D, Jayaraman A, and Cardona M 1980 *Phys. Rev. B* **22** 3877.

[3] Block S, Piermarini G J, Munro R G and Fuller E 1989 *Physica A* **156** 341.

[4] Kimmel G, Shimony Y, Raz O, and Dariel M P *Materials Structure* **6** 1999 149.

[5] Vaipolin A A, Goryonova N A, Kleshechinskii L I, Loshakova G V and Osmanov E O 1968 *Phys. Sat. Sol.* **29** 435.

[6] Zapol P, Pandey R, Ohmer M and Gale J 1996 *J. Appl. Phys.* **79** 671.

[7] Mintairov A M, Sadchikov N A, Sauncy T, Holtz M, Seryogin G A, Nikishin S A and Temkin H 1999 *Phys. Rev. B* **59** 15197.

[8] Shirakata S 1999 *J. Appl. Phys.* **85** 3294.

[9] Chaplot S L, Choudhury N, Ghose Subrata, Rao Mala N , Mittal R , and Goel P 2002 *Eur. J. Mineral* **14** 291.

[10] Miyauchi K, Minemura T, Nakatani K, Nakanishi H, Sugiyama M, and Shirakata S 2009 *Phys Stat Sol C* **6** 1116.

[11] Li K, and Xue D 2010 *Physica Scripta* **T139** 014072.

[12] Klotz S, Derollez P, Fouret R, Braden M, Hennion B, Hubert Ch., Meducin F, Lazewski J, Gonzalez J 2003 *Phys. Stat. Sol. B* **235** 331.

Journal of Physics: Conference Series **377** (2012) 012072 doi:10.1088/1742-6596/377/1/012072

# On high pressure β → ω phase transition in Ta and Zr-Nb

**Debojyoti Mukherjee, K.D. Joshi, and Satish C. Gupta**
Applied Physics Division, Bhabha Atomic Research Centre, Mumbai
India 400085

E-mail: debojyoti@barc.gov.in

**Abstract.** Transition from ω(hexagonal) to β(bcc) phase under pressure is well established in group IVB elements. It is interesting, however, to explore if pressure induced β to ω transformation is possible. In shock wave experiments ω phase has been observed in polycrystalline Ta samples retrieved from shock pressure of ~ 45 GPa (Hsiung and Lassila 1998), and also, in β stabilized Zr-Nb(Nb 20% by weight) quenched from shock pressure of ~ 15 GPa (Dey et al. 2004). In static pressure measurements, on the other hand, β to ω transition is neither found in Ta up to 170 GPa(Cynn et al. 1998), nor in Zr-Nb up to 8 GPa. Structural stability of β and ω phases has been examined under hydrostatic compression up to 243 GPa in Ta and up to ~ 18 GPa in Zr-Nb from first principles electronic band structure calculations. In agreement with experiments, present theoretical calculations suggest that the β phase is energetically favourable over ω phase at ambient conditions and remains stable throughout the compression range of calculations. Further, the calculation of activation barrier between β and ω phases at various pressures indicate that the ω phase is always unstable structure in Ta, however, the same phase is metastable Zr-Nb.

## 1. Introduction

The pressure induced ω → β phase transition in Group IV transition metals is well established [1-5]. The ω and β phases are simple hexagonal and bcc structure, respectively. Further, this observed structural transition under pressure has been associated to the pressure induced increase in *d*-electron population due to *sp-d* electron transfer [6, 7]. It is interesting, however, to explore if pressure induced β to ω transformation is possible. From electron microscopy measurements carried out on polycrystalline sample of elemental solid Ta retrieved from shock pressure of ~ 45 GPa, Hsiung and Lassila [8] have reported the occurrence of β → ω phase transition in this metal. However, in high pressure x-ray diffraction measurements carried out on Ta in diamond anvil cell (DAC) experiments, Cynn *et al.* [9] have reported no β → ω transition up to ~ 170 GPa. Subsequent, high pressure experimental studies using DAC by Dawaele *et al.* [10] up to 100 GPa and by Ling Yun *et al.* [11] up to 133 GPa also support experimental observation of Cynn *et al.* [9]. Similarly, in Zr-Nb(Nb 20% by weight) samples retrieved from shock pressure of ~ 15 GPa, Dey *et al.* [12] have reported the occurrence of β → ω phase transition in β stabilized phase of this alloy. However, this transition has not been observed under static loading at 8 GPa [12]. Later from theoretical calculations using WIEN97 code Gupta *et al.* [13] have demonstrated that β phase indeed remains a low energy structure as compared to ω phase under hydrostatic pressure.

In the present work, we have carried out *ab-initio* electronic band structure calculations on Ta and Zr-Nb alloy (i) to explore the possibility of pressure induced β to ω transition in these materials and (ii) to determine the equation state at 300 K. Further, in order to examine whether ω phase is thermodynamically unstable or metastable, we have determined the activation barrier between β and ω phases at various hydrostatic pressures. The calculations on Zr-Nb, though, are already been done by Gupta et al [13], we have further refined these by using more advanced version of WIEN code [14].

## 2. Theoretical Procedure

The total energy calculations have been performed as a function of hydrostatic compression on β and ω phase of Ta and Zr-Nb employing full potential linearized augmented plane wave (FP-LAPW) method (WIEN2K package) [14] within generalized gradient approximation (GGA) [15] for exchange - correlation energy. The plane wave cut-off parameter $RK_{max}$ is fixed at 8 for both the cases; however, a grid of 5000 k points and, 3000 k points respectively was used for sampling of the full Brillouin zone in Ta and Zr-Nb. The muffin tin radius used for both the structures was 2.0 a.u. in Ta and 2.2 a.u. in Zr-Nb. In Ta being a high Z metal; calculations have been performed both without and with spin orbit coupling (SOC). The total energy calculations for β and ω-phase in Ta were performed on $AlB_2$ type structure with one atom at the site 'Al' (0,0,0) and two atoms at site 'B' [(1/3,2/3,1/2+δ),(2/3,1/3,1/2-δ)]. It may be noted that δ= 1/6 for β- phase and δ = 0 for ideal ω-phase. For 0< δ< 1/6 the symmetry is trigonal and structures are termed as non-ideal ω. In case of Zr-Nb alloy the calculations were carried out on hexagonal super lattice (space group P6/mmm) and the c axis is doubled, *i.e.* the *c/a* ratio of the super cell is twice as compared that for the normal ω cell. The unit cell of the super lattice contain total six atoms located on three inequivelant sites '4h', 'a' and 'b' (international Table of Crystallography, Vol. 2). The site '4h' {(2/3, 1/3, 1/4-δ), (1/3, 2/3, 1/4+δ), (2/3, 1/3, $\overline{1/4-\delta}$), (1/3, 2/3, $\overline{1/4+\delta}$)} and 'a' (0, 0, 0) are occupied by Zr atoms, however, the site 'b' (0, 0, 1/2) is occupied by Nb atom. In this structure the Zr-Nb alloy contained 17.5% of Nb by weight, which is closer to the composition of Zr-20Nb sample used in the experiment. Again δ = 0 corresponds to ideal ω phase, and δ = 1/12 (*i.e.* displacement along c axis by 1/12 of $C_\omega$, the lattice parameter of super lattice) corresponds to β phase. Here, for 0< δ <1/12 the structures are non-ideal ω. For further clarity the ideal ω cells used for calculations on Ta and Zr-Nb are shown in figure 1.

The 300 K isotherm is determined by adding finite temperature corrections to 0 K calculations using the procedure followed by Wallace [16] and Joshi *et al.*[17].

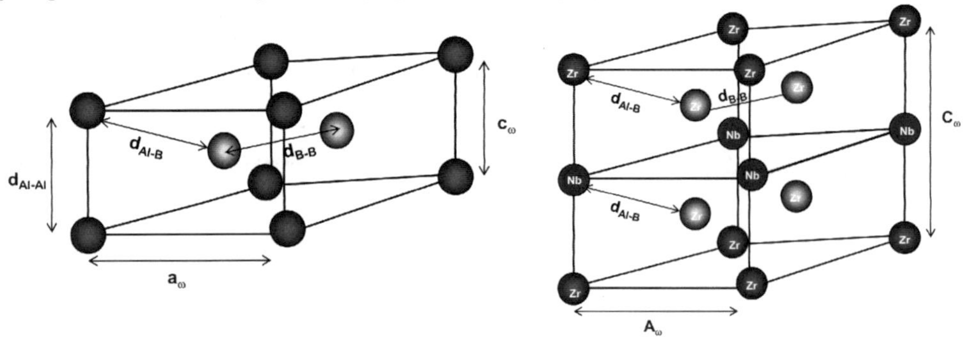


**Figure 1**. Omega lattice used for calculations on Ta (left) and Zr-Nb (right) alloy.

## 3. Results and Discussions

*Tantalum*

Figure 2 displays the total energy of β and ω phases relative to β phase for tantalum as a function of hydrostatic compression. The relative total energies have been plotted for both with and without considering SOC. It is clear from the figure that for both the cases the β phase is energetically

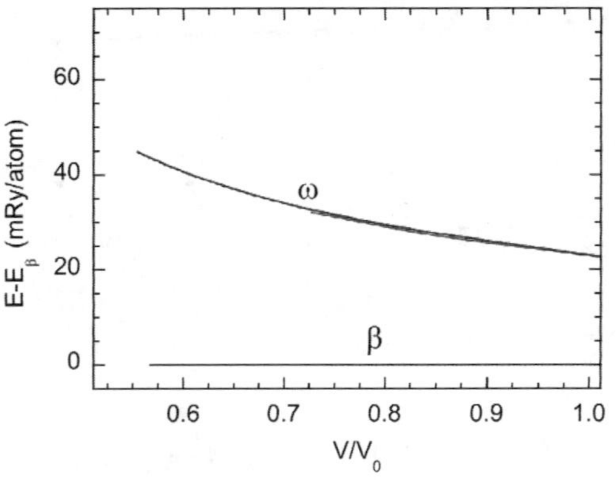


**Figure 2**. Total energy of β and ω phases relative to β phase for Ta. The red and green curves correspond to the energy of ω phase relative to β phase with and without considering SOC , respectively.

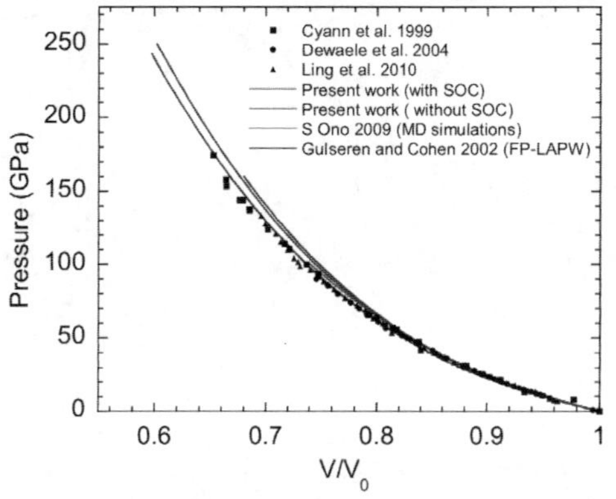


**Figure 3**. Comparison of theoretical 300 K isotherm with experimental data [9-11] and MD simulations [18, 19].

**Table 1**. The comparison of various physical quantities for β of Ta at ambient conditions. The values outside and inside the brackets for the present work are determined using calculations excluding and including SOC, respectively.

| | Theory other sources | Theory present work | Experimental |
|---|---|---|---|
| $V_0$(Å$^3$/atom) | 18.016 [18]<br>18.31 [19]<br>18.337 [23]<br>18.245 [24] | 18.238 (18.4) | 18.039 [9], 18.033 [10]<br>18.005 [11] |
| $B_0$(GPa) | 194 [18]<br>192 [19]<br>190 [23]<br>193 [24] | 193 (195.9) | 194.7 [9] 194 [10]<br>192.6[11], 196 [20] |
| $B_0'$ | 3.74 [18]<br>3.81 [23]<br>3.88 [24] | 3.63 (4.03) | 3.4 [9], 3.25 [10]<br>3.58 [11] |
| $\gamma_0$ | | 1.31 (1.51)*<br>1.65 (1.84)** | 1.69 [21], 1.63 [22] |

*Values determined using Dugdale- Macdonald relation; ** values determined using Slater's relation

favourable over ω phase throughout the compression range, indicating no β → ω transition up to ~ 243 GPa, which is in agreement with the experimental observations [9-11]. Further, difference in total energy of ω phase with β is more or less independent of SOC. In figure 3 we compare the theoretically determined equation of state with experimental data [9-11] and other theoretical works [18, 19]. The present theoretical 300 K isotherm determined from calculations carried out without SOC agrees well

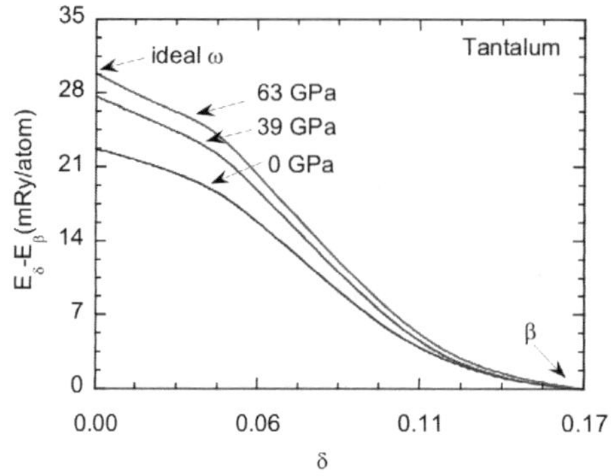


**Figure 4**. Total energy of ideal and non ideal ω structures relative to β phase in Ta as a function of δ, the displacement along c axis of B atoms from the 0. $5c_\omega$

with experimental data [9-11] and molecular dynamic (MD) simulations [18] throughout the compression range. However, that derived from calculations including SOC though compares well with experiments [9-11] and MD simulations [19] up to ~ 60 GPa, it starts deviating systematically above this pressure. Further, our theoretical isotherm determined from calculations without SOC is softer than that reported from earlier FP-LAPW calculations [19], however the same agrees well with the later if SOC is taken into account. In Table 1 we compare various physical quantities derived from theoretical isotherm to that obtained experiments[9-11, 20-22] and those available from other theoretical sources [18, 19, 23, 24]. The Gruneisen parameter ($\gamma_0$) is determined using $B_0'$ in both the Dugdale- Macdonald approximation and Slater's approximations.

Additionally, in order to examine whether ω phase is thermodynamically unstable or metastable, we have determined the activation barrier between β and ω phases at various hydrostatic pressures. In figure 4 we show total energy with respect to β phase along the path of transition in Ta at various pressures. It is clear from this figure that the ω phase is always an unstable phase. The occurrence of this transition under shock compression suggests that the shear stresses present in shock compression could stabilize even a phase unstable under hydrostatic conditions.

*Zr-Nb*

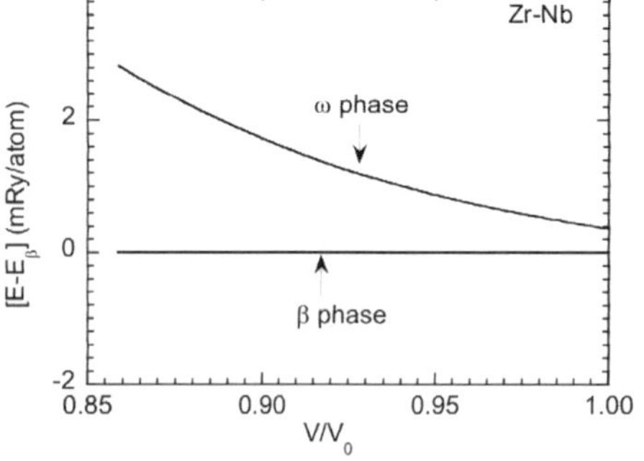


**Figure 5**. Total energy of β and ω phases relative to β phase for Zr-Nb.

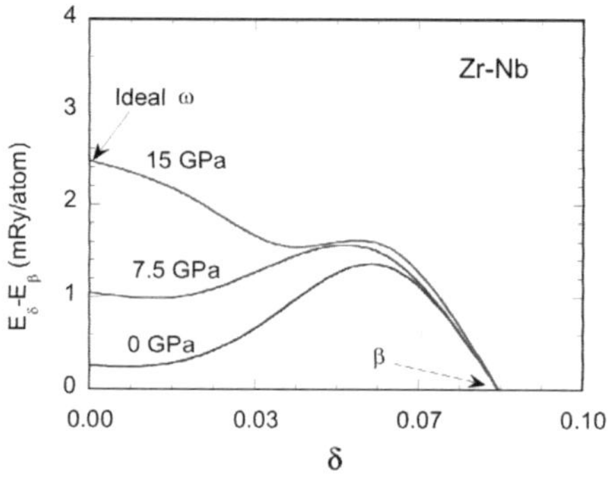


**Figure 6**. Total energy of ideal and non ideal ω structures relative to β phase in Zr-Nb as a function of δ, the displacement along c axis of B atoms from the 0. $25C_\omega$, where $C_\omega$ is the lattice parameter of super lattice as displayed in figure 1.

Figure 5 displays the total energy of ideal ω phase relative to that of β phase as a function of volume under hydrostatic compression for Zr-Nb alloy. As depicted in figure, like tantalum, in Zr-Nb also, the β phase is favourable over ω phase throughout the compression range under hydrostatic loading. The zero pressure volume and, bulk modulus and its pressure derivative at this volume for β phase was determined to be 22.06 $(Å)^3$/atom, 113 GPa and 5.64, respectively.

Finally, to determine the activation barrier between β and ω phase, we performed total energy calculations as a function of δ for various compressions. It may be noted that as mentioned in previous section, in case of Zr-Nb the displacement δ along c-axis is normalized with respect to the corresponding lattice parameter of super lattice. As shown in figure 6, for all compressions the total energy shows the two minima corresponding to stable β phase and metastable ω phase. At zero pressure, the local minimum occurs for $\delta = 0$, however, at 7.5 GPa and 15 GPa it shifts to rumpled ω with $\delta \sim 0.02$ and $\delta \sim 0.04$, respectively. The activation barriers of ~1.1 mRy/atom at zero pressure increases to ~ 1.5 mRy/atom at 15 GPa. The β to ω transition in Zr-Nb is possible, only if the available energy during compression is sufficient to overcome the activation barrier. The detection of ω phase in shock treated samples suggests that this activation barrier might have been overcome due to shear stress present in the shock loading.

**Summary**

To summarise, our analysis highlights the role of shear stresses in stabilizing the structures which, under hydrostatic condition are metastable or even unstable. It supports the viewpoint of Joshi *et al.*[5] who proposed that the occurrence of metastable γ and δ phases in Ti at Mbar pressures is related to the presence of shear stresses in DAC experiments.

**References**

[1] Gyanchandani J.S., Gupta S.C., Sikka S.K. and Chidambaram R. 1990 *High Press. Res.* **4** 472.
[2] Xia H., Dulcos S.J., Ruoff A.L. and Vohra Y.K. 1990 *Phys. Rev. Lett.* **64** 204.
[3] Xia Hui, Parthsarthy G., Luo H., Vohra Y.K. and Ruoff A. 1990 *Phys Rev B* **42** 6736.
[4] Ahuja R., Wills J. M., Johansson B. and Eriksson O. 1993 *Phys. Rev. B* **48** 16269.
[5] Joshi K.D., Jyoti G., Gupta Satish C., Sikka S.K. 2002 *Phys. Rev. B*, **65** 052106-1.
[6] Gyanchandani Jyoti S., Sikka S.K. and Chidambaram R., in *Recent Trends in High Press. Res., Proceedings of XIII AIRAPT International Conference on High Press. Science and Technology* (Ed: A.K. Singh), Oxford and IBH, New Delhi, 331, (1992).
[7] Joshi K.D., Jyoti G., Gupta Satish C. and Sikka S.K. 2002 *J. Phys. Cond. Matter* **14** 1.
[8] Hsiung L.M. and Lassila D.H., *Scripta mater* 1998 **39** 603.
[9] Cynn H. and Yoo C, *Phys. Rev B* 1998 **59** 8526.
[10] Dewaele Agnes, Loubeyre Paul and Mezouar Mohamed 2004 *Phys. Rev. B* **69** 092106.
[11] TANG Ling-Yun, LIU Lei, LIU Jing, XIAO Wan-Sheng, LI Yan-Chun, LI Xiao-Dong and BI Yan 2010 *Chin. Phys. Lett.* **27** 016402.
[12] Dey G.K., Tewari R., Banerjee S., Jyoti G., Gupta Satish C., Joshi K.D., Sikka S.K. 2004 *Acta Materialia* **52** 5243.
[13] Satish C. Gupta, K.D. Joshi and S. Banerjee 2008 *Metal. and Materials Trans* **39A**, 1593.
[14] Blaha P. Schwarz K. and Luitz J. *WIEN2K* (Technical University of Vienna).
[15] Perdew J.P. Burke S. and Ernzerhof M. 1996 *Phys. Rev. Lett.* **77** 3865
[16] Wallace D.C. *Thermodynamics of crystals* 1972 Wiley New York
[17] Joshi K.D., Gupta Satish C. and Banerjee S. 2009 *J. Phys. Cond. Mat.*, **21**, 415402.
[18] Ono Shigeaki 2009 *Int. J. Mol. Sci* **10**, 4342.
[19] Gulseren O. and Cohen R. E. 2002, *Phys. Rev. B* **65** 064103.
[20] Bolef D. I. 1961 *J. Appl. Phys.* **32** 100.
[21] Pearson W. B., *A Handbook of Lattice Spacings and Structures, of Metals and Alloys*

(Pergamon, New York, 19581).

[22] Isbell W. M., Christman D.R., Babcock S.G. 1971 *Measurements of Dynamic Properties of Materoals Volume VI Tantalum*; Report from Materiats and structures Laboratory, Manufacturing Devololument, General Motnrs Corporat'on, General Motors Technical Center. Warren, Michigan, USA

[23] Liu Zhong-Li, Li Xiao-Feng, Cai Ling-Cang, and Jing Fu-Qian 2011 *Phys. Status Solidi B* **248**, 175.

[24] Liu Zhong-Li, Cai Ling-Cang, Chen Xiang-Rong, and Jing Fu-Qian 2008 *Phys. Rev. B* **77** 024103-1.

# Electronic and Magnetic Properties of $CaS_{0.875}M_{0.125}$ (M = C, Si, Ge and Sn) by First Principles Theory

**M Yogeswari, R Umamaheswari and G Kalpana***
Department of Physics, Anna University, Sardar Patel road, Guindy, Chennai – 600 025, INDIA

E-mail: g_kalpa@annauniv.edu, g_kalpa@yahoo.com

**Abstract.** The total energy, energy bands, density of sates and half-metallic ferromagnetism of $CaS_{0.875}M_{0.125}$ (M = C, Si, Ge and Sn) have been studied using band structure calculation methods namely full potential linearized augmented plane wave (FP-LAPW) as well as tight binding linear muffin orbital (TB-LMTO) method. We find that within the generalized gradient approximation (GGA) all dopants induce half-metallic ferromagnetism in CaS, whereas in the case of local spin density approximation (LSDA) C- doping only induce half-metallicity in CaS. The calculated magnetic moment is found to be 2.00 $\mu_B$ per formula unit. The magnetism arises mainly from the p states of dopant atoms. The ground state properties such as lattice constant, total energy difference between non-magnetic and ferromagnetic state, total and partial magnetic moments have been calculated. Within the GGA calculation all dopants enhance the stability of ferromagnetic state.

## Introduction

It is highly desirable to explore satisfactory half-metallic (HM) ferromagnetic materials for the realization of high-performance spintronic devices [1]. HM ferromagnets are materials with one of their two spin channels metallic, whereas other has a semiconductor or insulating property. It shows 100% spin polarization around the Fermi level. Since its introduction by de Groot et al. [2] in $C1_b$ structure semi-Heusler alloy NiMnSb, several HM ferromagnets have been theoretically predicted such as half- and full Hesuler alloys, double perovskite compounds, magnetic oxides, several zinc-blende (ZB) phase transition metal (TM) pinictides and diluted magnetic semiconductors [3]. Few of them are synthesized experimentally [4]. In this case ferromagnetism substantially arises from 3d magnetic ion impurities. Recently Gao et al. [5] have reported sp HM ferromagnetism in rock salt and ZB MS (M = Li, Na and K) using FP-LAPW method. Moreover several work reveal that non-magnetic elements like B, C, N, Mg, Ca, and Cu as dopants can produce spin polarization in some semiconducting host materials [6]. More recently, unusual HM ferromagnetism at room temperature is found in insulators such as alkaline earth oxides doped with intrinsic non-magnetic elements like B, C and N. Recently electronic structure and trends of MgO systems doped with III, IV and V main group elements for a concentration of 25% was reported using plane wave ultra soft pseudo potential method. Based on density functional theory, Liu et al [7] has studied that the C- doping in alkaline earth chalcogenides (O and S) induce half-metallic ferromagnetism. In our previous work we have investigated the electronic structure and HM ferromagnetism in calcium chalcogenides doped with B, C and N [8]. Alkaline-earth chalcogenides are important wide bandgap semiconducting materials, which can produce excitons with large binding energy. Hence they have ample technological application in the areas of luminescent devices, catalysis, and microelectronics [9]. In this paper we study the possibility of HM ferromagnetism in calcium sulphide with 2p elements such as C, Si, Ge and Sn as dopant using full-potential linearized augmented plane wave method (FP-LAPW) and tight binding linear muffin tin orbital method (TB-LMTO). Realization of ferromagnetism in these compounds may improve the function of optical based spintronic devices.

## 2. Computational details

To study the electronic structure and possibility of HM ferromagnetism in $CaS_{0.875}M_{0.125}$ (M = C, Si, Ge and Sn), we have performed first principle calculations using all electron full potential-linearized augmented plane wave method based on density functional theory as implemented in Wien2k code [10]. Generalized gradient approximation of Perdew, Burke and Ernzerhof was used to treat exchange correlation potential. Relativistic effects are taken into account within the scalar approximation. We take $R_{mt}K_{max} = 8$ and the angular momentum expansion up to $l = 10$ in the muffin tins. We have used 4000 ***k***-points in the first Brillouin zone. The self-consistency calculations were considered to be converged only when the total energy deviation is less than $10^{-5}$ Ry./f.u. The calculations were performed with spin polarized potentials to analyse ferromagnetic phase of compounds. To study the doping effects 2x2x2 supercell was constructed, which contains 16 atoms with fcc lattice.

Furthermore we have used tight binding linear muffin tin orbital (TB-LMTO) method [11] within local spin density approximation (LSDA) to study the possibility of HM ferromagnetism in title compound. The von Barth and Hedin parameterization has been used for exchange and correlation potential. All relativistic effects have been taken into account except spin-orbit coupling. A mesh of 12x12x12 was taken in the irreducible wedge of Brillouin Zone. The **k** points and total energy convergence are checked. The 16 atom supercell was constructed to study the HM property in $CaS_{0.875}M_{0.125}$ (M = C, Si, Ge and Sn).

## 3. Results and discussions

### 3.1 Total energy and stability calculations

Total energy calculations are performed for binary host compound CaS, from which the equilibrium lattice constant was calculated. The calculated lattice constants are slightly overestimated by FP-LAPW and underestimated in TB-LMTO as compared with experimental value. Before doing the electronic and magnetic calculations, full atomic position and volume optimization was performed for $CaS_{0.875}M_{0.125}$ (M = C, Si, Ge and Sn). In order to determine the equilibrium lattice constant for doped systems, the total energies were calculated as a function of reduced volume for both non-magnetic and magnetic states and fitted into Murangaun equation of state. The predicted lattice constant, total energy difference ( $\Delta E = E_{NM} - E_{FM}$ ) are tabulated for both methods namely FP-LAPW and TB-LMTO (Table 1). The lattice constant of the doped systems decreases with size of the dopant, since bigger the radius of the dopant, the larger the lattice constant. The calculated total energy difference ( $\Delta E = E_{NM} - E_{FM}$ ) between the nonmagnetic and ferromagnetic states are positive (Table 1) except for $CaS_{0.875}Ge_{0.125}$ and $CaS_{0.875}Sn_{0.125}$, it shows that the ferromagnetic state is more stable than non-magnetic state. The total energy difference increases as atomic size of the dopant decreases. This is due to; when size of the dopant atom decreases the overlap with neighbouring atom decreases and the valence band become very narrow.

### 3.2 Electronic band structure and magnetic properties

The spin dependent band structure of $CaS_{0.875}M_{0.125}$ (M = C, Si, Ge and Sn) compounds were calculated at their equilibrium lattice constant along the selected high symmetry directions. The electronic band structure profile is same for all compounds, so for illustration, electronic band structure of $CaS_{0.875}Si_{0.125}$ calculated by FP-LAPW is shown in figure 1. From figure 1, in the majority

**Table 1.** Calculated equilibrium lattice constant 'a' (Å) for both nonmagnetic (NM) and ferromagnetic (FM) phases, total energy difference ΔE (meV/cell) and total magnetic moment ($\mu_B$) of $CaS_{0.875}M_{0.125}$ (M = C, Si, Ge and Sn).

| Compounds | 'a' (Å) | | | | ΔE (meV/cell) | | Total magnetic moment ($\mu_B$) | |
|---|---|---|---|---|---|---|---|---|
| | FP-LAPW | | TB-LMTO | | FP-LAPW | TB-LMTO | FP-LAPW | TB-LMTO |
| | NM | FM | NM | FM | | | | |
| $CaS_{0.875}C_{0.125}$ | 5.681 | 5.679 | 5.558 | 5.568 | 538 | 424 | 2.00 | 2.00 |
| $CaS_{0.875}Si_{0.125}$ | 5.772 | 5.769 | 5.636 | 5.639 | 223 | 73 | 2.00 | 1.87 |
| $CaS_{0.875}Ge_{0.125}$ | 5.778 | 5.774 | 5.642 | 5.641 | 157 | - 42 | 2.00 | 1.69 |
| $CaS_{0.875}Sn_{0.125}$ | 5.834 | 5.831 | 5.688 | 5.671 | 135 | - 36 | 2.00 | 1.30 |

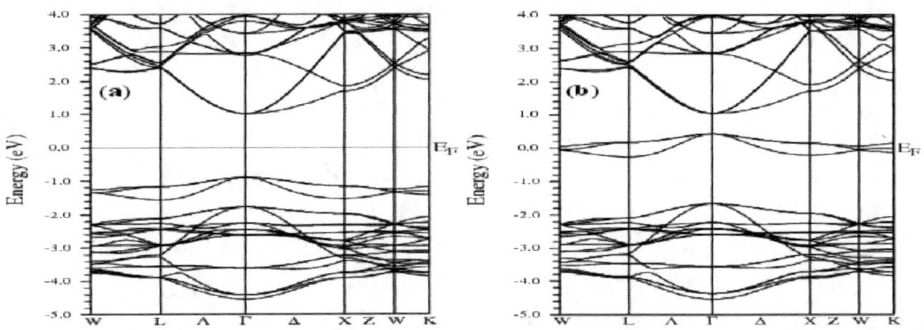

**Figure 1.** Band structure of (a) majority and (b) minority spin states of $CaS_{0.875}Si_{0.125}$ using FP-LAPW method (GGA scheme).

spin system 2p state of impurity atom is completely filled, so it substantially lies within the valence band by keeping semiconducting behaviour. While the minority 2p spin states (spin-down) has partially filled band, therefore it crosses the Fermi level and exhibit a metallic behaviour. It indicates that these materials are true half-metallic. The three bands nearer to the Fermi level are mainly from the 2p states of dopant (Si) atom. Spin polarized calculation based on FP-LAPW shows that substitution of C, Si, Ge and Sn at the anion site of CaS gives rise to total magnetic moment of 2.00 $\mu_B$ per formula unit. Whereas in TB-LMTO spin polarized calculation C- doping induce magnetism and exhibit half-metallicity, but Si-, Ge- and Sn- doping do not exhibit half-metallicity.

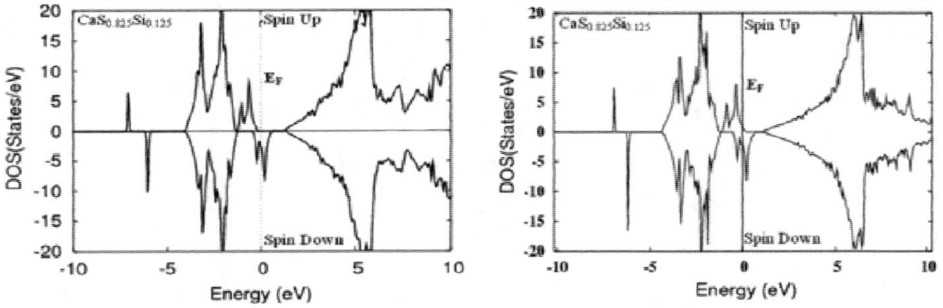

**Figure 2.** Spin polarized (a) GGA and (b) LSDA density of sates for $CaS_{0.875}Si_{0.125}$ at their equilibrium lattice constant. Fermi level is set to zero.

**3.3 Density of states calculations**

The dopant induced spin polarization in the host electrons is characterized by its density of states (DOS). The spin polarized DOS of $CaS_{0.875}M_{0.125}$ (M = C, Si, Ge and Sn) was calculated and for illustration the LSDA and GGA spin polarized total DOS of $CaS_{0.875}Si_{0.125}$ is shown in figure 2. In the minority spin channel in both approximations the spin-polarized DOS shows a metallic character. In the majority spin channel, there is a gap around the Fermi level in GGA DOS so it behaves like half-metallic ferromagnet. In the case of LSDA calculation it slightly crosses the Fermi level. Here it behaves like ferromagnet. Here the ferromagnetism mainly comes from 3p states of dopant. Replacement of anion by partially filled valence p shell elements introduce hole into the valence band. So these spin-polarized holes may mediate the ferromagnetic coupling through strong p-p interaction in these compounds.

**4. Conclusion**

To summarize, the possibility of magnetization in CaS with the presence of nonmagnetic sp impurities such as C, Si, Ge and Sn were studied based on both FP-LAPW and TB-LMTO-ASA method within the density functional study. Both LSDA and GGA lattice constant are follow the same trend. However, the exchange-correlation functions act differently on the orbital and from this the results about their half-metallic state and their ferromagnetism differ. Our calculation based on FP-LAPW illustrates that all dopants (C, Si, Ge and Sn) induce a total magnetic moment of 2.00 $\mu_B$ per f.u. on behalf of spin polarized p states of dopants in the bandgap. Our studies based on TB-LMTO indicate that C only can induce half-metallic ferromagnetic state in calcium chalcogenide. The strong p-p interaction between chalcogenide and impurity atom may be responsible for forming ferromagnetic coupling.

**Acknowledgement**

This work is supported by the Department of Science and Technology – Science and Engineering Research Council (DST-SERC), India. (Grant No. Sr/S2/CMP-29/2008).

**References**

[1] Wolf S A, Awschalom D D, Buhrman R A, Daughton J M, von Molnar S, Roukes M L, Chtchelkanova A Y and Treger D M 2001 *Science* **294** 1488
[2] de Groot R A, Mueller F M, van Engen P G and Buschow K H J 1983 *Phys. Rev. Lett.* 50 2024
[3] Kronik L, Jain M and Chelikowsky J R 2002 *Phys. Rev. B* **66** 041203
[4] Saeki H, Tabata H and Kawai T 2001 *Appl. Phys. Lett.* **79** 288
[5] Gao G Y, Yao K L, Song M H and Liu Z L 2011 *J.Mag. Mag. Mater.* **323** 2652
[6] Zhang C -W and Yan S -S 2010 *J. Appl. Phys.* **107** 043913
[7] Liu J, Chen L, Dong H -N and Zheng R -L 2009 *Appl. Phys. Lett.* **95** 132502
[8] Yogeswari M and Kalpana G 2011 *Mod. Phys. Lett. B* **25** 1
[9] Panday R and Sivaraman S 1991 J. Phys. Chem. Solids **52** 211
[10] Blaha P, Schwarz K, Madsen G K H, Kvasnicka D and Luitz J, **WIEN2k**, An Augmented Plane Wave + Local Orbital's Program for Calculating Crystal Properties (Karlheinz Schwarz, Techn. Universität Wien, Austria), 2001. ISBN 3-9501031-1-2
[11] Anderson O K and Jepsen O 1984 *Phys. Rev. Lett* **53** 2571

# Ground state study of Electronic and Magnetic Properties of Co2MnZ (Z = Ge, Sn) type Heusler Compounds : A first Principle Study.

**D P Rai, Sandeep, M P Ghimire and R K Thapa**

Department of Physics, Mizoram university, Aizawl, India 796004

E-mail: dibyaprakashrai@gmail.com

**Abstract.** The structural optimization was performed based on generalized gradient approximation (GGA) followed by the calculation of electronic structure and magnetic properties on $Co_2MnGe$ and $Co_2MnSn$. The calculation of electronic structure was based on local spin density approximation (LSDA) within full potential linear augmented plane wave (FPLAPW) method. We studied the structural, electronic and magnetic properties. The calculated density of states (DOS) and band structures shows the half-metallicity of $Co_2MnGe$ and $Co_2MnSn$.

## 1. Introduction

The term Heusler alloy is named after a German mining engineer and chemist Friedrich Heusler [1]. Half-metallicity attracted much attention [2], because of its prospective applications in spintronics [3]. Recently rapid development of magneto-electronics intensified the research on ferromagnetic materials that are suitable for spin injection into a semiconductor [4]. One of the promising classes of materials is the half-metallic ferrimagnets, i.e., compounds for which only one spin channel presents a gap at the Fermi level, while the other has a metallic character, leading to 100% carrier spin polarization at $E_F$ [5]. Ishida *et. al.* have also proposed that the full-Heusler alloy compounds of the type $Co_2MnZ$, where Z stands for Si and Ge, are half-metals [6]. Heusler alloys have been particularly interesting systems because they exhibit much higher ferromagnetic Curie temperature than other half-metallic materials [7]. The electronic and magnetic properties of $Co_2MnAl$ [8] and $Co_2CrSi$ [9] using LSDA shows the half-metallicity at the ground state. Rai and Thapa investigated the Electronic Structure and Magnetic Properties of $X_2YZ$ (X=Co, Y=Cr, Z=Al, Ga) type Heusler Compounds by using A first Principle Study and reported HMFs [10]. In half-metals, the creation of a fully spin-polarized current should be possible that should maximize the efficiency of magnetoelectronic devices [11]. Materials with high spin polarization can be used for tunnel magnetoresistance (TMR) and giant magnetoresistance (GMR) [5]. The Co-based Heusler alloys $Co_2YZ$ (Y: transition metal, Z: *sp* atom) are the most prospective candidates for the application in spintronics. This is due to a high Curie temperature beyond room temperature and the simple fabrication process such as dc magnetron sputtering in $Co_2YZ$ [12] . In the present paper, we systematically study the electronic and magnetic structure of Co based full Heusler alloys $Co_2MnZ$ (Z = Ge and Sn) to search for new halfmetallic ferromagnetic candidates. Among the systems studied $Co_2MnZ$ (Z = Ge and Sn) are predicted to be nearly half-metallic.

## 2. Crystal structure and Computational Methods

The considered full Heusler alloys $Co_2YZ$ adopt the ordered $L_{21}$-type structure (space group *Fm*-3-*m*) which may be understood as the result of four interpenetrating face-centered-cubic (fcc) lattices. The *Y* and *Z* atoms occupy two fcc sublattices with origin at (0, 0, 0) and (½, ½, ½) respectively. The Co atoms are located in sublattices with origins at (¼, ¼, ¼,) and (¾, ¾, ¾,).

**Fig. 1:** Structure of the $Co_2YZ$ Heusler alloys : Co (red), Mn(yellow) and Z (blue) atoms.

The first step, we optimized the parameters. The calculation was accomplished by using the WIEN2K code [13]. In the next step, we calculated the electronic structure and magnetic properties using FP-LAPW method. The accuracy is up to $10^{-4}$ Ry. The exchange–correlation potential is chosen in the local spin density approximation (LSDA) [14]. The selfconsistent potentials were calculated on a 20 × 20 × 20 **k**-mesh in the Brillouin zone, which corresponds to 256 *k* points in the irreducible Brillouin zone. The sets of valence orbitals in the calculations were selected as *3s, 3p, 4s, 4p, 3d* Co atoms, *3s, 3p, 4s, 4p, 3d* for Mn atoms, *3d, 4s, 4p* for Ge atoms and *4d, 5s, 5p* for Sn atoms. All lower states were treated as core states.

## 3. Results and Discussions

### *3.1. The equilibrium energy and lattice parameter*

The optimized lattice constant and isothermal bulk modulus are calculated by fitting the total energy to the Murnaghan's equation of state [15] as shown in Fig. 2. It is clear IVB column sp element (Ge or Sn) with four valence electrons state is the most stable. In table 1 we have tabulated the equilibrium lattice constants and the equilibrium energy. We observed $Co_2MnGe$ has a higher energy compare to $Co_2MnSn$.

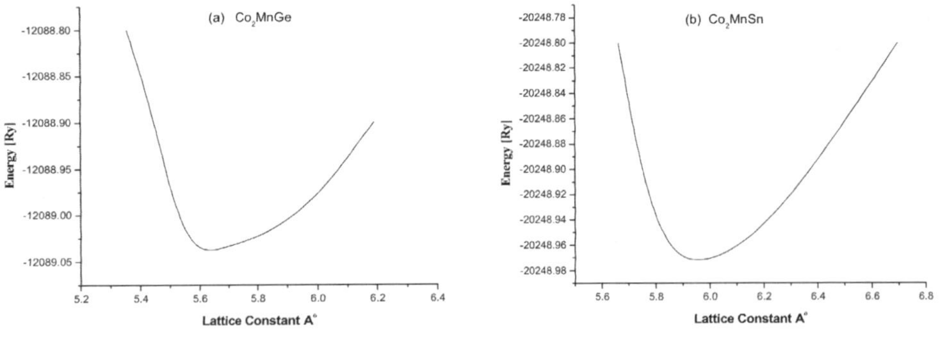


**Fig. 2:** Plot of Energy versus Lattice constant: (a) $Co_2MnGe$ and (b) $Co_2MnSn$

### *3.2 Density of states and magnetic moments*

The total DOS plots of $Co_2MnGe$ and $Co_2MnSn$ are shown in Fig. 3 and Fig. 4 respictively. From the DOS plots of $Co_2MnGe$ and $Co_2MnSn$, shown in Fig. 3 and Fig. 4, peaks are mostly due to *d* state electrons of Co atoms in the core, semi-core and the valence region below $E_F$ for both spin channels..

In spin up channel, the DOS intersects $E_F$ showing the metallic nature. For $Co_2MnGe$ we have observed from Fig. 3 that Mn-*d* electrons mainly contribute in the valence region with sharp peaks at -2.3 eV and -2.7 eV while, in case of $Co_2MnSn$ Mn-*d* electrons contribute with sharp peaks at -2.4 eV and -2.8 eV in spin-up configuration with a Fermi cut-off. In spin down channel, peaks are due to *d* states of Co atoms in the conduction region. In the conduction region of $Co_2MnGe$ peaks were observed at 1.0 eV, 2.0eV and for $Co_2MnSn$, peaks are observed at 0.9 eV and 1.8 eV. The *d* electrons of Co are found to strongly hybridized with Mn-*d* electrons [16]. The partial magnetic moments of Co, Mn and Ge atoms are $0.975\mu_B$, $3.097\mu_B$ and $-0.044\mu_B$ respectively for $Co_2MnGe$. The effective magnetic moment is $5.004\mu_B$ which is approximately an integer value 5.00 μB [16]. Similarly for $Co_2MnSn$, the partial magnetic moments of Co, Mn and Sn atoms are $0.950\mu_B$, 3.253 $\mu_B$ and -0.059 $\mu_B$ respectively. The effective magnetic moment is thus 5.016 $\mu_B$ which is approximately an integer value 5.03 $\mu_B$ [16]. The effective magnetic moments, energy gaps, optimized lattice are tabulated and compared with the previous results in table 1.

**Table 1:** The calculated equilibrium lattice parameters, spin magnetic moments in $\mu$B for the Co2MnZ ( Ge and Sn) compounds.

| | **Lattice constant $a_o$ Å** | | **Magnetic Moment $\mu_B$** | | **Equilibrium Energy (Ry)** |
|---|---|---|---|---|---|
| Compound | **Previous** | **Observed** | **Previous** | **Observed** | |
| $Co_2MnGe$ | 5.737[17]<br>5.743[19]<br>5.749[16] | 5.678 | 5.00[19]<br>5.00[16] | 5.004 | -12089.018 |
| $Co_2MnSn$ | 5.95[18]<br>5.984[16] | 5.970 | 5.02[17]<br>5.04[18]<br>5.03[16] | 5.016 | -20248.956 |

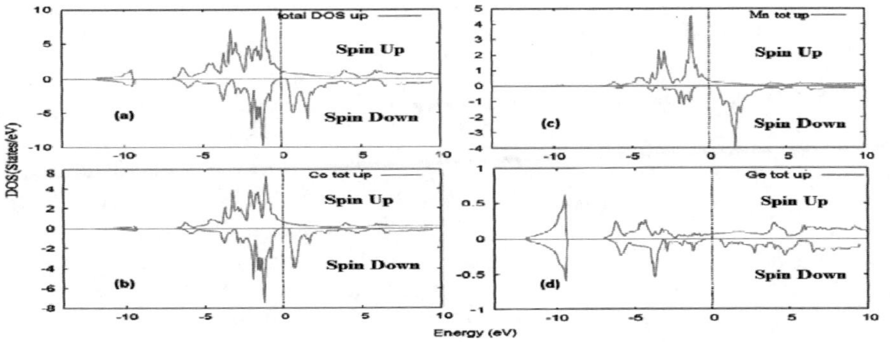


**Fig. 3:** (a) total DOS plots of $Co_2MnGe$, (b) partial DOS plots of Co (c) partial DOS plots of Mn (d) partial DOS plots of Ge.

For $Co_2MnGe$ an indirect energy gap obtained between symmetry point Γ and X is 0.60 eV which almost equals to 0.581 eV reported by Kandpal *et. al.* [16]. Similarly for $Co_2MnSn$, an indirect energy gap obtained between Γ and X is 0.40 eV which agrees with 0.411 eV reported by Kandpal *et. al.* [16]. The results of energy gap and bulk modulus are given in Table 2.

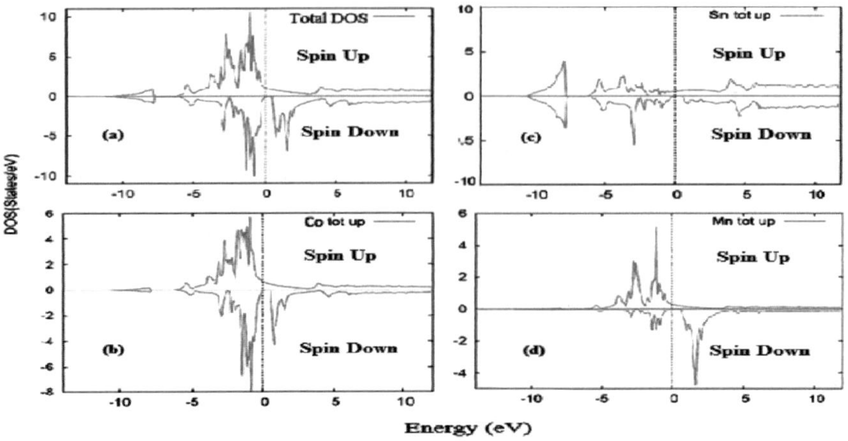


**Fig. 4:** (a) total DOS plots of $Co_2MnSn$, (b) partial DOS plots of Co (c) partial DOS plots of Sn (d) partial DOS plots of Mn.

**Table 2:** Energy gaps and bulk modulus.

| Compounds | Energy gap $E_g$ (eV) | | Bulk Modulus (GPa) |
|---|---|---|---|
| | **Previous** | **Observed** | |
| $Co_2MnGe$ | 0.581[16] | 0.600 | 409.33 |
| $Co_2MnSn$ | 0.411[16] | 0.400 | 212.07 |

## 4. Conclusions

We have studied the possibility of appearance of half-metallicity in the case of the full Heusler compounds $Co_2MnZ$, where Z is an sp atom belonging to the IVB column of the periodic table. We found that compounds $Co_2MnZ$ (Z=Ge, Sn) the ferromagnetic is stable at the equilibrium lattice constant. Although all compounds are not half-metallic at their equilibrium lattice constant, small expansion of the lattice pushes the Fermi level within the gap. Thus these compounds follow the Slater–Pauling behaviour and the 'rule of 24' [20]. The lighter the transition elements and the smaller the number of valence electrons, the wider is the gaps and the more stable is the half-metallicity. The results were in support of the HMF nature for $Co_2MnGe$ and $Co_2MnSn$ . The existence of energy gap in DOS for spin down of both systems is an indication of being a potential HMF. The calculated magnetic moment for $Co_2MnGe$ is 5.004 $\mu_B$ and for $Co_2MnSn$, it is 5.016 $\mu_B$.
:

**Acknowledgment**

DPR acknowledges MZU-UGC research fellowship, SD acknowledges a JRF and RKT a research grant from DAE (BRNS), Mumbai.

**References**

[1] Heusler F 1903 *Verh. Dtsch. Phys. Ges.* **12** 219.

[2] Katsnelson M I, Irkhin V Y, Chioncel L, Lichtenstein A I and de Groot R A 2008 *Rev. Mod. Phys.* **80** 315.
[3] Zutic I, Fabian J and Sarma S D 2004 *Rev. Mod. Phys.* **76** 323.
[4] Ohno H 1998 *Science* **281** 951.
[5] de Boeck, van Roy W, Das J, Motsnyi V, Liu Z, Lagae L, Boeve H, Dessein K and Borghs G 2002 *Semicond. Sci. Technol.* **17** 342.
[6] Ishida S, Akazawa S, Kubo Y and Ishida J 1982 *J. Phys. F: Met. Phys.* **12** 1111.
[7] Webster P J and Ziebeck K R A 1988 Alloys and Compounds of d-Elements with Main Group Elements. *Part 2(Landolt B¨ornstein New Series,* Group III, Vol. **19**, Pt.c) ed H R J Wijn (Berlin: Springer) pp 75–184.
[8] Rai D P, Hashemifar J, Jamal M, Lalmuanpuia, Ghimire M P, Sandeep, Khathing D T, Patra P K, Sharma B I, Rosangliana and Thapa R K 2010 *Indian J. Phys.* **84 (5)** 593.
[9] Rai D P, Sandeep, Ghimire M P and Thapa R K 2011 *Bull. Mat. Sc.* **34** 1219.
[10] Rai D P, A. Shankar, Sandeep, Ghimire M P and Thapa R K 2012 *Int. J. Mod. Phys.* B **26 8** 1250071.
[11] Yakushi K, Saito K, Takanashi K, Takahashi Y K, Hondo K 2006 *Appld. Phys. Lett.* **88** 2225041.
[12] Miura Y, Nagao K and Shirai M 2004 *Phys. Rev.* B **69** 144413.
[13] Blaha P., Schwarz K., Madsen G. K. H., Kvasnicka D., Luitz J., Schwarz K., 2008. An Augmented Plane Wave plus Local Orbitals Program for Calculating Crystal Properties:Wien2K User's Guide, *Techn. Universitat Wien, Wien.*
[14] Perdew J P and Wang Y 1992 *Phys. Rev.* B **45** 13244.
[15] Murnaghan F D *Proc. Natl. Acad. Sci. USA* **30** 244.
[16] Kandpal H C, Fecher G H and Felser C 2006 *J. Phys. D: Appl. Phys.* **40** 1507.
[17] Piccozi S, Continenza A, Freeman A J 2002 *Phys. Rev.* B **66** 094421
[18] Kubler J, Williams A R, Sommers C B 1983 *Phys. Rev.* B **28 4** 1745.
[19] Kurtulus Y and Dronskowski R 2005 *Phys. Rev.* B **71** 014425.
[20] Galanakis I, Dederichs P H and Papanikolaou N 2002 *Phys. Rev.* B **66** 174429.

23rd International Conference on High Pressure Science and Technology (AIRAPT-23) IOP Publishing
Journal of Physics: Conference Series **377** (2012) 012075 doi:10.1088/1742-6596/377/1/012075

# Anderson–Gruneisen parameter under high temperature in $(Fe,Mn,Co,Mg)_2SiO_4$

**S. Gupta[1] and S.C.Goyal[2]**
[1,2]Deptt. of Physics, Agra College, Agra-282002

Email:Seema_g_khushi@yahoo.co.in,

**Abstract**The Anderson–Gr ¨ uneisen parameter (δ) is of considerable importance to Earth scientists because it sets limitations on the thermo-elastic properties of the lower mantle and core. However, there are several formulations on the Gr ¨ uneisen parameter, which are in frequent use and predict varying dependence of d as a function of temperature. In this paper, the expressions for thermal expansion, thermal expansion coefficients and bulk modulus are obtained considering the anharmonic dependence on temperature and are applied to study these constants to $(Fe,Mn,Co,Mg)_2SiO4$. Using the derived expressions, we have shown that different parameters on which the Anderson–Gr ¨ uneisen parameter (δ) depends are temperature dependent, but above all the Anderson–Gr ¨ uneisen parameter (δ) is independent of temperature. The results obtained have been found to be comparable to experimental data.

## 1.Introduction

The behavior of thermo-elastic constants under the effect of temperature has attracted the attention of theoretical as well as experimental workers [1–5] because of their need in the study of technological problems. Though the theoretical attempts based on inter-ionic potential models [6,7] have been carried out by various workers, there are some weaknesses in the models adopted by them. These models involve not only various approximations but also heavy computational work in order to get the results. The purpose of the present study is to discuss a simple and straightforward method based on the Anderson–Gr ¨ uneisen parameter (δ) for the evaluation of thermo-elastic properties at different temperatures for $(Fe,Mn,Co,Mg)_2 SiO_4$.

The Anderson– Gr¨uneisen parameter is an important quantity in geophysics as it often occurs in equations, which describe the thermo-elastic behavior of materials at high pressures and temperatures. Anderson [1] introduced a parameter (δ), which has been found to be an important tool for investigating the temperature dependence of bulk modulus, describing microscopic behavior, internal structure and other related properties of solids [8–11] and also in geophysical understanding of Earth's interior [12–13].

Various phenomenological and empirical efforts [14-17] have been made to show the temperature dependence of the Anderson– Gr¨uneisen parameter (δ), assuming different forms for dependence of coefficient of volume thermal expansion (α) on temperature, and have explained the variation in bulk modulus with temperature. Many authors [18,19] have assumed that this parameter varies with temperature and thus it is temperature dependent. In the present study, we have tried to show that the different parameters of Anderson–Gr¨ uneisen parameter (δ) vary with temperature but the Anderson–Gr¨ uneisen parameter is independent of temperature.

It has been shown that different parameters on which the Anderson–Gr ¨ uneisen parameter (δ) depend are temperature dependent, but above all Anderson–Gr ¨ uneisen parameter (δ) is independent of temperature in case of (Fe,Mn,Co,Mg)$_2$SiO4.

## 2.Method of Analysis

The Anderson-Grüneisen parameter, which plays an important role in understanding the anharmonic properties of solids, is defined as follows [1]

$$\delta_T = \frac{1}{\alpha K_T}(\partial K_T / \partial T)_P \quad (1)$$

Where α is the coefficient of volume thermal expansion, $K_t$ is the isothermal bulk modulus and $(\partial K/dT)_P$ is the temperature variation of bulk modulus (B) at constant pressure.

A method to estimate the temperature dependence of the bulk modulus has been developed with approximation that α depends linearly on temperature [20]. The percentage difference increases with increasing temperature and becomes 9.75% at melting point. Thus, the approximation that α depends linearly on temperature is valid only for small expansions that are in the low temperature range.

Through comparing and analyzing some experimental data [1], we put forward that α depends quadratically on temperature, which can be put as follows:

$$\alpha = \alpha_0 + \alpha_0^{'}(T - T_0) + \frac{\alpha_0^{''}}{2}(T - T_0)^2 + \frac{\alpha_0^{'''}}{6}(T - T_0)^3 \quad (2)$$

where $\alpha_0$ is the value of α at $T_o$ $\alpha_o^{'}$, $\alpha_0^{''}$ and $\alpha_0^{'''}$ are respectively the first second and third order temperature derivatives of α at initial temperature $T=T_o$ =300K. If one assumes that the product $\alpha K_T$ remains constant [21] then we get from eq (1) the following approximate formulation:

$$\delta_T = \frac{1}{\alpha^2}\left(\frac{\partial \alpha}{\partial T}\right)_P \quad (3)$$

Asuming $\delta_T$ to be independent [20] of T, we get the following relations

$$\alpha_0^{'} = \left(\frac{\partial \alpha}{\partial T}\right)_P = \delta_T \alpha_0^2 \quad (4)$$

$$\alpha_0^{''} = \left(\frac{\partial^2 \alpha}{\partial T^2}\right)_P = 2\delta_{T0}^2 \alpha_0^3$$

And

$$\alpha_0^{'''} = \left(\frac{\partial^3 \alpha}{\partial T^3}\right)_P = 6\delta_{T0}^3 \alpha_0^4 \quad (5)$$

Where $\delta_{T0}$ is the value of $\delta_T$ at $T_o$ and at atmospheric pressure, i.e. at P=0,

Substituting the value of α from eq. (2) in eq. (1), we get

$$dK_T = -\delta_{T_0}[\alpha_0 + \delta_{T_0}\varepsilon_0^2(T - T_0) + \delta_{T_0}^2\alpha_0^3(T - T_0)^2 + \delta_{T_0}^3\alpha_0^4(T - T_0)^3]dT \quad (6)$$

### 3. Results and discussion

In recent years, temperature dependence of δ has been discussed many times. In our present work, we have computed different parameters on which δ depends and dependency of these parameters on temperature. By this computation we have tried to throw some light on the picture about dependence of δ on temperature in order to make it clearer.

**Table 1;** Values of input data from Ref. [22] at room temperature: α (in $10^{-5}$ $K^{-1}$), $\delta_0$ (dimensionless); bulk modulus ($B_0$; in GPa).

| | $Fe_2SiO_4$ | $Mn_2SiO_4$ | $Co_2SiO_4$ | $Mg_2SiO_4$ |
|---|---|---|---|---|
| $\delta^0$ | 7.34 | 8.19 | 7.32 | 5.94 |
| α | 2.61 | 2.27 | 2.27 | 2.72 |
| $B_0$ | 136.7 | 128.0 | 147.1 | 127.3 |

Journal of Physics: Conference Series **377** (2012) 012075 doi:10.1088/1742-6596/377/1/012075

It was found that eq. (2) improves the results as compared to earlier values [22]. Thus eq. (2) proposed seems to be suitable for analyzing the temperature dependence of α. From this, one can conclude that one parameter (α) on which δ depends is temperature dependent.

**Table2;** Computed values of the first order temperature derivative of bulk moduli

| T | $Fe_2SiO_4$ (K`$_T$) | $Mn_2SiO_4$ (K`$_T$) | $Co_2SiO_4$ (K`$_T$) | $Mg_2SiO_4$ (K`$_T$ |
|---|---|---|---|---|
| 300 | -0.02618817 | -0.02379686 | -0.02444272 | -0.02056761 |
| 400 | -0.02618816 | -0.02379686 | -0.02444272 | -0.02056761 |
| 500 | -0.02618812 | -0.02379683 | -0.0244427 | -0.02056759 |
| 600 | -0.02618795 | -0.02379669 | -0.02444261 | -0.02056752 |
| 700 | -0.0261875 | -0.02379633 | -0.02444237 | -0.02056734 |
| 800 | -0.02618656 | -0.02379557 | -0.02444187 | -0.02056696 |
| 900 | -0.02618485 | -0.02379419 | -0.02444097 | -0.02056628 |
| 1000 | -0.02618207 | -0.02379194 | -0.02443948 | -0.02056517 |
| 1100 | -0.02617784 | -0.02378853 | -0.02443723 | -0.02056347 |
| 1200 | -0.02617174 | -0.0237836 | -0.02443397 | -0.02056101 |
| 1400 | -0.02615205 | -0.02376768 | -0.02442343 | -0.02055307 |
| 1500 | -0.02613742 | -0.02375585 | -0.02441558 | -0.02054715 |
| 1600 | -0.02611885 | -0.02374082 | -0.02440559 | -0.02053961 |
| 1700 | -0.02609573 | -0.0237221 | -0.02439312 | -0.0205302 |
| 1800 | -0.02606744 | -0.02369917 | -0.02437783 | -0.02051865 |

**Table3;** Calculated values of Anderson–Gr¨ uneisen parameter (δ) up to 1800K along with others reported data.

| | $Fe_2SiO_4$ | | $Mn_2SiO_4$ | | $Co_2SiO_4$ | | $Mg_2SiO_4$ | |
|---|---|---|---|---|---|---|---|---|
| T | $\delta_T$ Present | $\delta_T$ [22 ] | $\delta_T$ Present | $\delta_T$ [22 ] | $\delta_T$ Present | $\delta_T$ [22 ] | $\delta_T$ Present | $\delta_T$ [22 ] |
| 300 | 7.34 | 7.34 | 8.19 | 8.19 | 7.32 | 7.32 | 5.94 | 5.94 |
| 400 | 7.34 | 6.85 | 8.19 | 7.35 | 7.32 | 6.56 | 5.94 | 5.58 |
| 500 | 7.34 | 6.62 | 8.19 | 6.96 | 7.32 | 6.19 | 5.94 | 5.49 |
| 600 | 7.34 | 6.50 | 8.19 | 6.76 | 7.32 | 6.01 | 5.94 | 5.48 |
| 700 | 7.34 | 6.45 | 8.19 | 6.63 | 7.32 | 5.88 | 5.94 | 5.49 |
| 800 | 7.34 | | 8.19 | | 7.32 | | 5.94 | 5.47 |
| 900 | 7.34 | | 8.19 | | 7.32 | | 5.94 | 5.46 |
| 1000 | 7.34 | | 8.19 | | 7.32 | | 5.94 | 5.47 |
| 1100 | 7.34 | | 8.19 | | 7.32 | | 5.94 | 5.46 |
| 1200 | 7.34 | | 8.19 | | 7.32 | | 5.94 | 5.49 |
| 1400 | 7.34 | | 8.19 | | 7.32 | | 5.94 | 5.44 |
| 1500 | 7.34 | | 8.19 | | 7.32 | | 5.94 | 5.37 |
| 1600 | 7.34 | | 8.19 | | 7.32 | | 5.94 | 5.38 |
| 1700 | 7.34 | | 8.19 | | 7.32 | | 5.94 | 5.40 |
| 1800 | 7.34 | | 8.19 | | 7.32 | | 5.94 | 5.42 |

It is clear from eq.(1) that δ depends on $\alpha K_T$ and K`$_T$ . As product of $\alpha K_T$ comes to be constant (i.e. independent of temperature) so dependence of δ on temperature fully dependent on values of B`$_T$ with temperature. The author obtained values of K`$_T$ with change of temperature and are calculated these values in Table (2). From Table (2), it is clear that K`$_T$ remains constant till some range of temperature and varies very slowly with temperature after definite temp range. This variation is so slow that it is not significant and hence we can say that K`$_T$ is also constant. Hence δ depends on which parameters all are temperature independent quantities so as whole δ is temperature independent quantity.

The values of δ are shown in Table (3). It is noted that the values of δ are nearly constants in all the four minerals, however the magnitude of this parameter is different in all the four minerals and it is increasing in magnitude in this order $Mg_2SiO_4 < Co_2SiO_4 < Fe_2SiO_4 < Mn_2SiO_4$.

Thus, the present study may be useful in analyzing the thermo elastic behavior, microscopic behavior, internal structure and other related properties of earth lower mantle solids. A study of this kind has become increasingly important over the last decade, as contributions from global seismic tomography, seismological investigations of geographically and radically localized regions, mantle discontinuities, analysis of normal modes of oscillations, and other types of studies have revealed the Earth's mantle in unprecedented. The present theory can be extended to other class of solids like earth's lower mantle solids, which are useful for geophysical studies.

**ACKNOWLEDGMENT:** One of the authors Seema Gupta is thankful to DST-Delhi and and S.C.G. is grateful to UGC for providing the financial assistance. The authors are also grateful to Principal, Agra College, Agra for providing the facilities.

## REFERENCES

[1] Anderson O.L. Isaak D.G.1995 in: Mineral Physics and Crystallography A Handbook of Physical Constants, T.J. Ahrens (Ed.), American Geophysical Union, Washington DC p. 64.
[2] Isaak D.G. Ohno I. 2003 Phys. Chem. Miner. **30** 430 .
[3] Sinogeikin S.V. Lakshtanov D.L Nicholas J.D.. Jackson J.M Bass J.D. 2005 J. Eur. Ceram. Soc. **25** 1313.
[4] Srivastava U.C.1980 Phys. Rev. B **21** 2602.
[5] Wolf R.J. Mansour K.A Lee M.W. Ray J.R 1992 Phys. Rev. B **46** 8027.
[6] Li X.F. Chen X.R. Meng C.M. Ji G.F. 2006 Solid State Commun. **139** 19.
[7] Zhao J.J. Winey J.M. Gupta Y.M. 2007 Phys. Rev. B **75** 094105.
[8] Shanker J. Singh B.P. Jitendra K 2008 Condens. Matter Phys. **11** 681.
[9] Vocˇ Adlo L. Poirer J.P. Price G.D. 2000 Am. Mineral. **85**, 390.
[10] Wachtman J.B.. Te!t W.E Lam D.G. Apstein C.S. 1961 Phys. Rev **122** 1754.
[11] Anderson O.L. 1966 Phys. Rev **144** 553.
[12] Jayaprakash R. Dixit S. Kushwah S.S. Shanker J. 1995 Indian J. Pure Appl. Phys. **33** 125 .
[13] Thomas L.M. Shanker 1996 J. Phys. Status Solidi B **195** 361.
[14] Kushwah S.S. Kumar P. Shanker J. 1997 Indian J. Pure. Appl. Phys. **35** 125.
[15] Kumar M. Bedi S.S. 1995 Phys. Status Solidi (b) **188** 645.
[16] Born M. Huang K. 1954 Dynamical Theory of Crystal Lattice, p. 50(Oxford university Press, Oxford.
[17] Anderson O.L.1998 Phys. Earth and Planet. Inter. **109** 179.
[18] Dinesh K.B. Singh P. 2005 Physica B **357** 525.
[19] Singh K.Y. Gupta B.R.K. 2003 Physica B **334** 266.
[20] Kumar M. 1995 Physica B **205** 175.
[21] Shanker J. Kumar M. 1993 Phys. Status Solidi B **179** 351.
[22] Anderson O.L. and Isaak G. 1995 Elastic Constanta of Mantle Minerals at High Temperature Mineral Phys. and Crystallography (A.G.U.,Handbook of Physical Constants).

23rd International Conference on High Pressure Science and Technology (AIRAPT-23) IOP Publishing
Journal of Physics: Conference Series **377** (2012) 012076 doi:10.1088/1742-6596/377/1/012076

# Pressure-induced structural phase transition and elastic properties in rare earth CeBi and LaBi

**V Mankad[1], S K Gupta[1], I Lukacevi´c[2] and P K Jha[1]***
[1]Department of Physics, Bhavnagar University, Bhavnagar-364022, India
[2]Department of Physics, University J. J. Strossmayer, Osijek, Croatia
E-mail: prafullaj@yahoo.com

**Abstract.** A comprehensive first principles study of structural, elastic, electronic and phonon properties of rare-earth CeBi and LaBi is reported within the density functional theory scheme. The ground state properties such as lattice constant, elastic constants, bulk modulus, and finally the phase transition and lattice dynamical properties of rare-earth CeBi and LaBi of rocksalt (B1) and CsCl (B2) structures are determined. The electron band structures for the two phases of both rare-earth crystals are presented. We have calculated phonon dispersion curve, showing that all phonon modes in phonon dispersion curves of B1 phase in CeBi and LaBi are positive, which indicates a stable phase for this structure.

## 1. Introduction

Pressure is one of the external parameters by which the interplay of the *f*-electrons with the normal conduction electrons may be varied. At ambient conditions, the rare-earth compounds are characterized by a fixed $f^n$ configuration of atomic-like *f*-electrons, but the decreased lattice spacing, resulting from the application of pressure, eventually leads to the destabilization of the *f*-shell [1-4]. A theoretical description of this electronic transition remains a challenge. They are considered to be part of the green technology industry, helping to improve energy efficiency in magnets, batteries, non-linear opto-electronics and computers. The volume change at the crystallographic transition is attributed to a decrease of the cerium valence or a lowering of the *p-f* hybridization due to the larger interatomic distances in both high pressure phases [4].

Leger *et al* [1] performed experiment of CeBi under high pressure and they showed that cubic and tetragonal phases coexist above the transition, with quite close volumes over the whole pressure range. These experiments give accurate structural information for pressures up to 100 GPa. On the other hand, enormous improvement in computational power and an existence of a number of new, efficient and easily accessible program packages [5], considerably expand the possibility to predict the physical properties of new phases at high pressures, as well as open an insight into the actual processes at extreme conditions. It is revealed from the literature, that role of *f*-electrons in Ce-ion, due to trivalent to tetravalent transition at high pressure on the electronic properties and other bulk properties, such as elastic constants and phonon spectra etc. is yet to be explored in details. In the present paper, we report structural, electronic and elastic properties of CeBi and LaBi rare earth compounds using first principles calculation. The present study reports a comprehensive study on structural, electronic band structures, elastic and lattice dynamical properties of rare earth monopnictides CeBi and LaBi,

[1]To whom any correspondence should be addressed.

using first-principles density functional calculations within the pseudopotential approximation. Both compounds possess NaCl (B1) structure at ambient pressure and transform either to CsCl or body centered tetragonal (BCT) structure at higher pressures [3-4]. Our results, concerning equilibrium lattice parameter and bulk modulus, agree well with the available experimental and previous theoretical data.

## 2. Computational Details

All calculations in the present study are based on the implementation of plane wave density functional theory (DFT) in the Kohn-Sham framework using ABINIT simulation package [5]. The wave functions describe only the valence and the conduction electrons, while the core electrons are taken into account for pseudopotentials. For the exchange-correlation functional, we have employed the local density approximation (LDA) and generalized-gradient approximation (GGA) functional developed by Perdew, Burke and Ernzerhof (PBE) [6]. The electron-ion interactions are described through the use of Troullier and Martins type pseudo-potentials [7]. A set of convergence tests have been performed in order to choose correctly the mesh of k-points and the cut-off kinetic energy for the plane wave basis, to start the ground state and linear response calculations. The kinetic energy cut-off is set to 50 Ha for both B1 and B2 structures of both compounds. The Brillouin zone is sampled by Monkhorst-Pack mesh of *k*-points for B1 and B2 phases for CeBi and LaBi. Convergence tests prove that the Brillouin zone (BZ) sampling and the kinetic energy cut-off are sufficient to guarantee an excellent convergence. The convergence is ensured to the total energy around 0.001 Ha, the phonon frequencies by 1-2 $cm^{-1}$ and the transition pressure by about 1 GPa. Phonon frequencies are subsequently obtained using the linear response approach [8] based on the density functional perturbation theory (DFPT). The expression is obtained from the second derivatives of the total energy with respect to the phonon displacement of atoms or an external electric field.

**Table 1.** Calculated values of lattice parameter, elastic constants (in GPa) and bulk modulus of rare-earth bismuthides.

| Structure | Lattice Constant (Bohr) | $C_{11}$ | $C_{12}$ | $C_{44}$ | Bulk Modulus (GPa) |
|---|---|---|---|---|---|
| CeBi (B1) Present | 12.45(LDA), 12.38 (GGA) | 109.70 | 5.97 | 9.94 | 40.54 |
| Others | 12.40[g] | 107.26[a] | 24.36[a] | 14.85[a] | 49.6 |
| Experiment | 12.26 | - | - | - | - |
| CeBi (B2) Present | 7.79 (LDA), 7.84 (GGA) | 46.80 | 37.39 | 10.30 | 40.52 |
| Experiment | 7.06 | - | - | - | - |
| LaBi (B1) Present | 12.04 | 172.35 | 17.65 | 6.65 | 69.19 |
| Other | 12.08[e], 12.35[c], 12.08[f], | 123.14[a], 119.7[b], 140.12[c], 120.63[d], | 32.17[a], 11.9[b], 58.15[c], 11.85[d] | 19.5[a], 17.1[b], 17.94[c], 20.52[d] | 65.18[e], 62.20[c], 65.18[f], 50.03[h] |
| Experiment | 12.40 | - | - | - | - |
| LaBi (B2) Present | 7.80 | 33.74 | 36.97 | 8.88 | 35.89 |

[a]Ref [10], [b]Ref [11], [c]Ref [3], [d]Ref [12], [e]Ref [13], [f]Ref [14], [g]Ref [2], [h]Ref [15]

### 3. Results and Discussion

As a first step, we have performed the structural optimization of B1 and B2 phases of CeBi and LaBi rare-earth crystals. The structural optimization has been performed under the minimum condition of the total energy. We computed the equilibrium lattice constant ($a_0$) and the bulk moduli ($B_0$) by fitting the energy-volume curves to the third order Birch-Murnaghan equation of states (EOS) [9]. The equilibrium lattice parameters for both compounds, along with available experimental and other theoretical data, are listed in Table I. It can be clearly seen that the calculated lattice parameter for the CeBi and LaBi compounds are in a satisfactory agreement with the previous calculations and available experimental data. Figure 1 presents the total energy vs lattice constant curves for CeBi using both LDA and GGA calculations. It is clear from the figure that there is a large difference between the total energies in both phases from both calculations. This may be due to the contribution of *f*- electron, which may be responsible for it.

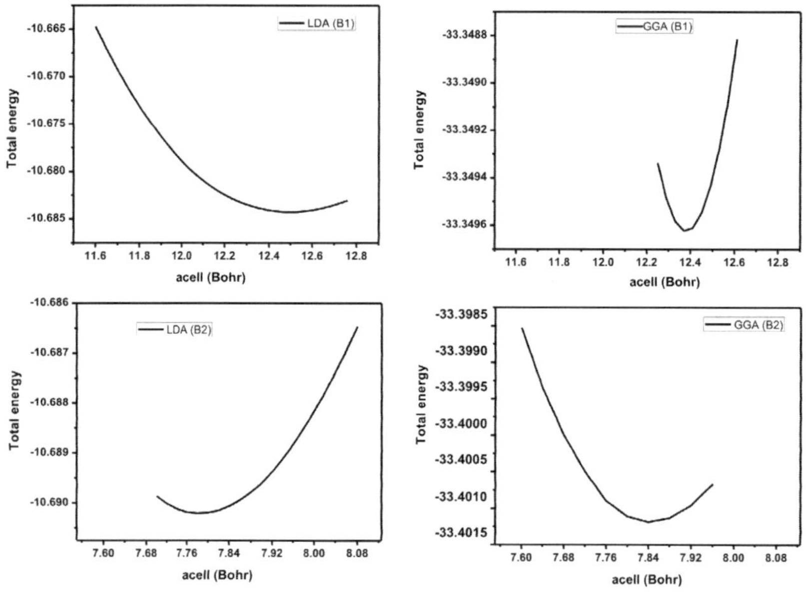


**Figure 1.** Total energy minimization of CeBi crystal in rocksalt and CsCl phase

Elastic constants which are associated with homogeneous strains, *i.e.* with macroscopic distortion of the crystals, can be calculated from the force constants. Most importantly, the knowledge of elastic constants is essential for many practical applications related to the mechanical properties of a solid, load deflection, inter-atomic bonding, internal strain, sound velocities and fracture toughness. The calculated elastic constants are in the range of earlier calculations and physically significant. However, it could not be compared with experiment due to non availability of data. The elastic or mechanical stability can be checked by the whole set of elastic stiffness constant $C_{ij}$ using the Born-Huang criterion. The key criterion for mechanical stability of a crystal is that a strain energy must be positive. It is clear that both CeBi and LaBi are mechanically stable in B1 and B2 phase. We have also calculated the electronic band structure for both phases of both compounds at different pressures and presented pressure dependent band structure for CeBi and LaBi in Fig. 2. As pressure increases, the conduction band width increases, because of the enhanced overlap of the wave function with the neighbouring atoms.

From the band structure, it is also seen that the lowest lying band in these compounds is mainly due to monopnictides '*s*'-like state and the band lying above mainly due to the monopnictides *p*-like state which hybridizes with Ce and La '*s*' and '*d*'-like states at Γ-points. The narrow bands originating above the Fermi level are mainly due to Ce and La '*f*'-like states, and the major contribution is mainly from Ce and La '*d*'-like states. Localization of the *f* electron has taken away the occupied *f*-electron weight of approximately one electron, with delocalized f electrons is situated in the vicinity of the Fermi energy (not shown in the figure). A trend of increasing localization of the *f* electron is observed with increasing nuclear charge of the ligand, which also causes a volume increase. We note that due to the increasing localization, the *f* electron in CeBi and LaBi remains localized in the B2 structure.

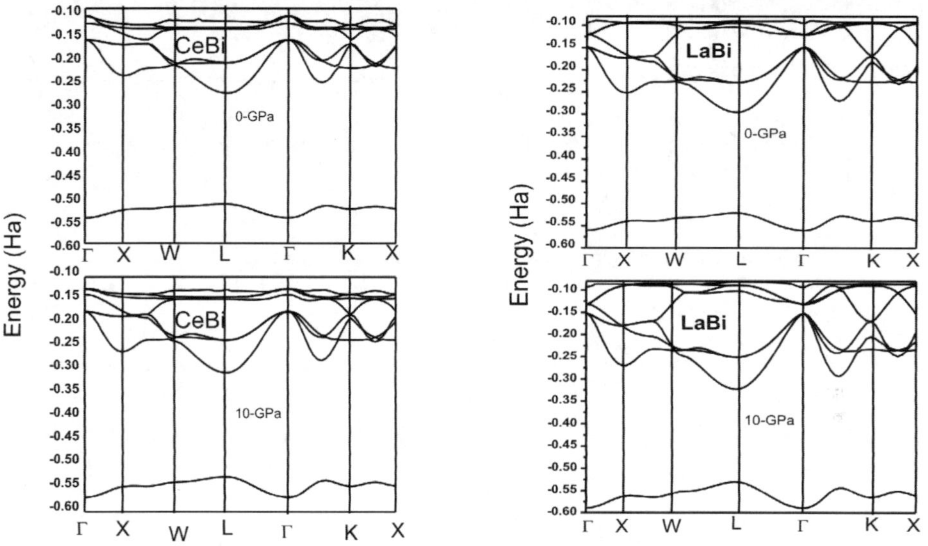


**Figure 2.** Calculated bandstructure of CeBi and LaBi at different pressure.

We have calculated phonon dispersion curve (presented elsewhere) gives impression that both compounds in rock salt structure are dynamically stable as the frequency of phonon modes throughout the Brillouin zone is positive.

## 4. Conclusion

In this study we have reported the results of *ab-initio* calculations on structural, elastic, electronic and vibrational properties for two CeBi and LaBi in their rocksalt and CsCl phase using plane wave method within GGA and LDA. There is a good agreement in the case of lattice parameters and elastic constant with previous reports. The band structure calculation shows the metallic nature. The energy difference between B1 and B2 phase of CeBi indicate that LDA and GGA are not suitable pseudopotential for the study of phase transition.

## 5. Acknowledgments

Authors are thankful to financial Department of Science and Technology, New Delhi, Department of Atomic energy, Mumbai Govt. of India. IL would like to acknowledge the usage of grid infrastructure

CRO-NGI (Croatian National Grid Infrastructure) available through SRCE (University Computing Center) organization, Zagreb, Croatia.

## 6. References

[1] Léger J. M, Oki K, Rossat-Mignod J and Vogt O 1985 *J. Physique* 46 89-894
[2] Svaney A, Szotekz Z,. Temmermanz W. M, Lægsgaardy J and Winter H 1998. *J. Phys.: Condens. Matter* 10 5309–5325
[3] Ciftci Y. O, Colakoglu K and Deligoz E 2008 *J. Phys.: Condens. Matter* 20 345202
[4] Vaitheeswaran G, Kanchana V, Rajagopalan M 2002 *Physica B* 315 64–73
[5] The ABINIT code is a common project of the University Catholique de Louvain, Corning Incorporated, and other contributions http://www.abinit.org.
[6] Perdew J. P, Burke K, Ernzerhof M 1996 *Phys. Rev. Lett* 77 3865–3868
[7] Troullier N, Martin J. L 1991 *Phys. Rev. B* 43 1993-2006
[8] Baroni S, Gironcoli S. de, Corso A. dal, Giannozzi P 2001 *Rev. Mod. Phys.* 73 515-562
[9] Mehl M. J, Barry B. M and Papaconstantopoulos D. A, in: Intermetallic Compounds: Principle and Practice, Principles, Vol. 1, edited by J. H. Westbrook and R. L. Fleischeir (John Wiely, London, 1995), pp. 195-210.
[10] Bhardwaj P, Singh S 2004 *Phys. Stat. Sol. (a)* 201 13
[11] Gokoglu G, Erkisi A 2008 *Solid State. Comm.* 147 221
[12] Khodja F. D, Boudali A, Amrani K, Kadoun A and Abbar B 2008 *Physica B* 403 4305
[13] Vaitheeswaran G, Kanchana V, Rajagopalan M 2002 *Physica B* 315 64
[14] Davidov D, Bucher E, Rupp L.W, Longinotti D, Rettori C 1974 *Phys. Rev.* 9 2879
[15] Pagare G and Sanyal S. P 2006 *Phase Transition* 79 935

# A first principles study of lattice dynamical and electronic properties of Yb-pnictides (YbX, X= N, P and As) compounds

**Himadri R. Soni**[1], **Nikita Rathod**[1,2], **Sanjeev K. Gupta**[1] **and Prafulla K. Jha**[1*]
[1]Department of Physics, Bhavnagar University, Bhavngar-364022, India
[2]Vadodara Institute of Engineering, Vadodara-391510, India

E-mail: prafullaj@yahoo.com

**Abstract.** The Yb-pnictides have drawn considerable interest in the field of solid state and material science due to it's diverge applications in the properties like electronic, magnetic, optical and phonon. These compounds have got keen interest due to its partially filled 4f bands. In the frame work of density functional theory (DFT), we determine structural, electronic, lattice dynamical and elastic properties of Yb-pnictides (YbX, X= N, P, As) in rocksalt (RS) phase using first-principles density functional theory (DFT) simulations based with a generalized gradient approximation (GGA) of the exchange correlation energy implemented in ABINIT package.

## 1. Introduction

The Yb-pnictides have drawn considerable interest in the field of solid state and material science due to it's diverge applications in the properties like electronic, magnetic, optical and phonon [1-3]. These compounds have got keen interest due to its partially filled 4f bands. Such rare earth (RE) compounds have electronic configuration such as $4f^{n-1}(5d6s)^m$, where this 4f band is partially filled. The special characteristic of these Yb-pnictides is the fluctuations in their valence electrons. The Yb ion in these compounds is predominantly trivalent with one *4f* hole. All YbX (X=N, P, As) compounds are semimetals with rocksalt (RS) structures. Because of the interesting properties which are associated with the question of the lanthanides (for instance, mixed and intermediate valency and heavy fermion behaviour). It is important to have a good understanding of what stabilizes a given electronic configuration. Furthermore, it is also observed that there is a strong electron-phonon coupling at the RE (Rare Earth) ion site due to 4f-5d hybridization giving rise to anomalies in their phonon dispersion curves [3-5]. The Yb-pnictides crystallize in rocksalt structure, while many more investigations are to be carried out to establish the physical properties of the Yb-pnictides in the present paper.

In the present paper we report the density functional theoretical study of lattice dynamical and electronic properties of such Yb-pnictides (YbX, X=N, P, As). We briefly outline the theory in section 2 and present our calculated results in section 3.

## 2. Computational Details

To determine the lattice dynamical and electronic properties of YbX (X=N,P and As) in RS phase, we have performed plane-waves and pseudopotentials (both norm-conserving and ultrasoft) calculations

Published under licence by IOP Publishing Ltd

using ABINIT code [6]. The generalized gradient approximation (GGA) [7-8] has been used for the exchange and correlation energy density function. A fully relativistic calculation is performed for core states, whereas the valence states are treated in a scalar relativistic scheme. In order to find an appropriate energy cutoff, the total energy as a function of the energy cutoff has been calculated for 23 different energy cutoff ranging from 10 Ry to 125 Ry. The total energy converges near the energy cutoff of 52 Ry, 32 Ry and 32 Ry for YbN, YbP and YbAs respectively. Here, for the self consistency, the initial potential for the next iteration is constructed using a convergence stabilization scheme. The number of sampling k-points used in the Brillouin zone (BZ) summation of the electronic density and total energy is increased until the total energy converges to the desirable tolerance. The self consistent calculations are considered to be converged when the total energy of the system is stable within convergence of $E_{tot} < 2.73\times10^{-3}$ eV and forces on atoms $F_i < 0.01$ eV/ Å. To generate the (6×6×6) mesh in the BZ, the scheme of Monkhorst-Pack [9] has been used. Further, the lattice dynamical study for YbX are performed using density functional perturbation theory [DFPT] [10]. In this method, the dynamical matrix which provides information on lattice dynamics of the system can be obtained from the ground state electron charge of the nuclear geometry. The kinetic energy cut off and numbers of k-points mentioned above are found to yield phonon frequencies converged to within 2-5 $cm^{-1}$.

| **Compound** | | **YbN** | **YbP** | **YbAs** |
|---|---|---|---|---|
| **Lattice parameter (Å)** | **Present work** | 4.8033 | 5.5545 | 5.7449 |
| | **Experimental** | 4.78[11] | 5.537[11] | 5.697[11] |
| | **Other work** | 4.787[15] | 5.552[14], 5.545[15] | 6.081[14], 5.694[15] |
| **Bulk modulus (GPa)** | **Present work** | 130.77 | 61.77 | 54.3363 |
| | **Experimental** | | | |
| | **Other work** | 130.55[12],136[13] | 70[13], 104±3[14] | 63[13], 85±4[14] |

**Table 1.** Calculated equilibrium lattice constants, bulk modulus for YbN, YbP and YbAs and compared with experimental and available other calculations.

## 3. Results and Discussion

The theoretical ground state constants such as lattice parameters and bulk modulus of the three pnictides YbN, YbP and YbAs are obtained using the method discussed in the above section and listed in Table 1, which also includes the available experimental [11] and other theoretical [12-15] data for comparison. It is clear that the present theoretical values agree reasonably well with the experimental data [11]. It is generally observed that GGA yields higher lattice constant. The total energy versus volume curve is fitted to the Murnaghan equation of state [16] to obtain the bulk modulus value as given below:

$$E(V) = E_0 - \frac{B_0 V_0}{B'_0 - 1} + \frac{B_0 V}{B'_0}\left[\frac{(V_0/V)^{B'_0}}{B'_0 - 1} + 1\right] \tag{1}$$

where $B_0$ and $B'_0$ are the bulk modulus and its derivative respectively, $E_0$ is the ground state total energy and $V_0$ is the volume. The bulk modulus so obtained is presented in Table I and compared with the available experimental [11] and other theoretical [12-15] data. There is an excellent agreement between the present and other theoretical data [12-15].

The total density of states (DOS) of these YbX (X=N, P, As) are presented in Fig. 1. One of the interesting features, which emerges from Fig. 1 is that for YbX (X=N, P, As) the Fermi level $E_F$

resides above the high dense peak which indicates instability. It is due to further filling up of *d*-orbitals. The additional *d* electrons for YbX (X=N, P, As) lead to a shift of the center of gravity of the total DOS towards lower energies. Furthermore, it is interesting to point out from Fig. 1 that, the major contribution in $N(E_F)$ comes from the Yb-4f states which is of course shifted a bit from the Fermi level due to hybridization.

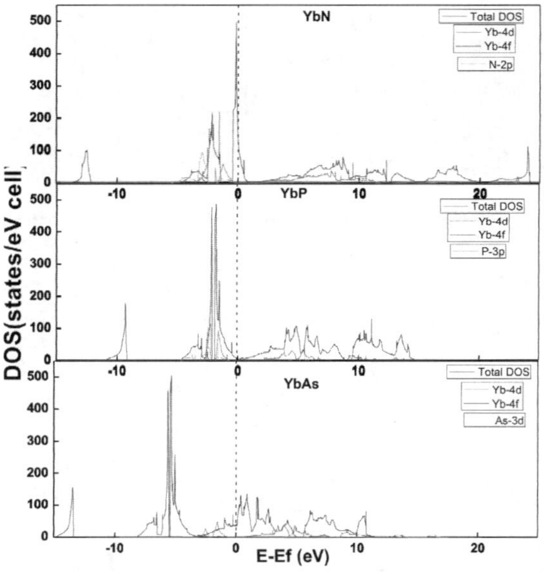


**Figure 1**. Electronics density of states of YbN, YbP and YbAs.

The phonon dispersion curves along with the phonon density of states for rocksalt YbX (Yb=N, P, As) are plotted in Fig. 2. These figures give impression that YbX compounds in rocksalt structure are dynamically stable as the frequency of phonon modes throughout the Brillouin zone is positive. It is clear from the figure that gap appears between the optical and acoustical phonon branches in the case of YbX, which is due to mass ratio of constituent atoms. We obtain the frequency of acoustic branches comparatively lower in magnitude and show a smooth variation along all the directions of the Brillouin zone. There is some anomalous behaviour for lowest TA phonon branch around zone centre in case of YbN. A systematic trend of variation of a LA phonon in all three compounds is observed. As the mass of the pnictides ion increases, a hump near the X-point from YbN to YbAs of the Brillouin zone gradually appears. This hump is maximum in the case of YbAs. No gap between LO and TO at Γ- point indicates metallic behaviour for YbX compounds in contrast to the rigid ion model calculation [3]. The rigid ion model around the ionic behaviour for these compounds thereby indicating strong coulomb interaction for crystal cohesion.

## 4. Conclusions

In the present study, we reported the results of ab-initio calculations on structural, electronic and phonon properties of YbX (X=N, P, As) in rocksalt structure. There is a good agreement in the case of lattice parameter and bulk modulus. Electronic density of states and phonon dispersion curves reveal metallic nature of YbX. The slight unusual behaviour of phonon dispersion curve is observed in the case of YbAs.

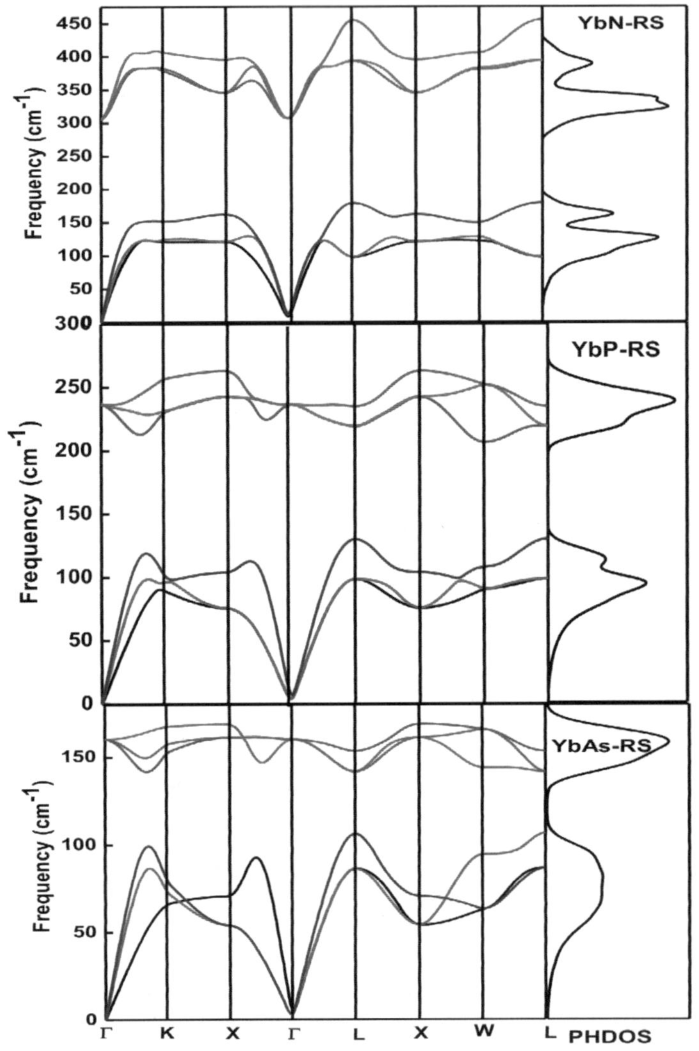


**Figure 2**. Calculated phonon dispersion curves along with the phonon density of states of YbN, YbP and YbAs.

**Acknowledgements**
Financial assistance from Department of Science and Technology, Govt. of India, New Delhi and Department of Atomic Energy, Govt. of India, Mumbai.

**References**

[1] Degiorgi L, Bacsa W and Wachter P 1990 *Phys. Rev. B* **42** 530
[2] Donni A, Furrer A, Fischer P, Hulliger F, Wachter P and Ott H R 1990 *J. Magn. Magn. Mater* **90&91** 143
[3] Jha P K and Sanyal S P 1995 *Phys. Rev. B* **52** 15898; Jha P K and Sanyal S P 1998 *Phys. State. Sol (b)* **205** 465

[4] Guntherodt G, Jayaraman A and Kress W 1981 *Valence Fluctuations in solids*, edited by Falicove L M, Hanke W and Maple M B (North Holland, Amstterdam).
[5] Jha P K and Sanyal S P 1993 *J. Pure Appl. Phys.* **31** 469
[6] The Abinit Code is a common project of the University Catholique de Louvain, Corning incorporated, and other contributions, <http://www.abinit.org>.
[7] Perdew J P, Burke K and Ernzerhof M 1996 *Phys. Rev. Lett.* **77** 3865
[8] Hua X, Chen X and Goddard W A 1997 *Phys. Rev. B* **55** 16103
[9] Monkhorst H J and Pack J D 1976 *Phys. Rev. B* **13** 5188
[10] Baroni S, Gironcoli S, De A, Dal C and Giannozzi P 2001 *Rev. Mod. Phys.* **73** 515
[11] Mendik M and Wachter P 1993 *Phys. Rev. B* **47** 6110
[12] Rukmangad A, Aynyas M and Sanyal S P 2010 *Indian J. Phys.* **84 (9)** 1183
[13] Winkler B and Milman V 2005 *Zeitschrift für Kristallographie* **220(8)** Phase Transitions, pp. 700-703
[14] Hayashi J, Shirotani I, Adachi T, Shimomura O and Kikegawa T 2004 *Philosophical Magazine* **84(34)** 3663
[15] Li D X, Oyamada A, Hashi K, Haga Y, Matsumura T, Shida H, Suzuki T, Kasuya T, Donni A, Hulliger F 1995 *J. Magn. Magn. Mater* **140-144** 1169
[16] Murnaghan F D 1944 *Proc Natl Acad Sci* **30(9)** 244 (USA)

# Pressure-induced vibrational and electronic properties of palladium per nitride

**Sanjay D. Gupta, Sanjeev K. Gupta and Prafulla K. Jha***
Department of Physics, Bhavnagar University, Bhavnagar 364002, India
*E-mail: prafullaj@yahoo.com

**Abstract.** We predict electronic and structural phase transition at 11GPa for crystalline $PdN_2$ in pyrite structure by employing first principles calculation using density functional theory implemented in Quantum Espresso code. The electronic band structure, equation of states and dynamical properties at zero as well as high pressure are calculated within the frame work of conventional GGA exchange correlation functional. We report the sharp change of the Pd-N and N-N bond length at phase transition pressure in support of isostructural phase transition driven by electronic phase transition for $PdN_2$ in pyrite structure. The calculated Raman frequencies near the phase transition pressure reveals the first order isostructural phase transition by the instability of longitudinal acoustical as well as optical branches, provides the mechanism for experimentally observed decomposition below 13GPa.

## 1. Introduction

Metallic nitrides are of great technological and fundamental importance due to extraordinary properties such as hardness, superconductivity, photoluminescence and many type of magnetism [1]. Traditional applications have taken advantage of the hard and refractory nature of many of these compounds, but numerous mere recent applications are based on their crystalline structure stability, electronic and vibrational properties [2]. Now a consensus has been reached concerning the observed crystal structure and stoichiometry i.e. one metal atom for every nitrogen dimmer [1-4]. However, it was found that pyrite $PdN_2$ structure decomposes at pressure below ~13GPa [3-5]. Therefore, the correct stoichiometry could not be determined using electron microprobe or X-ray photoelectron spectroscopy. Until now the crystalline structure of this new synthesized palladium per nitride has not been determined. All noble metal nitrides ($PtN_2$, $IrN_2$ and $OsN_2$) studied to date possess uncomplicated behaviour due to significant hybridization between 5d electrons and N-2p electrons at all hydrostatic pressure [6-8]. They are recoverable to ambient conditions after synthesis at high pressure and high temperature [in the range of 60GPa and 2000K]. In contrast to these compounds, the synthesized crystalline $PdN_2$ at high pressure and temperature does not appear to be recoverable at ambient conditions and decomposed below 13GPa [9]. Recently, Aberg et al. [4] have argued that the equation of states (EOS) cannot be accurately described within either the local density or generalized gradient approximations for $PdN_2$ and only the Heyd-Scuseria-Ernzerhof exchange-correlation functional (HSE06) provides very good agreement with experimental data. Motivated with this report, we have investigated the EOS, electronic band structure phase transition and vibrational properties using conventional generalized gradient approximations and analyzed with available experimental and theoretical data. Our analysis concludes that the conventional GGA is sufficient to predict the physical properties of $PdN_2$ comparable to the experimental observation. We are able to observe the decomposed below 13GPa similar to experiment.

## 2. Computational Methodology

We have investigated structural, electronic, and lattice dynamical properties using the plane wave pseudopotential within the GGA-XC functional as parameterized by Perdew, Burke, and Ernzerhof for treating the effects of exchange correlation implemented in PWSCF package. The Brillouin zone integrations were performed using $6\times6\times6$ Monkhorst-Pack k-point sampling for structural relaxations and the calculation of other properties. The plane-wave cutoff was set to 38Ry, and

Mathfessel-Paxton smearing with a width of 0.05eV was used to determine the eigenvalues. Atomic positions were relaxed with ionic forces converged to 0.001eV/Å. The computed equilibrium lattice constant and bulk modulus agree well with the experiment and other theoretical calculations.

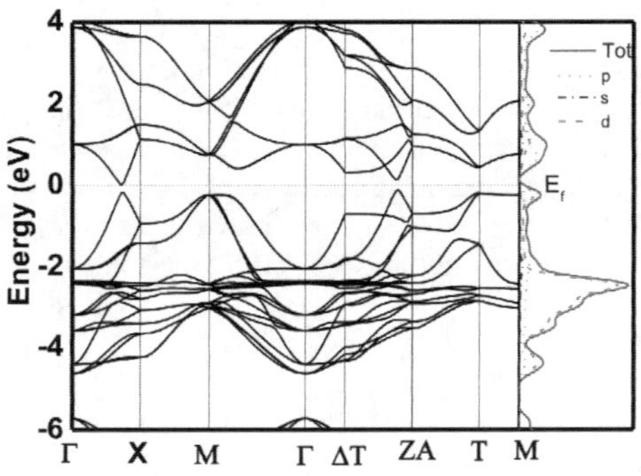


**Figure 1(a).** Electronic band structure for pyrite $PdN_2$ at zero pressure along with total and projected density of states

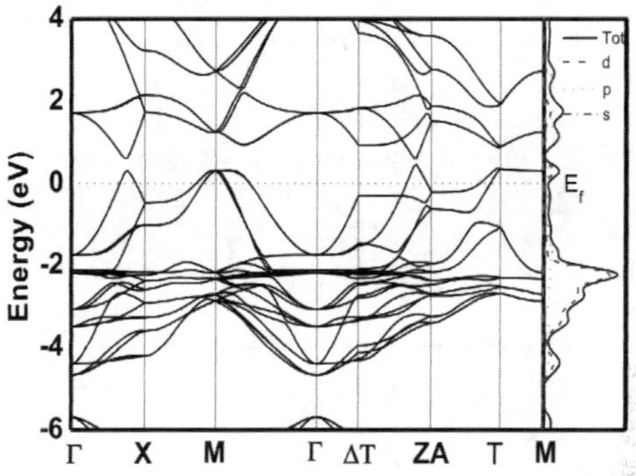


**Figure 1(b).** Electronic band structure for pyrite $PdN_2$ at 11GPa along with total and projected density of states

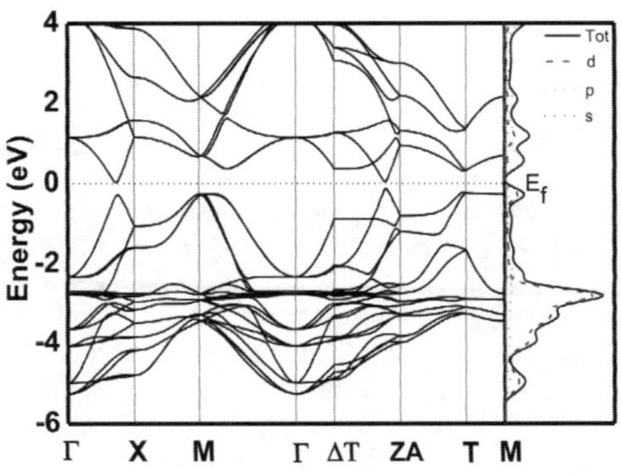


**Figure 1(c).** Electronic band structure for pyrite $PdN_2$ at 18GPa along with total and projected density of states

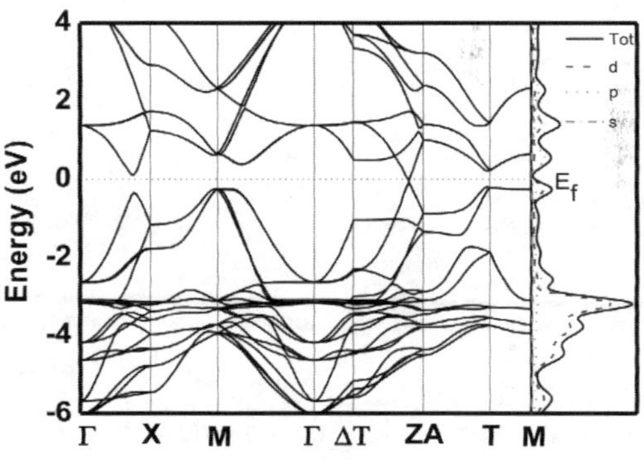


**Figure 1(d).** Electronic band structure for pyrite $PdN_2$ at 60GPa along with total and projected density of states

### 3. Results and Discussion

The structure of pyrite $PdN_2$ was optimized at zero and different high pressures (11, 18 and 60GPa). The optimized structural parameters are in agreement with earlier theoretical and experimental data. The calculated electronic band structures of pyrite $PdN_2$ at zero pressure, 11GPa, 18GPa and 60GPa are presented in Figs. 1(a), 1(b), 1(c) and 1(d) respectively along with total and partial electronic density of states. The Figs. 1(a) and 1(c) reveal a smaller direct band gap in Γ-X direction of the Brillouin zone at zero pressure as well as 18GPa. The projected density of states shown in the right panel of each figure at corresponding high pressure reveals that the valence and conduction orbital bands are determined by N-*p* and Pd-*d* electron states. With the application of pressure, the Fermi energy sharply shifts towards the lower energy region suggests a transformation from insulator to metallic crystal state. Our calculated electronic band structure and density of states are very well in agreement in reported by Chen et al. [5] with smaller band gap at 18GPa. At higher synthesized pressure the band again closes and makes $PdN_2$ as metallic form its insulator state as shown in Figs. 1(c) and 1(d). This electronic phase transition in pyrite crystalline $PdN_2$ can also lead to first order isostructural transition at applied higher static pressure near the decomposition pressure. Aberg et al. [4] had claimed that the EOS can not accurately describe within either the LDA or GGA approximations. Our calculated EOS data is in good agreement with the theoretically calculated EOS [4] and experimental X-ray diffraction measurements. The anomalous behavior of EOS at 11GPa corresponds to isostructural phase transition for $PdN_2$ pyrite crystal structure shown in Fig 2(a). Thus the electronic transition in the $PdN_2$ pyrite crystal structure leads to the isostructural phase transition at phase transition pressure 11GPa. This is due to the fact that pressure applied, bond length increases between nitrogen dimmers and transform from N≡N to N-N single bond character and forming thermodynamically stable pyrite $PdN_2$ crystal above 18GPa [5].

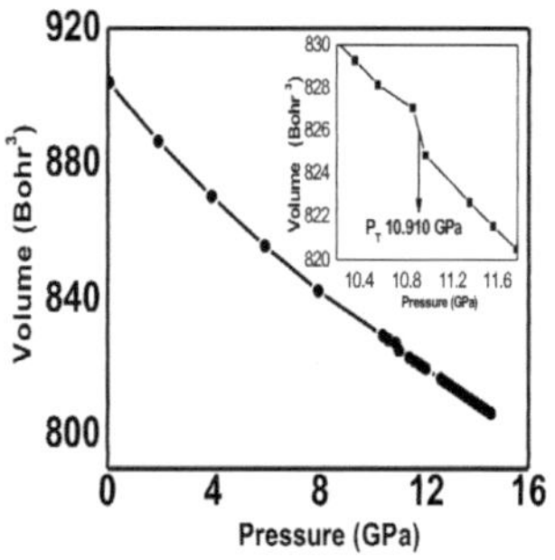


Figure 2(a). Equation of state as obtained from GGA-XC.

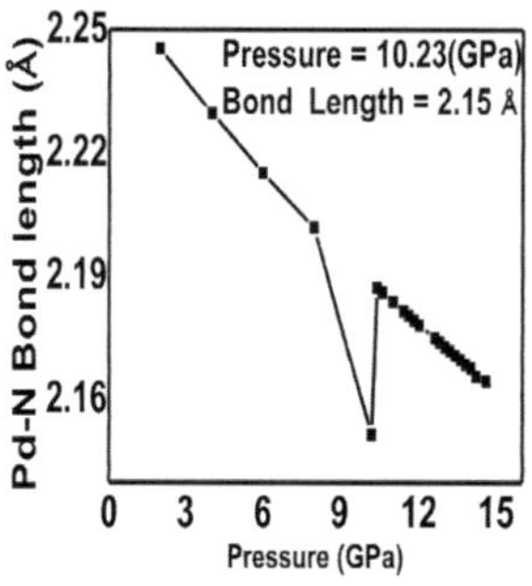


Figure 2(b). Bond length between palladium and nitrogen as function of pressure.

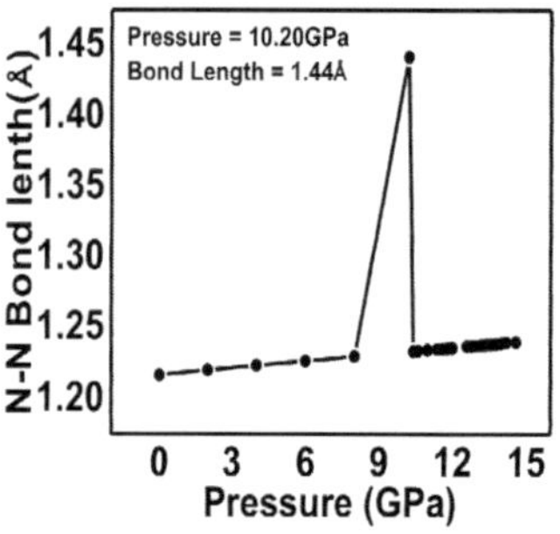


Figure 2(c). Bond length of nitrogen dimmers as a function of pressure.

We have obtained Pd-N and N-N bond length up to 15GPa by optimized at each pressure and shown in Figs. 2(b) and 2(c) respectively. The Figs. 2(b) and 2(c) clearly indicates sharp deep and rise in bond length near phase transition pressure (11GPa) between Pd-N and N-N bonding in nitrogen dimmer respectively. Thus the electronic phase transition leads to an exponential increase in free carrier concentration as well screening effect causes and anomalous behavior of the bond length. Thus the volume dependence bond length foreshadows of peculiar properties of this compound near the phase transition pressure. The palladium per nitride with pyrite structure is very soft and elastically stable at 0GPa, though thermodynamically unstable, become energetically favourable at high pressure but the lattice dynamical calculations reveal that the pyrite $PdN_2$ is dynamically stable at ambient and 60GPa [4]. The zone centre phonon frequencies 1042.9 $cm^{-1}$ and 1118.7 $cm^{-1}$ of $A_g$ and $T_g$ modes particularly at 60GPa, which are the Raman active phonons agree well with the experimental Raman data. The pressure variation of Raman active modes shows a linear variation; however, at higher pressure the optical frequencies decrease and acoustic frequencies increases [2-3]. From Fig. 3(b) shows that almost all the zone centre phonon branches behaves anomalously at calculated electronic phase transition provides clear support in the decomposition of $PdN_2$ pyrite structure at 11GPa and predict the quenching of crystal structure from high pressure to ambient pressure can be possible to recover crystalline pyrite $PdN_2$ in ambient pressure.

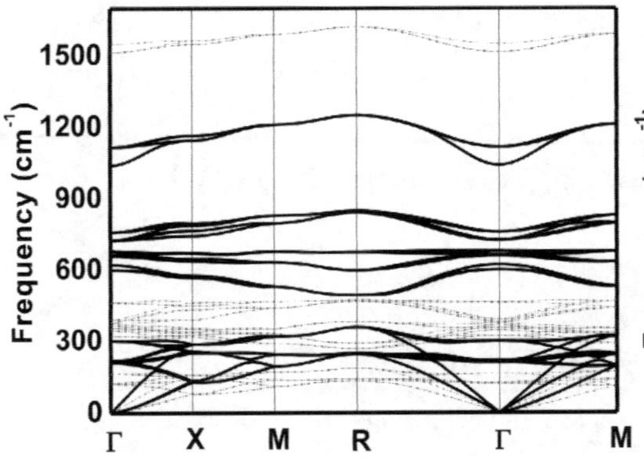


**Figure 3(a).** Phonon dispersion curve for pyrite structure of $PdN_2$ at ambient (dash curve) and 60 GPa (solid curve).

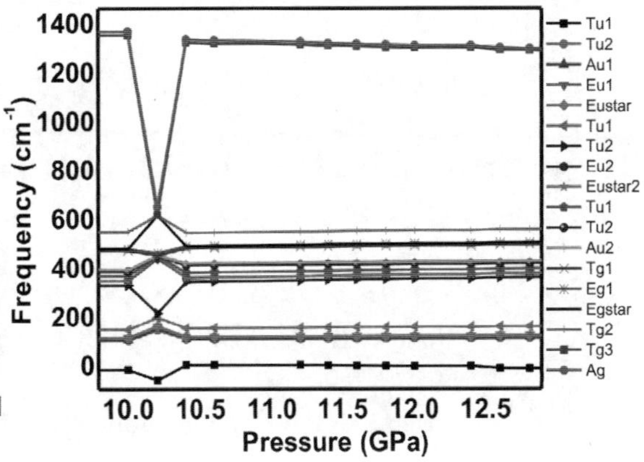


**Figure 3(b).** The high pressure behaviour of zone centre phonon frequencies for pyrite structure of $PdN_2$.

## 4. Conclusion

In conclusion, we have investigated structural electronic and phonon properties for $PdN_2$ in pyrite crystal structure up to 60GPa using first principles calculations. Our results show first time that all calculated properties reproduced accurately and consistent, with available experimental data calculated by GGA-XC functional. The electronic phase transition from insulator to metallic state, multivalue EOS, drastic change in Bond length of Pd-N and N-N, and dynamical instability as well as anomalous behavior of zone centre phonon frequencies at 11GPa clearly suggest that $PdN_2$ pyrite crystal structure decomposes at 11GPa sharply. The mechanical and dynamical stability clearly shows the $PdN_2$ pyrite crystal structure can be recovered in high crystalline form at ambient pressure.

## Acknowledgements

The financial support from the Department of Science and Technology, Govt. of India, New Delhi and Department of Atomic Energy, Govt. of India, Mumbai is highly appreciated for this work.

## References

[1] Toth L 1971 *Transition metal carbides and Nitrides* (New York: Academic Press)
[2] Samsonov G V and Vinittski I M, 1976 *Handbook of Refractory Compounds* (Metallurgiya, Moscow; IFZ/Plenum, New York, 1980)
[3] Aberg D, Sadigh B, Crowhurst J and Goncharov A F 2008 *Phys. Rev. Lett* **100** 095501
[4] Aberg D, Erhart P, Crowhurst J, Zaug J M, Goncharov A F and Sadigh B 2010 *Phys. Rev. B* **82** 104116
[5] Chen W, Tse J S and Jiang J Z 2010 *J. Phys.: Condens. Matter* **22** 015404
[6] Crowhurst J C, et al. 2006 *Science* **311** 1275
[7] Yu R, Zhan Q and Zhang X F 2006 *Appl. Phys. Lett.* **88** 051913
[8] Young A F, Montoya J A, Sanloup C, Lazzeri M, Gregoryanz E and Scandolo S 2006 *Phys. Rev. B* **73** 153102
[9] Soto G 2004 *Mater. Lett.* **58** 2178
[10] Nye J F 1985 *Physical Properties of crystals* (Oxford: Oxford University press)

Journal of Physics: Conference Series **377** (2012) 012079 doi:10.1088/1742-6596/377/1/012079

# High pressure effect on structural, elastic and thermal properties of DyP and DyAs

**Sanjay Bhajanker[1, 2], Vipul Srivastava[1, 2] and S P Sanyal[2]**

[1]Deptt of Physics, NRI Institute of Research & Technology, Bhopal-462021, India
[2]Deptt. of Physics, Barkatullah University, Bhopal – 462026, India

E-mail: san_bhajanker@yahoo.co.in

**Abstract.** We have performed an inter-ionic potential theory with modified ionic charge to investigate the structural, elastic and thermal properties of DyP and DyAs theoretically, at high pressure. We found a structural phase transition from NaCl ($B_1$) - to CsCl ($B_2$)-type structure at 57 GPa for DyP and 51 GPa for DyAs and other properties, such as lattice constant, bulk modulus, cohesive energy and second-order elastic constants were also calculated and compared with the available theoretical data. In order to gain further information, we have also predicted the Young's modulus (*E*), Poisson's ratio *(ν)*, anisotropy factor *(A)* and Debye temperature *($\theta_D$)*. Our results are in good agreement with theoretical data where available and provide predictions where they are not.

## 1. Introduction

The investigation on binary rare-earth (RE) monopnictides has been done in last decade with great interest because of these binary rare-earths (RE) compounds crystallize in simple NaCl-type structure. Most of these RE compounds show extensive structural, elastic and thermal properties [1]. Apart from this an interesting aspect of these compounds is the occurrence of localized strongly correlated 4f electrons which has been challenge for theoretian. There are various theoretical approaches to explain localized nature of the strongly correlated electron [2]. In recent years, dysprosium (Dy) monopnictides and monochalcogenides have been investigated as a suitable reference material for understanding the unusual properties of other rare-earth monopnictides and monochalcogenides [1, 3-6]. Shirotani et al. [3] have systematically investigated the pressure induced structural phase transition from NaCl-type ($B_1$) to CsCl-type ($B_2$) structure in some LnAs (Ln = Pr, Nd, Sm, Gd, Dy and Ho) compounds with NaCl-type structure has been studied up to 60 GPa at room. As far as some optical and electronic properties of Dy monopnictides are concerned Schoenes et al [4] have studied both experimentally and theoretically. Recently, Özı¸sık et al [1] have theoretically studied the structural, elastic and vibrational properties of DyX (X = P, As) compounds by using the VASP code.

In present paper, we have performed an inter-ionic potential theory [7] with modified ionic charge to investigate the structural, elastic and thermal properties of DyP and DyAs at high pressure, systematically. We predict a structural phase transition from NaCl ($B_1$) to CsCl ($B_2$)-type structure at pressure of 57 GPa for DyP and 51 GPa for DyAs and other properties such as lattice constant, bulk modulus, cohesive energy and second order elastic constants are calculated and compared with the available theoretical data. In order to gain further information, we have also predicted the Young's modulus ($E$), Poission's ratio ($v$), anisotropy factor ($A$) and Debye temperature ($\theta_D$) for these compounds.

## 2. Method of calculation

The interionic potential for DyP and DyAs compounds in the framework of the rigid ion model [7] is expressed as:

$$U(r)=\sum_{ij}\frac{Z_m^2e^2}{r_{ij}}+\sum_{ij}b\beta_{ij}\exp\left[\frac{r_i+r_j-r_{ij}}{\rho_{ij}}\right]+\sum_{ij}C_{ij}r_{ij}^{-6}+\sum_{ij}D_{ij}r_{ij}^{-8};\qquad i,j=1,2 \tag{1}$$

which includes long-range Coulomb (first term), Hafemeister and Flygare form of short-range repulsive (second term) and van der Waals multipole interactions (third and fourth terms). $Z_me$ is the modified ionic charge, which parametrically includes the effect of Coulomb screening by the delocalised *f*-electrons. $\beta_{ij}$ are the Pauling coefficients. $b$ and $\rho$ are the short range parameters, which can be determined from the bulk modulus and equilibrium condition:

$$\left|\frac{dU_{ij}(r)}{dr}\right|_{r=r_0}=0$$

The theory has been applied to the wide class of NaCl-type solids [8]

To understand the inter-atomic forces in cubic solids (NaCl-type) the second order elastic constants (SOEC) i.e. $C_{11}$, $C_{12}$ and $C_{44}$ are derived from the crystal potential energy by the method of homogeneous deformations as follows:

$$C_{11}=\left(\frac{e^2}{4r_0^4}\right)\left[-5.112Z_m^2+A_{12}+\frac{1}{2}(A_{11}+A_{22}+B_{11}+B_{22})\right] \tag{2}$$

$$C_{12}=\left(\frac{e^2}{4r_0^4}\right)\left[-0.225Z_m^2+B_{12}+\frac{1}{4}(A_{11}+A_{12}-\frac{5}{4}(B_{11}+B_{12}))\right] \tag{3}$$

$$C_{44}=\left(\frac{e^2}{4r_0^4}\right)\left[2.556Z_m^2+B_{12}+\frac{1}{4}(A_{11}+A_{12}+\frac{3}{4}(B_{11}+B_{12}))\right] \tag{4}$$

where, $A_{ij}$ and $B_{ij}$ (i,j = 1,2) are short-range force constants.

The bulk modulus is derived from elastic constants as, $B_0=\frac{1}{3}(C_{11}+2C_{12})$ (5)

We have estimated Debye temperature for DyP and DyAs using the following expressions[9,10]:

$$\theta_D=\frac{h}{K_B}\left[\frac{3n}{4\pi V_a}\right]^{1/3}S_m \tag{6}$$

where h is a Plank's constant, $K_B$ is Boltzmann's constant and $V_a$ is the atomic volume and $S_m$ is average sound velocity. The average sound velocity of the polycrystalline material is given by

$$S_m=\left[\frac{1}{3}\left(\frac{2}{S_T^3}+\frac{1}{S_L^3}\right)\right]^{-1/3} \tag{7}$$

where $S_T$ and $S_L$ are the longitudinal and transverse sound velocities.

## 3. Results and discussion

The calculated structural parameters of DyP and DyAs, under ambient conditions are listed in Table1. It is clear from the Table 1 that the calculated input and output parameter, phase transition pressure and relative volume change are in good agreement with the experimental and other theoretical values [1, 3-6].

**Table 1.** Input and out parameters, phase transition and relative volume change for DyP and DyAs compounds.

| Solids | $r_0$ (Å) | $B_0$(GPa) | $Z_m^2$ | b($\times10^{-19}$J) | ρ(Å) | $P_T$(GPa) | $\Delta V/V_0$ % |
|---|---|---|---|---|---|---|---|
| **DyP** | 2.83 | 82.7 | 3.00 | 6.743 | 0.293 | 57 | 17 |
| | 2.815[a] | 85.4[b] | - | - | - | 58.7[b] | - |
| **DyAs** | 2.91 | 73.7 | 2.95 | 7.248 | 0.297 | 51 | 18 |
| | 2.895[c] | 76.2[b] | - | -- | - | 52.6[b], 44[d], 48[e] | - |

[a]Ref [4], [b]Ref [1], [c]Ref [6], [d]Ref [5], [e]Ref [3]

The modified ionic charge parameter, $Z_m^2$, which has been derived from the measured value of bulk modulus shows decreasing trend in these compounds. In Fig. 1 graph is plotted between modified ionic charge parameter, $Z_m^2$, and lattice constants of these compounds. The large value of bulk modulus in DyP and DyAs shows its strong ionic character, whereas small value of $Z_m^2$ in DyP and DyAs indicates strong delocalization of f-electrons. Two other parameter b and ρ are the short range parameter which shows the increasing trend in these compound.

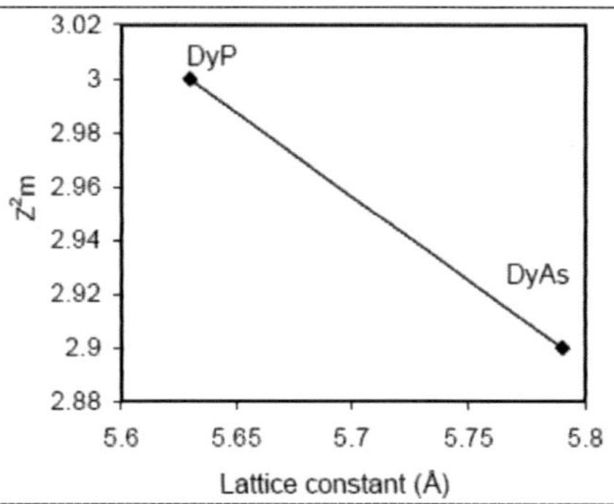


**Figure 1.** Variation of ionic charge parameter with lattice constants of DyP and DyAs.

The high pressure, along with ambient pressure studies on materials, can help in analyzing the nature of interaction around the atoms. It is found that under pressure; most of the rare earth compounds undergo a first-order phase transition from the NaCl-type to CsCl-type structure. The calculated phase transition pressure, $P_T$, for DyP and DyAs are 57 GPa and 51 GPa, respectively, which shows good agreement with the other value [1, 3, 5]. The equation of state for DyP and DyAs are given in Fig. 2 and Fig. 3 respectively.

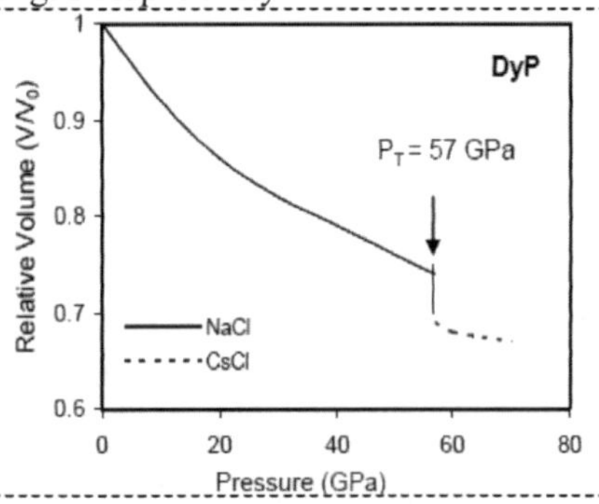


**Figure 2.** Equation of state for DyP.

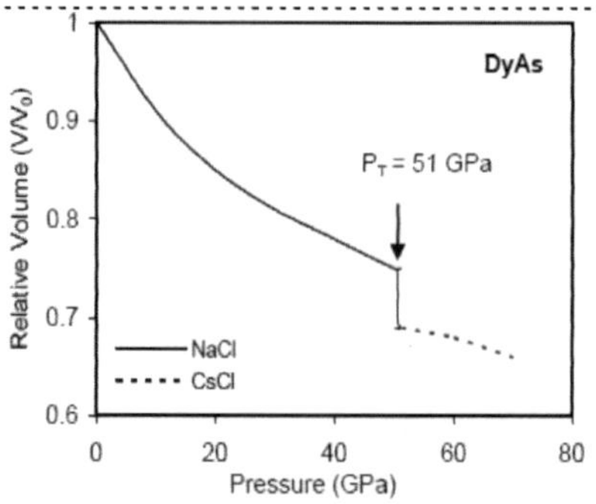


**Figure 3.** Equation of state for DyAs.

The elastic constants are useful for mechanical and dynamical behaviour of solids. Cubic structures have only three independent elastic constants namely $C_{11}$, $C_{12}$ and $C_{44}$. Therefore, the study of the second order elastic constants ($C_{11}$, $C_{12}$, and $C_{44}$) and their pressure derivatives at 0 k is quite important for understanding the nature of the interatomic forces. Recentally, Özı¸sık et al [1] calculated the elastic constants of DyP and DyAs and they concluded that these compounds are elastically stable and the values of elastic constants decrease with increasing the atomic number of pnictide atoms in $B_1$ phase. Our calculated results also follow the same trend [1], which is presented in Table 2. Mechanical properties such as Shear ($G$) and Young's modulus ($E$), Poisson ratio ($v$) and anisotropic factor ($A$) are also presented in Table 2 and compared with the available theoretical data [1].

**Table 2.** Elastic constants ($C_{11}$, $C_{12}$, $C_{44}$), Shear modulus ($G$), Young modulus ($E$), Poisson ratio ($v$), anisotropic ratio ($A$) and Debye temperature ($\theta_D$ in K) for DyP and DyAs compounds.

| Solids | $C_{11}$ | $C_{12}$ | $C_{44}$ | $G$ | $E$ | $v$ | $A$ | $\theta_D$ |
|---|---|---|---|---|---|---|---|---|
| **DyP** | 167.3 | 40.6 | 40.9 | 48.8 | 122.4 | 0.253 | 0.644 | 491.2 |
| | 208.9[a] | 28.1[a] | 46.4[a] | 64.0[a] | 147.4[a] | 0.21[a] | 0.51[a] | 214.7[a] |
| **DyAs** | 150.2 | 35.8 | 36.0 | 43.4 | 108.9 | 0.254 | 0.627 | 425.1 |
| | 187.2[a] | 24.6[a] | 38.8[a] | 52.4[a] | 127.9[a] | 0.22[a] | 0.47[a] | 182.5[a] |

[a]Ref[1]

The Debye temperature is an important physical quantity in characterizing lattice vibrations of a solid. It plays a key role in the discussions of the temperature dependence of the electrical resistivity in metals. Debye temperature is calculated by sound velocities, which are estimated by [9, 10]. We note that $\theta_D$ decreases when we move from DyP to DyAs. This is good agreement with the results reported in Ref. [1] but our calculated Debye temperature are higher than the results reported by Özı¸sık et al [1].

**Acknowledgments**
Authors are thankful to Madhya Pradesh Council of Science & Technology, Bhopal for Financial Assistance through Research Project (1904/CST/2009). One of the authors SPS acknowledges CSIR, New Delhi.

## 4. References

[1] Özişik H, Çiftci Y Ö, Çolakoğlu K and Deligöz E 2011 *Phys. Scr.* **83** 035601
[2] Hayashi J, Shirotani I, Tanaka Y, Adachi T, Shimomura O, Kikegawa T 2000 *Solid state Commun,* **114** 561
[3] Shirotani I, Yamanashi K, Hayashi J, Tanakde Y, Ishimatsu N, Shimomura O and Kikegawa T 2003 *Solid State Commun.* **127** 573
[4] Schoenes J, Repond P, Hulliger F, Ghosh D B, De S K, Kunes J and Oppeneer P M 2003 *Phys. Rev. B* **68** 085102
[5] Shekar N V and Sahu P Ch 2006 *J. Mater. Sci.* **41** 3207
[6] Duan C G, Sabirianov R F, Mei W N, Dowben P A, Jaswal S S and Tsymbal E Y 2007 *J. Phys.: Condens.Matter* **19** 315220
[7] Singh R K 1982 *Phys. Rep.* **85** 259
[8] Singh A, Srivastava V, Aynyas M, and Sanyal S P 2009 *Physica B* **404** 1852-1857
[9] Sun Z, Li S, Ahuja R, Schneide J M 2004 *Solid State Commun.* **129** 589
[10] Wachter P, Filzmoser and Rebizant J 2001 *Physica B* **293** 199

Journal of Physics: Conference Series **377** (2012) 012080 doi:10.1088/1742-6596/377/1/012080

# Mechanical and thermal properties of praseodymium monochalcogenides and monopnictides under pressure

**Sanjay Bhajanker [1, 3], Vipul Srivastava [1, 3], G Pagare [2] and S P Sanyal [3]**

[1]Deptt of Physics, NRI Institute of Research & Technology, Bhopal-462021, India
[2]Deptt. of Physics, Govt.MLB PG College, Bhopal - 462011, India
[3]Deptt. of Physics, Barkatullah University, Bhopal – 462026, India
E-mail: san_bhajanker@yahoo.co.in

**Abstract.** The mechanical and thermal properties of praseodymium monochalcogenides and monopnictides are calculated using elastic constants, those have been derived from two body inter- ionic potential theory. We have calculated the Young's modulus (*E*), Poisson's ratio ($\nu$), anisotropy factors (*A*), sound velocities and Debye temperature ($\theta_D$) for these compounds. The bulk modulus to shear modulus ratio (B/G) lies between 1.75-1.92 for all these compounds, which shows that all compounds are ductile in nature. Young's modulus shows that PrS and PrAs are stiffer than the other monochalcogenides and monopnictides of praseodymium. The variation of elastic constants ($C_{11}$, $C_{44}$) with pressure and the variation of Debye temperature with pressure are also presented for these compounds.

## 1. Introduction

In recent years, a great deal of interest has been focused on the rare earth chalcogenides and pnictides because they form an interesting and extensively studied series and have numerous applications in technologies [1]. The praseodymium chalcogenides have attracted great attention due to their potential applications in spintronics, spin filtering devices, hyperfine enhanced nuclear cooling, study of combined electron and nuclear ordering phenomenon at very low temperature [2] thus opened a new area for development of electronic devices such as metal base transistors. A few experimental and theoretical studies of praseodymium monochalcogenides and monopnictides are available in literature [1, 3-5]. Shirotani et al. [3] have systematically investigated the pressure induced structural phase transition from NaCl-type ($B_1$) to CsCl-type ($B_2$) structure in cerium and praseodymium monopnictides by measuring the powder X-ray diffraction pattern using synchrotron radiation. As far as, the electronic properties of these compounds are concerned, Vaitheeswaran et al. [4] have calculated the electronic structure of praseodymium monochalcogenides and monopnictides under ambient and high pressure. They found that the trivalent Pr configuration is the ground state in all these compounds and remains stable even under pressure up to ~50 GPa. Very recently, Kocak et al. [5] have investigated the structural, elastic and lattice dynamical properties of praseodymium chalcogenides (PrS, PrSe and PrTe) using the VASP code. A systematic structural study of praseodymium monochalcogenides (PrX: X = S, Se, Te) and praseodymium monopnictides (PrY: Y = P, As, Sb, Bi) is reported from our group [1] using the interionic potential theory with necessary modifications. These compounds undergo $B_1$ to $B_2$ phase transitions at various pressures [1]. Our further interest in these compounds is motivated for their mechanical and thermal properties. The knowledge of mechanical properties is essential for many practical applications such as load deflection, thermoelastic stress, sound velocities, and so on. The mechanical properties such as Young modulus (*E*), shear modulus (*G*), Poisson ratio (*v*) and anisotropic ratio (*A*) and thermal properties such as Debye temperature ($\theta_D$) of praseodymium monochalcogenides and monopnictides have not been sufficiently studied. This lack of information has motivated us to investigate the mechanical and thermal properties of these compounds.

In the present paper, we have investigated mechanical properties such as Shear modulus (*G*), Young modulus (*E*), Poisson ratio (*v*), anisotropic ratio (*A*) and thermal properties with respect to Debye temperature of praseodymium monochalcogenides and monopnictides incorporating data from previous work [1]. We have also presented the variation of elastic constants and of Debye temperatures with pressure for these compounds.

## 2. Method of calculation

### *2.1. Calculation of mechanical properties*

The mechanical properties such as Shear and Young modulus, Poisson ratio, and anisotropic ratio are calculated using standard formulism as given in equations 1-4. The incorporation of second order elastic constants (SOEC) in the formalism is calculated using classical inter ionic potential theory [6]. The study of SOEC $C_{11}$, $C_{12}$ and $C_{44}$ and their pressure derivatives at 0K is quite important for understanding of the inter-atomic force in solids. The bulk modulus is derived from elastic constants as

$$B = \frac{1}{3}(C_{11} + 2C_{12}) \tag{1}$$

In addition another physical quantity, namely the anisotropic ratio *A*, defined as: $A=2C_{44}/(C_{11}-C_{12})$ has also been calculated. For an ideal isotropic system, *A* is unity and deviation from unity measures the amount of elastic anisotropy. The Poission's ratio, *v* is calculated using the relation:

$$v = \frac{3B - 2G}{2(3B + G)} \tag{2}$$

where, *B* is bulk modulus and *G* is average shear modulus. As per Hill [7] average shear modulus, *G* is defined as arithmetic mean of Voigt, $G_V$ and Reuss, $G_R$ values, which can be expressed in terms of elastic constants as:

$$G_V = \frac{1}{5}(C_{11} - C_{12} + 3C_{44}) \quad \text{and} \quad G_R = \frac{5(C_{11} - C_{12})C_{44}}{3(C_{11} - C_{12}) + 4C_{44}} \tag{3}$$

The Young's modulus, *E* is calculated using the expression:

$$E = \frac{9BG}{3B + G} \tag{4}$$

### *2.2. Calculation of Debye temperature ($\theta_D$)*

The Debye temperature ($\theta_D$) is an important physical parameter of solids, which defines a demarcation line between quantum mechanical and classical behaviour of phonons. We have estimated Debye temperature of PrX and PrY compounds using the following expression [8, 9]:

$$\theta_D = \frac{h}{K_B}\left[\frac{3n}{4\pi V_a}\right]^{1/3} S_m \tag{5}$$

where h is a Plank`s constant, $K_B$ is Boltzmann`s constant and $V_a$ is the atomic volume and $S_m$ is average sound velocity.

## 3. Results and discussion

Praseodymium monochalcogenides and monopnictides crystallize in simple NaCl-type structure [10]. The ground state properties of these compounds have been studied by Soni et al. [1] from our group.

They have successfully calculated the lattice parameters, bulk modulus and elastic constant of these compounds. Incorporating all the above parameters we have further investigated Young modulus ($E$), shear modulus ($G$), Poisson ratio ($v$) and anisotropic ratio ($A$) for these compounds which are given in Table 1.

**Table 1.** Elastic constants ($C_{11}$, $C_{12} = C_{44}$) [1], bulk modulus ($B$), shear modulus ($G$), Young modulus ($E$), Poisson ratio ($v$), anisotropic ratio ($A$), calculated values of density ($\rho$ in Kg/m$^3$), the longitudinal, transverse, average sound velocity ($S_L$, $S_T$ and $S_M$ in m/s) and Debye temperature ($\theta_D$ in K) for praseodymium monochalcogenides and monopnictides.

| Solids | $C_{11}$ | $C_{12} = C_{44}$ | $B$ | $G$ | $E$ | $\nu$ | $A$ | $\rho$ | $S_L$ | $S_T$ | $S_M$ | $\theta_D$ |
|---|---|---|---|---|---|---|---|---|---|---|---|---|
| **PrS** | 233 | 41 | 105.8 | 58 | 148 | 0.266 | 0.435 | 6.01 | 5626 | 3245 | 3603 | 591 |
| **PrSe** | 211 | 31 | 91.5 | 48 | 123 | 0.275 | 0.349 | 6.87 | 4892 | 2821 | 3132 | 496 |
| **PrTe** | 137 | 20 | 59.5 | 31 | 80 | 0.274 | 0.354 | 7.00 | 3910 | 2255 | 2504 | 373 |
| **PrP** | 156 | 34 | 75.1 | 43 | 109 | 0.257 | 0.564 | 5.49 | 4968 | 2870 | 3186 | 508 |
| **PrAs** | 191 | 39 | 89.9 | 51 | 129 | 0.260 | 0.513 | 6.54 | 4977 | 2876 | 3192 | 500 |
| **PrSb** | 117 | 21 | 53.5 | 29 | 75 | 0.264 | 0.458 | 6.66 | 3804 | 2197 | 2439 | 360 |
| **PrBi** | 103 | 20 | 48.1 | 27 | 68 | 0.261 | 0.496 | 8.53 | 3184 | 1841 | 2043 | 298 |

We can also measure the brittle/ductile behaviour of these compounds by using empirical relation between the calculated bulk modulus and shear modulus. The empirical relation proposed by Pugh [11]. According to Pugh [11] a material is brittle if the B/G ratio is less than 1.75, otherwise it behaves in ductile manner. The present calculated values of B/G are found to be 1.82 for PrS, 1.91 for PrSe, 1.92 for PrTe, 1.75 for PrP, 1.76 for PrAs, 1.84 for PrSb and 1.78 for PrBi. One can clearly notice that these compounds show ductile nature. The Young's modulus, $E$ defines stiffness of the materials and it is clear from Table 1 that PrS is stiffer than other monochalcogenides of praseodymium; similarly PrAs is stiffer than the other monopnictides of praseodymium. The Poisson's ratio, $v$, defines the stability of solid against shear. The value of Poisson's ratio lies in between-1 and 0.5; the calculated values of $v$ for praseodymium monochalcogenides and monopnictides do not fall in this range and reveals a possibility of instability of structure. The anisotropic ratio, $A$, is a measure of the degree of elastic anisotropy in solid. For an ideal isotropic system, $A$ is unity and deviation from unity measures the amount of elastic anisotropy. The calculated anisotropic ratio for praseodymium monochalcogenides and monopnictides is less than 1 which indicates that these compounds are not elastically isotropic. We have also calculated the pressure dependence of the second order elastic constants, namely $C_{11}$ and $C_{44}$ for praseodymium monochalcogenides and monopnictides in Fig.1.

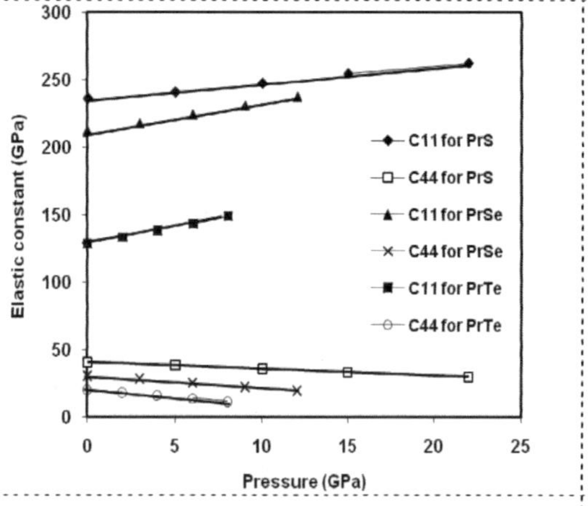


**Figure 1.** Variation of elastic constants with pressure for praseodymium monochalcogenides.

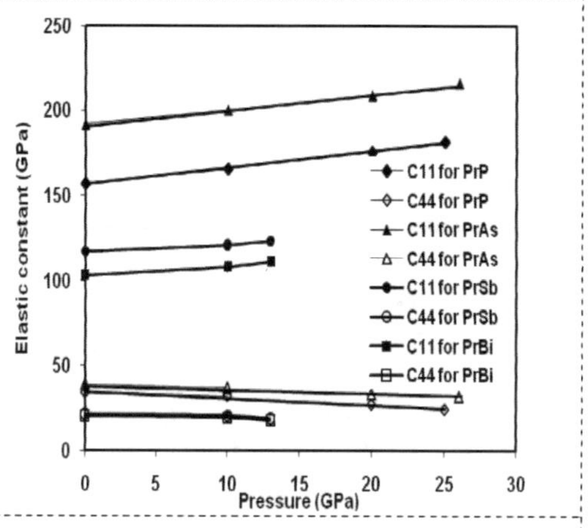


**Figure 2.** Variation of elastic constants with pressure for praseodymium monopnictides.

Fig.1 shows the variation of elastic constants with pressure for praseodymium monochalcogenides and in Fig. 2 the variation of elastic constants with pressure for praseodymium monopnictides. As the pressure increases in $B_1$ phase, $C_{11}$ increases linearly but $C_{44}$ decreases linearly this shows that $C_{44}$ is less sensitive to pressure as compared to $C_{11}$ [12]. There is no experimental data available in literature to compare our prediction.

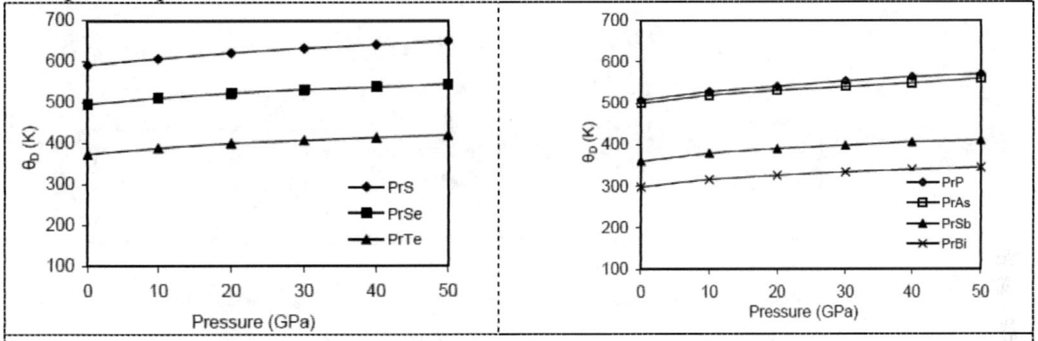


**Figure 3.** Variation of Debye temperature with pressure for (a) praseodymium monochalcogenides (b) praseodymium monopnictides.

The Debye temperature ($\theta_D$) is an important fundamental parameter closely related to many physical properties such as specific heat, melting temperature and thermal conductivity of solids. The Debye temperatures for praseodymium monochalcogenides and monopnictides are presented in Table 1. It is clear from Table 1 that the Debye temperature decreases with increasing lattice constant and atomic number of the praseodymium monochalcogenides and monopnictides. The calculated values of density ($\rho$ in Kg/m$^3$), the longitudinal, transverse and average sound velocity ($S_L$, $S_T$ and $S_M$ in m/s) for praseodymium monochalcogenides and monopnictides are also listed in Table 1.

We have also calculated the variation of $\theta_D$ with respect to pressure for praseodymium monochalcogenides and monopnictides and predicted them in Fig.3a and 3b. It is clear from the figure that $\theta_D$ increases as the pressure increases which shows quite normal behaviour for these compounds.

**Acknowledgments**

Authors are thankful to Madhya Pradesh Council of Science & Technology, Bhopal for Financial Assistance through Research Project (1904/CST/2009). One of the authors SPS acknowledges CSIR, New Delhi.

**4. References**

[1] Soni P, Pagare G, Srivastava V and Sanyal S P 2009 *Phase transitions* **82** 519-530
[2] Andres K and Bucher E 1971 *J. Appl. Phys.* **42** 1522
[3] Shirotani I, Hayashi J, Yamanashi K, Hirano K, Adachi T, Ishimatsu N, Shimomura O and Kikegawa T 2003 *Physica B: Condens. Matter* **334** 167-174
[4] Vaitheeswaran G, Petit L, Svane A, Kanchana V and Rajagopalan M 2004 *J. Phys.: Condens. Matter* **16** 4429-4440
[5] Kocak B, Ciftci Y O, Colakoglu K and Deligoz E 2011 *Comput. Mater. Sci.* **50** 1958-1964
[6] Singh R K 1982 *Phys. Rep.* **85** 259
[7] Hill R 1952 *Proc. Phys. Soc. London A 65* 349
[8] Sun Z, Li S, Ahuja R, Schneide J M 2004 *Solid State Commun.* **129** 589
[9] Wachter P, Filzmoser M, Rebizant J 2001 *Physica B* **293** 199
[10] Hullinger F 1979 *in Hand Book on the Physics and Chemistry of Rare Earths* Vol. 4, eds K A Gschneidner Jr and L Eyring (New York: North Holland) p 153
[11] Pugh S F 1954 *Philos. Mag.* **45** 823
[12] Singh A, Srivastava V, Aynyas M, and Sanyal S P 2009 *Physica B* **404** 1852-1857

Journal of Physics: Conference Series **377** (2012) 012081 doi:10.1088/1742-6596/377/1/012081

# Structural and electronic properties of Cd-rich lanthanide intermetallics

**Afroj A. Khan[1,2,*], Vipul Srivastava[2,3], M Rajagopalan[4] , Sankar P Sanyal[3]**

[1]Department of Engineering Physics, Sagar Institute of Science Technology & Engineering, Ratibad, Bhopal - 462044, India
[2]Department of Physics, Barkatullah University, Hoshangabad Road, Bhopal - 462 026, India
[3]Department of Engineering Physics, NRI Institute of Research & Technology, Raisen Road, Bhopal - 462 021, India
[4]Crystal Growth Centre, Anna University, Chennai - 600 025, India

[*]E-mail address: afroj8674@rediffmail,com

**Abstract.** The structural and electronic properties of B2-cadmium lanthanide (RE), CdLn (Ln=La, Ce and Pr) intermetallics have been calculated at T=0K at ambient and at high pressure by using tight binding linear muffin tin orbital method. We have estimated lattice parameter, bulk modulus, density of states for the three CdLn intermetallics. The *f*-electrons present in the –Ln, play important role. The variation in density of states under compression is calculated due to Pr-*f* effect in CdPr, which opens a possibility of structural instability.

## 1. Introduction.

Recently, in material design, intermetallics and alloys has a consideration for their significant application due to their ductile, high tensile strength and other properties [1-7]. The AB intermetallics, where 'A' is 'metal' and 'B' is eit her Lanthanide (Ln) or Actinide (Ac) or 'transition metal' (TM) are formed as unique class of materials. Particularly, Aluminium with -Ln and -TM systems have received a great attention due to their application in metallic glass technology [6,7]. The CdLn (Ln= La, Ce and Pr) intermetallics crystallize in body centered cesium chloride structure ($B_2$-phase, *Pm3m*, Space Group, *221*) [8, 9]. These intermetallics are very important with respect to their structural and electronic properties. One another interesting aspect of these materials is that they are having partially filled *f*- state. Those may get delocalized and give further possibility of phase change under pressure. Therefore, it becomes important to understand the electronic properties of this class of intermetallics. To the best of our knowledge the structural and electronic properties of CdLn intemetallics at normal and as well as at high pressure, have not been extensively studied and it is yet to be explored in details. We have, therefore, explored the structural and electronic properties of CdLn intermetallics under normal and compression by employing the non-spin polarized first-principles tight binding linear muffin tin orbital (TB-LMTO) method.

## 2. Method of calculations

The total energy, band structure and density of states for CdRE are calculated in a manner similar to our previous work [10-12] using TB-LMTO method [13,14] within the local-density approximation (LDA) [15]. von-Barth and Hedin [16] parameterization scheme was used for exchange correlation potential. The CdRE intermetallics crystallize in the $B_2$-type structure, and positioned at Cd: (0, 0, 0 ) and RE: (0.5, 0.5, 0.5). The tetrahedron method [17] of Brillouin zone integration was used to calculate the density of states. The total energy was computed by reducing the volume from 1.05 $V_0$ to 0.65 $V_0$, where $V_0$ is the equilibrium cell volume. The calculated total energy was fitted to Birch equation of state [18] to obtain the pressure volume relation. The pressure was obtained by taking volume derivative of the total energy. The bulk modulus, $B = -V_0\, dP / dV$ was also calculated from P-V relation.

## 3. Results and discussion

*3.1 Lattice parameter*

Total energy is calculated under compressions using TB- LMTO method for CdLn, intermetallics and is plotted in figure 1. The minimum of all the curves defines the equilibrium volume, $V_0$ corresponding to lattice parameter $a_0$ for CdLn. The present calculated values of lattice parameters are in good agreement with the Wycoff values [9], although the reported values of lattice parameters of CdLn intermetallics are slightly higher as compared to our results. This is due to usage of local density approximation (LDA) [15] in the present calculation. The bulk modulus for CdLa, CdCe and CdPr are calculated to be 64.77, 65.25 and 57.46 GPa, respectively.

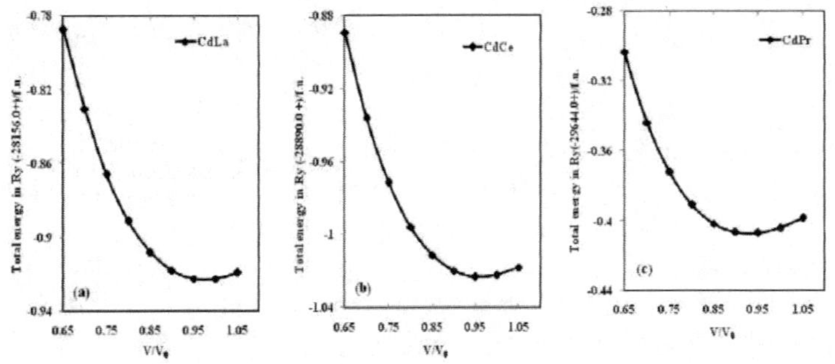


**Figure 1.** Variation of total energy as a function of relative volume for CdLn

*3.2 Band structure and density of states*

The band structures along the high symmetry directions for all the CdLn are shown in figure 2. The lowest energy bands are due "Cd-d" states. In C dLa, some of La-d states hybridize with Cd-p states, but small amount of La-*f* can be seen above Fermi level, but for Ce and Pr *f*-states can be seen at Femi level clearly and hybridize with 'Cd-*p*' states, which shows increased metallic character.

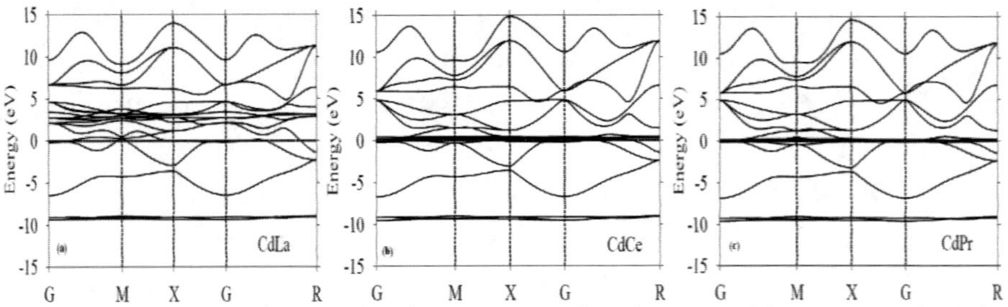


**Figure 2.** Band structure along the high symmetry directions at ambient pressure for CdLn intermetallics

To obtain clear picture of individual states, we have plotted partial and total density of states (DOS) for all CdLn intermetallics in figure 3. The 'Cd-*d*' states can be seen near energy value of -10eV. On the other hand, 'La-*f*' states can be seen in the region of 0-5eV in the figure 3. Localized 'Ce-*f*' and 'Pr-*f*' can be seen at Fermi level, which hybridizes with 'Cd-*p*' states.

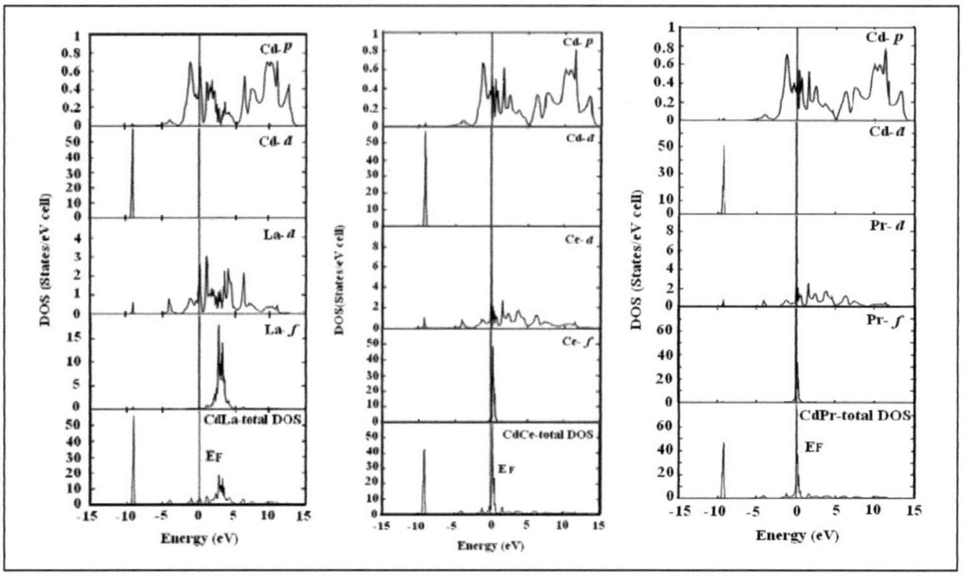

**Figure 3.** Partial density of states at Fermi level for CdLn intermetallics.

From the calculations, we have found that number of states increases as we go from –La to –Pr. In figure 4 we have plotted the variation in total DOS under compression for three CdLn intermetallics. From this figure one can notice a linear decrease in DOS at $E_f$ as we compress. The value of total DOS at $E_f$ for CdLa decreases and suddenly increases at $V/V_0 = 0.70$. In case of CdCe only decrease in DOS between compression value of $V/V_0 = 1.05$ to 0.65. As far as, such variation for CdPr is concerned, an increase in DOS is noticed at $V/V_0 = 1.05$ to 0.90 then decrease. Such interesting feature could be the indication of structural instability under compression. The variation of DOS under compression strongly related to the *f*-states of lanthanum atom. To understand such variation of DOS under compression in CdLa and CdPr, we have estimated the electronic DOS of *f* states and found variation in Pr-*f* states under compression. The reason can be understood that *f*-like electrons may get delocalized. The delocalization of the Pr-*f*-like states could be the reason of variation in total DOS at $E_F$ under compression, which has been justified by calculating electron localization function (ELF) ranging here 0.5-0.6.

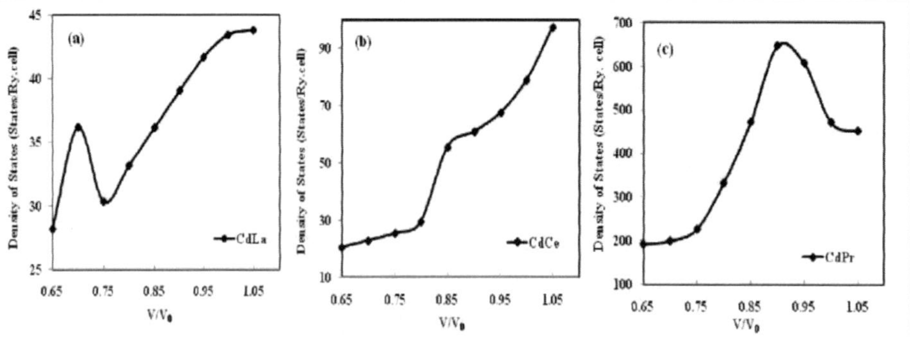

**Figure 4.** Variation of density of states at Fermi level with relative volume for CdLn

## 4. Conclusion

The CdLn intermetallics have been investigated theoretically with respect to their electronic aspect such as band structure, density of states, partial number of electrons, and other ground state properties in $B_2$ Phase also. It is found from present study that CdLn intermetallics are metallic in nature and metallicity increases from – La to –Pr. The variation in DOS number is noticed in CdLa and CdPr under compression, which may be due to the delocalization of 'f ' electrons under pressure.

**Acknowledgements**
AAK is thankful to the Chairman, Shri Sudhir Agrawal, SISTec-E and Secretary, Shri Siddharth Agrawal, SISTec-E Bhopal India for financial support. VS is thankful to MPCST, Bhopal, India for the award of research project (MPCST Project no. 1904/CST/R&D/2009) and for financial support. SPS gratefully acknowledges Council of Science and Industrial Research (CSIR), New Delhi, India. MR is thankful to CSIR, New Delhi, India for the award of emeritus Professor.

**References**
[1] Tao X, Ouyang Y, Liu H, Feng Y, Du Y and Lin Z 2008 *Solid State Commun.* **148** 314.
[2] Hu W, Xu H, Shu X, Yuan X, Gao B and Zhang B 2000 *J. Phys. D Appl. Phys.* **33** 711.
[3] Gschneidner Jr. K A, Ji Min, Wang C Z, Ho K M, Russell A M, Mudryk Ya, Becker A T, Larson J L 2009 *Acta Materialia* **57** 5876.
[4] Gschneidner Jr. K, Russell A, Pecharsky A, Morris J, Zhang Z, Lograsso T, Hsu D, Chester Lo C H, Ye Y, Slager A and Kesse D 2003 *Nat. Mater* **2** 587
[5] Divi M and Kuriplach J 1995 *Physica B* **205** 353.
[6] Tao X, Ouyang Y, Liu H, Zeng F, Feng Y, Jin Z 2007 *Physica B* **399** 27.
[7] Tao X, Ouyang Y, Liu H, Zeng F, Feng Y, Jin Z 2007 *Comp. Mater. Sci.* **40** 226 and references therein.
[8] Villars P and Calvert L D *1991 Pearson's Handbook of Crystallographic Data for Intermetallic Phases* vol. 1-4(ASM International, Materials Park /OH USA)
[9] Wyckoff R W G 1965 *Crystal Structures* 2nd edn (Inter-science)Vol **1-2** 196.
[10] Pagare G, Srivastava V, Sanyal S P and Rajagopalan M 2011 *Physica B* **406** 449.
[11] Srivastava V, Pagare G, Sanyal S P and Rajagopalan M 2009 *Physica Status Solidi B* **6** 206.
[12] Pagare G, Chohan S, Sanyal S P and Rajagopalan M 2011 *Physica B* **406** 339.
[13] Andersen O K 1975 *Phys. Rev. B* **12** 3060.
[14] Andersen O K, Jepsen O 1984 *Phys. Rev Lett.* **53** 2571.
[15] Kohn W, Sham L J 1965 *Phys. Rev A* **140** 1133.
[16] Barth U.van, Hedin L 1972 *J. Phys. C* **5** 1629.
[17] Jepsen O, Andersen O K 1971 *Solid State Commun* **9** 1763.
[18] Birch F., *J.* 1978 *Geophys. Rev.* **83**1257.

Journal of Physics: Conference Series **377** (2012) 012082 doi:10.1088/1742-6596/377/1/012082

# Effect of pressure on some physical properties of gallium based semiconductors

**P S Vyas[1], B Y Thakore[2], P N Gajjar[3] and A R Jani[2]**

[1] V. P. and R. P. T. P. Science College, Vallabh Vidyanagar 388 120, Gujarat, India
[2] Department of Physics, Sardar Patel University, Vallabh Vidyanagar, 388 120, Gujarat, India
[3] Department of Physics, University School of Sciences, Gujarat University, Ahmedabad 380 009, Gujarat, India

E-mail: p_s_vyas@yahoo.com

**Abstract**. The gallium based semiconductor compounds are very useful materials for optical spectroscopy and optoelectronic applications. we have studied the effect of pressure on various physical properties like total energy, static bulk modulus, energy band gap at the point X on the Jones-zone face, pressure derivative of bulk modulus and equation of state of gallium based binary compounds GaSb, GaAs, GaP and GaN using pseudopotential theory beyond second order with our well established single parametric model potential. We have incorporated Nagy's static local field correction function to include exchange and correlation effects. The results are compared with those obtained using few other local field correction functions. The present results agree satisfactorily with available experimental and other such theoretical data confirming the application.

## 1. Introduction

The gallium based semiconductor compounds and their ternary and quaternary alloys are very useful materials for optical spectroscopy and optoelectronic applications. Due to narrow energy gap, group III-V compounds are suitable to fabricate light emitting diodes, infrared detectors and infrared diode lasers, high speed switching devices and communication devices such as emitters, wave guides [1]. The second order perturbation theory based on the pseudopotential formalism, which has been successfully applied to simple metals, their alloys and glasses [2-5] is failed for covalent crystals. When the perturbation treatment is applying to the covalent semiconductors, it is seen that the pseudopotentials are large in these crystal compared with those in simple metals and there is an energy gap on the Jones-zone face corresponding to the Fermi surface. These facts are closely related to the covalent bonding character in covalent crystal, which comes from higher than second order terms and have important role for the stability of lattices of the crystals having zinc-blende structures. While in the second order perturbation theory the crystal are unstable for shear distortion.Therefore higher order terms are considered for covalent crystal to incorporate covalent bonding effect [6-8]. Soma [7, 8] used higher order perturbation theory to calculate total energy and bulk modulus of semiconductor elements and compounds. In the present paper, we have studied effect of pressure on various physical properties like total energy, static bulk modulus, energy band gap at the point X on the Jones-zone face, pressure derivative of bulk modulus and equation of state of gallium based binary compounds

GaSb, GaAs, GaP and GaN using pseudopotential theory beyond second order with our well established single parametric model potential [9, 10]. We have used Nagy's static local field correction function (N) [11] to include exchange and correlation effects. The results are compared with those obtained using few other local field correction functions due to Hartree (H) [12], Ichimaru et al (I) [13] and Sarkar et al (S) [14].

The paper is organized as follows. In section 1 we give a brief introduction. Section 2 contains a theoretical background for the calculation procedure. Section 3 covers discussion of the results. A summary of our conclusions is narrated in Section 4.

## 2. Theory

The bare-ion potential used in the present investigations is proposed in **r**-space [9, 10] is

$$W^{\text{ion}}(r) = -\frac{Ze^2}{2R_c}\left[2-\left(2-\frac{r}{R_c}-\frac{r^2}{R_c^2}\right)exp\left(\frac{-r}{R_c}\right)\right] \text{ for } r \le R_c$$
$$W^{\text{ion}}(r) = -\frac{Ze^2}{r} \text{ for } r \ge R_c \tag{1}$$

Here $Z$, $e$ and $R_c$ are valency, electronic charge and pseudo core radius respectively and $R_c$ is estimated through zero pressure condition. It is to be noted that to represent core-core repulsion usually Born-Mayer term is employed. We have also incorporated this feature in our potential through exponential factor. Inside the core, it is represented as the combination of linear and quadratic terms modulated by repulsive exponential factor, which tends to cancel Coulomb potential inside the core. While outside the core, it has the Coulombian tail. The smoothness of the bare-ion potential due to exponential term results in to faster conversions in **r**-space. It is continuous at the core.

For a covalent metallic crystal with zinc-blende structure, the total energy $E_T$ per atom using higher order perturbation theory based on pseudopotential method is given as [6, 7]

$$E_T = E_i + E_0 + E_1 + E_2 + E_{cov} \tag{2}$$

Here $E_i$ is the total Coulomb energy of positively charged pseudo metal ions in a uniformly negatively charged background; $E_0$ is the sum of kinetic, exchange and correlation energies of the uniform electron gas; $E_1$ and $E_2$ represent the first-order and second-order energies of the valence electron due to the pseudopotential, $E_{cov}$ is the covalent bond correction to the crystal energy corresponding to the third-order and fourth order terms.

To incorporate the exchange and correlation effects in the dielectric function, Nagy's static local field correction function [11] is used. It is given by,

$$f(q) = 1 - g(0,n) + \frac{cb}{c^2+q^2} - \frac{g(0,n)}{q}\tan^{-1} q/c \tag{3}$$

The detailed expressions of form factor, energy terms shown in equation (2) and Nagy's screening function (N) [11] are given elsewhere [6]. The local field correction functions proposed by H [12], I [13] and S [14] are also employed for comparative study with Nagy's screening function (N) [11].

The Jones zone of a zinc-blende type crystal is highly symmetrical with twelve {220} zone faces and does not deviate too much from a sphere. Heine and Jones [15] have shown that energy band gap at the point X ($k_x = (110)2\pi/a_0$) on the Jones-zone face is given for group III-V can be computed using

$$E_{gx} = 2\left|U_S(220) + \frac{a_0^2}{4\pi^2}\left[U_S(111)^2 + U_A(111)^2\right]\right| \tag{4}$$

**Table 1.** Input parameters

| Compound | Structure | Atomic Volume (a.u.)$^3$ |
|---|---|---|
| GaSb | Zinc-blende | 193.20 |
| GaAs | Zinc-blende | 152.50 |
| GaP | Zinc-blende | 136.70 |
| GaN | Zinc-blende | 76.90 |

## 3. Results and Discussion

The input parameters are given in the Table 1. In the present study we have considered only zinc-blende structures of GaSb, GaAs, GaP and GaN. The crystal energy per electron and bulk modulus computed at zero pressure (equilibrium) for the said compounds using N [11], H [12], I [13] and S [14] along with the experimental findings are shown in table 2. It is seen that H [12] produced higher values of energy and bulk modulus while I [13] produced lower values energy and bulk modulus in all screening functions. It suggests that inclusion of exchange and correlation effect is essential for energy calculations. The total energy and bulk modulus computed for all four compounds by N [11] are found in good agreement with the experimental data [8, 16].

**Table 2.** Total crystal energy and bulk modulus of GaAs, GaSb, GaP and GaN.

| Compound | $-E_T$ (Ryd / electron) | | | | | B (GPa) | | | | |
|---|---|---|---|---|---|---|---|---|---|---|
| | N [11] | H [12] | I [13] | S [14] | Exp. [8, 16] | N [11] | H [12] | I [13] | S [14] | Exp. [8, 16] |
| GaSb | 1.984 | 1.776 | 2.057 | 1.936 | 1.965 | 50.26 | 76.05 | 42.61 | 52.44 | 56.3 |
| GaAs | 2.155 | 1.932 | 2.232 | 2.107 | 2.157, 2.159 | 72.12 | 100.50 | 63.33 | 73.23 | 75.4, 74.8 |
| GaP | 2.237 | 2.010 | 2.316 | 2.189 | 2.221, 2.223 | 85.63 | 114.96 | 76.29 | 86.08 | 88.7 |
| GaN | 2.700 | 2.471 | 2.778 | 2.656 | 2.634, 2.635 | 215.48 | 247.42 | 202.32 | 210.19 | 210, 207 |

To study the effect of compression and expansion on the total energy, we have study volume dependency of total energy. The present results of total energy of all four gallium based binary compounds with relative volume using N [11] are shown in figure 1. It is seen that the variations in total energy with volume show same trend for all four compounds. GaN has produced lower value and GaSb has produced higher value of total energy per electron. The energy becomes minimum at equilibrium volumes. The pressure–volume relationship (equation of states) for all four compounds using N [11] is shown in figure 2. In the present study, we have considered compression up to 60%; which is now possible to achieve due to advancement in DAC techniques. It is found that on 60% compression, the pressure become 237 GPa, 316 GPa, 364 GPa and 813 GPa for GaSb, GaAs, GaP and GaN respectively. The pressure of GaN is largely influenced by the volume compression and GaSb showed minimum influence. Our findings of equation of state are compared with those obtained using Murnaghan [17] and Vinet [18] scheme and good agreement is observed.

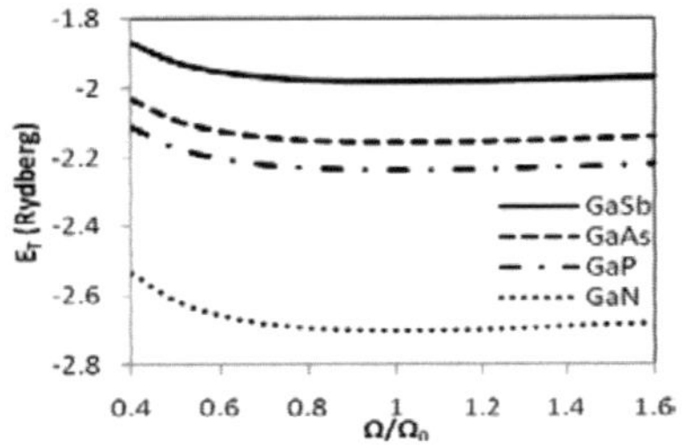


**Figure 1** Total energy as a function of relative volume for GaSb, GaAs, GaP and GaN.

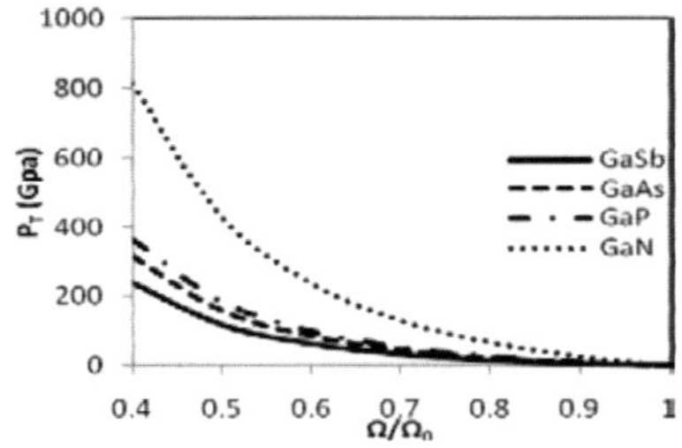


**Figure 2**Total pressure as a function of relative volume for GaSb, GaAs, GaP and GaN.

The variations of bulk modulus and energy band gap at point X on the Jones- zone face with pressure (up to 10 GPa) are shown in the figure 3 and 4 respectively. At zero pressure (equilibrium), present results of bulk modulus for all four compounds are agreed with the experimental findings [8]. The computed bulk modulus is increased linearly with the pressure show same trend for all four compounds and increased by 30-35 GPa on pressure changing from 0 to 10 GPa. The values of computed pressure derivatives of bulk modulus at zero pressure are 3.350, 3.193, 3.133 and 2.935 for GaSb, GaAs, GaP and GaN respectively. From figure 4, it is found that the energy band gap at point X on the Jones- zone face is increased by 0.003 to 0.005 eV as pressure is increased up to 10 GPa. Any noticeable change is not observed in $E_{gx}$ with pressure.

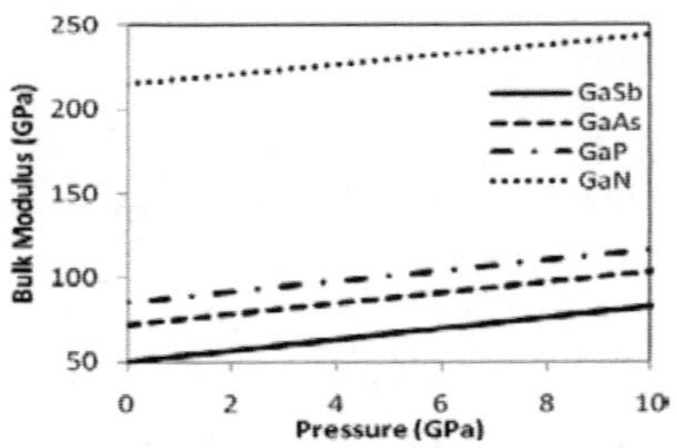


**Figure 3**Bulk modulus as a function of pressure for GaSb, GaAs, GaP and GaN.

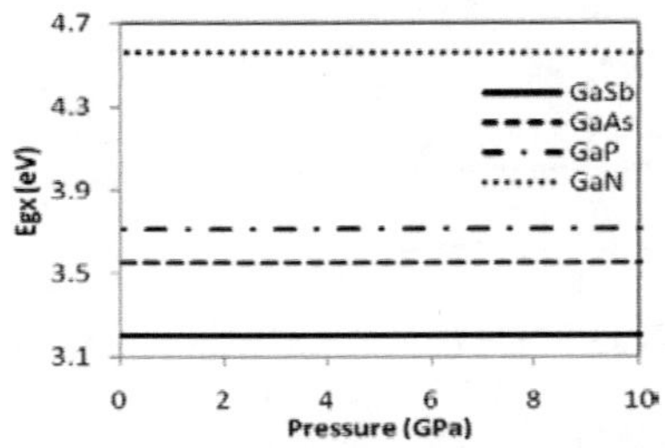


**Figure 4** Energy band gap at the point X on the Jones-zone face as a function of pressure for GaSb, GaAs, GaP and GaN.

## 4. Conclusions

In the present investigation it is found that total energies computed for all four gallium based binary compounds are matched with the experimental findings. GaN has produced lower value and GaSb has produced higher value of total energy per electron in all four gallium based binary compounds. The present findings of equation of state are found in good agreement with those computed by using Murnaghan [17] and Vinet [18] formalisms. The pressure of GaN is largely influenced by the volume compression in all four compounds. The bulk modulus is linearly increased with the pressure increased up to 10 GPa. But energy band gap at point X on the Jones-zone face is not much influenced by the pressure change up to 10 GPa. The pressure derivatives of bulk modulus are found matching with the experimental results with some deviation. Finally, we conclude that present results using our novel model potential are satisfactory and could be used for further comprehensive study of such system of semiconductor.

## Acknowledgements

The author P. S. Vyas is thankful to UGC, New Delhi, India for providing financial support under minor research project No. F. :47-651/08(WRO).

## References

[1] Bourissa N 2003 *Mater. Sci. and Eng.* B **100** 280
[2] Vyas P S, Thakore B Y, Gajjar P N and Jani A R 2008 *Commun. Theor. Phys.* **50** 763
[3] Gajjar P N, Thakore B Y and Jani A R 1998 *Acta Phys. Polo.* A **94** 33
[4] Shvets V T and Belov E V 1999 *Acta Phys. Polo.* A **96** 741
[5] Thakore B Y, Gajjar P N and Jani A R 2000 *Bull. Mater. Sci.* **23** 5
[6] Vyas P S, Thakore B Y, Gajjar P N and Jani A R 2010 *Commun. Theor. Phys.* **54** 573
[7] Soma T 1978 *Phys. Stat. Sol.* (b) **86** 263
[8] Soma T 1978 *J. Phys.* C **11** 2669
[9] Vyas P S, Thakore B Y, Gajjar P N and Jani A R 2010 *Am. Inst. Phys. Conf. Publ.* **1313** 379
[10] Vyas P S, Thakore B Y, Gajjar P N and Jani A R 2011 *Physica* B **406** 4412
[11] Nagy I 1986 *J. Phys.* C **19** 481
[12] Harrison W A 1998 *Elementary electronic structure* (Singapore World Scientific) pp 39
[13] Utsumi K and Ichimaru S 1980 *Phys. Rev.* B **22** 5203
[14] Sarkar A, Sen D, Haldar S and Roy D 1998 *Modern Phys. Lett.* B **12** 639
[15] Heine V and Jones R O 1969 *J. Phys.* C **2** 719
[16] Jivani A 2010 *Study of certain physical properties of semiconductors from group IV, III-V, II-VI, and their alloys using pseudopotential theory* (Vallabh Vidyanagar: Sardar Patel University Press)
[17] Murnaghan F D 1944 *Proc. Nat. Acad. Sci.* **30** 244
[18] Vinet P, Ferrante J, Smith J R and Rose J H 1986 *J. Phys.* C **19** L467

# Effect of uniaxial strain on plasmon excitations in graphene

**F. M. D. Pellegrino**[1,2], **G. G. N. Angilella**[1,2,3,4], **R. Pucci**[1,2]

[1] Dipartimento di Fisica e Astronomia, Università di Catania, Via S. Sofia, 64, I-95123 Catania, Italy
[2] CNISM, UdR di Catania, Via S. Sofia, 64, I-95123 Catania, Italy
[3] Scuola Superiore di Catania, Università di Catania, Via Valdisavoia, 9, I-95123 Catania, Italy
[4] INFN, Sez. di Catania, Via S. Sofia, 64, I-95123 Catania, Italy

E-mail: giuseppe.angilella@ct.infn.it

**Abstract.** Uniaxial strain is known to modify significantly the electronic properties of graphene, a carbon single layer of atomic width. Here, we study the effect of applied strain on the composite excitations arising from the coupling of charge carriers and plasmons in graphene, *i.e.* the plasmarons. Specifically, we predict that the plasmaron energy dispersion, which has been recently observed experimentally in unstrained graphene, is shifted and broadened by applied uniaxial strain. Thus, strain constitutes an additional parameter which may be useful to tune graphene properties in plasmaronic devices.

Graphene is a monoatomic carbon layer with honeycomb structure. After having been considered for decades as the ideal constituent of most compounds of carbon in the $sp^2$ hybridization state, it has been recently obtained in the laboratory [1], thus kindling intense research activity both on the experimental and on the theoretical side. Graphene is especially characterized by a quasiparticle band structure consisting of two bands, touching at the Fermi level in a linear, cone-like fashion at the so-called Dirac points $\pm\mathbf{K}$, and a linearly vanishing density of states (DOS) at the Fermi level [2, 3]. These peculiar electronic properties, along with the reduced dimensionality, have remarkable effects on the electromagnetic properties of graphene. These include, *e.g.,* the reflectivity [4], the optical conductivity [5, 6, 7, 8, 9], the plasmon dispersion relation [10, 11, 12, 13], as well as a newly predicted transverse electromagnetic mode [14], which is characteristic of a 2D system with a double band structure, such as graphene.

The composite elementary excitations arising from the coupling of charge carriers and plasmons, the so-called plasmarons, have been considered in a general context earlier on by Lundqvist [15, 16]. Recently, plasmarons have been experimentally observed in graphene by means of angular resolved photoemission spectroscopy (ARPES) [17], and their dispersion relation described theoretically within an improved version of the random phase approximation ($G_0W$-RPA) [18].

In $n$ doped graphene, a plasmaron mode with momentum $\mathbf{k}$ results from the relatively strong coupling of a quasihole with momentum $\mathbf{k}+\mathbf{q}$, and a plasmon with momentum $-\mathbf{q}$, the quasihole-plasmon coupling being stronger when the two excitations have the same group velocity [18].

At $\mathbf{k} = 0$, the plasmaron relative momentum modulus turns out to be

$$q = \frac{e^2}{8\pi\epsilon} \frac{\mu}{(\hbar v_F)^2}, \tag{1}$$

where $\mu$ is the chemical potential, $v_F$ the Fermi velocity, $\epsilon = \epsilon_0 \epsilon_r$ the dielectric constant. Therefore, the plasmaron binding energy with respect to the Fermi energy can be estimated, in first approximation, as the sum of the energies of the bare quasihole and plasmon, both having momentum modulus $q$, *viz.*

$$E_P = -\mu - \alpha \frac{c}{v_F} \frac{\mu}{2\epsilon_r}, \tag{2}$$

where $\alpha$ is the fine structure constant. In the realistic case of graphene on a $SiO_2$ substrate, Eq. (2) yields $E_P \simeq -1.25\mu$. A more accurate estimate, including the contribution of the quasihole-plasmon interaction at the $G_0W$-RPA level [18], yields $E_P \simeq -1.3\mu$, in better agreement with the experimental results [17].

We now consider the effect of uniaxial strain on the graphene sheet. It has been shown that this amounts to a shift of the Dirac points in reciprocal space, and to an anisotropic deformation of the Dirac cone centred at those points [19], with elliptic sections at constant energy. Such a strain-induced angular dependence may also be nonuniform, as in the case of coordinate-dependent strain [20]. The effect of applied uniaxial strain on the plasmon dispersion relation of graphene has been studied in Refs. [12, 13]. Besides, is has been shown [19] that strain may significantly modify the dispersion relation of a transverse plasmon mode, which has been recently predicted to occur in graphene [14]. Therefore, it can be expected, on general grounds, that strain affects the energy dispersion of the plasmaronic modes.

Denoting by $\varepsilon$ the modulus of applied strain, $\theta$ its direction (with $\theta = 0$ and $\theta = \pi/6$ referring to strain along the zig zag and armchair directions, respectively), and $\nu$ the Poisson's ratio of graphene ($\nu = 0.14$ [21], to be compared with the known experimental value $\nu = 0.165$ for graphite [22], and with $\nu = -1$, corresponding to the hydrostatic limit), following Ref. [19], the quasiparticle dispersion relation, to linear order in $\varepsilon$, reads

$$\epsilon_{\mathbf{q}} = \pm \hbar v_F q [(1 - \kappa(1-\nu)\varepsilon) - \kappa(1+\nu)\varepsilon \cos(2\theta + 2\varphi)], \tag{3}$$

where the + (−) sign refers to the conduction (valence) band, and while the plasmon dispersion relation under strain becomes

$$\hbar\omega_{\text{pl}} = \sqrt{\frac{e^2}{2\pi\epsilon}\mu} \,[1 - \kappa(1+\nu)\varepsilon \cos(2\theta - 2\varphi)] \sqrt{q}. \tag{4}$$

Deriving the corresponding group velocities from the above Eqs. (3) and (4), Eqs. (1) and (2) for the plasmaron momentum and energy, respectively, get modified into

$$q = [1 + 2\kappa(1-\nu)\varepsilon] \frac{e^2}{8\pi\epsilon} \frac{\mu}{(\hbar v_F)^2}, \tag{5a}$$

$$E_P(\varphi) = -\mu - \alpha \frac{c}{v_F} \frac{\mu}{2\epsilon_r} [1 + \kappa(1-\nu)\varepsilon - \kappa(1+\nu)\varepsilon \cos(2\theta - 2\varphi)], \tag{5b}$$

to linear order in the strain modulus $\varepsilon$. Here, $\kappa = (a/2t)|\partial t/\partial a| - \frac{1}{2} \approx 1.1$ is related to the logarithmic derivative of the nearest-neighbor hopping $t$ at $\varepsilon = 0$, $a$ is the carbon–carbon distance, and $\theta$ is the direction of the stress.

Eq. (5b) shows that, in the presence of applied uniaxial strain, the plasmaronic energy at $\mathbf{k} = 0$ acquires an explicit dependence on the angle $\varphi$ of the quasihole momentum $\mathbf{q}$. This is due to the anisotropy of both the electronic and the plasmon spectrum. Correspondingly, the plasmaron energy is characterized by a central value

$$E_P^c = -\mu - \alpha \frac{c}{v_F} \frac{\mu}{2\epsilon_r} [1 + \kappa(1-\nu)\varepsilon], \tag{6}$$

and a strain-induced energy spread

$$\Delta E_P = \alpha \frac{c}{v_F} \frac{\mu}{\epsilon_r} \kappa(1+\nu)\varepsilon. \tag{7}$$

Considering again the realistic case of graphene on a $SiO_2$ substrate, one can estimate the central plasmaron energy in the unstrained case as $E_P^c(\varepsilon = 0) = -1.25\mu = -125$ meV, for $\mu = 100meV$, with zero energy spread. Correspondingly, in the case of an applied strain $\varepsilon = 10$ %, one finds a central plasmaron energy of $E_P^c(\varepsilon = 10\ \%) = -127.4$ meV, with an energy spread $\Delta E_P(\varepsilon = 10\ \%) = 6.27$ meV.

In conclusion, the effect of applied uniaxial strain on graphene is therefore that of shifting and broadening the plasmaron energy, proportionally to the strain modulus. Therefore, by suitably applying uniaxial strain, one gains further control on the energy of the plasmaronic excitation, besides the possibility of tuning the relative dielectric constant $\epsilon_r$ [23]. This may be instrumental for the realization of 'plasmaronic' devices.

## References

[1] Novoselov K S, Jiang D, Schedin F, Booth T J, Khotkevich V V, Morozov S V and Geim A K 2005 *Proc. Nat. Acad. Sci.* **102** 10451
[2] Castro Neto A H, Guinea F, Peres N M R, Novoselov K S and Geim A K 2009 *Rev. Mod. Phys.* **81** 000109
[3] Abergel D S L, Apalkov V, Berashevich J, Ziegler K and Chakraborty T 2010 *Adv. Phys.* **59** 261
[4] Nair R R, Blake P, Grigorenko A N, Novoselov K S, Booth T J, Stauber T, Peres N M R and Geim A K 2008 *Science* **320** 1308
[5] Kuzmenko A B, van Heumen E, Carbone F and van der Marel D 2008 *Phys. Rev. Lett.* **100** 117401
[6] Wang F, Zhang Y, Tian C, Girit C, Zettl A, Crommie M and Shen Y R 2008 *Science* **320** 206
[7] Mak K F, Sfeir M Y, Wu Y, Lui C H, Misewich J A and Heinz T F 2008 *Phys. Rev. Lett.* **101** 196405
[8] Stauber T, Peres N M R and Geim A K 2008 *Phys. Rev. B* **78** 085432
[9] Pellegrino F M D, Angilella G G N and Pucci R 2010 *Phys. Rev. B* **81** 035411
[10] Hwang E H and Das Sarma S 2007 *Phys. Rev. B* **75** 205418
[11] Abedinpour S H, Vignale G, Principi A, Polini M, Tse W K and MacDonald A H 2011 *Phys. Rev. B* **84** 045429
[12] Pellegrino F M D, Angilella G G N and Pucci R 2010 *Phys. Rev. B* **82** 115434
[13] Pellegrino F M D, Angilella G G N and Pucci R 2011 *High Press. Res.* **31** 98
[14] Mikhailov S A and Ziegler K 2007 *Phys. Rev. Lett.* **99** 016803
[15] Lundqvist B I 1967 *Phys. kondens. Materie* **6** 193
[16] Lundqvist B I 1967 *Phys. kondens. Materie* **6** 206
[17] Bostwick A, Speck F, Seyller T, Horn K, Polini M, Asgari R, MacDonald A H and Rotenberg E 2010 *Science* **328** 999
[18] Polini M, Asgari R, Borghi G, Barlas Y, Pereg-Barnea T and MacDonald A H 2008 *Phys. Rev. B* **77** 081411(R)
[19] Pellegrino F M D, Angilella G G N and Pucci R 2011 *Phys. Rev. B* **84** 195407
[20] Pellegrino F M D, Angilella G G N and Pucci R 2011 *Phys. Rev. B* **84** 195404
[21] Farjam M and Rafii-Tabar H 2009 *Phys. Rev. B* **80** 167401
[22] Blakslee O L, Proctor D G, Seldin E J, Spence G B and Weng T 1970 *J. Appl. Phys.* **41** 3373
[23] Walter A L, Bostwick A, Jeon K J, Speck F, Ostler M, Seyller T, Moreschini L, Chang Y J, Polini M, Asgari R, MacDonald A H, Horn K and Rotenberg E 2011 *Phys. Rev. B* **84** 085410

Journal of Physics: Conference Series **377** (2012) 012084 doi:10.1088/1742-6596/377/1/012084

# Study of Bulk modulus, Volume, Energy, lattice parameters and magnetic moments in rare earth hexaborides using density functional theory

**Sandeep[1], M. P.Ghimire[2, 5], D. P. Rai[1], P. K. Patra[3], A. K. Mohanty[4] and R. K. Thapa[1]**
[1]Condensed Matter Theory Group, Department of Physics, Mizoram University, Aizawl, Mizoram, India 796009
[2]Faculty of Science, Nepal Academy of Science and Technology, Kathmandu, Nepal.
[3]Science Centre, North-Eastern Hill University, Shillong. Meghalaya, India.
[4]Nuclear Physics Division, Bhaba Atomic Research Centre, Mumbai, India.
[5]Condensed Matter Physics Research Center, Butwal, Rupendehi, Nepal.

E-mail: sndp.chettri@gmail.com

**Abstract.** We have calculated the theoretical lattice parameters, Bulk modulus, volume, energy, lattice parameters and magnetic moments for $RB_6$ (R=La, Ce, Pr and Sm) of $CaB_6$ type crystal structure with space group *Pm3m* using full potential linearized augmented plane wave (FP-LAPW) method. The bulk modulus was found to be 9.56 % higher for $LaB_6$ and 2.4% lower for $CeB_6$ compared to the experimental results Gupta *et al.* [4] and Ogita *et al* [5]. Magnetic moments for $LaB_6$, $CeB_6$ were found in qualitative agreement with the earlier reported results. The results based on generalized gradient approximation (GGA) were found and compared with local spin density approximation (LSDA) results for $CeB_6$ and $SmB_6$ as well.

## 1. Introduction

Rare earth hexaborides have unusual combination of properties with metallic conductivity and low work function as well as low volatility at temperatures providing technologically useful thermionic electron current density, micro-beam applications [1]. Hexaborides of the rare-earth elements are considered for applications as wear- and corrosion-resistant hard coatings for decoration of consumer products such as eye-glass frames and wristwatch casings etc [2]. We have calculated the theoretical lattice parameters, Bulk modulus, volume, energy, lattice parameters and magnetic moments for $RB_6$ (R=La, Ce, Pr and Sm) of $CaB_6$ type crystal structure using full potential linearized augmented plane wave (FP-LAPW) method.

## 2. Computational Details

We have performed our calculations using the theoretically optimized lattice parameters and the atomic positions Wyckoff (1965) [3] for $RB_6$. $RB_6$ have the $CaB_6$ type crystal structure and may be viewed as a CsCl-type lattice with the Cs replaced by a R ion and the chlorine by a $B_6$ octahedron. The atomic arrangement is described in terms of the space group *Pm3m*. Since R 4*f*-orbitals are rather localized, the 4*f*-electron correlations are expected to be strong. Consequently, the LSDA+U calculations [4] have been chosen to include the on-site Coulomb interaction. The values of on-site Coulomb energies (U) used are 5.00 eV for $LaB_6$ and $CeB_6$ [5], 6.00 eV for $PrB_6$ and $NdB_6$, 7.00 eV for $SmB_6$ [6] and the exchange parameter (J) 1.00 eV was chosen fixed as the usual value for R respectively. The density plane cut-off $R_{MT}*K_{MAX}$ is chosen to be 7.0, where $K_{MAX}$ is the plane wave cut-off and $R_{MT}$ is the muffin-tin radii. This amounts 419, 407, 408, 413 and 416 plane waves for $LaB_6$, $CeB_6$, $PrB_6$, $NdB_6$ and $SmB_6$ respectively. To provide a reliable Brillouin zone integration, a set of 120, 84, 165, 120 and 120 *k* points for $LaB_6$, $CeB_6$, $PrB_6$, $NdB_6$ and $SmB_6$ respectively in the irreducible wedge of the Brillouin zone (IBZ) were used. The self-consistency was better than 0.0001 e/a.u.$^3$ for charge density. The stability was better than 0.01 mRy for total energy per cell. For computations, WIEN2k code [7] is used.

## 3. Results and Discussions

The calculated lattice parameters are compared with the previous results in table 1.

Table 1. Lattice constant and total magnetic moment of rare earth hexaborides.

| | Lattice Constant (Å) | | |
|---|---|---|---|
| | Our Results | Experimental Results | Previous results |
| $LaB_6$ | 4.176 | 4.156 | 4.138[24]<br>4.185[24] |
| $CeB_6$ | 4.144 | 4.141 | 4.084[24] |
| $PrB_6$ | 4.151 | 4.121 | |
| $NdB_6$ | 4.157 | 4.128 | |
| $SmB_6$ | 4.172 | 4.133 | 4.111[24]<br>4.156[24] |

The theoretically calculated Bulk modulus, volume and energy with earlier results are tabulated for $RB_6$ in table 2.

Table 2 Bulk modulus, volume, Pressure derivative and energy of $RB_6$

| | Bulk Modulus (Present) | Bulk Modulus Theoretical LDA | Bulk Modulus Theoretical GGA | Bulk Modulus Expt. | Vol. | Energy |
|---|---|---|---|---|---|---|
| | (GPa) | (GPa) | (GPa) | (GPa) | (a.u.$^3$) | (Ryd) |
| $LaB_6$ | 189.5336 | 178.8[11] | 166.2 [11] | 164 [9]<br>173 [9] | 491.2956 | -17293.8 |
| $CeB_6$ | 162.1642 | 173 [11] | | 166 [10] | 480.2547 | -18029.2 |
| $PrB_6$ | 165.8493 | | | | 482.5787 | -18784.0 |
| $SmB_6$ | 146.6851 | 179.6[11] | 167.2 [11] | | 489.0972 | -21168.0 |

The volume versus energy curve was obtained by fitting the calculated values with the Murnaghan equation [8] to calculate the ground state lattice parameters for these compounds [Table 1]. The bulk modulus for $LaB_6$ was calculated to be 189.5336 GPa and was 9.56% higher than the experimental value of 173 GPa as determined by Ogita *et al.* [9] in the study of Raman scattering spectra for $RB_6$. We have observed a decrease of 2.4 % in the calculated Bulk modulus for $CeB_6$ (162.1642 GPa) when compared to the experimentally derived value by Leger *et al.* [10]. We have observed that the bulk modulus for $LaB_6$ was higher compared to the earlier experimental [9] as well as theoretical LDA and GGA results [11]. On the other hand we noticed found higher values of bulk modulus for $CeB_6$ [10, 11] and $SmB_6$ [11] compared to the earlier theoretical and experimental results.

Fig. 1 (a-d) shows the volume optimization curve for $LaB_6$, $CeB_6$, $PrB_6$ and $SmB_6$ respectively. The high value of the $LaB_6$ may be an indication of charge transfer from La to B atoms. It seems that the form of the GGA approach does not affect the results and they give closer values compared to those obtained by other. Therefore, we may conclude that the LSDA do not improve the quality of the structural properties, but rather it gives worse, especially for the bulk modulus. Thus the inclusion of the LSDA in the calculations of the structural properties of $RB_6$ is not recommended, even if it is known that the LSDA may play an important role in the accuracy of the electronic properties of heavy compounds. We have observed zero magnetic moment for $LaB_6$ as was reported by earlier studies of Gupta *et al.* [11] and Singh *et al.* [12]. The magnetic moment for $CeB_6$ was calculated to be 1.03 $\mu_B$ which is also in agreement with the results of Singh *et al.* [12]. This was found slightly on the higher side when compared with the value obtained by Kubo *et al.* [13], Biasini *et al.* [14] and Gupta *et al* [11]. The moment for $PrB_6$ was calculated to be 0.09% less than the value reported by Singh *et al* 12]. The $NdB_6$ moment was calculated to be 0.15% higher than the value reported by Singh *et al* [12]. The $SmB_6$ was found to have a moment of 5.87 $\mu_B$ which agrees well with the result obtained by Gupta *et al.* [11] and is higher than the moment calculated by Singh *et al* [12] by 4.2%.

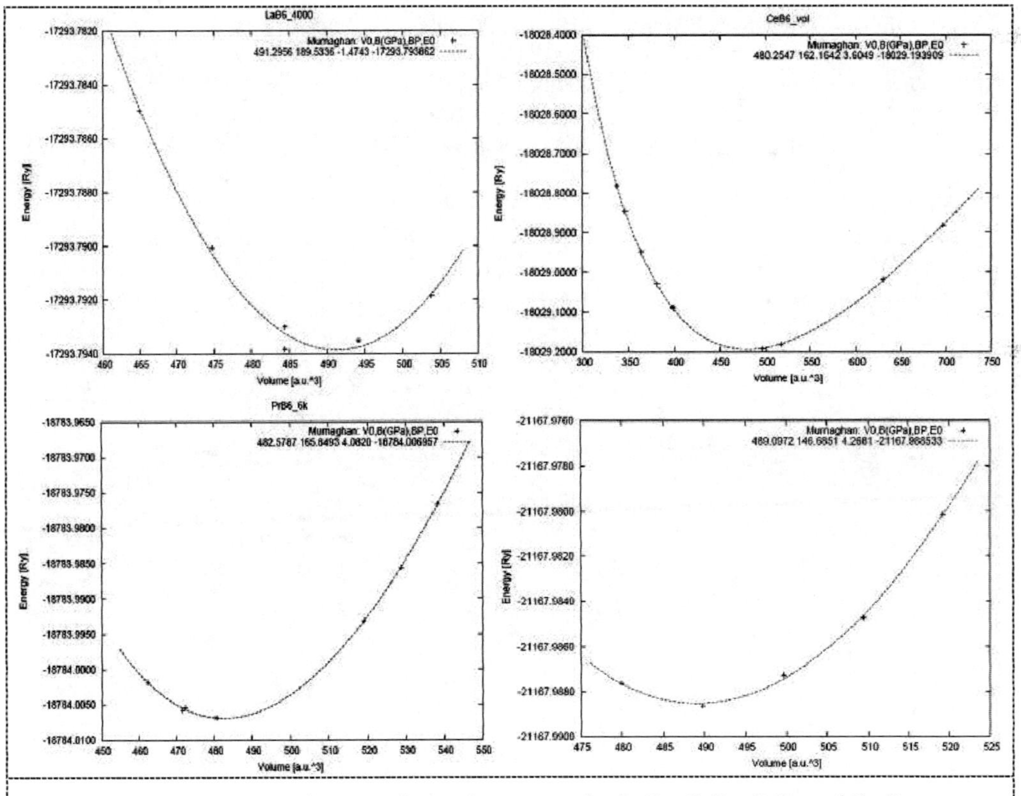


**Figure 1 (a-d)**: Volume optimization curves for $LaB_6$, $CeB_6$, $PrB_6$ and $SmB_6$.

**4. Conclusions** The spin polarized full potential linerized augmented plane wave method was employed for calculations of the bulk modulus, volume and total energy of $RB_6$. The bulk modulus was found to be 9.56 % higher for $LaB_6$ and 2.4% lower for $CeB_6$ compared to the experimental results Gupta *et al.*[11] and Ogita *et al* [9]. Magnetic moments for $RB_6$ were also found in qualitative agreement with the earlier reported results.

## 5. References

[1] Davis P R, Swanson L W, Hutta J J and Jones D L 1986 *J. Mater. Science* **21** 825.
[2] Mitterer C, Waldhauser W, Beck U and Reiners G 1996 *Surface and Coatings Technology*, **86-87** 715.
[3] Wyckoff R W G (ed) (1965) Crystal Structures vol 2 (Tucson, AZ: University of Arizona)
[4] Anisimov V I, Aryasetiawan F and Lichtenstein A I 1997 *J. Phys.: Condens. Matter* **9** 767
[5] Gschneidner K A Jr and Eyring L (ed) (1995) *Handbook on the Physics and Chemistry of Rare Earths* vol 20 (Amsterdam: North-Holland)
[6]Antonov V N, Harmon B N and Yaresko A N 2002 *Phys. Rev. B* **66** 165209
[7]Blaha P, Schwartz K, Madsen G K H, Kvasnicka D, Luitz J 2008: *An Augmented Plane Wave+Local Orbitals Program for Calculating Crystal Properties.* (Vienna University of Technology: Inst. of Physical and Theoretical Chemistry, Getreidemarkt 9/156, A-1060 2008, Vienna / Austria)
[8] Murnaghan F D 1944 *Proc. Natl. Acad. Sci. USA*, **30** 244.
[9] Ogita N, Nagai S, Okamoto N and Udagawa M 2003 *Phys. Rev. B* **68** 224 305.
[10] Leger J M, Rossat-Mignod J, Kunii S and Kasuya T 1985 *solid state communication* **54** 995
[11]Gupta H C and S. Kumar 2010 *AIP Conf. Proc. Xxii International Conference On Raman Spectroscopy* Vol 1267 p. 856-857
[12] Singh N, Saini S M, Nautiyal T and Auluck S 2007 *J. Phys.: condens. Matter* **19** 346226
[13] Kubo Y, Asano S, Harima H and Yanase A 1993 *J. Phys. Soc. Japan* **62** 205
[14] Biasini M, Fretwell H M, Dugdale S B, Alam M A, Kubo Y, Harima H and Sato N 1997 *Phys Rev. B* 56 10192.

Journal of Physics: Conference Series **377** (2012) 012085 doi:10.1088/1742-6596/377/1/012085

# Melting curve of metals using classical molecular dynamics simulations

**Karabi Ghosh**[1]
Theoretical Physics Division, Bhabha Atomic Research Centre,
Trombay, Mumbai-400085, India

E-mail: karabi@barc.gov.in

**Abstract.** The melting curves of Cu and Al have been generated using classical molecular dynamics simulation. The embedded atom method potential of Cai and Ye has been used to account for the interactions between atoms. The results obtained for both the metals agree well with Diamond Anvil Cell and with *ab initio* MD simulation results like the hysteresis or Z-method.

## 1. Introduction

The solid liquid phase boundary commonly known as the melting curve is an important part of the phase diagram and is widely used in high pressure physics, inertial confinement fusion, material science, astrophysics and geophysical sciences. Melting curves of metals can be obtained experimentally by the Diamond Anvil Cell (DAC) which is a static method or by the dynamic methods consisting of laser or ion beam driven ablation, etc. However it is difficult to design experiments because of the high pressures and temperatures involved. Various simulation methods like the Lindemann's law, dislocation mediated melting, classical and *ab initio* Molecular Dynamics (MD) etc. are in use to complement the experiments. The classical MD method [1] is very convenient as a large number of atoms compared to *ab initio* MD can be simulated using an empirical potential which properly describe the interatomic interactions. Pair potentials are insufficient for describing the properties of metals as the cohesive energy of a crystal is mainly described by the many-body interactions or by a volume dependent part [2]. Pure pair interactions also imply that the surface relaxes outwards instead of inwards as is observed in experiments. The embedded atom method (EAM) potential circumvents these problems by viewing each atom in a solid as embedded in a host comprising all the other atoms. The total energy is the sum of the embedding and pair potential energy [3]. EAM potentials are generated by fitting to experimental data or results of *ab initio* calculations. In this paper we use the EAM potential generated by Cai and Ye [4] which has the least number of fitted parameters.

## 2. Computational details

### *2.1. Classical Molecular Dynamics Simulation*

We have generated the melting curves of Cu and Al by classical molecular dynamics simulations employing the parallel molecular dynamics simulation package DL_POLY [5]. The interaction

[1] To whom any correspondence should be addressed.

between atoms is accounted by the EAM potential proposed by Cai and Ye for all the calculations. The potential parameters for both Cu and Al are reported in this reference. We considered cubic super cells consisting of 11*11*11 conventional FCC unit cells, which correspond to 5324 atoms. Three dimensional parallelopiped periodic boundary conditions were applied on the super cells to eliminate the surface effects and reproduce the bulk properties. The Berendsen isothermal-isobaric (NPT) ensemble was used to achieve constant temperature and pressure conditions [6]. The relaxation times for the thermostat and barostat were 1 and 3 ps respectively. We used a total of 10,000 time steps where the first 4000 steps were used for equilibration and the remaining 6000 for statistical averaging. The time step used in all the cases was 1 fs. We have used the leapfrog scheme for integrating the Newton's equations of motion [7]. In order to ascertain that not more than one image of a particle interacts with another particle, the cut off distances for force calculations have been chosen to be smaller than half the size of the supercell. The cutoff distances used for Cu and Al are 10Å and 15Å, respectively. The initial zero pressure FCC Cu supercell is generated using the program genlat.f in utility of DL_POLY. The origin of the Cartesian coordinate system is taken to be the centre of the supercell and the Cu atoms are placed one after the other starting from the (-,-,-) octant towards the (+,+,+) octant. The initial FCC structure shown in figure 1 (shown using the visualization software xcrysden [8]), evolves to that in figure 2 once melting occurs.

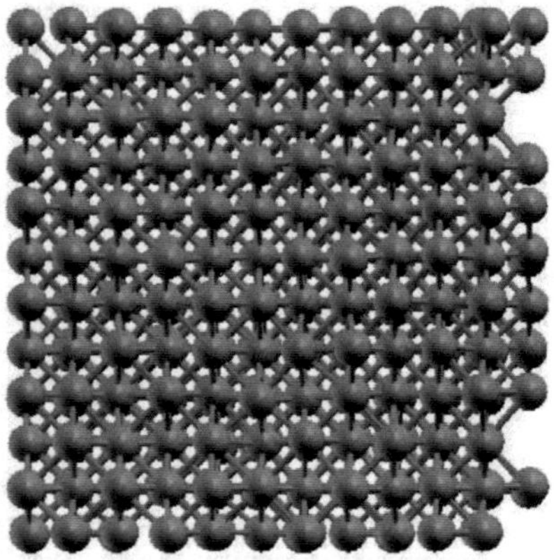

**Figure 1.** Initial FCC lattice structure for Cu

**Figure 2.** Configuration of atoms of Cu after melting.

The melting curve has been generated by determining the melting point at a specified pressure using the one-phase method [9] wherein the melting point is obtained from sharp increase in atomic volume, diffusion coefficient and energy as temperature is varied. Initially the temperature is increased by 50K until it melts and the range within which the jump occurs is observed. The structure obtained after equilibration at a certain temperature is used as the starting configuration for the subsequent higher temperature. The melting point for the specific pressure is next obtained by increasing the temperatures by 10K and observing the jump in the above mentioned quantities. As the long range order is lost once melting occurs, the height of the first peak of the radial distribution function (RDF) decreases drastically and the RDF~1 after 2-3 small humps as shown in figure 3. At a temperature of 1440K, the intermediate and long distance peaks slowly merge out indicating the onset of melting. The jump in diffusion coefficient for Cu is shown in figure 4 wherein the melting point at 1440K is shown using arrows. The whole melting curve is obtained within the range of pressure of interest by repeating the above steps.

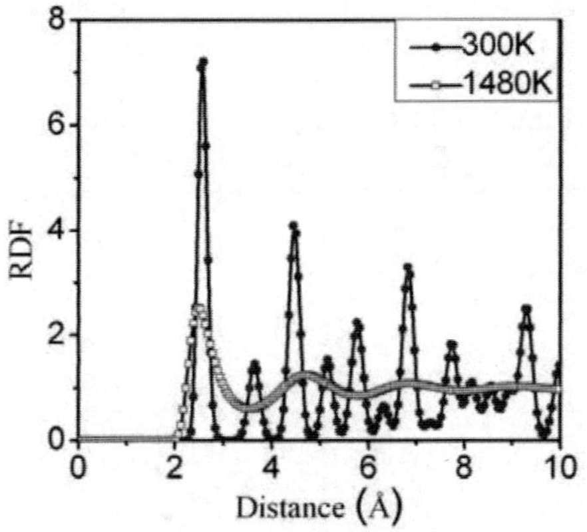


**Figure 3.** Radial distribution function of Cu for the initial FCC structure and that after melting.

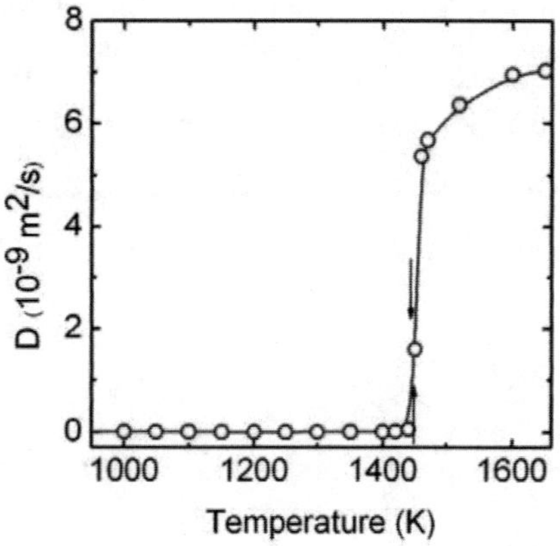


**Figure 4.** Jump in diffusion coefficient of Cu at the melting point.

*2.2. Embedded atom method (EAM) potential*

According to the EAM theory, the total energy of a system of N atoms is described as the energy required to embed these N atoms into the homogeneous electron gas caused by surrounding atoms plus a correction of energy from two-body interactions. Thus this total energy can be expressed as

$$E_{tot} = \sum_i F_i(\rho_i) + \sum_{i>j} \phi(r_{ij}) \quad [1]$$

where $E_{tot}$ is the total internal energy, $F_i(\rho_i)$ is the embedding energy required to place atom i in an electron density $\rho_i$, $\Phi(r_{ij})$ is the two-body potential between atoms i and j, $r_{ij}$ is the separation distance related to the specified pair of atoms i and j, and $\rho_i$ denotes the host electron density at atom i due to all other atoms and is given by

$$\rho_i = \sum_{j\neq i} f(r_{ij}) \quad [2]$$

where $f(r_{ij})$ is the electron density of the individual atom j which assumes an exponential form

$$f(r) = f_e \exp[-\chi(r - r_e)] \quad [3]$$

where $f_e$ is a scaling constant taken to be 1 for pure metals, $r_e$ is the equilibrium nearest neighbour distance and $\chi$ is an adjustable parameter. The embedding function is taken as the universal form suggested by Banerjea and Smith [10]

$$F(\rho) = -F_0[1 - \ln\left(\frac{\rho}{\rho_e}\right)^n]\left(\frac{\rho}{\rho_e}\right)^n + F_1\left(\frac{\rho}{\rho_e}\right) \quad [4]$$

where $\rho_e$ is the equilibrium host electron density, $F_0 = E_c - E^f_v$ is the difference between cohesive energy $E_c$ and vacancy formation energy $E^f_v$. Also, n=0.5 is a dimensionless constant and $F_1$ is a fitted parameter. The two-body potential is given by [11]

$$\phi(r) = -\alpha[1 + \beta(r/r_A - 1)]\exp[-\beta(r/r_A - 1)] \quad [5]$$

where $\alpha$,$\beta$ and $r_A$ are the fitted parameters. The fitted parameters $\chi$, $\alpha$, $\beta$, $r_A$ and $F_1$ are determined by minimizing the root square deviation ($X_{rms}$) between calculated and experimental thermodynamic data (ref [4] and references therein). For Cu and Al, the experimental data consists of the three elastic constants ($C_{11}$, $C_{12}$ and $C_{44}$), the vacancy formation energy ($E^f_v$), the equilibrium lattice constant ($a_0$), and the cohesive energy ($E_c$). During the fitting, the cutoff distance for the interaction is taken as

$r_{cut}$=1.65$a_0$ where $a_0$ is the lattice parameter. The fitting results and experimental data to which fitting is done along with the potential parameters for Cu and Al are displayed in table1.

**Table 1.** Values of the metal properties used in fitting and the potential parameters (Ref. [4]).

| | Cu | | Al | |
|---|---|---|---|---|
| | Calculated | Experimental | Calculated | Experimental |
| $a_0$ (Å) | 3.615 | 3.615 | 4.05 | 4.05 |
| $E_C$ (eV) | 3.52 | 3.54 | 3.32 | 3.36 |
| $C_{11}$ (ergs/cm$^3$) | 1.68 | 1.70 | 0.9 | 1.14 |
| $C_{12}$ (ergs/cm$^3$) | 1.263 | 1.225 | 0.702 | 0.619 |
| $C_{44}$ (ergs/cm$^3$) | 0.752 | 0.758 | 0.330 | 0.316 |
| $E^f_v$ (eV) | 1.31 | 1.30 | 0.73 | 0.75 |
| $X_{rms}$ (%) | 0.13 | - | 6.65 | - |
| | Potential | parameters | | |
| $\chi$ (Å$^{-1}$) | 3.0 | - | 2.5 | - |
| $\alpha$ (eV) | 0.3902 | - | 0.0834 | |
| $\beta$ | 6.0641 | - | 7.5995 | - |
| $F_1$ (eV) | 1.0241 | - | -0.1392 | - |
| $r_A$ (Å) | 2.3051 | - | 3.0169 | - |

This EAM potential explains the stability of the crystal structure of Cu and Al over a wide range. The Equation of state (EOS) of the total energy as a function of the lattice constant of Cu and Al obtained from this EAM potential is in good agreement with those from the EOS of Rose *et al.* which has been experimentally verified from static and shock wave experiments [11].

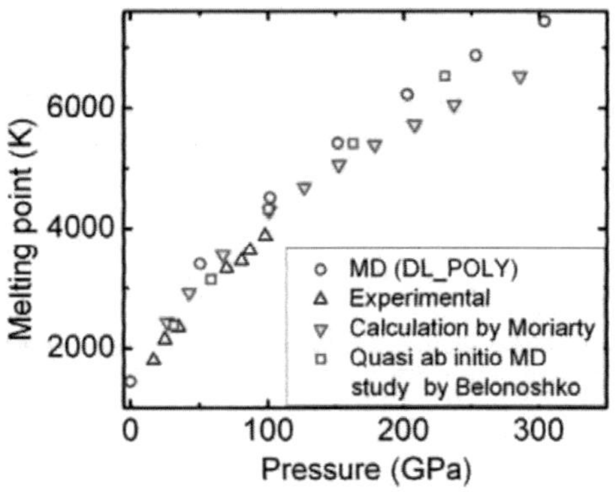


**Figure 5.** Melting curve for Cu.

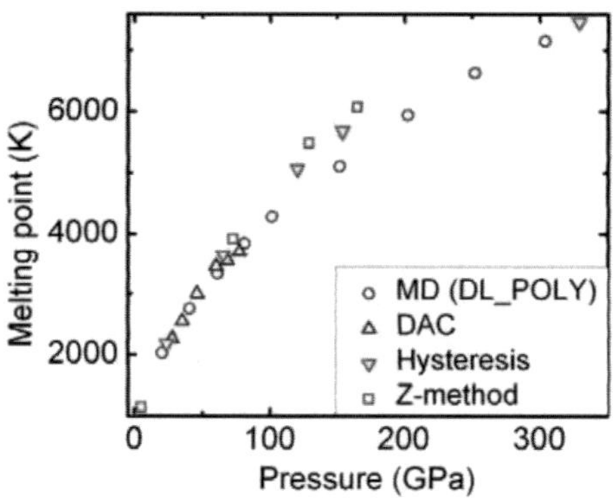


**Figure 6.** Melting curve for Al.

### 3. Results and discussions

The melting curves obtained for Cu and Al using MD simulations are plotted in figure 5 and figure 6 respectively. The melting curve obtained from our MD simulation for Cu is found to agree well with the quasi *ab initio* MD study by Belonoshko [12]. Belonoshko *et al.* have investigated the melting point of Cu using MD simulation employing the Sutton-Chen model for the interatomic interaction. This interaction has been fitted to reproduce results from first-principles self-consistent total-energy calculations within the local-density approximation using the full potential linear-muffin-tin-orbital method. The melting points obtained by Moriarty [13] using *ab initio* calculations agree well upto 100

GPa pressures and there is good agreement with the experimentally obtained laser-heated diamond cell results [14]. The steep slope of Cu confirms the key role of d-shell electrons in determining the temperature dependence of high pressure melting curve. The melting curve for Al agrees well with the experimental DAC [15] results. The results obtained from *ab initio* MD using the hysteresis and the Z-method [16] are found to be slightly higher at higher pressures.

## 4. Conclusion

Melting curves of Cu and Al have been generated using classical MD with the EAM potential as the interatomic potential. The EAM potential of Cai and Ye is found to produce results in good agreement with DAC and other *ab initio* MD results. The advantage of classical MD method is that it is computationally much less expensive compared to the *ab initio* methods and generates results of comparable accuracy.

## 5. References

[1] Allen M P and Tildesley D J 1989 *Computer Simulation of Liquids* (Oxford: Calendon Press)
[2] Holender J M 1990 *J. Phys.:Condens. Matter* **2** 1291
[3] Daw M S and Baskes M I 1984 *Phys. Rev.* B **29** 6443
[4] Cai J and Ye Y Y 1996 *Phys. Rev.* B **54** 8398
[5] Smith W, Forester T R, Todorov I T and Lesley M *DL_POLY 2 User Manual*, CCLRC, Daresbury Lab.
[6] Berendsen H J C, Postma J P M, van Gunteren W F, DiNola A, Haak J R 1984 *J. Chem. Phys.* **81** 3684
[7] Verlet L 1967 *Phys. Rev.* **159** 98
[8] Kokalj A 2003 *Comp. Mater. Sci.* **28** 155.Code available from http://www.xcrysden.org
[9] Mitrokhin Yu 2006 *Comp. Mater. Sci.* **36** 189
[10] Banerjea A and Smith J R 1988 *Phys. Rev.* B **37** 6632
[11] Rose J H, Smith J R, Guinea F and Ferrante J 1984 *Phys. Rev.* B **29** 2963
[12] Belonoshko A B, Ahuja R, Eriksson O and Johansson B 2000 *Phys. Rev.* B **61** 3838
[13] Moriarty J A 1986 *Shock Waves in Condensed Matter* ed Y M Gupta (New York: Plenum Press) p 101
[14] Japel S, Schwager B, Boehler R and Ross M 2005 *Phys. Rev. Lett.* **95** 167801
[15] Boehler R and Ross M 1997 *Earth and Planetary Sci. Lett.* **153** 223
[16] Bouchat J, Bottin F, Jomard G and Zerah G 2009 *Phys. Rev.* B **80** 094102

**Acknowledgements**
I would like to thank Dr. S V G Menon for his constant guidance and encouragement.

23rd International Conference on High Pressure Science and Technology (AIRAPT-23) IOP Publishing
Journal of Physics: Conference Series **377** (2012) 012086 doi:10.1088/1742-6596/377/1/012086

# Molecular Dynamics Simulation of Liquid-Vapor Coexistence Curves of Metals

**A. Sai Venkata Ramana**

Theoretical Physics Division, Bhabha Atomic Research Centre, Mumbai-400085, India

E-mail: asaivenk@barc.gov.in

**Abstract.** A Molecular dynamics implementation of Gibbs ensemble method is applied to determine liquid-vapor coexistence curves of metals using embedded atom model potentials. As an application of the code we developed, the liquid-vapor coexistence curves of Aluminum and Copper are simulated using Cai and Ye potential. The critical constants obtained were found to be slightly lower than the range of experimental results. The results show that the potentials parameters obtained by fitting to low temperature solid properties are not adequate to accurately determine the liquid-vapor phase diagram.

## 1. Introduction

The liquid-vapor phase diagram (LVPD) of metals finds important applications in a number of areas of physics and engineering. However, experiments to obtain LVPD of metals are difficult because of high pressures and temperatures involved. In such situations, computer simulations work as a reliable tool to complement the experiments. An elegant Monte-Carlo method for simulating the liquid vapor coexistence has been developed by Panagiotopoulos[1] and is called as Gibbs ensemble method. It has the advantage of direct calculation of phase coexistence properties of fluids from a single simulation. The method has been extensively used to obtain the LVPD of organic liquids using pair potentials. Later, many molecular dynamics versions of Gibbs ensemble have been proposed[2, 3, 4, 5, 6]. In this paper we use the Gibbs ensemble molecular dynamics (GEMD) method to obtain the LVPD of simple metals using Embedded Atom Model[7] Potentials. The method we use in the code we developed is close to that of Wen-ze et.al.[6] .

The EAM potential is introduced by Daw and Baskes for metals. It, in addition to the pair potential term contains an embedding term which takes into account the effect of de-localized electrons on the atoms or molecules in the system. The standard pair potentials which do not take this effect (metallic bonding) into account could not explain some characteristic properties of metals like the ratio of elastic constants $c_{12} : c_{44}$, the vacancy formation energy etc. The EAM potentials have been used in a wide variety of problems of materials science and excellent results have been obtained[8]. Many forms of EAM potentials have been formulated by fitting to data from experiments or ab-initio calculations for solids[8]. Of these, the one developed by Cai and Ye[9] is interesting as it has a smaller number of fitted parameters when compared to many other potentials. The Cai and Ye potential has been recently applied to study the properties of liquid nickel and good agreement has been found with experiments for zero pressure melting[10].

We choose this potential to study the liquid vapor phase diagram of aluminum and copper using molecular dynamics.

## 2. The Embedded Atom Model potential

The total energy $E$ of an N-particle interacting system in the EAM formalism can be written as

$$E[\rho] = \sum_{i=1}^{N} F(\rho_i) + \frac{1}{2}\sum_{i=1}^{N}\sum_{j(\neq i)=1}^{N} \phi(\vec{r}_{ij}) \quad (1)$$

where $F(\rho_i)$ is the embedding term which is equal to the energy required to place an impurity atom at the site $i$ and $\phi(\vec{r}_{ij})$ is the pair potential. In the Cai and Ye potential the embedding term is given by

$$F(\rho) = -F_0\left(1 - \ln\left(\frac{\rho}{\rho_e}\right)^n\right)\left(\frac{\rho}{\rho_e}\right)^n + F_1\left(\frac{\rho}{\rho_e}\right) \quad (2)$$

where $\rho_e$ is the equilibrium host electron density. $F_0 = E_c - E_v^f$ is the difference between cohesive energy $E_c$ and the vacancy formation energy $E_v^f$ and $n = 0.5$ is a dimensionless constant. The host electron density at $i^{th}$ site is given by

$$\rho_i = \sum_{i(\neq j)=1}^{N} f(\vec{r}_{ij}) \quad (3)$$

where

$$f(r) = f_e \exp(-\chi(r - r_e)) \quad (4)$$

$f_e$ is a scaling constant taken to be 1 for pure metals. $r_e$ is the equilibrium nearest neighbor distance.

The pair potential is given by

$$\phi(r) = -\alpha\left(1 + \beta\left(\frac{r}{r_a} - 1\right)\right)\exp\left(-\beta\left(\frac{r}{r_a} - 1\right)\right) \quad (5)$$

where $F_1, \alpha, \beta, \chi, r_a$ are fitted parameters. During the fitting, the cutoff distance for the interaction is taken as $r_{cut} = 1.65a_0$ where $a_0$ is the lattice parameter.The detailed simulation procedure is as follows.

## 3. The Gibbs Ensemble Molecular Dynamics Method

The basic idea of GEMD is to simulate the conditions of liquid-vapor coexistence. The system will contain two simulation boxes. Total number of particles in the system is kept constant. However the number of particles in individual boxes can vary. One box is assumed to be situated in an infinite medium of homogeneous liquid at a (given) constant temperature and the other is assumed to be situated in an infinite medium of homogeneous vapor at the same temperature. Since the idea is to get the thermodynamic properties of a macroscopic system, the interface effects are neglected. The coexistence conditions are simulated by evolving the boxes in such a way that they have same temperature, pressure and chemical potential after equilibration as required by the Gibbs phase rule. Initially, each box is given a guess density(i.e., no. of particles and volume). The equilibration would be faster if the guess densities are closer to the coexisting liquid and vapor densities. Periodic boundary conditions are applied to each box to ensure that they represent the bulk coexisting phases. Temperature fluctuations in each box are controlled by a Berendsen thermostat [11] so that they reach the given temperature. Pressures in both the boxes are equalized by controlling the volume fluctuations using a Berendsen barostat. This is done by adjusting the volume of each box such that the instantaneous pressure in one box becomes equal to the instantaneous pressure in the other. However the total volume of the two boxes is not restricted to be constant. The particle transfer step to equilibrate the chemical potential in both boxes is carried out after each five hundred time steps by comparing their chemical potentials. It is done as follows: A particle is chosen randomly from the box where chemical potential is more and is removed from it. Correspondingly a particle is introduced into the other with its potential energy calculated and the velocity taken from the Boltzmann distribution of corresponding temperature. However it is taken care that the introduced particle is not too close to any other particle in the box. With the above three procedures being done during the simulation, the two boxes evolve in time in such away that they have same temperature, pressure and the chemical potential after equilibration. For a given temperature, the system may phase separate into liquid in one box and gas in the other with proper choices of initial densities.

The method described above has been tested for the Lennard-Jones fluid initially. The phase diagram obtained exactly matched with that of the earlier simulations[1] validating the code we developed. Later the code has been used to simulate the liquid-vapor phase diagram of Aluminum and Copper using the Cai and Ye potential. Time step used for the simulations is 1 femto second and each simulation run has typically $5 \times 10^5$ equilibration and production steps. The chemical potential has been evaluated using Widom's test particle insertion method[12]. In each step 200 test particles are inserted and the chemical potential is calculated after each five hundred steps. Total number of particles used in the simulation are 1000. The initial densities have been chosen so that after equilibration, both the boxes contain a good number of particles (few hundreds) so that the averages are reliable and deviations are small. The parameters for Berendsen thermostat and barostat are taken from an earlier work[13].

## 4. Results and Discussion

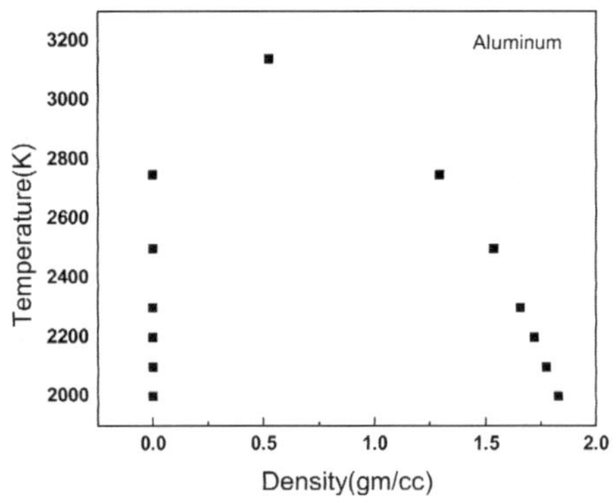


**Figure 1.** Temperature Vs Density for Aluminum.

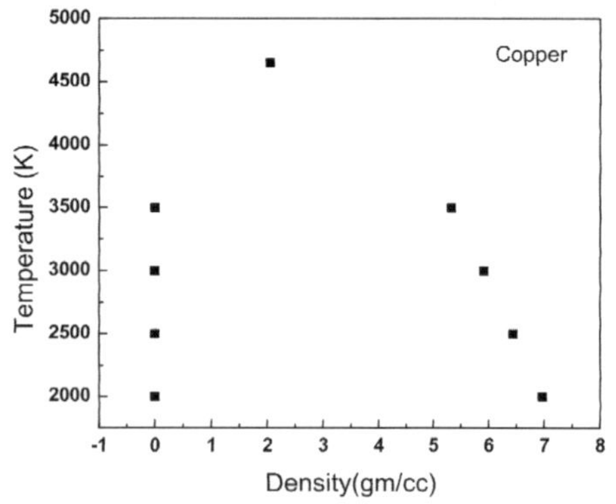


**Figure 2.** Temperature Vs Density for Copper.

The Temperature(T) Versus Density($\rho$) diagrams for Aluminum and Copper are shown in figures 1 and 2. The critical temperature($T_c$) and critical density ($\rho_c$) are obtained by fitting the simulation data using the law of rectilinear diameters

$$\frac{\rho_l + \rho_v}{2} = \rho_c + A(T - T_c) \qquad (6)$$

and the power law

$$\rho_l - \rho_v = B(T - T_c)^\beta \quad (7)$$

where $\rho_l$ and $\rho_v$ are liquid and vapor densities. $A$ and $B$ are fitting constants and $\beta = 0.33$

$T_c$ and $\rho_c$ for Aluminum are $3140 \pm 11$ K and $0.53 \pm 0.03$ gm/cc respectively. For Copper $T_c$ and $\rho_c$ are respectively $4651 \pm 70$ K and $2.05 \pm 0.022$ gm/cc. A view at the earlier literature data available for liquid-vapor phase diagram of Aluminum [14, 15, 16, 17] shows that there is a huge scatter in the value predicted for $T_c$ from around 5500 K to 9000 K. Whereas the predicted value for $\rho_c$ is in between 0.3 gm/cc and 0.7 gm/cc. Similarly, the literature data for Copper[14, 16, 18] for $T_c$ varied from 5100 and 8600 and $\rho_c$ has the range 1.7 to 3.6 gm/cc. The results show that for both Aluminum and Copper, the $T_c$ predicted by Cai and Ye potential is lesser than the minimum of the predicted value. However the $\rho_c$ predicted by it is within the range of predicted values from literature. To check the dependence of the results on particle number we have carried out simulations for few temperatures with 1728 particles. The results are seen to change negligibly.

The low $T_c$ predicted by the Cai and Ye potential may be attributed to low well depth of the pair potential and the embedding function. Also, the potential parameters are obtained by fitting to the solid data like vacancy formation energy, cohesive energy, elastic constants etc. Thus current simulation results show that the EAM potentials parametrized for solid state are not adequate to describe the fluid state far from melting. The potentials have to be re-parametrized using liquid state data like vapor pressure etc. Alternatively, ab-initio simulations may be done to obtain the liquid vapor phase diagrams. However, the enormous computational time required limits the total number of particles to a small number in order to obtain results in a reasonable period of time. But in the two phase coexistence regime, simulations with a small number of particles may not be reliable. Instead, the ab-initio simulations may be used to obtain the potential parameters and then the potential could be used in classical molecular dynamics simulations. Thus, with a good potential to describe the liquid metals, the gibbs ensemble method can be considered as a reliable tool to obtain the liquid-vapor phase diagram without going for experiments.

## References

[1] A. Z. Panagiotopoulos; N. Quirke; M. Stapleton; D. J. Tildesley Mol. Phys. 63, 527 (1988)
[2] M. J. Kotelyanskii and R. Hentschke Phys. Rev. E 51, 5116 (1995).
[3] Bruce J. Palmer and Chaomei Lo J. Chem Phys. 101, 10899(1994).
[4] Andras Baranyai and Peter T. Cummings Mol. Sim. 17, 21 (1996).
[5] Christoph Bratschi and Hanspeter Huber J. Chem Phys. 126, 164104 (2007).
[6] Wen-Ze Ou-Yang, Zhong-Yuan Lu, Tong-Fei Shi, Zhao-Yan Sun, and Li-Jia An J. Chem. Phys. 123, 234502(2005)
[7] M. S. Daw and M. I. Baskes, Phys. Rev. B 29, 6443 (1984); Phys. Rev. Lett. 50, 1285 (1983).
[8] M. S. Daw, S. M. Foiles and M. I. Baskes, Materials Science Reports 9, 251 (1993) and references therein.
[9] J. Cai and Y. Y. Ye, Phys. Rev. B 54, 8398 (1996).
[10] F. J. Cherne, M. I. Baskes and P. A. Deymier Phys. Rev. B 65, 024209 (2001).
[11] H. J. C. Berendsen, J. P. M. Postma, W. F. van Gunsteren, A. DiNola, and J. R. Haak J. Chem. Phys. 81, 3684 (1984)
[12] B. widom J. Chem. Phys. 39, 2803(1963)
[13] Howard E. Alper, Peter Politzer Int. J. Quantum Chem. 76, 670(2000)
[14] A. Ray, M.K. Srivastava, G. Kondayya, and S.V.G. Menon, Laser and Particle Beams 24, 437(2006)
[15] Gerald Faussurier, Christophe Blancard, and Pier Luigi Silvestrelli, Phys. Rev. B 79, 134202(2009)
[16] Jayant K. Singh , Jhumpa Adhikari , Sang Kyu Kwak, Fluid Phase Equilibria 246, 1 (2006)
[17] Divesh Bhatt, Ahren W. Jasper, Nathan E. Schultz, J. Ilja Siepmann and Donald G. Truhlar J. Am. Chem. Soc. 128, 4224 (2006)
[18] T. Aleksandrov, C. Desgranges, J. Delhommelle Fluid Phase Equilibria 287, 79 (2010)

Journal of Physics: Conference Series **377** (2012) 012087 doi:10.1088/1742-6596/377/1/012087

# On Dynamic and Elastic Stability of Lanthanum Carbide

**B.D. Sahoo, D. Mukherjee, K.D. Joshi and Satish C. Gupta**
Applied Physics Division, Bhabha Atomic Research Centre, Mumbai
India 400085

E-mail: bdsahoo@barc.gov.in

**Abstract.** The elastic stability up to 100 GPa and dynamic stability up to ~ 150 GPa of LaC have been investigated in NaCl (B1 phase) type and CsCl (B2 phase) structures, by first-principles calculations using the plane-wave pseudopotential method. The analysis of elastic moduli shows that the shear elastic modulus C′ (=(C11-C12)/2) of B1 phase is negative up to ~47 GPa, suggesting that the B1 phase is elastic unstable up to this pressure. The theoretically calculated phonon frequencies in various directions of the Brillouin zone are imaginary even up to 150 GPa, indicative of dynamic instability of B1 phase in this compound up to 150 GPa. These analyses in B2 phase of LaC suggest that this phase will also remain unstable elastically and dynamically up to 150 GPa.

## 1. Introduction

Transition metal carbides are potential candidates for technological applications due to their good mechanical properties such as extreme hardness, corrosive resistance and high melting points. Such properties make these compounds suitable for cutting tools and hard protective coating for magnetic storage devices etc. In past few years many studies have come out on their structural, elastic and thermodynamic properties [1-3].

Among other transition metal carbides, the carbides of group IIIB elements (Sc, Y and La) are also subject of current interest [4, 5]. In ScC and YC, stability of B1 phase at ambient conditions has been found out by Isaev *et al.* [4] using lattice dynamic simulations and Soni *et al.* [5] using full potential linearized augmented plane wave (FP-LAPW) calculations, which have been supported by experimental findings [6,7]. However, the stability of LaC has not been studied theoretically as well as experimentally. In fact, to our knowledge, the LaC has not been synthesized yet. From the *ab-initio* LMTO calculations Guillermet *et al.* [8] have determined the lattice parameters, enthalpy of formation, Debye temperature and cohesive energy for B1 phase of LaC.

In the present work, we have carried out first-principles theoretical analysis to understand the elastic and dynamic stability of B1 and B2 phase in LaC up to hydrostatic pressures of ~ 100 GPa and 150 GPa, respectively, employing plane-wave pseudopotential method

## 2. Theoretical Method

All the calculations have been performed employing plane wave pseudopotential method within the framework of Density Functional Theory (DFT) [9] implemented in Quantum Espresso package [10]. The elastic moduli of B1 and B2 phase have been determined from static lattice calculations. The *ab-initio* total energy calculations required for this purpose have been performed within generalized gradient approximation (GGA) for the exchange correlation potential [11]. The ultrasoft pseudopotential with valence configurations of $5d^1 6S^{1.5} 6p^{0.5}$ for Lanthanum with core radius of 2.2 a.u.

for s and d orbitals and 2.0 a.u. for semicore p orbitals has been used. For carbon valence configuration of $2s^2 2p^2$ with core radius 1.1 a.u. is used. Also, for C, non-relativistic calculations are performed and for La, non-linear core correction has been added with scalar relativistic calculations. Electronic wave functions are expanded in a plane wave basis set (PWs) with energy cut off, $E_{cutoff}$, of 60 Ry and charge density is represented in PWs with energy cut off of 400 Ry. A 16×16×16 Monkhorst–Pack [12] grid of k-points has been used for sampling of electronic Brillouin zone (BZ) with energy convergence criteria chosen to be $10^{-6}$ Ry. Further, we find that increasing the number of k points does not affect the phonon frequency by more than 5 $cm^{-1}$. The dynamic stability of B1 and B2 phase has been analysed from the phonon spectra determined from lattice dynamical calculations using linear response method [13], carried out within the framework of self consistent density functional perturbation theory with 4×4×4 q mesh in the first BZ for the interpolation of the force constants for phonon dispersion calculations and 12×12×12 k-mesh used for integration over the Brillouin zone.

To determine the isotherm for B1 and B2 phases of LaC, we have computed the total energy at 0 K as a function of volume. The theoretical isotherm is then determined by finding the negative volume derivative of total energy dependence on unit cell volume. The zero pressure bulk modulus ($B_0$) and its pressure derivative at zero pressure ($B_0'$) were obtained by fitting the P-V data so obtained to the third order Birch-Murnaghan equation of state.

Further LaC single crystal in B1 and B2 phase has only three independent elastic constants, $C_{11}$, $C_{12}$ and $C_{44}$ or equivalently three independent elastic moduli namely bulk modulus (B) and two shear moduli $C' = (C_{11}-C_{12})/2$ and $C_{44}$. To determine the shear moduli $C'$ and $C_{44}$ at a particular volume, the total energy is calculated as a function of different kinds of distortions [14,15] with the dependence of the total energy of the strained lattice on the deformation tensor expressed as [16]:

$$E_c(V,\underline{\varepsilon}) = E_c(V,0) + \frac{1}{2} V \sum_i \sum_j C_{ij} e_i e_j$$

The shear elastic moduli are then calculated from the second order derivative of the total energy with respect to strain $e_i$ (with $i,j$ = 1 to 6).

### 3. Results and Discussion

The zero pressure equilibrium volume ($V_0$) per formula unit, bulk modulus ($B_0$) and its pressure derivative ($B_0'$) for the B1 phase of LaC obtained by fitting theoretically determined pressure volume data to Birch-Murnaghan EOS are listed in the Table I. Our zero pressure volume, bulk modulus and its pressure derivative agree well with the available other theoretical values [4,8] confirming the accuracy of the calculation. In figure 1 we show the elastic constants of LaC single crystal in B1 phase as a function of pressure. The $C'$ and $C_{44}$ at zero pressure are determined to be -9.59 GPa and 47.81 GPa, respectively (Table I). Moreover, under pressure the $C'$ modulus first decreases with increasing pressure and reaches minimum value of ~ -16 GPa at ~ 21 GPa, thereafter, it increases continuously with further compression and becomes positive beyond 47 GPa. The $C_{44}$ however, decreases monotonically with increasing pressure but remains positive even at 100 GPa. The negative $C'$ modulus up to 47 GPa suggests that B1 phase in LaC is elastically unstable up to this pressure, however, it emerges as elastically stable phase above 47 GPa.

Additionally, we have determined the phonon dispersion relations (Fig. 2) for B1 phase of this carbide at zero pressure from lattice dynamic calculations. As is evident from figure 2, few phonon frequencies in various directions of Brillouin zone are imaginary even at highest pressure of ~ 150 GPa, indicating that the B1 phase in LaC is dynamically unstable up to ~ 150 GPa.

In B2 phase of LaC, the zero pressure equilibrium volume per formula unit, bulk modulus and its pressure derivative derived from the theoretical equation of state are 35.51 $Å^3$, 92.15 GPa and 3.379, respectively. Further, the elastic and dynamic stabilities of LaC have been tested for B2 phase also. As shown in figure 3, at zero pressure though the shear modulus $C'$ is positive (~ 16 GPa), the

**Table I:** Various physical quantities at ambient pressure for B1 phase of LaC

| | Present Work | Other Theoretical Works |
|---|---|---|
| $V_0$ (Å$^3$) | 40.07 | 40.11[4], 39.95[8] |
| $B_0$ (GPa) | 83.18 | 80 [4] |
| $B_0'$ | 3.502 | -- |
| $C'$ (GPa) | -9.591 | -- |
| $C_{44}$ (GPa) | 47.81 | -- |

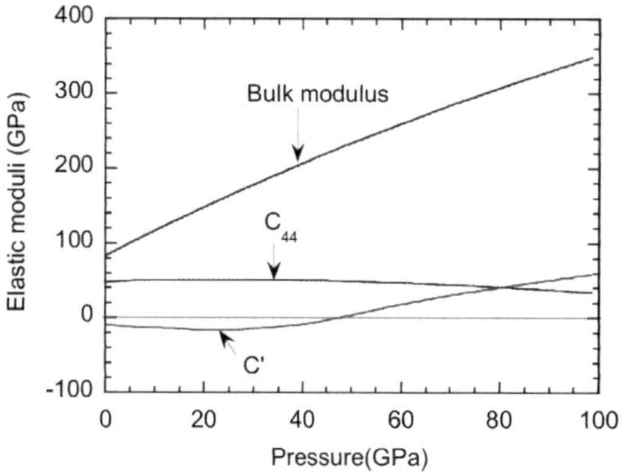


**Figure 1.** Elastic moduli of LaC as a function of pressure in B1 phase

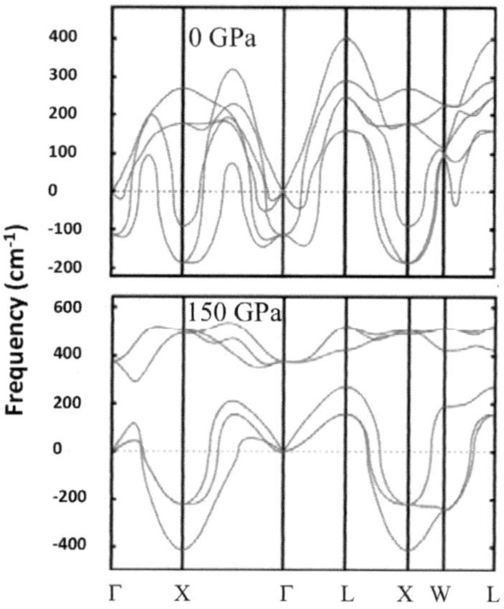


**Figure 2.** Phonon spectrum of LaC for B1 phase

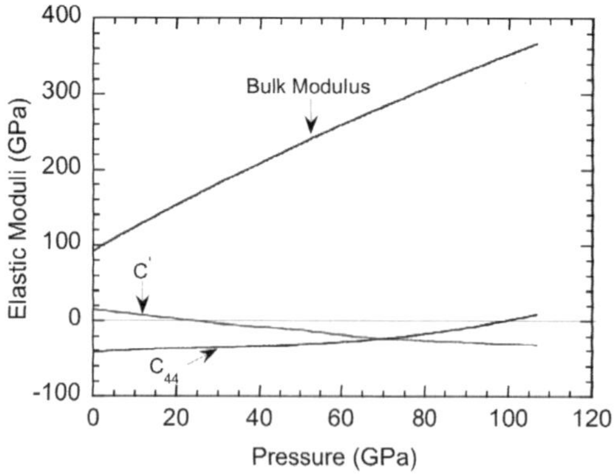


**Figure 3.** Elastic moduli of LaC as a function of pressure in B2 phase

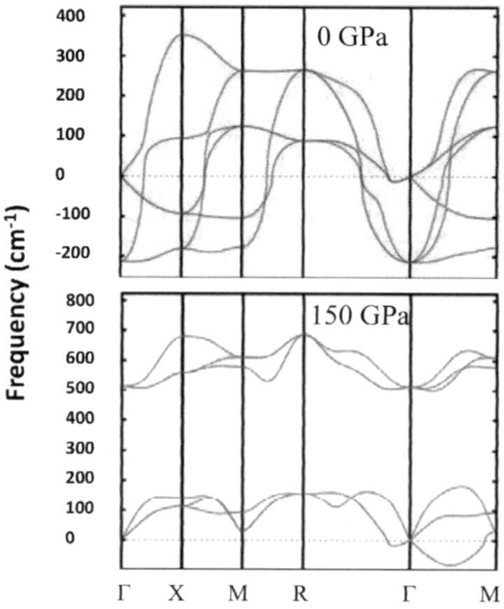


**Figure 4.** Phonon spectrum of LaC for B2 phase

$C_{44}$ modulus is negative (~ -40 GPa). Further, the $C_{44}$ modulus increases with increasing pressure and become positive above 100 GPa. On the contrary, the C′ modulus decreases monotonically with increasing pressure and vanish at ~ 25 GPa. These findings suggest that LaC in B2 phase is elastically unstable up to 100 GPa and expected to remain unstable even at still higher pressures as indicated by the monotonic decreasing trend of C′ modulus with increasing pressure. Finally, the phonon spectra of LaC in B2 phase (Fig. 4) shows that many phonon frequencies are imaginary at ambient as well as at 150 GPa, suggesting that the B2 phase will be also dynamically unstable up to 150 GPa.

## 4. Summary

To summarize, we have examined elastic and dynamic stability of B1 and B2 phase in single crystal of LaC. Analysis of elastic moduli and phonon spectrum as a function of pressure suggests that the B1 phase will be elastically unstable up to ~ 47 GPa, however, it remains dynamically unstable up to ~ 150 GPa. The B2 phase however, will be unstable dynamically as well as elastically up to ~ 150 GPa. This study suggests that the synthesis of LaC in B1 or B2 phase is not possible.

## Reference

[1] Aleksandra Vojvodic and Carlo Ruberto 2010 *J. Phys. Cond. Matter* **22** 3755011
[2] Zengtao Lv, Haiquan Hu, Cheng Wu, Shouxin Cui, Guiqing Zhang and Wenxia Feng 2011 *Physica B* **406** 2750
[3] Hui Li, Litong Zhang, Qingfeng Zeng, Haitao Ren, Kang Guan, Qiaomu Liu and Laifei Cheng 2011 *Solid. State Comm.* **151** 61
[4] Isaev E. I., Simak S. I., Abrikosov I. A., Ahuja R., Vekilov Yu. Kh., Katsnelson M. I., Lichtenstein A. I. and Johansson B. 2007 *J. Appl. Phys.* **101** 123519
[5] Soni,Pooja Pagare Gitanjali, Sanyal Sankar P. 2011 *J. Phys. and Chem. Solids* **72** 810
[6] Toth E., *Transition Metal Carbides and Nitrides*, Academic,NewYork,1971
[7] P.Villars and L.D.Calvet *Pearson's Handbook of Crystallographic Data for Intermetallic Phases*, American Society for Metals, Metals Park, OH, 1985
[8] Guillermet A. Fernandez 1993 *Phys. Rev. B* **48** 11673
[9] Baroni S, Giannozzi P and Testa A 1987 *Phys. Rev. Lett.* **58** 1861
[10] Baroni S, Dal Corso A, de Gironcoli S, Giannozzi P, Cavazzoni C, Ballabio G, Scandolo S, Chiarotti G, Focher P,Pasquarello A, Laasonen K, Trave A, Car R, Marzari N and Kokalj A http://www.pwscf.org
[11 Perdew J P and Burke K 1996 *Int. J. Quantum Chem.* **57** 309; Perdew J P, Burke K and Ernzerhof M 1996 *Phys. Rev. Lett.* **77** 3865
[12] Monkhorst H J and Pack J D 1976 *Phys. Rev.* B **13** 5188
[13] Giannozzi P, de Gironcoli S, Pavone P and Baroni S 1991 *Phys. Rev.* B **43** 7231
[14] Cohen R. E., Sixtrude L., and Wasserman E. 1997 *Phys. Rev. B* **56**, 8575
[15] Gupta S. C., Joshi K. D., and Banerjee S. 2008 *Metall. Mater. Trans. A* **39A**, 1593
[16] Nye J. F. *Physical Properties of Crystals* (Oxford University Press, Oxford, 1957)

23rd International Conference on High Pressure Science and Technology (AIRAPT-23) IOP Publishing
Journal of Physics: Conference Series **377** (2012) 012088 doi:10.1088/1742-6596/377/1/012088

# High pressure study on structural and vibrational properties of $NH_3BH_3$

**Ch. Bheema Lingam[1], K. Ramesh Babu[2], Surya P. Tewari[1,2], G. Vaitheeswaran[2*]**
[1] School of Physics and [2]Advanced Centre of Research in High Energy Materials (ACRHEM), University of Hyderabad, Hyderabad - 500046, India

*E-mail: gvsp@uohyd.ernet.in

**Abstract.** The structural and vibrational properties of potential hydrogen storage compound $NH_3BH_3$ have been studied through density functional theory at ambient as well as high pressures upto 10 GPa. The calculated lattice parameters at ambient pressure are in good agreement with experiments. The compressibility of $NH_3BH_3$ along the crystallographic *a*-axis is almost constant when compared to *b*- and *c*-axes, which is an indication of anisotropic compressibility. The zone center vibrational frequencies of $NH_3BH_3$ at ambient and 10 GPa pressures are calculated. The $A_2$ and $B_1$ symmetry modes are independent of pressure, whereas the $A_1$ symmetry modes increases and $B_2$ symmetry modes decreases with pressure.

## 1. Introduction

High pressure studies of many hydrogen storage compounds exhibit phase transitions, and the resultant high pressure phases exhibit superior hydrogen storage properties. Therefore it is important to understand the structural and dynamical behavior of these compounds under pressure [1]. Ammonia borane ($NH_3BH_3$) is one of the promising candidates for onboard hydrogen fuel cells. It has remarkable high gravimetric and volumetric hydrogen density and it releases $H_2$ at moderate temperatures [2]. The hydrogen can be released by thermolysis, starting at temperatures 80°C, but at high temperatures the process tends to release volatile gases such as borazine [3]. $NH_3BH_3$ related compounds or complexes containing alkali or other metals have been found to have significantly improved hydrogen release properties and rehydrogenation [3, 4]. The high pressure behavior and properties of $NH_3BH_3$ have been studied earlier [5, 6]. These studies are basically motivated by the anomalous lattice stability of $NH_3BH_3$. At room temperature, in contrast to the very similar compounds like diborane, $B_2H_6$ (liquid up to 4GPa [7]) and ethane, $C_2H_6$ (gas), $NH_3BH_3$ forms in solid phase. This stability in $NH_3BH_3$ is due to the existence of dihydrogen bonds between $NH_3BH_3$ molecules [8]. The crystalline phases of $NH_3BH_3$ are found to be orthorhombic (Pmn21) at 16K and tetragonal (I4mm) at 225K [9]. Upon pressure, $NH_3BH_3$ undergo to orthorhombic ($Cmc2_1$) phase at 1.22 GPa and further pressures it undergoes to triclinic (P1) phase at 8 GPa [10]. Recent studies suggest that the $Cmc2_1$ phase can be obtained from low temperature $Pmn2_1$ phase upon pressure [1]. In the present work we report the pressure studies of structure and vibrational properties of low temperature orthorhombic ($Pmn2_1$) phase of $NH_3BH_3$ up to 10 GPa.

## 2. Computational details

The first principles calculations were carried out on $NH_3BH_3$ compound using density functional theory as implemented in the Cambridge Sequential Total Energy Package (CASTEP) [11]. The electron-ion interaction can be seen as the electrons are interacting with the rigid ions. Valence electrons experience the core as screened by the core electrons. The pseudopotential is an effective potential constructed to replace the atomic all-electron potential such that core states are eliminated and the valence electrons are described by pseudo-wave functions. For $NH_3BH_3$, the basis orbitals used as valence states are N: $2s^2$, $2p^3$, B: $2s^2$, $2p^1$ and H: $1s^1$. We have used Vanderbilt-type ultrasoft Pseudo Potentials [12] for the electron–nuclei interactions together with the local density approximation (LDA) of Ceperley and Alder [13] parameterized by Perdew and Zunger (CA-PZ) [14] for exchange - correlation functionals for electron–electron interactions. A plane wave basis set with an energy cut-off of 520 eV has been used. For the Brillouin zone sampling, a 7 x 7 x 7 Monkhorst–Pack mesh [15] has been used, in which the forces on the atoms are converged to less than 0.0005 eV/Å. The maximum ionic displacement is within 0.005Å and the total stress tensor is reduced to the order of 0.02 GPa.

## 3. Results and Discussion

### *3.1. Structural properties at ambient pressure*

The optimized crystal structure of low-temperature orthorhombic phase of $NH_3BH_3$ contains three formula units in which each pair of B and N atoms are surrounded by three H atoms i.e., N is connected to one H1 and two H2, whereas B is connected to two H3 and one H4 [2].

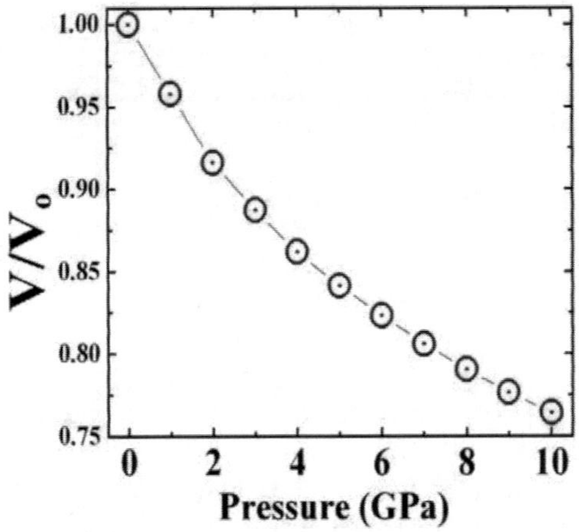


**Figure 1a.** The isothermal equation of states (EOS) for $NH_3BH_3$.

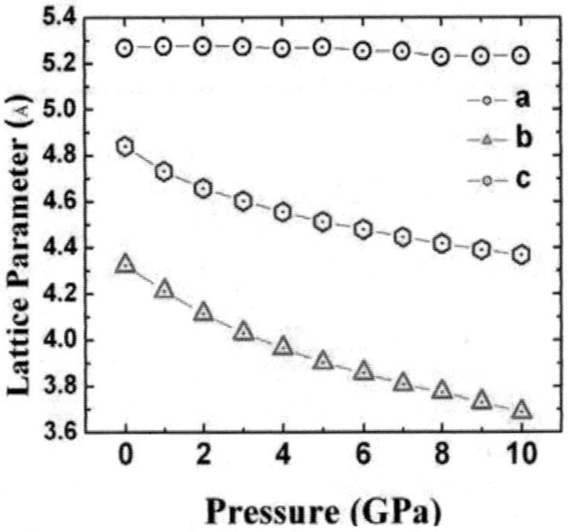


**Figure 1b.** Lattice constants of $NH_3BH_3$ as a function of pressure.

The calculated structural parameters and fractional coordinates are in good agreement with the other theoretical studies [2] and also compared well with experimental data [9]. The ground-state properties such as equilibrium lattice parameters, atomic positions, bond angles, and bond lengths are in agreement with the experiments with error ±10%, and they agree well with other theoretical calculations. This error is due to the inherent limitation of LDA for the exchange and correlation functional. The optimized structure for ambient pressure is found to agree well with the experiments. The external pressure was gradually increased by an increment of 1 GPa in each time. From the hydrostatic compression simulations, the isothermal equation of state (EOS) was obtained and is shown in Figure 1(a) as a function of the volume compression ratio, $V/V_o$, where V and $V_o$ are the compressed and uncompressed volumes, respectively. The volume compression of $NH_3BH_3$ is found to be linear with pressure. The crystallographic directions *a*-, *b*-, and *c-axis* are reported in Figure 1(b) as a function of pressure. The crystallographic directions *b*- and *c-axis* exhibit a relatively smooth monotonic behaviour with pressure, but the crystallographic direction *a*, is almost constant. It is very hard to compress $NH_3BH_3$ along the crystallographic direction *a-axis*. This is an indication of anisotropic compressibility. These results are in accord with the earlier studies [2], indicating that $NH_3BH_3$ is a soft material towards the compression.

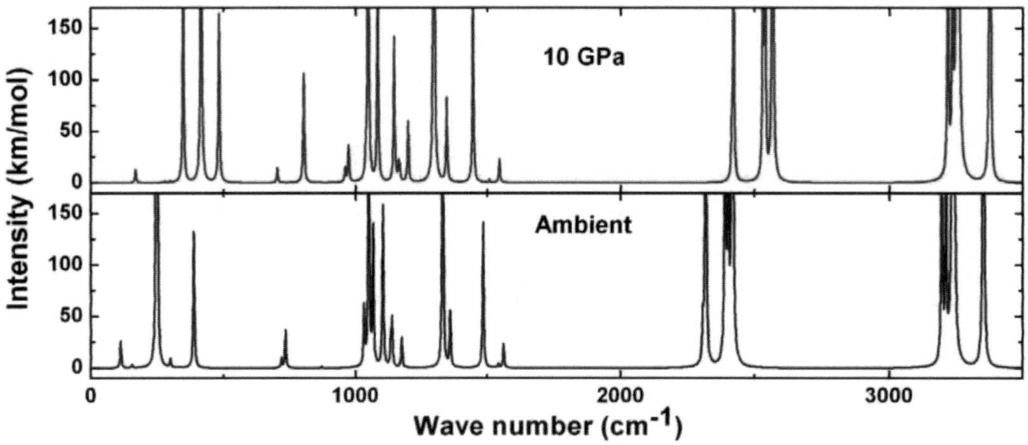


**Figure 2.** IR spectrum of $NH_3BH_3$ at ambient and 10 GPa pressures

### *3.2. Vibrational properties at ambient and high pressures*

We have also performed the density functional perturbation calculations to find the zone centre vibrational frequencies of $NH_3BH_3$ for ambient and 10 GPa pressures. The harmonic vibrational frequencies of the compound were computed from the matrix of Cartesian second derivatives (i.e. Hessian matrix). The Hessian elements were computed by displacing each atom, and finding a gradient vector thus building up a complete second derivative matrix. The infrared intensities are obtained from the atomic polar tensor, which is a second derivative of the total energy with respect to the Cartesian coordinates and dipole moments.

**Table 1.** Calculated Vibrational Frequencies (v) of $NH_3BH_3$ at Ambient and 10 GPa pressures

| Ambient | | 10GPa | | Ambient | | 10 GPa | |
|---|---|---|---|---|---|---|---|
| v ($cm^{-1}$) | Symmetry | v ($cm^{-1}$) | Symmetry | v ($cm^{-1}$) | Symmetry | v ($cm^{-1}$) | Symmetry |
| 15.65 | $B_2$ | 5.81 | $A_1$ | 1104.17 | $A_1$ | 1145.02 | $B_2$ |
| 19.06 | $A_1$ | 20.72 | $B_2$ | 1109.87 | $B_2$ | 1147.05 | $A_1$ |
| 27.20 | $B_1$ | 30.46 | $B_1$ | 1132.84 | $A_1$ | 1164.24 | $A_1$ |
| 95.00 | $A_2$ | 95.84 | $A_2$ | 1138.20 | $B_2$ | 1168.71 | $B_2$ |
| 98.99 | $A_1$ | 97.48 | $B_2$ | 1179.19 | $B_1$ | 1200.03 | $B_1$ |
| 155.48 | $B_2$ | 169.52 | $A_1$ | 1182.97 | $A_2$ | 1210.12 | $A_2$ |
| 218.84 | $A_2$ | 248.70 | $A_2$ | 1331.17 | $A_1$ | 1296.07 | $A_1$ |
| 227.67 | $A_1$ | 279.84 | $B_1$ | 1360.90 | $B_2$ | 1344.39 | $B_2$ |
| 245.68 | $B_1$ | 300.91 | $A_1$ | 1484.00 | $B_1$ | 1445.30 | $B_1$ |
| 248.56 | $B_2$ | 349.27 | $B_2$ | 1497.57 | $A_2$ | 1462.97 | $A_2$ |
| 274.05 | $A_2$ | 406.26 | $A_2$ | 1543.71 | $A_1$ | 1508.76 | $A_1$ |
| 296.74 | $B_1$ | 416.86 | $B_1$ | 1562.54 | $B_2$ | 1547.18 | $B_2$ |
| 395.90 | $B_1$ | 483.46 | $B_1$ | 2303.59 | $A_1$ | 2413.67 | $A_1$ |
| 416.79 | $A_2$ | 491.79 | $A_2$ | 2314.47 | $B_2$ | 2420.44 | $B_2$ |
| 720.84 | $B_2$ | 705.38 | $B_2$ | 2386.85 | $A_1$ | 2530.95 | $A_1$ |
| 729.09 | $A_2$ | 722.13 | $A_1$ | 2398.08 | $B_2$ | 2538.89 | $B_2$ |
| 736.00 | $A_1$ | 779.78 | $A_2$ | 2413.92 | $B_1$ | 2566.35 | $B_1$ |
| 739.71 | $B_1$ | 806.55 | $B_1$ | 2417.36 | $A_2$ | 2567.85 | $A_2$ |
| 877.13 | $B_2$ | 960.98 | $A_1$ | 3194.66 | $A_1$ | 3218.32 | $A_1$ |
| 877.15 | $A_1$ | 973.54 | $B_2$ | 3210.94 | $B_2$ | 3235.68 | $B_2$ |
| 1032.49 | $A_1$ | 1037.60 | $A_1$ | 3236.21 | $A_2$ | 3254.24 | $A_2$ |
| 1050.23 | $B_1$ | 1047.11 | $B_1$ | 3236.56 | $B_1$ | 3255.12 | $B_1$ |
| 1061.39 | $A_2$ | 1059.89 | $A_2$ | 3349.13 | $A_1$ | 3372.10 | $A_1$ |
| 1066.96 | $B_2$ | 1083.77 | $B_2$ | 3353.12 | $B_2$ | 3376.25 | $B_2$ |

The intensity of each mode can be evaluated as a square of all transition moments of the mode and expressed in terms of the atomic polar tensor matrix and eigenvectors of the mass-weighted Hessian. Figure 2 shows the infrared absorbance as a function of frequency. The zone centre vibrational frequencies are shown in Table 1. The group symmetry decomposition into irreducible representations of the Pmn21 space group of $NH_3BH_3$ yields sums $A_1+B_1+B_2$ for three acoustic modes and $13A_1+10A_2+9B_1+13B_2$ for 45 optical modes for ambient and 10 GPa pressures. Except $A_2$ symmetry modes, all other modes are found to be Raman active. At ambient pressure, the lowest frequencies of $A_1$, $A_2$, $B_1$ and $B_2$ are found to be 98.9 $cm^{-1}$, 95 $cm^{-1}$, 245.6 $cm^{-1}$ and 155.4 $cm^{-1}$, respectively. Whereas at 10 GPa, the lowest frequencies of $A_1$, $A_2$, $B_1$ and $B_2$ are found to be 169.5 $cm^{-1}$, 95.8 $cm^{-1}$, 279.8 $cm^{-1}$ and 97.4 $cm^{-1}$, respectively. The $A_2$ and $B_1$ symmetry modes are independent of pressure, whereas the $A_1$ symmetry modes increases and $B_2$ symmetry modes decreases with pressure. Higher bond orders results in higher frequencies i.e higher the frequency, stronger are the bonds. The frequencies around 2300-2400 $cm^{-1}$ are shifted to 2400-2500 $cm^{-1}$ at 10 GPa pressures, the lower frequencies upto 877 $cm^{-1}$ are shifted to 973 $cm^{-1}$ at 10 GPa pressures. A blueshift represents decrease in wavelength (increase in frequency); the opposite effect is referred to as redshift [16]. The frequencies around 2300-2400 $cm^{-1}$ are increased (shifted) to 2400-2500 $cm^{-1}$ at 10 GPa pressures, the lower frequencies upto 877 $cm^{-1}$ are increased (shifted) to 973 $cm^{-1}$ at 10 GPa pressures. This is an indication of blue shift. The blueshift indicates the contraction of bonds under pressure [16]. Rest of the frequencies are unchanged. This is a clear indication that, $NH_3BH_3$ under pressure becomes relatively hard material.

## 4. Conclusions

In summary, the structural and vibrational properties of $NH_3BH_3$ under hydrostatic compression up to 10 GPa have been studied using density functional theory. It is observed that $NH_3BH_3$ is a soft material towards the compression and it is becoming hard material.

## Acknowledgments

Ch BL thanks CAS-RFSMS and K RB thanks DRDO through ACRHEM for the financial support. Ch BL and K RB thank CMSD, University of Hyderabad for computational time.

## References

[1] Wang L, Bao K, Meng X, Wang X, Jiang T, Cui T, Liu B and Zou G 2011 *J. Chem. Phys.* **134** 024517
[2] Bheema Lingam Ch, Ramesh Babu K, Tewari S P and Vaitheeswaran G 2011 *J. Comp. Chem.* **32** 1734
[3] Staubitz A, Robertson A P M and Manners I 2010 *Chem Rev* **110** 4079
[4] Wu H, Zhou W, Pinkerton F E, Meyer M S, Srinivas G, Yildirim T, Udovic T J and Rush J J 2010 *J Mater Chem* **20** 6550
[5] Trudel S and Gilson D F R 2003 *Inorg Chem* **42** 2814
[6] Custelcean R and Dreger Z A 2003 *J Phys Chem B* **107** 9231
[7] Murli C and Song Y 2009 *J Phys Chem B* **113** 13509
[8] Crabtree R H, Siegbahn P E M, Eisenstein O, Rheingold A L and Koetzle T F 1996 *Acc Chem Res* **29** 348
[9] Bowden M E, Gainsford G J and Robinson W T 2007 *Aust J Chem* **60** 149
[10]Ravhi S K, Xuezhi Ke, Zhang J, Lin Z, Sven C V, Hartl M, Sinogeikin S, Daemen L, Andrew L C, Chen C and Zhao Y 2010 *Chem. Phys. Lett.* **495** 203
[11] Segall M, Lindan P, Probert M, Pickard C, Hasnip P, Clark S and Payne M 2002 *J. Phys.: Condens. Matter* **14** 271
[12] Vanderbilt D 1990 *Phys. Rev. B* **41** 7892
[13] Ceperley D M and Alder B J 1980 *Phys. Rev. Lett* **45** 566.
[14] Perdew J P and Zunger A 1981 *Phys. Rev. B* **23** 5048.
[15] Monkhorst H J and Pack J 1976 *Phys. Rev. B* **13** 5188.
[16] Joseph J, Jemmis E 2007 *J. Am. Chem. Soc*, **129** 4620.

23rd International Conference on High Pressure Science and Technology (AIRAPT-23) IOP Publishing
Journal of Physics: Conference Series **377** (2012) 012089 doi:10.1088/1742-6596/377/1/012089

# Pressure-induced phonon, electronic and thermal properties of antiperovskite $AlCNi_3$

**Prafulla K. Jha, Sanjay D. Gupta and Sanjeev K. Gupta**
Department of Physics, Bhavnagar University, Bhavnagar 364002, India
E-mail: prafullaj@yahoo.com

**Abstract.** First principle pseudopotential calculations were performed to investigate structural, electronic, vibrational, and electron phonon interaction of antiperovskite $AlCNi_3$. The stability of $AlCNi_3$ in the antipervoskite phase has been analysed on the basis of the results of electronic structure, electronic density of states and phonon calculation at ambient as well as hydrostatic 7.3 GPa high pressure. The dynamical stability, electron phonon interaction and superconductivity have been discussed at ambient as well as hydrostatic high pressure 7.3 GPa.

## 1. Introduction

The discovery of 8 K superconductivity in the non-oxide perovskite $MgCNi_3$ has attracted much attention among intermetallic non-oxide perovskite cubic structure. At first glance, $MgCNi_3$ should belong to the list of unconventional superconductors due to its high content of magnetic element Ni. However, the nature of its superconductivity is still controversial [1-2]. In spite of this, the occurrence of superconductivity through electron-magnon coupling spin fluctuations is found to be active in determining the properties of both superconducting and normal state [2-3]. As an alternative, it is significant to investigate the materials property which are closely related to $MgCNi_3$, i.e., $AlCNi_3$ (A=Zn, Al, Ga, In, Cd, and so on). As to Al compound, previous experimental studies indicate $AlC_xNi_3$ is nonmagnetic or PM [4-5], while a recent work shows a PM into weak FM transition at near 300 K in $AlCNi_3$ [6]. In order to eliminate the discrepancy, more efforts are needed. Several theoretical and experimental studies have been carried out on substitution of Mg with Al, Zn, in $MgCNi_3$, in order to determine the effect of such substitution on the superconducting transition temperature and dynamical stability. In recently the synthesis of $AlCNi_3$ in which Al replaces Mg sites of the superconducting intermettalic perovskite $MgCNi_3$. Even though $AlCNi_3$ isostructural with $MgCNi_3$ unfortunately the superconductivity has not been detected in these compound even when temperature lowered up to 4K. While a recent theoretical calculations shows that the calculated phonon modes are unstable along the [110] and [111] symmetry direction in the Brillouin zone [7]. Motivated with this fact the dynamical instability and softening of acoustic phonon branch at high symmetry direction leads to the high temperature superconducting material. It is known that the theoretical calculation on $AlCNi_3$ [8] shows that the density of states at Fermi level is smaller than that of $MgCNi_3$ and it has been argued that a strongly defined DOS at the Fermi level could be responsible for the absence of superconductivity in the $AlCNi_3$ non-oxide antiperovskite. It is therefore, worthwhile to examine theoretically the structural, electronic and vibrational properties of antiperovskite $AlCNi_3$. The results of this investigation can reveal some useful insights towards the behaviour and understanding for the properties of the $MgCNi_3$ superconductor, which is isostructural to $AlCNi_3$ and also can help in elucidating the property of $AlCNi_3$.

## 2. Computational Methodology

Calculations are performed using plane wave self consisted field (PWSCF) [9]. To present the interaction between ionic cores and valence electrons the electronic wave functions are presented in a plane wave basis set with a kinetic energy cut-off of 110 Ry. The Brillouin zone integration are carried out by the Gaussian smearing technique using 8x8x8 k-points mesh with shift from origin. We have also checked more dense grids up to 16x16x16 but the result did not affected considerably. Total energies converge to within $10^{-4}$ Ry/atom with response to energy cut-off from 30 to 120 Ry.

**Table 1** Calculated lattice parameter, bulk modulus, superconducting temperature and electron phonon coupling parameter for $AlCNi_3$

| $AlCNi_3$ | Lattice parameter (Å) | Bulk modulus ($B_0$) | (ω) log | λ | $N(E_f)$ | $T_c$ |
|---|---|---|---|---|---|---|
| Zero Pressure | 3.774, 3.80[a], 3.587[a], 3.697[b], 3.774[b] | 194.6 201.69[b], 237.35[b] | 86.188 | 1.232 | 16.634 | 7.969 |
| 7.3 GPa | 3.729 | - | 61.601 | 1.6935 | 15.770 | 7.818 |

[a]Ref[6], [b]Ref[8],

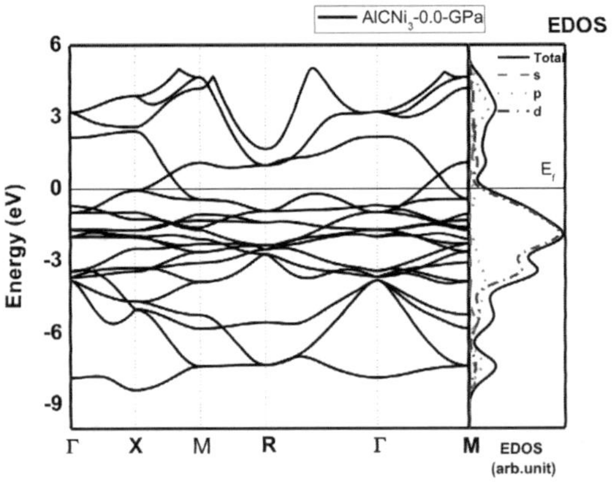


**Figure 1(a)** Calculated band structures of antiperovskite AlCNi3 at zero pressure along with DOS.

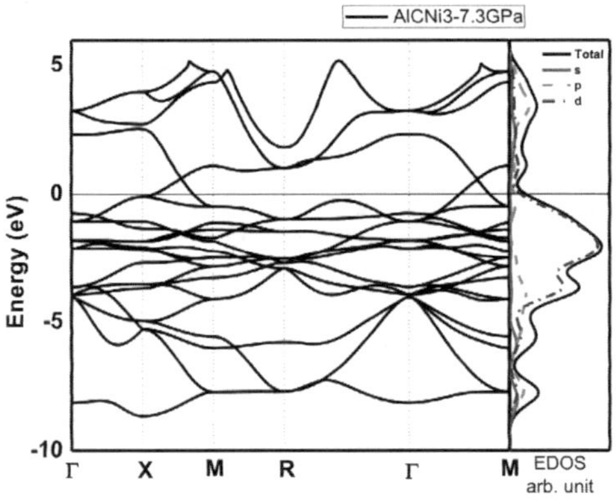


**Figure 1(b)** Calculated band structures of antiperovskite AlCNi3 at 7.3 GPa pressure along with DOS.

Lattice dynamical calculations (phonon dispersion curves and phonon density of states) are performed within the framework of the self-consistent density functional perturbation theory (DFPT). This allows the use of only the unit cell to calculate the phonons of any wave vector [10]. The dynamical matrices were calculated on a 4x4x4 grid of q-points sampling and all the structural parameters were fully relaxed.

## 3. Results and Discussion

The theoretically obtained equilibrium lattice constants a=3.774 Å, bulk modulus B0 =194.6 GPa and pressure derivative of the bulk modulus B′ =4.28 for the $AlCNi_3$ agree well with previous theoretical calculations [7-8] and experimental data [6] shown in table 1. The little higher value of lattice constant is inherent in the case of GGA calculation due to underestimation of total energy. The calculated electronic bandstructure of antiperovskite $AlCNi_3$ along the high symmetry direction of the simple cubic Brillouin zone is shown in Figure 1(a) and 1(b). The total and projected electronic density of states for ambient and high pressure have been calculated and shown along with electronic band structure in right side panel in figure 1(a) and 1(b) respectively.

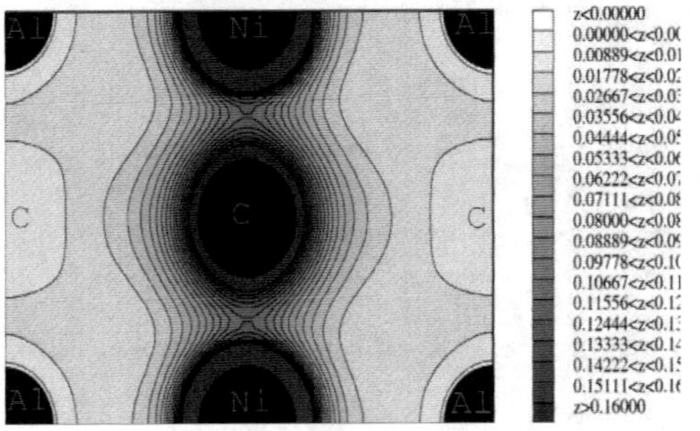


**Figure 2(a).** Charge density plot of $AlCNi_3$ in (110) direction at zero pressure.

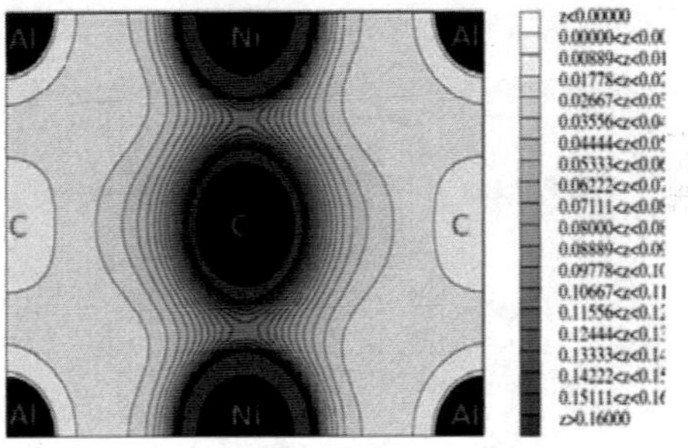


**Figure 2(b).** Charge density plot of $AlCNi_3$ in (110) direction at 7.3 GPa pressure.

Finite value of total electronic density of states at Fermi energy in both conditions clearly indicates the metallic nature of antiperovskite $AlCNi_3$. Our calculated total density of states for the compounds compares quite well in agreement with the density of states of the same compound calculated by the shim et al [11]. The d-Ni distributed mainly between -4.0 to 1.0 eV with band crossing the Fermi level between X and M as well as X-M and Γ-M direction in Brillouin zone. The width of valence band is 5eV. The contribution of d-Al band in valence portion is quite lowest compare to d-Ni band. The 2p-C band and d-Ni band hybridization gives the higher density of states just below the Fermi energy level. There is quite higher density of states contributed mainly due to d-Ni band at -2eV just below the Fermi level while comparing with the DOS of $MgCNi_3$, $AlCNi_3$ have lower density of states at Fermi energy level. While, the sharpen peak in DOS of $AlCNi_3$ quite below from Fermi energy comparing to $MgCNi_3$. The lowest broad band in the Figurers lies between -9 to -6 eV below the Fermi level is predominantly due to s-C states. In $AlCNi_3$, the contribution of d-Ni band to the density of states at the Fermi level N(Ef) is about 25% and total band of Ni atom is about 70% of the total DOS near the Fermi level. It has been generally observed that the total density of states of antiperovskite $AlCNi_3$ in narrow region around the Fermi level the magnitude of $N(E_f)$ decreases as the lattice constant increases. Further, to investigate the behaviour and nature of chemical bonding as well as the interaction between the ions in both conditions for antiperovskite $AlCNi_3$, we have calculated the charge density contour in the (110) planes shown in Figures 2(a) and (b) respectively. The aluminium atoms are positioned at the four corners, the C atom is at the centre of the charge contour plot while the Ni atoms are shown at on top and below position of the carbon atom. The charge density plot clearly indicates that the interaction between Ni and C band is quite strong at zero pressure as well as hydrostatic 7.3 GPa pressure. It can be seen from figures that Al-Ni bonding has strong covalent character. This shows that there is strong hybridization between d-Ni and p-C bands. It can be seen from Figures 2(a) and (b) that due to applied hydrostatic pressure applied C-Ni covalent bond become stronger and the strength of Al-Ni bonding becomes weaker. It is interesting to consider the effect of phonons and to determine the phonon modes involved in superconductivity. The theoretically obtained phonon spectra at zero pressure and 7.3 GPa pressure are shown in Figure 3. This figure shows different phonon spectra from its isostrcutural $MgCNi_3$ [7], for which all phonon modes are stable along the main symmetry directions [100], [100] and [111], except M high symmetry point in whole Brillouin zone.

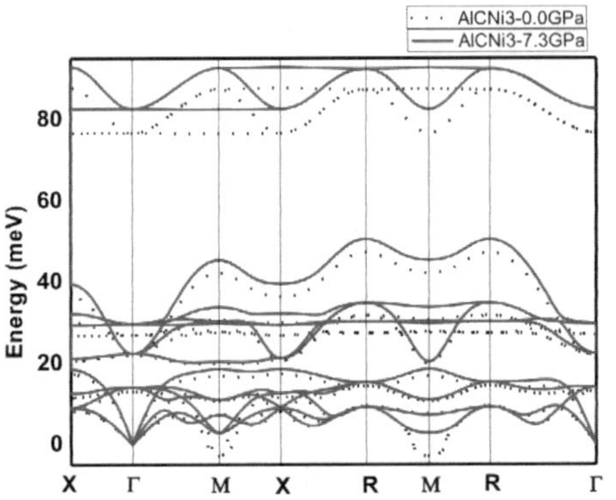


**Figure 3.** Calculated phonon dispersion curves for AlCNi3 at zero and 7.3GPa

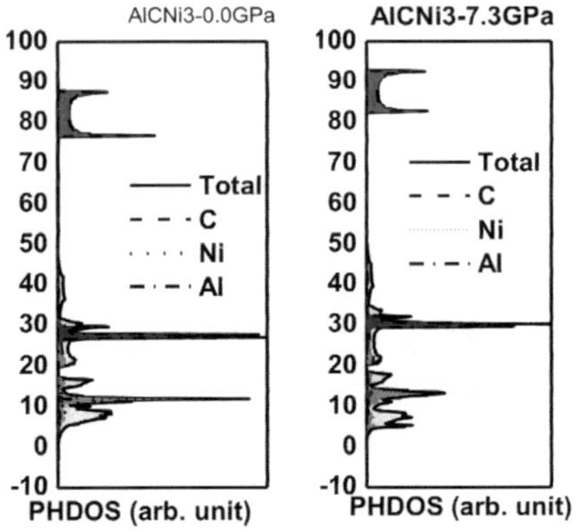


**Figure 4(a)** and **(b)** Calculated phonon DOS dispersion curves for $AlCNi_3$ at zero and 7.3GPa Pressure respectively.

The antiperovskite $AlCNi_3$ have five atoms per unit cell, from which derives total 15 phonon branches. From which, out of 15 there are 3 acoustical and 12 optical. The optical phonon at the gamma points corresponds to the following irreducible representations: $\Gamma\left(O_h^1\right) = F_{1u}^1 + F_{1u}^2 + F_{1u}^3 + F_{2u}$ The lowest acoustic phonon modes are due to the vibration of heavy Ni atoms. The $F_{1u}^1$ phonon modes is the resulting vibration from all atoms in the unit cell. We have described earlier all phonon modes are stable along the whole Brillouin zone except M points due to the large *q*-vector high symmetry M point in the Brillouin zone. The LA phonon branch lies lower in energy than the transverse acoustic branch (TA) for $AlCNi_3$ at zero pressure further the hydrostatic pressure applied by reducing the lattice constant. Furthermore, it has been already shown from experimentally and theoretically that decrease in lattice constant would lead to decrease in $N(E_f)$. It can be seen the $N(E_f)$ from density of states in figure 1(a) and figure 1(b) changes significantly. The applied hydrostatic pressure induces phonon hardening to all phonon branches and gives positive phonon spectra throughout the Brillouin zone. It is interesting to note that the applied hydrostatic pressure at 7.3 GPa the LA phonon branch having lower energy gains the energy and TA branch release the energy and become more softer with applied hydrostatic pressure. After presenting the outcome on dispersion curve, we now pay our attention to the *e-p* coupling. The mass of Al is higher than that of Mg, and the lighter Al-C layers are expected to give lower frequency phonon than Mg-C layers. This can reduce the magnitude of the average phonon frequency which appears in BSC strong coupling expression for Tc and leads to a lower $T_c$ in $AlCNi_3$. The e-p enhancement parameter and superconducting transition temperature ($T_c$) for $AlCNi_3$ are presented in table 1, which are reasonably in good agreement with the theoretical [7] and experimental investigation [12]. The superconducting transition temperature $T_c$ is calculated using modified Allen Dynes formula. Theoretically calculated e-p coupling parameter $\lambda$

increased at higher pressure and N(Ef) decreases with increasing pressure and resultant superconducting transition temperature decrease with applied pressure.

## 4. Conclusion

In conclusion, we have performed first principles investigation of structural electronic and phonon properties for $AlCNi_3$ at zero and external hydrostatic pressure 7.3 GPa using density functional theory. The results for structural electronic and phonon properties of $AlCNi_3$ at zero pressure are consistent with previously available experimental and theoretical data. The calculations reveal that this material remains metallic in both ambient and high pressure conditions. The applied hydrostatic pressure pushed the position of the density of states far from the Fermi level from its ground state. This unfavours for the formulation of cooper pars within the context of BCS theory of superconductivity. The phonon spectra clearly reveal that the superconductivity can be only due to soft phonon mode at high symmetry M points in the Brillouin zone. The hydrostatic pressure plays major role for the dynamical stability by increasing the bonding strength between N-C atoms and superconductivity reduces due to harden of low energy LA phonon mode at high symmetry M point in the Brillouin zone, which resulted in the decrement in predicting the high temperature superconductivity in the $AlCNi_3$ at 7.3 GPa high pressure.

## Acknowledgements

Financial assistance from Department of Science and Technology, Govt. of India, New Delhi and Department of Atomic Energy, Govt. of India, Mumbai.

## References

[1] Shan L, Liu Z Y, Ren Z A, Che G C, and Wen H H, Phys. Rev. B 71, 144516 (2005).
[2] Berk N F and Schrieffer J R, Phys. Rev. Lett. 17, 433 (1966).
[3] Jiji P, Joseph T and Singh P P, Phys. Rev. B 72, 064519 (2005).
[4] Huetter L J and Stadelmaier H H, Acta Metall. 6, 367 (1958).
[5] Hase I, Phys. Rev. B 70, 033105 (2004).
[6] Dong A F, Che G C, Huang W W, Jia S L, Chen H, and Zhao Z X, Physica C 422, 65 (2005).
[7] Tütüncü H M and Srivastava G P, J. Phys.: Condens. Matter 18, 11089 (2006).
[8] Okoye C M I, Solid State Communications 136, 605 (2005).
[9] Baroni S, Dal Corso A, Gironcoli S de and Giannozzi P, http://www.pwscf.org
[10] Baroni S, Gironcoli S de, dal Corso A, and Giannozzi P, Rev. Mod. Phys. 73, 515 (2001).
[11] Shim J H, Kwon S K, Min B I, J. Phys. Soc. Jpn. Suppl. 71, 341 (2002).
[12] Tong P, Sun Y P, Zhu X B and Song W H, Phys. Rev. B 74, 224416 (2006).

Journal of Physics: Conference Series **377** (2012) 012090 doi:10.1088/1742-6596/377/1/012090

# The role of pressure in elastic properties of cerium

**I Lukačević[1,4], V Mankad[2], S K Gupta[2], P K Jha[2] and D Kirin[3]**
[1] Department of Physics, University J. J. Strossmayer, Osijek, Croatia
[2] Department of Physics, Bhavnagar University, Bhavnagar, India
[3] Ruđer Bošković Institute, Zagreb, Croatia

**Abstract.** Cerium 4f electrons are peculiar in being spatially localized with a radial extent much smaller than that of 5s and 5p semicore states, yet having an energy in the region of valence 6s and 5f electrons. Two phases, α and γ, of cerium are under debate in terms of different properties like elastic properties or phase transition properties. The mechanical and elastic properties of cerium have generated substantial interest over the years both experimentally and theoretically. Theoretical studies, concerned with the lattice dynamics, which may be of use, as it is expected from the electron-phonon interaction, and plays a significant role during the $\gamma \rightarrow \alpha$ transition, are very limited.
The aim of the present work is to study the behavior of lattice dynamics and elastic constants under increasing pressure using first-principles methods within linear response approach for the face-centered cubic (fcc) γ-cerium and shed some light on the interrelationship among the elasticity and phonons and understand the role of phonons in the phase transition. Our calculations reproduce zero pressure lattice dynamical and elastic properties of cerium with a very good precision.

## 1. Introduction

The story about the one of the most interesting elemental crystals, cerium, and its isostructural phase transition from its γ to its α phase around 0.8 GPa, after several decades, finally closes in. Up to date, two mechanism are commonly used to describe the electronic nature of this phase transition: the Mott transition model and the Kondo-Volume-Collapse (see reference [1] and references therein for a more detailed description).

The question about the role of the lattice dynamics in this phase transition has been posed already in late 70's by Stassis et al. [2]. Based on their inelastic neutron scattering experiments, they have found that the phonon spectrum of γ-Ce is softer than it should be, when compared to the spectrum of thorium using the Lindemann homology rule [3]. These softening features were more pronounced in the phonon modes which are more sensitive to the influence of pressure in the elastic limit. More specifically, these are the modes which, in the elastic limit, involve the elastic constants $c_{11}$ and $c_{12}$, which in turn, determine the bulk modulus. And, not surprisingly, the bulk modulus in γ-Ce exhibits a dramatical decrease with increasing pressure [4,1]. On the ground of these results and the pressure dependence of the thermal displacements, measured by Jeong et al. [5], one can conclude that lattice dynamics and electron-phonon coupling play an important role in the explanation of the $\gamma \rightarrow \alpha$ phase transition and elastic instability in the elemental crystal of Ce [6].

---

[4]Author to whom any correspondence should be addressed (e-mail: ilukacevic@fizika.unios.hr).

A few theoretical studies have already been employed in order to illuminate the role of lattice dynamics. Huang and Chen [7] have calculated the phonon dispersion curves in the γ-Ce, but only at zero pressure. Still, they managed to reproduce, using the Haman scheme norm conserving pseudopotential method, the soft branches in the phonon dispersion curves. A very recent combined study by Krisch et al. [8] measured, using inelastic X-ray scattering, and calculated, using the *ab initio* projector augmented wave approach, the phonon dispersions at various pressures in both and _ phases of Ce. This way, they were, by comparison, able to note the significant changes in phonon dispersions through the phase transition, confirming the important role which lattice dynamics and electron-phonon interaction play in this case. This facts were the motivation for our work: to study in more detail the lattice dynamics and elastic properties across the γ → α phase transition in Ce. A detailed analyses of the electron-phonon interaction is currently in progress.

In a cubic lattice the bulk modulus depends on the shear modulus and the elastic constant $c_{11}$. Since it is known that the shear modulus is relatively insensitive to the applied hydrostatic pressure [4], the softening of the bulk modulus should be a consequence of the softening of $c_{11}$, which is connected with the longitudinal phonon modes. We have, thus, tried to calculate the elastic constants, in order to confirm that there indeed exists the softening of the $c_{11}$ constant. Longitudinal and shear moduli have been precisely measured by ultrasonic experiments by Decremps et al. [1], showing an anisotropic behavior of γ-Ce with respect to the propagation of elastic waves. This property was already predicted by Stassis et al. on the basis of Born-von Karman force constant model [2]. Furthermore, ultrasonic measurements show the softening of the longitudinal modulus with increasing pressure, while the shear modulus is almost constant in the whole stability interval of the phase of Ce crystal. All these elastic properties should be reproduced by our *ab initio* calculations.

## 2. Methods

In our study, we used standard *ab initio* methods based on DFT [9] and DFPT [10] as implemented in the ABINIT [11] code. We used pseudopotentials to substitute the all-electron potentials, plane waves to expand the wave functions and Monckhorst-Pack meshes [12] of k-points to represent the integrations over the reciprocal space. Of the available pseudopotential schemes, we used Troullier-Martins LDA [13] ones (4 valence electrons) for the elastic properties of Ce and Hartwigsen-Goedecker-Hutter LDA [14] ones (12 valence electrons) for the lattice dynamics. Each studied property was converged with respect to the number of the plane waves and k-points in the Brillouin zone: elastic constants under 1 GPa and phonon frequencies under 1 $cm^{-1}$. Above convergences were reached using the plane wave energy cut-offs of 60 and 80 Ha for the elastic constants and phonon frequencies, respectively, and using the 8×8×8 and 6×6×6 k-point meshes for the elastic constants and phonon frequencies, respectively. Marzari's "cold smearing" scheme was used for the occupance of electronic level with the smearing energy of 0.005 Ha. Elastic constants and phonon dispersions were calculated using the standard response function methods, as described in Refs. [15] and [10], respectively. Our equilibrium lattice constant of γ-Ce is 9.99 Bohr, which is in good agreement with experimental [2] and previous theoretical [7] results of 9.75 Bohr and 9.94 Bohr, respectively.

## 3. Results

### *3.1 Lattice dynamics*

Since the elastic instability in the phase of Ce is connected with the soft phonon branches, phonon dispersion curves should show lowered frequencies for the longitudinal modes and transversal modes at L and W points of the Brillouin zone. Figure 1, which represents the calculated full phonon dispersions of γ-Ce at zero pressure, shows clearly that it is so, although the calculated phonon branch at L point is not as soft as in experiments. As these phonon modes are connected with the $c_{11}$ and $c_{12}$ elastic constants in the elastic limit, we should expect to find the softening behavior of these elastic constants with increasing pressure (see subsection 3.2). No significant changes in the phonon dispersions of phase can be seen at the pressure of 0.8 GPa. More specifically, there are no soft modes which go to zero frequency at the phase transition, which, in turn, also indicates that it is the first order phase transition.

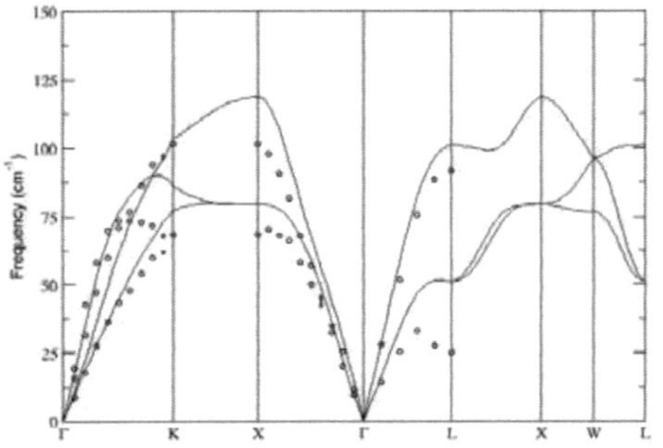


**Figure 1.** Phonon dispersion curves of γ-Ce at 0 Gpa. Circles represent the experimental data obtained by Stassis et al. (reference [2]).

*3.2 Elastic properties*

The influence of the hydrostatic pressure on the calculated elastic constants is shown in figure 2(a). It can be clearly seen that with used pseudopotential method, one cannot reproduce the softening of the $c_{11}$ constant as expected from experiments [2,5]. The zero pressure values of the elastic constants, on the other hand, are in an excellent agreement with the experimental values, which was not obtained in any previous theoretical study, up to our knowledge (see table 1).

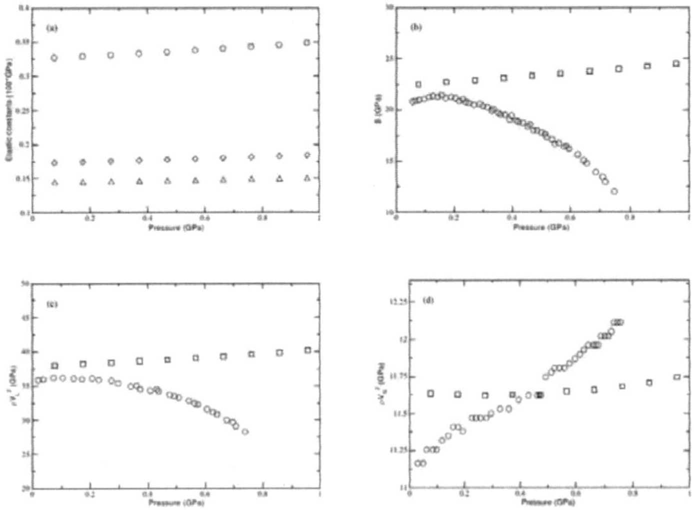

**Figure 2.** Pressure dependence of (a) elastic constants: $c_{11}$ (○), $c_{12}$ (◊) and $c_{44}$ (Δ). There is no expected softening of the $c_{11}$. (b) bulk modulus; (c) longitudinal modulus; (d) shear modulus. In figures (b), (c) and (d) calculated values are presented by □ and experimental [1] by ○.

Consequently, the calculated bulk modulus shows no softening with increasing pressure (Fig. 2(b)). Since longitudinal modulus depends on the bulk modulus, it should also show the softening trend with increasing pressure [1]. Our pseudopotential approach fails here, too (figure 2(c)). Oppositely, shear modulus is structured in such a way that the softening behavior of the $c_{11}$ constant and the bulk modulus is not apparent. It is almost insensitive to the pressure. Here our calculations correctly describe its dependence on pressure (figure 2(d)), because only the difference between the trends is important and not the trends themselves. Our calculations correctly predict the anisotropy in the propagation of the elastic waves, although not in the same measure as experiments.

**Table 1.** Elastic constants $c_{ij}$, bulk B, longitudinal $\rho V_{2L}^2$ and shear $\rho V_{2S}^2$ modulus at zero pressure.

| | | $c_{11}$ (GPa) | $c_{12}$ (GPa) | $c_{44}$ (GPa) | B (GPa) | $\rho V_{2L}^2$ (GPa) | $\rho V_{2S}^2$ (GPa) |
|---|---|---|---|---|---|---|---|
| Present | | 32.5 | 17.2 | 14.3 | 22.3 | 37.8 | 11.7 |
| Exp. | [2] | 24.1 | 10.2 | 19.4 | 14.8 | | |
| Exp. | [1] | | | | 20.6 | 35.6 | 11.2 |
| Exp. | [16] | | | | 21.0 | | |
| Theor. | [7] | 37.9 | 26.4 | 20.9 | 30.2 | | 5.8 |

It is altogether still not clear why the calculated $c_{11}$, B and $\rho V_{2L}^2$ agree so well at zero pressure, while at the same time do not follow the correct behavior with increasing pressure. It is the authors' presumption that the possible reason lies in the pseudopotential used. It has been checked, that the pressure does not increase the energy of the outer core electrons, which would make them more valence-like electrons, and which, thus, would had to be included into the valence bands.

## 4. Conclusions

Pseudopotential *ab initio* calculations of high pressure properties of γ-Ce should have had a part of elucidating the roles of lattice dynamics and elasticity in the phase transition occurring at about 0.8 GPa. Although our calculations correctly reproduced the softer phonon branches in the phonon dispersions and elastic constants (and consequently the bulk modulus) at zero pressure, they failed to reproduce the behavior of the elastic properties, as found in the experiments, at higher pressures. The underlying reason of this is still unknown to the authors. One of the possible reasons is the use of inadequate pseudopotential during the study of the elastic constants.

Nevertheless, the found characteristics of the lattice dynamical and elastic properties indicate that the electron-phonon interaction plays a significant role in the studied phase transition. The calculations of the pressure dependence of this interaction are currently under way.

## Acknowledgments

IL would like to acknowledge the usage of grid infrastructure CRO-NGI (Croatian National Grid Infrastructure) available through SRCE (University Computing Center) organization, Zagreb, Croatia. VM, SKG and PKJ are thankful to the financial Department of Science and Technology, New Delhi, Department of Atomic energy, Mumbai Govt. of India.

## References

[1] Decremps F, Antonangeli D, Amadon B and Schmerber G 2009 *Phys. Rev.* B **80** 132103-6
[2] Stassis C, Gould T, McMasters O D, Gschneider, K A Jr and Nicklow R M 1979 *Phys. Rev.* B **19** 5746-53
[3] Lindemann F A 1910 *Phys. Z.* **11** 609
[4] Voronov F F, Vereshchagin L F and Goncharova V A 1960 *Dokl. Akad. Nauk, SSSR* **135** 1104
[5] Jeong I-K, Darling T W, Graf M J, Proffen Th and Heffner R H 2004 *Phys. Rev. Lett.* **92** 105702-5
[6] Bergmann D J and Halperin B I 1976 *Phys. Rev.* B **13** 2145

[7] Huang Li and Chen Chang-An 2007 *J. Phys.:Condend. Matter* **19** 476206-17
[8] Krisch M *et al.* 2011 *PNAS* **108** 9342-9345
[9] Hohenberg P and Kohn W 1964 *Phys. Rev.* **136** B864
[10] Sham L J and Kohn W 1965 *Phys. Rev.* **140** A1133
[11] Gonze X 1997 *Phys. Rev.* B **55** 10337
Gonze X and Lee C 1997 *Phys. Rev.* B **55** 10355
Baroni S, de Gironcoli S, Dal Corso A and Giannozzi P 2001 *Rev. Mod. Phys.* **73** 515
[12] Gonze X *et al.* 2009 *Computer Phys. Commun.* **180** 2582
[13] Monkhorst H J and Pack J D 1976 *Phys. Rev.* B **13** 5188
[14] Troullier N and Martins J L 1991 *Phys. Rev.* B **43** 1993
[15] Hartwigsen C, Goedecker S and Hutter J 1998 *Phys. Rev.* B **58** 364162
[16] Hamann D R, Wu X, Rabe K M and Vanderbilt D 2005 *Phys. Rev.* B **71** 035117
[17] Shick A B, Pickett W E and Liechtenstein A I 2001 *J. Electron Spectrosc. Relat. Phemon.* **753** 114-6

Journal of Physics: Conference Series **377** (2012) 012091 doi:10.1088/1742-6596/377/1/012091

# *Pressure-induced Structural Phase Transitions in $Ti_{1-x}Zr_xC$ Solid Solution*

**Bhoopendra Dhar Diwan**[1] **and Anurag Srivastava**[2]
[1]MPBOU, Bhopal 462016 M.P., India, [2]Advanced Material Research Lab.; ABV-Indian Institute of Information Technology & Management, Gwalior 474010 M.P., India.
E-mail: bddiwan@gmail.com

**Abstract.** In present paper we have analyzed the relative Structural stability and pressure dependent B1→B2 structural phase transformations in $Ti_{1-x}Zr_xC$ solid solution. For the present work, we have used three-body force potential model which includes the long range as well as short range interactions, where Coulomb and three body interactions are the contents of long range part and the van der Waals multipole interactions and the Hafemeister and Flygare overlap repulsive interactions (effective up to the next nearest neighbors) makes the short range part. The calculated transition pressures are 490, 461, 431, 402 and 372 GPa for TiC, $Ti_{0.75}Zr_{0.25}C$, $Ti_{0.50}Zr_{0.50}C$, $Ti_{0.25}Zr_{0.75}C$, and ZrC compounds respectively.

## 1. Introduction

In the last few years, the refractory metal compounds, particularly the carbides are promising engineering materials application due to their scientific, technological, and commercial importance. On account of their effective use in wide range of industries such as in petroleum, electronics, aerospace, high power, high-temperature engineering, mechanical engineering, nuclear industry, chemical industry [1,2], in cutting tools, information storage technology, optoelectronic deceives, hard coatings, thin films for electronic devices and low temperature fuel cells [3]. These materials have exceptional characteristic, such as high melting temperature, extreme hardness, electrical resistivity, magnetic susceptibility, superconductivity and chemical inertness etc. [4] and are brittle, show metallic lustre, metallic conductivity and few of them are of superconducting in nature. Few experimental and theoretical works confirms that the refractory metal compounds combine covalent, metallic as well as ionic type bonding due to the charge transfer and shows unique properties as ultra hard in nature because of the great degrees of covalency [5-7].

The structural, electronic, mechanical, thermal properties and pressure induced phase transition in some transition metals and their carbides have been investigated through different theoretical approximation [8-15]. Ahuja et al. [16] have predicted NaCl (B1) to CsCl (B2) type phase transitions in TiC, TiN, and TiO using FPLMTO method. Srivastava et al. [17] have recently reported the structural properties and pressure induced structural phase transition in transition metal carbides (TMCs) by using ab initio methods. In an another report by Singh et al. [18] by using rigid ion model the prediction of

[1] bddiwan@gmail.com.
[2] profanurag@gmail.com

phase transition needs revisit of the process when they use unreasonable assumptions as they have considered the same values of inter ionic separation ($r_o$) [24, 25, 30] and bulk modulus $B_o$ (or elastic constants) [24, 25, 30] in both the phases, as input, is not justified. The different properties of ternary compounds of these carbides have also been studied by many workers through different approximations [19-22] in view of these findings, importance of these methods and few challenges in model calculations we thought it pertinent to perform the present study thought revisit the model calculations.

## 2. Theory and Method of Computations

The phase transition is a phenomenon, where the parental structure transforms to the most stable structure at elevated pressure and/or temperature and the stability of any system can be explained on the basis of minimization of the total energy of the system which in terms of its Gibbs free energy can be written as

$$G = U + PV - TS \tag{1}$$

Where, S is the vibrational entropy at absolute temperature T, V is the volume of the unit cell at pressure P, and U is the internal energy, which at $T = 0$ K equivalent to the cohesive energy due to mutual interaction of the ions. Internal energies of the system can be expressed as

$$U(r) = -\frac{\alpha_m e^2 z}{r}[z + 2nf(r)] - \frac{C}{r^6} - \frac{D}{r^8} + nb\beta_{+-}\exp\frac{(r_1 + r_2 - r)}{\rho}$$
$$+ n'\frac{b}{2}\beta_{++}\exp\frac{(2r_1 - kr)}{\rho} + n'\frac{b}{2}\beta_{--}\exp\frac{(2r_2 - kr)}{\rho} \tag{2}$$

where the first two terms are the potential energies due to long range coulomb and three body interactions [23], third term is due to van der Waals interaction governed by dipole-dipole (d-d) and dipole-quadrupole (d-q) interactions [24-25] and, the last three terms are the Hafemeister and Flygare (HF) type overlap repulsive interactions [26] effective up to the next nearest neighbour ions, k is the structure factor for the B1and B2 type phases. C and D are the overall van der Waals constants represented in terms of the vdW coefficients $c_{ij}$ and $d_{ij}$ weighted by the appropriate lattice sums ($S_{ij}$ and $T_{ij}$) and reported by Tosi [24-25] for the B1and B2 type phases, respectively. The values of these coefficients $c_{ij}$ and $d_{ij}$ have been evaluated by the relationship given in Slater and Kirkwood variational (SKV) approach [23, 27] and $\beta_{ij}(i = +; j = -)$ is the Pauling coefficients [28]. $Ze$ is the ionic charge with e as the electronic charge; $b$, $\rho$ are the range and hardness parameters; r is the nearest neighbor separation; $r_1(r_2)$ the ionic radii of the ion $i(j)$ and $f(r)$ is the three body interaction parameter defined by Cochran [29]. Cohesive energies as expressed in equations 2 contains three material dependent model parameters $b$, $\rho$ and $f(r)$, calculated by making use of the equilibrium condition, i.e. first derivative of lattice energy equal to zero, the isothermal bulk modulus and the Cauchy discrepancy $C_{12} - C_{44} \neq 0$ from the knowledge of lattice constant ($2r_o$).

## 3. Results and Discussion

The model parameters hardness parameters (ρ), range parameter (b) and three body interaction parameter ($f(r)$) are calculated using the measured values of the equilibrium inter ionic spacing $r_o$, second order elastic constants $C_{ij}$, reported in Table 2 and following the equilibrium conditions, for TiC, VC and their compound $Ti_{1-x}Zr_xC$ in its each phase B1 (B2). Our calculated the cubic type structure considered to be stable at the anion to cation radius ratio between 0.41 to 0.73 for NaCl-type structure and above 0.73 to 1.00 for CsCl-type structure [24, 25, and 30]. Therefore, the assumption made by Singh et al. [18] in their artical, that the cation and anion radii (as inter ionic separation $r_o$) have the same value in both the phases is not justified. Another reason might be the lattice parameter $a$ is inversely correlated with the bulk modulus $B_o$ [31], so we cannot take the same values of bulk modulus (or elastic constants) as an input to compute the model parameters for

both the phases B1 and B2. Same is with another assumption made by Singh et al. [18] that the elastic constants $C_{ij}$ (bulk modulus $B_o$) has the same value in both phases is also not justified. In high pressure structural phase transition from B1 type phase to B2 type phase, the inter-ionic potential reduces considerably and the coordination number increases. In our calculation, we have used the coordination numbers to calculate the input parameter for B2 type phase instead of taking same values.
The values of model parameters are depicted in Table 2 and utilized to compute the internal lattice energies (U) for binary and ternary carbides. The U-V relationship curve defines the stability of compound and give the equilibrium inter ionic separation $r_o$ associated with the minimum energies for B1 and B2 type phases, respectively. The preference for the rock salt structure (the cubic symmetry) may be traced back to the ionic character of the C-atoms, seems that the phase transition occurs from B1 type phase to B2 type phase.
The phase transition between two phases occurs, when in both the phases the Gibbs free energy $G_{B1}$ ($G_{B2}$) is equal. In order to get information about the transition pressure $P_T$, we have calculated their respective cohesive energy $U_{B1}$ and $U_{B2}$ at T = 0 K as a function of corresponding volume $V_{B1}$ and $V_{B2}$ in B1 and B2 phase. The calculated results in the B1 and B2 phases are plotted in Figure 2 and 3 and draw by the common tangent of the both energy–volume curves phase transition pressure can be determined. TiC, ZC and their compound $Ti_{1-x}Zr_xC$ at ambient pressure crystallize in the NaCl type structure and undergo a transition to the CsCl structure upon compression. Our computed phase transition range from 372 GPa to 490 GPa and we found that under compression, the transition pressure decreases as we increase the Zr composition in TiC shown in figure 1. Table 2 shows the present calculated value of phase transition pressure $P_T$ whereas is close the other one theoretical report [16] but deviate from other reported values [17, 18]. This deviation in results might be due to the type of interaction taken into consideration and the approach of generating the model parameter, particularly at model calculation.

**Table 1.** Input parameter for TiC and ZrC.

| TMCs | $r_o$ (Å) | $C_{11}$ (Mbar) | $C_{12}$ (Mbar) | $C_{44}$ (Mbar) |
|---|---|---|---|---|
| TiC | 2.165[a] | 5.13[c] | 1.06[c] | 1.78[c] |
| ZrC | 2.349[b] | 4.70[c] | 1.00[c] | 1.60[c] |

[a]Ref [32], [b]Ref [33], [c]Ref [8]

**Table 2.** Model parameters, inter ion spacing $r_o$, cohesive energy U and transition pressure $P_T$ for $Ti_{1-x}Zr_xC$ Solid Solution in its phases B1and B2.

| X | $\rho$ (Å) | | $b$ ($10^{-19}$J) | | $f(r)$ ($10^{-2}$) | | $r_o$ (Å) | | U (eV) | | $P_T$ (GPa) |
|---|---|---|---|---|---|---|---|---|---|---|---|
| | B1 | B2 | B1 | B2 | B1 | B2 | B1 | B2 | B1 | B2 | |
| 0 | 0.621<br>0.200[a] | 0.431 | 9.764<br>0.534[a] | 3.74 | 2.111 | 4.046 | 2.164<br>2.170[a]<br>2.165[b] | 2.326<br>2.270[a] | -25.80<br>-25.69[a] | -22.20<br>-24.95[a] | 490<br>490[c]<br>691[d]<br>709[d]<br>68[a] |
| 0.25 | 0.615 | 0.422 | 9.814 | 4.196 | 2.135 | 4.032 | 2.21 | 2.37 | -24.68 | -20.98 | 461 |
| 0.50 | 0.61 | 0.414 | 9.864 | 4.653 | 2.159 | 4.019 | 2.26 | 2.42 | -23.55 | -19.75 | 431 |
| 0.75 | 0.604 | 0.405 | 9.914 | 5.109 | 2.182 | 4.005 | 2.31 | 2.47 | -22.43 | -18.53 | 402 |
| 1.0 | 0.598<br>0.202[a] | 0.396 | 9.964<br>0.978[a] | 5.565 | 2.206 | 3.991 | 2.36<br>2.350[a]<br>2.349[e] | 2.529<br>2.450[a] | -21.30<br>-21.51[a] | -17.30<br>-19.91[a] | 372<br>525[d]<br>98[a] |

[a]Ref[18], [b]Ref [32], [c]Ref [16], [d]Ref [17], [e]Ref [33]

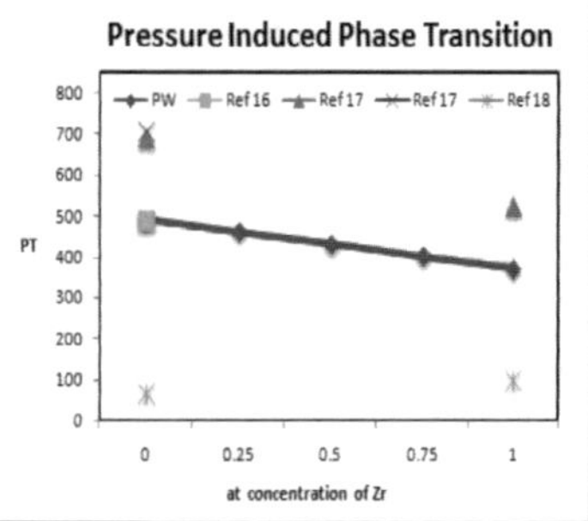


**Fig. 1** Pressure Induced Phase Transition of $Ti_{1-x}Zr_xC$; as a function of composition (Zr).

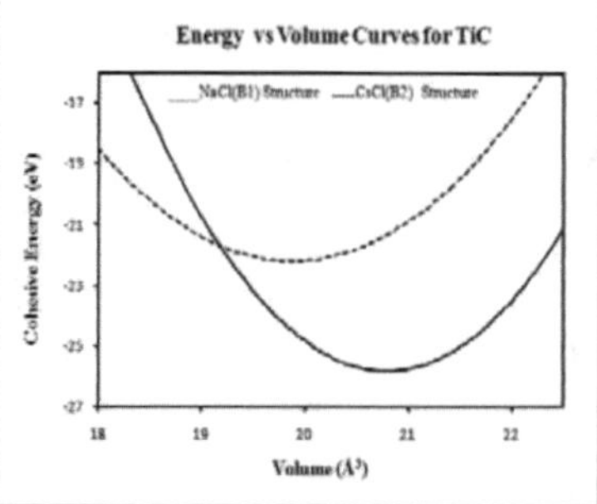


**Fig. 2.** Variation of cohesive energy as a function of corresponding volume of TiC for both structures B1 and B2.

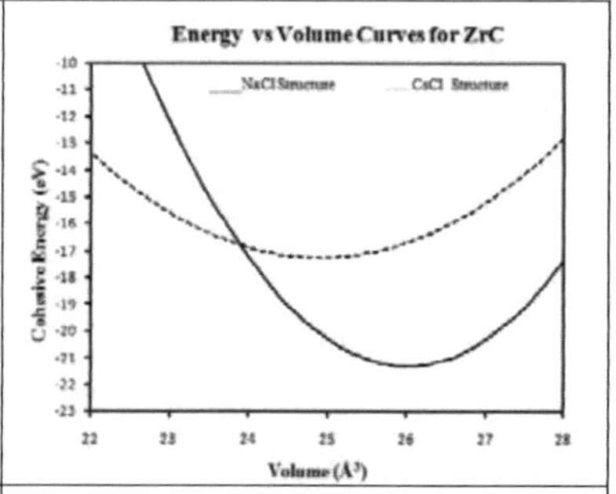


**Fig.3.** Variation of cohesive energy as a function of corresponding volume of ZrC for both structures B1 and B2.

## 4. References

1. Shveikin G P, Alyamovskii S I, Zainulin Yu G, et al. 1984 (UNTs AN SSSR, Sverdlovsk,)
2. Gusev A I, Rempel A A, and Magel A J 2001 Springer, New York, 607
3. Ham D. & Lee J S 2009 Energies **2**, 873–899
4. Toth L E Transition 1971 Metal Carbides and Nitrides Academic Press, New York
5. Blaha, P Redinger J and Schwarz K 1985 Phys. Rev. B **31**, 2316
6. Wijeyesekera S 1984 Organometallics **3**, 2179-2189
7. Samsonov G 2008 Powder Metallurgy and Metal Ceramics **47**, 13–20
8. Weber W 1973 Phys. Rev. B **8** 5082–5092
9. Upadhyaya K S et al. 2005 Sciences-New York **64** 299-304
10. Isaev E et al. 2007 Journal of Applied **101** 123519
11. Haglund J, Grimvall G, Jarlborg T, and Guillermet A F 1991 Phys. Rev. B 43 14400–14408.
12. Guillermet A F, Haglund J and Grimvall G 1992 Phys. Rev. B 45 11557–11567
13. Liermann H 2004 Int. Journal of Refractory Metals and Hard Materials **22** 129-132.
14. Ogorodnikov V and Rogovoi Y I 1993 Powder Metallurgy and Metal Ceramics **32** 532–538.
15. Zhao Z et al. 2010 Materials Science arXiv: **1011**. 3579
16. Ahuja R, Ericsson O, Wills J M and Johansson B 1996 Phys. Rev. B **53** 3072–3079
17. Srivastava A, Chauhan M and Singh R K 2011 Phase Transitions **84** 58-66
18. Singh A, Aynyas M and Sanyal S 2009 Phase Transitions **82** 576-586
19. Zaoui A, Bouhafs B and Ruterana P 2005 Materials Chemistry and Physics **91, 1** 108-115.
20. Adjaoud O, Burton B P and Walle A V D 2009 Phys RevB.**80** 134112 32-34.
21. Chen K and Zhao L 2007 Journal of Physics and Chemistry of Solids **68** 1805-1811
22. Ramasubramanian S, Rajagoplan M, Thangavel R and Kumar J 2009 The European Physical Journal B **69** 265-268
23. Slater J C and Kirkwood K G 1931 Phys. Rev. **37** 682-697
24. Tosi M P and Fumi F G J. 1962 Phys. Chem. Solids **23** 359
25. Tosi M P 1964 Solid State Physics **16** 1-120
26. Hafemiester D W and Flygare H F 1965 J. Chem. Phys. **43** 795-800
27. Narayan R J 1977 Phys. Chem. Solids **38** 1097
28. Pauling L Z 1928 Kristallogr **67** 377-404
29. Cochran W 1971 C.R.C. Crit. Rev. Solid State Sci.**2** 1-44
30. Alan J Cohen and Roy G Gordon, Phys. Rev. B **12**, 3228 (1975)
31. Bajorath J 2001 J. Chem. Inf. Comput.Sci. **41** 233-45
32. Zhukov V P et al. 1988J. Phys. Chem. Solids **49** 841–849.
33. Krmpter C P and Fries R J 1960 Anal. Chem. **32** 570

Journal of Physics: Conference Series **377** (2012) 012092 doi:10.1088/1742-6596/377/1/012092

# Pressure effects on band structures in dense lithium

**Naoyuki Goto**
Division of Frontier Materials Science, Graduate School of Engineering Science, Osaka University, 1-3 Machikaneyama-cho, Toyonaka, Osaka, 560-8531, Japan

**Hitose Nagara**[1]
Graduate school of Engineering, Tottori University, 4-101 Koyamacho-minami, Tottori, Tottori 680-8552, Japan

E-mail: nagara@damp.tottori-u.ac.jp

**Abstract.** We studied the change of the band structures in some structures of Li predicted at high pressures, using GGA and GW calculations. The width of the 1s band coming from the 1s electron of Li shows broadening by the pressurization, which is the normal behavior of bands at high pressure. The width of the band just below the Fermi level decreases by the pressurization, which is an opposite behavior to the normal bands. The character of this narrowing band is mostly p-like with a little s-like portion. The band gaps in some structures are really observed even by the GGA calculations. The gaps by the GW calculations increase to about 1.5 times the GGA values. Generally the one-shot GW calculation (diagonal only calculations) gives more reliable values than the GGA, but it may fail to predict band gaps for the case where band dispersion shows complex crossing near the Fermi level. There remains some structures for which GW calculations with off-diagonal elements taken into account are needed to identify the phase to be metallic or semiconducting.

## 1.Introduction

Lithium has interesting properties under high pressure. When it is compressed to tens of gigapascal, it shows superconductivity with high $T_c$ and the $T_c$ reaches to 20K, which was the highest record among elements before the 29K in Calcium was found[1-3]. Another interesting property is the metal to semiconductor transition at high pressure[4-8]. At ambient pressure it is a metal as is well known, but when the pressure is applied its conductivity decreases and finally it becomes a semiconductor, which is an opposite behavior to that in the normal substances. Semiconductor-to-metal transition is usually observed at high pressure, which is due to the increased overlapping of the wave functions of neighboring atoms by the decrease of the interatomic distance. To understand this peculiar behavior in lithium at high pressure, we studied the change of the band structures in lithium under high pressure using first principles calculations.

## 2.Structures

Lithium undergoes successive structural phase transitions under high pressure. The bcc structure at ambient pressure transforms to the fcc structure and, around 40GPa, to the *cI16* structure through the narrow region of *hR1* structure. Up to the phase with the *cI16* structure, Li remains metallic. The *cI16* structure prevails above 70GPa where another structural phase transition has been experimentally reported[8,9]. The structures above the *cI16* have not yet been identified experimentally. Theoretical

[1] To whom any correspondence should be addressed.

Published under licence by IOP Publishing Ltd

studies have predicted further structural phase transitions at higher pressures, as well as many low energy structures[10,11]. In some of the predicted structures the band gap openings have been reported theoretically, which may correspond to the experimentally reported semiconductor phase. We studied the change of the band structures in some of the structures predicted at high pressures using GGA and GW calculations. The band structures studied in this paper are of *cI16* , of *Aba2-24*, of *C2*, and of *Cmca-24,* all of which have been determined or predicted as the structures at higher pressures[10,11]. The calculations are performed using the packaged software Quantum-Espresso[12] and VASP[13].

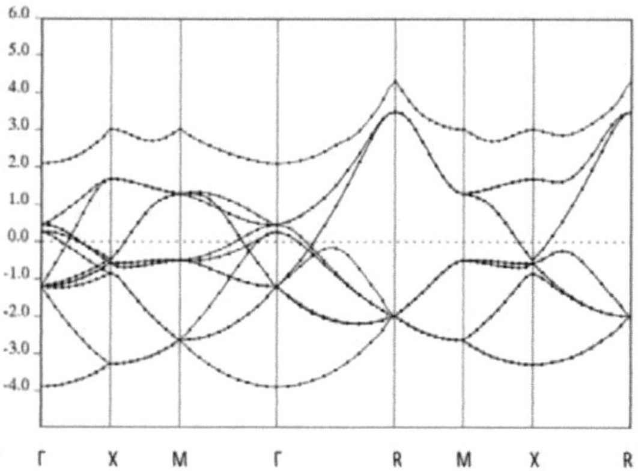


**Fig.1** Band structure of *cI16* at 38GPa.

### 3.Results

Fig.1 is the band structure of the *cI16* structure at 38GPa which clearly shows a metallic band with band dispersion curves crossing the Fermi level. Fig.2 shows the band structure of the *Aba2-24* structure calculated by the GGA at 89GPa with corrections by the one-shot GW approximation (diagonal only). The top of the valence band is shifted a little from the bottom of the conduction band which is located at Γ point in the Brillouin zone. The gap in this structure is the nearly direct-gap type as shown in the figure. The GGA band gap is estimated to be 0.1717 *eV.* The band gap decreases with decreasing pressure around this pressure. The correction by the one-shot GW approximation, where only diagonal matrix elements of the self-energy are taken into account, shows no apparent change of the feature of the band structure. It only shifts the one electron energies and the band gap near the Γ point is increased to about 1.6 times the GGA value. Fig.3 shows the similar band structure of *C2* structure at 89GPa with corrections by the GW approximation. The band gap of this structure is a indirect type. The top of the valence band is located at B and the bottom of the conduction band at Γ. The indirect band gap in this structure is 0.2243*eV*, which decreases with decreasing pressure and vanishes around 70 GPa. The direct band gap at Γ point is 0.496*eV* which is increased to 1.4 times bigger value by the GW correction. For the above two structures where the band dispersion curves do not

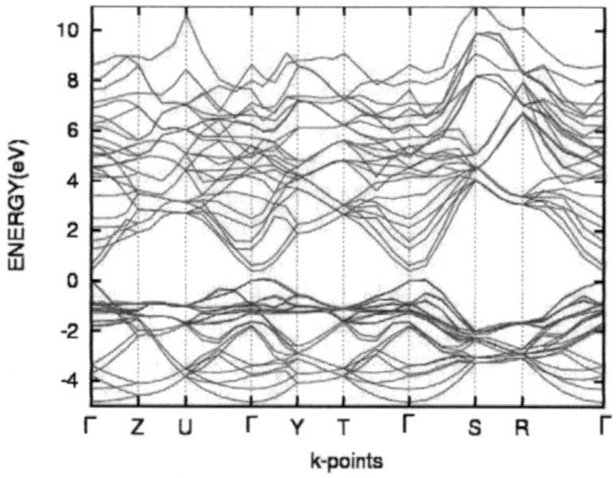


**Fig.2** Band structure of the *Aba2-24* at 89GPa with correction by the GW (red lines).

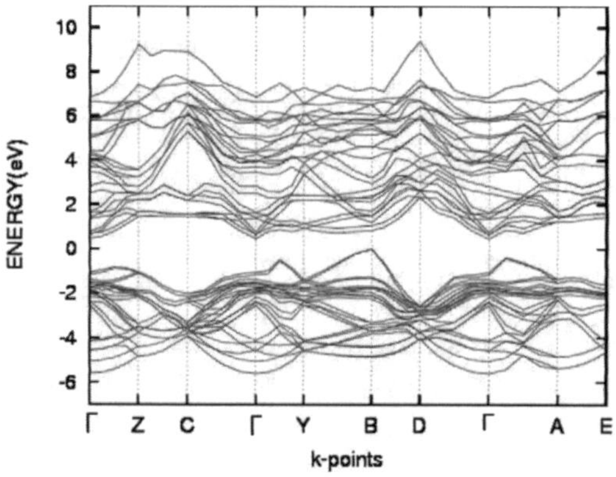


**Fig.3** Band structure of *C2* at 89GPa with correction by the GW(red line).

cross the Fermi level, located at 0 *eV,* the band gap corrections can be rather accurately obtained by the one-shot GW calculation (diagonal only calculation). Fig.4 shows the band structure of the *Cmca-24* structure obtained by the GGA. The GGA band structure shoes metallic behavior as is observed. What if the correction by the GW approximation is made? The one-shot GW calculations are known to fail when the band dispersion curves show complicate crossing near the Fermi level. Apparent example of the failure is the band structure of Ge at ambient pressure[14]. Ge is well known to be a semiconductor but the GGA band structure is a metallic character in which the dispersion curves complicatedly cross the Fermi-level. The one-shot GW calculation, however, do not correct this problem and Ge remains to be metal. To solve the problem we need to take off-diagonal elements of the self-energy into account [14].

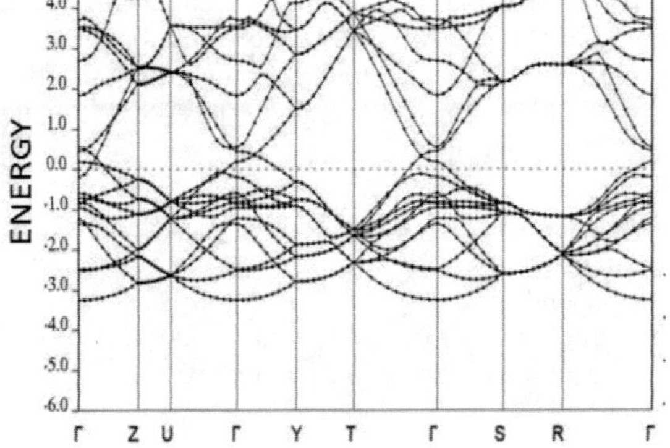


**Fig.4** Band structure of *Cmca-24* structure at 80GPa calculated by the GGA.

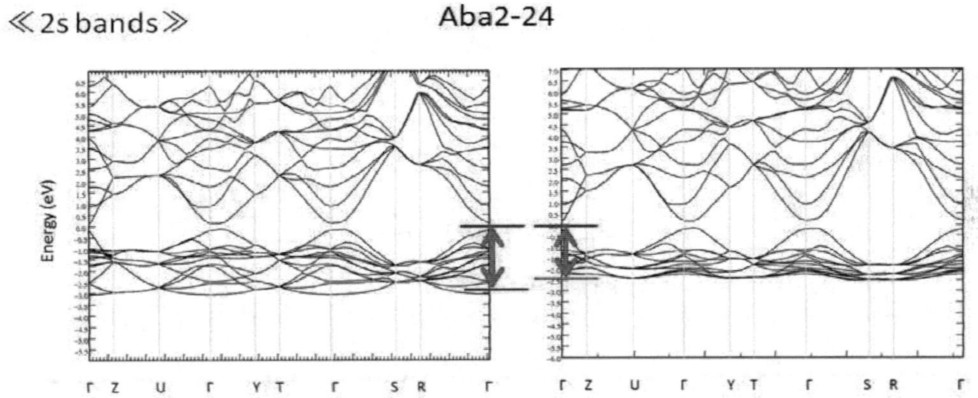


**Fig.5** Change of the 2s band width.

We have not yet tried the GW calculation taking the off-diagonal elements into account, but there remains a possibility that the band structure turns into semiconductor type. In many structures predicted so far, there may remain some structures with semiconductor type and more extensive studies using GW approximation with off-diagonal elements taken into account are necessary.

Another interesting feature of the pressure effect on the band structures in the above studies is the narrowing of band width in the band below the gap. In many semiconductors the band width increases with

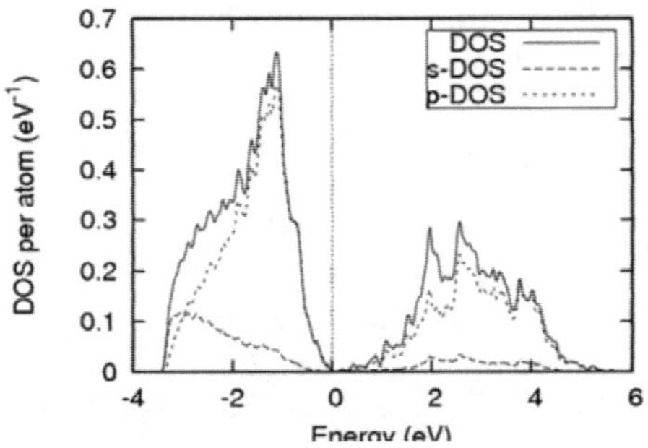


**Fig.6** P-DOS of the band in *Aba2-24* at 55GPa.

increasing pressure. Fig.5 shows the narrowing of the band width just below the gap. This narrowing shows some localization of the electrons constituting these bands. Fig.6 shows the projected DOS of those bands. The electrons forming this band originally come from the 2s electron of Li. The character, however, has changed into p-character with s-character remains as a small portion in the DOS. The band of the 1s electron shows broadening with increasing pressure, which is the normal behavior of the band width when the overlap of the wave function increases. The increase of the band width is shown in Fig.7. The 1s band is located more than $40eV$ below the Fermi level. But the overlap of the wave function is observed with increased band width. In the band structure calculations on Li at high pressure the 1s electron is usually treated as the band electron, not the core electron. This comes from the fact that significant overlap of the 1s wave functions and the resulting broadening of the band are observed in Li at high pressure.

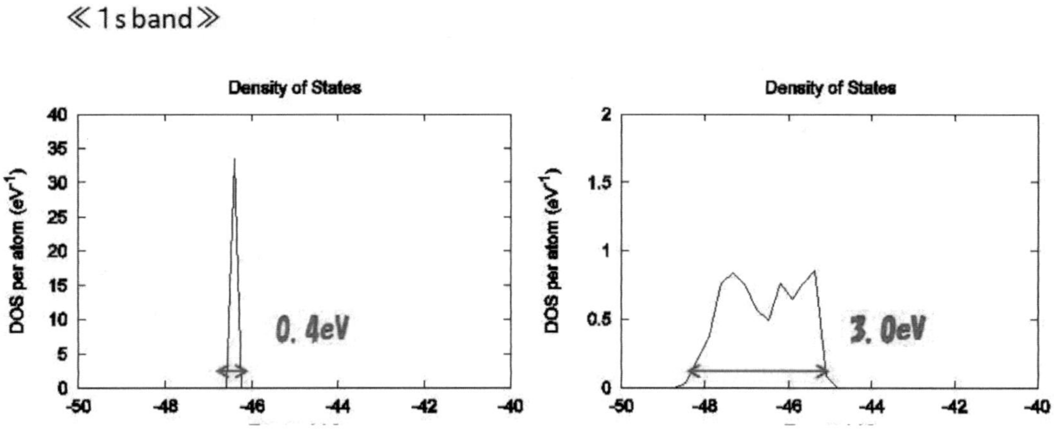


**Fig.7** Broadening of the 1s bands in Li at high pressure

**References:**

[1] N. E. Christensen and D. L. Novikov, Phys. Rev. Lett. **86**, 1861 (2001).
[2] K. Shimizu et al. Nature **419**, 597 (2002).
[3] M. Sakata et al., Phys. Rev. B**83**, 220512(R) (2011).
[4] J. B. Neaton and N. W. Ashcroft, Nature **400**, 141 (1999).
[5] A. Bergara, J. B. Neaton, and N. W. Ashcroft, Phys. Rev. B**62**, 8494 (2000).
[6] F. Siringo, R. Pucci, and G.G.N. Angilella, High Press. Res. **15**, 255 (1997).
[7] G.G.N. Angilella, F. Siringo, and R. Pucci, Eur. Phys. J. B**32**, 323 (2003).
[8] T. Matsuoka and K. Shimizu, Nature **458**, 186 (2009).
[9] M. Hanfland, K. Syassen, N. E. Christensen, and D. Novikov, Nature **408**, 174 (2000).
[10] Y. Yao, J. Tse, and D. Klug, Phys. Rev. Lett. **102**, 115503 (2009).
[11] J. Lv, Y. Wang, L. Zhu, and Y. Ma, Phys. Rev. Lett. **106**, 015503 (2011).
[12] P. Giannozzi et al. J. Phys. Condens. Matter **21**, 395502 (2009)
[13] G. Kresse and J. Hafner, Phys. Rev. B**47**, 558(1993).
[14] M. van Shilfgaarde et al PRL **96**, 226402(2006).

Journal of Physics: Conference Series **377** (2012) 012093 doi:10.1088/1742-6596/377/1/012093

# Electron band structure of the high pressure cubic phase of $AlH_3$

**Hongliang Shi[1], Niliffar Zarifi[2], Wai-Leung Yim[1] and J.S. Tse[2*]**
[1] Institute of High Performance Computing,
1 Fusionpolis Way, #16-16 Connexis, Singapore 138632
[2] Department of Physics and Engineering Physics,
University of Saskatchewan, Saskatoon SK, Canada S7N

E-mail: John.Tse@usask.ca

**Abstract.** The electronic band structure of the cubic Pm3n phase of $AlH_3$ stable above 100 GPa is examined with semi-local, Tran-Blaha modified Becke-Johnson local density approximation (TB-mBJLDA), screened hybrid density functionals and GW methods. The shift of the conduction band to higher energy with increasing pressure is predicted by all methods. However, there are significant differences in detail band structure. In the pressure range from 90 to160 GPa, semi-local, hybrid functional and TB-mBJLDA calculations predicted that $AlH_3$ is a poor metal. In comparison, GW calculations show a gap opening at 160 GPa and $AlH_3$ becomes a small gap semi-conductor. From the trends of the calculated band shifts, it can be concluded that the favourable conditions leading to the nesting of Fermi surfaces predicted by semi-local calculation have disappeared if the exchange term is included. The results highlight the importance of the correction to the exchange energy on the band structure of hydrogen dominant dense metal hydrides at high pressure hydrides and may help to rationalize the absence of superconductivity in $AlH_3$ from experimental measurements.

Since the suggestion of possible superconductivity with high critical temperature ($T_c$) in dense hydrogen dominant metal hydrides at high pressure [1], there have been enormous experimental and theoretical efforts to investigate the structural and electronic properties on a wide variety of-hydride compounds. Experimentally, superconductivity was discovered in $SiH_4$ at high pressure [2]. However, the nature of $SiH_4$ and the origin of the superconductive behaviour have been a subject of current debate [3-5]. On the other hand, very optimistic predictions of $T_c$ ( > 80 K) have been reported in a number of theoretical studies [6,7]. Without exception, all the theoretical studies employed the "standard "Kohn-Sham method of density function theory (DFT) with the ground state electronic structures and vibrational spectra obtained from semi-local (generalized gradient corrected) density functionals using the pseudopotential plane wave method [8]. The electronic and vibrational information were then used in subsequent electron-phonon calculations [9]. Therefore, the reliability of the prediction of superconductivity depends critically on the accuracy of the theoretical band structure.

Unlike $SiH_4$, there is no dispute on the crystal structure of the predicted superconductive phase of $AlH_3$ stable beyond 100 GPa [10]. Both experimental x-ray diffraction and theoretical studies show it has a cubic Pm-3n structure with a body-center arrangement of Al atoms and the H atoms occupy the interstitial sites [10]. The band structure obtained from DFT calculations using the Perdew-Burke-Ernzerhof (PBE) functional [11] revealed two electronic bands cross the Fermi level forming an electron pocket at R and an electron hole at M symmetry points (Fig. 1). The theoretical results show the dispersions of the two bands are similar (parallel to

* To whom any correspondence should be addressed.

each other) and the two pieces of Fermi surfaces can be nested in the M → R direction. As the result, a large mass enhancement parameter $\lambda$=0.74 was obtained from electron-phonon calculations and the cubic phase is expected be superconductive. Using the extended McMillan equation based on the Migdal-Eliashberg theory, a $T_c$ of 24 K was predicted at 110 GPa [10]. Experimental resistance measurement of have been performed down to 4 K and up to 164 GPa [10]. Although the results confirm $AlH_3$ is a poor metal, no superconductivity was found. Usually, with a judicial choice of a nominal Coulomb repulsion parameter ($\mu^*$, *ca.* 0.1 – 0.15) the calculated $T_c$ is often in reasonable agreement with experiment. In this case, the discrepancy between the theory and experiment is puzzling and has motivated several theoretical investigations [13,14]. It was shown that the lattice vibrations in $AlH_3$ are highly anharmonic [13]. If the renormalized frequencies were used in the calculations of the phonon linewidths, a smaller mass-enhancement parameter was obtained [13]. On the other hand, anharmonicity may enhance the electron-phonon couplings. In fact, this is one of the basic assumptions of the Garland-Benneman-Mueller theory on the superconductivity in disordered solids [15]. More recently, calculations on the optical spectrum of cubic $AlH_3$ has predicted a highly damped low energy transitions below 1 eV that may induce an abrupt edge in the reflectivity [14]. This observation is significant as the electronic excitation energy is very close to the H-H vibration frequencies thus raising the possibility of strong coupling between them. In this study, we took a different viewpoint. Since the large electron-phonon coupling in $AlH_3$ was attributed to the nesting of the Fermi surfaces predicted from semi-local PBE calculations, it is important to investigate possible changes in the band structure calculated with improved DFT methods proposed recently that may help to explain the lack of superconductivity.

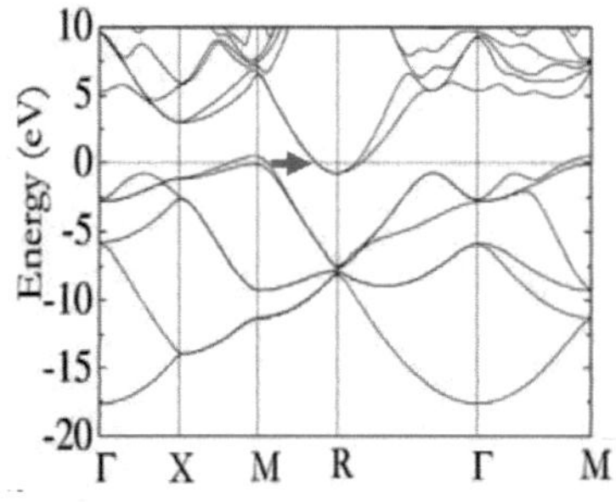


**Figure 1.** Electronic band structure of the cubic *Pm-3n* phase of $AlH_3$ calculated with PBE functional at 100 GPa (adapted from ref.10). Two bands cross the Fermi level at M and R. The red arrow indicates the nesting vector connecting the two pieces of Fermi surfaces.

The Kohn-Sham (KS)-DFT method is computationally efficient and can be used on a wide variety of complex materials. The calculated structural and chemical properties are often in very good agreement with experiments. However, there are several known shortcomings with the DFT approach using the "standard" functionals. Relevant to the discussion to be presented below, is that KS-DFT calculations often underestimate the energy gaps between occupied valence and unoccupied conduction band and some cases, even the general characteristic of the band structure. This deficiency can be traced back to the failure of DFT on describing the electron self-interaction and/or exchange interaction correctly. Several practical remedies to this problem have been proposed recently. Within the local density approximation (LDA) framework, an approximate functional designed to mimic the gap structure in the exact exchange of an atom was introduced by Becke-Johnson (BJLDA) [16] and later extended to solids by Tran and Blaha (TB-mBJLDA) [17,18]. An alternative is to add a small amount of non-local Hartree-Fock (HF) exchange to the semi-local density functional. The PBE0 hybrid functional is a combination of the popular PBE functional with a 25% Hartree-exchange [19]. Taking advantage of the fast spatial decay of the short range HF exchange, a screened Coulomb hybrid functional separating the short-range and long range HF exchange was introduced by Heyd-Scuseria-Erhzerhof (HSE) [20]. The HSE functional offers a highly efficient computational strategy for extended periodic systems. A different approach based on many-body perturbation theory (GW) is often used to correct for the band gap problem. The GW method [21,22] goes beyond the local exchange mean field approximation and includes many-body electron-electron interactions through the screening of the exchange term with inverse dielectric matrices. Comparisons of the different methods have been made in several recent publications [23-27] together discussions on their strength and weakness and will not be repeated here. In general, all the above mentioned methods have succeeded to ameliorate the underestimated band gap energies, albeit each has their own limitations and problems.

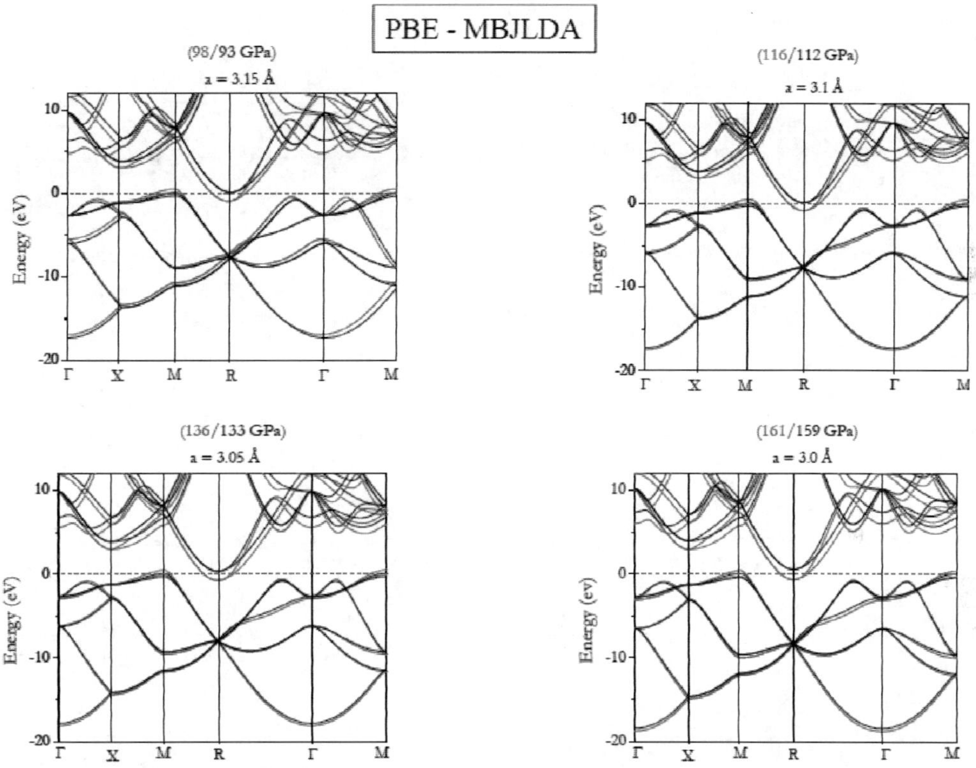


**Figure 2.** Electronic band structure of $AlH_3$ calculated from PBE (red) and TB-mBJLDA functionals using the WIEN2*k* code (black). The pressures are obtained from the equation of states calculated by the respective functionals.

In this study, electronic band structures of of $AlH_3$ in the pressure range from 90 to 160 GPa calculated from First-principles methods PBE [11], TB-mBJLDA [17], HSE [26,27], single shot G0W0 and self-consistent GW (SCGW) [23,24] were investigated to investigate the . The pressure was obtained from the calculated equation of state (EOS, i.e. energy vs volume) by the respective methods. The exception is that the pressure for G0W0 and SCGW were taken from the PBE and HSE EOS. The pseudopotential planewave code VASP [28] and the all electron full-potential linearized augmented plane wave (FLAPW) WIEN2k program [29] were used for the PBE calculations. The WIEN2k code was also used for TB-mBJLDA calculations using PBE optimized wavefunctions as the initial estimates. HSE, G0W0 and SCGW calculations were performed with the VASP code. Converged FPLAPW results can be achieved with a $RK_{max}$ value of 5 and a plane wave cutoff, $g_{max}$, of 20 Ryd. Projected augmented wave (PAW) potentials [30] for Al and H were used in the PAW planewave calculations. The energy for the planewave cutoff is 710 eV. A 12×12×12 Monkhorst-Pack k-point mesh was used for Brillouin integration. For G0W0 and SCGW, calculations were performed with both PBE and HSE

SCF optimized reference wave functions. Test calculations show that 186 empty bands and 80 frequency grid points were necessary to achieve convergence in the calculation of the response function. For the SCGW procedure, convergence on the quasi-particle energies on consecutive iterative step was carefully monitored. Quasi-particle correction to the all the k-points of the 12×12×12 Monkhorst-Pack mesh were computed. Since the Wannier interpolation routine was not available VASP 5.2 [28], the full band structure was not computed and the discussions of the GW results will focus only on the R and M symmetry points.

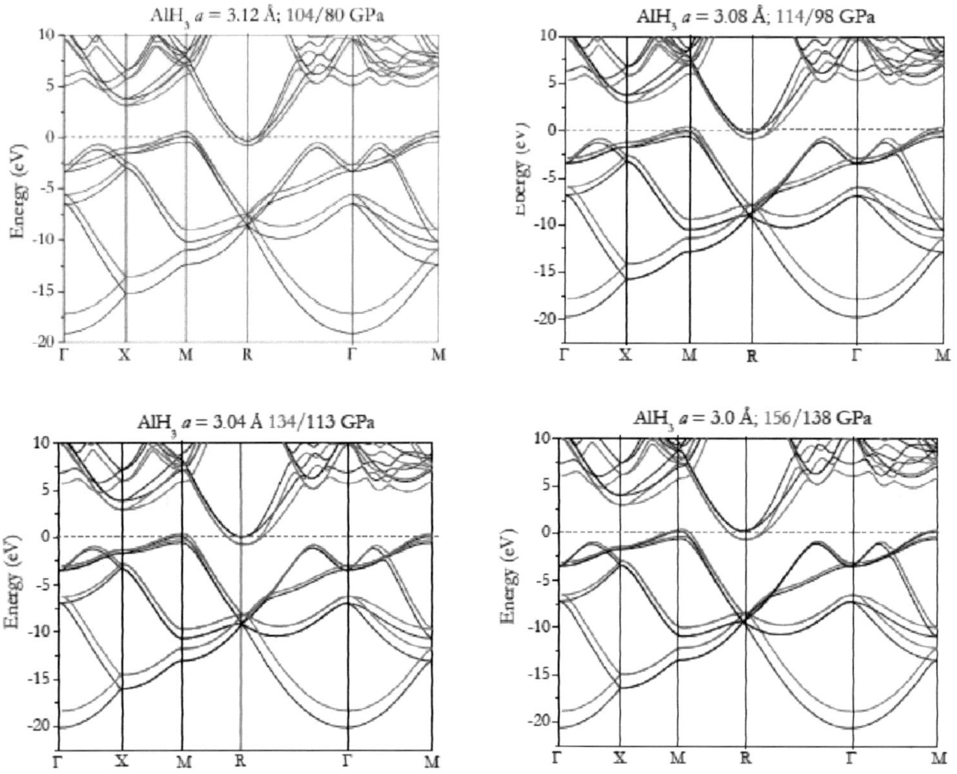


**Figure 3**. Electronic band structure of $AlH_3$ calculated from PBE (red) and HSE functionals using the VASP code (black). The pressures were obtained from the equation of states calculated from the respective functionals.

The high pressure cubic $AlH_3$ has shown to be a metal from DFT calculations. It is useful to compare with the result of TB-mBJLDA on the electronic band structure. Standard DFT calculations introduce a large error in the proper description of the H atom due to an incomplete cancellation of the Hartree and exchange-correlation self-interaction energies [25]. The PBE and TB-mBJLDA band structures for $AlH_3$ computed using WIEN2k at selected pressures are compared in Fig. 2. The band structure profiles are the same as previous PBE results (Fig. 1) [10]. The focus of ensuing discussion will be on the energy profile of the two bands crossing the Fermi level at the M and R point. At M, the electronic band is an inverted parabola while the band dispersion at R resembles a parabolic free-electron band. At first glance, there is little difference in the band structure of the occupied bands obtained from the two methods. In comparison, the vacant TB-mBJLDA conduction bands were shifted to higher energies. This observation is consistent with previous reports which TB-mBJLDA exhibits a similar trend in band shift. At 93 GPa ($a$=3.15 Å), the maximum at R and the minimum at M cross the Fermi level forming an electron hole and pocket, respectively. As the pressure increases, the energy minimum at R near the Fermi level also increases. Fig. 3 compares the band structures obtained from PBE and HSE functionals using the VASP code at different pressures. There are significant differences between the results obtained by the two methods. Relative to the Fermi level, the HSE valance bands below the Fermi level shifted to lower energies whilst the opposite trend is observed for the conduction bands. The energy maximum at M and the energy minimum at M stay very close to the Fermi level. Even though, the "pseudo-gap" between M and R also increases with pressure, it is not at the same extent as predicted by TB-mBJLDA.

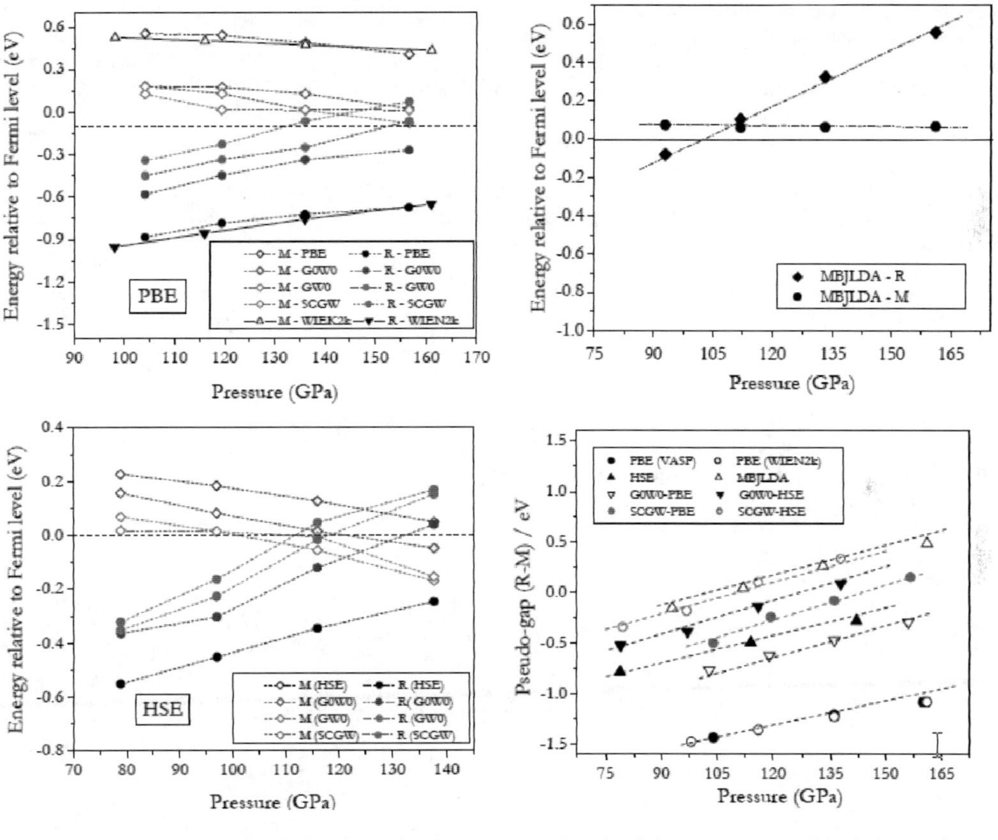


**Figure 4.** Comparison of calculated maximum energy at *R* and minimum energy at *M* relative to the Fermi energy using (a) PBE functional with VASP and Wien2*k* code and *GW* corrections; (b) TB-MBLJDA; (c) HSE functional with VASP and *GW* corrections; (d). pseudo-gap energy between the *R* and *M* using PBE (VASP and WIEN2k), TB-mBJLDA, HSE, *G0W0* (PBE and HSE) and *SCGW* (PBE and HSE). The lines are intended to guide the eye and do not indicate

Variation of the energy with pressure at R and M relative to the Fermi level from PBE, TB-mBJLDA, HSE, G0W0 and SCGW calculations are analyzed in detail and the results are presented in Fig. 4. Two points are relevant to the ensuing discussion. If the energy at M is positive, the electronic band crosses the Fermi level. If the energy at R is positive, the band does not cross the Fermi level. The corollary is, if the energy at M is negative and the energy at R is positive then the system is a semiconductor (insulator). As long as the energy at M remains positive, the system is metallic. First we examine the consistency between the pseudopotential and all electron calculations using the PBE functional. As shown in Fig. 4a, the results are almost identical. This observation confirms the choices of various computational parameters. The PBE results show a steady decrease in the maximum energy at the M point with increasing pressure which almost touches the Fermi level at pressure above 160 GPa. The minimum energy at R shows exactly the opposite. Since M is positive and R is negative

within this pressure range, $AlH_3$ is metallic and the nesting of two pieces of Fermi surfaces at R and M is still possible. This conclusion is the same as reach in a previous study using the PBE functional [10].
Although the trends on the variation of the band energy at M and R are similar to PBE, the TB-mBJLDA results show a different band structure profile. At 93 GPa, the band structure is broadly similar to that of the PBE functional. The maximum energy of the band at M is not very sensitive to pressure and remains above the Fermi level showing that the system is metallic up to the highest pressure studied here. The energy minimum at R, however, moves to higher energy at high pressure and no longer crosses the Fermi level. Therefore, nesting of the two Fermi surfaces is no longer possible.

The HSE results show a similar pattern as PBE (Fig. 4c). The difference is that the energy at R and M relative to the Fermi level are substantially lower. For example, at 105 GPa the TB-mBJLDA maximum energy at M of 0.52 eV and minimum energy at R -0.87 eV differs quite significantly from the corresponding HSE values of 0.32 eV and -0.25 eV. AlH3 is still metallic within the pressure range studied, however, since the HSE bands were shifted, the curvatures of the two Fermi surfaces will no longer be same as predicted by PBE and nesting of the Fermi surfaces may not be feasible (Fig. 3).

One shot G0W0 corrections employing reference ground state PBE and HSE wave functions are reported in Fig. 4a, 4c and 4d. The variation of the energy at M and R with increasing pressure follows the trend observed in the corresponding PBE and HSE calculations. G0W0 corrections to the quasi-particle energy are noticeably larger with the PBE wave function. Relative to the PBE, G0W0 (PBE) calculations show an almost uniform shift of +0.3 eV of the energy at R and -0.4 eV for the energy at M over the entire pressure range. In comparison, the G0W0 (HSE) corrected quasi-particle energies at R are much smaller. The quasi-particle corrections, however, are more significant at M where the G0W0 (HSE) energies are ca. 0.2 eV lower than the corresponding HSE values. There is one important difference between the PBE and HSE G0W0 results. At pressure higher than 135 GPa, G0W0 (HSE) energy of M becomes lower than the Fermi energy and is higher than the Fermi level at R. This observation suggests that $AlH_3$ may become a small gap insulator. As it was not able to compute the full band structure using the GW method with VASP, this speculation cannot be established unambiguously based on the quasi-particle corrections on only two k points.

Before presenting results of SCGW calculations using PBE and HSE ground state reference wave functions, the convergence of GW corrected quasi-particle energies at M and R point calculated at a = 3.08 Å are compared in Fig.5. The convergence rate was found to be faster for the HSE wave function. Differences in the quasi-particle energies between successive iteration step drop below 10-3 eV just after the second step. In comparison, although the quasi-particle energy correction at M converged quickly with SCGW (PBE), the energy at R only converged after the fourth iterations. The converged SCGW (PBE) and SCGW (HSE) quasi-particle energies at M are found to be practically identical; whilst the converged SCGW (PBE) quasi-particle energy at R approaches that of the SCGW (HSE) value within 0.04 eV. The results also show the SCGW quasi-particle energies are quite close to that obtained from single shot G0W0 (HSE) calculation. An observation has already been made in previous studies [27].

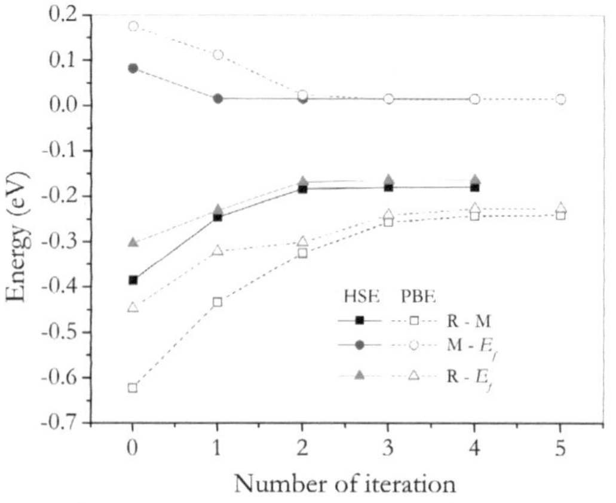


**Figure 5**. Convergence of *SCGW* calculations.

Results of G0W0, GW0, and SCGW corrections on the PBE and HSE wave functions are compared in Fig. 4a and 4b, respectively. Both the PBE and HSE quasi-particle energy differences relative to the Fermi energy at M and R are reduced with increasing level of correlation treatment. For example, for the HSE corrections at ca. 100 GPa, the energies at M relative to the Fermi level decrease respectively from 0.18 eV (G0W0), to 0.12 eV (GW0) and then to 0.02 eV (SCGW). Similarly, the energies at R relative to the Fermi level change steadily from -0.46 eV (G0W0), to- 0.32 eV (GW0) and finally to 0.16 eV (SCGW). The GW0 corrections to the quasi-particle energies are close to of the SCGW values but the absolute differences are still fairly significant ( < 0.1 eV). Shifts in the band energies lowered the possible "metal $\rightarrow$ insulator" transition pressure from the G0W0 (HSE) predicted value of 135 GPa to 120 GPa for SCGW. A similar trend in the energy shift is also observed on the quasi-particle GW corrections to the PBE eigenvalues at different level of approximations. The SCGW energies of the M and R points so as the pseudo R-M gaps calculated from the PBE wave functions are still noticeably different from the corresponding results using the HSE wave functions as references.

Since the original proposal of possible superconductivity in $AlH_3$ was attributed to the nesting of two pieces of Fermi surfaces cutting the Fermi level at M and R [10], the critical parameter to be examined is the pseudo-gap between the M and R point. The main results obtained from the present investigation are summarized in Fig. 5. All the methods studied here produced the same general trend. Within the pressure range studied by PBE, TB-mBJLDA and HSE calculations, show $AlH_3$ is shown to be metallic although there is gradual opening of the gap between M and R. In contrast, GW calculations show the gap opens much faster with pressure and AlH3 may become an insulator. It is interesting to note that even though the TB-mBJLDA calculations predicted the energy change in at M is relatively insensitive to pressure. The calculated R-M pseudo-gap energies are very close to the SCGW (HSE) results. Contrary to previous observation [27], there are significant differences between the pseudo-gap energies in $AlH_3$ computed from HSE, G0W0 (HSE) and SCGW. The substantial changes in the conduction band structure can be attributed to the large self-interaction for the H atoms in the hydrides [25, 32]. Therefore, great care must be exercised in the study of hydrogen dominant dense metal hydrides particularly in the situation with long H-H distance where the property of H is closer to the atom. Owing to a lack of experimental data, it is not possible to ascertain the accuracy of the band structures computed from the different methods. Experimental resistance measurements [10] show the resistivity increases from 100 to 160 GPa and the temperature profiles resembled that of a bad metal. Present results obtained from more refined density functionals suggest there is an increase in the pseudo-gap of between M and R with pressure. This band shift will affect curvatures of the bands at the Fermi level and the possibility of extensive nesting might diminish. The prediction from PBE calculations that the two electronic bands crossing the Fermi level near M and R led to nesting of Fermi surfaces may be not be realized in reality. It can be concluded that $AlH_3$ will be a poor metal or even a low band gap semiconductor under pressure up to 160 GPa, and it is not superconductive. In passing, it is noteworthy in spite of many successful applications of standard DFT calculations in reproducing structures and structural transitions for solids at high pressure, in some systems, accurate band structures are critical for the determination of the phase transitions. An example is that DFT calculations underestimated the insulator to metal transition in $\varepsilon$-$O_2$ transition pressure by almost 50 GPa [33,34]. The discrepancy with experiment is removed once the quasi-particle corrections were made with G0W0 calculations [35]. In a recent calculation of the electron-phonon coupling in $C_{60}$ molecules [36], hybrid functional although did not alter the structure and vibrational properties significantly, it increased the contribution of the exchange energy to the total electron-phonon coupling by almost 40% bringing the theoretical prediction in better agreement with experiment.

## References

[1] Ashcroft N.W. 2004 *Phys. Rev. Lett.* **92** 187002.

[2] Eremets M.I., Trojan I.A., Medvedev S.A., Tse J.S. and Yao Y. 2008 *Science* **319** 1506.

[3] Strobel T.A., Goncharov, A.F., Seagle, C.T., Liu, Z., Somayazulu, M., Struzhkin, V.V. and Hemley, R.J., 2011 *Phys. Rev. B* **83** 144102.

[4] Hanfland M., Proctor J.E., Guillaume C.L., Degtyareva O., and Gregoryanz E. 2011 *Phys. Rev. Lett.* **106** 095503.

[5] Degtyareva O., Proctor J.E., Guillaume C.L., Gregoryanz E. and Hanfland M. 2009 *Solid State Comm.* **149** 1583.

[6] Tse J.S., Yao Y. and Tanaka K. 2007 *Phys. Rev. Lett.* **98** 117004.

[7] Jin X., Meng X. He, Z., Ma, Y., Liu, B., Cui, T., Zou, G. and Mao, H-k. 2010 Proc. Natl. Acad. Sci., **107** 9969.

[8] Burke K., Perdew, J.P. and Ernzerhof M. 1998 *J. Chem. Phys.* **109**, 3760.

[9] Baroni S., de Gironcoli S., Dal Corso A. and Giannozzi P. 2001 *Rev. Mod. Phys.* **73** 515.
[10] Goncharenko I,. Eremets M.I., Hanfland M., Tse J.S., Amboage M., Yao Y. and Trojan I.A. 2008 *Phys. Rev. Lett.* **100** 045504.
[11] Perdew J.P., Burke K. and Ernzerhof M. 1996 *Phys. Rev. Lett.* **77** 3865.
[12] Allen P.B. and Dynes R.C. 1975 *Phys. Rev. B* **12** 905.
[13] Rousseau B. and Bergara A. 2010 *Phys. Rev. B* **82** 104504.
[14] Gurtubay I.G., Rousseau B. And Bergara A. 2010 *Phys. Rev. B* **82** 085113.
[15] Garland J.W., Bennemann K.H. and Mueller F.M. 1968 *Phys. Rev. Lett.* **21** 1315.
[16] Becke A.D. and Johnson E.R. 2006 J. Chem. Phys. **124** 221101.
[17] Tran F. and Blaha P. 2009 *Phys. Rev. Lett.* **102** 226401.
[18] Koller D., Tran F. and Blaha P. 2011 *Phys. Rev. B* **83** 195134.
[19] Ernzerhof M. and Scuseria G.E. 1999 *J. Chem. Phys.* **110** 5029.
[20] Heyd J., Scuseria E.E. and Ernzerhof M. 2003 *J. Chem. Phys.* **118** 8207.
[21] Aryasetiawan F. and Gunnarsson O. 1998 *Rep. Prog. Phys.* **61** 237.
[22] Onida G., Reining L. and Rubio A. 2002 *Rev. Mod. Phys.* **74** 601.
[23] Fuchs F., Furthmüller J., Bechstedt F., Shishkin M. and Kresse G. 2007 *Phys. Rev. B* **76** 115109.
[24] Shishkin M. and Kresse G. 2006 *Phys. Rev. B* **74** 035101.
[25] Singh D.J., 2010 *Phys. Rev. B* **82** 205102.
[26] Gerber I.C., Ángyán J.G., Marsman M. and Kresse G. 2007 *J. Chem. Phys.* **127** 054101.
[27] Paier J., Marsman M., Hummer K., Kresse G., Gerber I.C. and Ángyán J.G. 2006 *J. Chem. Phys.* **124** 154709.
[28] Kresse G. and Hafner J. 1993 *Phys. Rev. B* **48** 13115; Kresse G. and Furthmüller J. 1996 *Comput. Mater. Sci.* **6** 15.
[29] Blaha P. Schwarz K., Madsen G.K. and Luitz J. 2009 Computer Code WIEN2k, Technical University of Vienna, Vienna.
[30] Kresse G. and Joubert D. 1998 *Phys. Rev. B* **59** 1758.
[31] Hamann D.R, and Vanderbilt D. 2011 *Phys. Rev. B* **83** 195134.
[32] Ruzsinszky A., Perdew J.P., Csonka G.I., Scuseria G.E. and Vydrov O.A. 2008 *Phys. Rev. A* **77** 060502R.
[33] Ma Y., Oganov A.R. and Glass C.W. 2007 *Phys. Rev. B* **76** 064101.
[34] Kim D.Y., Lebègue S., Araújo C.M., Arnaud B., Alouani M. and Ahuja R. 2008 *Phys. Rev. B* **77** 092104.
[35] Tse J.S., Klug D.D., Yao Y. and Desgreniers S. 2008 *Phys. Rev. B* **78** 132101.
[36] Janssen J.L. Côté M., Louie S.G. and Cohen M.L. 2010 *Phys. Rev. B* **81** 073106.

Journal of Physics: Conference Series **377** (2012) 012094 doi:10.1088/1742-6596/377/1/012094

# High Pressure Phase Transitions in Barium Tungstate

**Prabhatasree Goel[1], R. Mittal and S. L. Chaplot**
Solid State Physics Division, Bhabha Atomic Research Centre, Trombay, Mumbai 400085, India

E-mail: knp@barc.gov.in

**Abstract.** Alkaline-earth tungstates exhibit an interesting phase diagram with respect to pressure. $BaWO_4$ is found to undergo a pressure-driven phase transition at 7.1 GPa from tetragonal scheelite structure to monoclinic fergusonite structure. It is observed to become amorphous beyond 45 GPa. In this work, we report molecular dynamics simulation studies on transformations of the scheelite phase in $BaWO_4$ with increasing pressure at ambient temperature. Our calculated equation of state is in very good agreement with reported experimental data and first principles calculations. In our calculations there is no apparent volume discontinuity at the scheelite to fergusonite transformation in $BaWO_4$ around 7.5 GPa. The tungstate transforms to an amorphous phase beyond 35 GPa with discernable volume collapse. The pair correlation functions show that there are subtle changes in the arrangement of the $BaO_8$ polyhedra with increasing pressure. However, the $WO_4$ tetrahedra remain unperturbed with increasing pressure until amorphization occurs.

## 1. Introduction

Alkaline–earth tungstates, $MWO_4$ (M= Ba, Ca, Sr, Pb) are important for their scientific and technological applications. These compounds find their use in solid- state scintillators and optoelectronic devices. Tungstate divalent scheelite compounds find use in solid state laser applications, geological aspects, optoelectronic devices and various other technological applications [1-6]. This has motivated extensive interest in the fundamental physical properties of the $AWO_4$ tungstates. $MWO_4$ compounds under normal conditions crystallize in the tetragonal scheelite structure with ($CaWO_4$ mineral structure). The space group is $I4_1/a$ and the number of formula units per crystallographic unit cell is 4. These compounds exhibit a rich phase diagram. Recent works on angle dispersive X-ray diffraction (ADXRD), x-ray absorption near edge structure measurements (XANES) [2] have observed that upon compression $BaWO_4$ undergoes a scheelite-to fergusonite phase transition at room temperature. $BaWO_4$ is found to undergo a further change to another monoclinic phase at around 10 GPa at room temperature. While at high temperature and high pressure, it undergoes a transition to a distorted scheelite phase (*I2/a*). Despite several experimental [1,2,3,6,7] and theoretical efforts [2,4,5], a complete thorough understanding of the phase diagram of the compound is still elusive. Extensive Raman and infrared scattering techniques [4,7] have been used to study the zone centre phonon modes in several of these tungstates.

$BaWO_4$ is the heaviest member of the tungstate family. The scheelite structure adopted by $BaWO_4$ at low pressures has a body centered unit cell, with barium atoms occupying $S_4$ symmetry

[1] To whom any correspondence should be addressed.

while the sixteen oxygen atoms occupy $C_1$ sites. Each W is surrounded by four equivalent oxygen atoms in a tetrahedral symmetry and Ba is surrounded by eight oxygen atoms. Along a-axis the $WO_4$ units are directly aligned whereas along the c-axis there is a Ba cation between two $WO_4$ tetrahedra. Hence, different arrangement of hard $WO_4$ tetrahedra along the **c** and **a**-axis accounts for the difference in compressibility of the two unit-cell axes. In this paper we have carried out molecular dynamics simulations to understand the mechanism of pressure driven phase changes. We have tried to understand the behavior of various polyhedra under compression.

## 2. Theoretical formalism

Molecular dynamics simulations have been used to study the response of the tungstate to increasing pressure. The parameters used for the studies were obtained from the previous lattice dynamical and inelastic neutron studies on orthotungstates [8]. A supercell of dimension 8**a** × 8**b** × 4**c** was used to study the evolution of the room temperature ambient pressure scheelite phase ($I4_1/a$) with increasing pressures up to 60 GPa. We have used an interatomic potential comprising of Coulomb and short range forces in addition to a van der Waal interaction for the oxygen atoms. We have also included a stretching potential for the W-O bond. All the calculations have been done using the molecular dynamics program MOLDY developed at Trombay, Mumbai [9].

## 3. Results and Discussions

Alkaline-earth scheelites exhibit an interesting phase diagram with respect to increasing pressure. $BaWO_4$ have been found to undergo a displacive phase transition from ambient scheelite phase to a monoclinic fergusonite phase at around 7 GPa.

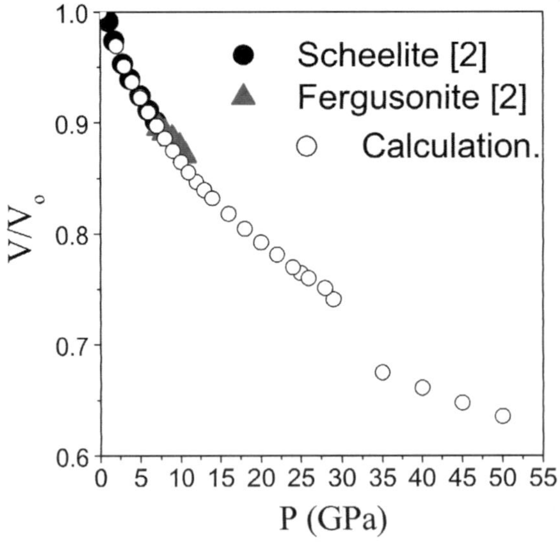


**Figure 1**: Equation of state in $BaWO_4$(open circles-our work) compared with reported data [2] (solid symbols)

This transformation is caused by small displacements of the Ba atoms from their high symmetry positions and due to changes in the O positions which lead to polyhedra distortion. Figure 1 gives the equation of state in $BaWO_4$. Our calculations of the equation of state are in very good agreement with reported experimental and first principles results [2,5,6]. The new monoclinic fergusonite phase is achieved by slight distortion of the original scheelite phase. The lattice parameters in the two phases are very close, their difference being around 0.03 Å and the monoclinic angle β is around 90.09°. Similarity between fergusonite and scheelite is so high that the phase transition could only be inferred from very subtle differences in peaks of the ADXRD spectra. In our calculations the difference between **a** and **b** lattice parameters and β angle is within the statistical errors of molecular dynamics simulations.

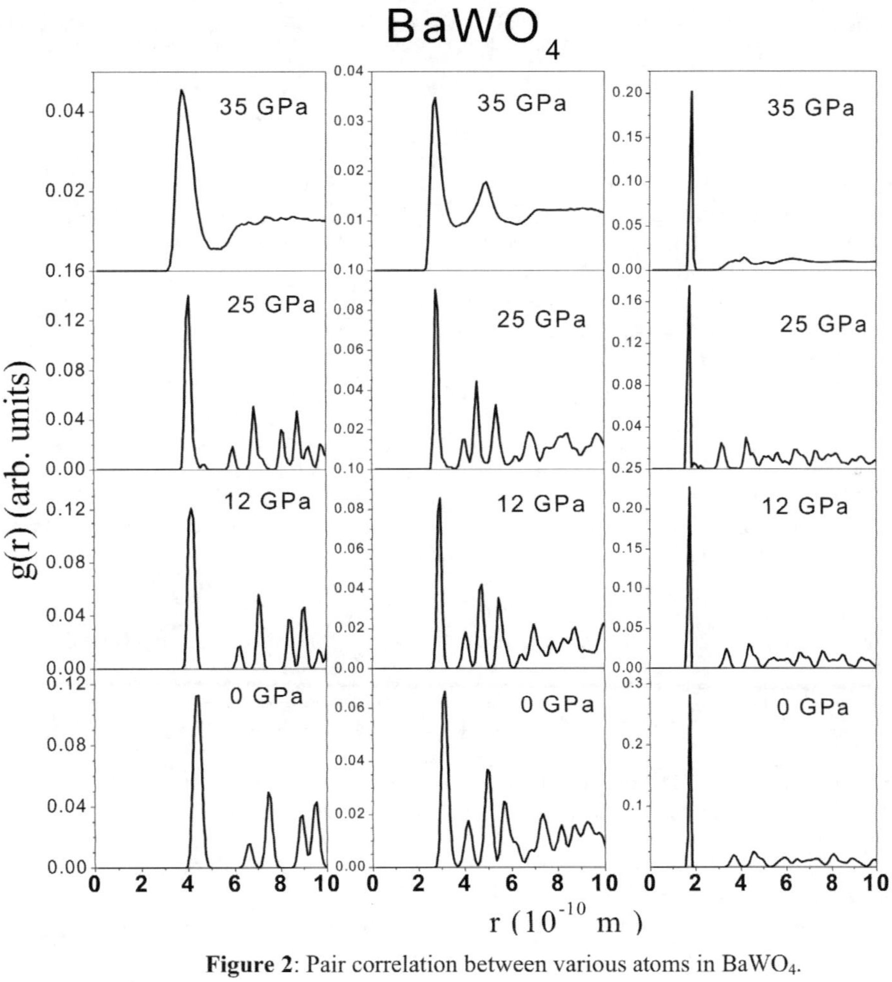


**Figure 2**: Pair correlation between various atoms in $BaWO_4$.

Scheelite phase becomes unstable with respect to a fergusonite-like distortion with increase in pressure. But the distortion is very small to be observed explicitly within our calculations. We calculated the pair correlation function to look at the microscopic changes occurring in the lattice. Figure. 2 gives the pair correlation function in $BaWO_4$ at various pressures. The pair correlations up to 35 GPa look similar. There are very subtle differences between the correlations for Ba-O with increasing pressure. Experimental studies report [2] a transition to another monoclinic structure $BaWO_4$-II at around 10 GPa. Some of the peaks broaden as pressure is further increased beyond 20

GPa. But at all these pressures the W-O correlations remain unchanged. All the peaks corresponding to Ba-W, Ba-O and W-O broaden at 35 GPa. This implies that the system has amorphized. Experimental XANES and ADXRD results [2] indicate amorphization of $BaWO_4$ around 47 GPa. There is a loss of translational periodicity as can be seen from the pair correlations of Ba-O and Ba-W atoms. The experimental studies indicate that there is very little difference in between the scheelite and fergusonite structure.

## 4. Conclusions

Molecular dynamics simulations have been carried out on barium tungstate using pair potentials. We have calculated the pair correlation functions of the Ba-W, Ba-O and W-O pairs with increasing pressure. The pressure-induced phase changes occur due to the distortion of the $BaO_8$ bisdisphenoids. $WO_4$ tetrahedra remain rigid and do not show any change with increasing pressure, depicting the high covalency of the bonding. There are slight changes in the values of the **a** and **b** parameters, which leads to the distortion of the scheelite structure to a monoclinic phase. The compressibility observed in this compound is due to the shortening of the Ba-O bonds. Owing to closeness between the scheelite and fergusonite phases, there is no apparent volume discontinuity observed as can been seen from the equation of state of the compound. Barium tungstate is found to amorphize around 35 GPa which is in agreement with reported experimental works. These studies may have important geophysical and geochemical implications, since scheelite orthotungstates occur in various kinds of rocks in earth's upper mantle. Further, their behavior under different pressure-temperature conditions would help to understand structural stabilities and distortions in these compounds, which could in turn be useful in the optimization of their various technological applications.

## References

[1] Nakamura T, Katagiri M, Ukibe M, Ikeuchi T and Ohkubo M. 2004 Nucl. Instrum. Methods Phys. Res. A **520** 6
[2] Errandonea D, Pellicer-Porres J, Manjon F J, Segura A, Ch. Ferrer-Roca, Kumar R S, Tschauner O, Lopez- Solano J, Rodríguez-Hernandez P, Radescu S, Mujica A, Munoz A, and Aquilanti G 2006 Phys. Rev. B **73** 224103
[3] Lopez-Solano J, Rodriguez-Hernandez P, Radescu S, Mujica A, Munoz A, Errandonea D, Manjon F J, Pellicer-Porres J, Garro N, Segura A, Ch. Ferrer-Roca, Kumar R.S, Tschauner O and. Aquilanti G 2007 Phys. Stat. Sol. B **244** 325
[4] Manjon F J, Errandonea D, Garro N, Pellicer- Porres J, Lopez-Solano J, Rodríguez-Hernandez P, Radescu S, Mujica A, and Munoz A 2006 Phys. Rev. B **73** 144112
[5] Rodriguez-Hernandez P, Lopez-Solano J, Radescu S, Mujica A, Munoz A, Errandonea D, Manjon F. J, Pellicer-Porres J, Segura A, Ch. Ferrer-Roca, Kumar R.S, Tschauner O and Aquilanti G, 2006 J Phys. Chem. of solids, **67** 2164
[6]. Grzechnik A, Crichton W. A, Hanfled M, 2005 Phys. Stat. Solidi, **242** 2795
[7] Miller P J, Khanna R K and Lippincott E R 1973 J. Phys. Chem. Solids **34** 533
[8] Prabhatasree Goel, Mittal R, Chaplot S L and Tyagi A K, 2008 Pramana J. Phy. **71** 1135
[9] Chaplot S L and Rao K R, 1987 Phys. Rev. B **35** 9771

# Thermomechanical approach to the modeling of HP–HT material processing

**O O Lyeshchuk, S B Polotniak and M V Novikov**
V. M. Bakul Institute for Superhard Materials NAS of Ukraine,
2, Avtozavodska St., Kyiv 04074, Ukraine

E-mail: leshchuk@ism.kiev.ua

**Abstract.** The mathematical models, algorithms and software for finite element simulations both superhard materials spontaneous synthesis in high pressure apparatus, and martensitic phase transformation of material at compression in diamond anvil cell have been developed. The phase transformation processes in the reaction volume of high pressure apparatus for diamond synthesis and in the sample of polycrystalline potassium chloride when deformed in the diamond anvil cell have been numerically simulated.

## 1. Introduction

In this study we demonstrate the possibility of applying methods of continuum mechanics for computer modeling and simulation of material processing at high pressure and temperature (HP–HT). This thermomechanical approach allows us to simulate both the processes of spontaneous synthesis of superhard materials in high pressure apparatuses (HPA), and processes of phase transformations in diamond anvil cells (DAC). This results in determination of the fields of thermodynamic parameters in the transformation zone. The present numerical technique is used to optimize the existing technologies of the HP–HT synthesis of materials and to develop new ones.

## 2. Modeling the diamond spontaneous synthesis in HPA

Spontaneous crystallization of diamond particles results from joint action of several physico-mechanical processes, namely, resistance heating, heat transfer, thermoelastoplastic deformation and phase transition. Prior to heating, the reaction cell contains a compressed (up to 4–5 GPa) powder mixture of graphite and fine particles of carbon solvent alloy. Then, the cell is heated by passing the electric current. During heating, the stress field inside HPA is evolving and, after a certain pressure and temperature level was reached, carbon in the reaction volume undergoes phase transition. The diamond crystals begin to grow in the metal melt on carbon clusters being the crystallization centres.

Two-level theoretical model of the diamond spontaneous crystallization in HPA has been developed [1]. At the macro level, the model describes the coupled processes of resistance heating, heat transfer, thermoelastoplastic deformation and graphite-to-diamond phase transition at the reaction volume scale. At the diamond grain (micro) level, it describes the thermoelastic fields and conditions of phase transitions in the local diamond–metal melt–graphite system.

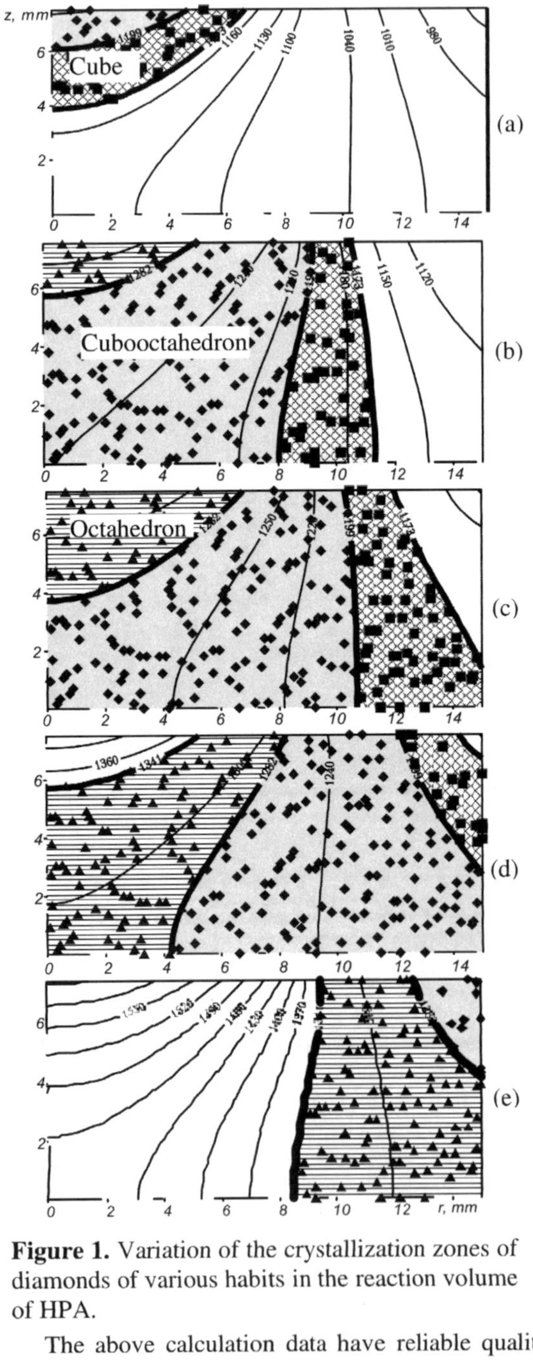


**Figure 1.** Variation of the crystallization zones of diamonds of various habits in the reaction volume of HPA.

The thermomechanical state of the HPA in the process of diamond crystallization is described by the following closed system of coupled equations of continuum mechanics:

- steady-state current conduction,
- transient heat conduction,
- plasticity theory,
- graphite-to-diamond phase transition criterion,
- kinetic equation of the graphite-to-diamond phase transition,
- equations of thermoelasticity for the micro level.

By applying the methods of continuum thermomechanics and computer simulation, the temperature, stress and mass fraction fields in the reaction volume of HPA and in the local diamond–melt–graphite system under diamond synthesis conditions have been determined. The calculated variation in the configuration of the crystallization zones of diamonds of various habits in the reaction volume of recessed anvil-type HPA as the heating current power $W$ increases from 2.10 to 2.74 kW is given in figure 1. It has been found that diamonds start to crystallize in the central part of the reaction volume. In a rather small volume of the crystallization zone, diamonds of the cubic and cubooctahedral habits crystallize simultaneously. This fact is due to the high temperature gradient in the region under study. With increase in the $W$ the boundary of the diamond crystallization zone shifts towards both the centre and the periphery of the reaction volume. When $W$ = 2.33 kW (figure 1b), an additional crystallization zone of octahedral diamonds develops and the whole morphological series of diamond crystals (from a cube to an octahedron) is observed. With further increase in $W$ when its value exceeds 2.48 kW (figure 1d), the crystallization zone starts to narrow and shifts from the centre to the periphery of the reaction volume. With $W \geq$ 3 kW $p$ and $T$ parameters in the reaction volume are beyond the diamond thermodynamic stability region and crystallization does not occur.

The above calculation data have reliable qualitative and quantitative experimental verifications. Firstly, at the optimal pressure the temperature conditions required for diamond crystallization are not arranged in the whole of the reaction volume. Secondly, the volume of the diamond crystallization zone (all other things being the same) depends on the heating current power whose optimal value coincides with the calculated value of 2.5 kW. Diamond synthesis at this heating current power offers the greatest amount of single crystals mostly of cubooctahedral habit with well-developed faces.

## 3. Numerical simulation of the B1 → B2 phase transformation in a potassium chloride sample when deformed in a gasket of DAC

Phase transformations (PT) in crystalline materials when deformed in DAC with the use of gaskets and without a pressure medium have an "instantaneous" nature and are accompanied by an abrupt change of the volume and all physico-mechanical properties of the material. Such PT are classified as martensitic diffusionless or as polymorphous first-order ones. At a PT the crystalline lattice of the new phase forms through the ordered displacement of atoms of the initial phase and the relative displacements of neighbouring atoms are small as compared to the atomic spacing. A high nucleation rate of the new phase crystals is defined by a small deformation, low energy barrier for the uniform PT as compared to the bond energy in a crystal, the ordered structure, and a high mobility of the phase boundaries. With some deviation from the equilibrium state, the excess energy required for the nucleation of the new phase crystals becomes comparable with the energy of defects presented in the initial phase, and the nucleation of martensitic crystals occurs at a higher rate.

The aim of this study was to investigate the kinetics of the B1 → B2 martensitic PT in KCl crystals when deformed in DAC at a constant temperature. To study the regularities of the PT progress in materials deformed in DAC a finite element procedure [2] has been developed. The procedure is based on the thermomechanical theory of PT in inelastic materials, the thermodynamically justified criterion for the PT in elastoplastic materials and extreme principle for the determination of all unknown parameters [3].

Potassium chloride is one of the well characterized alkali–halide crystals. In KCl crystals under the external action (pressure, temperature) one can observe the PT from the B1 stable low-pressure phase, which has the structure of NaCl type, to the B2 metastable high-pressure phase with the structure of CsCl type. The PT is accompanied by an abrupt decrease in the volume (volumetric strain of about 12%) and variation of physico-mechanical properties.

The process of the compression of a polycrystalline KCl sample placed into a stainless steel gasket in a DAC at room temperature has been considered. The following assumptions were taken:

- diamond anvils and the B2 phase of KCl (after the PT) deform elastically, while a deformable gasket from a stainless steel and the B1 phase deform elastoplastically;
- elastic properties of the B1 and B2 phases of KCl are equal;
- at the constant temperature and a specified external loading a PT with a volumetric phase deformation of –0.12 occurs in the sample material;
- phase boundary is coherent, i.e. the displacements at the boundary are continuous;
- the PT pressure was taken to be 1.96 GPa.

The compression of a KCl sample in a DAC has been simulated step-by-step. At each step of loading a certain value of the anvil axial displacement was set and the distribution of fields of displacements, deformations, and stresses in the region under study was calculated.

The analysis shows that starting from the first step of loading the sample deforms plastically, the maximum value of the accumulated plastic strain at the 18th step of loading (the latter step without the PT) reaches 1.5%. The distribution of accumulated plastic strain over the sample volume is characterized by a considerable nonuniformity. The pressure and the volumetric strain distribution at the first steps of loading are also of the nonuniform nature. At the further loading, regions of the highest values of these parameters are considerably widened in the direction to the sample symmetry plane, their distribution becomes less nonuniform and by the 18th step the gradients of the pressure and volumetric strain variations become negligible. At the 19th step the small volumes of the material, for which the PT conditions start to be fulfilled, appear in the sample.

The PT in FE 496 brings about a substantial redistribution of physico-mechanical fields: in the cases of plastic and volumetric strains, the local redistribution takes place in some neighbourhood of this element, but for the pressure and the work integral, the redistribution occurs in the whole volume of the sample. A decrease in the pressure and integral force applied to the anvils is observed. A check of the fulfilment of the PT conditions after that PT has shown that the further progress of the transformation is possible only after an increase in the force applied to the anvils. After the 20th loading step, the PT conditions start to be fulfilled for FE 495 and 486. A nucleus of the new phase forms. However, in the subsequent loading in parallel with the existing B2 phase nucleus growth, the formation of a new nucleus is observed (figure 2). Such behaviour of the sample is caused by the local redistribution of stress and strain tensors components in the elastoplastically deforming B1 phase of the sample. Because of the plastic relaxation of stresses, the formation of a new nucleus becomes more preferable than the growth of the old one. The PT process runs in the direction of the force loading, and is characterized by a considerable nonuniformity of the local stresses and strains distribution, which facilitates the PT. At the initial stage of the PT these changes are local in nature, but as the new phase concentration exceeded 5%, the disturbance caused by the element, in which the PT has occurred, becomes essential, produces considerable stress and strain concentrations and contributes to the PT in the sample.

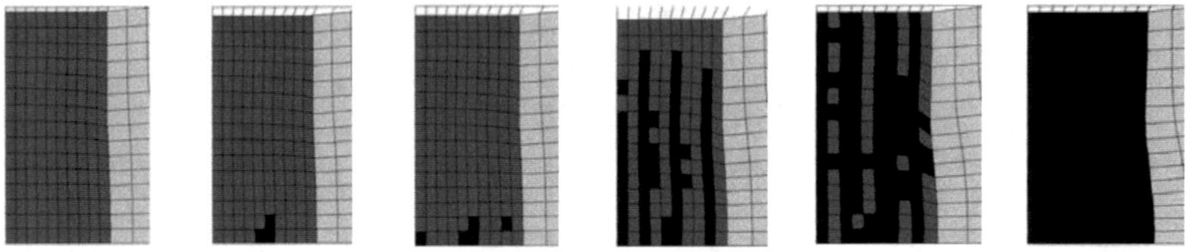

**Figure 2.** Stages of the B1 → B2 phase transformation in the polycrystalline KCl sample: □ – diamond, ▩ – stainless steel, ▩ – B1 KCl phase, ■ – B2 KCl phase.

Figure 3 gives the photo of the KCl sample made when studying the regularities of the material straining in DAC. The distribution of the B1 and B2 phases in the sample agrees with the calculated one (figure 2). The calculated and experimental pressure dependences of the B2 phase concentration and the normalized volume of the polycrystalline KCl sample are shown on figures 4 and 5. Some disagreement of the data on the concentration stems from the fact that in the experiment the PT started at 2.09 ± 0.03 GPa and terminated at 2.82 ± 0.10 GPa, while in the calculation the pressure of the PT onset was 1.96 GPa. In addition, in the experiment the sample was loaded by stages with a holding time at the constant pressure for several min in the course of the PT, while in the calculation after an increase of loading, a decrease in the pressure in the sample during the PT was allowed.

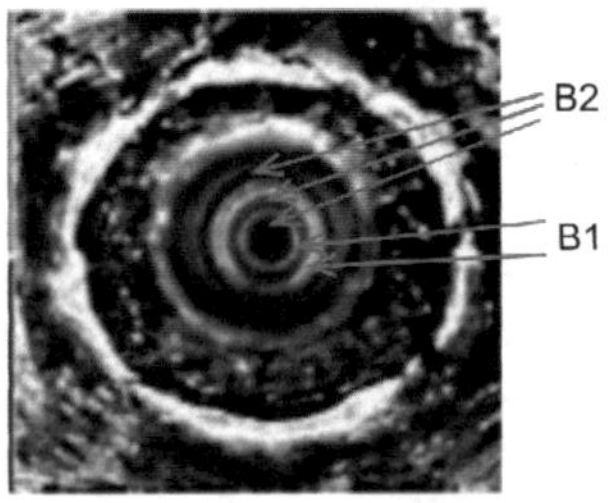


**Figure 3.** PT in KCl sample when deformed in a DAC.

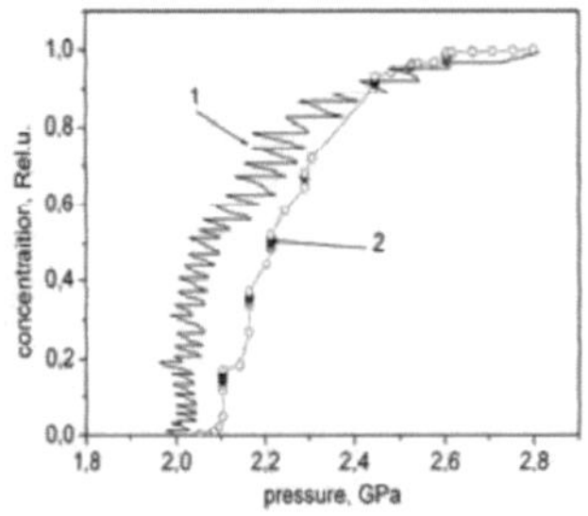


**Figure 4.** Pressure dependence of the B2 phase concentration in the KCl sample: (1) calculated data, (2) experiment.

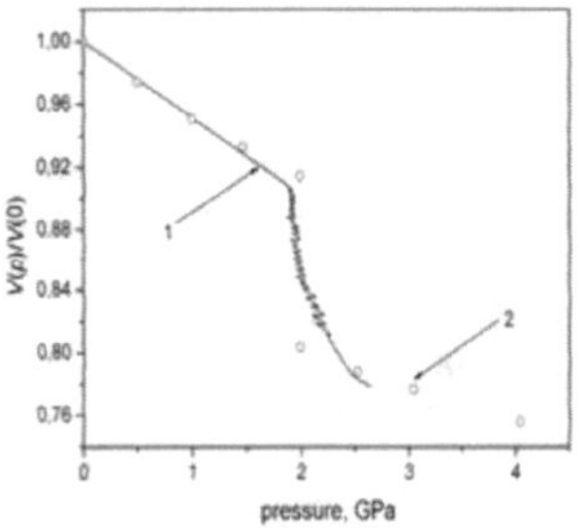


**Figure 5.** Pressure dependence of the normalized volume $V(p)/V(0)$ of the KCl sample in loading and PT: (1) calculated data, (2) experiment.

[1] Lyeshchuk O 2010 *Comp. Mater. Sci.* **49** S85–S94
[2] Polotnyak S B 2008 *J. Superhard Mater.* **30** No. 2 82–93
[3] Levitas V I 1998 *Int. J. Solids Struct.* **35** 889–940

Journal of Physics: Conference Series **377** (2012) 012096 doi:10.1088/1742-6596/377/1/012096

# Uniaxial compression influence on valence sub-bands energy spectrum and electroluminescence in n-AlGaAs/GaAsP/p-AlGaAs diode structures

**N Ya Minina, E V Bogdanov and S S Shirokov**

Physics Faculty, Lomonosov Moscow State University, 119991 Moscow, Russia

E-mail: min@mig.phys.msu.ru

**Abstract.** Numerical calculations of the valence band and conduction band size quantized levels in a strained $p$-$Al_xGa_{1-x}As/GaAs_{1-y}P_y/n$-$Al_xGa_{1-x}As$ (y = 0.16) double heterostructure were performed for different values of the external uniaxial compression along [110] direction. They indicate that the two upper levels in the valence band merge at pressure about 4 kbar and a strong state mixing develops around the merging point. The results of calculations explain the nonlinear character of the photon energy shift and electroluminescence intensity increase that were experimentally observed in these structures under uniaxial compression up to 5 kbar.

## 1. Introduction

For a long time hydrostatic pressure and uniaxial compression were used as a strong external influence on the band structure and energy spectrum of charge carriers in solids leading to their quantitative and qualitative reconstruction. Recently, the possibility to change the wavelength of laser diodes by means of external uniaxial stress [1] and uniform compression in hydrostatic pressure chambers [2] has been successfully demonstrated. The method of uniaxial compression used in [1] up to pressure P = 4 - 5 kbar is more convenient for optical measurements, because it allows a free withdrawal of radiation from a sample for a spectral analysis.

It has been shown [1] that under uniaxial compression in [110] and [1-10] directions electroluminescence (EL) spectra of strained $p$-$Al_xGa_{1-x}As/GaAs_{1-y}P_y/n$-$Al_xGa_{1-x}As$ (y = 0.16) double heterostructures usually used in TM emitting 808 nm high-power diode lasers demonstrate a blue shift, while the electroluminescence intensity shows 2 - 3 times increase (figure 1). Moreover, the effect of uniaxial stress influence on EL polarization (figure 2) indicates the relative decrease of TE-mode under compression in respect to TM-mode. The photon energy shift in respect to the applied stress is not linear (see insert in figure 1) and determined by increase of the energy gap in quantum well (QW) material under compression, but the enhancement of the EL intensity remained uncertain and can not be qualitatively explained either by decrease of nonradiative recombination under compression or by arising piezoelectric field or a potential barrier lowering.

The present paper is aimed on the numerical calculations and analysis of the energy spectrum in $p$-$Al_xGa_{1-x}As/GaAs_{1-y}P_y/n$-$Al_xGa_{1-x}As$ (y = 0.16) diode structure under uniaxial compression, since the EL data described above [1] (see figures 1, 2) are representative of the significant change in the energy band structure and size quantized level states in the QW under uniaxial stress.

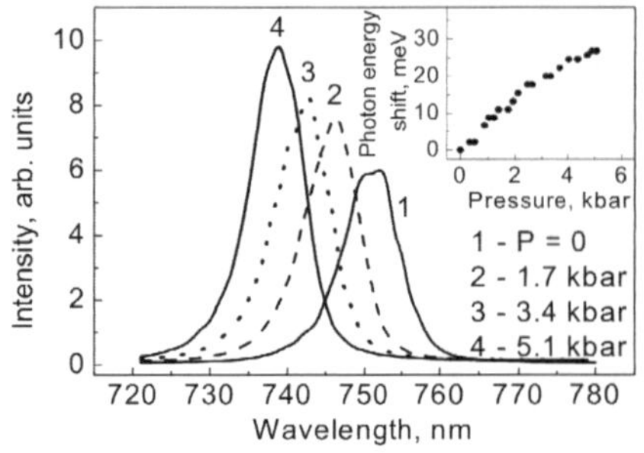


**Figure 1.** EL spectra measured at 77 K under uniaxial compression along [110] axis in *p*-$Al_xGa_{1-x}As/GaAs_{1-y}P_y$/*n*-$Al_xGa_{1-x}As$ (y = 0.16) heterostructure.

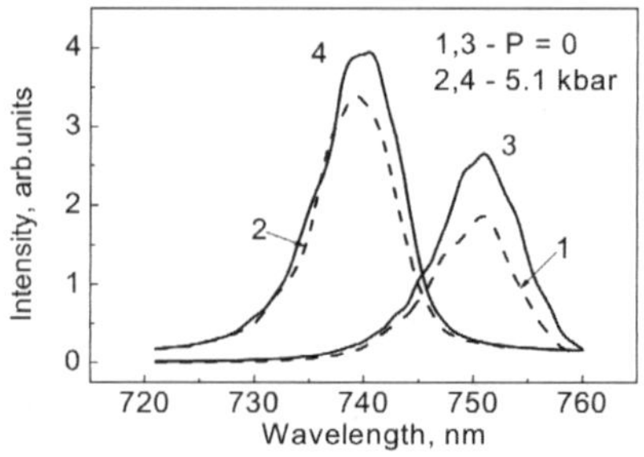


**Figure 2.** Spectra of TM-mode (1, 2) and TE-mode (3, 4) measured at a temperature of 77 K under uniaxial compression along [110] direction.

## 2. Numerical calculations and discussion

The *p*-$Al_xGa_{1-x}As/GaAs_{1-y}P_y$/*n*-$Al_xGa_{1-x}As$ (y = 0.16) structures under investigations were grown on silicon doped (001) GaAs substrates by metal organic vapor-phase epitaxy. A $GaAs_{0.84}P_{0.16}$ QW of a 14 nm width is surrounded on both sides by $Al_{0.45}Ga_{0.55}As$ waveguide barrier layers of 1 micron thickness on the whole with various *p*- and *n*-types levels of doping, which, starting from $1 \times 10^{17}$ cm$^{-3}$ near the QW, reaches $2 \times 10^{18}$ cm$^{-3}$ in the peripheral areas of the structure.

The valence and conduction band profiles, as well as size quantized levels and wave functions of electrons and holes in the investigated $GaAs_{0.84}P_{0.16}$ QW, were numerically calculated for different values of the external uniaxial compression along [110] direction. The Luttinger-Kohn Hamiltonian with strain terms [3] was self-consistently solved together with Poisson's equation for the electrostatic potential using the finite-difference **k·p** method. Numerical calculations were performed in the framework of the program "Heterostructure Design Studio 2.1" in the vicinity of the zone centre at the Γ point. There was used the model developed in [4], which considers the Luttinger-Kohn Hamiltonian 6×6 and describes the conduction band, light and heavy hole subbands but doesn't take into account the valence split-off subband lying about 300 meV below the top of the valence band in $GaAs_{0.84}P_{0.16}$ [5]. Necessary for calculations parameters were taken from literature [5].

According to the calculations, in the strained $GaAs_{0.84}P_{0.16}$ QW under investigation at P = 0 the LH1 level is the ground state in valence band while the heavy hole HH1 level is the next one in the energy scale (figure 3). Optical transitions between the lowest electron level e1 and the highest hole level h1 in the valence band determine an optical gap that is equal to the experimentally observed emitted photon energy and increases with applied stress. Under uniaxial compression levels h1 and h2 move toward each other, and their merging with the following repulsion takes place at P ≈ 4.5 kbar (figure 3). As soon the transitions from the electron level e1 to the uppermost hole level h1 dominate in EL, the h1-h2 exchange in the character of the pressure dependence after the merging can explain the observed nonlinearity in the EL photon energy shift (see figure 4).

In calculations, the hole wave functions were expanded in basis functions of Luttinger-Kohn representation with the total angular momentum J = 3/2, and its projections $m_J = \pm 1/2$ and $m_J = \pm 3/2$ correspond to light holes and heavy holes respectively [4]. Applied uniaxial stress reduces the symmetry and causes the hole state mixing. Analysis of the envelope functions related to the two upper levels in the valence band QW permits to evaluate the contribution of the basis functions with different total angular momentum projections into the wave functions on h1 and h2 levels under uniaxial compression. The results depicted in figure 5 indicate that at P = 0 light holes (LH1 state, h1 level) are described only by basis functions with $|m_J| = 1/2$, while heavy holes (HH1 state, h2 level) – by basis function with $|m_J| = 3/2$. The picture demonstrates the development of light hole – heavy hole states mixing under compression and LH1-HH1 states merging at P ≈ 4 kbar.

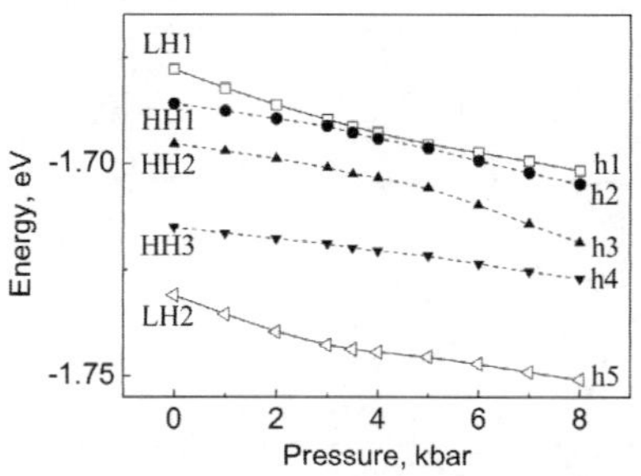


**Figure 3.** Calculated energy shifts of five upper (h1-h5) hole levels at Γ point under uniaxial compression along [110] direction.

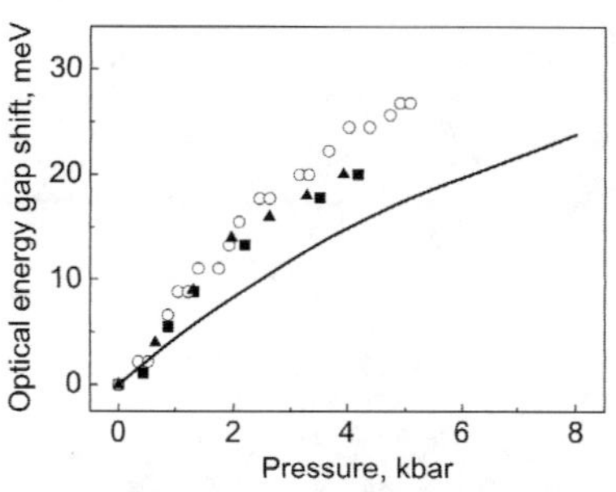


**Figure 4.** Calculated pressure dependence of the optical energy gap shift. Dots – experimental data obtained for different samples.

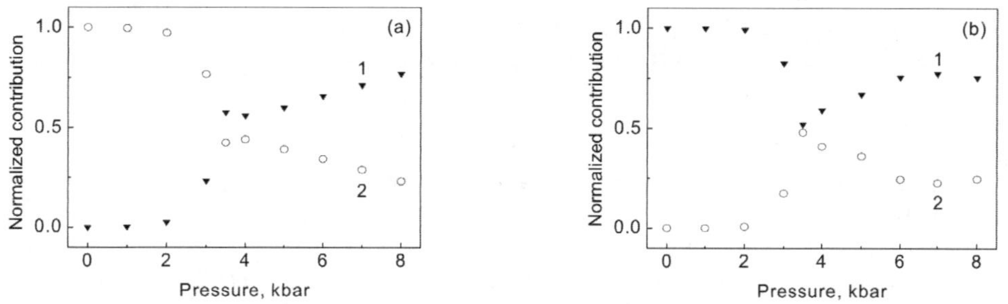


**Figure 5.** Relative contribution of basis functions with different angular momentum projections $|m_J| = 3/2$ (1) and $|m_J| = 1/2$ (2) into wave functions of holes on h1 (a) and h2 (b) levels under uniaxial compression.

From the electron and hole wave functions matrix elements of electron-photon interaction operator for interband transitions and, further, absorption coefficient and optical gain may be calculated [6]. Optical gain spectra for TM- and TE-modes of polarized light are represented in figure 6 for different values of applied stress. In calculations, electron and hole concentration in QW was taken to be $n = p = 2 \times 10^{12}$ $cm^{-2}$ that is characteristic for the used range of currents. The TM-mode optical gain represented in figure 6 demonstrates significant increase under compression up to 8 kbar in a wide photon energy range. Transformation of TE-mode optical gain is complicated: it increases at low stress and drops at $P > 4$ kbar. These results of numerical calculations are qualitatively consistent with experimentally observed effects: (1) 2 - 3 times EL intensity increase under compression (see figure 1) and (2) decrease of relative light polarization at high pressures (see figure 2).

The increase of the optical gain under uniaxial compression is evidently connected with the LH1-HH1 merging at $P \approx 4$ kbar that determines growing of the joint hole density of states due to the higher value of the heavy hole effective mass in comparison with the light hole one. Since the strong mixing of heavy hole and light hole states develops in a definite pressure interval (figure 5), in this region of pressure there exists a noticeable mixing of their wave functions and, as a result, selection rules "softening" and optical transition probability enhancement. The degree of mixing depends on the h1 and h2 levels splitting Δ. This situation leads to a smooth EL intensity increase under compression in the vicinity of $\Delta \approx 0$. At $P \geq 4$ kbar EL intensity increment is still possible due to the high density of heavy hole states. It is remarkable that non-linear behavior of optical energy gap starts to be evident at pressure $P \approx 2$ kbar (figure 4) where the strong LH1-HH1 state mixing begins (figure 5).

It should be noted, that the valence band mixing was studied previously mainly in a series of experiments with different samples, and the shift of light and heavy hole levels in QW in dependence on the internal biaxial strain was regulated by QW and barrier constitution [7]. In [1] the development

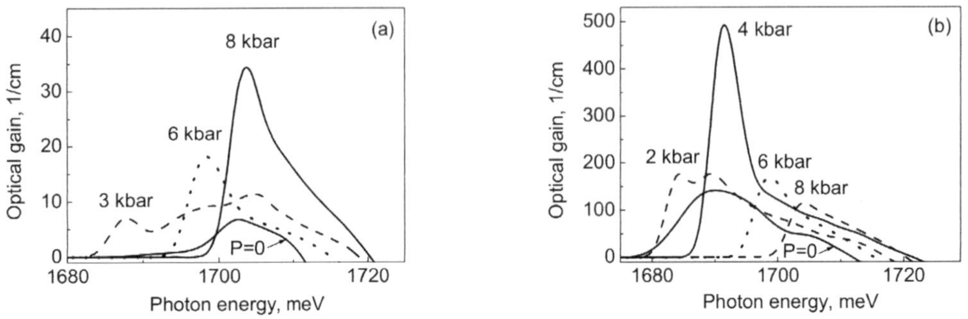


**Figure 6.** Optical gain spectra of TM-mode (a) and TE-mode (b) under compression along [110] direction.

of light and heavy hole wave functions mixing process was investigated on the sample in the course of one experiment that makes the observed phenomenon of EL intensity increase under uniaxial compression single-valued and highly reliable.

## 3. Conclusion

The experimental data on EL spectra of strained $p$-$Al_xGa_{1-x}As/GaAs_{1-y}P_y/n$-$Al_xGa_{1-x}As$ (y = 0.16) diode structures under compression up to 4 - 5 kbar [1] demonstrated a nonlinear blue photon energy shift, 2 - 3 times increase of EL intensity and relative decrease of TE-mode in respect to TM-mode under compression, that are representative of a significant change in the energy band structure.

Numerical calculations of the $p$-$Al_xGa_{1-x}As/GaAs_{1-y}P_y/n$-$Al_xGa_{1-x}As$ (y = 0.16) band structure and size quantized levels in QW under uniaxial stress performed in the present paper indicate a merger of two uppermost hole levels (LH1-HH1 states) at pressure about 4 kbar that determines nonlinear increase of the optical gap magnitude and correspondently EL photon energy. Development of strong LH1-HH1 mixing and transition of the hole ground state in the QW from LH1 nature ($|m_J| = 1/2$) to HH1 nature ($|m_J| = 3/2$) are shown. Matrix elements of the electron-photon interaction Hamiltonian and optical gain were calculated for different polarizations (TE- and TM-modes). The result indicates increase of these modes intensity in the vicinity of LH1–HH1 merging as well as its relative change under condition of LH1–HH1 mixing.

## Acknowledgements

We are grateful to our colleagues Prof. A.E. Yunovich from Moscow State University and Dr. H. Kissel from DILAS Diodenlaser GmbH in Mainz for their cooperation.

## References

[1] Berman I V, Bogdanov E V, Kissel H, Minina N Ya, Shirokov S S and Yunovich A E 2009 *Phys. Status Solidi (b)* **246** 522
[2] Trzeciakowski W, Bercha A, Dybala F, Bohdan R, Adamiec P and Mariani O 2007 *Phys. Status Solidi (b)* **244** 179
[3] Bir G L and Pikus G E 1974 *Symmetry and Strain-Induced Effects in Semiconductors* (New York: Wiley)
[4] Kolokolov K I, Savin A M, Beneslavski S D, Minina N Ya and Hansen O P 1999 *Phys. Rev.* B **59** 7537
[5] Vurgaftman I, Meyer J R and Ram-Mohan L R 2001 *J. Appl. Phys.* **89** 5815
[6] Wu S, Cao Y, Tomic S and Ishikawa F 2010 *J. Appl. Phys.* **107** 013107
[7] van der Poel C J, Ambrosius H P M M, Linders R W M, Peeters R M L, Acket G A and Krijn M P C M 1993 *Appl. Phys. Lett.* **63** 2312

# Analysis of equations of state and temperature dependence of thermal expansivity and bulk modulus for silicon

**Tushar C Pandya[1], Nilesh A Thakar[2] and Apoorva D Bhatt[1]**
[1] St.Xavier's College, Navrangpura, Ahmedabad-380009, India
[2] K. K. Shah Jarodwala Maninagar Science College, Rambaug, Maninagar, Ahmedabad-380008, India

E-mail: pandyatc@gmail.com

**Abstract.** In the present paper an attempt has been made for the comparative study of different equations of state for silicon (Phase-1, cubic diamond structure) in the pressure range of 0-11 GPa. We compare the results of different equations of state (EOS) with available experimental data. The Kwon and Kim EOS is found to give far better agreement with the available experimental data. Results obtained by Poirier-Tarantola, Vinet, Tait and Suzuki's equations of state are not giving satisfactory agreement with the available experimental data. In the present study simple methods based on thermodynamic functions are presented to investigate the temperature dependence of thermal expansivity and bulk modulus for silicon. The results are reported for silicon. The calculated values of thermal expansivity are in good agreement with experimental data.

## 1. Introduction

Compression behaviour of the diamond-type semiconductors has been a topic of experimental and theoretical study for a long time [1, 2]. Large amount of theoretical [3-5] and experimental work [6-9] have been devoted to the high pressure studies of Group IV element silicon. Angilella G.G.N [10] et.al have summarized some studies, both theoretical and experimental, relating to x-ray diffraction for diamond –lattice semiconductors under compression. Silicon undergoes a series of structure phase transition with pressure: cubic diamond [CD] to β-Sn, to simple hexagonal, to hexagonal closed packed, to face centred cubic [11]. Since the CD phase has the lowest minimum total energy, it is the most stable phase of silicon at ambient pressure [11].

Studies on equations of state [EOS] are very useful to understand the high pressure behaviour of silicon. In the present paper we report the analysis of equations of state for the CD-phase of silicon in the pressure range of 0-11 GPa. Attempt has been made to identify the better equation of state that gives satisfactory agreement with the available experimental data. In the present study simple methods based on thermodynamic functions are presented to investigate the temperature dependence of thermal expansivity and bulk modulus for the CD phase of silicon.

## 2. Method of analysis

### *2.1. Equations of state*

We consider a number of equations of state representing the relationship between pressure and volume at a fixed temperature. These are briefly described as follows along with the basic assumptions implied in their derivations.

Kwon and Kim [12, 13] presented the following method for the determination of equation of state

$$dV = -V\frac{dP}{B} \tag{1}$$

Or,

$$V - V_0 = -V_0\left(\frac{P - P_0}{B_0}\right) \tag{2}$$

Where, $P_0$ is the atmospheric pressure referred as zero, $V_0$ and $B_0$ are the volume and isothermal bulk modulus at zero pressure. Equation (2) thus gives,

$$P = B_0\left(1 - \frac{V}{V_0}\right) \tag{3}$$

Equation (3) seems to be very simple for the determination of pressure (P) at different values of compression ($V/V_0$) [12, 13] in a number of papers.

A theoretical method based on thermodynamic analysis for EOS has also been presented by Kumar [14]. The form of Shankar EOS is written as follows:

$$P = \frac{B_0}{A}\left[\exp\left\{A\left(1 - \frac{V}{V_0}\right)\right\} - 1\right] \tag{4}$$

The derivation of equation (4) is based on the volume dependence of Anderson-Gruneisen parameter $\delta_T$ as proposed by Anderson et.al [15] in the following form,

$$\frac{\delta_T + 1}{V/V_0} = A \tag{5}$$

Where A is the parameter equal to $\delta_T^0 + 1$. Here, it should be mentioned that if we take $\delta_T$ is independent of V, equation (4) becomes identical to the well known Murnaghan EOS [12].

It has also been discussed in the literature [15] that for some solids $\delta_T = B_0'$, the first order pressure derivative of bulk modulus. If we consider this idea and determine $A = B_0' + 1$ then the form of equation (4) resembles with so called Tait equation of state as quoted by MacDonald [16]. Tait equation [17] which is widely used in recent literature for studying P-V data is given as,

$$\frac{\Delta V}{V_0} = \frac{1}{B_0'+1}\ln(1 + \beta P) \tag{6}$$

Here, $\beta = (B_0' + 1)/B_0$

Vinet et.al [18] have proposed a new universal equation of state [UEOS] which is found to be extensive use for all kinds of solids subjected to high pressure.

Vinet UEOS is written as follows

$$P = 3(1 - x)B_0\frac{\exp[\eta(1 - x)]}{x^2} \tag{7}$$

Where, $x = (V/V_0)^{1/3}$ and $\eta = 3/2(B_0' - 1)$

The natural strain equation of state developed by Poirier-Tarantola EOS [19] reads as follows.

$$P = B_0 \frac{V_0}{V}\left[\ln\frac{V_0}{V} + \frac{(B_0 - 2)}{2}\left(\ln\frac{V_0}{V}\right)^2\right] \tag{8}$$

The EOS based on the Suzuki's theory of thermal expansivity reads as follows [20]

$$P = B_0\left[\left(1-\frac{V}{V_0}\right) + \left(\frac{B_0'-1}{2}\right)\left(1-\frac{V}{V_0}\right)^2\right] \tag{9}$$

*2.2. Temperature dependence of thermal expansivity*

Following the approximation given by Anderson et.al [21-23], that the product of the coefficient of volume thermal expansion [α] and the isothermal bulk modulus [B] remains constant under the effect of temperature (T). We can write;

$$\alpha B = const \tag{10}$$

$$\alpha\left(\frac{dB}{dT}\right)_P + B\left(\frac{d\alpha}{dT}\right)_P = 0 \tag{11}$$

Which gives,

$$\delta_T = -\frac{1}{\alpha B}\left(\frac{dB}{dT}\right)_P = \frac{1}{\alpha^2}\left(\frac{d\alpha}{dT}\right)_P \tag{12}$$

Where, α is defined as,

$$\alpha = \frac{1}{V}\left(\frac{dV}{dT}\right)_P \tag{13}$$

Using equations (12) and (13) we get,

$$\delta_T = \frac{V}{\alpha}\left(\frac{\partial\alpha}{\partial V}\right)_P \tag{14}$$

At constant pressure integration of equation (14) gives,

$$\left(\frac{\alpha}{\alpha_0}\right) = \left(\frac{V}{V_0}\right)^{\delta_T} \tag{15}$$

Now equation (12) can be written as,

$$\delta_T = \frac{1}{\alpha^2}\left(\frac{d\alpha}{dT}\right)_P \tag{16}$$

The integration of equation (16) gives,

$$\frac{\alpha}{\alpha_0} = \left[1 - \alpha_0\delta_T(T - T_0)\right]^{-1} \tag{17}$$

Equation (17) gives directly the values of α at required temperature and thus does not require the knowledge of $V/V_0$ as a function of temperature.

*2.3. Temperature dependence of bulk modulus*

The isothermal Anderson-Gruneisen parameter $\delta_T$ is defined as:

$$\delta_T = -\frac{1}{\alpha B}\left(\frac{dB}{dT}\right)_P \tag{18}$$

From equations (13) and (18) we get,

$$\delta_T = -\frac{V}{B}\left(\frac{dB}{dV}\right)_P \tag{19}$$

Thus at constant pressure we get,

$$\frac{dB}{B} = -\delta_T \frac{dV}{V} \tag{20}$$

Integrating equation (20) within the limits $B = B_0$ at $V = V_0$, assuming $\delta_T$ constant, we get

$$\frac{B}{B_0} = \left(\frac{V}{V_0}\right)^{-\delta_T} \tag{21}$$

This equation may be used to calculate B at different temperatures provided that $V/V_0$ versus T are known. At constant pressure equation (18) may be rewritten as,

$$\frac{dB}{dT} = -(\alpha B \delta_T) \tag{22}$$

Integrating equation (22) gives,

$$\int_{B_0}^{B} dB = -(\alpha_0 B_0 \delta_T)\int_{T_0}^{T} dT \tag{23}$$

$$\frac{B}{B_0} = [1 - \alpha_0 \delta_T (T - T_0)] \tag{24}$$

This equation (24) is based on the assumption that $(\alpha B \delta_T)$ is a constant under the effect of temperature. We make the use of equation (24) to calculate B at different temperatures.

## 3. Results and Discussion

The input parameters required for present calculations are compiled from literature [24]. The values of $B_0$, $B_0$' and $\alpha_0$ are taken as 97.8 GPa, 4.1 and $8.76\times10^{-6}$ $K^{-1}$ respectively. The isothermal Anderson-Gruneisen parameter for silicon is 3.11 [15].The results for silicon based on the various EOS for pressure (P) versus compression ($V/V_0$) are given in Figure 1 along with the experimental data. It is found that the results obtained by Kwon-Kim equation are in very good agreement with the X-ray diffraction data [25]. A good agreement between theory and experiment demonstrates the validity of simple form of the Kwon-Kim EOS, applicable for the entire range of pressure, from atmospheric pressure up to the structural transition pressure of Silicon. The results obtained by Poirier-Tarantola, Vinet, Suzuki and Tait EOSs are not giving satisfactory agreement with the available experimental data [25]. The results of temperature dependence of thermal expansivity based on equation (17) are reported in Figure. 2. The calculated values of thermal expansivity are in good agreement with available experimental data [24]. Equation (17) gives directly the values of $\alpha$ at required temperature and thus does not require the knowledge of $V/V_0$ as a function of temperature. Thus it is clear that application of equation (17) is simpler as compared with equation (15). The results of pressure dependence of bulk modulus obtained using equation (24) are shown in Figure 3 in the absence of experimental data. These results may be of some interest to the researchers engaged in the experimental work.

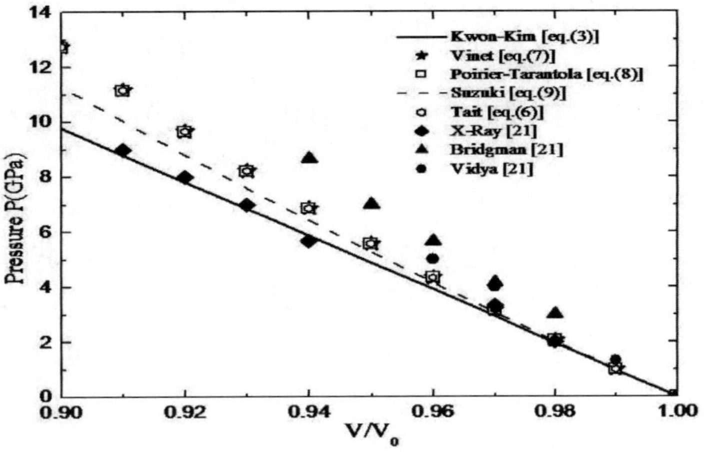


**Figure 1.** Compression behaviour of silicon using different EOS models along with available experimental data.

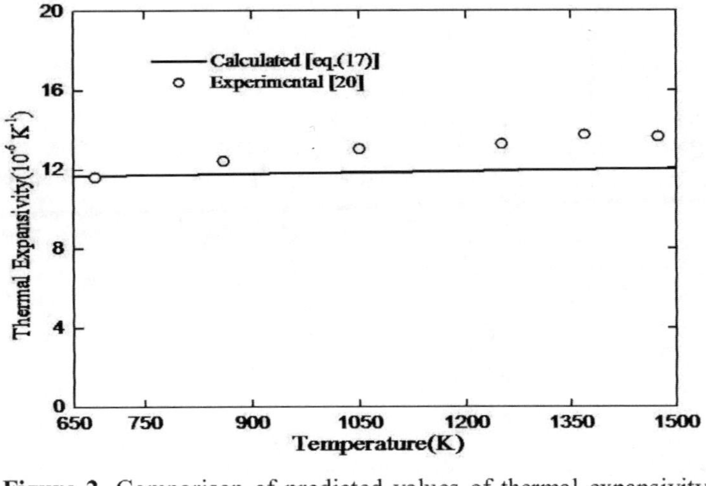


**Figure 2.** Comparison of predicted values of thermal expansivity at different temperatures through equation (17) with the experimental data.

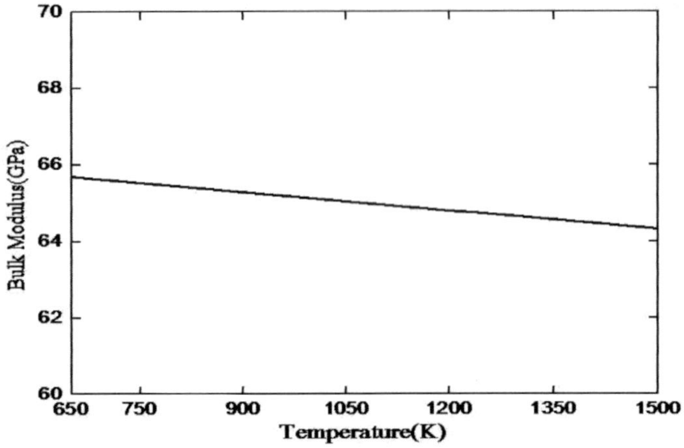


**Figure 3.** Predicated values of bulk modulus of silicon at different temperature through equation (24).

## 4. References

[1] Paul W and Warschauer D M 1963 *Solids Under Pressure* (New York: Mc Graw-Hill)
[2] Sharma S M 2006 *Pramana J. Phys.* **67** 141
[3] Olijnyk H, Sikka S K, and Holzapfel W B 1984 *Phys. Lett.* **103A** 137
[4] Hu J Z and Spain I L 1984 *Solid State Commun.* **51** 263
[5] Duclos S J, Vohra Y K and Ruoff A L 1987 *Phys. Rev. Lett.* **58** 775
[6] Yin M T and Cohen M L 1982 *Phys. Rev.* B **26** 5668.
[7] Chang K J and Cohen M L 1985 *Phys. Rev.* B **31** 7819
[8] Dacorogna M M, Chang K J and Cohen M L 1985 *Phys. Rev.* B **32** 1853
[9] Ovsyannikov S V, Shchennikov V V, Misiuk A and Shchennikov Jr V V 2004 *Solid State Commun.* **132** 545
[10] Angilella G G N, March N H, Howard I A and Pucci R 2008 *J. Phys.: Conf. Ser.* **121** 032006
[11] Liu A Y, Chang K J and Cohen M L 1988 *Phys. Rev.* B **37** 6344
[12] Kwon T H and Kim S C 1995 *Physica* B **205** 331
[13] Kwon T H 1995 *Solid State Commun.* **95** 255
[14] Kumar M 1995 *Physica* B **212** 391
[15] Anderson O L, Isaak D G and Oda H 1992 *Rev of Geophys* **30** 57
[16] MacDonald J R 1969 *Rev Mod Phys* **41** 316
[17] Thomas L M and Shanker J 1995 *Phys. Status Solidi* (b) **189** 363
[18] Vinet P, Smith J R and Rose J H 1987 *J Phys* C **19** L467
[19] Poirier J and Tarantola A 1998 *Phys. Earth Planet. Inter.* **109** 1
[20] Suzuki I 1975 *J. Phys. Earth* **23** 145
[21] Anderson O L and Zou K 1989 *Phys Chem Minerals* **16** 642
[22] Anderson O L, Isaak D G and Oda H 1991 *J. Geophys Res* **96** 1803
[23] Kumar M 1998 *Indian J Pure & Appl Phys* **36** 119
[24] Martienseen W and Warlimont H(Eds.) 2005 *Springer Handbook of Condensed Matter and Data Materials* (New York: Springer)
[25] Sennu M, Mii H and Fujishiro I 1976 *J. of Phys. Soc. of Japan* **41** 1562

Journal of Physics: Conference Series **377** (2012) 012098 doi:10.1088/1742-6596/377/1/012098

# Modelling and Analysis of High Pressure Peaking Switch

**Bindu S** [1], **Mrunal Parekh** [2], **H A Mangalvedekar** [2], **Archana Sharma** [3], **D P Chakravarthy** [3]
[1,2] Department of Electrical Engineering, Veermata Jijabai Technological Institute, Matunga,
[3] APPD, BARC,Trombay, Mumbai-400085

E-mail: bindubalu@rediffmail.com

**Abstract.** This paper presents modelling and analysis of peaking switch used in Marx generator, such that the rise time of the pulse produced by the Marx generator is reduced substantially. Towards this FEMM ( Finite Element Methods Magnetics) software is used for the field modelling of the switch and MATLAB for circuit modelling to observe the rise time. The switch has to produce pulse with sub-nanosecond rise time, hence the electrode distance has to be minimum. This switch can withstand high voltage only under high pressure. A mathematical model is simulated in MATLAB to see the performance under high pressure.

## 1 Introduction

The high power pulses are used in many industrial, societal and defense applications. Marx generator is one of the source used to generate the short time high energy pulse. The output of the Marx generator is coupled to a peaking capacitor and peaking switch to decrease the rise time of the generated pulse. The short rise time plays an important role in the pulsed systems. This short rise time, high power pulse is fed to various loads like plasma gun, ion gun, electron gun, vircator / antenna etc. The applications include high power radar, plasma ion implantation, diamond cutting, food irradiation, surface coating, waste water processing, medical electronics, Ultra Wide Band(UWB) systems, defence etc.[1-5]. The goal of an Ultra Wide Band(UWB) system is to interact with the increasing combination of soft and hardware functions in electronics.

This paper deals with the design and modeling of peaking switch to modify the output pulse of a Marx generator to achieve the required rise time to develop UWB signals. The pulse from the Marx generator forms the input to an antenna which will generate and radiate the UWB signal. The radiated signal from the antenna depends upon the pulse and the rise time of the pulse generated by the peaking switch. The output pulse depends on the gap inductance of the switch, the fast discharge capacity of the peaking capacitor, the dielectric medium in the switch and the gas pressure [6-13]. This paper discusses the circuit model of the switch with a detailed analysis carried out using the software FEMM and MATLAB [14]. The inter-electrode distance in the switch is very small and high voltage pulse is applied under very high pressure to the switch. The high pressure is needed in the switch to withstand the high voltage of the pulse. Therefore an attempt has been made in this paper to formulate a mathematical model for the switch under pressure.

Journal of Physics: Conference Series **377** (2012) 012098 doi:10.1088/1742-6596/377/1/012098

## 2. Aspects of ultra fast switching

The Peaking switch is a major component of an UWB system. A fast pulse rise time decides the high frequency components of the resulting spectrum. To sharpen the rise time of a pulse, a spark gap configuration, called a peaking switch is used. The crux of the peaking switch is the establishment of very high electric fields in the inter electrode spacing. A uniform electric field is generally achieved using a Rogowski electrode profile [12], hence this profile is used for modelling the switch. The uniform field is characterised by constant electric field through out the inter electrode region associated with linear voltage profile. The uniform field between the electrodes is produced by applying high voltage between the electrodes, under high pressure. The electric fields produced in the inter electrode region is in the range of MV/cm. The velocity of propagation of the electron avalanche is proportional to the electric field applied to electrodes.

The peaking switch is said to have closed when the potential difference between the electrodes is sharply reduced and current becomes circuit limited. The development of the pre breakdown phase of the spark is related to the minimum closure time of the spark gap, which in turn determines the rise time [7]. This rise time depends on electric field intensity, gas pressure and type of gas medium. This could be modelled as an R-L-C network.

## 3. Development of Equivalent circuit model of the switch

Figure.1 shows a peaking switch with Rogowski electrode profile. The input to the switch is given to the electrode-I and load is connected to the electrode-II, $r_{arc}$ is the radius of the arc , $r_{inner}$ is the radius of the electrode, $r_{outer}$ is the radius of the outer enclosure of the switch and 'd' is the inter electrode distance. In this section we develop an equivalent R-L-C circuit model of the switch during the arcing phenomena.

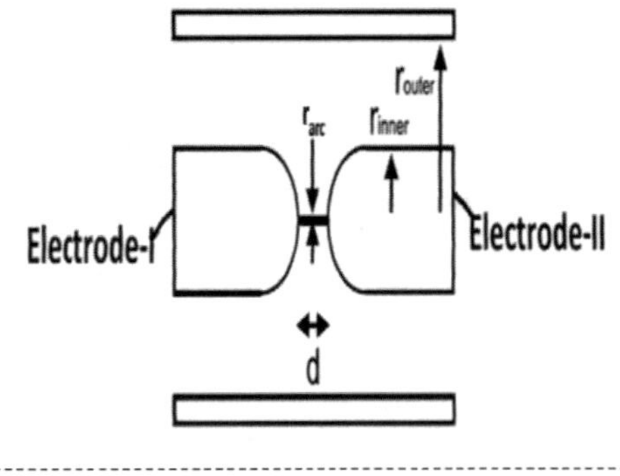


**Figure 1.** Switch model with Rogowski profile electrode

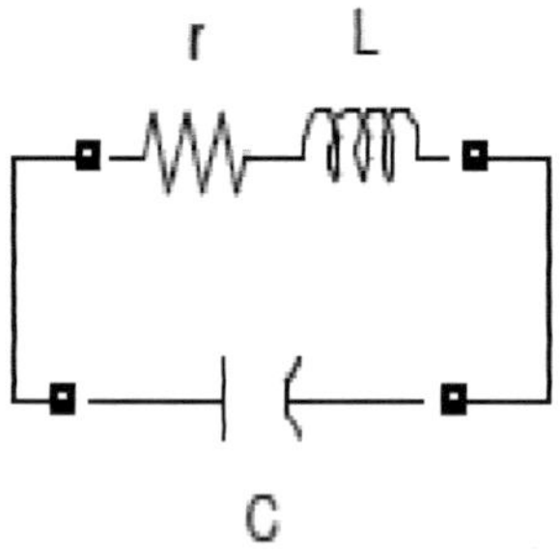


**Figure 2.** Switch equivalent circuit

### *3.1 Switch Inductance under arcing condition*

The inductor, '$L_s$' of the switch can be considered as a co-axial transmission line with $r_o$ and $r_{arc}$ as the outer and inner radius of the transmission line[6].

$$L_s = d\frac{\mu_r\mu_o \ln[\frac{r_o}{r_{arc}}]}{2\pi} \qquad (1)$$

Where 'd' is the channel length in cm,$\mu_0$ is the permeability and $\mu_r$ is the relative permeability
The inductance per unit length $L_s$' of the spark gap is then given by

$$L_{s}' = \frac{\mu_r \mu_o \ln \left[\frac{r_o}{r_{arc}}\right]}{2\pi} \tag{2}$$

The $L_s$ can be also written as[13]

$$L_s = 2 * d * \ln\left(\frac{r_o}{r_{arc}}\right) \tag{3}$$

Since $r_o >> r_i$, the rise time of few nano second the log term is about seven. Hence

$$L_s \approx 14 * d \tag{4}$$

The channel inductance limits the achievable rise time of the switch by inhibiting the rate of rise of voltage. Hence to achieve minimum rise time the channel length should be as small as possible.

*3.2 Switch Capacitance under arcing condition:*

During the switch breakdown the gap is closed by the streamer movement. There will be some small current flowing due to the increasing capacitance at the gap associated with the smaller distance between positive and negative charges in the gap region.

The small inter electrode distances, to achieve low channel inductance, yield high spark gap capacitances. The switch capacitance can be evaluated by

$$C = \frac{2\pi \epsilon_r \epsilon_o}{ln\frac{r_o}{ri}} \tag{5}$$

where $\epsilon_o$ the absolute permittivity of the medium and '$\epsilon_r$' is the relative permittivity of the medium.

*3.3 Switch resistance under arcing condition*

Since we are modeling the switch for breakdown of electrode gap for the duration when the arc occurs, we consider the arc resistance as the resistance of switch Apart from the above two parameters arc resistance has to be considered. The arc is nonlinear in nature. At the time of switching the value of resistance is very less. For the first part of the simulation arc is considered a constant and it is of the order of 0.015 to 0.13Ω in air [11].

The switch has to operate under high pressure to withstand very high voltage. The circuit parameter which is changing is the arc resistance. Under high pressure the resistance is represented by Rompe and Weizel arc resistance model given in the equation [9,10]

$$R_{channel} = d.\frac{P}{2a\int i(t)^2 dt} \tag{6}$$

where *d* is the gap length, *p* is the gas pressure, *i*(*t*) is the current flowing through the channel, and *a* is a coefficient characterizing the nature of the gas . It was obtained that a = 0.8-1 (atm-cm$^2$)/(s-V$^2$) for air and nitrogen and 30 (atm-cm$^2$)/(s-V$^2$) for argon[10].

*3.4 Peaking Capacitor*

The entire experiment consists of the pulse generator which is a Marx generator to which a peaking capacitor and peaking switch is connected as in 'figure.3'. The rise time of the output pulse can be improved by using peaking capacitor. The peaking capacitor will hold the energy for a short time and discharge much faster than the erected Marx generator. The value of the peaking capacitor is evaluated by considering the load impedance '$R_{load}$' and the source impedances $L_{marx}$ and $C_{marx}$, of Marx generator.

$$C_p = \frac{\frac{L_{marx}}{R_{load}^2}}{1+\frac{L_{marx}}{R_{load}^2 C_{marx}}} \tag{7}$$

The peaking capacitor is selected to achieve exponential decay through the load. It is placed after all the Marx spark gap switches and before the peaking switch as in 'figure.3'

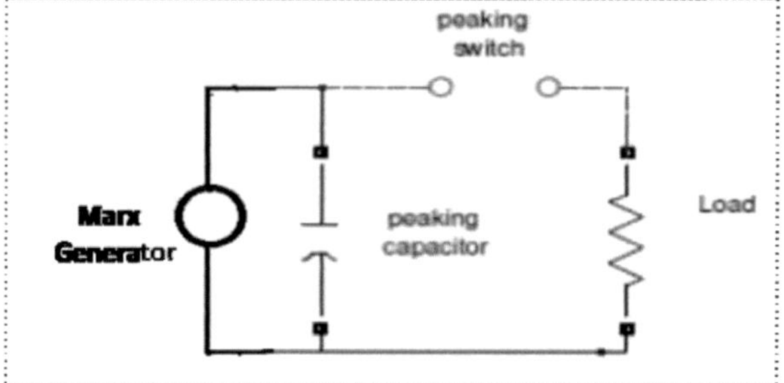


**Figure 3.** Marx generator and peaking capacitor connected to the load

## 4. Electrostatic modelling

The 'figure 4' shows the coaxial spark gap model of the peaking switch with electrodes having Rogowski profile modeled using FEMM. One of the conductors is charged to a high voltage (300 kV) and other one is at ground potential. The switch geometry taken for simulation is having electrode diameter 2.63cm and gap distance 2mm. The diameter of the whole cylindrical structure of the switch is 8.1cm and its volume is 289.46 cubic cm. This analysis gives the electric field distribution, field density, energy stored in the switch and total switch capacitance. 'figure. 5' shows the field intensity plot and it is uniform throughout the distance. A comparative study of different electrode geometries like plane-plane, point-plane, partially uniform electrode and Rogowski profile are carried out using this software. The modelling helps to detect non-uniformities in the field intensity. The energy stored in the switch and its capacitance is given in Table 1. The table shows that plane-plane and Rogowski electrode profiles, which have uniform field intensity have higher energy storage. Therefore the selection of Rogowski profile for the shape of the electrodes in the switch is perfectly justified.

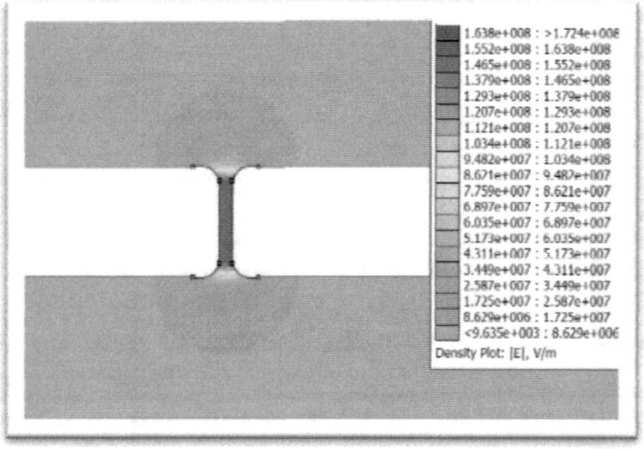


**Figure 4.** FEMM switch model shows density plot Of a Rogowski profile electrode

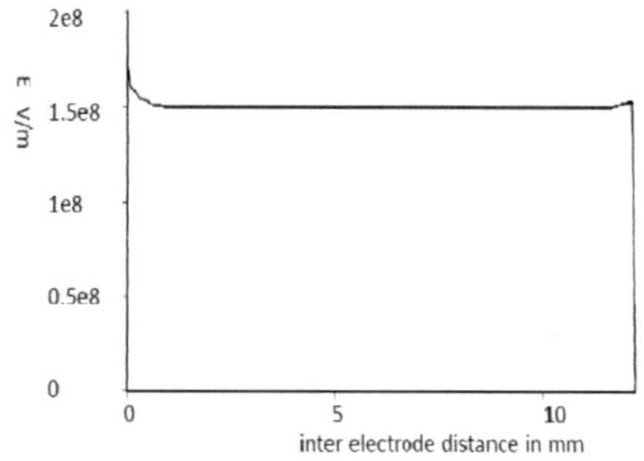


**Figure 5.** Electric field intensity plot for a Rogowski profile

**Table 1** Energy stored and capacitance for different geometries

| Electrode configuration | Energy stored (Joules) | Capacitance of the gap (pico farads) |
|---|---|---|
| Plane-Plane | 0.00538726 | 0.1197 |
| Point plane | 0.00362011 | 0.0804 |
| Partly uniform | 0.00353064 | 0.0784 |
| Rogowski | 0.00404199 | 0.0898 |

## 5. Circuit simulation

The circuit simulations are helpful for investigating the principle of operation of the overall system performance. The accuracy of this method depends on the value of the equivalent circuit parameters evaluated from the geometry. The simulations are carried out with and without pressure

*5.1 Simulation without pressure*

Switch is modelled as a series combination of inductor and resistor, in parallel with a capacitor [8]. This switch is connected to the peaking capacitor and load resistance. The inductance of the switch calculated using equation(2) is 25nH. At the time of switching the value of resistance is kept constant and its value is 0.13Ω [11]. The switch capacitance is 0.0898pF and peaking capacitor is 40pF . The circuit simulation is done with MATLAB Simulink. The peaking capacitor is initially charged to 300kV; which discharges through the switch and the load.The rise time obtained is 1ns and peak voltage is 220kV as shown in 'figure.6'. This simulation is not taking care of gas inside the switch and its pressure. It only gives the effect of gap distance and geometry on the output pulse.

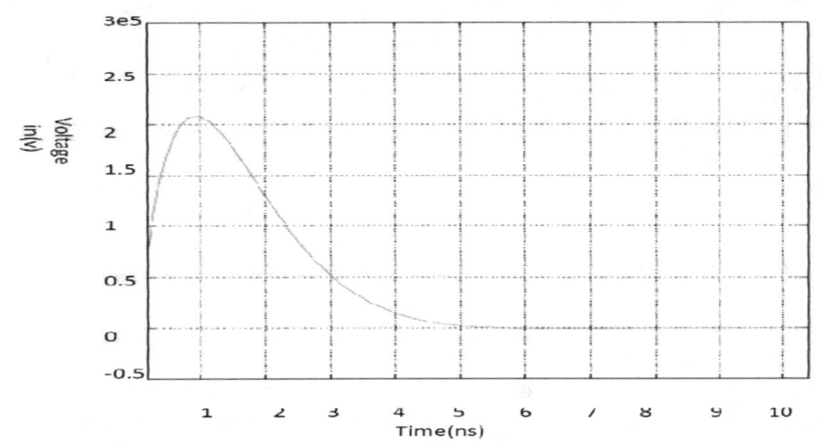


**Figure 6.** Output pulse for a switch with 2mm gap distance and air as the medium without pressure

*5.2 Simulation under pressure*

The switch is expected to withstand the voltage of 300kV for a gap distance of 2mm under pressurized conditions. The gas pressure under various inter electrode distances could be calculated from Paschen's law. The breakdown voltage under uniform field in air is given by equation(8)[12].

$$V = 6.72\sqrt{Pd} + 24.36Pd \tag{8}$$

voltage V as a function of Pd value is given by

$$V = \frac{B*Pd}{lnPd+C} \tag{9}$$

where B and C are constants given in reference [13]. 'Figure 7' Shows the breakdown voltage vs gap distance for air and $SF_6$ gas under 5bar and 10 bar pressure.

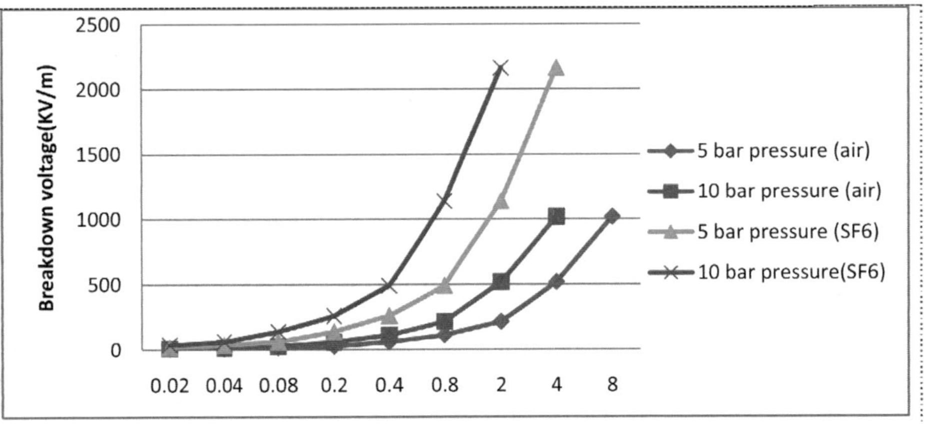


**Figure 7.** Graph showing the gap distance Vs breakdown voltage under different pressure for air and $SF_6$

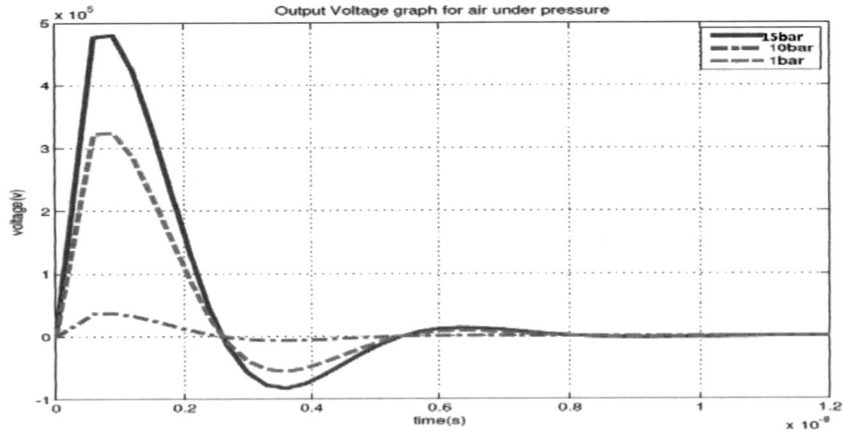


**Figure 8.** Output pulse for a switch with 2mm gap distance and air as the medium under high pressure

The objective of this paper being the reduction in rise time, it is essential to minimize the channel inductance. The inductance reduces when the electrode gap distance is small. Once the gap distance reduces the operating pressure has to increase to withstand high voltage. The lumped element model of the switch is developed using Rompe and Weizel arc resistance model. The non linear differential equation is solved by finite difference method to find the rise time and peak voltage of the output pulse generated by the switch under different pressure.

Now the fixed arc resistance as shown in 'figure.2' could be made variable by incorporating the resistance as a function of pressure as in equation (6). This gives the model of the switch under pressured conditions. The result obtained from such incorporation is shown in 'figure. 8' which gives the typical output pulse from a peaking switch with air as the medium under various pressures. For 1 bar pressure the voltage is 40kV and rise time is 0.06ns. As the pressure increases the peak voltage also increases. When the pressure is 15 bar the peak voltage is 450kV. The result show that as the pressure is increased the breakdown voltage increase, which means that switch gets closed for higher voltage. This modelling show that the rise time can be reduced by reducing the switch inductance and the peak applied voltage can be increased by increasing the pressure in the switch.

**6. Conclusion**

The paper has modelled a peaking switch and brought out the salient points towards reducing the rise time of the switch. It has been shown that reduction in switch inductance reduces the rise time of the pulse output. The switch needs to be applied high voltage under high pressure. It has been shown that the high voltage could be achieved by increasing the pressure in the switch.

The switch model for the desired rise time has been made and analyzed in FEMM software. Switch equivalent circuit model is simulated using MATLAB to observe the pulse rise time. The generation of sub-nanosecond pulse is possible only under high pressure. High pressure switch model is made by considering the arc resistance as a pressure dependent variable. These modelling techniques will be useful in implementing the high pressure peaking switch for UWB application

**References**

[1] Martin John Christoher 1992 *Proceeding of the IEEE* **80** 934
[2] Carey W.J , Mayes J.R 2002 *Power modulator symposium* 236
[3] Grishin D. M et al 2008 *Technical Physics Letters* **34** 822
[4] Pecasting Laurent et al 2006 *IEEE transaction on plasma scienc* **43** 1822
[5] Cadilhon B, Pecasting Laurent, Thierry Reess, Alain Gibert, 2008 *IET* **2** 248
[6] Lehr J.M, Baum C.E, Prather W.D and Agee F. J, 1997, *Pulsed Power conference* **2** 1033
[7] Pulsed power technology-May.2010 *Lecture delivered at the school*
[8] Carl E. Baum, Prashanth Kumar, Serhat Altunc , Apr.2009,*Switching Notes,* Note 35
[9] Xingxin Wang, Haiyun Luo, Yuan Hu 2002 *IEEE transaction on Plasma.***35**. 1916-1922
[10] Mesytas G 2005 *Pulsed Power* (New York:Springer)
[11] Engel T. G 1989 *IEEE Transaction on Plasma Sci.* **17** 323
[12] Ravindra Arora and Wolfgang Mosch *High voltage Insulation engineering* (New age international (p) limited)
[13] Heylen A. E. D 2006 *IEEE Electrical Insulation Magazine* **22** 25
[14] Frost C.A *et al.* 1993 *Proceeding of the IEEE International Pulsed Power Conference* **1** 491

Journal of Physics: Conference Series **377** (2012) 012099 doi:10.1088/1742-6596/377/1/012099

# Simulation of impact/explosive driven spallation in metals: comparative study of damage and dynamic strength models

**V R Ikkurthi, S Chaturvedi**
Computational Analysis Division, Bhabha Atomic Research Centre, Visakhapatnam – 530 012, India

E-mail: ramana468@gmail.com

**Abstract.** Spallation refers to fracture occurring in materials due to tensile loading. This paper presents an overview of our extensive one- and two-dimensional simulations done to study spallation in impact/explosive loaded Copper or Mild Steel targets, using an Arbitrary-Lagrangian-Eulerian hydrocode. Three methods of computing the spall strength and spall thickness have been employed. The computed spall parameters have been compared with Russian spall experiments. In impact loaded targets, due to a square shock wave, only one scab has been observed. In the case of explosive loading, as the rarefaction associated with the incoming shock wave is triangular, there occur many high-tension regions leading to the formation of multiple scabs (typically 2-3 scabs). Effect of flier velocity and target temperature on spallation has been studied. These trends match with experiments. Edge effects due to finite diameter of the target attenuate the incoming shock wave by lateral release, resulting in more marked damage near the axis than on the periphery. Four damage models and three dynamic strength models have been examined to determine the best-suited model for spallation studies. It is found that the Void Growth (VG) damage model and Zerilli-Armstrong strength model yield spall parameters close to experiments.

## 1.Introduction

"Spall fracture"' is damage that occurs in a body when two rarefaction waves interact and produce enough dynamic tension to fracture it. Spallation is a rate-dependent process in which fracture occurs simultaneously over an area by the nucleation, growth and coalescence of many micro cracks or voids. The schematic of impact and explosive loaded spallation experiments is shown in figure 1. In case of impact loading, two one-dimensional shock waves will be generated. One propagates into the target and the other into the flyer plate. These shock waves reflect as rarefaction waves from the free surfaces of the flyer and target plates respectively. With a thin flyer, these rarefaction waves interact inside the target, producing a state of tension in some region. If this tensile stress level exceeds the dynamic yield strength of the material, fracture takes place, producing a scab (``spall element") from that section of the target. After fracture has taken place, stress or spall signals are produced that propagate to, and accelerate, the free surface of the target. The velocity and the thickness (``spall thickness") of this scab depends upon the shape of the pressure profile at the free surface and the yield strength of the target. Here, ``spall strength" refers to the relative resistance of the material to spallation.

Many theoretical damage and strength models, both empirical and micro-physical in nature, have been developed to describe and predict high strain phenomena such as spall fracture [1]. These models compute the void volume in the region of fracture to quantify the damage level.

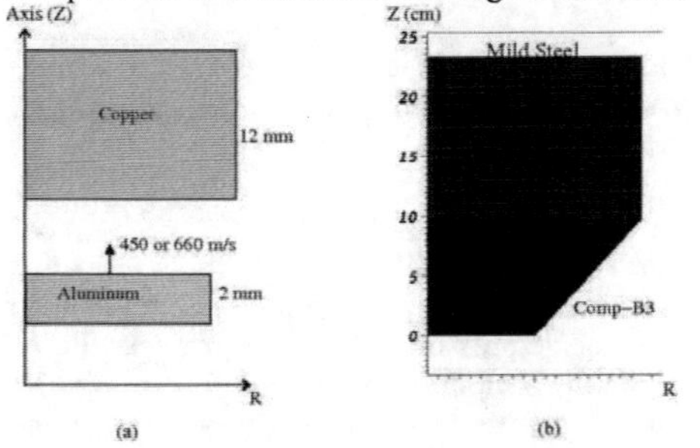


**Figure 1**. Schematic of (a) Impact, and (b) Explosive loaded spallation experiments.

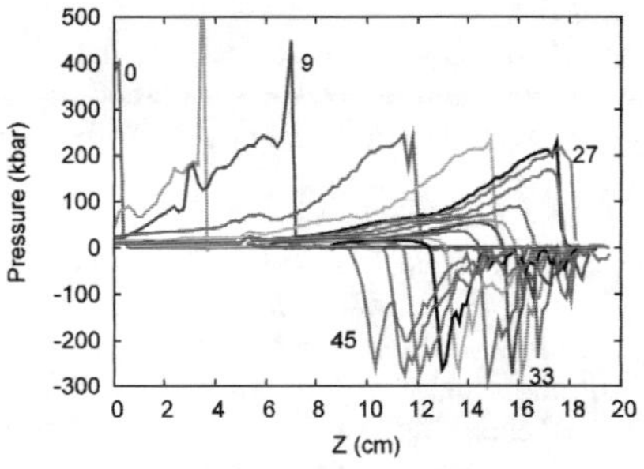


**Figure 2.** Temporal variation of pressure along axis for explosive-loaded MS (fracture process accounted for).

There are a variety of practical problems involving high-velocity impact, and spallation frequently plays an important role. It is, therefore, of great interest to determine the effect on spallation of parameters like the impact pressure, the duration of the pressure pulse and the initial temperature of the target material. It is also of interest to assess the relative performance of different fracture and dynamic strength models in predicting spall strength and thickness for the flyer and target materials of interest. Finally, edge effects can also critically affect spallation, and their importance must be examined. Using two dimensional simulations, the above problems have been studied in detail and the results are presented in the following sections.

## 2. Computational Method

An Arbitrary-Lagrangian-Eulerian hydrocode has been used for these simulations. An Aluminum metallic flyer or a cylindrical charge of Composition-B3 (density = 1.717 g/cc) explosive drives a shock into a Copper or Mild Steel (MS) target plate. The schematic of the spallation problem is shown in figure 1. A programmed-burn model is used to compute burn propagation in the explosive. The JWL equation of state (EOS) is used for Composition-B3 and a six-parameter EOS [2] is used for Al/Cu/MS. The microphysical mechanism of ductile damage is nucleation, growth and coalescence of voids in the tension region. As the voids increase, the material softens, i.e., yield strength (Y) and shear modulus (G) decrease. Four damage models [1] and three dynamic strength models have been examined in this study. The damage models compute the void volume fraction (VVF) and also compute Y and G values as functions of VVF. Dynamic strength models compute variation of Y and G with strain, strain rate, pressure, temperature, etc. In our simulations, we monitor the free-surface velocity ($v_{fs}$) of the target and also the spatio-temporal evolution of Y, G and VVF.

We consider three methods of computing the spall strength and spall thickness, based on $v_{fs}$, VVF profiles and degradation of Shear Modulus. The first method, which is routinely used in analyzing experimental data, uses the pullback velocity and the period T of the oscillations in $v_{fs}$, based on the following equations

$$P_{sp}=\frac{1}{2}\rho_0 C_b \Delta w \quad \& \quad t_s=\frac{1}{2} C_b T_p$$

where $\rho_0$ and $C_b$ are density and bulk sound speed at zero pressure. The second method uses a criterion following Eftis [3] that when VVF in a computational cell approaches a critical value (~ 0.3), spallation takes place at that point. The third method is where the shear modulus drops to half its normal value. Using these methods, we have computed the spall thickness of the scabs generated in impact/explosively-loaded targets and compared with the experiments.

## 3.Results

Extensive 1-D and 2-D simulations have been performed to study spallation and edge effects. This is an overview of our work on spallation done over many years. Due to limit on the length of the paper, only some results are presented with illustrations or numbers and others are simply listed.

### *3.1. Multiple spallation*

Two-dimensional hydrodynamic simulations have been done to study spallation in explosive-loaded MS target. Here simulations show that multiple scabs emerge from the target. The physical mechanism of multiple spallation is due to two reasons: Very high shock pressures and nonsquare profile of incoming shock wave. This can be explained in detail using the temporal variation of pressure profile along the axis, as shown in figure 2. The rarefaction associated with the incoming shock wave can be assumed triangular and when this incoming rarefaction wave meets the rarefaction wave generated due to interaction of shock front with the free-surface, a tension region is created. As the interaction proceeds the tension strength gradually increases. Spallation or fracture occurs at the point or region where the tension strength exceeds material strength. This creates a free-surface and the new situation is like a comparatively lower strength triangular shock wave interacting with the free-surface. Hence, the whole process repeats and if the initial shock strength is very high two or more spallation regions can occur. This leads to the formation of multiple scabs. The number and size of scabs emerge from the target agree well with the experimental observations.

### *3.2. Comparison of Damage models*

We have done spallation simulations using either of the four damage models [1]: Void growth (VG), Nucleation and Growth (DFRACT), Cochran-Banner (CB), Johnson-Cook (JC). Figure 3 shows the measured and computed free-surface velocity profiles of the spalled element in a copper target, for the parameters of the experiment labelled B-62 [4]. Our simulations show that both VG and DFRACT models give results that match fairly well with experiments, with the VG model performing somewhat better.

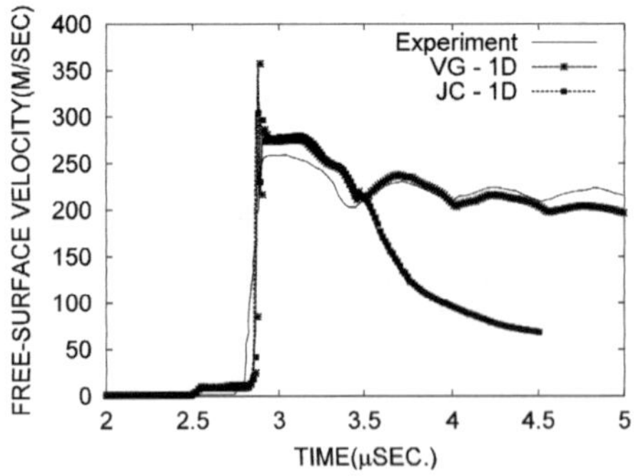


**Figure 3.** Experimental and computed free-surface velocity profiles for experiment labelled B-62 using VG and JC damage models, with copper target (aluminum flyer).

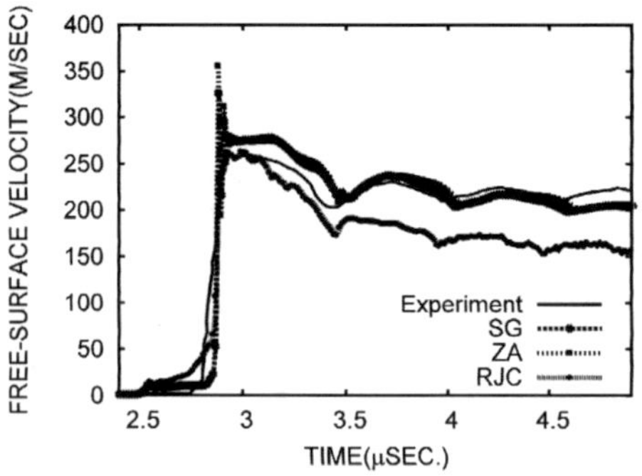


**Figure 4.** Experimental and computed free-surface velocity profiles for experiment labelled B-62 using SG, ZA and RJC models, with copper target (aluminium flyer).

### *3.3. Comparison of Strength models*

We have repeated spallation simulations using either of the three dynamic strength models: Steinberg-Guinan (SG) [5], Zerilli-Armstrong (ZA) [6], Revised Johnson-Cook (RJC) [7]. Figure 4 shows the experimental and computed free-surface velocity profiles of the spalled element in a Cu target, for the parameters of the experiment labelled B-62 [4]. The results show that ZA model has the best performance for the experiments examined, with SG also doing fairly well. The RJC model generally does not yield a good match.

### *3.4. Effect of impact pressure and target temperature*

We find that $t_s$ monotonically decreases with increase in flyer velocity, for both copper and MS targets. For a copper target (B62), $P_{sp}$ first decreases marginally, but then increases monotonically with the impact pressure. For MS targets (B16), the initial drop is quite large, although it is then followed by a monotonic increase. This can be explained using dislocation mechanics which has been accounted through strength models. With increase in strength of compressive pulse dislocations pile-up and hardens the material resulting in an increase of $P_{sp}$. We see that the spall strength of copper increases only slowly with $T/T_m$, with a maximum increase of only ~20%. These trends agree with experimental observations reported for copper. For MS, the behaviour is more erratic, and no clear trend can be seen.

### *3.6. Edge effects on Spallation*

In situations where the thickness and diameter of the target material are of the same order, the incident shock is attenuated not only by the axial unloading waves but also by lateral release. Due to this extra attenuation, it is only near the axis that the tensile stresses and stored energy are sufficient to produce complete fracture. In case of explosive-loaded metals, heavy damage (3-5 scabs) on the axis has been observed. But, far from axis, due to edge effects, reduced damage (1-2 scabs) has been observed. Hence, the net result is that 2-3 scabs emanate from the target.

## 4. Conclusion

One- and two-dimensional simulations have been done to study spallation in impact/explosive loaded targets. The computed spall parameters have been compared with Russian spall experiments. In impact loaded targets, due to a square shock wave, only one scab has been observed. In the case of explosive loading, as the rarefaction associated with the incoming shock wave is triangular, multiple scabs (typically 2-3 scabs) emerge. In impact problems, the computed spall thickness decreases with increase in impact pressure. The spall strength is found to be increasing with impact pressure. These trends match with experiments. Edge effects due to finite diameter of the target attenuate the incoming shock wave by lateral release, resulting in more marked damage near the axis than on the periphery. It is found that the Void Growth (VG) and Nucleation and Growth models yield spall parameters close to experiments, with the VG model doing marginally better. In case of strength models, the spall parameters obtained using Zerilli-Armstrong (ZA) and Steinberg-Guinan models show reasonable agreement with experiments. ZA model is a micro-structural model and takes into consideration the crystal structure and hence performing better.

## 5. References

[1] Ikkurthi V R and Chaturvedi S 2004 *International J. Impact Engg.* **30** 275, and references therein.

[2] Eliezer J S, Ghatak A, Hora H and Teller E 1986 *An introduction to equations of state, Theory and applications* (Cambridge University Press)

[3] Eftis J 1996 Constitutive modelling of spall fracture *High pressure shock compression of solids II* ed L.Davison, D.E.Grady, S.Mohsen (Berlin: Springer)

[4] Antoun T H, Seaman L and Curran D R 1998 Dynamic failure of materials, vol.2- Compilation of Russian spall data *Tech. Rep. No. DSWA-TR-96-77-V2* DSWA, Alexandria, VA.

[5] Steinberg D J, Cochran S G and Guinan M W 1980 *J. Appl. Phys.* **51** 1498.

[6] Zerilli F J and Armstrong R W 1987 *J. Appl. Phys.* **61** 1816.

[7] Rule W K and Jones S E 1998 *International J. Impact Engg.* **21** 609.

Journal of Physics: Conference Series **377** (2012) 012100 doi:10.1088/1742-6596/377/1/012100

# Interatomic potential parameters for molecular dynamics simulations of RDX using a reactive force field: A validation study

**M Warrier, P Pahari and S Chaturvedi**

Computational Analysis Division, BARC, Visakhapatnam, India - 530012

E-mail: Manoj.Warrier@gmail.com

**Abstract.** The parameter sets of the ReaxFF potential distributed with the open source, Large-scale Atomic/Molecular Massively Parallel Simulator (LAMMPS) code, is validated for simulating crystal RDX. These parameters are used to model crystal RDX and obtain its unit cell size and bulk modulus. It is seen that the parameters supplied with LAMMPS (5-April, 2011 release) do not reproduce the unit cell size and bulk modulus of crystal RDX as reported by experiments and by other simulations using the ReaxFF potential. The simulation method and relevant parts of the LAMMPS code implementing the method has been earlier validated for Cu. We conclude that either the parameter sets provided with the LAMMPS distribution or its implementation of the ReaxFF potential are not suitable for modeling crystal RDX.

## 1. Introduction

LAMMPS [1] is a classical molecular dynamics code distributed as an open-source code and is widely used. It has a wide set of interaction potentials and shows a very good parallel scaling due to which several tens of millions of atoms can be simulated on a modest 128 node compute cluster. Materials like RDX, TATB, HMX, PETN are widely used in defense and in industry [2, 3]. These materials are composed of the elements Carbon, Hydrogen, Nitrogen and Oxygen. Using first principle methods are desirable for simulating reactions due to changes in electron sharing amongst the constituent atoms. This is however not practical due to severe computational time limitations. Van Duin, et al. have developed a reactive force field, ReaxFF [4], which is fitted to first principles generated data which has the capability to model large molecules and reactions between its constituent atomic species. LAMMPS has three sets of parameters for the ReaxFF force field to simulate C, H, N and O interactions [5–7] and therefore can be used to simulate RDX. We hereafter to these parameter sets as the general purpose hydrocarbon parameterization (GPHP) [5], reactive force field for Nitramines (RFFN) [6] and reactive force field for PETN (RFFPETN) [7]. These parameter sets are validated by carrying out MD simulations of crystal RDX to extract the unit cell sizes and the Bulk Modulus of crystal RDX.

We show that none of the above parameter sets reproduce either the reported values for unit cell sizes, or, the bulk modulus of crystal RDX. It should be emphasized that ReaxFF has been earlier used to to obtain the unit cell sizes and the bulk modulus of RDX with other MD simulations [8] which show a good match with experiments. We have earlier validated our method and the parts of the LAMMPS code involved in our calculation by using it to obtain the correct bulk modulus and unit cell size for Copper using the embedded atom method potential [9].

In the next section, the simulation method to obtain unit cell size and bulk modulus of a crystal is discussed. The simulations of crystal RDX using the various parameter sets are described in the third section and the results are presented. Finally we present our conclusions.

## 2. Method to obtain bulk modulus of crystal RDX

In order to determine the bulk modulus and equilibrium unit cell dimensions of crystal RDX the parameter sets provided with LAMMPS (RFFN, RFFPETN) are used to carry out 20 NVT simulations at +10% to −10% of the experimentally reported equilibrium volume in steps of 1% of equilibrium volume. A single unit cell with periodic boundary conditions is used. An MD time step of 5 femto-sec is used. Average values of the pressure and total energy of the system are sampled once the system attains equilibrium.

The variation of energy with volume from the MD simulations were fitted with the Murnaghan equation of state [10].

$$E_V = E_o + \frac{B_o V}{B'_o}\left[\frac{(v_o/V)^{B'_o}}{B'_o - 1} + 1\right] - \frac{B_o V_o}{B'_o - 1} \tag{1}$$

where, $E_o$ and $V_o$ are the energy and volume at zero pressure respectively, $E_V$ is the energy corresponding to a volume $V$, $B_o$ is the equilibrium bulk modulus at constant temperature and $B'_o$ is the pressure derivative of $B_o$. We obtain the bulk modulus and the equilibrium unit cell volume of RDX by the Murnaghan EOS fit.

## 3. Molecular Dynamics Simulations of crystal RDX

20 NVT simulations at 300 K as described in Section.2, for the RFFN and RFFPETN parameter sets are carried out. A single unit cell with periodic boundary conditions (PBC) is used. The MD time step is 0.5 femto-sec. A steepest descent (SD) minimization procedure is used to initially relax the RDX crystal prior to the MD simulations. The SD method was carried out for 100 iterations till the energy tolerance was $4.47 \times 10^{-5}$.

The variation of energy of RDX with volume for the RFFN and RFFPETN cases are shown in Fig.1 and Fig.2 respectively. Note that the variation is neither smooth nor monotonous and does not have a well defined minima as seen for copper [9]. There seems to be non-monotonic shifts in energy. The MD data is fitted with the Murnaghan Equation of State (Eqn.1). From the fit, the bulk modulus of RDX at 300 K is 5.905 GPa ($B'_o = 8.96$) for the RFFN case and 7.22 GPa ($B'_o = 6.88$) for the RFFPETN case. These are smaller than the experimentally reported values of 13.0 GPa [11] and results (13.7 GPa) from other MD simulations [8] using ReaxFF. The equilibrium volume from the fit is 1774.2$\AA^3$ for the RFFN case and 1795.9$\AA^3$ for the RFFPETN case. The corresponding unit cell sizes for these and for the NPT simulations that follow are compared in Table.1.

Atomic visualization software show no obvious structural changes in the RDX crystal. The SD minimization method in LAMMPS, used to relax the RDX crystal does not allow volume change, thereby not really taking the RDX crystal to a minimum energy state and also disallowing molecular adjustments that may occur with a varying volume. This, coupled with the above mismatches with experimental and simulation results motivates us to carry out NPT simulations of crystal RDX. This would also yield the unit cell size parameters which can be verified with the above results.

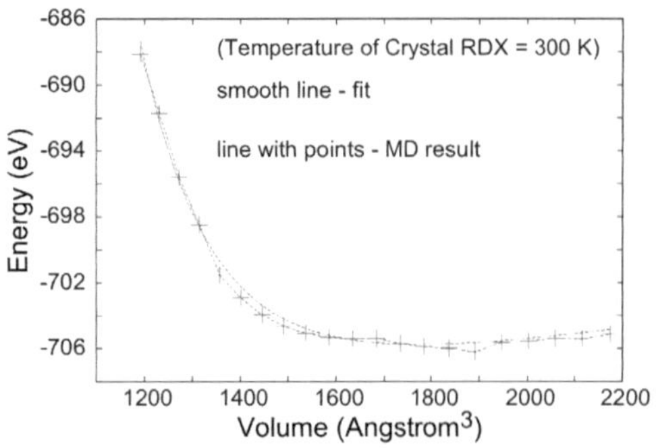


**Figure 1.** Total energy variation with volume for RDX for the RFFN case

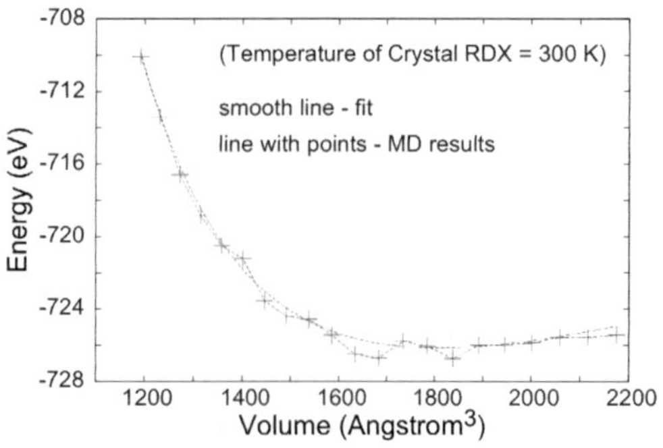


**Figure 2.** Total energy variation with volume for RDX for the RFFPETN case

**Table 1.** Comparison of unit cell sizes from our NVT and NPT simulations of crystal RDX using the RFFN and RFFPETN parameter sets, with reported experimental and simulation results

| | NVT RFFN | NVT RFFPETN | NPT RFFN | NPT RFFPETN | Expt. [11] | Sim. [8] |
|---|---|---|---|---|---|---|
| a | 13.55 Å | 13.60 Å | 14.04 Å | 14.60 Å | 13.18 Å | 13.78 Å |
| b | 11.89 Å | 11.95 Å | 12.36 Å | 12.84 Å | 11.57 Å | 12.03 Å |
| c | 11.01 Å | 11.05 Å | 11.43 Å | 11.88 Å | 10.71 Å | 10.96 Å |

*3.0.1. NPT simulations of crystal RDX* NPT simulations of a $2 \times 2 \times 2$ unit cells of RDX are carried out by varying the crystal temperature from 50 K to 300 K over 100 pico-seconds using a Nose-Hoover thermostat and barostat. The system is then allowed to relax at 300 K for 200 pico-seconds. A time step of 0.1 femto-sec is used with PBC along all the three axes. The simulations are carried for all the parameter sets (GPHP, RFFN, RFFPETN). Note that in these simulations the volume of the simulation box will change to accommodate stresses that develop within the molecules, thereby allowing molecular reorientation and structural changes within the molecule.

Table.1 shows the average unit cell sizes along X, Y and Z for the three parameter sets and also from the equilibrium volume values obtained from the NVT simulations. Fig.3 shows the initial configuration of the $2 \times 2 \times 2$ unit cells of RDX. Fig.4 shows the final configuration of the RDX molecules at the end of the NPT simulation. Structural changes in the molecular crystal structure of RDX is seen around 180 K as the crystal temperature is changed from 50 K to 300 K. The RFFPETN and GPHP results for RDX crystal structure also do not preserve the RDX crystal structure, but are not displayed here due to lack of space. From these results it is evident that none of the parameter sets preserves the crystal structure of RDX.

## 4. Conclusions and discussion

We have explored the possibility of simulating crystal RDX with three parameter sets (GPHP, RFFN and RFFPETN) supplied with LAMMPS. We see that the RFFN and RFFPETN parameter sets yield wrong results for the bulk modulus. We see from Table.1 that the unit

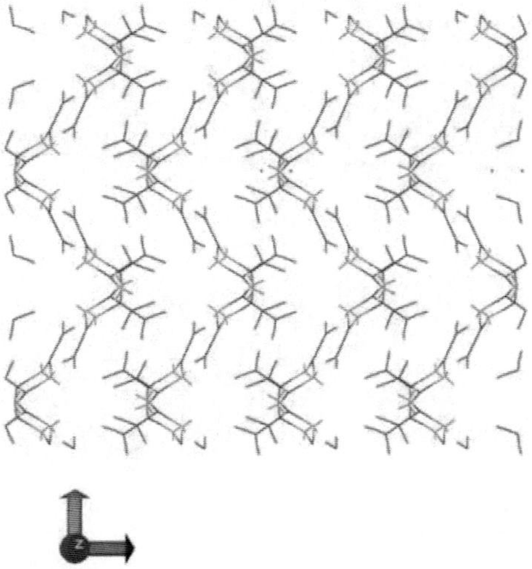

**Figure 3.** Initial atomic configuration of the $2 \times 2 \times 2$ RDX crystal used in the NPT simulations

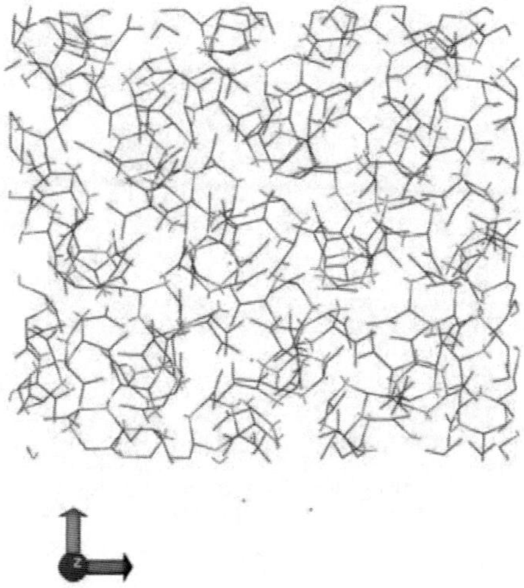

**Figure 4.** Final atomic configuration of the $2 \times 2 \times 2$ RDX crystal with RFFN parameter set after the NPT simulations

cell sizes of crystal RDX obtained from the NVT simulations with the RFFN and RFFPETN parameter sets are close to the NVT simulation results [8] and Experiments [11]. However the NVT simulations constraints the molecular readjustments by imposing a constant volume. We see that NPT simulations of RDX crystal, which allows more scope for molecular position adjustments due to variable volume, do not preserve the RDX crystal structure. The average value of the unit cell sizes in the NPT case do not match the reported NVT simulation [8] and experimental [11] results (Table.1). Therefore we conclude that the ReaxFF implementation in LAMMPS is not suitable for simulating RDX. In the future we will explore the applicability of the provided parameter sets to simulate single RDX molecule.

**Acknowledgments**
All the simulations in this article were carried out on the tserver cluster at BARC-Vizag. M. Warrier thanks S. Balasubraminian and S. K. Reddy from JNCASR-Banglore for discussions

**References**

[1] Plimpton S J 1995 *Jnl. Comput. Phys.* **117** 1–19 http://lammps.sandia.gov
[2] Bennet C M and Thomas D 1998 *LANL Report* **No. LA-UR-98-3820**
[3] Hegde G, Pathak J, Jagadeesh G, Oomen C, Arunan E and Reddy K 2009 *Proceedings 26$^{th}$ Int. Symposium on Shock Waves, Göttingen, Germany*, Ed. Klaus Hannemann, Friedrich Seiler 377–382
[4] Van Duin A C T, Dasgupta S, Lorant F and Goddard W A 2001 *J. Phys. Chem. A* **105** 9396–9409
[5] Mattsson T R, Lane J M D and K R Cochrane et al. 2010 *Phys. Rev. B* **81** 054103
[6] Zhang L, Van Duin A C T, Zybin S and Goddard W A 2009 *Journal of Physical Chemistry A*, **113** 10770–10778
[7] Budzien J, Thompson A P and Zybin S V 2009 *J. Phys. Chem.* **113** 13142
[8] Strachan A, van Duin A C and Debashis Chakraborty et al. 2003 *Phys. Rev. Lett.* **91** 098301
[9] Warrier M, Rawat S and Chaturvedi S 2011 *Proceedings of the 55th DAE Solid State Physics Symposium 2010, AIP Conf. Proc.* **1349** 95–96
[10] Murnaghan F 1944 *Proceedings of the National Academy of Sciences* **30** 244–247
[11] Olinger B, Roof B and Cady H 1978 *Commissariat a lEnergie Atomique Centre dEtudes de Vajours, Sevran, France* Symposium International Sur Le Comportement Des Milieux France, 3

Journal of Physics: Conference Series **377** (2012) 012101 doi:10.1088/1742-6596/377/1/012101

# Comparisons between Fast Shock Tube Simulations and Tests

**V Mehra[1], V Mishra, Sijoy CD and S Chaturvedi**
Computational Analysis Division
Bhabha Atomic Research Centre
IDA 'B' Block, 4th cross-road, Autonagar, Visakhapatnam 530 012, A.P., India

E-mail: vmehra10@yahoo.com
**Abstract.** The experiments of Menikoff *et al* on a projectile hypervelocity launcher using a fast shock tube (FST) are modelled using smooth particle hydrodynamics (SPH) technique. In a FST, the progressive detonation of a co-axial HE cylinder induces a cumulative shock in the liquid-filled core. This shock hits a thin flyer and accelerates it to hypervelocity. The comparisons are made on flyer velocity profile, peak pressure and shock speed in liquid core. The SPH reproduces the qualitative and quantitative aspects of the FST and is well-suited to the high strain-rate feature of this experiment.

## 1. Introduction

The design of hypervelocity launch systems requires understanding physics of projectile acceleration at high driving pressures (megabar range) and short acceleration times (few $\mu$s). The Fast Shock Tube (FST) has been proposed as a hypervelocity launcher [1] and as an injector for an electromagnetic rail gun [2]. The FST consists of a cylinder of high-explosive (HE) that drives a flat, axial shock in a fluid-filled core inside the HE . First experiments on FST were carried out by Menikoff *et al* [1]

Drive conditions much higher than direct HE drive can be obtained in FST. The core contains propellant material such as polystyrene foam or hexane. The HE is initiated at the left hand of the FST assembly in a cylindrically symmetric way and the detonation wave acts like a virtual piston that drives a strong axial wave in the core. The shocked core hits the flyer and accelerates it to a very high velocity. The choice of the design parameters is vital and these include the diameters and lengths of HE tube and core, type of HE and core density. The proper choice of these parameters achieves the flat shock and uniform flow conditions behind the shock in the core [2]. The experimental parameters are so set that the axial shock velocity in the core equals the detonation velocity of the HE otherwise the axial shock would overtake the detonation front and send a precursor wave into the HE leading to a loss in the core shock intensity.

## 2. The FST tests

The schematic of FST experiments that were carried out by Menikoff *et al* is shown in fig.1 The core of the shock tube is 6.746 cm long by 0.856 cm inner diameter and is surrounded by a HE cylinder of diameter 5 cm. The HE chosen is PBX-9501 that has a Chapman-Jouget (CJ) detonation velocity of 8.8 km/s and a CJ pressure of 370 kbar. The working fluid in the inner tube is n-hexane with density

[1] To whom correspondence should be addressed.

0.657 g/cm³. The parameters were chosen so that the test is close to the energy limit for propagating an axial shock---with a lower working fluid density or higher HE energy, the axial shock would get ahead of the detonation front and send a precursor back into the HE and thus complicating the analysis and weakening the axial shock that propagates in the hexane tube [1].

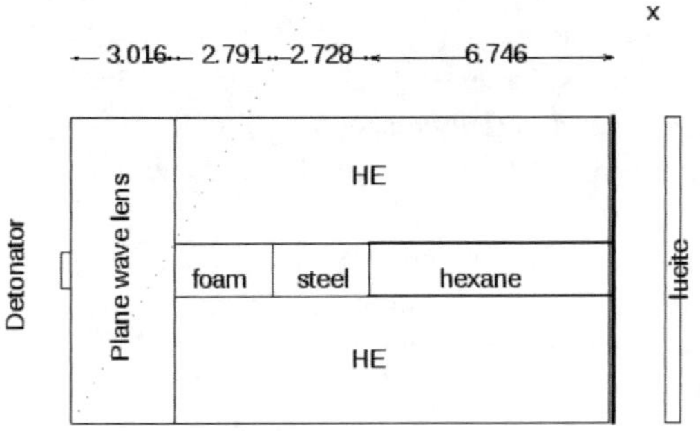


**Figure 1**:Schematic of Menikoff experiment. Distances are in cm.

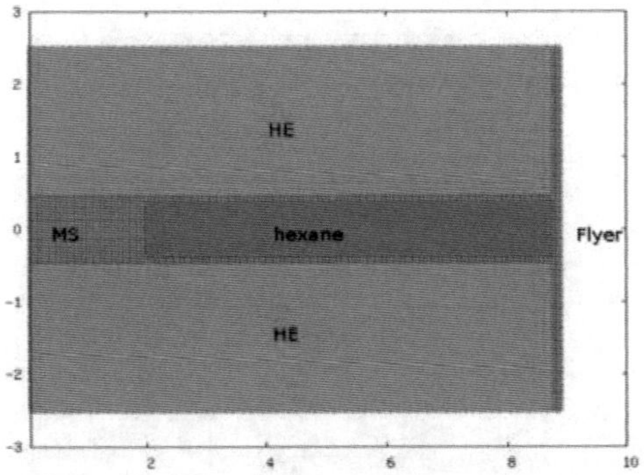


**Figure 2**: The SPH model of the FST experiment. The HE nodes a shown as green, hexane nodes as blue, and MS (mild steel) nodes that form end plug, core casing and flyer as red. Reflecting boundary is put at $x$=0 and free boundary at $y$=±2.5. A total of 9087 nodes are used. Distances are in cm.

A mild steel (MS) flyer plate of 0.79 mm thickness is placed at the end of hexane tube., followed by a 1.5 cm air gap that separates it from a lucite block. The axial shock in the hexane sends a shock into the flyer plate. The shock arriving at the rear free surface changes the surface roughness and thus the surface reflectance is also changed. The rear flyer surface is illuminated with a flash lamp. The arrival of the shock at the free surface is thus marked with the change in intensity of the reflected light. Then the air gap between the flyer rear surface and the lucite block is compressed and heated to white-hot by the motion of the flyer plate and the impact of the flyer on the lucite is marked by a bright flash. The flyer plate velocity is measured by the time delay between the reflectance change and the flash. The shock strength in the hexane is obtained by impedance matching [1]. The arrival times of the shock at flyer surface and the flyer impact at the lucite are *radially resolved.*. The radial profile of the flyer plate velocity is obtained from the radially resolved flyer plate timings.

### 3. The Simulations

The FST experiments of Menikoff *et al* are modelled using the Smooth particle hydrodynamics (SPH) , a meshless technique often used to model high-deformation events such as hypervelocity impact and penetration, jet formation in shaped charges detonation *etc.* [3]. The SPH equations are obtained from the continuum equations by discretization onto a set of co-moving nodes. The fields of continuum mechanics--- pressure, density, temperature stress and strain tensors are defined on the nodes. The node-node interaction is obtained by the discretization of continuum equations and is mediated by a kernel function $W$. The function $W$ depends upon the inter-node distance and a parameter called the smoothing length $\boldsymbol{h}$ that measures the zone of influence of a SPH node. For the specific case of cubic spline kernel function that is commonly used in SPH, the node-node interaction vanishes for inter-node distance that is greater than $2\boldsymbol{h}$. The kernel $W$ are normalized and reduce to delta function in the limit $\boldsymbol{h}\rightarrow 0$ [3].

The SPH model of the experimental situation is built by filling up the geometry with SPH nodes. For FST model, we have PBX-9501 high explosive (HE) , n-hexane working fluid and mild steel as the flyer and the casing material for HE and n-hexane. The nodes are initially placed on a regular lattice that may be square or triangular. The ratio of smoothing length $\boldsymbol{h}$ to average initial inter-node spacing $\boldsymbol{d}$ is an important SPH parameter relevant to the stability of the SPH algorithm. Initially the smoothing lengths are set by scaling the inter-node spacing $\boldsymbol{d}$, thus $\boldsymbol{h/d}$ is taken as a global ratio, between 1 and 2 [4]. The local resolution is adaptively varied by evolving the smoothing lengths in space and time according to the Benz prescription [5].

Simulations in this paper have been carried out in two-dimensional ($x$-$y$) Cartesian geometry. The initial placement of SPH nodes is shown in fig.2. There are three types of nodes corresponding to HE, n-hexane and mild steel flyer. The HE is modelled with Jones-Wilkins-Lee (JWL) equation of state (EOS), hexane and MS are modelled with EOS of Gruneisen type[1]. Von Mises plastic flow criterion is used for MS.

The detonation is simultaneously initiated at $y=\pm\ 2.5$ *i.e.* along the periphery of HE charge. The detonation produces an axial shock in the hexane core that moves along the $x$-axis The nodal configurations and pressure distribution are shown in Fig.3 at 6 μs and 9 μs after HE initiation.

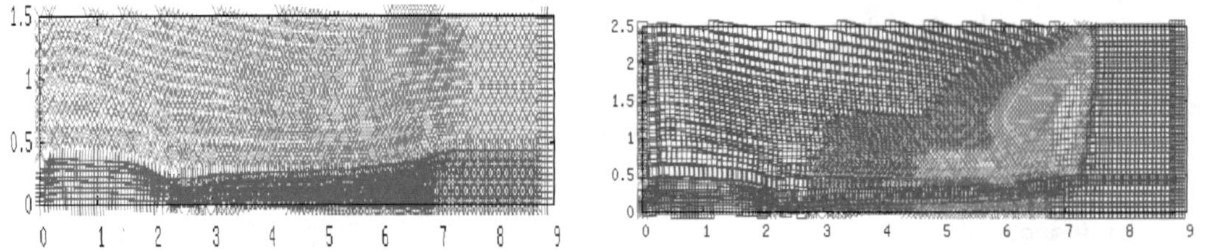

**Figure 3:** SPH nodal configurations (left) and pressure distribution (right) at 6 μs (top) and 9 μs (bottom) after the detonation is initiated. The axial shock velocity is determined from the visual inspection of these plots. On left, HE nodes are shown green, hexane nodes are blue and MS nodes are red. On right, red marks nodes with pressure > 200 kbar; green marks nodes with pressure >100 but <200 kbar; blue marks nodes with pressure >50 kbar but <100 kbar and pink marks pressure <50 kbar. The axis of the shock tube is along $x$-axis. Distances are in cm.

Shock velocity of 9.8 km/s is determined from the relative positions of the axial shock at different times. Experimentally, the shock speed is 8.9 km/s [1]. The SPH simulations yield 330 kbar as the peak pressure generated within hexane as the axial shock hits the flyer (experimentally 290 kbar). The simulations is ended at 12 μs from the HE initiation and the final nodal configuration is shown in Fig 4 below.

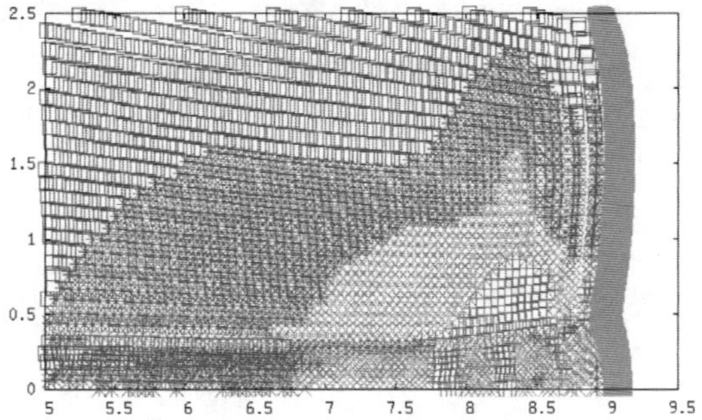

**Figure 4**: The pressure distribution and flyer deformation as the axial shock hits the flyer (shown in light blue). The red marks nodes where pressure >200 kbar, green marks nodes with pressure >100 kbar but <200 kbar. Blue marks nodes where pressure >50 kbar but <100 kbar and pink are nodes with pressure <50 kbar. Distances are in cm.

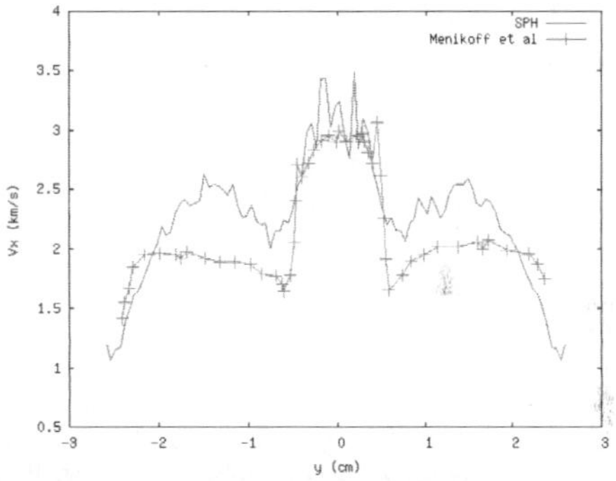


**Figure 5**. The axial velocity ($v_x$) profile of the flyer along the *y*-axis. The red line is SPH result and green is from Menikoff experiment. Experimentally, the central part of the flyer reaches 3 km/s and the outer part 2 km/s [1]. The corresponding values from our 2D simulations are 3.5 km/s and 2.5 km/s resp. Along *x*-axis is radial distance from core axis in cm and *y*-axis is axial velocity in km/s.

The fig.5 shows the comparison between the experimental velocity profile and the velocity profile produced through SPH. The peak flyer velocity from SPH is 3.5 km/s (exp. 3 km/s). The agreement is satisfactory considering the 2D nature of our simulations.

## 4. Conclusions

The experiments of Menikoff et al on a projectile hypervelocity launcher using a fast shock tube (FST) are modelled using smooth particle hydrodynamics (SPH) technique. In a FST, the progressive detonation of a co-axial HE cylinder induces a cumulative shock in the liquid-filled core. This shock hits a thin flyer and accelerates it to hypervelocity. The comparisons are made on flyer velocity profile, peak pressure and shock speed in liquid core. The SPH reproduces the qualitative and quantitative aspects of the FST quite well and is well-suited to the high strain-rate feature of this experiment.

## References

[1] Menikoff R, Lackner KS, Johnson NL, Colgate SA, Hyman JM and Miranda GA 1991 *Phys. Fluids.* A **3(1)** 218
[2] Tian Y, Mai M, Zheng P and Wang Y 2005 *IEEE Transactions on magnetics* **41 (1)** 365
[3] Mehra V and Chaturvedi S 2006 *J. Comp. Phys.* **212** 318
[4] Balsara DS 1995 *J. Comp. Phys.* **121** 357
[5] Benz W 1988 *Comp.. Phys. Comm.* **48** 97

# Tensile Instability and Artificial Stresses in Impact Problems in SPH

**Vishal Mehra, Sijoy CD, Vinayak Mishra and Shashank Chaturvedi**

Computational Analysis Division Bhabha Atomic Research Centre IDA 'B' Block, 4th cross-road, Autonagar, Visakhapatnam 530 012, A.P., India

E-mail: vmehra10@yahoo.com

**Abstract.** The smooth particle hydrodynamics (SPH) is a meshless computational technique that is popular in the modeling of impact and penetration problems. However, SPH is liable to a tensile instability that manifests itself as a bunching of nodes and formation of artificial voids and no generally accepted formulation exists to counter this instability. We examine the performance of two methods that have been proposed to deal with the tensile instability— the Monaghan artificial stresses and the Godunov-type SPH. The impact and penetration of 0.5 cm radii steel spheres on 2 mm thick aluminium plate at 3.1 km/s is chosen for comparison. We show that the artificial void formation in St-Al impact is suppressed but not eliminated by using Monaghan stresses while the void formation is entirely eliminated by using Godunov-type formulation of SPH that was proposed by Parshikov and Medin.

## 1. Introduction

The meshless Smooth Particle Hydrodynamics (SPH) technique is widely used to model high-velocity impact and penetration problems since it handles material deformation and erosion without needing a special logic, unlike mesh-based Eulerian or Lagrangian codes [1]. However, the SPH suffers from a numerical instability in tension and consequently the regions subject to large tension exhibit 'numerical fracture' and artificial voids [2]. The underlying cause is the lack of formal consistency in SPH: it can not reproduce exactly any class of functions on a defined set of points. As a consequence , a numerical clumping instability manifests itself when nodes are mutually attracted. Essentially, the SPH kernel function $W$ is unable to keep the nodes apart once they are sufficiently close to one another. This can be a significant concern in impact and penetration problems since the elastic dynamics involves appreciable attractive forces. The instability is much less pronounced in purely hydrodynamic flows. There is no universally accepted technique for countering this instability. In this paper, we compare and contrast two different stabilizing methods that have been proposed and we apply them to a high velocity impact problem.

## 2. Tensile Instability

The SPH solves the continuum equations of motion by representing the continuum as a set of co-moving and disconnected nodes [3]. The nodes have mass and carry the relevant fields such as velocities, density, pressure, temperature, stresses and strains. The nodes have overlapping regions of influence that leads to node-node interaction. The equation of motion for the nodes

Published under licence by IOP Publishing Ltd

Journal of Physics: Conference Series **377** (2012) 012102 doi:10.1088/1742-6596/377/1/012102

are obtained by discretization and interpolation. A kernel function $W$ with required properties handles the interpolation As the nodes are disconnected, the SPH does not require a special logic to handle topological changes that might occur in a penetration event. The discretization procedure yields the following set of equations [1]:

$$\begin{aligned} \frac{d\rho_i}{dt} &= \rho_i \sum_j \frac{m_j}{\rho_j}(\mathbf{v_i} - \mathbf{v_j}) \cdot \nabla W_{ij} \\ \frac{d\mathbf{v}_i}{dt} &= -\sum_j m_j (\frac{\sigma_i}{\rho_i^2} + \frac{\sigma_j}{\rho_j^2} + \Pi_{ij}) \cdot \nabla W_{ij} \\ \frac{dE_i}{dt} &= -\sum_j m_j (\mathbf{v_i} - \mathbf{v_j}) \cdot (\frac{\sigma_i}{\rho_i^2} + \frac{1}{2}\Pi_{ij}) \cdot \nabla W_{ij} \end{aligned} \quad (1)$$

The summations run over neighboring particles, the neighborhood being defined by the support of $W$ Here, $\mathbf{v_i}$ is the velocity, $\rho_i$ is the density, $E_i$ is the internal energy per unit mass, and $\sigma_i$ is the stress tensor associated with the $i$th node. $\Pi_{ij}$ is an artificial viscous pressure that is required to control unphysical oscillations around shocks. The kernel function $W_{ij} = W(|x_i - x_j|, h)$ depends upon the smoothing length $h$ that measures the zone of influence of a SPH node. For the cubic spline kernel used in this work, the kernel vanishes for $|x_i - x_j| > 2h$ [1,3] where $x_i, x_j$ are the positions of $i$th and $j$th nodes resp. The smoothing length $h$ is related to the average inter-node spacing $d$ by $h_{fac} = h/d$. The parameter $h_{fac}$ is important for the stability of SPH and should be between 1 and 2 [4]. The smoothing length $h$ may vary spatially and in that case we speak of $h_i$–the smoothing length associated with the node $i$ and $h$ in above equations is replaced by $h_{ij} = (h_i + h_j)/2$.

This set of SPH equations is liable to the tensile instability. An example may be seen in the high-velocity impact of mild steel spheres of 0.5 cm radii onto 0.2 cm thick aluminium plate at 3.1 km/s (St-Al impact)[1,5] (fig. 1 shows the initial setup of the nodes). The fig. 2 shows target and projectile 8 $\mu$s post-impact. Voids are seen between the target and the front part of the projectile and also inside the projectile. These voids are not seen in the free-Lagrange simulations of the impact that were done by Howell and Ball [5] and thus may be a numerical artifact.

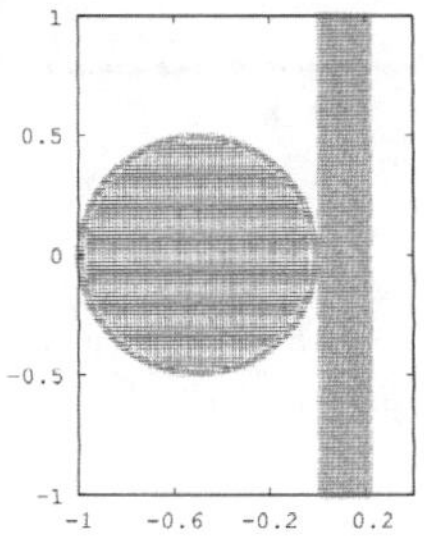


**Figure 1.** Geometry and the initial node placement for St-Al impact. The circular projectile (red nodes), on the left, has a radius of 0.5 cm. On the right is the target plate (green nodes) having a thickness of 0.2 cm.

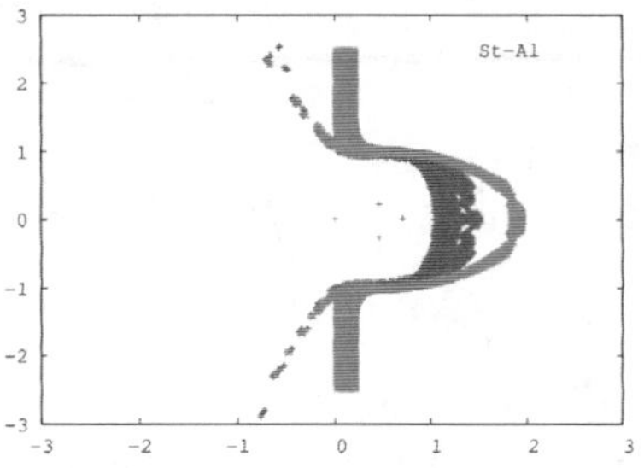


**Figure 2.** The St-Al impact by conventional SPH showing void formation and tensile instability. The projectile is shown in red and target in green. The distances are in cm. The configuration is shown 8 $\mu$s post-impact at 3.1 km/s.

## 3. Monaghan Stresses

Monaghan [2] proposed to deal with the tensile instability by adding certain artificial stresses $Rf_{ij}$ to the momentum equation:

$$\frac{d\mathbf{v}_i}{dt} = -\sum_j m_j \Big(\frac{\sigma_i}{\rho_i^2} + \frac{\sigma_j}{\rho_j^2} + \Pi_{ij} + Rf_{ij}\Big) \cdot \nabla W_{ij} \tag{2}$$

where

$$R = -0.15 \left(\frac{\sigma_i}{\rho_i^2} + \frac{\sigma_j}{\rho_j^2}\right)$$

and

$$f_{ij} = \left[\frac{W(|x_i - x_j|, h_{ij})}{W(h_{ij}/h_{\text{fac}}, h)}\right]^4.$$

Since $h_{ij}/h_{\text{fac}}$ is just the average nodal spacing at the neighbourhood of particles $i, j$, the term $f_{ij}$ expresses a force that increases as the particle separation decreases. As it seeks to correct a numerical instability, the $f_{ij}$ is expressed in terms of the kernel function.

The Monaghan stress have been shown to work for a variety of low-velocity problems [2] but have not been tried on high-velocity impact and penetration problem. The fig. 3 shows the St-Al impact configuration produced using the Monaghan stresses. The voids have reduced in size but they are still very prominent. Thus the Monaghan stresses do not eliminate tensile instability in the high-velocity impacts.

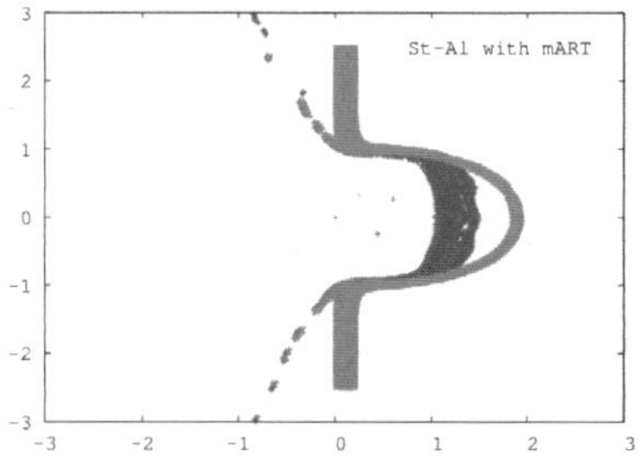


**Figure 3.** The St-Al impact simulated using Monaghan artificial stresses shown 8 $\mu$s post-impact. The projectile is shown in red and target in green. The distances are in cm.

## 4. Godunov-type SPH

Implementation of Godunov-type scheme into SPH [6] (CON) offers an alternative way to tackle the tensile instability. CON casts the problem of computing the derivatives $\nabla \mathbf{v}, \nabla \sigma$ , that occur in the momentum and energy balance equations of SPH, into the solution of a Riemann problem [7]. The Riemann problem directly solves the conservation equations at a discontinuity that is supposed to lie between two given nodes. The solution of the Riemann problem yields the values of hydrodynamic fields at the contact surface. In conventional SPH, the field values at the contact surface are just the mean values of the fields at interacting nodes. The CON procedure does not require artificial viscosity. Godunov-type methods are commonplace for gas dynamic applications. However, the applications to materials with strength has been rare, due to increased complexity of formulation. Shown below is the St-Al impact with CON. In figure. 4, there are no visible voids, either in projectile front or inside. Thus the tensile instability in SPH is dealt better with CON than with the Monaghan artificial stresses at high impact velocities. CON also offers freedom from setting arbitrary artificial viscosity parameters, a subject of some controversy in conventional SPH [8].

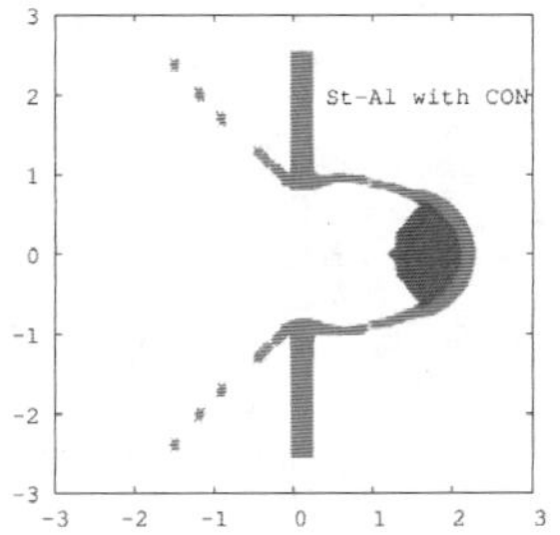


**Figure 4.** The St-Al impact simulated using CON formulation of SPH shown 8 $\mu$s post-impact. The projectile is shown in red and target in green. The distances are in cm.

The absence of visible tensile instability in CON simulation may merely indicate a slower growth of instability that has not had sufficient time to get itself manifest in this particular case. Futher studies with CON on different impact scenarios with different dimensions of projectile and target and impact velocities are needed to definitively rule out tensile instability in CON.

## 5. Conclusion

Impact and penetration of 0.5 cm radii steel spheres on 2 mm thick Aluminium plate at 3.1 km/s (St-Al impact) has been studied through meshless smooth particle hydrodynamics (SPH) technique. Conventional SPH is liable to a tensile instability that manifests itself as a bunching of nodes and formation of artificial voids. The St-Al impact as modeled through conventional SPH shows evidences of tensile instability. Monaghan proposed adding certain artificial stresses to SPH equations to counter tensile instability. We show that the artificial void formation in St-Al impact is suppressed but not eliminated by using Monaghan stresses while the void formation is entirely eliminated by using an alternative Godunov-type formulation of SPH that has been proposed by Parshikov and Medin. More studies with different impact scenarios are required to definitively rule out the presence or absence of tensile instabilty in the Godunov-type SPH formulation

[1] Mehra V and Chaturvedi S 2006 *J. Comp. Phys.* **212** 318
[2] Monaghan JJ 2000 *J. Comp. Phys.* **159** 290
[3] Gingold RA and Monaghan JJ 1977 *Mon. Not. R. Astronom. Soc..* **181** 375
[4] Balsara DS 1995 *J. Comp. Phys.* **121** 357
[5] Howell BP and Ball GJ 2006 *J. Comp. Phys.* **175** 128
[6] Parshikov A and Medin SA 2002 *J. Comp. Phys.* **180** 358
[7] Dukowitz J 1985 *J. Comp. Phys.* **61** 129
[8] Shaw A and Reid SR 2009 *Comput. Meth. Appl. Mech. Engg* **198** 3962

23rd International Conference on High Pressure Science and Technology (AIRAPT-23) IOP Publishing
Journal of Physics: Conference Series **377** (2012) 012103 doi:10.1088/1742-6596/377/1/012103

# Equation of state and electrical conductivity of expanded Al

**V. Mishra V. Mehra and S. Chaturvedi**

Computational Analysis Division, Bhabha Atomic Research Centre, Autonagar, Visakhapatnam - 530012, A.P. India

E-mail: v.mishra.11@gmail.com

**Abstract.** Ab-initio molecular dynamics (AIMD) simulations have been performed for generating equation of state (EOS) data as well as atomic configurations of expanded states of Al. The generated atomic configurations have been used for the calculation of electrical conductivity. AIMD simulations have been performed using the ABINIT code and electrical conductivity has been calculated using the Kubo-Greenwood formula as implemented in ABINIT. The generated equation of state data have been fitted to a three-term EOS model using adjustable parameters of ionic and electronic Gruneisen parameters. This three-term EOS model has been used to estimate critical density, temperature and pressure in the liquid-vapour region. The calculated values of critical density, temperature and pressure show good agreement with results available in the literature.

## 1. Introduction

Experiments and simulations of electrically exploded conductors are of great current interest because of their applications [1]. In such experiments, a large current is passed through a thin foil/wire of some conductor, such as aluminum or copper. This causes intense Joule heating of the conductor on a time-scale that is short in comparison with sound transit time through the conductor. The conductor material thus passes rapidly through the solid, liquid and vapour phases and may even enter the plasma phase. Hydrodynamic simulations of such systems require, as input, electrical conductivity and equation of state (EOS) data of the conductor material for densities ranging from normal density to a fraction of the normal density and temperatures ranging from room temperature to a few thousand Kelvin.

Ab-initio molecular dynamics (AIMD) is a powerful simulation technique for calculating properties of matter such as EOS and electrical conductivity in the density-temperature regime of exploding conductor experiments [2, 3].

In the present work, we have used the AIMD method to generate EOS and electrical conductivity data for expanded states of Al. Using the generated EOS data, we have estimated the critical temperature, density and pressure of Al in the liquid-vapour region.

This paper is organized as follows. In section 2 we have described computational details of generating EOS data and atomic configuration data using AIMD simulations. We also describe the method of computing DC electrical conductivity using the atomic configurations generated by AIMD simulations. In section 3.1 we have compared calculated electrical conductivity results with the reported results for the validation of AIMD simulations. In Section 3.2 we have presented EOS results of AIMD simulations and their fitting to a three-term EOS model. In

Section 3.3, we describe the procedure for estimating critical density, temperature and pressure and compare these results with those available in the literature. Concluding remarks are presented in Section 4.

## 2. Computational Details

We used the ABINIT code [4, 5] for performing AIMD simulations, in isokinetic ensemble. The simulations are first run until equilibration is achieved, i.e., until the variations in pressure stabilize to an oscillatory pattern, without any long-term trend. Following equilibration, the pressure and energy are calculated by averaging over the next 500-1000 time steps of 1 femtosecond (fs) each. The exchange correlation interactions of electrons are treated in generalized gradient approximation (GGA), using Perdew-Burke-Ernzerhof (PBE) parametrization [6]. Ion-electron interactions are described by the pseudo-potential generated by Troullier and Martins method [7].

All calculations have been done at the $\Gamma$ point, i.e., with number of k-points taken as one. A few runs were also done with a higher number of k-points but with no significant effect on the output pressure. In ABINIT, the energy cut off *ecut* is an important parameter which controls the number of plane waves and can have an enormous effect on the quality of a calculation. We performed several calculations at various values of *ecut* to investigate the convergence needed for reliable results and found that 300 eV is a good value for Al. Runs were done with 108 atoms in an appropriate cubic box to set the required density.

For lower densities ($< 1.0\ g/cc$) runs were done with 32 atoms, as large volume slows down the speed of calculations. For checking the accuracy with only 32 atoms, a few runs were done with 108 atoms. The results were not found to change significantly. We ensured that there were always some unoccupied levels for better convergence of all runs. The convergence criterion for energy was taken as $10^{-4}eV$. The pressure contribution of ions was calculated by using the ideal gas equation.

Out of the atomic configurations generated by AIMD after stabilization of pressure oscillations, 10-15 statistically-independent configurations are used for the calculation of electrical conductivity. Frequency-dependent electrical conductivity of Al is calculated using the Kubo-Greenwood formula as implemented in ABINIT code. Quantities like Fermi-Dirac occupation for each band and each k point, weight of each k point, electronic eigenstate and eigenvalues for each electronic state, and the derivative of the Hamiltonian with respect to the wave vector of the three directions, are required as input for Kubo-Greenwood formula. These quantities are calculated by using ABINIT code. For selected statistically independent atomic configurations of 108 (or 32) atoms, a self-consistent ground state calculation is performed with the ABINIT code to get all these quantities. For conductivity calculations, we found that four k points are necessary for better convergence. We ensured that there were always some unoccupied levels for better convergence of all runs. For calculating the DC electrical conductivity, we extrapolated Kubo-Greenwood results using the Drude formula.

## 3. Results and discussion

### *3.1. Validation of AIMD simulations*

The electrical conductivity as a function of temperature is plotted in Figure 1. These results are the average of 10-15 calculations, one for each statistically-independent configuration. We can notice that the electrical conductivity as a function of temperature shows a break (change of slope) near 6000 K. The cause for this can be explained using Drude theory. Following the Drude theory, the electrical conductivity can be expressed as

$$\sigma = \frac{ne^2\tau}{m} \tag{1}$$

Here, $n$ is the number density of electrons near Fermi surface, $e$ is the electronic charge, $\tau$ is the relaxation time and $m$ is the mass of electron. Initially conductivity decreases very fast with temperature (up to 6000 K) as shown in Figure 1. This can be understood in a simple manner. The ions vibrate about their equilibrium positions which leads to the thermal vibration of the crystal lattice. The amplitude of the vibration increases with the temperature. The electrons are scattered by collisions with the lattice ions. Consequently with the increase in the amplitude of vibration, the collision frequency increases. This leads to a decrease in the relaxation time $\tau$ and hence to a decrease in the conductivity. At very high temperature (above 6000 K), though the relaxation time $\tau$ is still decreasing with the temperature leading to decrease in the conductivity, but at the same time the number density of electrons near Fermi surface also starts increasing causing increase in the conductivity. These both phenomenas viz. the decrease in the relaxation time $\tau$ and the increase in the number density of electrons near Fermi surface jointly lead to less decrease in the electrical conductivity above the temperature 6000 K as compared to below this temperature. Hence, we see a break in the slope near 6000 K.

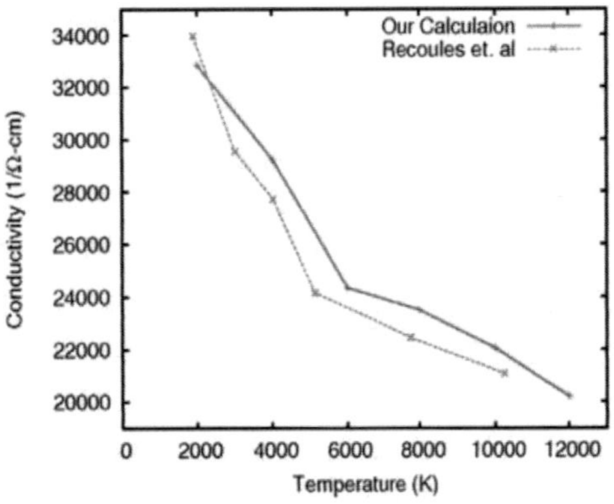


**Figure 1.** DC conductivity of Al at density 2.35 g/cc. For comparison results from [3] are also shown.

For comparison we have also given electrical conductivity data from [3] in Figure 1. Our results show good agreement with the reported results. This confirms the validity of our AIMD simulations.

### *3.2. EOS results of AIMD simulations and their fitting to three-term EOS model using adjustable parameters*

After confirming the validity of our AIMD simulations, we calculated EOS by averaging energy and pressure outputs over 500-1000 steps after equilibration. The calculated pressure as a function of density for temperatures 10,000 K and 15,000 K are shown in Figure 2 with symbols. A three-term EOS model, that was fitted to these AMID data, is given by:

$$P_{TOT}(\rho,\ T) = \rho\left(f_1 f_2 \rho^{f_2} - f_3 f_4 \rho^{f_4}\right) + \frac{\rho k T}{A M_P}\left(1 + \left[3\left(b_1 + b_2\rho/\rho_0\right) - 1\right] w^{\frac{1}{3}}\right) + \frac{1}{2}\gamma_E \beta T^2 \rho \quad (2)$$

Here the first term is cold term $P_C$ which have been represented by a soft sphere function in this model [9], second term is ion-thermal term $P_{IT}$ [8, 9] and third term is electron thermal term $P_{ET}$ [10, 11]. $w = T_m(\rho)/T$, $T_m$ is melting temperature, $3\left[b_1 + b_2\rho/\rho_0\right] - 1] = \gamma_F$ is lattice Gruneisen parameter and $\gamma_E$ is electronic Gruneisen parameter. The function $P_{TOT}(\rho,\ T)$ is fitted to AIMD results using adjustable parameters $b_1$, $b_2$ and $\gamma_E$.

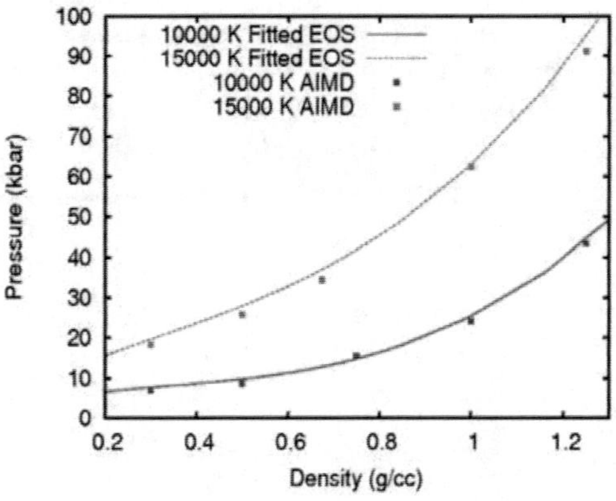


**Figure 2.** Isotherms for temperatures 10000 K and 15000 K calculated by AIMD simulations and three-term EOS fitting. Points represent results of AIMD simulations and lines represent three-term EOS fitting

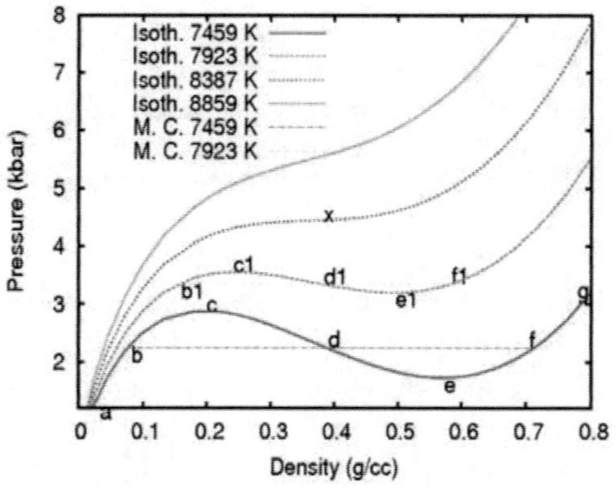


**Figure 3.** Isotherms for temperatures 7459 K, 7923 K 8387 K and 8859 K calculated by three term EOS model, fitted to AIMD simulation results. Straight lines $b-d$ and $b1-d1$ are the Maxwell constructions (M. C.).

### *3.3. Estimation of critical point*

We used the fitted three-term EOS model to estimate density temperature and pressure at critical point. In Figure 3 we have shown isotherms for temperatures 7458 $K$, 7923 $K$, 8387 $K$ and 8851 $K$ for expanded states of Al. We can see that isotherms of temperatures below 8387 $K$ show an oscillatory behaviour (wavelike structure). For example, the 7458 $K$ isotherm shows oscillatory behaviour from point $b$ to point $f$. In this range of the 7458 $K$ isotherm, the oscillating part can be divided in four parts $b-c$, $c-d$, $d-e$ and $e-f$. In the first part $b-c$ of the oscillating isotherm, the pressure first starts increasing from a value of 2.24 bar at $b$ and reaches a maximum value of 2.89 bar at $c$. In the second part $c-d$, pressure decreases from this maximum value to 2.24 bar at point $d$. In the third part $d-e$ of the oscillating isotherm, pressure starts decreasing below 2.24 bar and reaches a minimum value 1.7 bar at $e$, and in the fourth part $e-f$, pressure starts increasing from the minimum value of third part to the minimum value of the first part i.e. 2.24 bar. Thus it can be readily seen that, for a certain pressure 2.24 bar, there are three values of density marked as $b$, $d$ and $f$. In the density range below starting point of the first part (from point $a$ to $b$) and in the density range above the maximum density of the fourth part (from point $f$ to $g$) the expanded state of Al is stable because in these density ranges $\partial P/\partial \rho > 0$. The state in the second and third parts of the oscillating isotherm is unstable because in these states $\partial P/\partial \rho < 0$. First and fourth part of the oscillatory isotherm are metastable states [12]. These represent super-saturated vapour and super heated liquid respectively.

We notice that on 7458 K isotherm, for a pressure 2.24 bar, there are three values of density marked as $b$, $d$ and $f$. If the temperature is decreased the isotherms would move down, loops of the wavelike structure of the isotherms would become larger and larger and the points corresponding to $b$, $d$ and $f$ move farther from one another. As the temperature is raised, the isotherms move up, the loops become smaller and the points corresponding to $b$, $d$ and $f$ get closer to one another. At a certain temperature, the wave-like structure disappears and these three points become identical. The point where these three points merge is shown by the point $x$.

The temperature of this particular isotherm at which critical point lies is the critical temperature, and the density and pressure corresponding to this point $x$, are the critical density and critical pressure. $\partial P/\partial\rho$ and $\partial^2 P/\partial\rho^2$ are equal to zero at this point. In Table 1, we have given results of estimated critical density, temperature and pressure. For comparison, we have also given results reported in the literature. We can see that our results agree reasonably well with these results.

**Table 1.** Critical density, temperature and pressure obtained in this work and reported results

| Reference | $\rho_c(g/cc)$ | $T_c(K)$ | $P_c(bar)$ |
|---|---|---|---|
| This work | 0.385 | 8387 | 4.45 |
| Ref [13] | 0.28 | 8860 | 4.68 |
| Ref [14] | 0.43 | 8944 | 4.726 |
| Ref [15] | 0.785 | 8472 | 5.094 |
| Ref [16] | 0.66 | 7917 | 4.67 |

## 4. Conclusion

AIMD simulations have been performed for generating EOS and atomic configuration data. Generated atomic configurations have been used for calculating electrical conductivity. Generated EOS data have been used to estimate critical temperature, density and pressure of Al in liquid-vapour region. The electrical conductivity, critical density, critical temperature and critical pressure results show good agreement with the reported results.

**Acknowledgment**

Vinayak Mishra thanks Dr. A. K. Verma of High Pressure and Synchrotron radiation Physics Division, BARC, for useful discussions.

## References

[1] S A Chaikovsky, V I Oreshkin G A Mesyats N A Ratakhin I M Datsko and B A Kablambaev 2009 *Phys. Plasmas* **16** 042701
[2] M P Desjarlais, J D Kress and L A Collins 2002 *Phys. Rev. E* **66** 025401
[3] Vanina Recoules and Jean-Paul Crcombette 2005 *Phys. Rev. B* **72** 104202
[4] X Gonze et. al. 2002 *Comput. Mat. Sci.* **25** 478
[5] X Gonze et. al. 2009 *Computer Phys. Commun.* **180** 2582
[6] J P Perdew S Burke M Ernzerhof 1996 *Phys Rev Lett* **77** 3865
[7] N Troullier and J L Martins 1991 *Phys Rev B* **43** 1993
[8] R M More, K H Warren, D A Young, and G B Zimmerman 1988 *Phys. of Fluids* **31** 3059
[9] David A Young and Ellen M Corey 1995 *J. Appl. Phys.*, **78**, 3748
[10] S. Eliezer, A. Ghatak and H. Hora, 1986 *An Introduction to Equations of State: Theory and Applications* (Cambridge: Cambridge University Press) chapter 13 pp 211-212
[11] B K Godwal, T C Kaushik and S K Sikka 1997 *Laser and Particle Beams* **15** 353
[12] G Muller 2011 *Statistical Physics I* www.phys.uri.edu/ gerhard/PHY525/tln31.pdf
[13] A A Likalter 1996 *Phys. Rev. B* **53** 4386
[14] H Hess and Z Metallkd 1998 as given in the review paper V Morel, A Bultel and B G Cheron 2009 *Int. J. Thermophy* **30** 1853
[15] J K Singh, J Adhikari and S. K. Kwak 2006 *Fluid Phase Equilib.* **248** 1
[16] D G Gordeev, L F Gudarenko, M V Zhernokletov, V G Kudelkin and M A Mochalov 2008 *Combust. Expl. Shock Waves* **44** 177

# Fracture during high-velocity impact of copper plates: a molecular dynamics study

**S Rawat[1,2], M Warrier[2], S Chaturvedi[2] and V M Chavan[1]**

[1] Refueling Technology Division, Bhabha Atomic Research Centre, Mumbai 400 085, India
[2] Computational Analysis Division, Bhabha Atomic Research Centre, Visakhapatnam 530 012, India

E-mail: alig.sunil@gmail.com

**Abstract.** High velocity impact of copper plates has been studied using the molecular dynamics code LAMMPS. The impact of copper plates at 1100 m/s impact velocity shows that spall of the material takes place at 160 kbar tensile pressure. Void nucleation is not observed at lower impact velocities where the peak tensile pressure stays below 160 kbar. No significant void nucleation occurs in the neighbouring regions where the peak tensile pressure stays below 160 kbar. For impact velocities close to the threshold, we observe stochastic behaviour of the spall process with respect to small changes in the initial atomic coordinates.

## 1. Introduction

Solid fracture under high strain-rate deformation is of interest for high velocity impact and penetration problems. The impact of a flyer onto the target produces compressive stress waves which travel toward the respective free surfaces of the flyer and target. When they reach the free surfaces, they are reflected back as release waves and interact at some location in the target. The interaction of release waves produces tension in the target material. If the tension created exceeds the tensile strength of the material, nucleation, growth and coalescence of the voids take place, leading to spall. This process is called spallation.

We have simulated high velocity impact of copper plates using the LAMMPS [1] code. In the present work, we validate the void nucleation threshold (160 kbar) obtained under triaxial strain conditions [2] by studying the impact of copper plates at 1100 m/s. We also show the stochastic behaviour of spall process for the impact velocities close to the spall threshold of the material. The summary of work done by other workers is presented in [3].

## 2. Computational Method

The classical molecular dynamics code LAMMPS [1] has been used for this study. We have used embedded-atom method potential parameters generated by Foiles et al [4]. As a first step, the simulations have been validated by reproducing the Hugoniot for crystal copper [5]. The simulation domain contains $1.08\times10^6$ atoms corresponding to $300\times30\times30$ unit cells. Free boundary conditions are applied along the impact direction (x-direction) and periodic boundary conditions are applied along the transverse directions (y- and z- directions). The thickness of the target plate is taken as twice that of the flyer. The target is divided into small bins and pressure is computed in each bin of the target using binning analysis. Impact is done with an impact velocity of 1100 m/s. The typical strain-rate in this impact simulation is ~O($10^9$/s), comparable with the triaxial strain simulation [2].

## 3. Results and Discussion

### *3.1. Free surface velocity*

The free surface velocity of the target for an impact velocity of 1100 m/s is shown in figure 1 (a). The impact at 1100 m/s produces compressive stress waves at the point of impact. These waves propagate toward the respective free surfaces of the flyer and target. When they reach the free surfaces, they reflect as release waves. In this process, the free surface of the target is accelerated to approximately twice the particle velocity behind the shock [6]. The shock arrives at the free surface of the target at 14 ps as shown in figure 1(a). The interaction of release waves at some location in the target creates tension in the material. The centro-symmetry parameter analysis [3] shows that defects are created only during the tension. This means that plastic deformation of the material takes place only under the tension not under the compression state of the material. If the tension created exceeds the tensile strength of the material, nucleation, growth and coalescence of the voids take place leading to spall of the material. It is seen in the figure 1 (a) that there is a pullback in the free surface velocity which is the signature of spall [7]. Snapshots of the domain at different time points are shown in the side panel of figure 1(a).

### *3.2. Pressure-time history*

The pressure-time history in the region where spall occurs is shown in figure 1 (b). Spall region here refers to the region where failure of the material takes place. It is different from other regions where only voids are created and material does not fail. It is seen in figure 1 (b) that spall of the material takes place at 160 kbar tensile pressure. No significant void nucleation is observed in the neighbouring regions where tensile pressure stays below 160 kbar. This validates the void nucleation threshold (160 kbar) obtained for crystal copper under triaxial strain conditions [2]. As soon as spall of the material takes place, pressure in the region rapidly drops to zero. The flat portion of the curve around zero beyond 40 pico-second (ps) is the signature of spall in the material.

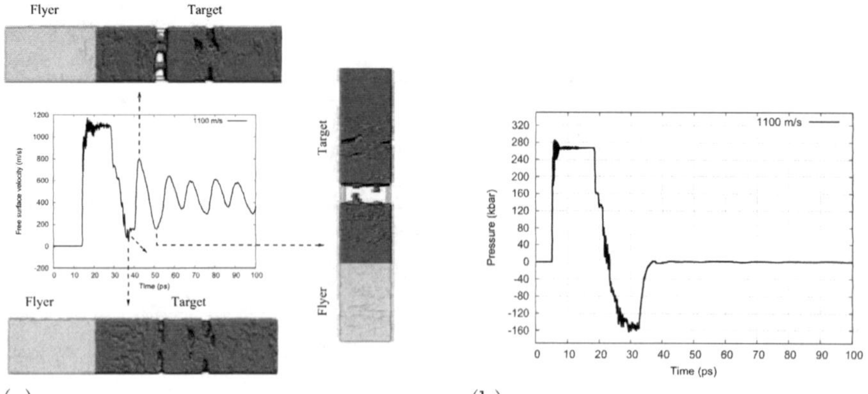


(a) (b)

**Figure 1.** (*a*) Free surface velocity of the target for an impact velocity of 1100 m/s (*b*) Pressure as a function of time in the spall region

## 4. Stochastic effect on the spallation process

To study the stochastic effect on the spallation process, we have performed following two exercises

- Impact at 1000 m/s without perturbing initial atomic coordinates
- Impact at 1000 m/s by perturbing initial atomic coordinates

*4.1. Impact at 1000 m/s without perturbing initial atomic coordinates*
*4.1.1. Free surface velocity* The free surface velocity of the target as a function of time is shown in figure 2 (a). In figure 2 (a), it is seen that there is no pullback in the free surface velocity of the target corresponding to first traversal of the shock. Note that the pullback in the free surface velocity is the signature of the spall [7]. This means that there is no spall corresponding to first traversal of the shock. The pullback in the free surface velocity occurs corresponding to second traversal of the shock which is due to the creation of voids, not due to spall of the material. In this case, voids grow and coalesce leading to incipient spall like condition [7]. However, spallation does not occur as shown in the side panel of the figure 2 (a).

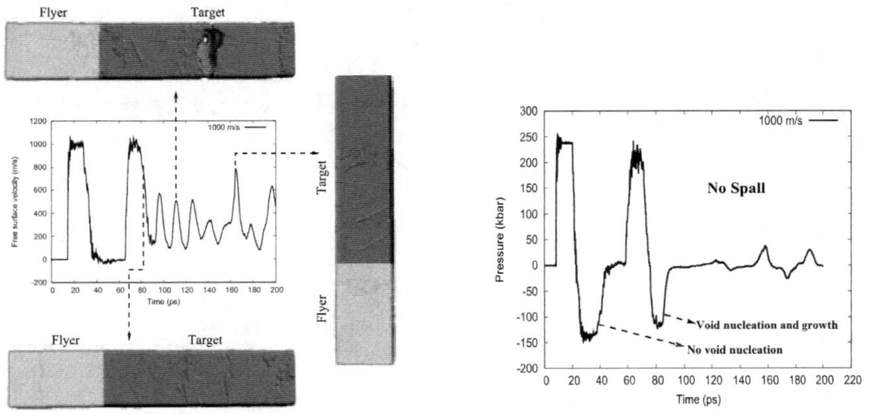


(a) (b)

**Figure 2.** (*a*) Free surface velocity of the target for an impact velocity of 1000 m/s (*b*) Pressure as a function of time in the spall region

*4.1.2. Pressure-time history* The pressure-time history in the region where voids nucleate is shown in figure 2 (b). It is seen that there is no void nucleation during the tension corresponding to first traversal of the shock. Void nucleation and growth occur during the tension created by second reverberation of the shock at lower value of tensile pressure. The tension created by first traversal of the shock creates many defects (dislocations, stacking faults, etc) in the material. These defects accumulate and become void nucleation sites for the tension created by second traversal of the shock leading to void nucleation at lower value of tensile pressure.

*4.2. Impact at 1000 m/s by perturbing initial atomic coordinates*
We perturb the initial x-coordinates of the atoms randomly between $\pm$ 0.001 $\AA$. No perturbation is applied to the y and z coordinates.

*4.2.1. Free surface velocity* The free surface velocity of the target for this case is shown in figure 3 (a). In figure 3 (a), there is no pullback in the free surface velocity of the target corresponding to first traversal of the shock. The pullback in the free surface velocity occurs after 80 ps due to nucleation and growth of the voids. In this case, growth and coalescence of the voids lead to the spall of the material. The compressive wave resulting from the stress relaxation due to spall of the material reverberates between the free surface and spall surface of the scab. The ringing in the free surface velocity after 90 ps (figure 3 (a)) corresponds to the reverberations of the trapped compressive wave. The snapshots at different times points in the free surface velocity are shown in the side panel of the figure 3 (a).

*4.2.2. Pressure-time history* The pressure as a function of time in the spall region is shown in figure 3 (b). No void nucleation occurs during the tension created by first traversal of the shock. However, various defects (partial dislocations, stacking faults, etc) are created. These defects become void nucleation sites for the tension corresponding to second traversal of the wave and hence lead to the nucleation and growth of voids at very low tensile pressure. This, in turn, results in spall. Therefore we see that, close to the spall threshold of the material, a small change in the atomic coordinates affects the spallation process of the material

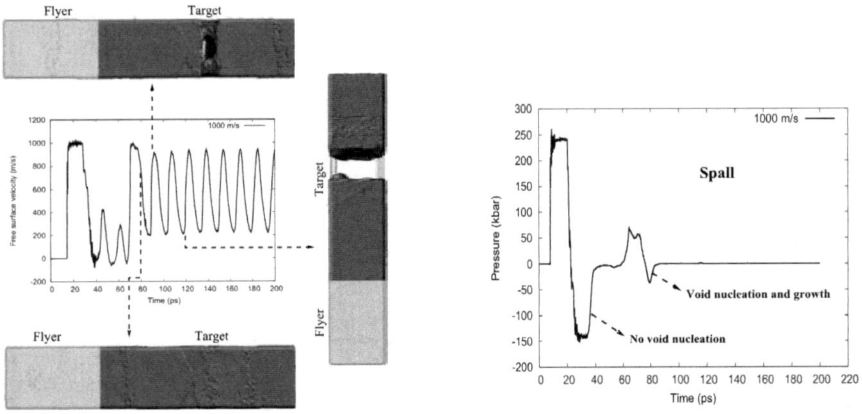


(a) (b)

**Figure 3.** (*a*) Free surface velocity of the target for an impact velocity of 1000 m/s (*b*) Pressure as a function of time in the spall region

## 5. Conclusions

We have performed molecular dynamics simulations of high velocity impact of copper plates to validate the void nucleation threshold obtained under triaxial strain conditions and to study the stochastic effect on the spallation process. For the impact of copper plates at 1100 m/s impact velocity, the spall of the material takes place at 160 kbar pressure. This validates the void nucleation threshold (160 kbar) obtained under triaxial strain conditions [2].

The stochastic effect on spallation process has been studied by performing impact of copper plates at 1000 m/s impact velocity. We find that a small change in the initial atomic coordinates affects the spallation process. This means that for impact velocities close to the spall threshold, spallation process is stochastic.

## References

[1] Plimpton S 1995 *J. Comp. Phys.* **117** 1-19
[2] Rawat S, Warrier M, Chaturvedi S and Chavan V M 2011 *modell. Simul. Mater. Sci. Eng.* **19** 025007
[3] Rawat S, Warrier M, Chaturvedi S and Chavan V M 2012 *modell. Simul. Mater. Sci. Eng.* **20** 015012
[4] Foiles S M, Baskes M I and Daw M S 1986 *Phys. Rev.* B **33**(12) 7983-91
[5] Bringa E M et al 2004 *J. Appl. Phys.* **96**(7) 3793-99
[6] Antoun T et al 2003 *Spall Fracture* (New York: Springer-Verlag)
[7] Minich R W, Cazamias J U, Kumar M and Schwartz A J 2004 *Metall. Mater. Trans.* A **35 A** 2663-73

Journal of Physics: Conference Series **377** (2012) 012105 doi:10.1088/1742-6596/377/1/012105

# A comparison of quotidian equation of state of aluminium with ab-initio calculations

**V. Mishra, Sijoy C. D., P. Pahari and S. Chaturvedi**

Computational Analysis Division, Bhabha Atomic Research Centre, Autonagar, Visakhapatnam - 530012, A.P. India

E-mail: v.mishra.11@gmail.com

**Abstract.** We present results of theoretical calculation of the equation of state of Al covering several orders of magnitude of density and temperature. The data is generated using a QEOS model. The QEOS data have been used to calculate $U_s - U_p$ and $P - V$ hugoniots. The calculated hugoniots show good agreement with the experimental hugoniots. The cold curve generated by QEOS model has been compared with FP-LAPW results for compressed states and good agreement has been found. The QEOS data has also been compared with AIMD simulation results for expanded states – agreement is not good. These comparisons confirm that QEOS results are accurate for compressed states and less accurate for low temperature expanded states of Al.

## 1. Introduction

The equation-of-state (EOS) which defines the pressure and energy as functions of density and temperature is important in high pressure studies. It depends on the nature of the inter-atomic interactions and thus provides a test of fundamental condensed matter theories. At the same time, it can be used to determine thermodynamic properties. The metal EOS is an essential input to the hydrodynamic simulations of pulse power experiments such as metallic liner acceleration [1] and exploding wires/foils [2].

The EOS of metals have three contributions [3, 4]:

$$E_{TOT}(\eta, T) = E_C(\eta, T) + E_{IT}(\eta, T) + E_{ET}(\eta, T) \qquad (1)$$

$$P_{TOT}(\eta, T) = P_C(\eta, T) + P_{IT}(\eta, T) + P_{ET}(\eta, T) \qquad \eta = \rho/\rho_0 \qquad (2)$$

Here the first term (T = 0 or cold term) $P_C$ and $E_C$ are the electronic pressure and specific internal energy at zero Kelvin (cold-curve). Second term $P_{IT}$ and $E_{IT}$ are the contribution of ionic vibrations to the pressure and energy at finite temperatures, and third term $P_{ET}$ and $E_{ET}$ are the electron thermal contributions. This representation of EOS is known as three-term EOS model. In QEOS, Thomas - Fermi (TF) model is used for both contributions of electrons viz cold as well as thermal [5]. Cowan model is used for ion thermal contribution [5, 6]. TF model does not consider quantum mechanical structure of atom therefore it does not give correct results near cold solid. TF theory predicts a few megabar pressure for the cold solid where the actual pressure should be zero due to attractive forces. In order to obtain an approximately correct equation of state in this region, an empirical bonding correction is applied to remove this inaccuracy. This correction is applied in such a way that pressure and bulk modulus are made to match the experimental value under normal conditions and the correction term has little effect away from the region near the cold solid. Hence the total energy and pressure are given by the following equations in QEOS [5, 6]

$$E_{TOT}(\eta, T) = E_B(\eta, T) + E_{IT}(\eta, T) + E_E(\eta, T) \tag{3}$$

$$P_{TOT}(\eta, T) = P_B(\eta, T) + P_{IT}(\eta, T) + P_E(\eta, T) \tag{4}$$

$E_B$ is the binding correction term and $E_E$ is the electron contribution to EOS which is calculated by using Thomas - Fermi model.

## 2. Computational Details

We have used More's methodology for performing QEOS calculations to generate EOS data of Al. For comparing QEOS data for T=0, we have performed first principles calculations using Full Potential Linearized augmented plane wave (FP-LAPW) method [7]. For comparing QEOS data in expanded states we have performed Ab-Initio Molecular Dynamics (AIMD) simulations. The details of QEOS, FP-LAPW and AIMD calculations are described in this section.

### *2.1. QEOS code using More's methodology*

We have developed a quotidian equation of state (QEOS) code, based on the methodology of More, for generating EOS tables, specific heat and sound speed over a wide range of temperatures and densities, including compressed as well as expanded states. This yields the pressure, energy, entropy and free energy, taking into account density-dependent melting temperature, Debye temperature and Gruneisen parameter for any material. Cold and electron thermal terms are given by the TF model itself, a binding correction is applied to correct the cold - curve. Ion-thermal term is calculated separately using Cowan model as described in [5]. In this model energy and pressure are calculated by using following expressions:

$$E_{IT}(\eta, T) = \frac{3kT}{2AM_P}\left(1 + w^{1/3}\right), \quad w = T_m(\rho)/T \tag{5}$$

$$P_{IT}(\eta, T) = \frac{\rho kT}{AM_P}\left(1 + \gamma_F w^{1/3}\right) \tag{6}$$

Here $T_m$ is melting temperature, $\gamma_F$ is Gruneisen parameter.

Using More's technique, we have incorporated a thermodynamically consistent interpolation scheme for calculating EOS, and density and temperature derivatives of pressure and energy.

### *2.2. Full Potential Linearized Plane Wave (FPLAPW) Calculations*

For validation of the QEOS results, we have performed first principles calculations to generate cold-curve for Al. For this purpose we used WIEN-2k [7] code which is based on full potential linear augmented plane wave (FP-LAPW) method within the framework of density functional theory. The exchange correlation potential within GGA is calculated using the scheme of Perdew - Burke- Ernzerhof [8]. These calculations have been done using a constant muffin-tin radius $R_{mt}$ of 1.7 a.u. In the WIEN2k code, it is very important to select a good basis set and k mesh size for getting accurate results. We start by taking a low value $R_{mt}K_{max}$ of = 7.0. For purposes of determining a suitable k -mesh size, we then calculate the total energy as a function of the size of the k -mesh increasing the size by 1000 at each step. We find that 10,000 k-points are sufficient. Having determined a good k-mesh, we now vary $R_{mt}K_{max}$ from 7 to 12, and find that 9.5 is a good choice. Selected k-mesh size and $R_{mt}K_{max}$ correspond to an energy accuracy of $10^{-4}$ $Ry$. We have selected an energy cut-off of 6.0 Rydberg to separate the core from the valence states.

### 2.3. Ab-Initio Molecular Dynamics (AIMD) Simulations

We used the ABINIT code [9] for performing AIMD simulations, in isokinetic ensemble. The simulations are first run until equilibration is achieved, i.e., until the variations in pressure stabilize to an oscillatory pattern, without any long-term trend. Following equilibration, the pressure and energy are calculated by averaging over the next 500-1000 time steps of 1 femtosecond (fs) each. The exchange correlation interactions of electrons are treated in generalized gradient approximation (GGA), using PBE parametrization [8]. Ion-electron interactions are described by the pseudo-potential generated by Troullier and Martins method [10]. All calculations have been done at the $\Gamma$ point, i.e., with number of k-points taken as one.

## 3. Results and Discussion

### 3.1. QEOS-Results

In figure 1 we have shown the TF as well as QEOS pressures as functions of density at T=200 K. The TF pressure is approximately 1 megabar at normal density. The binding correction term makes pressure zero at normal density. Binding correction has little effect at higher densities Hence QEOS and TF models give similar representation of electron contribution at very high density. TF model is accurate in high density limit. In order to get zero pressure at standard conditions, the cold energy should have a minimum at normal density. However, as shown in figure 2, the TF cold energy does not exhibit a minimum at normal conditions. The binding correction removes this discrepancy.

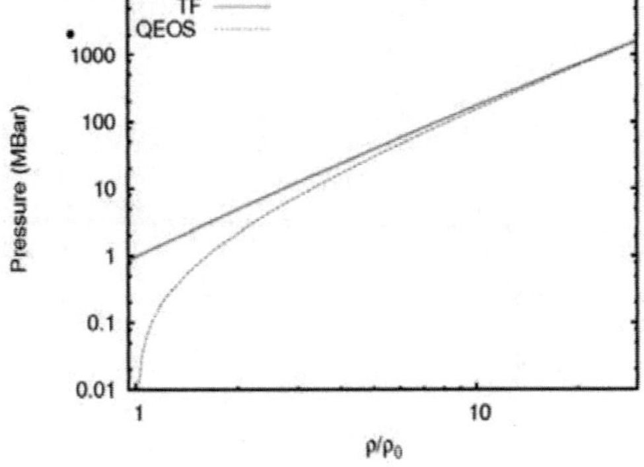


**Figure 1.** TF and QEOS cold pressures as function of density at T=200 K.

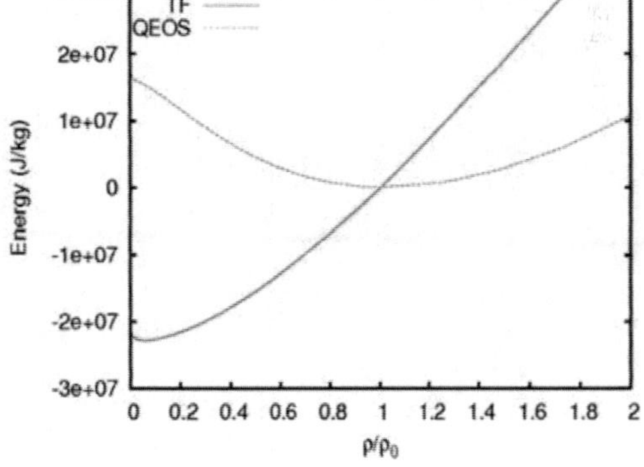


**Figure 2.** TF and QEOS cold energies as function of density.

The EOS generated using QEOS model is combined with the three hugoniot equations viz., $\rho_0 U_s = \rho_1 (U_s - U_p)$, $P_1 - P_0 = \rho_0 U_s U_p$ and $E_1 - E_0 = \frac{1}{2}(P_1 + P_0)(V_0 - V_1)$ to get the $P - V$ and $U_s - U_p$ hugoniots [4]. In these equations, $\rho$, $E$, $P$ and $V$ have same meaning as in equations 1, $U_s$ and $U_p$ are shock and particle velocities respectively. The suffixes 1 and 0 represent the quantities in the shocked and the un-shocked regions. The calculated $U_s - U_p$ and $P - V$ hugoniots are plotted in figure 3 and figure 4 respectively. For comparison experimental shock compression data [11, 12, 13] are also shown. Good agreement with the experimental hugoniot can be seen.

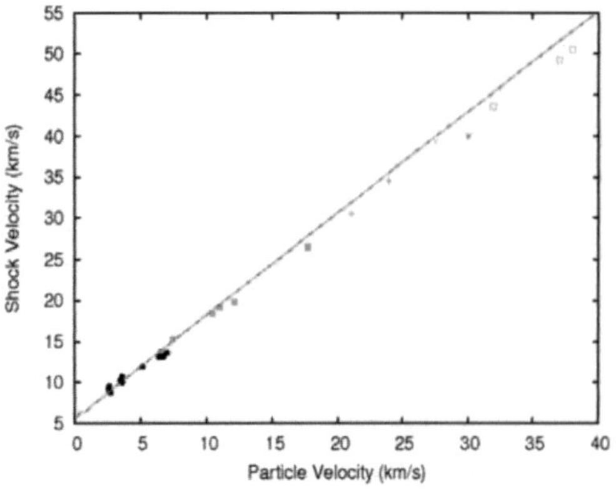


**Figure 3.** $U_s - U_p$ hugoniot. Solid line represents QEOS results and points represent experimental results [11, 12]

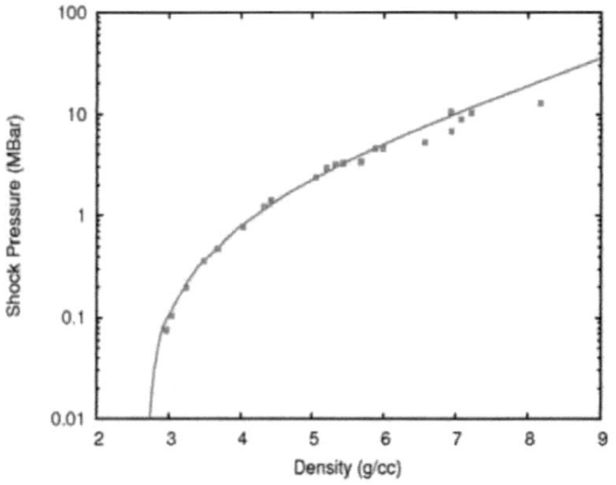


**Figure 4.** P-$\rho$ hugoniot. Solid line represents QEOS results and points represent experimental results [12, 13]

### *3.2. Comparison of QEOS results with ab-initio calculations*

We next compare the cold curve generated by QEOS by first-principles calculations using the FP-LAPW method within the framework of density functional theory. For generating cold curve using FP-LAPW method we have assumed Al to exist in FCC structure at all pressures. Although pressure-induced FCC-HCP and HCP-BCC structural transformations occur at high pressures, the differences in the energies of FCC and HCP structures and HCP and BCC structures are so small near the transition pressure that these structures coexist over a wide range of pressures [14, 15, 16]. This has been confirmed experimentally using X-ray powder diffraction experiments performed by Akahama [14]. In our calculations, we observed that differences in the energies of three structures were very small. and density of states curves of FCC and HCP structures at Fermi level were similar near the transition pressure [15, 16]. Therefore for the calculation of cold contribution we ignored structural transformation and assumed Al to exist in the FCC structure. In figure 5 we can see that there is a very good agreement between first principles results and QEOS results. These results match well with reported experimental data obtained by powder x-ray diffraction experiments [14]. The QEOS cold curve also agrees well with the cold-curve deduced from shock wave experiments [17, 18].

In figure 6 we have shown a comparison of QEOS results with AIMD simulation results for temperature 10000 K and expanded states. AIMD simulations are known to yield an accurate EOS in this regime, especially for $(T < 5eV)$ [19, 20]. It can be seen from figure 6 that QEOS results do not show good agreement with the results of AIMD simulations. Hence, QEOS results are less accurate for expanded metals at low temperatures.

## 4. Conclusion

We presented equation of state results of Al using quotidian equation of state model and from first principles calculations. QEOS calculations have been performed using More's methodology. Full potential linearized augmented plane wave (FPLAPW) calculations have been performed within the framework of density functional theory for accurate calculations of cold curve. Ab-initio molecular dynamics simulations have been performed for calculating EOS for expanded metals. FPLAPW and AIMD simulation results have been compared with the QEOS results. The cold curves generated by using QEOS and FPLAPW methods for compressed states show good agreement, implying that QEOS results are accurate for compressed Al. QEOS results

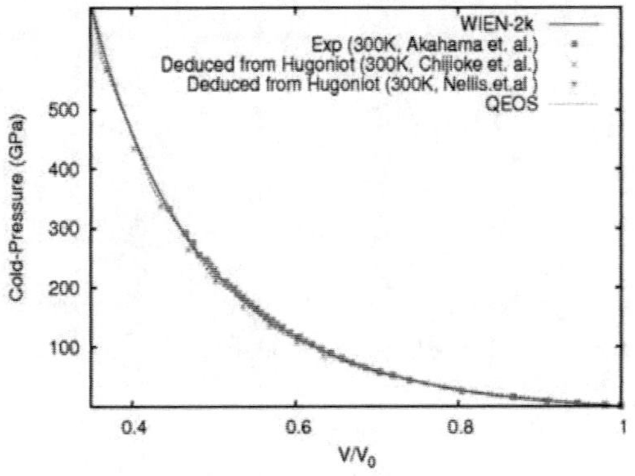


**Figure 5.** Cold curve generated by QEOS and FP-LAPW calculations and their comparison with experimental results [14, 17, 18]

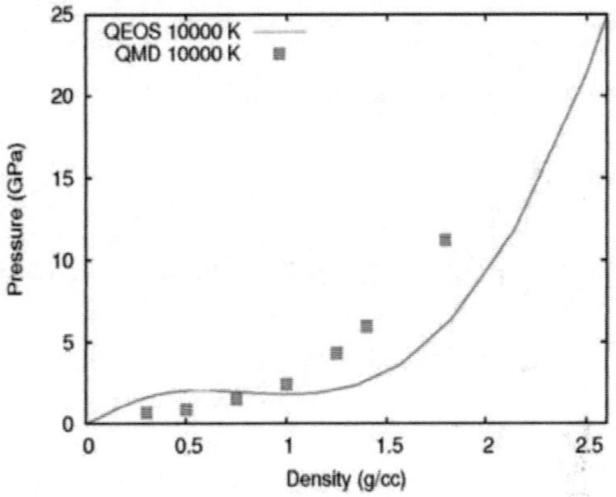


**Figure 6.** Isotherm (10000 K) generated by QEOS and AIMD simulations

do not show good agreement with the AIMD simulations for expanded states, indicating that QEOS method gives less accurate results for the expanded states and low temperatures.

## References

[1] Degnan et. al. *Phys. Rev. Lett.* 1995 **74** 1
[2] A V Majalee, V Mishra, P Pahari, V R Ikkurthi and S. Chaturvedi Nov. 2008 *Proceedings of Indo Russian Workshop on High Energy Density Physics Dynamics* I2IT Pune India
[3] B K Godwal, T C Kaushik and S K Sikka 1997 *Laser and Particle Beams* **15** 353
[4] S. Eliezer, A. Ghatak and H. Hora, 1986 *An Introduction to Equations of State: Theory and Applications* (Cambridge: Cambridge University Press)
[5] R M More, K H Warren, D A Young, and G B Zimmerman 1988 *Phys. of Fluids* **31** 3059
[6] David A Young and Ellen M Corey 1995 *J. Appl. Phys.*, **78**, 3748 2676
[7] P Blaha, K Schwarz, G K H Madsen, D Kvasnika, J Luitz 2001 *WIEN 2k, An Augmented Plane Wave Plus Local Orbital Program for Calculating Crystal Properties, Karlheinz Schwarz, Technical University, Wien, Austria* ISBN 3-9501031-1-2
[8] J P Perdew, S Burke, M Ernzerhof 1996 *Phys Rev B* **77** 3865
[9] X Gonze et. al. 2002 *Comput. Mat. Sci.* **25** 478
[10] N Troullier and J L Martins 1991 *Phys Rev B* **43** 1993
[11] I V Lomonosov 2007 *Laser and Particle beams* **25** 567
[12] A V Bushman I V Lomonosov and K V Khishchenko *K V Shortwaves Database* http://teos.fisp.ac.ru/rusbank
[13] Michel Penicaud 2009 J. phys. : Cond. Matter **21** 0954091
[14] Y Akahama, M Nishimura, K Kinoshita and H Kawamura 2006 *Phys. Rev. Lett.* **96** 045505
[15] V Mishra and S. Chaturvedi 2007 *Physica-B*, **393** 278
[16] V Mishra and S. Chaturvedi 2008 *Physica-B*, **403**
[17] A D Chijioke, W J Nellis, and Isaac F Silvera 2006 *J. Appl. Phy.* **98** 075261
[18] W J Nellis, J A Moriarty, A C Mitchel, M Ross, R G Dandrea, N W Ashcroft, N C Holmes and G R Gathers, Phys. Rev. Lett., 60 (1988) 1414-1417.
[19] M P Desjarlais, J D Kress and L A Collins 2002 *Phys. Rev. E* **66** 025401
[20] Vanina Recoules and Jean-Paul Crcombette 2005 *Phys. Rev. B* **72** 104202

# A VOF based multi-material method to study impact and penetration problems

**C.D. Sijoy, Vishal Mehra, Vinayak Mishra and Shashank Chaturvedi**
Computational Analysis Division, Bhabha Atomic Research Centre, Visakhapatnam, India
E-mail: cjoy_davis@yahoo.co.in

**Abstract.** A 2D axisymmetric Eulerian method is developed for studying multi-material elastic-plastic problems involving large material deformations. It consists of a Lagrangian plus remap strategy using a volume-of-fluid (VOF) based material interface tracking. The multi-material formulation used in this method allows one to update the energy and stress components of each material in a mixed cell independently. This assumes common strain-rates to all materials present in a mixed cell. The equivalent pressure and stress in a mixed cell are determined using volume weighted average. The present scheme, therefore, eliminates the commonly used pressure relaxation method, mixed EOS evaluation and energy partition schemes. The stress components of each material are transported using a second-order monotonic upwind scheme (MUSCL) due to van Leer. The capability of the proposed method is demonstrated by applying to various impact and penetration problems involving large material deformations. Reasonable agreement with experimental results are observed.

## 1. Introduction

The difficulties associated with mesh based pure Lagrangian methods for problems involving large material deformations can be minimized using an Eulerian or ALE method with a specialized technique called interface reconstruction [1], where the materials are assumed to be separated by a sharp macroscopic material interface in a mixed cell. These interfaces are constructed and tracked by using VOF method [2]. The objective of this work is to develop a simple and robust algorithm using VOF method for multi-material elastic-plastic problems involving large material deformations and strong shock waves, e.g. high velocity impact and penetration problems. Also, to include a relatively simple multi-material strength model which avoids commonly used stress/pressure iteration procedures in mixed computational cells.

## 2. Physical model

Most of the multi-material hydrodynamic algorithms assume a common velocity field and pressure for all materials present in a mixed cell. This pressure can be determined in different ways. One method is the pressure relaxation method, where a common pressure to all materials in a computational cell is found by using an iteration procedure. In each iteration step, the volume fraction or mass fraction of each material is adjusted until the pressures of those materials are converged to a common pressure. Another method (the simplest approach) is to use the volume weighted average of pressures in a mixed cell. The assumption of local pressure equilibrium in multi-material cells and associated iteration procedures may lead to pressure oscillations near material interfaces [3]. Also, physically, the pressure equilibrium in a computational cell cannot take place in a time shorter than that required for a sound wave to propagate through the cell at least once. For numerical stability, most of the hydrodynamic calculations are performed with a numerical time-step much smaller than this time (Courant time-step condition). Similarly, the assumption of same temperature and pressure to all materials in a mixed cell may lead to unrealistic energy flow from a hot material to a cold material after iteration procedure.

In this work, therefore, we have used the ideas given in Refs. [3, 4] for basic multi-material hydrodynamic formulation in which individual materials can possess different energies and temperatures. This leads to an evaluation of $m$ different energy equations in each mixed cell; where $m$ is the total number of materials in a mixed cell. The mean pressure in a multi-material cell is determined using a simple volume weighted average and thus avoids the iteration procedure. The same approach is extended to determine the equivalent stress in a mixed cell i.e. by using volume weighted average of individual stress components, where the individual stress components for each material in a mixed cell is updated with an assumption of common strain rates to all materials present in a cell [5]. The governing equations can be found in Ref. [3, 4].

## 3. Numerical method

The physical variables are updated using a 'directional split' method. A fixed Eulerian mesh is used with a virtual Lagrangian calculation. The effect of Lagrangian deformations are remapped on to the original mesh during Eulerian advection and remap step at each time-step. All the solution variables including the material stress components for each material are advected in a conservative manner with the help of VOF algorithm [2] and a MUSCL scheme due to van Leer. However, there is no guarantee that the individual materials will satisfy the yield criterion after the advection step. This discrepancy is removed by applying Von-Mises yield criterion prior to next time-step calculations.

The details of VOF method used in this work can be found in Ref. [2]. The interface line is represented by $\vec{n}\cdot\vec{x}-\alpha = 0$; where, $\vec{n}$ is the exterior normal to the line and $\alpha$ is the perpendicular distance from a local origin in a cell. The local origin in a cell for each material is decided by checking the slope of the interface line [2]. The slope of a material interface line and hence the normal vector are determined using the gradient of volume fractions from neighboring cells. The material interfaces in a mixed cell are constructed using 'onion-skin' model with the help of a 'material order list' determined dynamically [2].

## 4. Results and discussions

The numerical experiments described in following sections are performed with a time-step calculated using an expression given in Ref. [6] with CFL coefficient equal to 0.67.

### *4.1. Impact and penetration of metal sphere on thin metal plate*

First, we examine the case of 2D-axisymmetric high velocity impact of metal sphere on thin metal plate [8]. A detailed numerical analysis of the same problem have been reported in Ref. [9] by Mehra et. al using their 2-D and 3-D SPH codes. Two different cases from Ref. [8] are considered here. The initial parameters for both cases are summarized in Table 1. The material strength is modeled using parameters given in Ref. [10] and a Mie-Gruneisen EOS [10] is used. The computational domain consists of 600 $\times$ 240 cells in $z$ and $r$ directions with a mesh-size of 0.1667 mm in each direction. The background material (vacuum) is modeled with an extremely low density ideal gas with $\rho_0 = 10^{-12}$ kg/m$^3$ and $\gamma = 1.4$. Flux limited artificial viscosity (quadratic plus linear) is used.

The sequence of material deformation (for case 1 in Table 1) at different times are depicted in Fig. 1. The Al sphere completely penetrates through the thin Al slab. Both the impactor and

**Table 1.** Parameters used for impact problem. In the table, Y and $\mu$ are the yield and shear modulus respectively (in GPa). The velocity $v_c$ is given in km/s. $R_c$ is the sphere radius (cm) and $t_{slab}$ the target thickness (cm).

| Case | $R_c$ | $v_c$ | $t_{slab}$ | Impactor | Slab | $Y_{Al}$, $Y_{Cu}$ | $\mu_{Al}$, $\mu_{Cu}$ |
|---|---|---|---|---|---|---|---|
| 1 | 0.5 | 6.18 | 0.4 | Al | Al | 0.29, 0.12 | 27.6, 47.7 |
| 2 | 0.5 | 5.75 | 0.15 | Al | Cu | 0.29, 0.12 | 27.6, 47.7 |

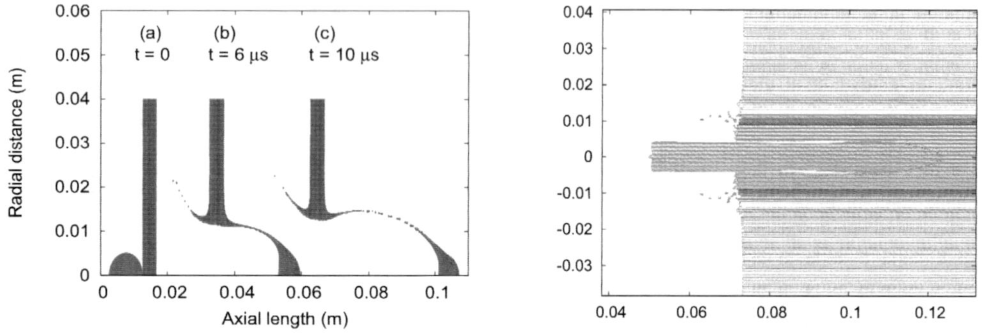


**Figure 1.** The left side plot show the material interfaces at different times for sphere impact problem described in Sec. 4.1. Sub-plots (b) and (c) are conveniently shifted along the axis by 2 and 5 cm, respectively. The right side plot show the projectile shape and depth-of-penetration at 390 $\mu$s for penetration problem described in Sec. 4.2.

**Table 2.** Comparison of crater diameter, $d_{cr}$, with experimental [8] and simulation values given in [8, 9] with our results for impact problem at 20 $\mu$s. The values are given in cm.

| Set | Our Sim. | Expt. [8] | Sim. [8] | 2D-Sim. [9] | 3D-Sim. [9] |
|---|---|---|---|---|---|
| 1 | 2.7 | 2.75-3.45 | 3.5 | 3.2−3.3 | 2.4−3.2 |
| 2 | 2.1 | 2.12 | 2.3 | 2.1−2.5 | 2.0−2.3 |

target undergo severe deformation, See Fig. 1. The nature of debris cloud expansion and the deformation of both the target and the impactor are in good agreement with the results given in Refs. [8, 9].

A comparison of simulated crater diameter ($d_{cr}$) and length to width ($l/w$) ratio of the debris cloud against experimental results is summarized in Table 2 and Table 3. The comparisons are made 20 $\mu$s after the impact, the same time point where experimental results are reported in Ref. [8]. The crater size from simulation show reasonable agreement with experimentally observed values. Inspecting the values given in Table 3, it is clear that our simulation results are close to the 3-D SPH simulation results. The reason for the difference in our simulation results with experimental values may be due to the assumption of a constant yield value for materials in our simulations. Perhaps, a rate dependent material strength model may give better match with experimental results. However, the present algorithm is capable of handling high material deformations with fairly accurate prediction of overall dynamics.

**Table 3.** Comparison of debris length to width ratio, $l/w$, with experimental [8] and simulation values given in [8, 9] with our results for impact problem.

| Set | Our Sim. | Expt. [8] | Sim. [8] | 2D-Sim. [9] | 3D-Sim. [9] |
|---|---|---|---|---|---|
| 1 | 1.58 | 1.39 | 1.33 | 1.31−1.44 | 1.53−1.9 |
| 2 | 1.51 | 1.39 | 1.11 | 1.41−1.51 | 1.7−2.3 |

### *4.2. Impact and penetration of an ogive nosed metal projectile in to a metal slab*

Next, we have studied the penetration of metal targets by ogive nose metal projectile [11]. The experiments in Ref. [11] were conducted for a range of velocities from 0.5 to 3 km/s. We have compared the depth-of-penetration for a particular case reported in Ref. [11] with our simulation result. The dimensions of the projectile and target are taken from Ref. [11]. A steel projectile is used with an ogive shaped nose with an axial length of the nose equal to 11.8 mm. The projectile diameter and total length (including ogive nose) are 7.11 mm and 71.1 mm respectively. The aluminum target is $\sim$ 250 mm in diameter. The calculations were performed with a variable mesh consists of 550 $\times$ 75 cells. The material parameters are taken from Ref. [11]. Also, both the projectile and target were modeled with Mie-Gruneisen EOS. The initial projectile velocity was 570 m/s. The right side plot in Fig. 1 shows the shape of the projectile and target. The penetration has stopped at about 390 $\mu$s. The projectile shows no permanent deformation (globally), which is consistent with the experiment [11]. However, a small amount of local material erosion is observed (towards the end of penetration for both the projectile and target, which was not reported in [11]. A contact algorithm such as *boundary-layer* or *free-slip* algorithm may mitigate this problem. The depth-of-penetration from simulation, 52 mm, matches well with the experimental value of 55 mm [11].

## 5. Summary and conclusions

We have presented an axisymmetric multi-material algorithm with material strength model and volume of fluid (VOF) method for material interface tracking. The present formulation allows one to have different EOS for each material in a mixed cell and eliminates the pressure equilibration method and mixed EOS evaluation. Numerical experiments show that the approximation of equivalent stress and pressure in a multi-material mixed cell using a simple volume-weighted average works well to approximate physical solutions reasonably. Also, the proposed algorithm can handle multi-material penetration problems with large material distortions.

## References

[1] D J Benson 2002 *Appl. Mech. Rev.* **55** pp 151
[2] C D Sijoy and Shashank Chaturvedi 2010 *J. Comp. Phys.* **229** pp 3848-3863
[3] D de Niem, E Kuhrt and U Motschmann 2007 *Comp. Phys. Comm.* **176** pp 170-190
[4] G H Miller and E G Puckett 1996 *J. Comp. Phys.* **128** pp 134-164
[5] D J Benson 1997 *Comput. Meth. Appl. Mech. Eng.* **140** pp 59-86
[6] D J Benson 1998 *Comput. Meth. Appl. Mech. Eng.* **67** pp 191-205
[7] G T Camacho 1997 *Comput. Methods. Appl. Mech. Eng.* **142** pp 269-301
[8] S Hiermaier et. al. 1997 *Int. J. Impact Eng.* **20** pp 363-374
[9] Vishal Mehra and Shashank Chaturvedi 2006 *J. Comp. Phys.* **212** pp 318-337
[10] M L Wilkins 1998 *Computer Simulation of Dynamic Phenomena* Scientific Computation (Springer-Verlag)
[11] A J Piekutowski et. al. 1999 *Int. J. Impact Engng.* **23** pp 723-724

# Excitation of characteristic modes of a crystal during solid fracture at high tensile pressure

**S Rawat[1,2], M Warrier[2], D Raju[3], S Chaturvedi[2] and V M Chavan[1]**

[1] Refueling Technology Division, Bhabha Atomic Research Centre, Mumbai 400 085, India
[2] Computational Analysis Division, Bhabha Atomic Research Centre, Visakhapatnam 530 012, India
[3] Institute for Plasma Research, Gujarat 382 428, India

E-mail: alig.sunil@gmail.com

**Abstract.** We have performed uniform triaxial deformation of single crystal copper at high strain rate using molecular dynamics code LAMMPS. The best-fit void nucleation and growth parameters are obtained using a macroscopic nucleation and growth (NAG) model. The detailed analysis of the data shows that voids nucleate at specific locations in the domain, and subsequently grow and coalesce. We explain the spatial location of first void nucleation in terms of the excitation and interaction of characteristic modes in the crystal domain using singular value decomposition analysis.

## 1. Introduction

High-velocity impact involving metals produces high tensile pressures in a small region. The resulting nucleation, growth and coalescence of voids lead to spall. Existing fracture models at high strain-rates, such as the Nucleation and Growth (NAG) model [1], involve several parameters, best-fit values of which are required for different materials. These best-fit parameters are available in the literature only for a few materials. If some of these parameters can be obtained from atomistic simulations, others can be determined by comparison with the experiment, e.g. spall produced by plate impact. In our earlier work [2], we have performed uniform triaxial deformation of crystal copper at $5\times10^9$/s strain rate and best-fit void nucleation and growth parameters are obtained for NAG model.

In the present work, we perform singular value decomposition analysis to determine the characteristic modes excited in the system, including their frequencies and associated distortions and attempt to explain the locations where nucleation first takes place.

## 2. Computational Method

Uniform triaxial deformation of single crystal copper has been performed at $5\times10^9$/s strain-rate using molecular dynamics (MD) code LAMMPS [3]. We have used embedded-atom method potential parameters generated by Foiles et al [4] for this study. Periodic boundary conditions are applied along all three directions. Simulation domain contains 360 atoms corresponding to $10\times3\times3$ unit cells. The system is first relaxed to 300 K and 1 bar pressure using an NPT simulation. The equations of motion are integrated with a time step of 1 femto-second using the velocity-Verlet algorithm.

## 3. Singular Value Decomposition (SVD) Analysis

### *3.1. Introduction*

Details of the singular value decomposition analysis are described in [5]. A brief description of the method is presented here. In this technique, a matrix $A_{m\times n}$ is decomposed into three new matrices:

$$A_{m\times n} = U_{m\times m}\, S_{m\times n}\, V^T_{n\times n} \tag{1}$$

where $UU^T = I$, $VV^T = I$ (i.e., U and V are orthogonal matrices); the columns of U are the orthogonal eigen vectors of $AA^T$, the columns of V are the orthogonal eigen vectors of $A^TA$ and S is a diagonal matrix containing the square roots of eigen values from U or V in descending order.

### *3.2. Application to MD output*

Matrix $A_{MN}$ generated from MD output is given as follows:

$$A_{MN} = \begin{bmatrix} A_{11} & A_{12} & A_{13} & \cdots & A_{1N} \\ A_{21} & A_{22} & A_{23} & \cdots & A_{2N} \\ A_{31} & A_{32} & A_{33} & \cdots & A_{3N} \\ \cdots & \cdots & \cdots & \cdots & \cdots \\ \cdots & \cdots & \cdots & \cdots & \cdots \\ A_{M1} & A_{M2} & A_{M3} & \cdots & A_{MN} \end{bmatrix} \tag{2}$$

where M is the number of time steps for which atomic positions are output and $N = d\, N_{atoms}$, where $N_{atoms}$ is the number of atoms used in the MD simulation and d is the dimension of the system. Therefore, we can write

$$A_{11} = x_{11}, A_{12} = y_{11}, A_{13} = z_{11}; A_{14} = x_{12}, A_{15} = y_{12}, A_{16} = z_{12}, \ldots\ldots$$

Rewriting the above matrix (Eq. 2) as follows:

$$A_{MN} = \begin{bmatrix} x_{11}\, y_{11}\, z_{11} & x_{12}\, y_{12}\, z_{12} & \cdots & x_{1N}\, y_{1N}\, z_{1N} \\ x_{21}\, y_{21}\, z_{21} & x_{22}\, y_{22}\, z_{22} & \cdots & x_{2N}\, y_{2N}\, z_{2N} \\ x_{31}\, y_{31}\, z_{31} & x_{32}\, y_{32}\, z_{32} & \cdots & x_{3N}\, y_{3N}\, z_{3N} \\ \cdots & \cdots & \cdots & \cdots \\ \cdots & \cdots & \cdots & \cdots \\ x_{M1}\, y_{M1}\, z_{M1} & x_{M2}\, y_{M2}\, z_{M2} & \cdots & x_{MN}\, y_{MN}\, z_{MN} \end{bmatrix} \tag{3}$$

here x, y and z are the atomic positions. Decomposition of this matrix using SVD gives information on temporal variations of atoms (U), information on spatial correlations amongst the atoms (V) and the amplitude (S) of the modes excited in the system. As discussed above, the singular values (S values) which are the representative of the amplitude of modes are arranged in descending order which means that even fairly weak modes excited in the system can also be identified.

## 4. Results and Discussion

Using the atomic coordinates output every 5 femto-seconds, we have performed the SVD analysis [6] to determine the characteristic modes excited in the system during deformation.

*4.1. Excited characteristic modes before and after the void nucleation*
For 360 atoms in the system, all 1080 modes are obtained using SVD. The variation of amplitude of modes excited in the system uniformly deformed at $5\times10^9$/s strain-rate before the void nucleation and after the first void nucleation is shown in figure 1 (a) and figure 1 (b), respectively.

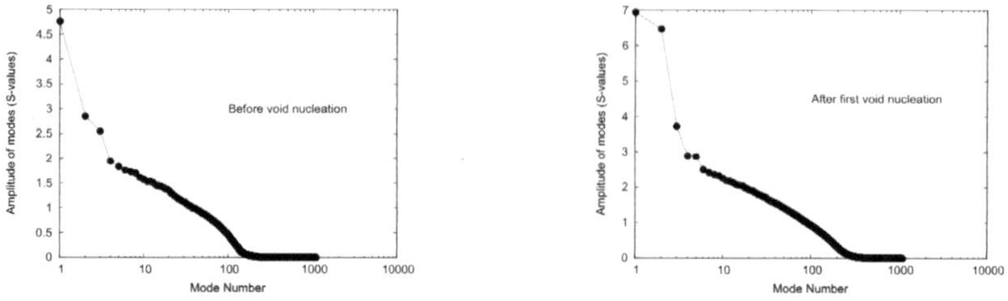

(a) (b)

**Figure 1.** Amplitude of modes (singular values 'S') as a function of mode number (*a*) before the nucleation of a void (*b*) after the nucleation of a first void

*4.2. Frequency Spectrum before and after the void nucleation*
The comparison of frequency spectrum for first and third modes before and after the nucleation of first void is shown in figure 2 (a) and figure 2 (b), respectively. It is seen in figure 2 (b) that a shift in the frequency occurs after the nucleation of a void. In addition to this, some new frequencies also appear after the first void nucleation. The creation of new frequencies and a shift in frequencies are the signature of the nucleation of a void.

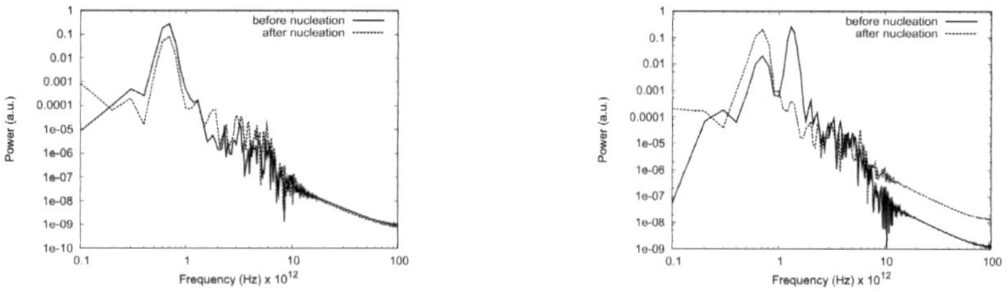

(a) (b)

**Figure 2.** Frequency spectrum of (*a*) mode 1 (*b*) mode 3

*4.3. Temporal variation after the void nucleation*
The temporal variation of the first strong mode after the first void nucleation is shown in figure 3. A sudden growth in the amplitude of the mode (figure 3) refers to the nucleation of a void in the system.

*4.4. Spatial variation before and after the void nucleation*
The spatial variation of first four modes excited in the system before and after the first void nucleation is shown in figure 4 (a) and figure 4 (b), respectively. The first strong mode before and after the void nucleation corresponds to the wavelength of $\lambda$. Other weak modes correspond to the wavelengths of $3\lambda/2$ and $5\lambda/2$. Note that hierarchy of the modes has been changed after the nucleation of a first void.

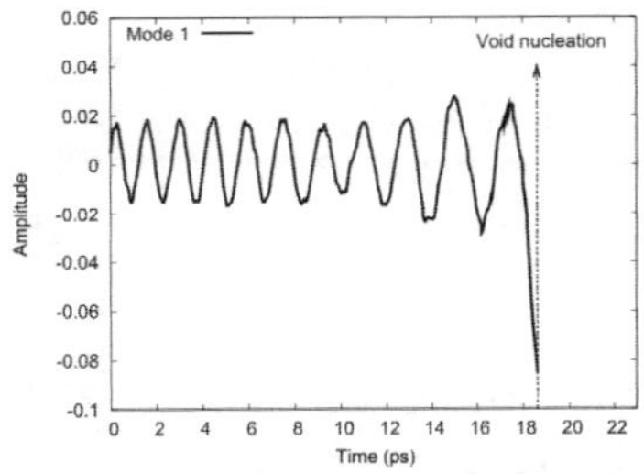


**Figure 3.** Temporal variation of the first mode after the first void nucleation

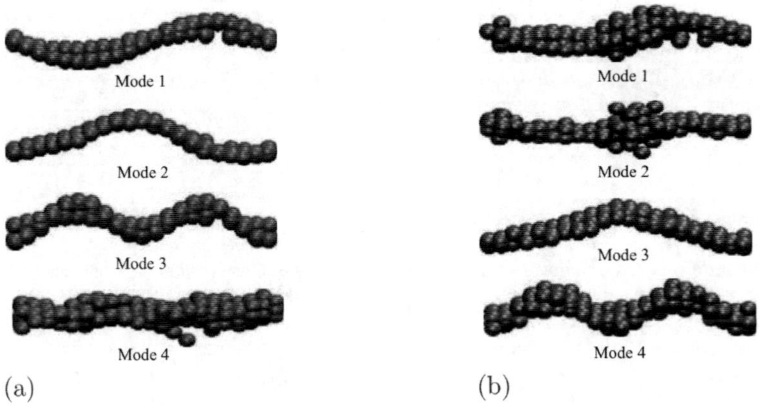


**Figure 4.** Spatial variation of modes (*a*) before the void nucleation (*b*) after the void nucleation

## 5. Conclusions

We have performed uniform triaxial deformation of single crystal copper at $5\times10^9$/s strain-rate and data is analysed using singular value decomposition (SVD) technique. We find that after the nucleation of first void

- there is a shift in the frequencies of modes
- new frequencies are created
- strong modes correspond to the wavelength of $\lambda$, $3\lambda/2$ and $5\lambda/2$.
- hierarchy of modes changes after the void nucleation

In future work, we plan to quantify the linkage between the change in mode behaviour (before and after void nucleation) and the nucleation process itself.

## References

[1] Curran D R 1987 *Phys. Rep.* **147** 253-388
[2] Rawat S, Warrier M, Chaturvedi S and Chavan V M 2011 *modell. Simul. Mater. Sci. Eng.* **19** 025007
[3] Plimpton S 1995 *J. Comp. Phys.* **117** 1-19
[4] Foiles S M, Baskes M I and Daw M S 1986 *Phys. Rev.* B **33**(12) 7983-91
[5] Press W H et al 1998 *Numerical Recipes in Fortran* (Cambridge University Press)
[6] Negi S and Chaturvedi S 2010 *Int. J. Nanosci.* **5** 471-86

23rd International Conference on High Pressure Science and Technology (AIRAPT-23) IOP Publishing
Journal of Physics: Conference Series **377** (2012) 012108 doi:10.1088/1742-6596/377/1/012108

# Time-Resolved Synchrotron X-ray Diffraction on Pulse Laser Heated Iron in Diamond Anvil Cell

**Choong-Shik Yoo,[1]* Haoyan Wei,[1] Ranga Dias,[1] Guoyin Shen,[2] Jesse Smith,[2] Jing-Yin Chen,[3] William Evans[3]**
[1]Institute for Shock Physics, Washington State University, Pullman, WA 99164
[2]HPCAT/APS, Carnegie Institute of Washington, Argonne, IL 60439
[3]Lawrence Livermore National Laboratory, Livermore, CA 94550

E-mail: csyoo@wsu.edu

**Abstract.** We present time-resolved synchrotron x-ray diffraction to probe the ε-δ phase transition of iron during pulse-laser heating in a diamond anvil cell. The system utilizes a monochromatic synchrotron x-ray beam, a two-dimensional pixel array x-ray detector and a dual beam, double side laser-heating system. Multiple frames of the diffraction images are obtained in real-time every 22 ms over 500 ms of the entire pulse heating period. The results show the structural evolution of iron phases at 17 GPa, resulting in thermal expansion coefficient $1/V(\Delta V/\Delta T)_p = 7.1*10^{-6}$ /K for ε-Fe and $2.4*10^{-5}$ /K for γ-Fe, as well as the evidence for metastability of γ-Fe at low temperatures below the ε−γ phase boundary.

## 1. Introduction

Understanding the dynamic response of solid under extreme conditions of pressure, temperature and strain rate is a fundamental scientific quest and a basic research need in materials science. Specifically, obtaining the atomistic/molecular level description of structural and chemical changes of solid under rapid heating and/or compression over a large temporal, spatial and energy range is critical to understanding material stability or metastability, transition mechanism, crystal order or disorder, and novel physical and chemical properties of solids.

The high brightness of third-generation synchrotron x-rays has revolutionized the field of high-pressure material science and geoscience in recent years. High-pressure synchrotron research activities to date have, however, concentrated largely on probing static properties of materials that are stable near minimum energy configurations and governed by thermodynamic constraints. Only a few studies have acquired dynamic properties of materials such as metastable structures, chemical mechanisms, transition dynamics, and mechanical deformation, much of which require the time-resolved structural information of solid that undergoes rapid phase or chemical changes. In this regard, developing time-resolved x-ray diffraction capabilities is significant for high-pressure sciences and synchrotron technology developments. It is also timely with the proposed development of dynamic compression sector (DCS) at the APS and other similar efforts at the NSLS-II and LCLS.

The machine characteristics of third-generation synchrotron are ideally suitable for probing dynamic single-event phenomena such as shock-induced phase changes and highly exothermic metal

* To whom any correspondence should be addressed. csyoo@wsu.edu; (509) 335 - 2742

combustions. As the first step, we have recently demonstrated a novel time-resolved x-ray diffraction capability probing structural and chemical evolutions of metallic solids that undergo rapid phase and chemical changes [1]. Briefly, these experiments were performed using monochromatic synchrotron x-rays at 16IDD beamline of the HPCAT/APS. The system utilized monochromatic synchrotron x-rays and a two-dimensional (2D) pixel array x-ray detector (PILATUS) in combination of a fast-rotating diffraction beam chopper, providing a time (in azimuth) and angle (in distance)-resolved x-ray diffraction image, continuously recorded at a time resolution of ~30 μs over a time period of 3 ms, as shown in Fig. 1. Multiple frames of the TARXD images were obtained with time resolutions between 30 and 300 μs over three to several hundreds of ms, depending on the dynamics of experiments in interest. This method is applicable to a wide range of dynamic experiments to study both single event phenomena of solid under thermal, electric or mechanical impact conditions and non-single event changes using dynamic-DAC and high frequency pulse lasers.

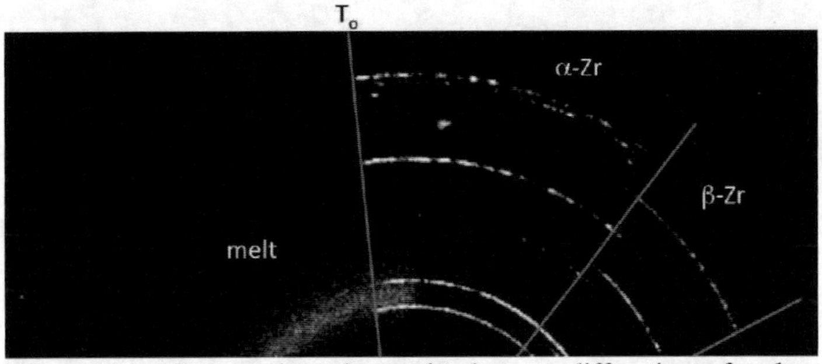


**Figure 1.** Time- and angle-resolved x-ray diffraction of pulse-heated Zr, showing the early time structural phase transition and melting, followed by the late time highly exothermic combustion of the melt (not shown). This 2D diffraction pattern was recorded for ~30 μs x-ray exposure over ~3 ms. Time runs clockwise from $T_o$ - concentric to the Debye-Scherrer's diffraction rings.

The ultimate goal of our efforts is to perform similar time-resolved structural studies at high pressures and temperatures using diamond anvil cells (DAC), dynamic-DAC, and pulse laser-heated DAC. In this paper, we describe a preliminary result showing the time-resolved diffraction change associated with the ε−γ phase transition in pulse laser-heated iron in DAC.

## 2. Experimental Method

All diffraction experiments were carried out using monochromatic synchrotron x-rays at 14.413 keV (or 0.8602 Å in wavelength) from an undulator at the 16IDD of the Advanced Photon Source (APS). We used a focused x-ray beam of ~50 (Horizontal) x 30 (Vertical) μm at the sample.

Figure 2 shows the actual experimental setup for time-resolved x-ray diffraction using laser-heated DAC in the hutch. The system consists of (i) a 2D pixel array x-ray detector (PILATUS), (ii) a dual beam, double side laser-heating system, and (iii) a set of electronics for time synchronization and controls of the experiments, which includes a power supply, a function generator, a delay generator, and a digital scope analyser. The power output of the laser is adjusted to ramp up and down by sending a saw-shape analogue input signal to the laser power supply, as shown in Fig. 3 (the blue curve). The temperature of the sample is measured from thermal emission of laser-heated area during this ramp up/down period [2], while the real-time thermal emission signal is also recorded on a fast photodiode through an optical fibre (Fig. 3, green). Note that the photodiode signal is offset from the laser power about 10-20 ms both at the start and the end. This is due to low thermal emission intensity proportional to the fourth power of temperature, which makes difficult to measure the intensity at low temperatures at the beginning and ending of pulse heating. For the same reason the measured temperature is actually close to the "maximum" temperature rather than "average" temperature. In this particular experiment, the maximum temperature of ~2400 (±100) K is achieved using about 15% of our laser power. A total of twenty diffraction images were collected during the entire period of one heating and cooling cycle for 500 ms, by acquiring 22 ms x-ray exposure per frame with 3 ms readout (or blockout) time between the frames, as illustrated in Fig. 3, purple. To avoid multiple heating of sample, we gate the laser only for time long enough for one ramp heating cycle (Fig. 3, red).

**Figure 2.** The experimental setup employed for time-resolved x-ray diffraction on double-side, pulse laser-heated iron in DAC at the 16IDD/APS.

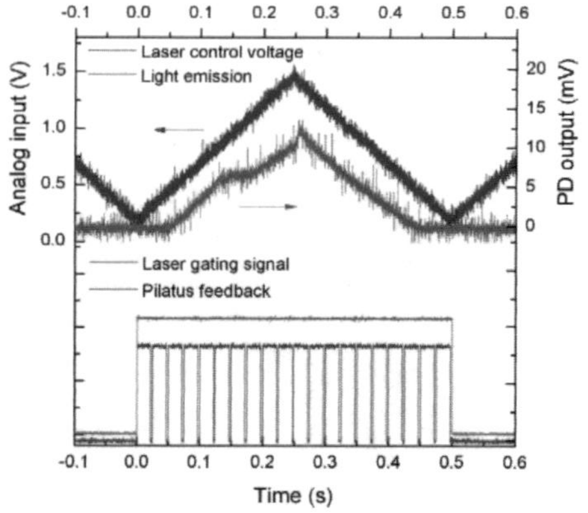


**Figure 3.** An input signal for pulse laser power, a photodiode output of thermal emission, and a PILATUS feedback of x-ray records, illustrating the time-synchronization of the experiments.

## 3. Results

In this paper, we describe an experiment of iron in $Al_2O_3$ pressure medium performed at 17 GPa and the temperature ramped up to ~2400 K - near the melting temperature of iron [3-5]. Note that the photodiode record in Fig. 3 exhibits a knee at ~120 ms after the heating starts, signifying the occurrence of the ε−γ phase transition during heating. Interestingly, such knee is absent during cooling. These results are consistent with the time-resolved diffraction data shown in Figs. 4-6.

Figure 4 shows a selected set of time-resolved diffraction images of pulse laser heated iron at 17 GPa, showing the reflections of ε(hcp)-Fe (white dots), γ(fcc)-Fe (green dots), and $Al_2O_3$ (organge dots) used as a pressure medium. The quality of the diffraction images is rather poor, largely due to a short x-ray exposure time for each image (22 ms) and, more importantly, high level of the structure-less background x-ray intensity from diamonds and strong diffraction lines from $Al_2O_3$. Nevertheless, it does depict the evidence of the ε−γ phase transition that occurs at ~100-120 ms after the initial pulse heating, as mentioned in Fig. 3. Note that the diffraction lines of $Al_2O_3$ appear relatively strong with respect to those of iron. This is likely due to the low energy x-ray (15 keV) used in this experiment, which also limits the time-resolution of the present experiment. In fact, we found that it was possible to obtain a 1 ms time-resolution using nitrogen pressure medium.

By integrating the time-resolved diffraction images in Fig. 4, we obtain the time- and angle-resolved diffraction patterns as shown in Fig. 5. Again, the signal-to-noise ratio of the spectra is not good, but sufficient to discern three diffraction lines from ε-Fe (010, 002 and 101, marked in blue), two lines from γ-Fe (111 and 002 in green), and four lines from $Al_2O_3$ (014, 110, 113, and 024, marked in orange). Note that the 002 line of ε-Fe shows relatively strong intensity in comparison with the

100, probably representing a preferred orientation of the sample. The γ-phase appears at ~100 ms during heating and then remains over the entire cooling period, while the ε-phase disappears during heating and reappears during cooling. This indicates the metastability of γ-Fe at low temperatures well below the γ/ε phase boundary at this pressure, confirming the photodiode record in Fig. 3. A similar metastable behaviour of γ-Fe was often observed in the previous laser-heating experiments [3-5]. The diffraction data also shows subtle changes in the peak positions of diffraction lines with time, apparently a signature for thermal expansion (or contraction). The thermal expansion of ε and γ-phases can be seen more clearly in Fig. 6.

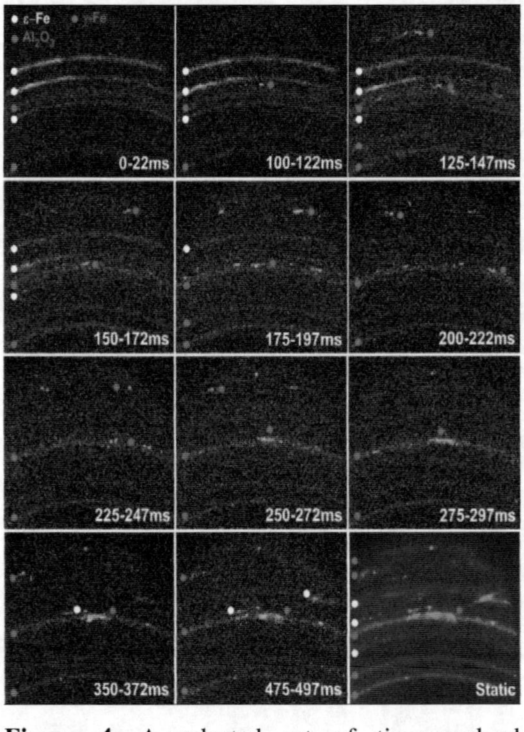


**Figure 4.** A selected set of time-resolved diffraction images of pulse laser heated iron at 17 GPa, showing the ε(hcp, white dots)-γ(fcc, green dots) phase transition together with the features from $Al_2O_3$ (organe dots).

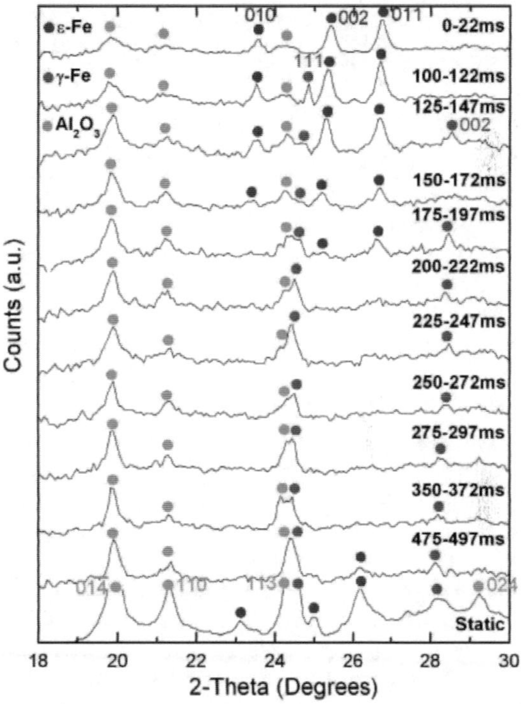


**Figure 5.** Time- and angle-resolved x-ray diffraction patterns of iron at 17 GPa, showing the ε–γ phase transition during pulse laser heating. The blue, green, orange dots signify the diffraction lines of ε-Fe, γ-Fe and $Al_2O_3$, respectively.

Figure 6 plots the time (or temperature)-dependent volume changes of the ε and γ- phases, obtained from the data in Fig. 5. Again, the plot depicts the coexistence of two ε and γ phases between 120 ms and 200 ms. The ε–γ phase transition accompanies about ~3% volume expansion occurring abruptly at the transition, 120 ms. Based on the measured "maximum" temperature 2400 K and the ε-γ transition temperature of 1000 K at 17 GPa, we estimate the thermal expansion coefficient $1/V(\Delta V/\Delta T)_p$ = $7.1*10^{-6}$ /K for ε-Fe and $2.4*10^{-5}$ /K for γ-Fe at 17 GPa. These values are compared with those measured at ambient pressure, $6.5*10^{-5}$ /K for γ-Fe and $3.3*10^{-5}$ /K for α(bcc)-Fe [6]. Based on these data, we find the decrease of thermal expansion coefficient for γ-Fe with increasing pressure in accordance with $\{\delta(\ln \alpha)/\delta(\ln V)\}_T = 6.1$, which agrees well with the previously reported value of 6.5 (±0.5) determined by a wire-heating technique in DAC [7].

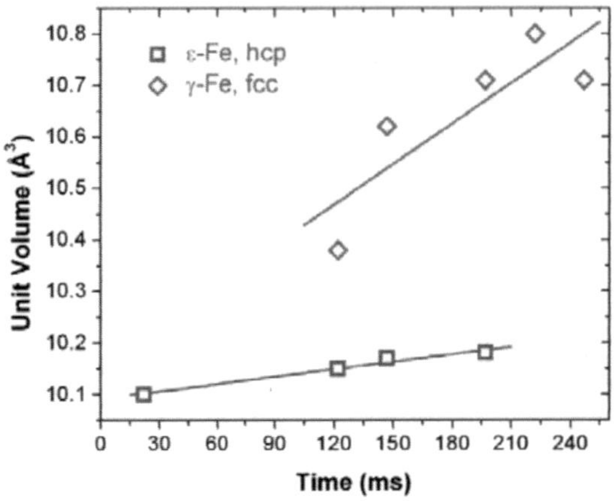


**Figure 6.** Time (or temperature)-dependent volume changes of the ε(hcp) and γ(fcc) phases, obtained from the data in Fig. 5. The phase transition occurs at around 120 ms – consistent with the knee observed in the PD record in Fig. 3.

## 4. Conclusion

We have shown the feasibility of time-resolved x-ray diffraction applied to pulse laser-heated samples in DAC. The results performed on Fe at 17 GPa show real-time evidences for the ε–γ phase transition in time-resolved thermal emission (or temperature) and diffraction records. Based on the diffraction data, we obtain the thermal expansion coefficient of $1/V(\Delta V/\Delta T)_p = 7.1*10^{-6}$ /K for ε-Fe and $2.4*10^{-5}$ /K for γ-Fe at 17 GPa.

The present method employs a high-resolution 2D PILATUS detector, providing a convenient way of obtaining time-resolved powder x-ray diffraction images in a ms resolution, or in a μs time resolution when it is used in combination with a fast recording chopper as shown in Fig. 1 [1], with high sensitivity nearly comparable to an intensified CCD but substantially lower noise. The time resolution of the present experiment can be enhanced to 10-100 μs permitting a signal averaging over a reasonable experimental period of ~5 min. The spatial-resolution can also be increased by use of a larger detector such as PILATUS 300K or 2M of which active area can be 10-100 times the present one. The main limitation, on the other hand, is relatively low x-ray energy used in the present experiments. This is because the quantum efficiency of PILATUS drops rapidly at high x-ray energies, from 99% at 12 keV to 55 % at 15 keV or ~20% at 20 keV, which limits the use of thick diamond anvils and pressure medium.

**Acknowledgement**

We appreciate Drs. P. Chow, Y. Xiao, and E. Rod for their technical assistance at the beamline. The present study has been supported by the NSF-DMR under Grant No. 0854618 and U.S. DHS under Award Number 2008-ST-061-ED0001. The views and conclusions contained in this document are those of the authors and should not be interpreted as necessarily representing the official policies, either expressed or implied, of the U.S. Department of Homeland Security. The synchrotron x-ray diffraction studies were done at the HPCAT 16IDD beam line at the Advanced Photon Source. The use of the HPCAT facility was supported by the DOE-NNSA, CDAC, and DOE-BES. The work at LLNL was funded by the Laboratory Directed Research and Development Program under project tracking code 11-ERD-046.

**References**

[1] Yoo C S, Wei H, Chen J Y, Shen G, Chow P, and Xiao Y 2011, submitted.
[2] Shen G, Rivers M L, Wang Y, and Sutton S R 2001 *Rev. Sci. Inst.* **72** 1273
[3] Boehler R 1993 *Nature* **363** 534
[4] Yoo C S, Akella J, Cambell A, Mao H K, and Hemley R J 1995 *Science* **270** 1473

[5] Shen G, Mao H K, Hemley R J, Duffy T S, and River M L 1998 *Geophy. Res. Lett.* **25** 373
[6] Wei H *et al.* 2011, to be published
[7] Boehler B, von Bargen N and Chopelas A 1990 *J. Geophys. Res.* **95** 736

23rd International Conference on High Pressure Science and Technology (AIRAPT-23) IOP Publishing
Journal of Physics: Conference Series **377** (2012) 012109 doi:10.1088/1742-6596/377/1/012109

# High Pressure Kinetics Studies of Water Solidification in *Dynamic*-DAC

**Jing-Yin Chen[1*], Choong-Shik Yoo[1]**
[1]Institute for Shock Physics and Chemistry Department, Washington State University, Pullman, WA 99164

E-mail: chen72@llnl.gov or csyoo@wsu.edu

**Abstract.** The phase diagram of water is both unusual and complex, exhibiting a wide range of polymorphs including proton-ordered or disordered forms. In addition, a variety of stable and metastable forms are observed. The richness of $H_2O$ phases attests the versatility of hydrogen bonded network structures that include kinetically stable amorphous ices. In addition, natural gas hydrates are observed upon cooling and/or pressurization. The structures of gas hydrates are quite different from ices, yet guest molecules are enclathrated inside hydrogen bonded water cages. Hence, the presence of small impurity molecules can affect strongly the behaviours of hydrogen bonding and alter the phase diagram and phase boundaries of $H_2O$. In this paper, we report the effect of methane impurity on solidification and melting of $H_2O$ under dynamic loading conditions. The presence of impurity mainly affects the growth rate of metastable ice VII, slowing down from 1.4 m/s in pure $H_2O$ to 0.26 m/s in impure $H_2O$, neither the occurrence of metastable ice VII nor the melting of ice VI.

## 1. Introduction

Abundant in nature, water is a major constituent of planets and living organisms alike. The phase diagram of water exhibits a large number of polymorphs (Fig. 1) with great diversity in crystalline structures, chemical bonding, and collective interactions [1-3]. The hydrogen bond (OH-bond) angles and topology of relatively weak OH-bonds (with respect to covalent OH-bonds) are subject to large distortions, leading to proton and structural disorders and myriad phases—both stable and metastable (including amorphous). In addition to the fifteen known solid phases of $H_2O$, there are many metastable phases. These include several metastable phases of ice IV near the melt line [4] and VII observed in the stability field of ice VI [5], high- and low-density amorphous (HDA and LDA) at low temperatures [6-10], and high- and low-density water (HDW and LDW) [11]. Furthermore, liquid water under dynamic conditions is observed beyond the equilibrium phase boundaries. Therefore, the stability limit conjecture of liquid water [2], shown as the dotted broken line, is estimated from the homogeneous nucleation temperatures of the super-cooled liquid. Recently, HDA [12] ice is reported at room temperature under rapid compression, forming from metastable ice VII in the stability field of ice VI. This is in addition to a whole series of intermediate structures arising from amorphization, dipole ordering transitions, and symmetrization of hydrogen bonding [6,13]. The strength of OH-bonds varies in these metastable structures, as does the transition dynamics that is not well understood.

* To whom any correspondence should be addressed. Chen's current affiliation is High Pressure Physics Group, Lawrence Livermore National Laboratory, Livermore, CA 94550.

Recent advances in *dynamic*-Diamond Anvil Cell (*d*-DAC) [5,12,14-16] enable the studies of metastable/intermediate states that are path-dependent and kinetically controlled. For example, dynamic pressure-induced dendrites have been observed and explained by Lee et al. In addition, metastable ice-VII, previously observed in the stability field of ice VI, is accounted for a lower interfacial surface energy between ice VII and super-compressed water [5]. Moreover, HDA ice [12] was observed at room temperatures under rapid compression. HDA is formed from metastable ice VII in the stability field of ice VI, well above the crystalline temperature of 150 K. It also arises from the structural similarity between HDA and ice VII. The kinetic data indicates that the formation follows an interfacial growth mechanism unlike the melting process. Nevertheless, the occurrence of HDA along the extrapolated melt line of ice VII (Fig. 1) resembles the ice Ih-to-HDA transition [6], indicating that the structural instabilities of parent ice VII and Ih drive the pressure-induced amorphization. The reported results strongly show that the *d*-DAC is a strong method to study the kinetics of materials and furthermore enables to study thermodynamics by varying pressures at different time scales.

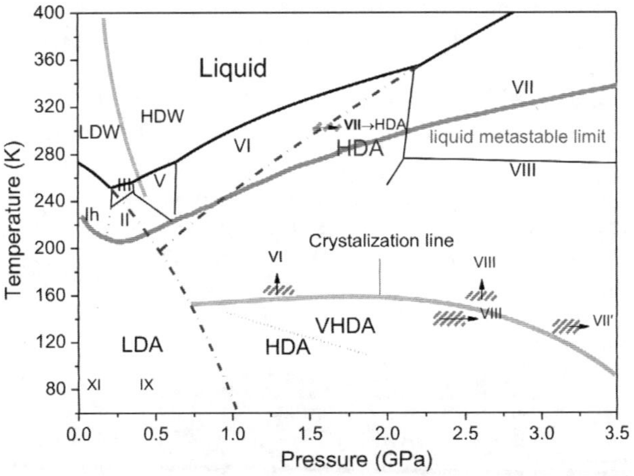


**Figure 1.** The phase diagram of $H_2O$ showing a wide range of metastable and disordered phases of water and ice. The present data signifies the appearance of metastable ice VII (hatched in baby blue) in the stability field of ice VI at room temperature – well before the equilibrium transition pressure. The blue dashed-dotted curves are the extrapolated melting lines of ice Ih and ice VII. The grey lines signify the proposed phase boundary of LDW and HDW[11] and the suggested crystallization line of amorphous ices[6-7,10-11]. The green curve is the calculated liquid metastable limit[2].

In the presence of small gaseous molecules, water form gas hydrate clathrates at low temperatures and low pressures, adding complexity in the phase diagram of water. Natural gas hydrates are made of hydrogen-bonded cages of water molecules containing a wide range of guest molecules ($CH_4$, $H_2$, $CO_2$, *etc.*). The existence of gas hydrates strongly show impurities play an important role on altering the phase diagram of $H_2O$ via changing hydrogen bonding. Hence, the influence of impurity (methane) on the phase boundaries of metastable ice VII and ice VI under dynamic conditions is investigated in this study.

## 2. Experiments

Using a micro-syringe with a stainless steel needle, a drop of pure distilled water (< 10 nl) was loaded into a small (0.08-0.13 mm) hole of a pre-indented stainless steel gasket mounted between two opposed diamond anvils with 0.3 or 0.5 mm flats, together with a few small ruby chips for pressure calibrations. For dirty water samples, methane is used as the impurity. Highly pure methane gas (>99.999%, Advanced Specialty Gases) was used without future purification. A small bubble was then introduced in the sample chamber for the space of methane. Methane gas was then loaded into the sample chamber using a high-pressure gas loader developed at Washington State University. The confocal micro-Raman spectroscopy with a 0.8 $cm^{-1}$ spectral resolution is used to identify the status of water and ice phases. The Raman spectra were obtained in a back scattering geometry and using an $Ar^+$ ion laser (Spectra Physics) at the 514.5 nm excitation wavelength. The formation of ice was recorded *in-situ*, using a high-speed camera (Photron FASTCAM APX RS). The sizes of growing or melting ices are measured in micro-photographic images by scaling to the known size of Ruby ball.

Our *d*-DAC incorporates three piezo-actuators (APC International, Ltd.) into a conventional DAC. The peak and modulating pressures and the compression rate of *d*-DAC are controlled by the shape and magnitude of the input electric signal to these actuators. We used a function generator and a power amplifier to tailor the input signal of the piezo-actuators. A digital delayed generator was used to synchronously operate all associated instrument, which includes a *d*-DAC, a CCD detector, a high-speed camera, and an oscilloscope. In this study, we typically used different time scales of symmetry trapezoidal waveforms. During these rise and fall periods, the images and pressures of the sample were measured *in-situ* using the high-speed camera and time-resolved Ruby luminescence, respectively. The Raman spectra were obtained before, at the top and after the waveforms. Figure 2 illustrates a 2-D image plate of time-resolved pressure measurement and the corresponding microphotographs as an example.

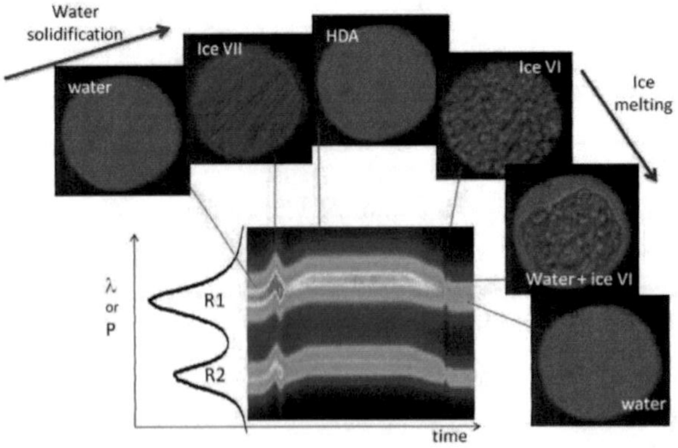


**Figure 2.** The time-resolved ruby luminescence recorded on a 2D CCD as a function of wavelength (or pressure) and time, shown together with the selected real-time micro-photographic images distinctive to different phases of water and ice.

## 3. Results and Discussions

The metastable ice VII with a "needle-like" morphology has been observed to form from super-compressed water at rapid compressions in the stability field of ice VI [5,12]. In addition, the pressure-induced amorphization of metastable ice VII at room temperature, HDA, is resulted from the structural instability of ice VII [12]. Interestingly, the metastable ice VII is also formed from super-compressed "dirty"-water. Surprisingly, the evolutions of metastable ice VII are different in the pure and dirty waters. Here, solidification of water, the evolutions of ices and melting of ice VI would be investigated in the pure and dirty water.

### *3.1. The water-to-metastable ice VII transition*

At higher compression rate (> 0.1 GPa/s) to the final pressure above 1.3 GPa, the metastable ice VII is crystallized from super-compressed water. Figure 3 shows the representative micro-photographic images of the sample during rapid compression in pure (A) and dirty (B) water. The measured lengths of metastable ice VII as a function of time are also illustrated on the right. Firstly, pure water is solidified to metastable ice VII within one frame (less than 67 μs) and turns into HDA in couple of milliseconds. The detailed discussion regarding the Raman spectra of pure water, ice VII and HDA and the VII-HDA transition mechanism can be found in our recent paper [12]. The impure water sample shows well-separated regions of fluid methane (bubbles in Fig. 3B) and liquid water (featureless area in Fig. 3B). The metastable ice VII grows in one-dimensional direction, according to the real-time micro-photographic images. The Raman spectra of fluid methane and dirty water (including broad water Raman peaks and weak methane peak) are taken before the pressure modulation. The metastable ice VII with needle-like texture is proven at the peak pressure by ice VII Raman signature [12], not shown here.

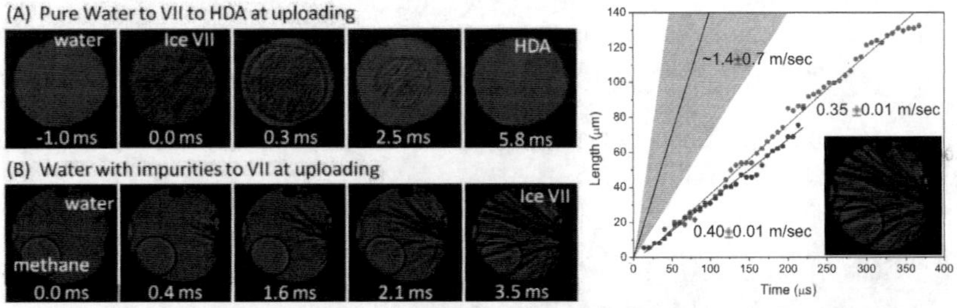


**Figure 3.** High-speed microphotographic images of the sample during rapid compression showing (A) pure water solidification to ice VII and to HDA in comparison with (B) dirty water solidification to ice VII. The growth rates of metastable ice VII are illustrated on the right, ~ 1.4 m/s (black line) from pure water and ~ 0.35/0.40 m/s (red and blue lines) from dirty water.

At both pure and dirty water conditions, the metastable ice VII observed exhibits a needle-like texture. According to the images during ice VII forms in Fig. 3(B), it is reasonable to assume that needle-like ice VII grows one-dimensionally from nucleation sites. Thus, the growth rate can be obtained by directly measuring the length of ice VII in a sequence of images. Two representative lengths of metastable ice VII (red and blue dashed lines) are illustrated in the right panel of Fig. 3, which yield an average growth rate of ice VII, 0.26±0.14 m/sec, in the dirty water samples. Since the metastable ice VII in the pure water case forms so fast (< 67 μs across ~80 μm) that only one or two frames are caught by our high-speed camera even at the fastest recording rate of 30000 frames per second. Therefore, the growth rate of ice-VII in pure water samples of ~ 1.4±0.7 m/sec is only estimated with a large error bar (hatched area), based on the diameter of the entire sample, ~100-210 μm, and the one or two frame time of 70-140 μs. Furthermore, the growth rate of ice VII may also depend on the compression rate and other factors. Clearly, additional experimental data are needed for more meaningful discussions.

It is very interesting that metastable ice VII forms even in dirty water. First of all, impurities seem to make almost no contribution on changing the stability limit conjecture of liquid water, shown in Fig.

1. In other words, impurities would not influence homogeneous nucleation mechanisms of super-cooled or super-compressed liquid. This implies that the assumption regarding one-dimensional growth is suitable for this study. However, based on the observed growth rates of ice VII from pure and dirty waters, it is clear that the presence of impurity slows down the growth rate of metastable ice VII. The fact that metastable ice VII forms from super-compressed water indicates a similar local bcc-like structure between ice VII and super-compressed water. The observed slower growth rate in the dirty sample is then due to the perturbed local order by the impurity. In this regard, it would be interesting to find at which point (or level of impurities) the local order of ice VII and water become different.

Note that no HDA is observed in dirty water samples. In contrast, in pure water samples, metastable ice-VII converts into HDA, signifying the structure similarity between the two phases. This difference is either due to the higher transition pressure (close to the stability field of 2.1 GPa) to form ice VII in dirty water that that of pure (1.4-2.1 GPa) or to the formation of methane hydrates that can be formed in the water-methane mixtures [16].

### *3.2. The ice VI to water transition*

Figure 4 shows the representative sequences of micro-photographic images of pure water (A) and dirty water (B), taken during pressure unloading. In pure water, HDA turns into ice VI and ice VI melts to liquid water upon decompression. In dirty water, ice VII transforms into ice VI and ice VI melts. It both cases, ice VI with a gland-like texture melts close to the equilibrium transition pressure, ~0.9 GPa, independent of the decompression rates. The time zero is defined by the frame right before water starts to appear. The measured area of ice VI directly from the images is scaled by the whole sample in order to compare among different samples. In dirty water, the interaction between $H_2O$ and methane during ice VI melts is ignored.

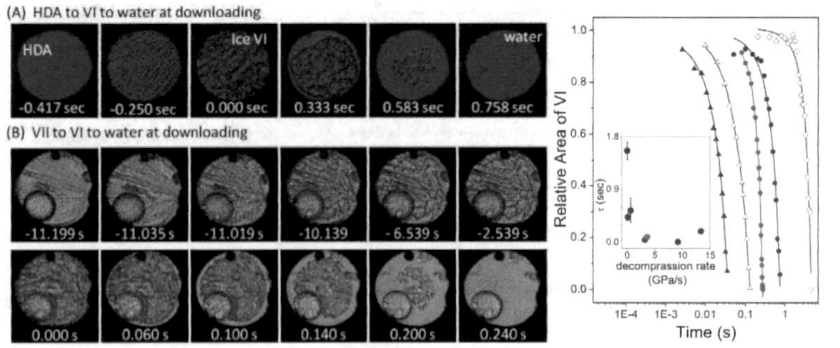


**Figure 4.** High-speed microphotographic images of the sample during decompression showing (A) pure water: HDA to ice VI and to liquid in comparison with (B) dirty water: ice VII to ice VI and to liquid. The relative areas of ice VI at different melting rates are illustrated on the right. The inset shows the melting time constant as a function of decompression rate.

In dirty water, ice VII directly transforms into ice VI, in contrast to HDA transforming into ice VI in pure water. The branches of ice VII grow wider and wider and turn into grand-like ice VI. In both cases, it is difficult to define the conversion rate to ice VI, either from HDA or ice VII. Nevertheless, we estimate the conversion rate based on the relative area changes of ice VI in pure (blue symbols) and dirty (red markers) water with time as plotted in Fig. 4 right. The different symbols signify the different decompression rates. Note that all melting behaviours of ice VI follow an exponential decay as illustrated in the well-fitted lines. Again, the melting of ice VI occurs always at the equilibrium transition pressure, regardless the initial pressure of ice VII. The melting time constants, τ, extracted from the exponential fits (Fig. 4 Inset) indicate that the slower decompression results in the slower melting as well as that the impurity has no bearing on the melting process of ice VI in terms of the transition pressure or the dynamics of melting.

## 4. Conclusions

Using *dynamic*-DAC, we have demonstrated that metastable ice VII can be crystallized directly from super-compressed water in both pure and dirty water systems. This transition results from the similar bcc-like local structure between ice VII and super-compressed water. The presence of methane impurity seems to perturb the interfacial structure between the two, lowering the growth rate of ice VII from 1.4±0.7 m/sec in pure water to 0.26±0.14 m/sec in impure water. In pure water, HDA transforms from metastable ice VII in the pressure range of 1.4-2.1 GPa, again because of the similar local structure. This transition is absent in dirty water, either because the transition pressure to ice-VII is higher - close to the stability field or due to the formation of methane hydrates perturbing the local structure. Upon decompression, HDA or Ice VII converts to ice VI, which melts at the equilibrium transition pressure, *always*, independent of the impurity level. Thus, the result indicates little effect of impurity on melting. Ice VI melts faster at higher decompression rate, following an exponential rate dependence.

## Acknowledgments

The present study was performed in support of the National Science Foundation Division of Materials Research (0854618), and American Chemical Society Petroleum Research Fund (49207-ND10). Acknowledgment is made to the Donors of the American Chemical Society Petroleum Research Fund for partial support of this research.

## References

[1] Hemley RJ, Jephcoat AP, Mao HK, Zha CS, Finger LW, and Cox DE 1987 *Nature* **330** 737
[2] Soper AK 2002 *Science* **297** 1288
[3] Nelmes RJ, Loveday JS, Strassle T, Bull CL, Guthrie M, Hamel G, and Klotz S 2006 *Nat. Phys.* **2** 414
[4] Chou IM, Blank JG, Goncharov AF, Mao H-K, and Hemley RJ 1998 *Science* **281** 809
[5] Lee GW, Evans WJ, and Yoo C-S 2006 *Phys. Rev. B* **74** 134112
[6] Mishima O, Calvert LD, and Whalley E 1984 *Nature* **310** 393; 1985 *Nature* **314** 76
[7] Klotz S, Hamel G, Loveday JS, Nelmes RJ, Guthrie M, and Soper AK 2002 *Phys. Rev. Lett.* **89** 285502
[8] Tse JS, Klug DD, Tulk CA, Swainson I, Svensson EC, Loong CK, Shpakov V, Belosludov VR, Belosludov RV, and Kawazoe Y *Nature* **400** 647
[9] Finney JL, Bowron DT, Soper AK, Loerting T, Mayer E, and Hallbrucker A 2002 *Phys. Rev. Lett.* **89** 205503
[10] Yoshimura Y, Mao H-k, and Hemley RJ 2006 *Chem Phys Lett* **420** 503
[11] Saitta AM and Datchi F 2003 *Phys. Rev. E* **67** 020201
[12] Chen J-Y and Yoo C-S 2011 *Proc. Natl. Acad. Sci. USA* **108** 7685
[13] Knight C, Singer SJ, Kuo J-L, Hirsch TK, Ojamae L, and Klein ML 2006 *Phys. Rev. E* **73** 056113
[14] Evans WJ, Yoo C-S, Lee GW, Cynn H, Lipp MJ, and Visbeck K 2007 *Rev. Sci. Instrum.* **78** 073904
[15] Lee GW, Evans WJ, and Yoo C-S 2007 *Proc. Natl. Acad. Sci. USA* **104** 9178
[16] Chen J-Y and Yoo C-S 2011 *MRS Proc.* **1262**-W04-03

Journal of Physics: Conference Series **377** (2012) 012110 doi:10.1088/1742-6596/377/1/012110

# Transition of strong, wide band-gap oxide insulators to metallic glasses at 100 GPa (Mbar) pressures, etc

**W J Nellis**
Harvard University, Department of Physics
Cambridge, Massachusetts 02138, USA

E-mail: nellis@physics.harvard.edu

**Abstract.** This paper is a review of experimental results observed for shock and statically compressed sapphire (single-crystal $Al_2O_3$) and implications of these results. Many of these were unexpected prior to experiments: (i) Hugoniot and 300-K theoretical isotherm are nearly coincident to 340 GPa, (ii) shock-compressed sapphire probably achieves minimum metallic conductivity (MMC) at ~300 GPa, as does fluid H at 140 GPa, (iii) Both crossovers are driven by shock dissipation and both probably occur by a bonds to bands transition (BtBT), (iv) the BtBT is driven by shock-induced damage which breaks bonds at such high densities that wave functions of atoms so formed hybridize into energy bands; (v) disorder caused by sluggish phase transitions in a DAC suggest sapphire also reaches MMC in a DAC at ~300 GPa; (vi) the structural ground states of all solids at very high pressures might be metallic glasses; (vii) Super Earths can have magnetic fields caused only by convecting conducting oxides without the need for Fe; (viii) alumina powders in a DAC might attain MMC at ~300 GPa and perhaps at pressures as low as ~200 GPa. $Al_2O_3$ is probably amorphous in a DAC at 200 GPa

## 1. Introduction

Extreme conditions of density and temperature produced by high pressures achieve novel states of matter. Dynamic pressure achieves compression and temperature simultaneously and adiabatically. The latter results from the short experimental lifetimes, typically of order 1 to 100 ns, which are too brief for significant thermal transport in or out of a sample. Thus, heating produced by dynamic compression is retained during experimental lifetimes. The experimental lifetimes are also sufficiently brief to preclude mass diffusion out of a sample holder. The fast time scale also means that dynamic compression is irreversible and dissipative. Dissipation energy goes into something other than compression, in this case heating and disorder. Consider, for example, fast compression of an ordered crystalline solid. If compression is sufficiently fast, an ordered crystal structure cannot survive that fast application of pressure, nor can the initial order be regained on the associated fast release (~100 ns) of dynamic pressure. Fast dynamic compression is commonly known as shock or multiple-shock compression. Shock wave velocities are supersonic. In general, a shock-compressed fluid is in thermal equilibrium and thus shock compression produces a shock temperature T and entropy S in bulk fluid. However, a shock-compressed strong solid is not generally in thermal equilibrium up to a stress of a few multiples of its elastic strength under shock, called the Hugoniot elastic limit (HEL). In this case shock heating and damage are heterogeneous. Densities and temperatures achieved dynamically can be tuned to a significant degree by tuning the temporal shape of the pressure pulse applied to a sample.

Static compression is isothermal because experimental lifetime is sufficiently long to enable heat produced by compression to be transported out of a pressure cell prior to making measurements. Temperature is varied independently by heating or cooling the sample or the entire cell. Experimental lifetimes for static compression range from seconds to years. Such long lifetimes enable light, mobile elements, such as H, to rapidly diffuse out of a static-pressure cell.

Work on strong oxide insulators discussed herein was motivated initially by experiments that synthesized metallic fluid hydrogen by reverberating a shock wave in liquid $H_2$ at 20 K contained between two *c*-cut sapphire (single-crystal $Al_2O_3$) disks [1]. The reverberating shock compresses the hydrogen with multiple shocks, which tunes the final temperature to ~1/10 what would be achieved with a single shock to the same final pressure. At 140 GPa, 10-fold compressed liquid density, and ~3000 K, fluid hydrogen achieves minimum metallic conductivity (MMC). In the process, however, transparent colorless sapphire becomes opaque because of shock-induced damage. Because sapphire becomes opaque, thermal emission from the dense hydrogen cannot exit the sample holder, and so its temperature cannot be determined from gray body spectra. In order to determine if some direction of shock propagation, other than along the *c* axis, might produce a more transparent anvil, shock propagation in seven crystallographic directions was investigated experimentally [2]. The nature of those shock waves in sapphire proved to be contrary to expectations prior to those measurements and showed that strong sapphire damages substantially at stresses above its HEL [3].

One widely accepted view is that shock compression produces shock heating, which produces thermal pressures in addition to pressures produced by "cold compression" at ambient temperature. As a result shock pressures are widely assumed to be much greater than pressures achieved by static compression to the same density, particularly at 100 GPa pressures. However, extreme conditions produce unusual and unexpected results. In the case of oxides, dynamic and static compression produce similar results at 100 GPa pressures. This paper reviews a number of unexpected results discovered for sapphire at ~100 GPa pressures. Similar results have been observed for $Gd_3Ga_5O_{12}$.

## 2. Virtual coincidence of Hugoniots and isotherm of oxides at high pressure and implications

The Hugoniot of sapphire was measured up to 340 GPa [4], no phase transitions or cross overs are apparent on the Hugoniot, and the Hugoniot is virtually identical to the theoretical 300-K isotherm up 340 GPa [3,5]. (A crossover is a transition that occurs over some range of pressure, density, and temperature, as opposed to a sharp first-order phase transition with a discontinuity in density.) The Hugoniot of sapphire does not deviate significantly from the theoretical 300-K isotherm until shock pressure exceeds ~400 GPa [6]. This fact suggests dissipation on the sapphire Hugoniot is dominated by disorder up to ~400 GPa and once disorder is complete additional dissipation goes into temperature above ~400 GPa. Electrical conductivities of shocked sapphire have been measured up to 220 GPa [7]. Measured conductivities of shocked sapphire extrapolate to MMC at ~300 GPa. Sapphire was not expected to have metallic conductivity at such a relatively low pressure. Remarkably, sapphire might attain MMC under static compression at ~300 GPa as well.

In a diamond anvil cell (DAC) the as-compressed phase of sapphire is only corundum ($\alpha$-$Al_2O_3$) up to a static pressure of 180 GPa. Predicted phase transitions occur but only with laser-heating and thermal quenching at high pressures. The predicted corundum ($\alpha$-$Al_2O_3$) to $Rh_2O_3$ (II)-type transition occurs at 103 GPa and the $Rh_2O_3$(II)-type to $CaIrO_3$-type transition occurs at 130 GPa and persists to 180 GPa. X-ray spectra of laser-heated and thermally quenched samples consist of broad individual diffraction peaks superimposed on a significant broad background, indicative of disordered structures with short-range order. Above 180 GPa, even with laser heating, sapphire is apparently amorphous because no crystal-structure type has been reported above 180 GPa [8]. $Al_2O_3$ in a DAC is substantially disordered, as is shocked sapphire. Since both shock and statically compressed sapphire are highly disordered and their pressure-density curves are quite similar up to 180 GPa [9], it is quite possible that amorphous $Al_2O_3$ in a DAC achieves MMC at ~300 GPa, as shocked $Al_2O_3$ might do. That is, $Al_2O_3$ in a DAC at ~300 GPa might well be a metallic glass in the solid at 300 K.

Analysis of a large amount of data on dynamically compressed soft fluids ($H_2$, etc) and strong oxides ($Al_2O_3$, etc) lead to the realization that these two apparently different types of materials probably become poor metals by essentially the same mechanism at ~100 GPa pressures. Remarkably, the H-H and Al-O bond strengths are both 4.5 eV. This common mechanism is probably the bonds to bands transition (BtBT). That is, once strong bonds in $Al_2O_3$ and $Gd_3Ga_5O_{12}$ break under dynamic compression, electronic wave functions of resulting atoms overlap at sufficiently high density to form a disordered metal or metallic glass [9-11]. At some high pressure bonds in all insulators break to form a poor metal. Material strength determines the pressure at which the BtBT occurs but not whether it occurs. Because of shock-induced disorder, maximum electrical conductivity is MMC. Sluggish phase transitions induce disorder into sapphire compressed in a DAC [8,10] so that sapphire probably attains MMC under static pressure as well as dynamic.

## 3. Ground-state structure at very high pressures and 0 K

The prominence of disorder at high pressures discussed above raises a fundamental question. At very high pressures, densities are sufficiently high for all materials to be metals. So one question is whether the structural ground state of all solids is well-ordered metal or a metallic glass at sufficiently high pressures and densities? A metallic glass is an amorphous metal, which is a poor metal with MMC because of strong electron scattering caused by disorder. To my knowledge there is no known *a priori* reason to assume that solids at very high pressures are perfectly-ordered crystals. Calculations of high-pressure structures assume perfectly ordered structures, total energy is calculated for each, the one with least total energy is declared the likely equilibrium structure, and pressure is the mathematical derivative of energy with respect to volume. Amorphous structures have yet to be included in such comparisons. To my knowledge, the mechanical stability of such structures under pressure is not calculated. That is, what happens dynamically if a bond is tilted slightly from its direction in a perfectly ordered crystal at high pressures? Such a disturbance might happen, for example, as atoms in one structure start moving in a transition to another structure. The key question is whether high-pressure structures are in stable equilibrium or not?

## 4. Can Super Earths have magnetic fields caused only by convecting molten oxides, without Fe?

Yes, it is possible and thus probable. A Super Earth is a planet with a mass up to ten times greater than that of Earth. Planetary magnetic fields are caused by dynamos, which are convecting conducting material. The magnetic fields of Earth and Jupiter are caused by convection of liquid Fe and metallic liquid hydrogen, respectively with MMC [12].

The highest pressure and temperature of rock (oxides) inside Earth are 130 GPa and 3000 K at the core-mantle boundary between the solid-rock mantle and the liquid Fe outer core. In planets larger than Earth interior pressures and temperatures are larger than in Earth. In some of them temperatures in oxides will exceed melting temperatures of rock at pressure and then that molten rock can convect. Because of very high pressures those molten oxides will probably decompose into a mixture of atoms. Wave functions of outermost electrons of those atoms could overlap to form metallic liquid metal with MMC, as suggested above occurs in $Al_2O_3$. Such a liquid metallic alloy would have a conductivity of MMC, no matter what the exact chemical composition of the oxide might be. Thus, molten rock could convect and conduct and so produce an external magnetic field by dynamo action. The actual magnitude of magnetic field would depend on details of conductivity and convecting hydrodynamic flows [13].

Solid rock deep in Super Earths might be metallic glasses at pressures less than required for melting. Structural defects, dislocations, grain boundaries, sluggish phase transitions, and chemical impurities cause such bond distortions in ideal ordered structures. Under high physical pressures bonds might creep into an amorphous phase. While planetary interiors are at high temperatures, annealing effects depend sensitively on details of temperature, density, electronic structure (strong chemical bonds or metallic bonding) and creep. Within a given Super Earth it is not possible to say, but in principle metallic-glass oxides probably exist inside some Super Earths.

**5. Does an oxide exist that is an electrical insulator above ~200 GPa?** Alumina powder is used as an electrical insulator in a DAC at ~100 GPa pressures. Given the possibility that sapphire might become a poor metal at ~300 GPa in a DAC and given that sapphire is highly disordered in a DAC at 180 GPa [8], it would seem prudent to measure electrical conductivities of powders used as electrical insulators in a DAC at high pressures to be sure such powders are insulators at pressures in which they are to be used. If the BtBT occurs at 300 K in a DAC for some reason, it is quite possible such powders could become poor metals. This question can only be answered by experiment.

## 6. References

[1] Nellis W J, Weir S T and Mitchell A C 1999 *Phys. Rev. B* 59 3434
[2] Kanel G I, Nellis W J, Savinykh A S, Razorenov S V and Rajendran A M 2009 *J. Appl. Phys.* 106 043524
[3] Nellis W J, Kanel G I, Razorenov S V, Savinykh A S and Rajendran A M 2010 *J. Phys.: Conf. Ser.* 215 012148
[4] Erskine D 1994 *High-Pressure Science and Technology-1993*, ed S C Schmidt, J W Shaner, G A Samara and M Ross (New York: AIP Press) p 141
[5] Umemoto K and Wentzcovitch R M 2008 *Proc. Natl. Acad. Sci. USA* 105 652
[6] Hicks D G, Celliers P M, Collins G W, Eggert J. H and Moon S J (2003) *Phys. Rev. Lett.* 91 035502
[7] Weir S T, Mitchell A C and Nellis W J 1996 *J Appl. Phys.* 80 1522
[8] Ono S, Oganov A R, Koyama T and Shimizu H 2006 *Earth Planet. Sci. Lett.* 246 326
[9] Nellis W J 2010 *Phys. Rev. B* 82 092101
[10] Nellis W J 2011 Europ. Phys. J.: Special Topics 196 121
[11] Nellis W J 2011 *Materials* 4 1168
[12] Stevenson D J 2010 *Space Sci. Rev.* 152 651
[13] Nellis W J 2011 to be published in Proceedings of 2011 Conference of Shock Compression of Condensed Matter in *J. Phys.: Conf. Ser.*

# Crystallographic phases in heavy rare earth metals under megabar pressures

**G K Samudrala and Y K Vohra**

Department of Physics, University of Alabama at Birmingham, Birmingham, AL 35294 USA

E-mail: gopi@uab.edu

**Abstract**. Experiments aimed at understanding the crystallographic phases of heavy rare earth metals were carried out in a diamond anvil cell at the Advanced Photon Source, Argonne National Laboratory. Heavy rare earth metals dysprosium (Dy), holmium (Ho), erbium (Er) and thulium (Tm) were compressed to multi-megabar pressures. The rare earth crystal sequence hcp→Sm-type→dhcp→distorted-fcc (dfcc) is observed in all four elements. Upon further compression, a structural transformation to a monoclinic C2/m phase has been observed. We summarize the results from these experiments and present Rietveld structural refinements on high pressure phases for the specific case of dysprosium.

## 1. Introduction

Temperature and pressure are two key parameters which allow us to access the rare earth crystal structure sequence hexagonal close packed (*hcp*) →samarium type (*Sm type*)→double hexagonal close packed (*dhcp*)→face centered cubic (*fcc*) in the individual elements of lanthanide series[1, 2]. Consequently, the behavior of 4f-shell rare earth metals under extreme conditions of pressure is of interest because pressure can cause electronic transfer between various energy sub-bands during compression and hence drive structural phase transitions in these materials. A further increase in pressure beyond the s–d electron transfer can cause Kondo volume collapse or f-shell delocalization, leading to participation of f-shell electrons in bonding at extreme compressions [3, 4]. The appearance of low-symmetry crystal structures with an accompanying volume collapse under high pressures in f - elements is generally attributed to f-shell delocalization. In light rare earth metals cerium (Ce), praseodymium (Pr), and neodymium (Nd); low-symmetry crystal structures like body centered tetragonal (bct), orthorhombic alpha-uranium ($\alpha$-U), and monoclinic *C2/m* crystal structures have been observed under high pressures. Similarly, low-symmetry crystal structures and volume collapse phase transformations have been reported in heavy actinide metals under pressure [5–8]. These low-symmetry structures signal the participation of f-electrons in bonding. The study of lanthanides at ultra high pressures is of interest as researchers try to find if the orthorhombic *Pnma* phase is stable in these elements, as this phase is known to be stable in actinides under high pressures [5 – 8]. We have examined the lanthanide elements dysprosium, holmium, erbium and thulium under multi-megabar pressures in order to understand the low symmetry crystal structures of these elements under extreme conditions.

## 2. Results and discussion

Matched pairs of beveled diamond anvils with central flats ranging from 35μm to 70μm, with bevel angle of 7 to 8.5 degrees and culet size of 300 to 350 μm in diameter were utilized to achieve megabar pressures in studying heavy rare earth metals dysprosium, holmium, erbium and thulium. In all experiments, copper has been placed along with the rare earth metals in the sample chambers of diamond anvil cells and has been utilized as pressure marker. All four elements have been studied at room temperature. In our experiments, dysprosium has been compressed to 35% of its initial volume at 210 GPa, holmium has been compressed to 41% of its initial volume at 137 GPa, erbium has been compressed to 40% of its initial volume at 151 GPa and thulium has been compressed to 38% of its initial volume at 195 GPa. Figure 1 shows the distorted fcc to monoclinic phase transition pressures for rare earth metals which have been known to adapt this structure at high pressures.

Detailed discussions on the phase transitions of holmium, erbium and thulium could be found in [9, 10 & 11] respectively. A discussion on the phase transitions of dysprosium follows. At ambient conditions, dysprosium is in hexagonal close packed structure and the *hcp* → *Sm-type* phase transition is observed starting at 7 GPa at ambient temperature. This is followed by a phase transformation from *Sm type* → *dhcp* starting at 18 GPa. The *dhcp* phase is observed between 18 GPa till 41 GPa. Above 42 GPa, *dhcp* phase transformed to the distorted fcc phase (*dfcc*, hR24 phase) proposed by Hamaya et al [12, 13]. The *dfcc* phase is observed between 42 and 80 GPa. Beyond 82 ± 2 GPa, a new yet unreported phase transition was observed that was attributed to a monoclinic (*C2/m* phase). Figure 2 shows all five phases observed in dysprosium at different pressures.

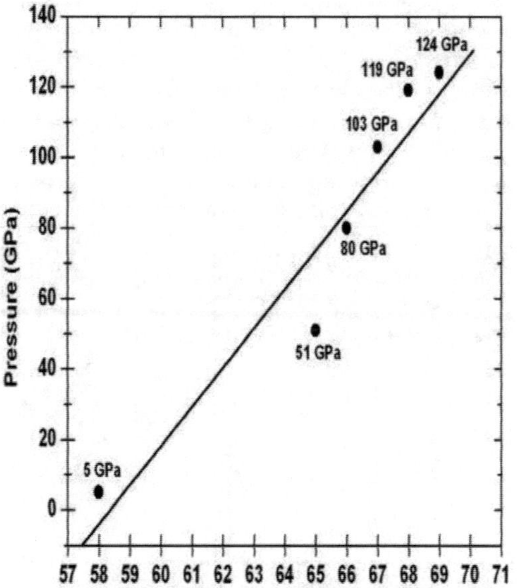


**Figure 1:** The pressure at which a post-dfcc phase *C2/m* is observed to occur in rare earth metals is plotted against the atomic number for elements. The pressure increases from 5 GPa for cerium to the highest pressure of 124 GPa for thulium. The solid line is merely a guide to the eye.

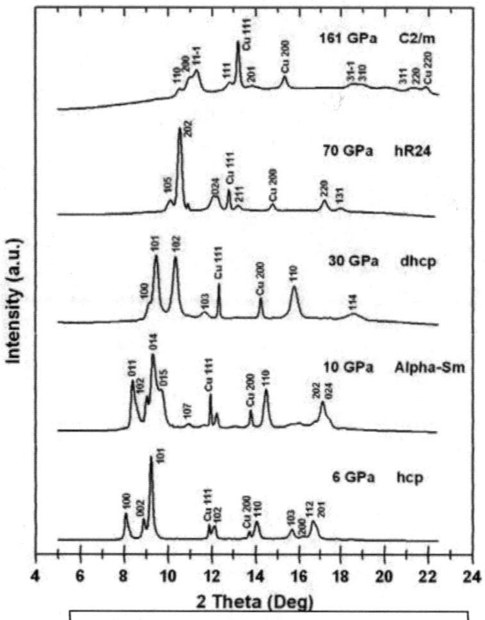


**Figure 2:** The integrated x-ray diffraction patterns for five different phases of dysprosium recorded at various pressures. The patterns shown are hcp phase at 6 GPa, α-Sm-type phase at 10 GPa, dhcp phase at 30 GPa, hR24 phase at 70 GPa, and C2/m phase at 161 GPa.

The unit cell of this hexagonal *hR24* phase consists of 24 atoms sited to produce periodic distortion of the *fcc* unit cell. The space group is *R-3m* with the *6c* and *18h* Wyckoff sites occupied *6c* (0, 0, $z_1$) and *18h* ($x$, $-x$, $z_2$). Fig. 3 (a) shows the integrated x-ray diffraction profile intensity versus 2θ at a pressure of 70 GPa where the sample is in the distorted-fcc phase. The diffraction pattern at 70 GPa was refined using the Rietveld method and yielded the lattice parameters a = 5.6673±0.0012 Å and $c$ = 13.8837±0.0064 Å and structural parameters $z_1$= 0.2789±0.0009, $x$= 0.5071±0.0018, $z_2$= 0.2497±0.0009 with volume = 16.091 $Å^3$ /atom. It should be added that these structural parameter values may have been affected by some preferred orientation effects in the polycrystalline samples. Hence, these structural parameters should be regarded as approximate until a full analysis of preferred orientation in the sample is performed. If $z_1 = z_2 = 0.25$, and $x = 0.5$ then an undistorted *fcc* structure is obtained provided the hexagonal axial ratio c/a = √6. The structural distortion from the *fcc* phase was found to increase in the *hR24* phase when the pressure was increased between 42 GPa and 80 GPa.

At 82 ± 2 GPa, we observe major changes in the x-ray diffraction pattern, as a major doublet in the *hR24* phase at 2θ values between 9-11 degrees changes to a triplet peak signaling a phase transition. This new phase can be indexed to the monoclinic *C2/m* phase with 4 atoms per unit cell and was previously reported in cerium under pressure [14]. This monoclinic phase is also similar to one reported in other heavy rare earth metals terbium [15], holmium [9], and thulium [11]. In a study on erbium [10], we have also examined orthorhombic *Ibam* proposed for praseodymium [16] but found the monoclinic *C2/m* to be the better fit for the low symmetry phase.

Fig. 3 (b) shows the angle-dispersive x-ray diffraction pattern of dysprosium at a pressure of 167 GPa and the Reitveld refinement of this pattern. The fit to monoclinic C2/m lattice parameters at this pressure are $a$ = 4.4992±0.0024 Å, $b$= 2.6273±0.0010 Å, $c$= 4.1349±0.0043 Å, β= 106.09±0.054 degrees with volume = 11.74 $Å^3$/atom. This monoclinic phase can be derived from an *fcc* lattice $a$ (monoclinic) = $c$ (monoclinic) = √1.5 $a$ (*fcc*) and $b$ (monoclinic) = $a$ (*fcc*)/√2 with β (*fcc*) = 109.5 degrees [15]. The four atoms in this *C2/m* structure occupy Wyckoff sites *4i* that are given by ($x$, 0, $z$) and if $x = z = 0.25$ then the structure reduces to a body cantered monoclinic phase with space group *I2/m*. The GSAS refinement gives $x$= 0.3974±0.0072, $z$= 0.4522±0.0066 indicating a structure quite distorted from the body cantered monoclinic phase.

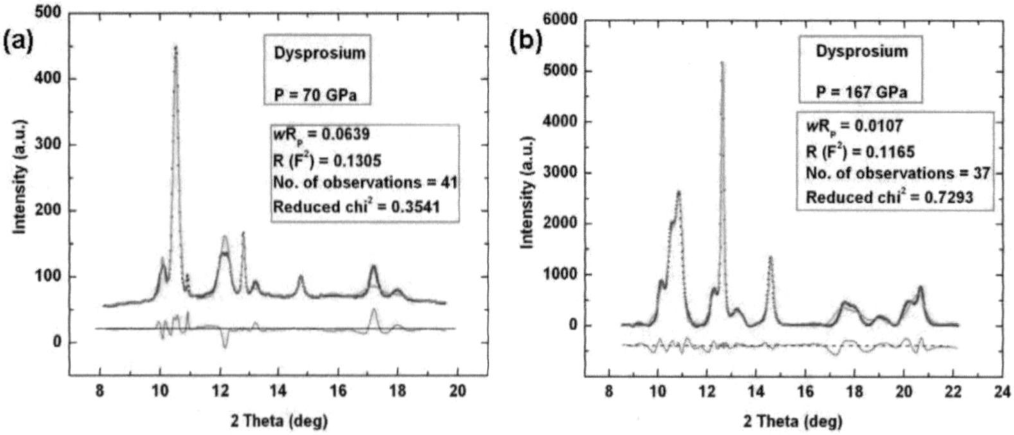


**Figure 3:** (a) and (b) show the Reitveld refinements of dysprosium at 70 GPa and 167 GPa respectively. The lattice parameters obtained are given in text. The symbols represent the observed intensity profile using FIT2D and the solid line represents the calculated values using the Rietveld refinements assuming (a) hexagonal *hR*24 phase and (b) monoclinic *C2/m* phase. The difference profile *I* (observed) – *I* (calculated) is shown at the bottom.

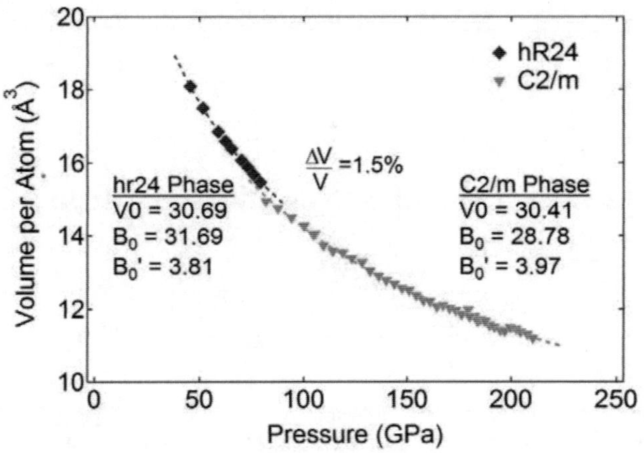


**Figure 4:** The measured pressure-volume curve of dysprosium for the two high pressure phases. Dashed curve is Birch–Murnaghan fit.

In summary, we have studied four rare earth metals dysprosium, holmium, erbium and thulium at megabar pressures and found the low symmetry phase preferred by these elements at ultra high pressures can be indexed as a monoclinic *C2/m* structure. Further, first principle calculations of heavy rare earth metals under extreme conditions are needed to confirm the eventual 4f-delocalization and its participation in bonding.

**Acknowledgments**
This material is based upon work supported by the Department of Energy (DOE) – National Nuclear Security Administration (NNSA) under Grant No. DE-FG52-10NA29660. Portions of this work were performed at HPCAT (Sector 16), APS, Argonne National Laboratory.

**References**

[1] Donohue J 1982 *The Structure of the Elements* (Robert E. Krieger, Malabar, FL).
[2] Beaudry B J and Gschneidner Jr. K A 1978 *Handbook on the Physics and Chemistry of Rare Earths*, Vol. 1, eds. Gschneidner Jr K A and Eyring L (North-Holland, Amsterdam p.173.
[3] Hewson A C 1993 *Kondo Problem to Heavy Fermions* (Cambridge: Cambridge University Press).
[4] Held K, Huscroft C, Scalettar R T and McMahan A K 2000 *Phys. Rev. Lett.* **85** 373.
[5] Lindbaum A, Heathman S, Litfin K, Meresse Y and Haire R G 2001 *Phys. Rev.* B **63** 214101.
[6] Penicaud M 2002 *J. Phys.: Condens. Matter* **14** 3575.
[7] Lindbaum A, Heathman S, Le Bihan T, Haire R G, Idiri M and Lander G H 2003 *J. Phys.: Condens. Matter* **15** S2297.
[8] Heathman S, Haire R G, Le Bihan T, Lindbaum A, Idiri M, Normile P, Li S, Ahuja R, Johansson B and Lander G H 2005 *Science* **309** 110.
[9] Vohra Y K, Sangala B R, Stemshorn A K and Hope K M 2008 *Mater. Res. Soc. Symp. Proc.* **1104** 1104-NN01-04.
[10] Samudrala G K, Thomas S A, Montgomery J M and Vohra Y K 2011 *J. Phys.: Condens. Matter* **23** 315701.
[11] Montgomery J M, Samudrala G K, Tsoi G M and Vohra Y K 2011 *J. Phys.: Condens. Matter* **23** 155701.
[12] Hamaya N, Sakamoto Y, Fujihisa H, Fujii Y, Takemura K, Kikegawa T and Shimomura O 1994 *AIP Conf.* **309** 457.
[13] Hamaya N, Okabe N, Yamakata M, Kagi T and Shimomura O, 1996 *High Press. Res.* **14**, 287.
[14] McMahon M I and Nelmes R J 1997 *Phys. Rev. Lett.* **78** 3884.
[15] Cunningham N C, Qiu W, Hope K M, Liermann H and Vohra Y K 2007 *Phys. Rev.* B **76** 212101.

# Pulsed particle beam high pressure/shock research in India

**Anurag Shyam**[1], **Rohit Shukla**
Energetics & Electromagnetics Division, Bhabha Atomic Research Centre, India
E-mail:anurag.shyam@gmail.com

**Abstract.** We have two major facilities for particle beam driven shock wave/high pressure generation. One being AMBA and the other being 1.2MJ capacitor bank RUDRA. Apparatus for Mega-Ampere Beam Application which is known as AMBA is now with India and the experiments are being planned from the facility for the shock wave and high pressure studies using the AMBA for intense light-ion beam generation and then bombarding them on a flyer target. To enhance the AMBA machine to double the output current is also under consideration. AMBA is a pulsed power source which delivers 50kJ of energy in 50ns with 1.7 MV minimum peak voltages maintained as an average of various shots in the case of positive polarity output in a suitable ion-diode. The output impedance of the AMBA machine is 2.25ohms and hence it is a 1.5 TW machine. With peak power densities up to ~1TW/cm$^2$, and proton ranges in condensed matter of 10 to 20 μm, specific energy depositions of several MJ/g at deposition rates of the order of 100 TW/g are obtained. This way the AMBA system can be used as a shock wave generator in both, direct drive and impact experiments. We also have 1.2MJ capacitor bank capable of delivering 3.6MA peak current at 44kV charging voltage to be used for Magnetized target fusion based on z-pinch regime of target material compression. The related diagnostics for the system, which are currently being developed, are mentioned in the present paper. Both the systems and the high pressure experiments to be conducted are described in the paper. A brief detail on the plasma focus devices, which also produce shock waves using particle beams, is also presented in the paper.

## 1. Introduction

The shock wave studies are mostly done using gas-gun, rail-gun and electric-gun [1] or high-power lasers [2]. Similarly the particle beams have also shown the potential of driving the projectiles to a velocity as high as 10km/s using electron and ion-beams. AMBA setup in India will be one of this kind of facility for the particle beam generated shock/high pressure studies. The other major facility RUDRA which is 1.2 Megajoule capacitor bank is also the largest capacitor bank facility of its kind for the similar experiments.

## 2. AMBA pulsed power machine:

The AMBA generates a high-voltage output pulse and applies to the ion-diode working as load in the following steps:
(1) A capacitor bank of 50 energy storage capacitors, connected in a Marx generator circuit, is DC charged to a level of up to 200KV (using a $\pm$100 KV);
(2) The Marx generator is electrically triggered and generated impulse charges a water-insulated intermediate storage capacitor in a time period of about 800 ns;

[1] To whom any correspondence should be addressed.

(3) The energy is transferred for the second time to a 2.25-Ω PFL in about 175 ns;
(4) The PFL is discharged by a low inductance switch generating an output pulse that is finally conducted to the load by a matching 2.25-Ω transmission line.

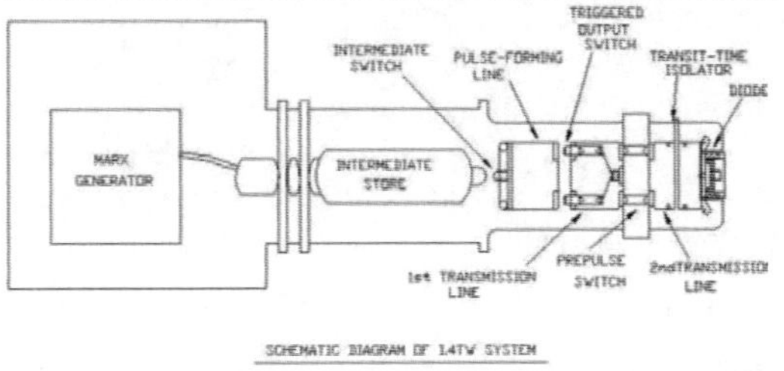


**Figure 1.** AMBA component assembly scheme

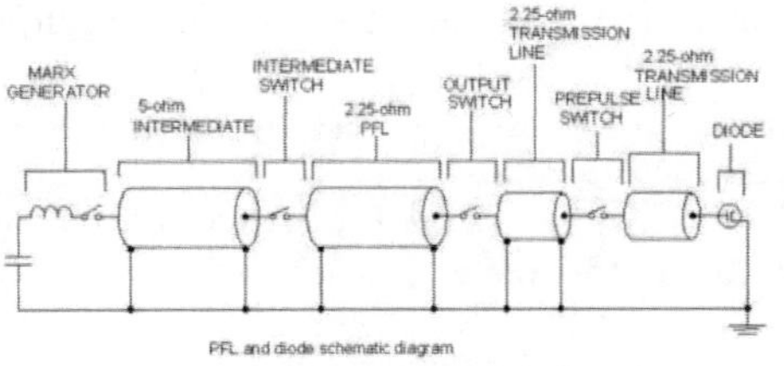


**Figure 2.** AMBA electrical Schematic

Figure 1 shows AMBA components assembly scheme and Figure 2 shows the electrical Schematic of the Machine. The system components are explained below.

**2.1 The AMBA machine: Marx Generator**

The primary energy storage is a 25-stage Marx generator, which can store 250 KJ at maximum charge level. Each stage of the Marx generator consists of two numbers of energy storage capacitors of 1microfarad charged from a bipolar power supply up to ± 100KV. In each stage a spark gap switch is also present which is pressurized with sulfur hexafluoride gas. All the 25 stages are arranged in a hybrid Marx circuit. This scheme derives its name from the fact that it combines the features of the capacitance Marx circuit and the resistively triggered Marx circuit.

The mechanism referring to these designations causes the spark gaps to successfully break down and discharge the generator into its load. In the capacitance Marx circuit, high-voltage transients generated by the firing of previous stages are coupled through stray capacitances to succeeding spark gaps, causing them to break down. In the resistively triggered Marx circuit, each spark gap contains a triggering electrode connected to successive stages through a coupling resistor. As each stage fires, one trigger pulse is generated that is applied to the unfired spark gap. The hybrid circuit uses both of these mechanisms to achieve very reliable Marx triggering during the operation at less than 50% of the self-firing voltage. Components of the hybrid Marx are arranged in such a way that the current discharge path doubles back on itself in zigzag fashion and, by so doing, maintain a relatively low generator inductance.

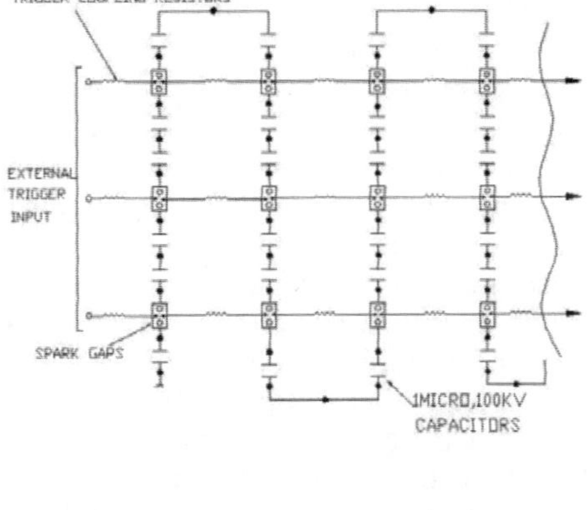


**Figure 3** AMBA; MARX erection scheme

**Figure 4.** AMBA; Marx bank assembly

The basic schematic of a representative portion of a hybrid Marx generator is shown in Figure 3. The circuit in this figure is simplified by omitting the charging circuits. The Marx circuit is suspended in the insulating oil in seven columns of three stages and in two columns of two stages. The nylon suspension straps are attached to the capacitors with special quick-release mechanisms to facilitate ease of installation and removal. The generator is positioned off-center within the tank to reduce electric fields in the vicinity of the capacitor bushings and to allow space for servicing. Figure 4 is the photograph of the Marx generator. A small two-stage Marx trigger circuit suspended in the Marx tank triggers the Marx generator. The trigger circuit is dc charged to 90kV (±45 kV) and is triggered by another trigger generator. The output of the Marx trigger circuit is resistively connected to the trigger electrodes of the first three spark gaps of the main Marx generator, while the remaining spark gaps are resistively coupled to each other.

## 2.2 Pulse Forming lines

The output pulse of the pulsed power system is generated in three stages: (a) energy is transferred form the Marx generator to a water-insulated intermediate storage capacitor in about 800 ns: (b) it is then transferred a second time to the PFL in less than 200 ns: (c) the charged PFL then is discharged to the load through a matching transmission line. This three-stage process allows for a more efficient design of the pulse-forming system by taking advantage of the time dependence of the electrical break down of water.

**Figure 5.** AMBA Ion-Diode and Marx Capacitors

The intermediate storage capacitor is a coaxial transmission line with dimensions chosen to minimize the electric stress in the water. The minimum electrical stress occurs for an infinite line when the ratio of the diameters of the two conductors is equal to the constant e. This ratio would represent an impedance of 6.7 Ω with water which is used as the dielectric. The optimum value is more on the order of 4.5 to 5 Ω because of local field enhancements at the ends of lines and the polarity effects. The intermediate store impedance for the system is chosen at about 5 Ω, and length is chosen so that the capacitance is about 16 nF when fringing fields at the ends are taken into consideration. Transfer of energy form the intermediate store to the PFL is accomplished with a self-closing water spark gap on the axis of the machine between the two lines. This switch has one fixed electrode and one that is adjustable by means of a hydraulic actuator controlled from outside of the machine. The operating voltage then is raised or lowered by remotely increasing or decreasing the distance between the electrodes. The out put impedance of the PFL is 2.25 Ω and the electrical length is 50 ns. The PFL is also charged at nominal power to 3.6 MV, the same level as that of the intermediate store. A low inductance switch at the output end of the PFL generates the output pulse by discharging the PFL into the first transmission line. Triggering electrodes are located on the output switch between the main electrodes, which are connected to a master trigger switch. Each trigger electrode is maintained at a potential that reduces field enhancement at its edge during the time that the PFL is being charged. When the desired voltage level is reached on the PFL, the master-trigger switch closes, abruptly

changing the potential of each of the trigger electrodes. A large field enhancement is produced at the edges of the trigger electrodes, thereby initiating rapid closure of the output switch.
The output pulse is conducted from the PFL to the diode through two lengths of matching 2.25-Ω transmission lines that are separated by a pre pulse switch. This pre pulse switch consists of 11 gas-pressurized spark gaps that are adjusted to withstand the pre pulse voltage levels, but rapidly synchronously break down when the main output pulse arrives. Following the pre pulse switch is the second length of transmission line, which connects to the diode assembly.
The system with similar parameters as that of AMBA is used as a shock wave generator in both direct drive and impact experiments [3,4] , delivering up to 40 kJ of proton beam energy to a 7- to 10-mm-diameter focal spot. With peak power densities up to ~1 TW/cm$^2$, and proton ranges in condensed matter of 10 to 20 μm, specific energy depositions of several MJ/g at deposition rates of the order of 100 TW/g are obtained. Aluminum foils of 10 to 30 μm thickness had been accelerated to velocities beyond 12 km/s. Within an uncertainty of ~5%, the threshold pressure was found to be 64 GPa for aluminum, 136 GPa for copper, 86.5 GPa for titanium, and 252 GPa for molybdenum. We will use this system as a shock wave generator for further study in the field of hypervelocity impacts and equations of state. The experiments that will be done include the Hugoniot measurements and measurements at unloading of shock-compressed state down to the vaporization region. Figure 5 presents the images of ion-diode and Marx capacitor bank of the AMBA machine. To generate high-power focused ion beam pulses, electrical energy is converted into particle beam energy in ion diodes. With peak proton energies of ~1.7 MeV, specific power densities of up to 200 TW/g and energy densities of several MJ/g can be realized at ~40 ns FWHM. The massive energy deposition in a zone of 5-10 mg/cm$^2$ leads to fast vaporization and ablation of the material and causes intense compression waves to propagate into the residual part of the target. In this way thin foil plates with a thickness of several tens of micrometers can be accelerated to velocities larger than 10 km/s.
We are also trying to upgrade AMBA machine to AMBA+ which will have two AMBA machines driving a single load simultaneously and thereby increasing the output power.

## 3. RUDRA 1.2Megajoule Capacitor Bank:

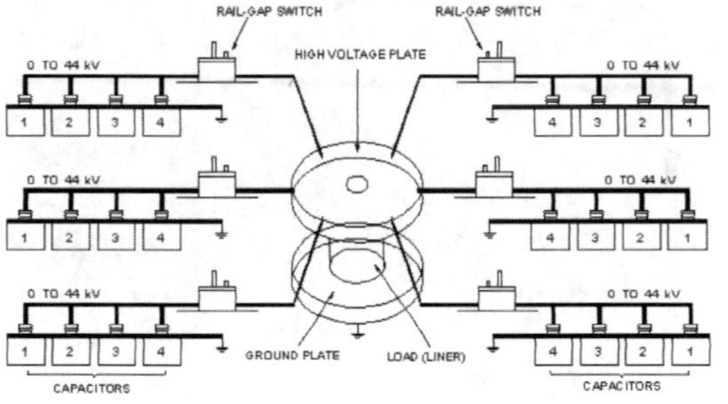


**Figure 6.**Layout of 1.2MegaJoule Capacitor Bank

**Figure 7** RUDRA 1.2MegaJoule Capacitor Bank

A capacitor bank named RUDRA is installed and commissioned in our laboratory The 1.2MJ capacitor bank consists of 24 nos. of capacitors with stored energy of 50kJ each. At the charging voltage of 44kV, each capacitor can deliver a peak current of 150kA and the voltage reversal is limited to 10%. The equivalent series inductance (ESL) of the capacitor is rated as 70nH. In order to segregate the coulomb transfer loading on switches and to minimize the effective circuit inductance, the capacitor bank is subdivided in to modules. There are six modules and four nos. of capacitors are connected to each module by parallel plate transmission line assembly (which further couples to Railgap switch). The individual Railgap switches are capable of transferring maximum 10 coulombs of charge at a peak current of 750kA with a jitter of <2ns. The output of all modules terminates to centralized collector plate through RG218 cables. The schematic of implemented configuration of 1.2MJ Capacitor Bank and the actual modular layout of bank has been shown in Fig. 6 and Fig.7 respectively. Other than RUDRA capacitor bank we also have small energy capacitor banks. RUDRIKA is another capacitor bank of 136kJ and is faster bank capable of delivering 3.6MA in short circuit configuration. It is shown in figure 8.

**Figure 8** RUDRIKA: 136kJ capacitor bank

CHANDI (80kJ) and CHANDI-II (160kJ) capacitor banks are shown in figures 9 and 10 respectively.

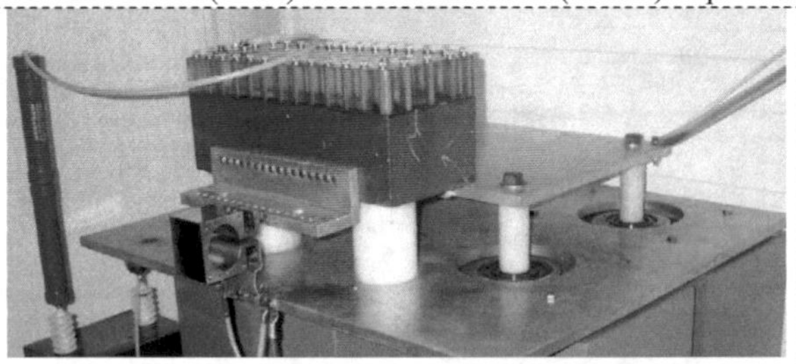

**Figure 9.** CHANDI (80kJ) Capacitor Bank

**Figure 10.** CHANDI-II (160KJ) Capacitor Bank.

We are upgrading the RUDRA Capacitor bank to RUDRA-II to make a 2.4 MJ capacitor bank which will have double the energy as compared to RUDRA bank and will be capable of delivering near double the currents. The application of this capacitor bank for high pressure studies is mainly in the fields of high energy plasma focus and in the z-pinch implosion studies. In the z-pinch implosion by high current capacitor banks, a cylindrical metallic target is compressed using the interaction of z-direction current with the theta-direction magnetic field. The targets are usually thin metal cylinders or a metallic coating on some cylindrical insulator. The current is passed through these cylinders. As the z-current interacts with the theta- magnetic field, the liner starts imploding. At the end of this radial motion, the plasma collides with itself in small radius column and kinetic energy gets converted to thermal energy increasing the temperature and pressure in the interaction region.

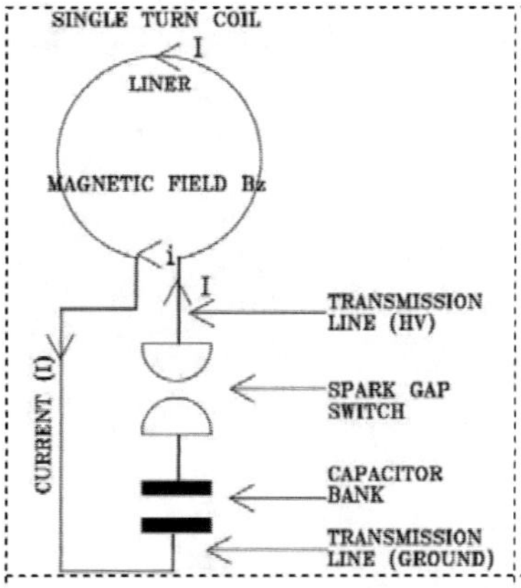


**Figure 11.** Scheme of single turn coil experiment

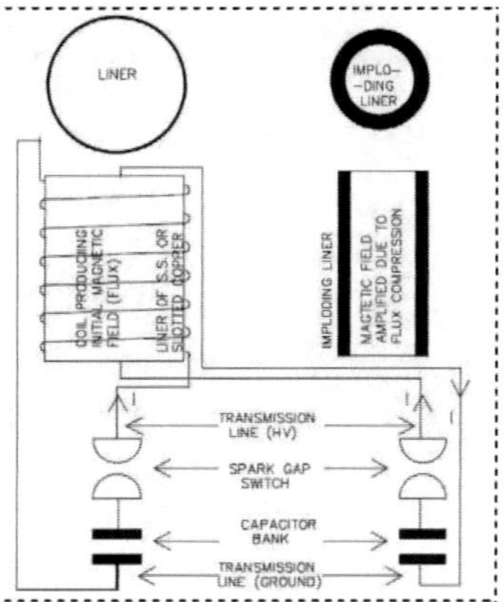


**Figure 12.** Scheme of flux compression scheme using two capacitor banks.

We are following two schemes for theta-pinch type high magnetic field/pressure generation which are depicted in figures 11 and figure 12 respectively. With RUDRA capacitor bank, we presently have 500Tesla pulsed magnetic field facility and we aim to extend it to 700 Tesla by making RUDRA-II capacitor bank in case of directly driven single turn coil configuration. Using Flux compression scheme we expect to get 1000Tesla by using RUDRA-II capacitor bank. Presently we are getting uneven compression, as is shown in figure 13, and working for achieving uniformity in compression. Other application of the capacitor bank is MTF/MAGO [5] like experiment which involves particle driven shock and high pressure research. In the direction of plasma focus also the capacitor bank is aimed to be used for generation of ion sheath, acceleration of it to create a plasma focus spot. In the both phases of acceleration namely axial acceleration phase and radial compression phase, the velocities of plasma sheath are very high and shock waves propagation takes place during the event.

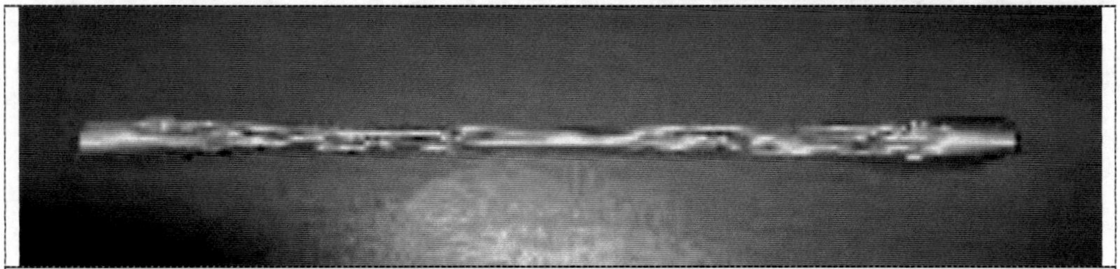

**Figure 13** Uneven collapse of liner

## 4. References

[1] T.C. Kaushik, A.K. Saxena, S.C. Gupta AIRAPT-23 Mumbai, India (2011)
[2] P.A.Naik, AIRAPT-23 Mumbai, India (2011)
[3] K.Bamung, H.J. Blume, P.Hoppe,H.U.Karow, D.rusch, O.Stoltz, J.Singer, G.I.Kanel, A.V. Utkin,S.V. Razoranov, V.E.Fortov Int. J. Impact Engng. 17, pp37-46 (1995).
[4] G.I.Kanel, J.R.Asay, K.Baumung, H. Blume, L.C. Chhabildas, V.E. Fortov, B.Goel, P. Hoppe, T.Mehlhorn, S.V. Razorenov, D. Rusch and A.V Utkin, Int. J. Impact Engng. 23, pp 421-430 (1999).
[5] S.F. Geranin, IEEE Trans. on Plasma Science 26 (4), 1230-1238, (1998).